Solid-Phase Synthesis

Solid-Phase Synthesis

A Practical Guide

edited by

Steven A. Kates
Consensus Pharmaceuticals, Inc.
Medford, Massachusetts

Fernando Albericio
University of Barcelona
Barcelona, Spain

CRC Press
Taylor & Francis Group
Boca Raton London New York

CRC Press is an imprint of the
Taylor & Francis Group, an **informa** business

CRC Press
Taylor & Francis Group
6000 Broken Sound Parkway NW, Suite 300
Boca Raton, FL 33487-2742

First issued in paperback 2019

© 2000 by Taylor & Francis Group, LLC
CRC Press is an imprint of Taylor & Francis Group, an Informa business

No claim to original U.S. Government works

ISBN-13: 978-0-8247-0359-2 (hbk)
ISBN-13: 978-0-367-39871-2 (pbk)

Visit the Taylor & Francis Web site at
http://www.taylorandfrancis.com

and the CRC Press Web site at
http://www.crcpress.com

Foreword

New biologically active peptides continue to be discovered at a rapid rate and the remarkable advances in sequencing the human genome promise to provide a near infinite number of new peptide and protein sequences. These will require chemical synthesis to fully understand their functions and mechanisms of action and to provide suitable agonists or antagonists for medical purposes. Solid-phase peptide synthesis will play an important role in the development of these areas. To aid in realizing this vast opportunity there is a clear need for a new book on the subject, and this volume fills that need. It is especially useful because it describes the historical development of various aspects of the field and continues with descriptions of the most recent and important advances. Equally importantly it provides detailed experimental procedures describing how to implement the methodology.

Both Fernando Albericio and Steven Kates, as well as most of the authors of individual chapters, are well known to members of the synthetic peptide community and the information they provide can be considered authoritative and reliable. Extensive, but careful, referral to the primary literature is exceedingly valuable. The general principles of the various approaches to solid-phase peptide synthesis with their advantages and disadvantages are expertly covered. Similarly, the very large numbers of polymeric solid supports, side-chain protecting groups, and temporary α-amine protecting groups and the many options for activation and coupling reactions are handled thoroughly. Purification and analytical methods for the products are also described in detail with appropriate examples and references. In addition to stepwise solid-phase synthesis the fragment or convergent approach and the new ligation methods are covered. The book also has chapters on various peptide derivatives of intense current interest, including glyco-, phospho-, and sulfopeptides, peptide mimetics, disulfides, cyclic peptides, peptide-nucleic acids, and more. Alternative approaches to combinatorial synthesis of peptide and nonpeptide libraries are well presented and will be of great interest to the reader.

While writing these words I had occasion to look at the manuscript for answers to two specific questions and quickly found both. I intend to keep a copy of this book on my desk for ready reference and I can recommend that every peptide chemist and many classical organic chemists do the same.

R. B. Merrifield
The Rockefeller University
New York, New York

Preface

Oganic chemistry in the last half of the 20th century has evolved to a level of extreme sophistication in which complex macromolecules once thought to exist only in nature can now be prepared in a laboratory hood, using glassware and chemicals. The process typically involves manually adding reagents and an organic solvent to a vessel that can be temperature controlled. Reactions are monitored and, upon completion, an aqueous solution is added for "workup." The biphasic liquid is transferred to a separatory funnel and the layers are divided. The organic layer is extracted from the aqueous layer and removed in vacuo to give either an oil or a solid that may require purification for a future application. To accelerate the purification process, flash chromatography was developed in which a crude product is passed through a column of silica gel under nitrogen pressure. The identity of the product is typically determined using infrared and nuclear magnetic resonance spectroscopy as well as mass spectrometry.

Bruce Merrifield was the first to recognize an alternative approach for the preparation of organic compounds. He applied this method to synthetic peptides and was awarded the Nobel Prize in 1984 for this discovery. The concept was to perform the chemistry proved in solution but add a covalent attachment step that links the target to an insoluble polymeric support. Other advantages of the solid-phase technique include simple filtration and washing without manipulative losses and the fact that the process lends itself to automation. Merrifield conceived this idea primarily because of a recognized need for new techniques, deriving it from the fundamentals of chromatography. In order to purify the peptides that were assembled in solution, the biomolecules were passed through gel filtration columns. Merrifield realized that because molecules bind to the stationary support, permanently attaching a peptide would presumably allow the synthesis to be performed more effectively, since reactions may be driven to completion with excess soluble reagents. Merrifield initially selected cellulose as the solid support but concluded that the polymer was not stable under synthesis conditions. Finally, copolymers of styrene and divinylbenzene similar to those used to prepare the ion-exchange Dowex-50 resins were chosen. This support formed small uniform spheres that could be readily solvated, washed, and filtered. Merrifield considered these beads the key to the development of solid-phase methods.

Solid-phase peptide synthesis (SPPS) strategies incorporate the following concepts: (1) polymeric support (resin) (see Chapter 1), (2) anchoring linkages (handles) (see Chapter 5), (3) coupling reagents (activators) (see Chapter 6), and (4) protection schemes (see Chapters 2, 3, and 4). As methods advanced for SPPS, more complex sequences were prepared. Cyclic peptides were one of the initial targets for strategies to be developed (see Chapters 7 and 8). Cyclic structures are of interest because of their potential for improved metabolic stabilities, increased potencies, better receptor selectivities, and more controlled bioavailabilities.

The repetitive process in peptide chemistry makes it possible to prepare ~30-residue sequences in high yield. Although the two-step process (deprotection and coupling) is extremely efficient, the chemistry unfortunately is still not fully effective for the ready preparation of small proteins (~70 residues or more). An alternative approach, referred to as convergent solid-phase peptide synthesis (CSPPS), relies on the same principles as those for natural product synthesis (see Chapter 9). In CSPPS, fragments of a desired sequence are synthesized and may be further purified and then attached to each other.

As the field of peptide synthesis matured, researchers applied the concepts and strategies to assemble oligodeoxynucleotides and carbohydrates. Historically, deoxyribonucleic acid (DNA) was constructed using the phosphate di- and triester methods. However, in an effort to apply the solid-phase approach as well as the advantages of automation, chemistry based on phosphorus (III) was developed. In the most widely used method, deoxynucleoside 3'-phosphoramidites are incorporated in the phosphite triester as the building blocks. This strategy is more commonly referred to as the phosphoramidite approach (see Chapter 11). Contrary to the case of peptide synthesis, oligonucleotides can be constructed rapidly (less than 8 min per nucleotide addition) and efficiently (coupling yield per cycle is ~98–99% in automated instrumentation). Thus, 50 mers can routinely be prepared in 1 day and many sequences exceed 100 nucleotides. The nucleic acid analogue peptide nucleic acid (PNA) has use as a diagnostic or molecular biology tool. The pseudopeptide backbone with the four standard nucleic acid bases as side chains is also amenable to solid-phase construction (see Chapter 13). The third major class of biopolymers found in eukaryotes, polysaccharides, has recently been adapted for synthesis on a solid phase (see Chapter 14). Oligosaccharides mediate a variety of biological functions such as inflammation, immunological response, metastasis, and fertilization. Solid-phase strategies to assemble these biomolecules are more complex than those for proteins and nucleic acids because similar functionalities and reactions must be differentiated and stereospecific, respectively.

Glycopeptides are an important class of biomolecules because of their involvement in biological recognition and transport processes. These complex natural products have a carbohydrate attached to a peptide sequence via an ether or amide linkage. For the former, the attachment occurs at a hydroxyl function (typically Ser, Thr, or Tyr) to produce an *O*-glycopeptide. For the latter, an amino sugar reacts with a carboxylic acid side-chain function (typically Asp or Glu) to provide an *N*-glycopeptide (see Chapter 10). Conjugates of peptides and nucleic acids have also been prepared because the combination of these two biopolymers can produce enhanced biological and biophysical properties relative to the individual sequences (see Chapter 12).

Before the concept emerged of small molecule libraries prepared on solid phases, chemists such as Clifford Leznoff, Josep Castells, and Henry Rappoport in the 1970s examined polymeric supported reactions for modification of symmetrical diamines and alcohols, Wittig reactions, and Dieckmann cyclizations. However, it required another twenty years for organic chemists to realize that solid-supported organic chemistry could be applied and expanded to medicinally relevant compounds. A great number of organic reactions are known in solution and would provide greater diversity because there was no limitation to 20 building blocks (amino acids). With proven success in peptide synthesis libraries, the field of combinatorial solid-phase organic chemistry developed. Amino acids and linear peptide sequences anchored to a solid support provide excellent precursors for the construction of heterocyclic compounds (see Chapter 15). Alternatively, researchers have prepared peptidomimetics as potential therapeutic agents (see Chapter 16). These derivatives incorporate a modified peptide backbone to increase oral bioavailability and resistance to enzymatic degradation.

The techniques in solid-phase organic chemistry complement those from solid-phase peptide assembly. Synthesis is performed on resins using anchoring linkages and orthogonal protection schemes. Molecules are cleaved from the solid support, purified (see Chapter 18), and then analyzed by high-performance liquid chromatography or mass spectrometry (see Chapters 19 and 20). Reactions can be performed in a simple apparatus that includes a syringe fitted with a polypropylene frit connected to a vacuum aspirator and to a complex automated instrument and workstation (see Chapter 17).

The chapters in this volume represent a broad range of solid-phase applications and combine recent advances as well as historical perspectives, addressing many of the current and future themes related to drug discovery. Finally, we would like to acknowledge the contributors to this volume, who have succeeded in highlighting the emerging field of solid-phase synthesis.

Steven A. Kates
Fernando Albericio

Contents

Contributors

Fernando Albericio   Department of Organic Chemistry, University of Barcelona, Barcelona, Spain

Nicholas P. Ambulos, Jr.   Department of Microbiology and Immunology, University of Maryland School of Medicine, Baltimore, Maryland

David Andreu   Department of Organic Chemistry, University of Barcelona, Barcelona, Spain

George Barany   Department of Chemistry, University of Minnesota, Minneapolis, Minnesota

Laurent Bellon   Department of Oligonucleotide Chemistry, Ribozyme Pharmaceuticals, Inc., Boulder, Colorado

Lisa Bibbs   Protein and Nucleic Acids Core Facility, The Scripps Research Institute, La Jolla, California

Christopher Blackburn   Drug Discovery, Millennium Pharmaceuticals, Cambridge, Massachusetts

Lynda F. Bonewald   Department of Medicine/Endocrinology, The University of Texas Health Science Center at San Antonio, San Antonio, Texas

Ralph A. Casale   Core PNA Research and Development, PE Biosystems, Framingham, Massachusetts

Scott B. Daniels   Instrument Development, PE Biosystems, Framingham, Massachusetts

Vern de Biasi   Analytical Sciences, SmithKline Beecham Pharmaceuticals, Essex, England

Amanda L. Doherty-Kirby   Department of Chemistry, University of Waterloo, Waterloo, Ontario, Canada

Michael Egholm   Core PNA Research and Development, PE Biosystems, Framingham, Massachusetts

Ramon Eritja   Instituto de Biologia Molecular de Barcelona, Consejo Superior de Investigaciones Cientificas, Barcelona, Spain

Gregg B. Fields   Department of Chemistry and Biochemistry, Florida Atlantic University, Boca Raton, Florida

Pilar Forns   Department of Chemistry and Biochemistry, Florida Atlantic University, Boca Raton, Florida

Carlos García-Echeverría   Oncology Research, Novartis Pharma Inc., Basel, Switzerland

Ernest Giralt   Department of Organic Chemistry, University of Barcelona, Barcelona, Spain

Gilles Guichard   Laboratory for Immunological Chemistry, UPR 9021 CNRS, Institut de Biologie Moléculaire et Cellulaire, Strasbourg, France

Richard A. Houghten   Department of Chemistry, Torrey Pines Institute for Molecular Studies, San Diego, California

Steven A. Kates   Department of Chemistry, Consensus Pharmaceuticals, Inc., Medford, Massachusetts

Ashok Khatri   Endocrine Unit/Medicine, Massachusetts General Hospital, Boston, Massachusetts

Gilles A. Lajoie   Department of Chemistry, University of Waterloo, Waterloo, Ontario, Canada

Paul Lloyd-Williams   Department of Organic Chemistry, University of Barcelona, Barcelona, Spain

John S. McMurray   Department of Neuro-Oncology, The University of Texas M. D. Anderson Cancer Center, Houston, Texas

Katalin Fölkl Medzihradszky   Department of Pharmaceutical Chemistry, University of California San Francisco, San Francisco, California

Adel Nefzi   Department of Chemistry, Torrey Pines Institute for Molecular Studies, San Diego, California

Ernesto Nicolás   Department of Organic Chemistry, University of Barcelona, Barcelona, Spain

Andrew Organ   Computational and Structural Sciences, SmithKline Beecham Pharmaceuticals, Essex, England

John M. Ostresh   Department of Chemistry, Torrey Pines Institute for Molecular Studies, San Diego, California

Jacques Y. Roberge   Early Discovery Chemistry, Bristol-Myers Squibb, Princeton, New Jersey

Paolo Rovero   Dipartimento di Scienze Farmaceutiche, Università di Salerno, Fisciano, Italy

John D. Wade   Howard Florey Institute, University of Melbourne, Parkville, Victoria, Australia

Susan T. Weintraub   Department of Biochemistry, The University of Texas Health Science Center at San Antonio, San Antonio, Texas

Francine Wincott   Department of Oligonucleotide Chemistry, Ribozyme Pharmaceuticals, Inc., Boulder, Colorado

T. Scott Yokum   Department of Chemistry, University of Minnesota, Minneapolis, Minnesota

Introduction: The Impact of Genome Sequencing on Drug Discovery

The third millennium will usher in major changes in medical and pharmaceutical research. Soon, the complete sequence of every human gene (and protein) will be available and accessible to all. Completion of the mouse genome will not be far behind. Mapping and identification of human disease genes will be dramatically accelerated, as will the development of mouse models of these diseases. Complete genomes of pathogens will become routinely available. These advances will create an unprecedented wealth of new validated drug targets for the pharmaceutical industry. However, they will also create a need for new technologies to deal with the overwhelming information. The ability to make sense out of gene sequences is paramount if the information is to be used efficiently. Scientists will rely increasingly on bioinformatics programs that tease out similarities and patterns from vast beds of information.

How will this new age of information affect drug discovery, beyond the obvious increase in validated targets? Five arenas in which utilization of this information will drive new technologies for drug development relevant to the theme of this book come to mind based on national discussions on functional genomics: (1) the identification of protein targets of existing drugs or lead compounds will be dramatically accelerated, thereby facilitating development of improved drugs; (2) global transcriptional responses to drugs or lead compounds will be utilized to identify additional downstream targets for intervention and potential harmful side effects of lead compounds; (3) high-throughput screens of combinatorial libraries at the cellular level will be feasible using reporters for genes in pathways of interest; (4) the prediction of protein structure and function and the role played by individual proteins in cellular regulation on the basis of primary sequence will become a reality, decreasing the interval from gene identification to functional assay; and (5) combinatorial libraries will be designed for families of related proteins and it will be possible to screen every member of the family to verify the specificity of lead compounds.

I. IDENTIFICATION OF DRUG TARGETS

Until recently, virtually all drugs were discovered by a relatively random process in which natural products or derivatives of natural products were added to microbes, human cell cultures, or animals (sometimes humans) in hope of some desired effect. The age of molecular biology has now made it possible to validate a protein as a drug target long before the availability of a lead compound. However, this technology has not eliminated natural product screening at the cellular level. Indeed, the development of improved techniques for generation of combinatorial libraries (discussed in this book) and the improved availability of reporter genes have increased screening at the cellular level. Thus, the emergence of lead compounds with unknown targets will accelerate.

Fortunately, the knowledge of complete genome sequences will accelerate identification of the drug targets. For example, it will be possible to identify a protein that has been affinity labeled or affinity purified with a lead compound from as little as 25 fmol of material. Mass spectrometric analysis of tryptic fragments, when compared with the complete genome, will, in most cases, provide unambiguous identification. This would not be possible without the genome sequence. Alternatively, targets of lead compounds will be identified by screening libraries of protein fragments expressed in bacterial vectors or phage. Once a binding fragment (domain) is identified, the complete clone will be identifiable from databases and available from complementary DNA libraries. This process will be considerably enhanced with the eventual development of "normalized" libraries in which all human proteins (or domains) are present at similar levels. These libraries will be contained within vectors that allow rapid shuttling into mammalian, bacterial, or insect cell expression systems. Thus, the time required to advance from a lead compound that induces a desired effect at the cellular level to an in vitro assay with a purified drug target will be considerably diminished.

II. GENE MICROARRAYS FOR GLOBAL EFFECTS OF DRUGS

The completion of the human genome sequence allows unprecedented possibilities for assessing cellular (or tissue) responses to drugs or lead compounds. It is already feasible to obtain microarrays of oligonucleotide probes that cover the complete genomes of microbes. Microarrays bearing a small but significant proportion (6000–10,000 genes) of the human genome are also available. Ultimately, all ~90,000 human genes will be available on these arrays. With this technology, it will be possible, in theory, to assess fully all changes in messenger RNA expression that occur in response to activation of a signaling pathway or addition of a drug that affects a pathway.

Although many acute responses to drugs will not be directly detected by this technology (e.g., protein degradation, phosphorylation, protein relocation), secondary responses to such events are likely to be detected, allowing one to infer the acute events. For example, from the induction of a set of genes known to be regulated by cyclic AMP, one would infer that protein kinase A was activated. Even acute morphological changes such as actin rearrangement or detachment from adjacent cells or from the extracellular matrix induce identifiable transcriptional responses. In fact, because it is becoming apparent that signaling networks invariably branch and regulate families of genes with particular temporal responses, it is likely that one need only follow a subset of well-chosen reporter genes to infer which upstream pathways have been affected by a cellular activator or drug. Computers will be trained to recognize these characteristic patterns and report back to the investigator. At one level, the identity or function of the genes that are regulated is less important than the pattern that implies the pathway affected. Undesirable effects of lead compounds related to lack of specificity for the desired target will become immediately available and interpretable.

III. HIGH-THROUGHPUT SCREENS AT THE CELLULAR LEVEL

The identification of reporter genes for specific pathways downstream of disease genes or other pathways relevant for drug intervention facilitates development of high-throughput screens at the cellular level. Such assays often rely on a visual response (such as expression of enhanced green fluorescent protein) that can be detected in a small number of cells or a single cell. In this way it is possible to grow cells in thousands of micropools into which distinct combinatorial variants of molecules have been deposited. Automated readings of color development in individual pools reveal hits.

Although the expense of current microarray gene chip technology is prohibitive for use in large-scale screening, one could imagine a scenario in which individual members of a library of compounds are simultaneously assessed for all possible effects on gene regulation rather than effects on a single reporter. Computer evaluation could reveal the likely target (or pathway), and this information would be stored for future evaluation. In this way, the probability that a given drug will hit a target will be multiplied by ~90,000-fold (the number of human genes). Of course, the relevance of the hit for human disease may not be immediately obvious, but subsequent identification of a disease gene in the same pathway will stimulate one to recall the hit and do further evaluation. This technology is not here today but is likely to arrive in the not so distant future.

IV. PREDICTION OF PROTEIN STRUCTURE AND FUNCTION

Perhaps the most frustrating barrier between the discovery of disease genes or other validated drug targets and an efficient and rational assay for drug development is ignorance of the structure and function of the protein. As a consequence, the human genome project is creating a backlog of validated targets that must be investigated by tedious processes that involve guessing at function on the basis of cellular responses to overexpression or deletion of the gene. Approximately 20% of newly discovered genes have a domain with significant similarity to a protein whose structure and/or function has been determined. A larger subgroup have domains that can be assigned on the basis of significant homology to a family of related proteins of unknown structure and/or function. With the completion of genomes of humans and *Drosophila*, it is expected that most genes will be assigned into such families. Crystallographers and nuclear magnetic resonance (NMR) researchers will concentrate their efforts on the novel domains that emerge from these sequences.

The availability of families of related domains considerably accelerates structure determinations by allowing more reliable predictions of the domain borders. By expressing the precise domain rather than a partial domain or a domain with floppy extensions, the probability of obtaining a crystal or a soluble protein for NMR analysis is enhanced. Thus, the interval between defining a domain and knowing the structure of a family member is becoming amazingly short. The structure of a single member of a family will facilitate alignment of other family members, especially when a large number of family members are identified. Improved threading programs will facilitate the modeling of unsolved structures based on the structure of a single family member. Thus, although reliable prediction of de novo protein folding in the absence of a template is still well in the future, improved software for aligning families of proteins, increased numbers of available sequences, and improved threading programs will make protein structure predictions increasingly accurate.

Ultimately, one would like to read the structure and function of a protein from the primary sequence the way one reads a paragraph by deducing the meaning of a linear collections of letters (words). The metaphor is attractive because the number of amino acids (20) is similar to the number of letters in the English alphabet (26). A motif is about the size of a word and a domain the size of a sentence. For example, this paragraph has 543 characters and four sentences and thus has information content similar to that of a 60-kd, four-domain protein.

Thus, the problem at hand is analogous to deciphering a foreign language or breaking a secret code. The problem is that even if we correctly deduce the structure of a protein, this does not necessarily define its function. Ultimately, key motifs within the protein reveal function just as key words within sentences reveal intent. These motifs may be the set of amino acids that form the catalytic center of an enzyme or they may be short stretches of amino acids that confer modification by another enzyme (myristylation, farnesylation, phosphorylation, ubiquination, proteolysis, acetylation, glycosylation) or confer interaction with another protein or domain (motifs for binding to SH2, SH3, PTB, PDZ, and WW domains). Ultimately, knowledge of the location of the motif in the context of the overall structure of the protein is critical for understanding function just as location of a key word in the context of a sentence or paragraph is critical for understanding meaning.

Unfortunately, we do not have a Rosetta Stone for translation of protein motifs and must rely on experimentation. However, considerable progress is being made in defining motifs through the use of peptide libraries or protein mutations. The rules have been deduced for a handful of protein kinases and protein interaction domains. It is not unreasonable to expect that computer programs will soon be able to make reliable predictions not only of protein function but also of likely sites of modification and likely in vivo partners. This information will considerably shorten the time required to go from the discovery of a gene for a potential drug target to placing the gene product in a defined pathway. Knowledge of protein function and regulation will suggest screens for drug discovery.

V. DESIGN OF COMBINATORIAL LIBRARIES FOR MEMBERS OF PROTEIN FAMILIES

Perhaps one of the greatest impacts of complete genome sequences will be the knowledge of the existence of highly related proteins that are likely to bind a given drug and result in unintended (and potentially avoidable) side effects. Rather than approach a potential drug target with blinders, hoping that inadvertent inhibition of homologous proteins in other tissues will not pose a problem, one can collect all homologues and screen libraries for specificity at the outset. Pharmaceutical companies have already begun such strategies based on partial genome sequences. This approach has the added advantage that one is likely to uncover drugs specific for other members of the family that may be useful for different diseases. In addition, this approach couples well with combinatorial library techniques. A library designed from a lead compound for a single family member is likely to produce high-affinity ligands for other family members.

In summary, genome sequencing and improved techniques for protein structure determination, protein structure prediction, combinatorial library synthesis, and drug screening will all have impacts on drug discovery. The acceleration of screening techniques and increased number of targets will create demands for libraries of compounds of increasing diversity and bioavailability. Still, the greatest barrier to drug development is failure in clinical trials because of problems of toxicity or bioavailability. In theory, increased knowledge of proteins related to the drug target of interest and improved techniques for assessing cellular responses (discussed before) could lower the probability of toxic side effects. However, prediction of bioavailability is still an inexact science. Thus, it remains to be seen whether the 1990s revolution in new technologies will significantly lower the ultimate temporal and financial barrier in drug discovery.

Lewis C. Cantley
Harvard Medical School
and Beth Israel Deaconess Medical Center
Boston, Massachusetts

1

The Solid Support

Pilar Forns and Gregg B. Fields
Florida Atlantic University, Boca Raton, Florida

I. INTRODUCTION

The term "solid support" is commonly used to denote the matrix upon which chemical reactions are performed. However, in reality solid-phase reactions are carried out not merely on the surface of the "solid" polymer but also inside these particles. The term solid support is more appropriate in describing the insolubility of the polymer, which allows filtration or centrifugation and hence separation of reactants from products.

A variety of supports have been developed since the initial introduction of polystyrene cross-linked with 2% divinylbenzene (DVB) by Merrifield [1] for solid-phase peptide synthesis (SPPS). The development of novel supports has had its maximum impact in the past decade because of problems found in the synthesis of difficult peptides and the evolution of solid-phase organic chemistry (SPOC). As the use of one particular support may not always provide a reasonable yield of a desired biomolecule, an obvious need developed for additional supports. Some of the desired properties of new resins include enhanced swelling or rigidity (mechanical strength) and chemical inertness. The recent interest in small-molecule combinatorial libraries has led to rebirth of SPOC and subsequent need for both new types of resins and linkers with varied physical properties to accommodate a wide range of reagents and reactions. Traditional SPPS utilizes linker resins that, after cleavage, result in peptide acids or amides. Linkers with different functionalities are thus required for SPOC as well as production of unusual peptides by SPPS (see Chapter 6). Advances in linker development include the use of activatable linkers [2–4], photolabile linkers [5], and silicon linkers [6,7]. Good leaving groups such as Br or I have been incorporated into the resins used for SPPS [8].

The characteristics of an efficient solid support are as follows:

1. It must be in particles of physical size and shape that permit ready manipulation and rapid filtration from liquids. It must be physically stable.
2. It must be inert to all reagents and solvents used during the synthesis. SPOC may require a resin that is in contact with a wide range of reagents (strong acidic or basic conditions; radical, carbene, carbanion, or carbenium ion chemistry; reducing and oxidizing conditions; or conditions of nitrations or halogenations) and subject to a variety of reaction conditions (variation of temperature and pressure.)
3. It must swell extensively in the solvents used for synthesis, allowing all reagents to penetrate readily throughout the particles of the polymer. Interactions of the dry polymeric matrix with appropriate solvents often lead to dramatically expanded structures. The extent of swelling may be quantitated by a variety of techniques, including volumetric analysis [9–13], gravimetric analysis after centrifugation or rapid filtration [14,15], direct microscopic examination [16], and exclusion of high-molecular-weight substances [17].
4. It must be readily modified to allow attachment of the first entity (i.e., an amino acid in SPPS or a variety of organic molecules in SPOC) by a covalent bond. The introduction of functional groups suitable for further chemical elaboration is usually accomplished by derivatization of preformed polymers, although, in a number of cases, a functionalized monomeric unit may be copolymerized in the course of preparation of the support. A variety of anchoring strategies have been used. One could link the anchoring group to a protected amino acid or peptide and then react with the functionalized polymer; alternatively, the anchoring group may have its origin on the support. The initial anchored unit will be the limiting reagent for all subsequent reactions. The selection of an appropriate initial "loading" level of the first reactant will determine the maximum possible yield of a synthesis. It has generally been assumed in SPPS that too high a level of growing chains dispersed throughout the polymeric matrix may result in decreased efficiencies of individual chemical steps because of steric crowding, intersite side reactions, decreased diffusion of reagents, and/ or changes in the solvation properties of the system. It has been demonstrated that high loadings (0.8–2 mmol/g polystyrene resin) can result in synthetic failures after 10 to 20 cycles. Lower loadings (0.3–0.5 mmol/g) have thus been more commonly used [18].

To achieve adequate reaction rates, it is required that all functional sites be located in well-solvated, accessible regions of the solid support, minimizing diffusion or steric barriers. Conventional wisdom states that supports should have the minimal level of cross-linking consistent with stability, resulting in well-solvated gels within which solid-phase chemistry takes place [12,19]. Alternatively, porous but rigid supports with a high degree of cross-linking can be used [20,21]. The following classification of solid supports, based on physical properties, has been proposed [22]:

1. *Gel-type supports.* Gel-type supports are most often used for solid-phase synthesis and feature equal distribution of functional groups throughout a highly solvated and inert polymer network ideal for the assembly of large molecules. The support capacity can be adjusted to afford problem-free synthesis and a high yield per volume of resin. The polymer network is flexible, and the resin can expand or exclude solvent to accommodate the growing molecule within the gel. There are four types of gel resins:

    a. Polystyrene (PS) resins. The hydrophobic polystyrene resins are produced from styrene and cross-linked with 1 or 2% DVB. They may be substituted with various functionalities as described later in this chapter.

    b. Polyacrylamide resins. A hydrophilic alternative to the polystyrene resins, polyacrylamide resins are obtained by cross-linking poly-N,N-dimethylacrylamide, bis-N,N'-acyloylethylene-diamine, and N-acryloylsarcosine methyl ester [23]. The ester is converted to a functional group after polymerization. Also, polyamide resins can be obtained by free radical–initiated copolymerization of N,N-dimethylacrylamide, N,N'-biacrylyl-1,3-diaminopropane, and N-(2-(methylsulfonyl)ethyloxycarbonyl)-allylamine (MSC-allylamine) or N-acrylyl-1,6-diaminopropane hydrochloride or N-methacrylyl-1,3-diaminopropane hydrochloride [24].

    c. Polyethylene glycol (PEG) grafted resins. PEG-PS was developed as a mechanically more stable resin that spaces the site of synthesis from the polymer backbone. PEG-PS is formed either by graft polymerization on polystyrene beads (Tentagel) [25] or by reaction of preformed oligooxyethylenes with aminomethylated polystyrene beads [26,27].

    d. PEG-based resins. PEG-based resins [28–31] are composed either exclusively of a PEG–polypropylene glycol (PPG) network or of a combination of PEG with a small amount of polyamide or polystyrene. The resins have been obtained by partial derivatization of PEG epichlorohydrin, chloromethylstyrene, or acryloyl chloride (using bis-aminopropyl-PEG). The resulting mixture of non-, mono-, and difunctionalized PEG can be mixed with other monomers to vary the polymer composition, and polymerization results in a highly cross-linked polymer with long cross-linkers of PEG.

2. *Surface-type supports.* Many different materials are used for surface functionalization, including beads made from sintered polyethylene, cellulose fibers (cotton, paper, Sepharose, and LH-20), porous highly cross-linked polystyrene or polymethacrylate, controlled pore glass, and silicas.

3. *Composites and supported gels.* To increase their mechanical stability, gel-type polymers can be supported by rigid matrices. An early example used kieselguhr as the matrix support by polymerizing a polyamide monomer solution after absorption into granules of the inorganic matrix. Later a highly cross-linked polystyrene sponge was used, and in its interior the polyamide was grafted by covalent attachment. Teflon membranes have also been modified by polymethylacrylate to produce a flow-stable gel.

4. *Brush polymers.* A linear component, such as polystyrene, is grafted onto a polyethylene film or tubing. The polymer chains extend from the rigid surface in a brushlike fashion. Functionalization can be achieved along the extending chains, and the functional groups can be further modified with spacers prior to solid-phase synthesis.

II. POLYSTYRENE RESINS

A. Low Cross-Linked Polystyrene

Cross-linked polystyrene is obtained by suspension polymerization [32–34] of styrene and DVB (Fig. 1). The product is a hydrophobic bead that is solvated by nonpolar solvents such as toluene and dichloromethane (DCM) [11,32]. The bead size can be controlled [35,36] such that virtually mono-sized beads (usually from 10–200 μm) have been obtained [33,34]. The solvation of pure polystyrene–divinylbenzene beads is not very efficient [37–39]. As peptide synthesis proceeds, resin solvation properties may change dramatically as a result of solvation of the growing peptide chains. Depending on the peptide composition, careful selection of solvent mixtures may be required to match the solubility parameters of both the peptide and the resin [12].

Functionalization of the resin can be performed by copolymerization, in which one of the monomers carries the functional group (Table 1). This

(*Text continues on pg. 26*)

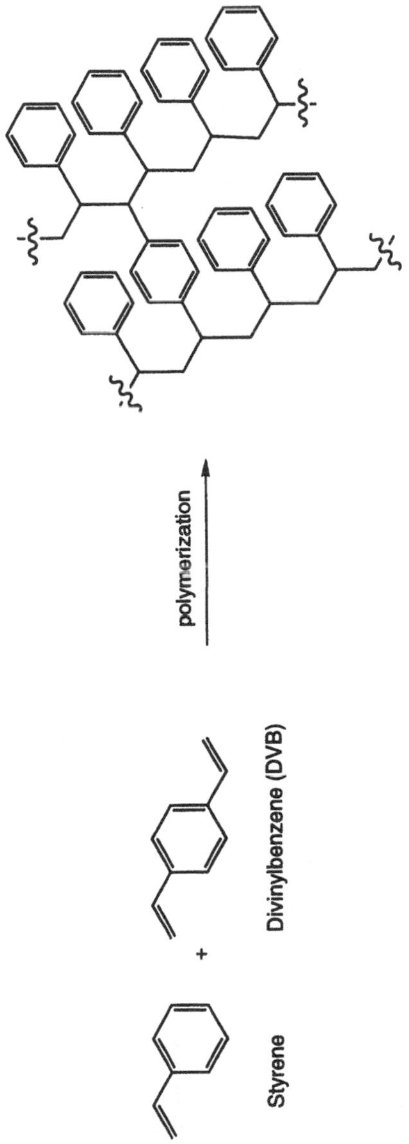

Figure 1 Synthesis of copoly(styrene-divinylbenzene) resin.

Table 1 Preparation of Polystyrene Resins

Amino resin

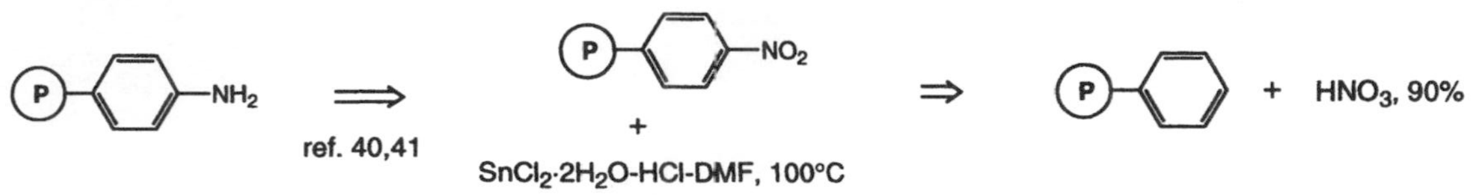
P
NH2
ref. 40,41
NO2
SnCl2·2H2O-HCl-DMF, 100°C
HNO3, 90%

Aminomethylated resin

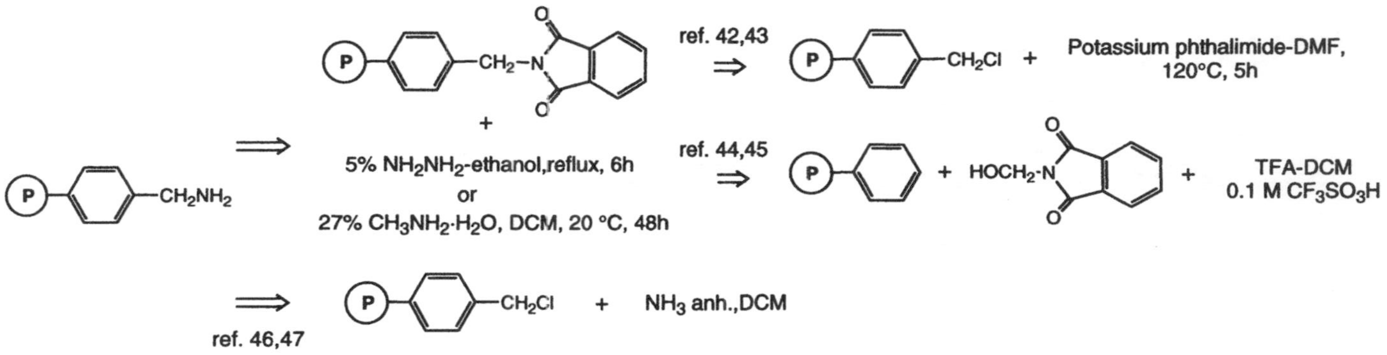
P
CH2NH2
CH2–N
5% NH2NH2-ethanol,reflux, 6h
or
27% CH3NH2·H2O, DCM, 20 °C, 48h
ref. 42,43
CH2Cl
Potassium phthalimide-DMF, 120°C, 5h
ref. 44,45
HOCH2-N
TFA-DCM
0.1 M CF3SO3H
ref. 46,47
NH3 anh.,DCM

4-Aminomethyl-3,5-dimethoxybutyryl-MBHA resin (PAL-MBHA) resin

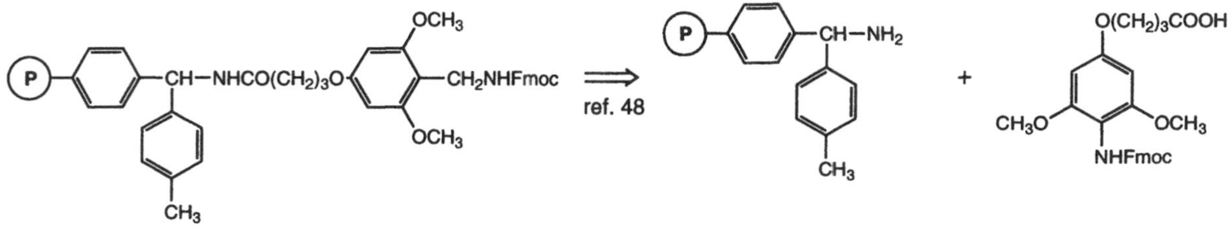
P
CH–NHCO(CH2)3O
OCH3
CH2NHFmoc
CH3
ref. 48
CH–NH2
O(CH2)3COOH
CH3O
NHFmoc

4-(Aminomethyl)-3-nitrobenzoylaminomethyl resin

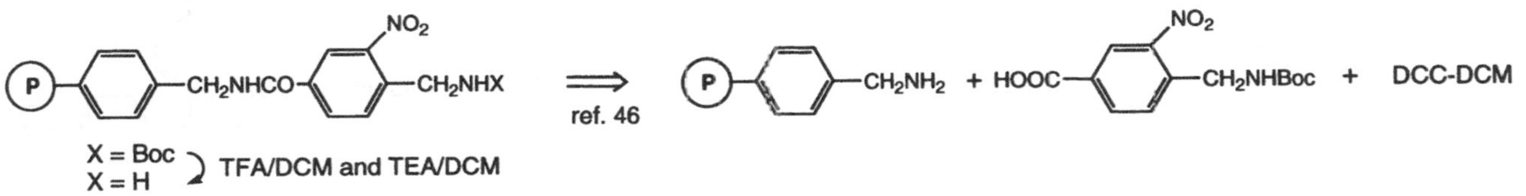

P—CH₂NH₂ + HOOC— (3-NO₂-benzene) —CH₂NHBoc + DCC-DCM

⟹ ref. 46

P—CH₂NHCO— (3-NO₂-benzene) —CH₂NHX

X = Boc
X = H } TFA/DCM and TEA/DCM

Aminomethylphenoxymethyl resin

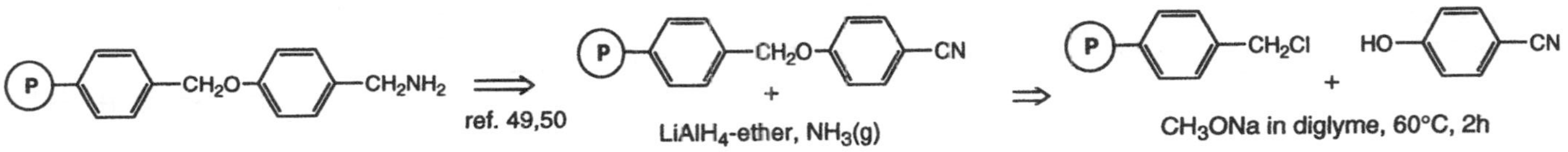

P—CH₂Cl + HO— (benzene) —CN

CH₃ONa in diglyme, 60°C, 2h

P—CH₂O— (benzene) —CN

+ LiAlH₄-ether, NH₃(g)

⟹ ref. 49,50

P—CH₂O— (benzene) —CH₂NH₂

4-(4-Aminooxymethyl-3-methoxyphenoxy)butyryl-MBHA resin

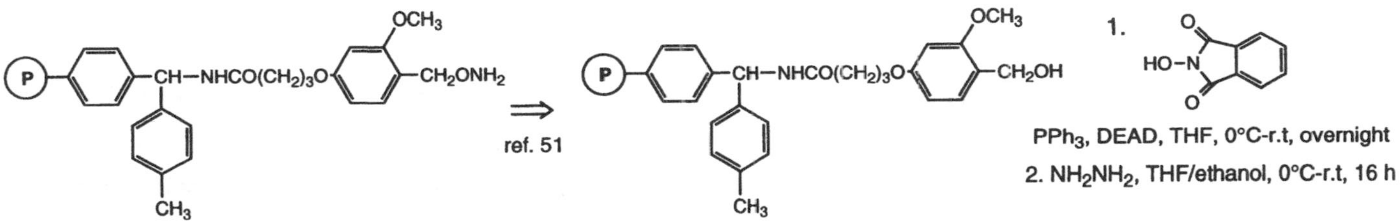

P— (benzene) —CH—NHCO(CH₂)₃O— (benzene, OCH₃) —CH₂OH
                |
         (benzene)—CH₃

1. (phthalimide N—OH) PPh₃, DEAD, THF, 0°C-r.t, overnight
2. NH₂NH₂, THF/ethanol, 0°C-r.t, 16 h

⟹ ref. 51

P— (benzene) —CH—NHCO(CH₂)₃O— (benzene, OCH₃) —CH₂ONH₂
                |
         (benzene)—CH₃

Table 1 Continued

Benzhydrylamine resins (Y_1=H, BHA resin; Y_1=OCH_3, MBHA resin)

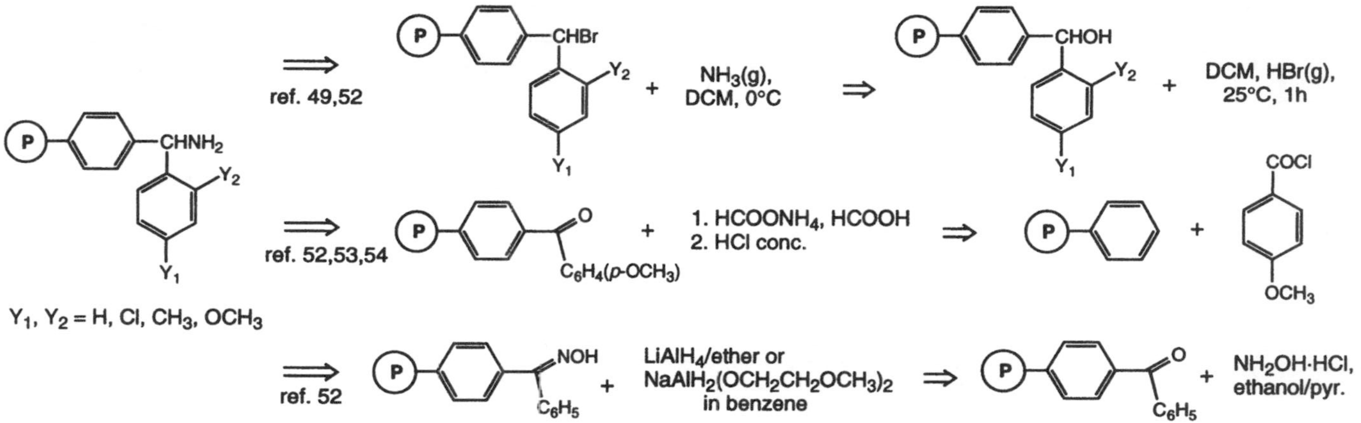
P
CHNH2
Y2
Y1
Y1, Y2 = H, Cl, CH3, OCH3
ref. 49,52
CHBr
NH3(g), DCM, 0°C
CHOH
DCM, HBr(g), 25°C, 1h
ref. 52,53,54
O
C6H4(p-OCH3)
1. HCOONH4, HCOOH
2. HCl conc.
COCl
OCH3
ref. 52
NOH
C6H5
LiAlH4/ether or NaAlH2(OCH2CH2OCH3)2 in benzene
NH2OH·HCl, ethanol/pyr.

Benzhydrylazide resin

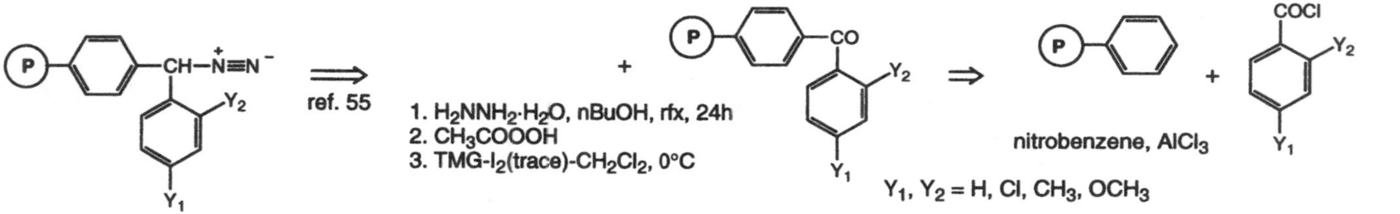
P
CH—N≡N
Y2
Y1
ref. 55
1. H2NNH2·H2O, nBuOH, rfx, 24h
2. CH3COOOH
3. TMG-I2(trace)-CH2Cl2, 0°C
CO
nitrobenzene, AlCl3
COCl
Y1, Y2 = H, Cl, CH3, OCH3

Benzhydrylhalo resin

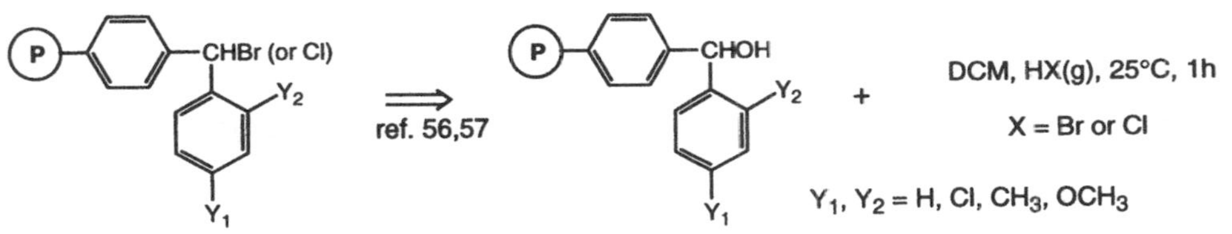

ref. 56,57

CHBr (or Cl), Y$_2$, Y$_1$ (P) $\Leftarrow$ CHOH, Y$_2$, Y$_1$ (P) + Y$_1$, Y$_2$ = H, Cl, CH$_3$, OCH$_3$

DCM, HX(g), 25°C, 1h

X = Br or Cl

Benzhydrylhydroxy resin

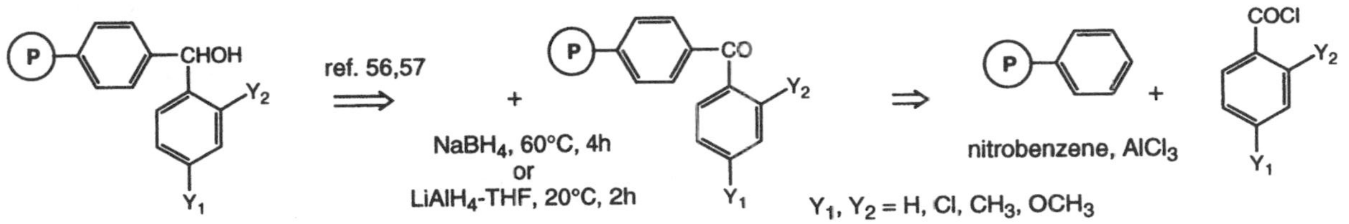

ref. 56,57

CHOH, Y$_2$, Y$_1$ (P) $\Leftarrow$

NaBH$_4$, 60°C, 4h
or
LiAlH$_4$-THF, 20°C, 2h

CO, Y$_2$, Y$_1$ (P) $\Leftarrow$ (P) + COCl, Y$_2$, Y$_1$

nitrobenzene, AlCl$_3$

Y$_1$, Y$_2$ = H, Cl, CH$_3$, OCH$_3$

3-(p-Benzyloxyphenyl)-1,1-dimethylpropyloxycarbonylhydrazide resin

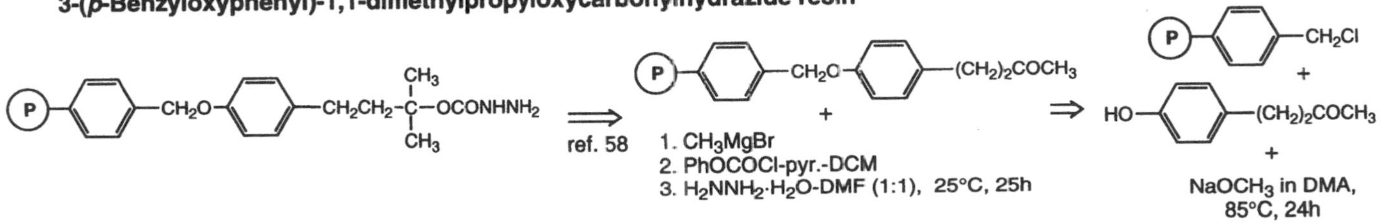

(P) CH$_2$O— —CH$_2$CH$_2$—C(CH$_3$)$_2$—OCONHNH$_2$

ref. 58

1. CH$_3$MgBr
2. PhOCOCl-pyr.-DCM
3. H$_2$NNH$_2$-H$_2$O-DMF (1:1), 25°C, 25h

(P) CH$_2$O— —(CH$_2$)$_2$COCH$_3$ + (P) CH$_2$O— —(CH$_2$)$_2$COCH$_3$

$\Leftarrow$ (P) CH$_2$Cl + HO— —(CH$_2$)$_2$COCH$_3$

NaOCH$_3$ in DMA, 85°C, 24h

9

Table 1 Continued

Benzylchloroformiate resin

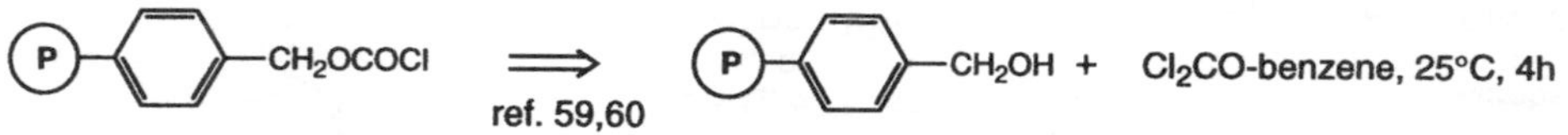
P
CH2OCOCl
ref. 59,60
P
CH2OH +
Cl2CO-benzene, 25°C, 4h

4-Bromo-2-butenoylaminomethyl resin (HYCRAM resin)

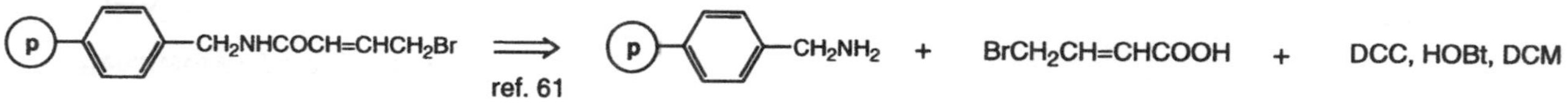
p
CH2NHCOCH=CHCH2Br
ref. 61
p
CH2NH2
+
BrCH2CH=CHCOOH
+
DCC, HOBt, DCM

4-Bromo-3-cyanophenyl(dimethyl)silylbenzyloxymethyl resin

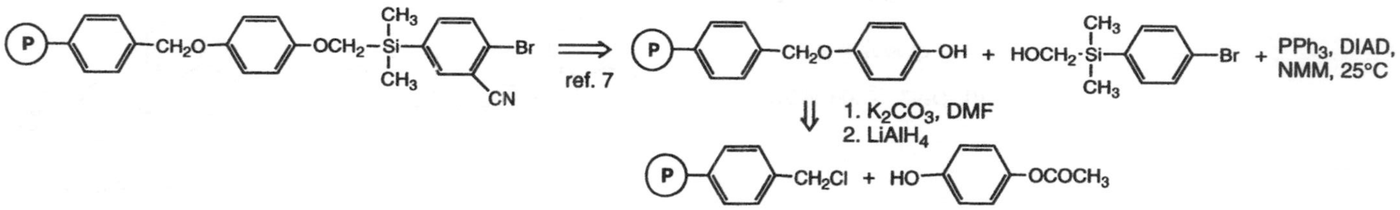
P
CH2O
OCH2–Si
CH3
CH3
Br
CN
ref. 7
P
CH2O
OH
+
HOCH2-Si
CH3
CH3
Br
+
PPh3, DIAD,
NMM, 25°C
1. K2CO3, DMF
2. LiAlH4
P
CH2Cl
+
HO
OCOCH3

Bromomethyl resin

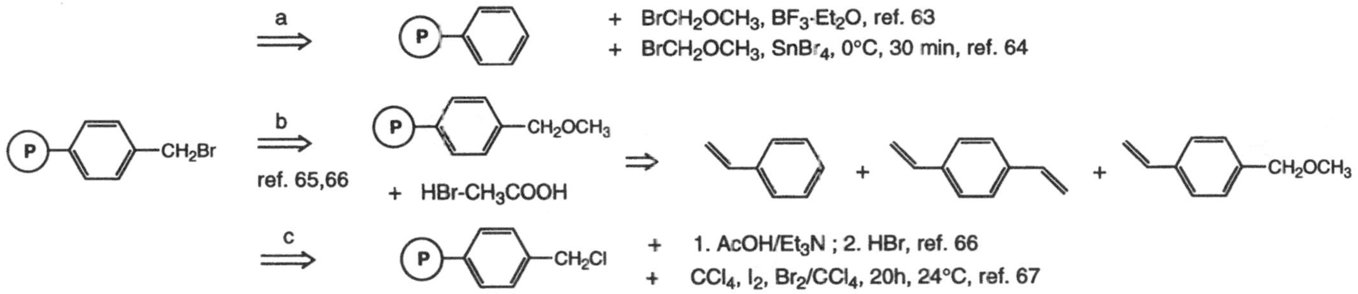
CH2Br
a
P
+ BrCH2OCH3, BF3·Et2O, ref. 63
+ BrCH2OCH3, SnBr4, 0°C, 30 min, ref. 64
b
ref. 65,66
CH2OCH3
+ HBr-CH3COOH
+
CH2OCH3
c
CH2Cl
+ 1. AcOH/Et3N ; 2. HBr, ref. 66
+ CCl4, I2, Br2/CCl4, 20h, 24°C, ref. 67

4-Bromomethyl-3-nitrobenzoylaminomethyl resin

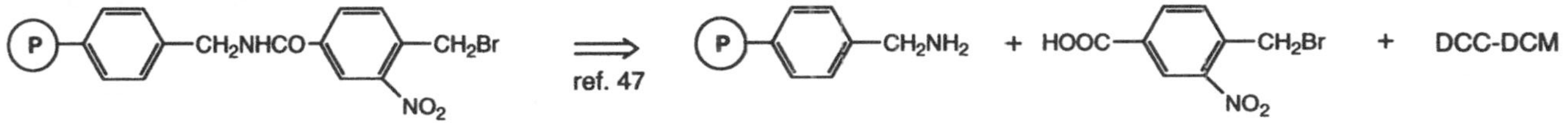
P
CH2NHCO
CH2Br
NO2
ref. 47
CH2NH2
+ HOOC
+ DCC-DCM

3-Bromomethyl-3-nitrobenzoyl-*N*-propylaminomethyl resin

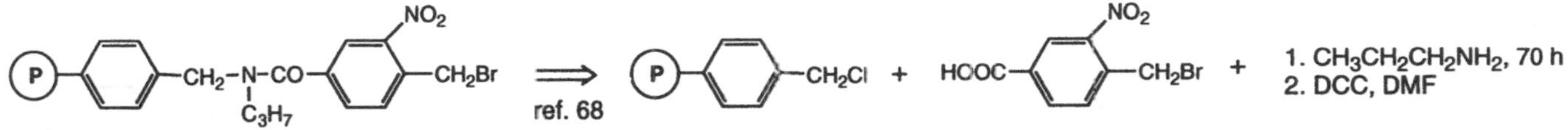
P
CH2–N–CO
C3H7
NO2
CH2Br
ref. 68
CH2Cl
+ HOOC
+
1. CH3CH2CH2NH2, 70 h
2. DCC, DMF

4-Bromophenyl(dimethyl)silylmethoxybenzoyloxymethyl resin

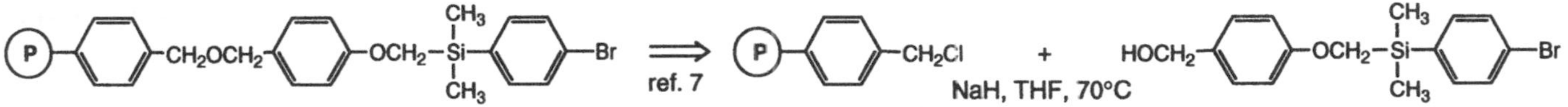
P
CH2OCH2
OCH2–Si
CH3
Br
ref. 7
CH2Cl
+
HOCH2
NaH, THF, 70°C

Table 1 Continued

2-Bromopropionyl resin (Bromo Wang resin)

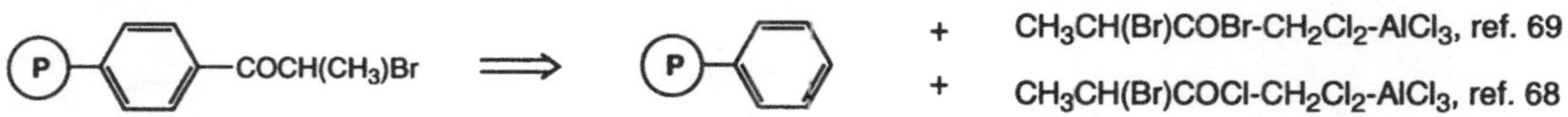

P
COCH(CH3)Br
P
+
CH3CH(Br)COBr-CH2Cl2-AlCl3, ref. 69
+
CH3CH(Br)COCl-CH2Cl2-AlCl3, ref. 68

Carbonylchoride resin

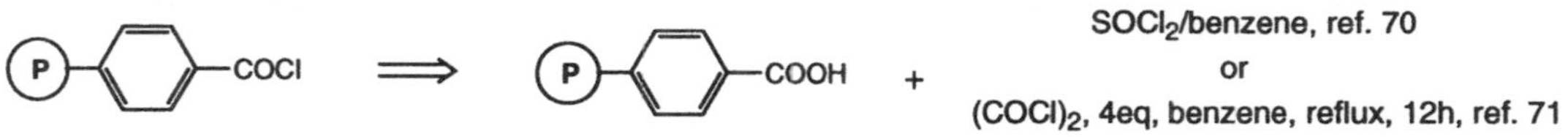

P
COCl
P
COOH
+
SOCl2/benzene, ref. 70
or
(COCl)2, 4eq, benzene, reflux, 12h, ref. 71

Carboxy resin

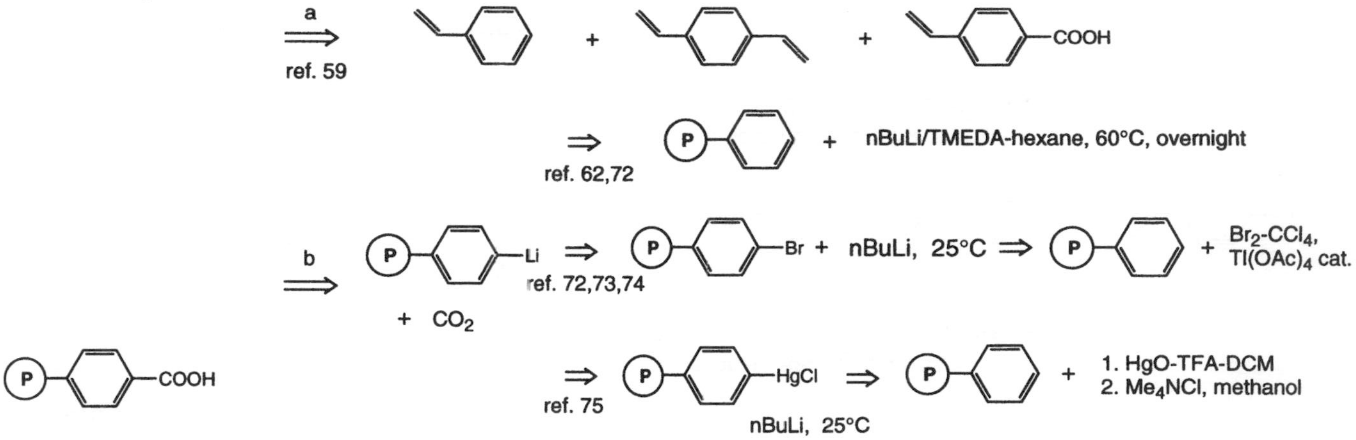

a
ref. 59
+
+
COOH
ref. 62,72
P
+
nBuLi/TMEDA-hexane, 60°C, overnight
b
P
Li
+ CO2
ref. 72,73,74
P
Br + nBuLi, 25°C
P
+
Br2-CCl4,
Tl(OAc)4 cat.
P
COOH
ref. 75
P
HgCl
nBuLi, 25°C
P
+
1. HgO-TFA-DCM
2. Me4NCl, methanol

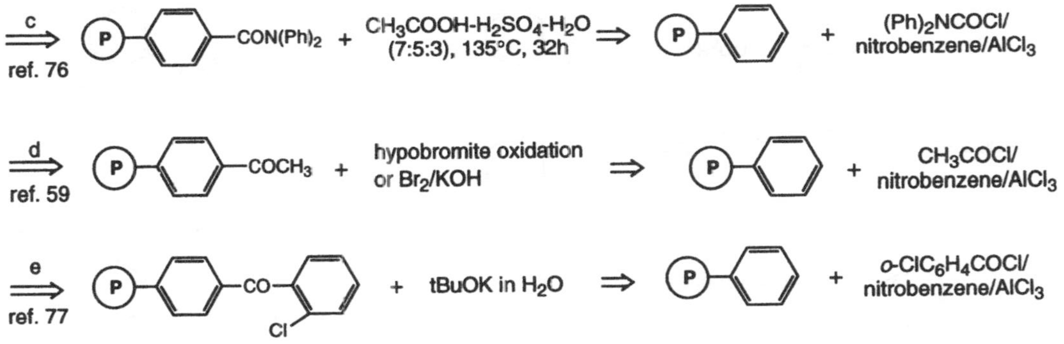

c ref. 76

P—CON(Ph)$_2$ + CH$_3$COOH-H$_2$SO$_4$-H$_2$O (7:5:3), 135°C, 32h ⟸ P—⟨benzene⟩ + (Ph)$_2$NCOCl/ nitrobenzene/AlCl$_3$

d ref. 59

P—COCH$_3$ + hypobromite oxidation or Br$_2$/KOH ⟸ P—⟨benzene⟩ + CH$_3$COCl/ nitrobenzene/AlCl$_3$

e ref. 77

P—CO-(o-Cl-phenyl) + tBuOK in H$_2$O ⟸ P—⟨benzene⟩ + o-ClC$_6$H$_4$COCl/ nitrobenzene/AlCl$_3$

Chloride triphenylphosphoniummethyl resin

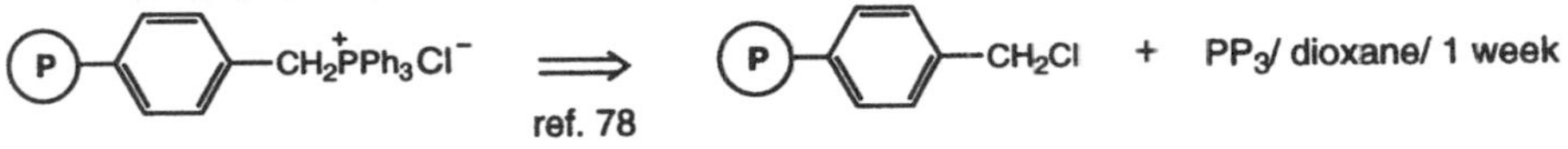

P—CH$_2$$\overset{+}{P}Ph_3$ Cl$^-$ ⟸ P—CH$_2$Cl + PP$_3$/ dioxane/ 1 week

ref. 78

Chloromethyl resin (Merrifield resin)

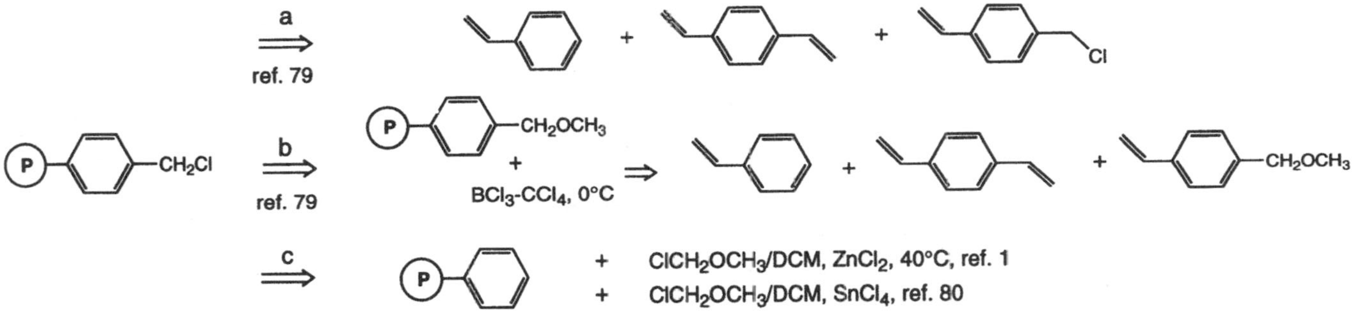

a ref. 79

P—CH$_2$OCH$_3$ + BCl$_3$-CCl$_4$, 0°C ⟸ P—⟨styrene⟩ + ⟨styrene⟩ + ⟨divinylbenzene⟩ + ⟨divinylbenzene⟩

b ref. 79

P—CH$_2$Cl ⟸ P—CH$_2$OCH$_3$ + ⟨styrene⟩ + ⟨divinylbenzene⟩ + ⟨chloromethyl-vinylbenzene⟩

c

P—⟨benzene⟩ + ClCH$_2$OCH$_3$/DCM, ZnCl$_2$, 40°C, ref. 1 + ClCH$_2$OCH$_3$/DCM, SnCl$_4$, ref. 80 ⟸ ⟨styrene⟩ + ⟨divinylbenzene⟩ + ⟨CH$_2$OCH$_3$-styrene⟩

13

Table 1 Continued

4-(Chloromethyl)phenylacetyl-(*N*-hexyl)aminomethyl resin

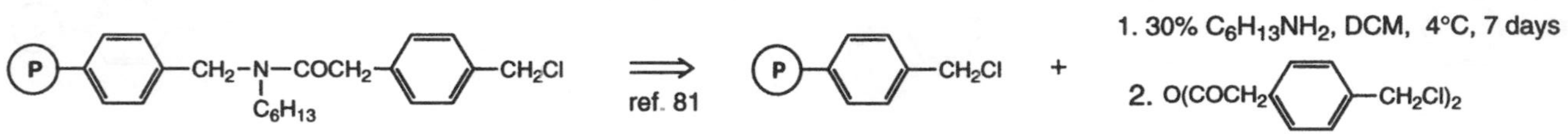
P
CH2-N-COCH2
C6H13
CH2Cl
ref. 81
P
CH2Cl
+
1. 30% C6H13NH2, DCM, 4°C, 7 days
2. O(COCH2
CH2Cl)2

Chlorosulphonyl resin

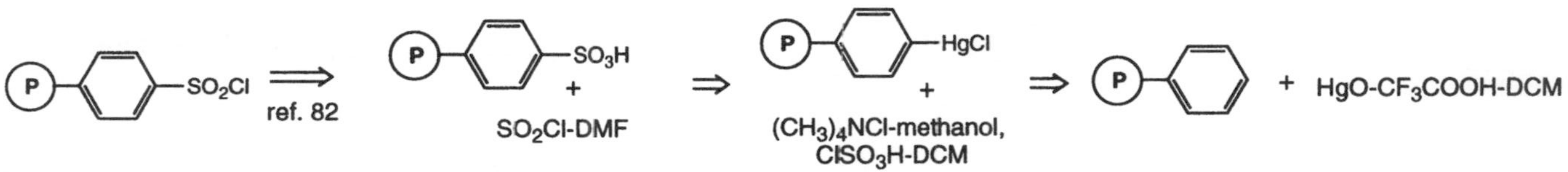
P
SO2Cl
ref. 82
P
SO3H
+
SO2Cl-DMF
P
HgCl
+
(CH3)4NCl-methanol,
ClSO3H-DCM
P
+ HgO-CF3COOH-DCM

4-[1-(1,2-dihydroxy-1-phenyl)ethyl]benzyl(acetylated)aminomethyl resin

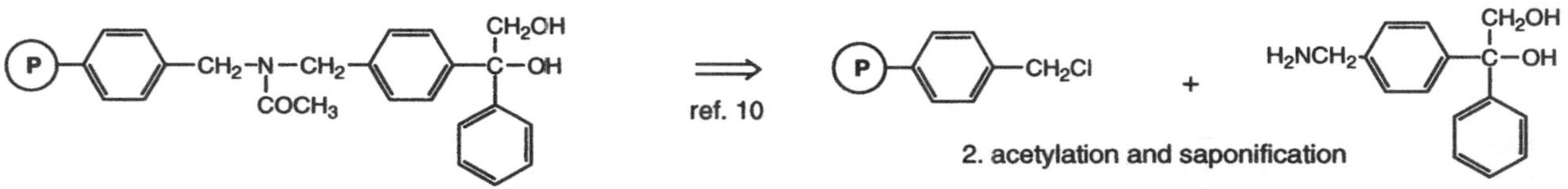
P
CH2-N-CH2
COCH3
CH2OH
C-OH
ref. 10
P
CH2Cl
+
H2NCH2
CH2OH
C-OH
2. acetylation and saponification

4-(2',4'-Dimethoxy-*N*-Fmoc-aminomethyl)phenoxyacetyl-2-chlorotrityl resin

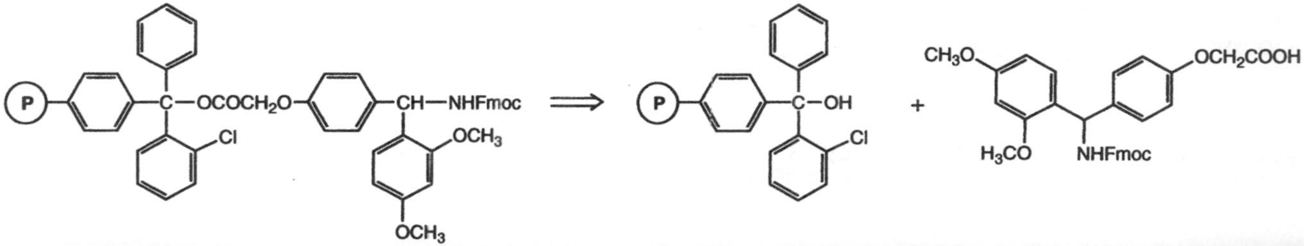
P
C-OCOCH2O
CH-NHFmoc
Cl
OCH3
OCH3
P
C-OH
Cl
+
CH3O
OCH2COOH
H3CO
NHFmoc

4-(2',4'-Dimethoxy-N-Fmoc-aminomethyl)phenoxyacetyl-MBHA resin

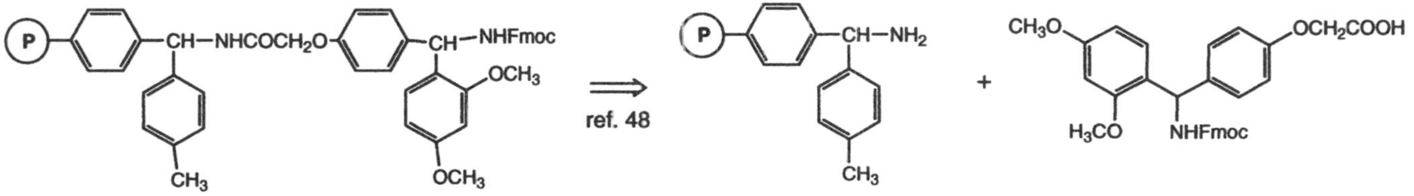

OCH$_2$COOH

NHFmoc

CH$_3$O

H$_3$CO

+

P — CH-NH$_2$ — CH$_3$

ref. 48

P — CH-NHCOCH$_2$O — CH-NHFmoc — OCH$_3$, OCH$_3$

CH-NHFmoc — CH$_3$

4-(2',4'-Dimethoxyphenylhydroxymethyl)phenoxymethyl resin (Rink acid resin)

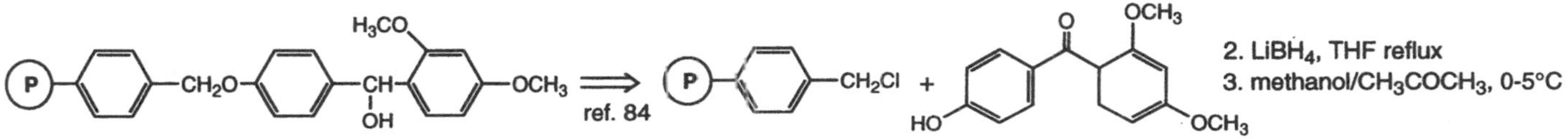

2. LiBH$_4$, THF reflux

3. methanol/CH$_3$COCH$_3$, 0-5°C

OCH$_3$

O

HO

P — CH$_2$Cl +

ref. 84

P — CH$_2$O — CH-OH — H$_3$CO, OCH$_3$

4-(2',4'-Dimethoxyphenyl-N-Fmoc-aminomethyl)phenoxymethyl resin (Rink amide resin)

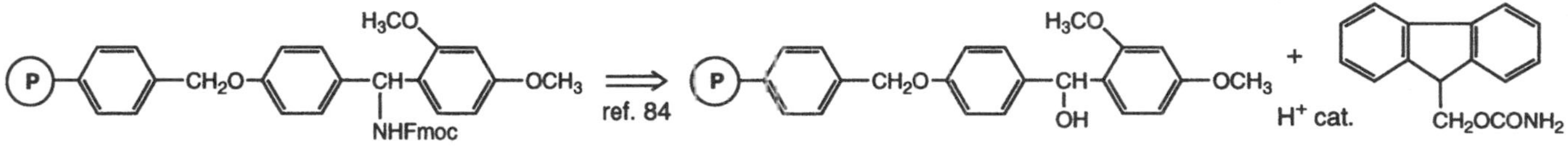

CH$_2$OCONH$_2$

+

H$^+$ cat.

OCH$_3$

P — CH$_2$O — CH-OH — H$_3$CO

ref. 84

P — CH$_2$O — CH-NHFmoc — H$_3$CO, OCH$_3$

1,1-Dimethyl-3-phenylpropyloxycarbonylhydrazide resin

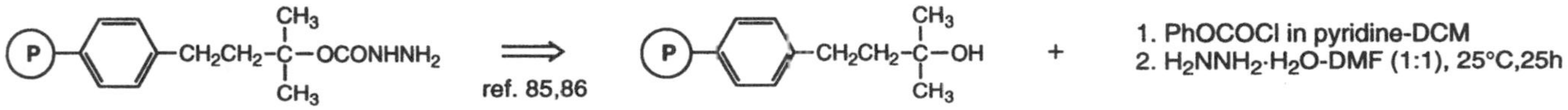

CH$_3$

P — CH$_2$CH$_2$-C-OH — CH$_3$

+

1. PhOCOCl in pyridine-DCM

2. H$_2$NNH$_2$·H$_2$O-DMF (1:1), 25°C, 25h

ref. 85,86

CH$_3$

P — CH$_2$CH$_2$-C-OCONHNH$_2$ — CH$_3$

Table 1 Continued

(5-Fluoro-2,4-dinitrophenyl)aminoacetylhydroxymethyl resin

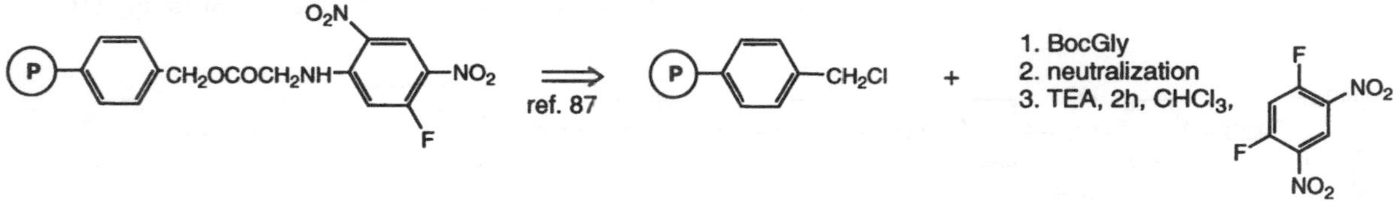
O2N
CH2OCOCH2NH
NO2
F
ref. 87
CH2Cl
1. BocGly
2. neutralization
3. TEA, 2h, CHCl3,
F
NO2
F
NO2

9-Fmoc-amino-xanthen-3-yloxymethyl resin (Sieber resin)

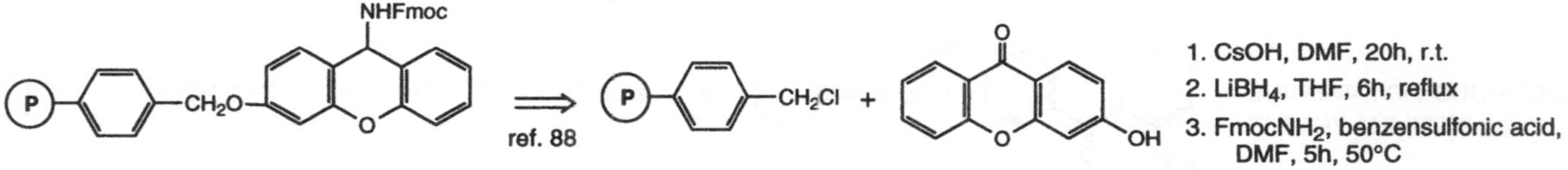
NHFmoc
CH2O
O
ref. 88
CH2Cl
O
O
OH
1. CsOH, DMF, 20h, r.t.
2. LiBH4, THF, 6h, reflux
3. FmocNH2, benzensulfonic acid, DMF, 5h, 50°C

Formyl resin

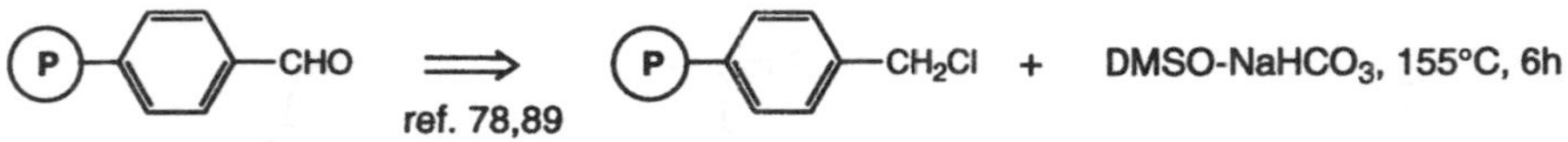
CHO
ref. 78,89
CH2Cl
DMSO-NaHCO3, 155°C, 6h

Haloacetyl resin

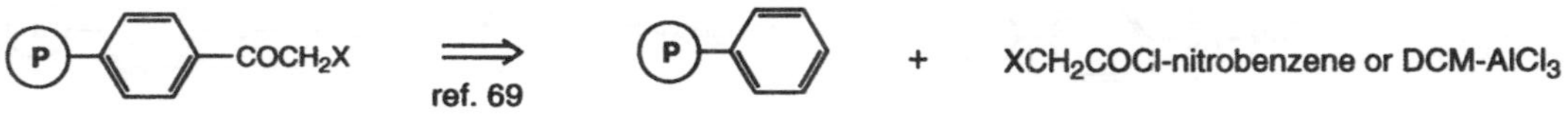
COCH2X
ref. 69
XCH2COCl-nitrobenzene or DCM-AlCl3

X = Cl or Br

Hydroxyalkyl(acetyl)aminomethyl resin

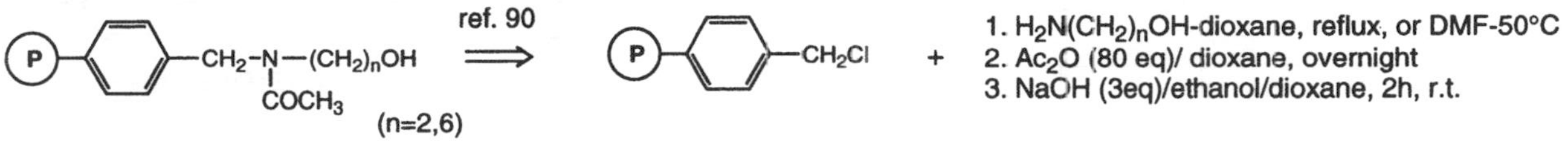

$$\text{(P)}-\text{CH}_2\text{Cl} \quad + \quad \xrightarrow{\text{ref. 90}}$$

1. $H_2N(CH_2)_nOH$-dioxane, reflux, or DMF-50°C
2. Ac_2O (80 eq)/ dioxane, overnight
3. NaOH (3eq)/ethanol/dioxane, 2h, r.t.

$$\text{(P)}-\text{CH}_2-\text{N}-(\text{CH}_2)_n\text{OH}$$
$$\quad\quad\quad\quad | $$
$$\quad\quad\quad\text{COCH}_3 \quad (n=2,6)$$

4-Hydroxymethylbenzoylaminomethyl resin

$$\text{(P)}-\text{CH}_2\text{NHCO}-\text{CH}_2\text{OH} \quad \xLeftarrow{\text{ref. 91}} \quad \text{(P)}-\text{CH}_2\text{NH}_2 \quad + \quad \text{HOOC}-\text{CH}_2\text{OH}$$

2-Hydroxy-2-methylbutyl resin

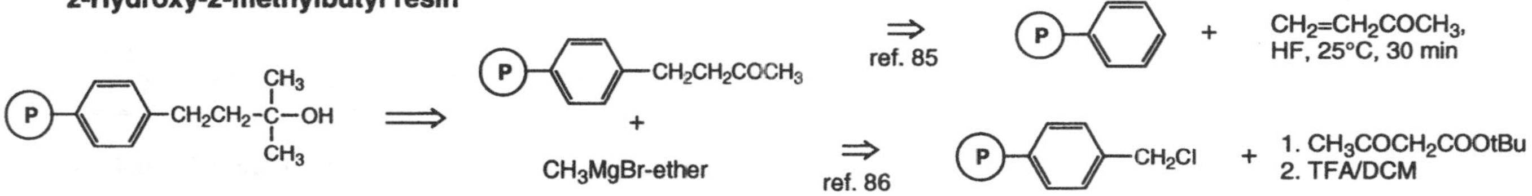

$$\text{(P)}-\text{CH}_2\text{CH}_2-\overset{\text{CH}_3}{\underset{\text{CH}_3}{\text{C}}}-\text{OH} \quad \xLeftarrow{\text{ref. 85}} \quad \text{(P)}-\text{CH}_2\text{CH}_2\text{COCH}_3 \quad + \quad \text{CH}_3\text{MgBr-ether}$$

$$\xLeftarrow{\text{ref. 86}}$$

$$\text{(P)}- \quad + \quad \text{(P)}-\text{CH}_2\text{Cl} \quad + \quad \text{CH}_3\text{COCH}_2\text{COOtBu}$$

$$CH_2=CH_2COCH_3,$$
$$HF, 25°C, 30 \text{ min}$$

1. $CH_3COCH_2COOtBu$
2. TFA/DCM

4-Hydroxymethyl-3-methoxyphenoxybutyryl-benzydrylamino resin (X=H, MPPB-BHA; X=CH₃, HMPPB-MBHA)

$$\text{(P)}-\text{CH}-\text{NHCO(CH}_2)_3\text{O}-\text{CH}_2\text{OH (OCH}_3)$$

X = H, CH₃

$$\xLeftarrow{\text{ref. 92}}$$

$$\text{(P)}-\text{CH}-\text{NH}_2 \quad + \quad \text{HOCO(CH}_2)_3\text{O}-\text{CH}_2\text{OH (OCH}_3) \quad + \quad \text{DCC-DCM}$$

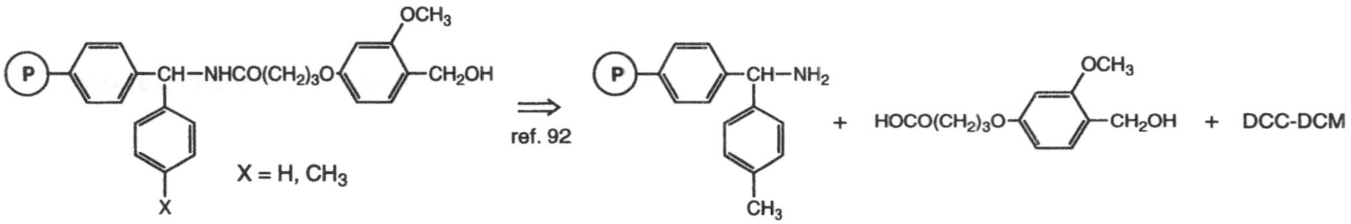

Table 1 Continued

4-Hydroxymethyl-3-methoxyphenoxymethyl resin (Sasrin resin)

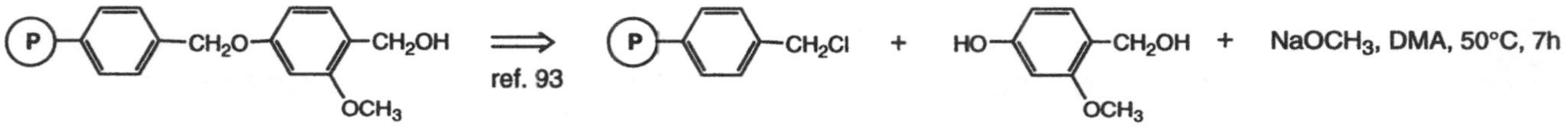

4-Hydroxymethylphenylacetylaminomethyl resin

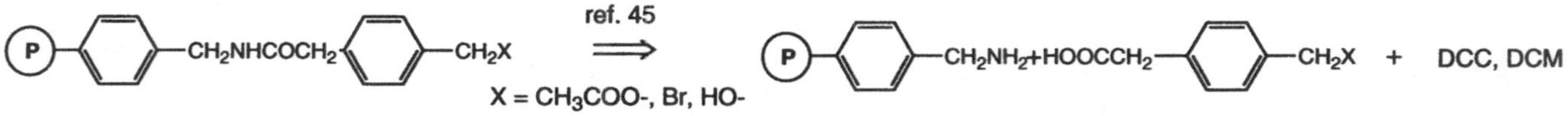

Hydroxy resin

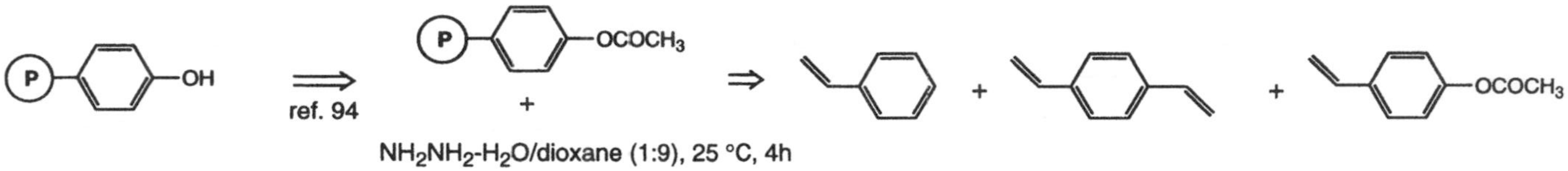

Hydroxybutyryl resin

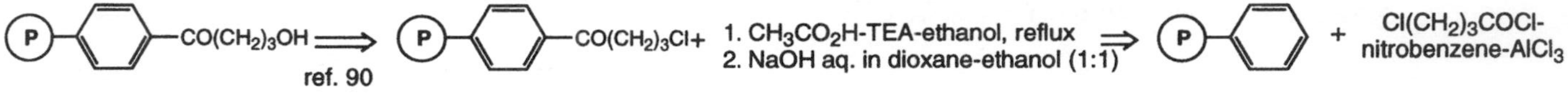

Hydroxyethyl resin

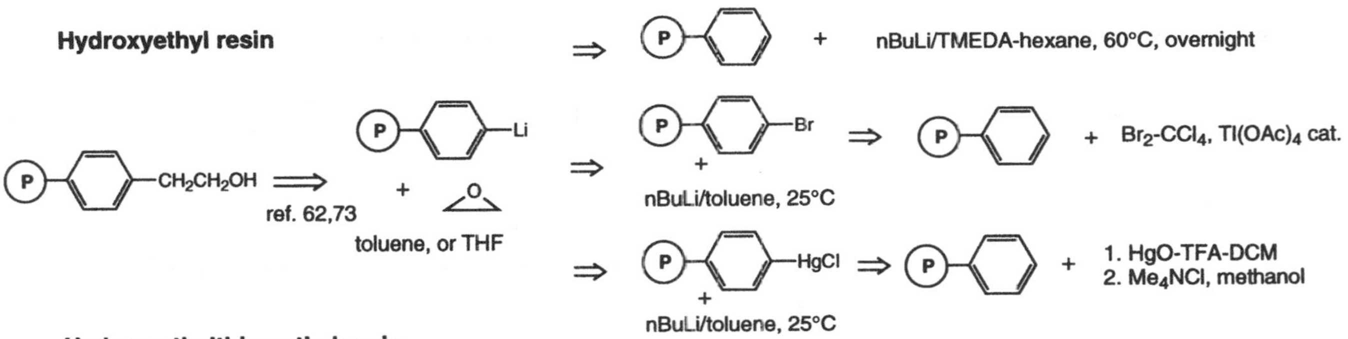

P⟶CH$_2$CH$_2$OH ⟸ P⟶ + ⟵

ref. 62,73

toluene, or THF

P⟶Li + O (epoxide)

P + ⟸ P⟶ +

nBuLi/TMEDA-hexane, 60°C, overnight

P⟶Br + + Br$_2$-CCl$_4$, Tl(OAc)$_4$ cat.

nBuLi/toluene, 25°C

P⟶HgCl + + 1. HgO-TFA-DCM
2. Me$_4$NCl, methanol

nBuLi/toluene, 25°C

Hydroxyethylthiomethyl resin

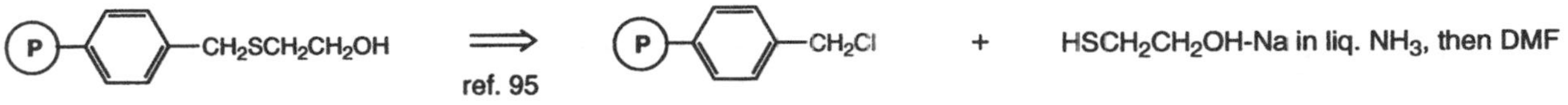

P⟶CH$_2$SCH$_2$CH$_2$OH ⟸ P⟶CH$_2$Cl + HSCH$_2$CH$_2$OH-Na in liq. NH$_3$, then DMF

ref. 95

Hydroxyhexyl resin

P⟶(CH$_2$)$_6$OH ⟸ P⟶(CH$_2$)$_5$COOH + LiAlH$_4$-ether

ref. 96

P⟶(CH$_2$)$_5$COOH ⟸ P⟶CO(CH$_2$)$_4$COOCH$_3$ + NH$_2$NH$_2$·H$_2$O-KOH in hot PEG

P⟶CO(CH$_2$)$_4$COOCH$_3$ ⟸ P + ClCO(CH$_2$)$_4$COCH$_3$, AlCl$_3$

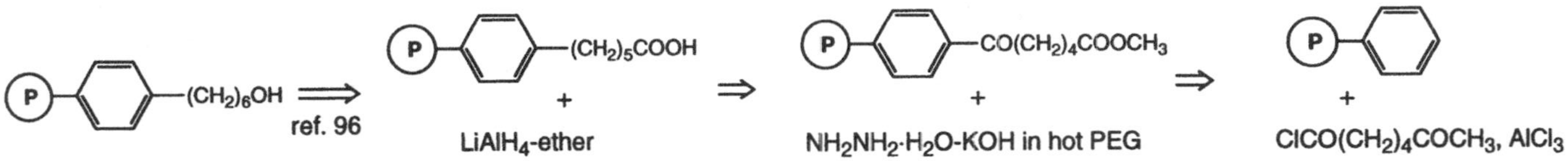

Hydroxylamine-Wang resin

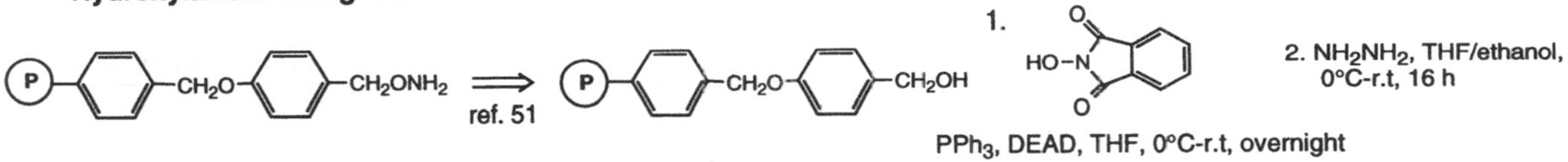

P⟶CH$_2$O⟶CH$_2$ONH$_2$ ⟸ P⟶CH$_2$O⟶CH$_2$OH

ref. 51

1. HO-N (phthalimide), PPh$_3$, DEAD, THF, 0°C-r.t, overnight

2. NH$_2$NH$_2$, THF/ethanol, 0°C-r.t, 16 h

19

Table 1 Continued

Hydroxymethyl resin

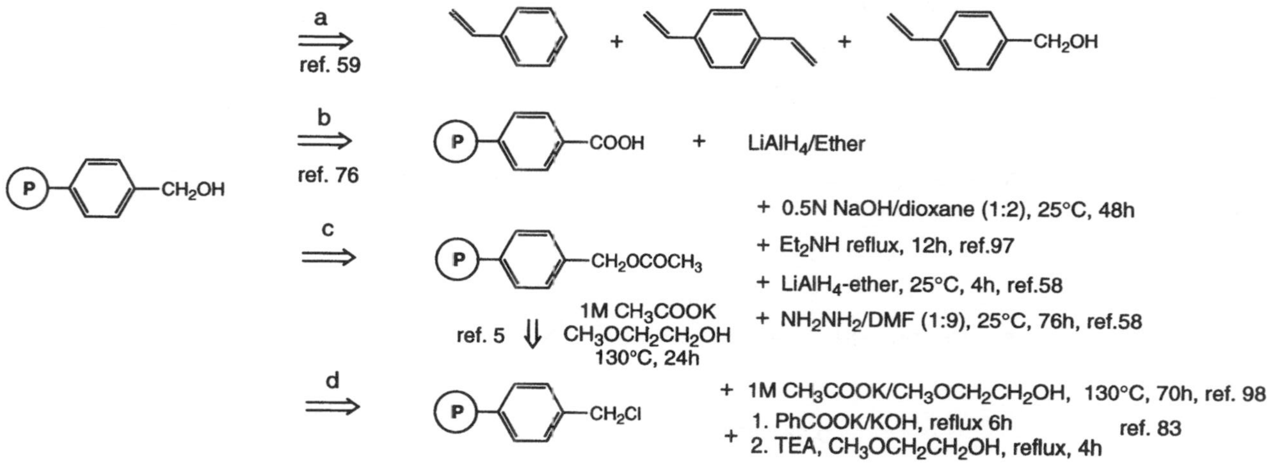
CH2OH
a
ref. 59
+
+
CH2OH
b
ref. 76
P
COOH
+
LiAlH4/Ether
c
P
CH2OCOCH3
+ 0.5N NaOH/dioxane (1:2), 25°C, 48h
+ Et2NH reflux, 12h, ref.97
+ LiAlH4-ether, 25°C, 4h, ref.58
+ NH2NH2/DMF (1:9), 25°C, 76h, ref.58
ref. 5
1M CH3COOK
CH3OCH2CH2OH
130°C, 24h
d
P
CH2Cl
+ 1M CH3COOK/CH3OCH2CH2OH, 130°C, 70h, ref. 98
1. PhCOOK/KOH, reflux 6h
2. TEA, CH3OCH2CH2OH, reflux, 4h
ref. 83

Hydroxymethylphenoxymethyl resin (Wang resin)

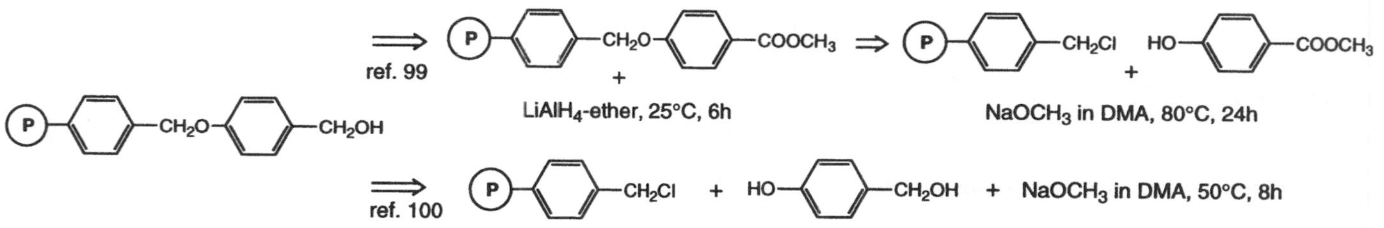
P
CH2O
CH2OH
ref. 99
CH2O
COOCH3
CH2Cl
HO
COOCH3
+
LiAlH4-ether, 25°C, 6h
NaOCH3 in DMA, 80°C, 24h
ref. 100
CH2Cl
+
HO
CH2OH
+ NaOCH3 in DMA, 50°C, 8h

Hydroxyphenylacetoxymethyl resin

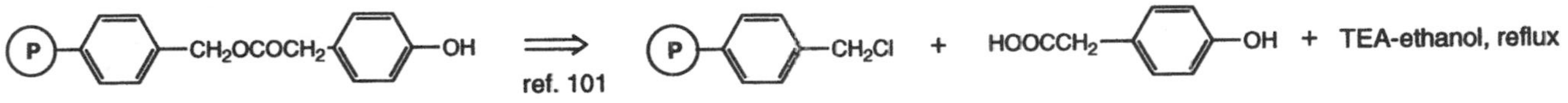
P
CH2OCOCH2
OH
ref. 101
P
CH2Cl
+
HOOCCH2
OH
+ TEA-ethanol, reflux

Hydroxyphenylthiomethyl resin

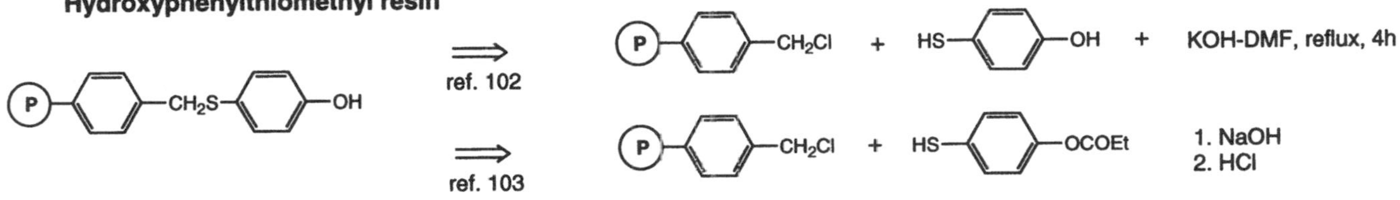
P
CH2S
OH
ref. 102
ref. 103
P
CH2Cl
+
HS
OH
+
KOH-DMF, reflux, 4h
P
CH2Cl
+
HS
OCOEt
1. NaOH
2. HCl

***o*-Bromo-chloromethyl resin**

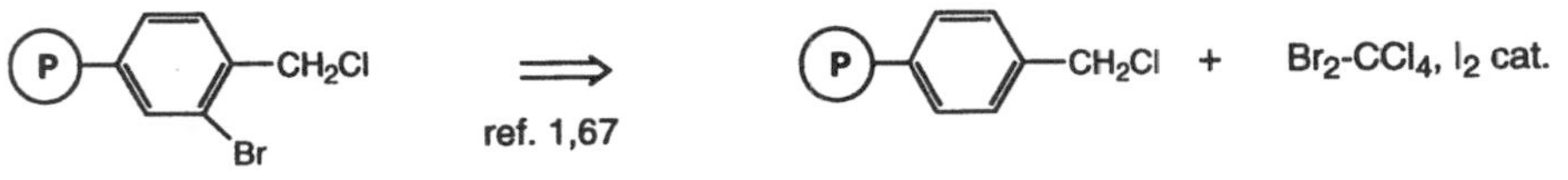
P
CH2Cl
Br
ref. 1,67
P
CH2Cl
+
Br2-CCl4, I2 cat.

***o*-Nitro-chloromethyl resin**

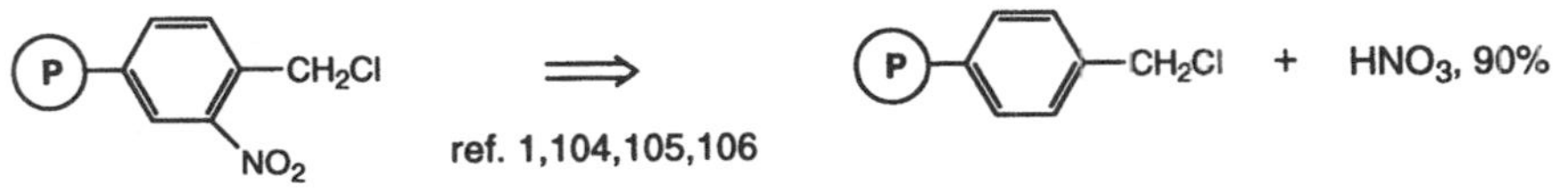
P
CH2Cl
NO2
ref. 1,104,105,106
P
CH2Cl
+
HNO3, 90%

Table 1 Continued

p-Nitrobenzophenone oxime resin (Kaiser oxime resin)

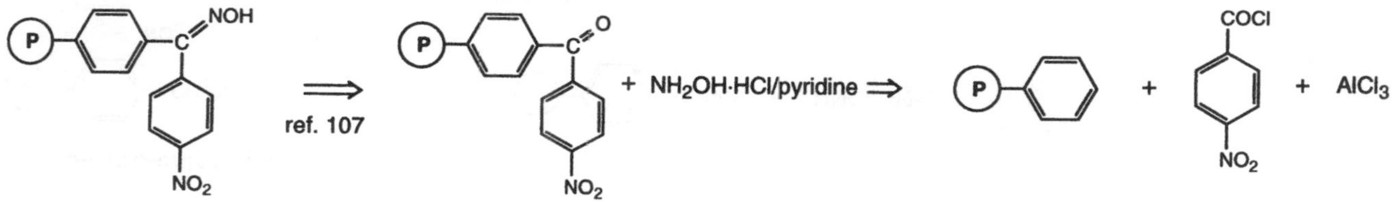

4-Nitrophenylcarbonatemethyl resin

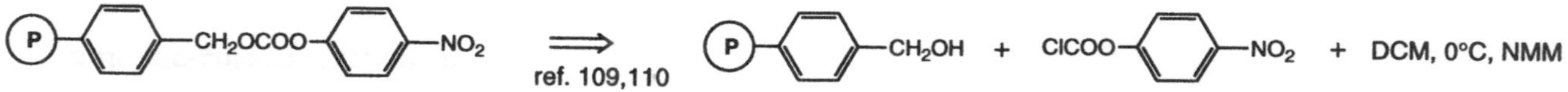

p-Nitrophenylcarbonate-Wang resin

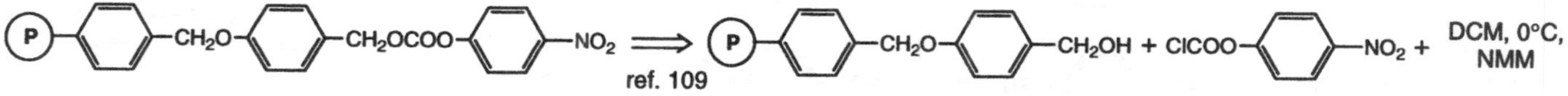

p-Benzyloxyphenylmethyloxycarbonylhydrazide resin

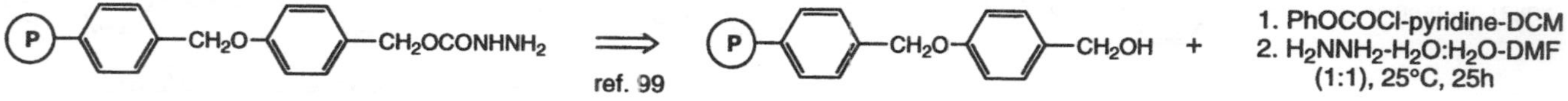

Succinyl aminomethyl PS resin carbamate

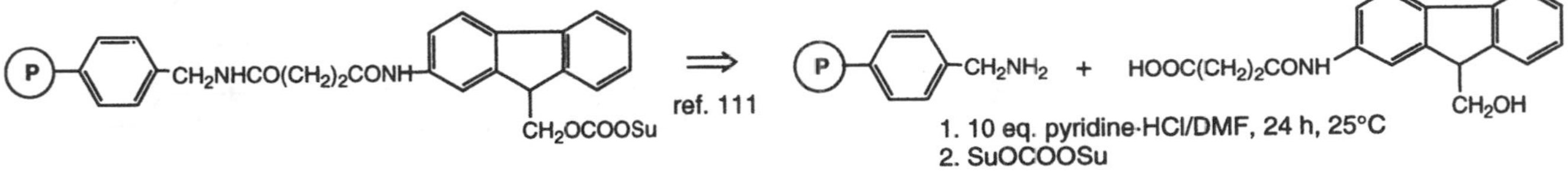

$CH_2NHCO(CH_2)_2CONH$—

ref. 111

CH_2NH_2 + $HOOC(CH_2)_2CONH$—

CH_2OH

$CH_2OCOOSu$

1. 10 eq. pyridine·HCl/DMF, 24 h, 25°C
2. SuOCOOSu

Succinimiyl 4-(benzylaminocarbonylethoxy)phenylmethylcarbonate resin

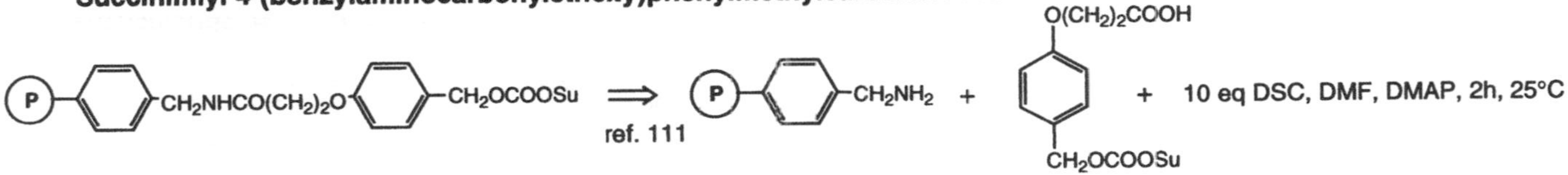

$CH_2NHCO(CH_2)_2O$—$CH_2OCOOSu$

ref. 111

CH_2NH_2 + $O(CH_2)_2COOH$

+ 10 eq DSC, DMF, DMAP, 2h, 25°C

$CH_2OCOOSu$

4-Sulfamylbenzoyl-4'-methylbenzhydrylamine resin

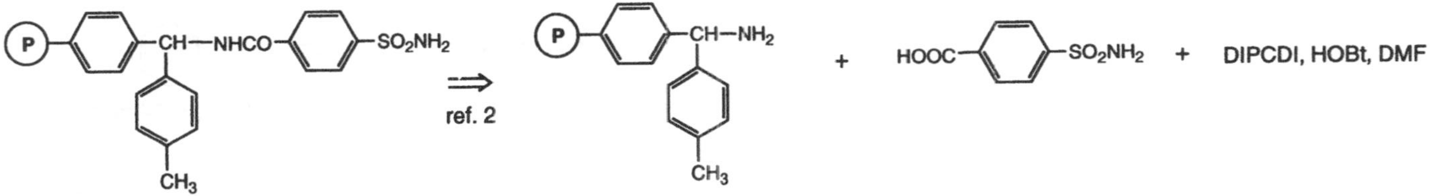

CH—$NHCO$—SO_2NH_2

CH_3

ref. 2

CH—NH_2

CH_3

+ $HOOC$—SO_2NH_2 + DIPCDI, HOBt, DMF

4-Sulfamylbenzoylaminomethyl resin

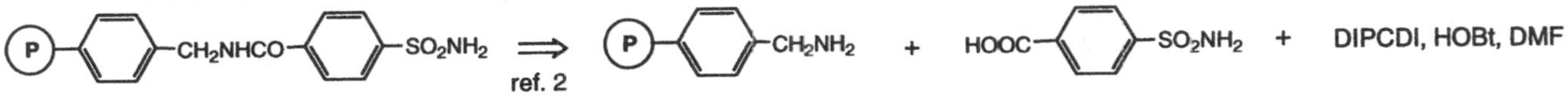

CH_2NHCO—SO_2NH_2

ref. 2

CH_2NH_2 + $HOOC$—SO_2NH_2 + DIPCDI, HOBt, DMF

Table 1 Continued

Sulfonamide resin

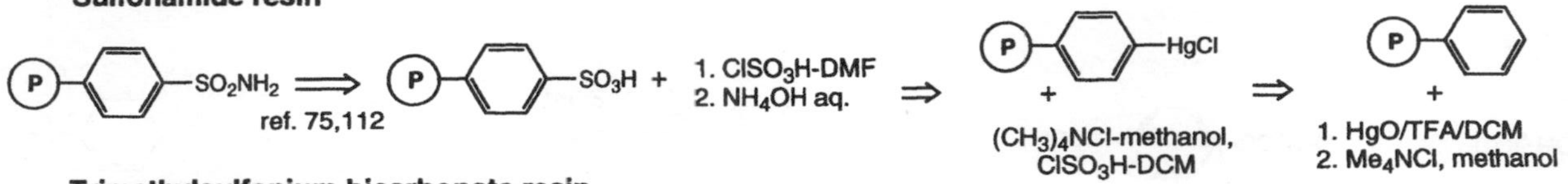
P
SO₂NH₂
ref. 75,112
SO₃H +
1. ClSO₃H-DMF
2. NH₄OH aq.
HgCl
+
(CH₃)₄NCl-methanol,
ClSO₃H-DCM
+
1. HgO/TFA/DCM
2. Me₄NCl, methanol

Trimethylsulfonium bicarbonate resin

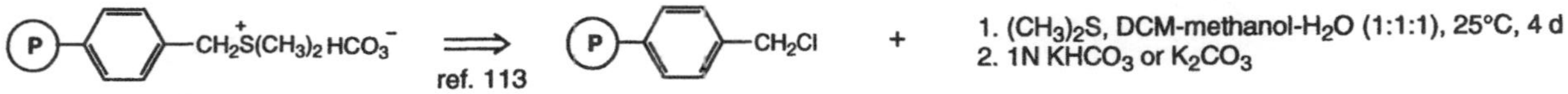
P
CH₂S⁺(CH₃)₂ HCO₃⁻
ref. 113
CH₂Cl
+
1. (CH₃)₂S, DCM-methanol-H₂O (1:1:1), 25°C, 4 d
2. 1N KHCO₃ or K₂CO₃

Tritylalcohol resin

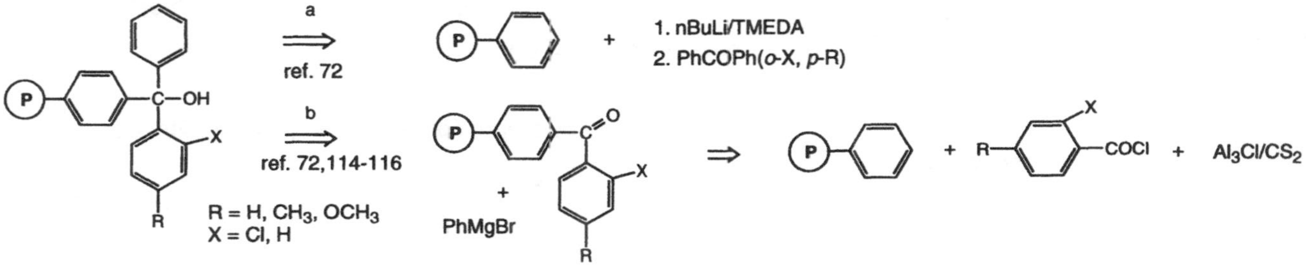
P
C–OH
X
R
a
ref. 72
b
ref. 72,114-116
R = H, CH₃, OCH₃
X = Cl, H
+
1. nBuLi/TMEDA
2. PhCOPh(o-X, p-R)
PhMgBr
COCl
+
Al₃Cl/CS₂

Tritylchloride derivatives resins

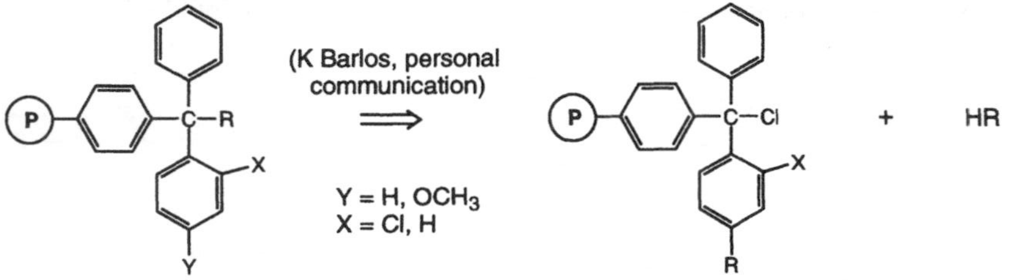
P
C–R
X
Y
(K Barlos, personal
communication)
Y = H, OCH₃
X = Cl, H
C–Cl
R
+
HR

R = piperazine, $HN(CH_2)_nNH_2$, $S(CH_2)_2NH_2$, $O(CH_2)_2NH_2$, $NHNH_2$, $OCO(o,p\text{-}NO_2C_6H_3)$, SH, $O(CH_2)_2OH$

Tritylchloride resin

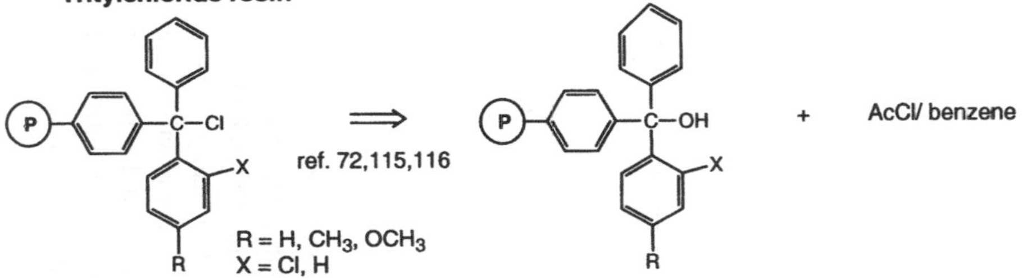

+ AcCl/ benzene

ref. 72,115,116

R = H, CH$_3$, OCH$_3$
X = Cl, H

Vinylcarbonyloxymethyl resin (REM resin)

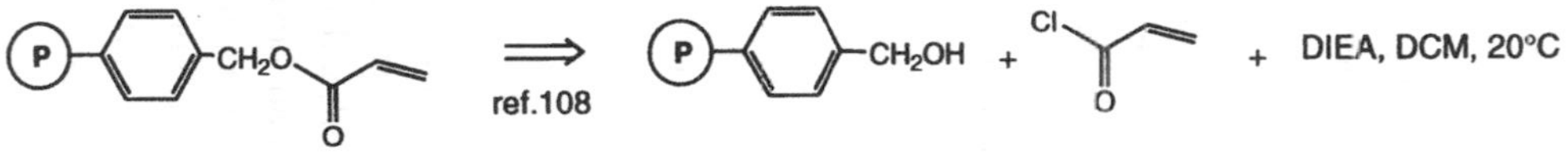

+ DIEA, DCM, 20°C

ref.108

[4-(2-Bromopropionyl)phenoxy]acetylaminomethyl resin

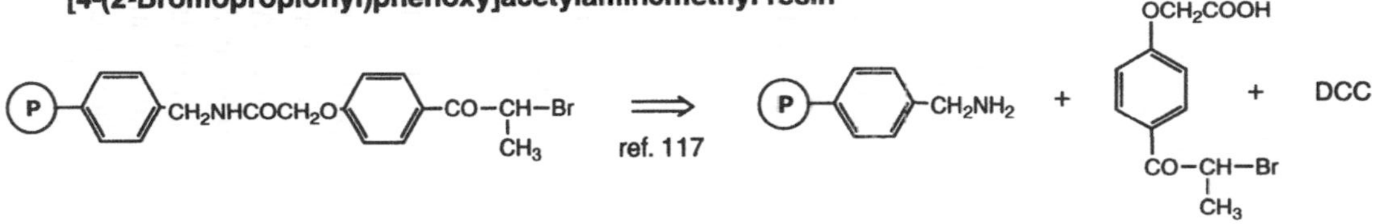

DCC

ref. 117

[4-(2-Hydroxypropionyl)-2-methoxy-5-nitrophenoxy]butyrylaminomethyl resin

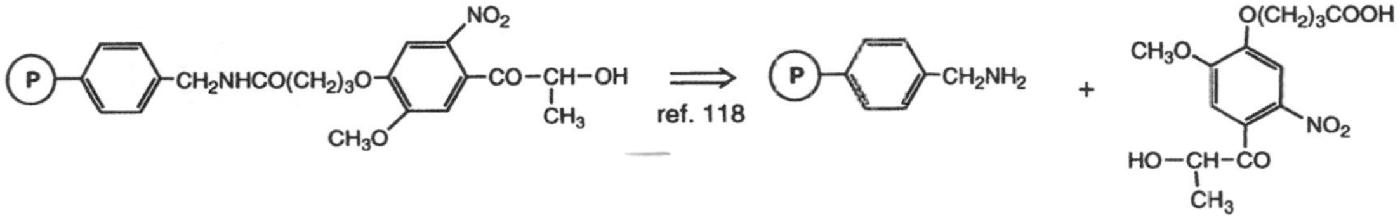

ref. 118

route has not been widely applied, because functionalized monomers are occasionally difficult to synthesize in pure form and few laboratories have the experience to carry out the actual polymerization. The substitution level and the extent of cross-linking in the product polymer can be controlled by the composition of the starting polymerization mixture. When the reactivity ratio of the monomers is near unity, the functional groups are anticipated to be randomly distributed throughout the linear structure of the polymer. In this way, the hypothetical "local clustering" of sites that might occur during the course of direct derivatization can be avoided. On the other hand, it is conceivable that some of the functional groups will be found in sterically unfavorable positions of the three-dimensional polymeric matrix, i.e., adjacent to cross-links, and thus will subsequently react more slowly.

The direct derivatization approach has been most commonly used for solid-phase methodology. More specifically, aromatic substitutions of the styrene ring (Table 1) and Friedel–Crafts alkylations (Table 1) or acylations (Table 1) have served as the principal approaches for the initial introduction of functional groups into the polystyrene resins most commonly used in solid-phase synthesis. A major problem with many of these methods is the potential for introduction of new cross-links in the polymeric matrix, such as methylene or sulfone bridges during chloromethylation or sulfonation, respectively. This problem is expected to be accentuated by longer reaction times, more active Friedel–Crafts catalysts, more reactive functional groups (e.g., bromomethyl > chloromethyl), and higher overall levels of functionalization. Additional cross-linking can also arise during further transformations, such as conversion of chloromethyl sites to aminomethyl sites by treatment with ammonia [46].

Table 1 lists some of the polystyrene resins that have been used in solid-phase synthesis and also includes a schematic retrosynthesis. Before we describe these syntheses in detail [119], it is important the proper methods for handling resins be reviewed.

1. General Procedure for Handling Resins

Beads of 1% cross-linked polystyrene are fragile and can easily be broken into fragments. Never use any form of stirring in which resin beads are rubbed between moving surfaces. An overhead stirrer with a Teflon blade is the best option. Large batches of resin can best be dried by using suction on a filter only until free solvent is removed, as further suction will cause the resin to become very cold because of solvent evaporation and hence wet from moisture in the air. Allow the resin to air dry overnight or until there is no perceptible DCM odor. The resin may then be dried in a vacuum desiccator for several hours.

The degree of cross-linking of polystyrene beads can be estimated by measuring the volume to which the beads swell in DCM [11]. Place 200 mg of resin in a 10-mL graduated cylinder and add 8 mL of DCM. Cover and allow to stand for 30 min. Resin that is 1% cross-linked should occupy a 2-mL layer at the top of the solvent. More highly cross-linked resins do not swell as efficiently.

2. General Resin Washing Procedure

Place the resin in a fritted glass Büchner funnel and wash it with a succession of solvents adequate to remove all reactants and by-products. Eight to ten aliquots should be used for each solvent, using a glass rod to stir each aliquot with the resin and allowing the solvent to equilibrate at least 1 min before removing it by suction filtration.

3. Removal of Linear Polystyrene

Place 100 g of resin in a 1 L round-bottom flask equipped with an overhead stirrer and a Teflon blade that does not scrape the bottom of the flask. Add 500 mL trifluoroacetic acid (TFA) and stir overnight. An alternative method is to use a three-neck flask with a reflux condenser and heat with stirring in an 80°C oil bath for 1 h, then cool. Wash the resin with TFA and DCM and air dry overnight.

4. Removal of Fine Resin Particles

Very fine particles of resin (often produced by fragmentation of beads by stirring or grinding) can plug the fritted disk in some synthesis reaction vessels. These fines can be removed by resin flotation. A homogeneous resin, when suspended in DCM, rises to the top, leaving a sharp line of demarcation at the bottom of the floating resin. If the resin contains fine particles, they will rise more slowly and the line of demarcation will not be sharp. Swell the resin in DCM, using about 20 mL/g resin in a separatory funnel. Allow the funnel to stand until most of the resin has risen to the top and a line of demarcation appears at the bottom of the resin. Drain the solvent, carrying the suspended fine particles of resin, out of the funnel and discard. Repeat this flotation three more times, each time mixing and allowing the resin to stand until a fairly sharp demarcation line appears at the bottom of the floating resin. When no particles remain below rising more slowly, the resin is homogeneous. Return the resin to the Büchner funnel, filter, and dry. If the resin has much higher density than the DCM, CCl_4 can be used.

The resin is now ready to be used for further derivatization or for synthesis. What follows is a description of the most commonly used methods for *functionalization* of polystyrene resins. Further derivatization of the resins, such as attachment of handles, is usually performed on these functionalized resins. Several of these derivatizations are also described.

5. Preparation of Chloromethyl Resin

Swell 50 g of copoly(styrene–1% divinylbenzene) resin beads by stirring at 25°C for 1 h in 200 mL of $CHCl_3$ in a three-neck round-bottom flask; then cool to 0°C. To the stirring mixture add a cold solution of 3.8 mL anhydrous $SnCl_4$ in 100 mL of chloromethyl methyl ether from a dropping funnel. Continue stirring for 30 min at 0°C. Filther the mixture on a fritted glass Büchner funnel and wash with 1 L of water–dioxane (1:3) followed by 1 L of 3 N HCl–dioxane (1:3), allowing brief periods for the wash solvent to soak into the beads. Wash the beads thoroughly with DCM, allowing time for solvent to penetrate the beads. Spread out the resin to air dry overnight; then dry over $CaCl_2$ under high vacuum. **Warning:** Chloromethyl methyl ether is carcinogenic and must be used with extreme care. It should be handled only in a well-ventilated hood. Avoid inhalation of vapor and all contact with liquid or solutions.

To determine the degree of chloromethylation, heat an aliquot of the resin (200 mg) in 3 mL pyridine in a test tube for 2 h at 100°C. Transfer the mixture quantitatively to a 125-mL Erlenmeyer flask with 30 mL of 50% acetic acid (HOAc), and add 5 mL of concentrated HNO_3. Analyze the liberated chloride by the modified Volhard method (see Section VII), omitting the addition of HNO_3 and adding only 5 mL of $AgNO_3$ (0.1 N). The preceding conditions for chloromethylation have yielded 0.9 to 2.0 mmol Cl/g resin. The degree of chloromethylation can be controlled by changing the amount of $SnCl_4$ used or the time and temperature of the reaction; 0.75 mmol Cl/g resin may be ideal for most solid-phase syntheses.

6. Preparation of Bromomethyl Resin

Stir 5 g chloromethyl resin (1.0 mmol Cl/g) in 40 mL CCl_4 for 10 min. Add 165 mg iodine and a solution of 8 mL bromine in 17 mL CCl_4 and stir the mixture for 20 h at 24°C in the dark. Filter the resin; wash with 150 mL each of dioxane, water, 1 M $NaHCO_3$, water, *N,N*-dimethylformamide (DMF), methanol, and DCM; and dry. The resin weight gain should be ~65%. The theoretical weight gain for monobromination of all aromatic rings is 71%.

7. Preparation of Hydroxymethyl Resin

Cover chloromethyl resin (0.5 to 1 mmol Cl/g resin; 1 equivalent Cl) and potassium acetate (10 equiv.) with methyl cellosolve (approximately 6 mL/

g resin) and heat with stirring in an oil bath at 125–135°C for 24 h. Carry out the reaction in a 24/40 round-bottom flask fitted with a 50-cm water condenser carrying a $CaCl_2$ drying tube. Filter the resulting acetoxy resin and wash thoroughly with water and methanol. The yield of acetoxy groups can be measured by titration of chloride in the combined filtrate and washes (modified Volhard method). Chloride analysis of the resin should show essentially complete conversion to the acetoxy form. This series of reactions can also be monitored qualitatively by infrared (IR) spectroscopy. KBr pellets may be made of the polymer beads (about 10 mg) directly with or without grinding.

Although the acetoxy resin can be converted to the hydroxymethyl resin by saponification, hydrazinolysis is much more satisfactory. Conversion with $LiAlH_4$ is also much more laborious. Stir 25 g acetoxy resin in 250 mL DMF containing 30 mL anhydrous hydrazine (50 equiv. hydrazine per equivalent of acetoxy groups) at room temperature for 72 h. Transfer the resin to a fritted glass Büchner funnel; wash with DMF, methanol, and DCM; and dry. The IR spectrum of these beads should show complete absence of the ester carbonyl band.

8. Preparation of p-Methoxybenzhydrylamine (MBHA) Resin

a. Acylation of Resin. Place 25 g copoly(styrene–1% divinylbenzene) resin beads, 200–400 mesh, and 250 mL $ClCH_2CH_2Cl$ in a 500-mL three-neck round-bottom flask fitted with a thermometer, a $CaCl_2$ drying tube, and a 200-mL dropping funnel. Preform the acylating reagent in the dropping funnel by mixing 50 mmol (7.73 g) p-toluyl chloride and 50 mmol (6.67 g) $AlCl_3$ in 150 mL $ClCH_2CH_2Cl$. After chilling the resin suspension to −15°C with an ice bath, add the acylating reagent carefully, with stirring, so that the temperature of the resin suspension remains below 5°C. After addition is complete, remove the ice bath and continue stirring for 4 h at room temperature. Collect the resin in a fritted glass funnel and wash successively with ethanol, DCM, ethanol, methanol, and DCM. Typically toluyl (and benzoyl) resins exhibit a strong IR band at 1650 cm^{-1} with an intensity approximating that of the 1600 cm^{-1} band in polystyrene. The same procedure may be used for the synthesis of BHA resin by using an equivalent amount of benzoyl chloride instead of p-toluyl chloride in the first step.

To determine the degree of acylation of polymers, heat 500 mg ketone resin, 500 mg hydroxylamine · HCl, and 2 mL pyridine overnight at 105°C in a closed tube. Collect the resin on a fritted filter; wash successively with DCM, ethanol, water, ethanol, methanol, and DCM; and dry. The IR spectrum should show complete loss of the C=O band. Combustion analysis for N gives the degree of acylation.

b. Reductive Amination of Acetylated Polymers by the Leuckart Reaction. Mix 24 g ketone resin and 75 g ammonium formate with 90 mL formamide, 60 mL 88% formic acid, and 300 mL nitrobenzene in a 1 L three-neck round-bottom flask fitted with an overhead stirrer, a thermometer, and a Dean–Stark trap. Heat the contents of the reaction flask to an inner temperature of 165°C for 22 h using an electric mantle. Approximately 40 mL of aqueous phase should be *collected* in the trap and removed during the initial heating period. The aqueous phase is colorless; nitrobenzene is yellow. Because the density of nitrobenzene changes dramatically with temperature, the water layer may be on top or on the bottom; it must be removed. Nitrobenzene removed concomitantly (if the water is on top) may be added back through the condenser. Unless the aqueous phase is removed, the reaction temperature will not reach 165°C; very little additional distillate is obtained once an inner temperature of 165°C is reached. When the reaction is finished, cool the mixture and collect the resin by filtration on a fritted funnel. Wash with ethanol, DCM, and ethanol. Deformylate the resin by hydrolysis for 60 min in 300 mL of 12 M HCl–ethanol (1:1) at refluxing temperature. Collect the amine hydrochloride resin by filtration; wash with ethanol, DCM, methanol, and DCM; and dry. The amine hydrochloride resin shows a decreased IR carbonyl band and a new band at 3600 cm^{-1}. When subjected to the ninhydrin test, the amine resin gives a strong blue color after 60 s at 155°C. The direct Volhard chloride procedure may be used on the amine·HCl resin to obtain quantitative substitution data.

9. Preparation of Aminomethyl Resin

a. Preparation of Phthalimidomethyl Resin. Place 200 g of copoly (styrene–1% divinylbenzene) resin and 117 g (1.0 mol) *N*-(hydroxymethyl)phthalimide in a three-neck round-bottom flask (5 L) equipped with an overhead stirrer. Add 2 L of TFA–DCM (1:1, v/v) and suspend the resin by rapid stirring. Slowly add 18 mL (0.20 mol) trifluoromethanesulfonic acid with rapid stirring and continue stirring at room temperature. The amidoalkylation reaction is followed by IR of KBr pellets of washed [i.e., TFA–DCM (1:1), DCM followed by aspiration on filter funnel] resin samples (~10 mg). The substitution of the resin can be estimated using the formula ([IR intensity at 1720 cm^{-1}]/[IR intensity at 1601 cm^{-1}]) × 0.17 = mmol/g. Stop the reaction at the desired substitution level (in this case at 10 h), filter the resin, and wash with 4 L TFA–DCM (1:1, v/v) followed by 8 L DCM. Dry the resin under vacuum overnight. This specific preparation yielded a product containing 0.28% N (0.20 mmol N/g) by elemental analysis. The concentration of reagent and catalyst and the reaction time have been varied to yield phthalimidomethyl resins containing substitution levels of 0.05–3.60 mmol/g.

b. Preparation of Aminomethyl Resin. Reflux 180 g of phthalimido resin (36 mmol) without stirring for 16 h in 2 L ethanol containing 5% hydrazine (Eastman 95+%). Filter the resin while hot and wash four times with 2 L hot ethanol and four times with 2 L methanol. Dry the resin under vacuum. These specific conditions resulted in 0.26% N (0.19 mmol N/g) by elemental analysis, 0.22 mmol NH_2/g by picric acid tritation, and no carbonyl groups by IR analysis. Examination of the DCM-solvated resin under the microscope shows the beads to be identical in appearance to the starting polystyrene resin.

10. Preparation of 4-(Hydroxymethyl) Phenylacetamidomethyl (Pam) Resin

a. Preparation of 4-(Acetoxymethyl) Pam Resin. Shake 1 g (2.48 mmol) of aminomethyl resin for 10 min with a solution of 0.52 g (2.48 mmol) 4-(acetoxymethyl)phenylacetic acid in 10 mL DCM. Add 0.51 g (2.48 mmol) *N,N'*-dicyclohexylcarbodiimide (DCC) in 2 mL DCM and shake the suspension for 20 h at room temperature. Filter the resin, wash with DCM, and shake in 20 mL of pyridine–acetic anhydride (1:1) for 2 h. Filter the resin; wash with DCM, DCM–HOAc (1:1), HOAc, isopropanol, and DCM; and vacuum dry. The IR spectrum of the resin shows carbonyl bands at 1739 (ester) and 1650 (amide) cm^{-1}. This method yielded 1.90 mmol acetoxy/g resin.

b. Preparation of 4-(Hydroxymethyl) Pam Resin. Shake the acetoxymethyl Pam resin with 10% hydrazine in DMF for 23 h at room temperature. Filter the resin; wash with DMF, isopropanol, and DCM; and vacuum dry. The resin should show no ester carbonyl in the IR spectrum and should contain <0.02 mmol acetoxy/g.

11. Preparation of *p*-Alkoxybenzyl Alcohol Resin

Stir 45 g (40.5 mmol) chloromethyl resin suspended in 250 mL dimethylacetamide gently with 16 g (105 mmol) methyl 4-hydroxybenzoate and 5.8 g (107 mmol) sodium methoxide at 80°C for 24 h. Collect the resin and wash with DMF, dioxane, and methanol to give 49.5 g of 4-methoxycarbonylphenoxymethyl resin containing 0.8 mmol/g methoxy groups and showing an ester carbonyl band at 1712 cm^{-1} in the IR spectrum (KBr pellet). Suspend 45 g of this methyl ester resin in 600 mL anhydrous ether and add 4.6 g $LiAlH_4$ in small portions over a 20-min period. Stir for 6 h; then collect the resin and wash with ethanol, methanol, dioxane, and methanol. To remove the slightly grayish color, stir in 2 L of 1 N H_2SO_4–dioxane (1:1) for 45 h. One should obtain white *p*-alkoxybenzyl alcohol resin containing 0.3% OCH_3 (0.1 mmol/g). The ester band in the IR spectrum should disappear completely. The same *p*-alkoxybenzyl alcohol resin can also be prepared by treating 5.1 g (4.6 mmol) chloromethyl resin with 0.74 g (5.9 mmol) 4-hydroxybenzyl alcohol and 0.32 g (6.1 mmol) sodium methoxide under similar conditions.

12. Preparation of 3-(*p*-Benzyloxyphenyl)-1,1-dimethylpropyloxycarbonylhydrazide Resin

Treat 10 g (7.3 mmol) chloromethyl resin suspended in 70 mL DMF with 1.64 g (10 mmol) 4-(*p*-hydroxyphenyl)-2-butanone in the presence of 0.54 g (10 mmol) sodium methoxide at 85°C for 24 h. Filter the dark brownish reaction mixture and wash the resin with DMF, DCM, and methanol to yield 10.9 g of buff-colored material. The IR spectrum (KBr) shows a band at 1730 cm^{-1}. Suspend the resin in 200 mL benzene and treat with Grignard reagent freshly prepared from 0.54 g Mg turnings in 300 mL ether bubbled with dry CH_3Br. After 60 min of additional stirring, wash the resin with benzene and dioxane and stir in 1 N H_2SO_4–dioxane (1:2) for 120 min. Collect the tertiary alcohol resin and wash with dioxane–water (1:1), dioxane, DMF, and DCM. Treat the resin with 7.9 mL pyridine and 9.8 mL phenyl chloroformate in 120 mL of DCM at 0°C for 16 h. Pour the reaction mixture into 100 mL ice water and filter to collect the resin. Wash the mixed carbonate resin with ice water, dioxane, and DMF. Stir resin with 120 mL DMF that contains 10 mL anhydrous hydrazine for 6 h to produce the desired 3-(*p*-benzyloxyphenyl)-1,1-dimethylpropyloxycarbonylhydrazide resin.

B. High Cross-Linked Polystyrene

Polystyrene containing 50% divinylbenzene provides a nonswelling, rigid support that possesses the attractive features of rapid reaction kinetics, efficient washing with organic solvents, and mechanical stability during oligonucleotide synthesis. This support has been derivatized to provide a primary amino functionality (Fig. 2) by the same procedure described for aminomethylated polystyrene resin (see Section II.A).

C. Polystyrene Grafted onto Polyethylene Film or Tubing

Linear chains of polystyrene may be grafted onto the rigid surface of polyethylene. Long chains can be obatined by γ-ray–initiated graft polymerization of 440% linear polystyrene in methanol on non–cross-linked polyethylene [120]. The rigid polyethylene surface imposes the restriction that the swelling of the polystyrene chains be one-dimensional. Thus, depending on the number of grafts per surface area, there is an inherent problem with

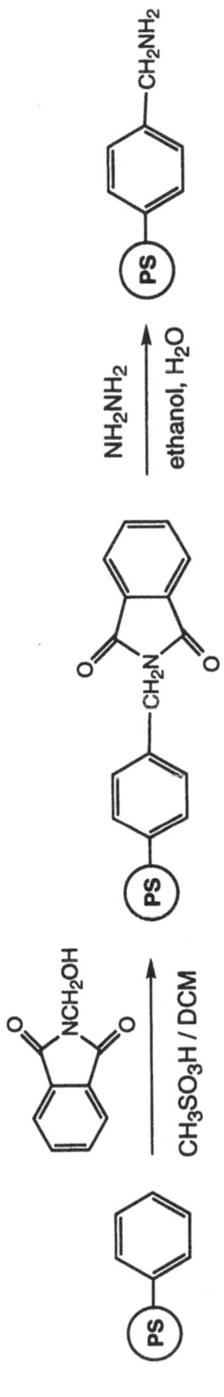

Figure 2 Preparation of aminomethyl polystyrene resin.

the geometry of the brush polymer in the case of syntheses of large and/or bulky molecules.

SPOC has been performed primarily using copoly(styrene–1% divinylbenzene) or copoly(styrene–2% divinylbenzene) resins (Table 2) [121,122]. The 2% DVB polymer is preferred for reactions at higher temperatures or for reactions with organometallic reagents. However, polystyrene resins have some limitations, such as mechanical stability and large interior surfaces. Other poylmers, such as polyethylene glycol grafted onto polystyrene resin, have been developed. This polymer has a lower loading capacity than traditional polystyrene resins but swells better in polar solvents such as methanol and water.

III. POLYETHYLENE GLYCOL RESINS

A. Polyethylene Glycol Grafted Resins

The properties of polystyrene supports have been modified by grafting polyethylene glycol (PEG) to the functional groups of the polystyrene. PEG itself is often used as a soluble support for synthesis [123,124]. Polymerization of oxyethylene onto functional hydroxymethyl groups afford PEG grafting with a chain length of 7–10 ethylene glycol residues or more. The PEG chain acts as a spacer for peptide attachment [25,125]. The PEG chains may also be grafted by attachment of long polyethylene glycol chains through an oxycarbonylglycine bond to a functional aminomethyl group of polystyrene [27,126]. The weight ratio of PEG to polystyrene is often about 1:1 in these resins, and the presence of PEG results in a complete change in the solvation behavior. The beads swell in both nonpolar and polar solvents except water. More important, these resins assume a state of complete solvation from the initiation of the synthesis. PEG-PS resins are thus very useful in SPOC, where a wide variety of solvent conditions may be required for optimal results, and in solid-phase synthesis of large peptides. PEG-PS resins are also physically stable in flow systems. The uniformity of bead size (typically 90–130 μm) makes these resins ideal for synthesis of libraries using the "one bead–one compound" approach. The PEG-PS graft polymeric supports may be further functionalized (see Fig. 3 and Table 3).

B. Polyethylene Glycol–Based Resins

Optimal support properties can be achieved by using PEG as the major component of the resin. The PEG molecule has an amphipathic nature and is solvated well by both polar and nonpolar solvents [22]. At the same time, the mobility and reorganization of the PEG molecules create a highly dy-

Table 2 Resins Used in SPOC

Reaction type	Polystyrene	PEG	Others
Amide formation and related reactions	Benzoic acid resin Aminomethyl resin Chloromethyl resin Wang resin 2-Chlorotrityl resin Rink amide resin Kaiser oxime resin	TentaGel S OH TentaGel S NH_2 TentaGel S RAM	CPG
Aromatic substitution	Aminomethyl resin Chloromethyl resin Wang resin 2-Chlorotrityl resin Rink amide resin Hydroxymethl resin Sasrin resin	TentaGel TentaGel S NH_2 TentaGel S RAM TentaGel S PHB	
Condensation	Aminomethyl resin Chloromcthyl resin Wang resin 2-Chlorotrityl resin Rink amide resin Hydroxymethyl resin Sasrin resin Sieber resin Aldehyde resin PAL amide resin	TentaGel S OH TentaGel S NH_2	
Cycloaddition	Chloromethyl resin Wang resin Rink amide resin Benzoic acid resin Sasrin resin	TentaGel S AC	
Deprotection–protection	Aminomethyl resin Chloromethyl resin Wang resin 2-Chlorotrityl resin Benzoic acid resin HMPB-MBHA resin	TentaGel	
Grignard and related reactions	Chloromethyl resin Chlorotrityl resin Wang resin		
Heterocycle formation	Aminomethyl resin Chloromethyl resin Wang resin 2-Chlorotrityl resin Rink resin Rink amide resin Hydroxymethyl resin Sasrin resin Kaiser oxime resin	TentaGel S NH_2 TentaGel S RAM TentaGel S HMB	

Table 2 Continued

Reaction type	Polystyrene	PEG	Others
Michael addition	Hydroxyethylene resin HMPA resin Chloromethyl resin Hydroxymethyl resin Tritylchloride resin Rink amide resin Wang resin REM resin		
Multiple component reaction	Aminomethyl resin Chloromethyl resin Wang resin Rink amide resin PAL amide resin MBHA resin		
Olefin formation	Chloromethyl resin Chlorotrityl resin Wang resin PAM resin	PEG-PAL resin	
Oxidation	Polystyrene resin Chloromethyl resin Chlorotrityl resin Wang resin Rink amide resin Sasrin resin	TentaGel S OH	
Reduction	Polystyrene resin Chloromethyl resin Benzoic acid resin Wang resin Trityl resin 2-Chlorotrityl resin PAL resin Rink amide resin MBHA resin REM resin Hydroxymethyl resin	TentaGel S NH$_2$	Spheron Ara 1000[a] PS-PEG
Substitution, nucleophilic–electrophilic	Polystyrene Aminomethyl resin Chloromethyl resin 2-Chlorotrityl resin Hydroxyethyl resin Rink amide resin Wang resin MBHA resin Sieber resin	TentaGel TentaGel S RAM TentaGel S OH TentaGel S NH$_2$	PINS (Rink handle)

Table 2 Continued

Reaction type	Polystyrene	PEG	Others
Cleavage, nonstandard	Aminomethyl resin	TentaGel S OH	Spheron Ara 1000[a]
	Chloromethyl resin	TentaGel S PHB	
	Benzoic acid resin		
	Hydroxymethyl resin		
	Rink amide resin		
	Wang resin		
	PAL amide resin		
	MBHA resin		
	Kaiser oxime resin		
	REM resin		
Nonstandard immobilization reactions	Aminomethyl resin	TentaGel S PHB	
	Carboxy resin	TentaGel S OH	
	Chloromethyl resin	TentaGel S NH$_2$	
	2-Chlorotrityl resin		
	Hydroxymethyl resin		
	Wang resin		
	HMHA resin		
	HMPA resin		
	HMPB-HMBA resin		
	Rink amide resin		
	Kaiser oxime resin		
	Sasrin resin		
Enzymatic			Polyacrylamide
			CPG
			Aminopropyl silica
			PEGA resin
Miscellaneous reactions	Chloromethyl resin (enolization, iodoetherification, ene rearrangement, bromination)		
	Rink amide resin (hydrolysis, esterification)		
	Wang resin (dehydration, transamidation, α-diazoketo formation)		
	MBHA resin (amination)		
	Hydroxyethyl resin (deoxygenation)		

[a]Poly(2-hydroxyethylmethacrylate) cross-linked to poly(HEMA) containing aryl-amino groups.
Source: Refs. 121 and 122.

Figure 3 Functionalization of PEG-PS resin.

namic process. The four types of cross-linked PEG-based polymers currently available are discussed in the following.

1. PEGA Resins

Partial acroylation of bis-2-aminopropyl PEG with acryloyl chloride followed by radical-initiated inverse suspension polymerization of the resulting macromonomer mixture yields resins that are well suited for continuous-flow peptide synthesis [28,127,128]. PEGA resins contain only polyether backbones and polyacrylamide backbones linked together by amide bonds (see Fig. 4, compound **5** or **6**) that are inert and stable under all peptide synthesis conditions including cleavage with strong acids or aqueous base. The family of PEGA resins are flow stable and beaded, and they can be used as solid supports in split synthesis of libraries. The resins can easily be modified by addition of small amounts of monomers such as N,N-dimethylacrylamide, acrylamide, acrylates, or acrylonitrile. The capacity of the resin can be increased to 0.8 mmol/g by addition of small functional group monomers (compound **4** in Fig. 4).

a. Preparation of PEGA Resin. Copolymerization can be performed using monomers **1, 2,** and **3** or **1, 3,** and **4.** The latter method (Fig. 4) proceeds as follows: acryloyl sarcosine ethyl ester **4** is prepared by reaction of acryloyl chloride (0.59 g, 6.5 mmol) with sarcosine ethyl ester hydrochloride (1 g, 6.5 mmol) in the presence of triethylamine (2 equiv.). Inverse polymerization of **4** as a functionalizating agent along with N,N-dimethylacrylamide **3** (1.49 g, 15.03 mmol) and bis-2-acrylamidoprop-1-yl-PEG$_{1900}$ **1** (10.54 g, 5 mmol) results in PEGA polymer **5** with a loading of 0.44 mmol/g. The capacity can be increased to 0.8 mmol/g.

b. Functionalization. The PEGA resin is treated with ethylenediamine (see aminoethylamidation for the functionalization of polyamide resin in Section VI.A). Free amino group capacity is determined by incorporating Fmoc-Gly (0.44 mmol/g and 0.8 mmol/g) onto the support and measuring the ultraviolet (UV) absorption of the benzofulvene–piperidine adduct formed by treatment with piperidine–DMF (1:4).

c. Attachment of the Linkage Agent. The 4-hydroxymethylbenzoic acid linker is quantitatively attached to the amino functionalized resin using 2 equiv. of the linker (for the experimental procedure see attachment of the linkage agent in Section VI.A).

2. Polyoxyethylene–Polystryene Resins (POEPS)

Partial alkylation of PEG (**7** in Fig. 5) with chloromethylstyrene gives a mixture of non-, mono- (**8**), and bisalkylated (**9**) PEG that can be polymer-

Table 3 Functionalized PEG-PS Resins

Bromo resin

PS — PEG — OCH₂CH₂Br

Amino resin

PS — PEG — OCH₂CH₂NH₂

Alcohol resin

PS — PEG — OCH₂CH₂OH

Carboxy resin

PS — PEG — OCH₂CH₂COOH

Thiol resin

PS — PEG — OCH₂CH₂SH

5-(4-Hydroxymethyl-3-methoxyphenoxy)pentanoylamino resin

PS — PEG — NHCO(CH₂)₄O — (aryl, OCH₃) — CH₂OH

4-Hydroxymethylbenzoylamino resin

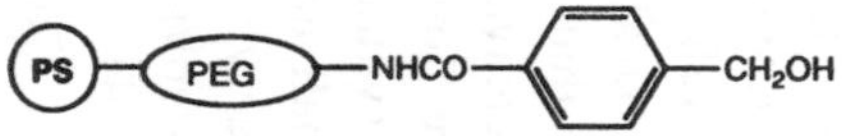

PS — PEG — NHCO — (aryl) — CH₂OH

4-(Hydroxymethyl)phenoxyacetamido resin

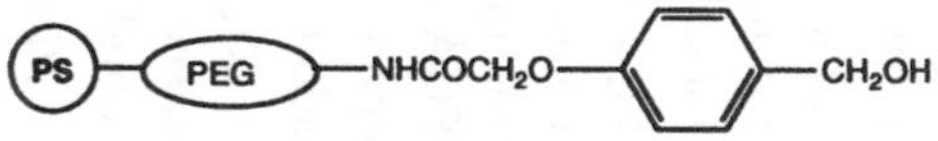

PS — PEG — NHCOCH₂O — (aryl) — CH₂OH

Succinamido resin

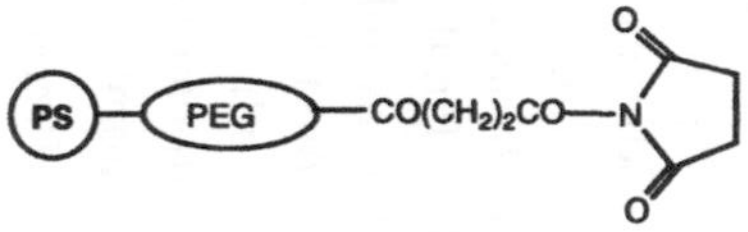

PS — PEG — CO(CH₂)₂CO — N (succinimide)

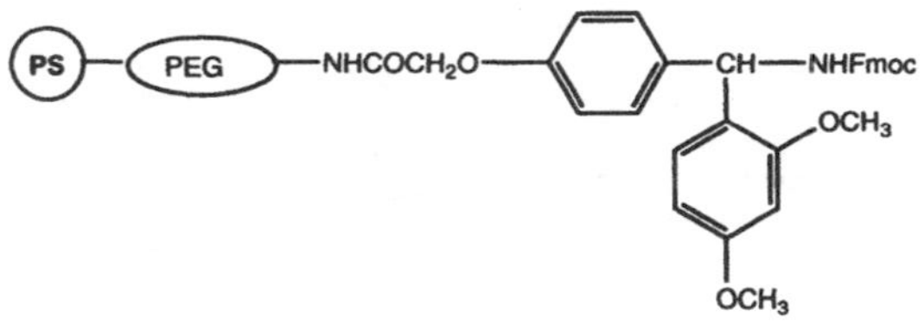

Sieber resin

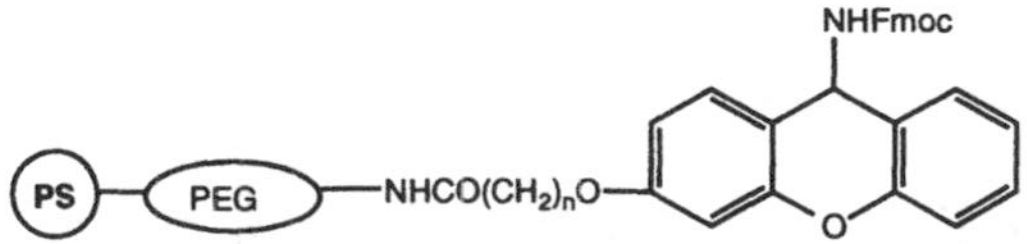

Hydroxytrityl resin

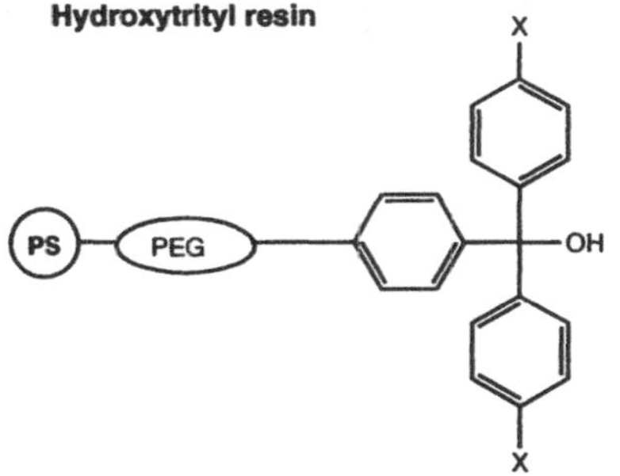

X = H, Cl

Hydroxytrityl resin

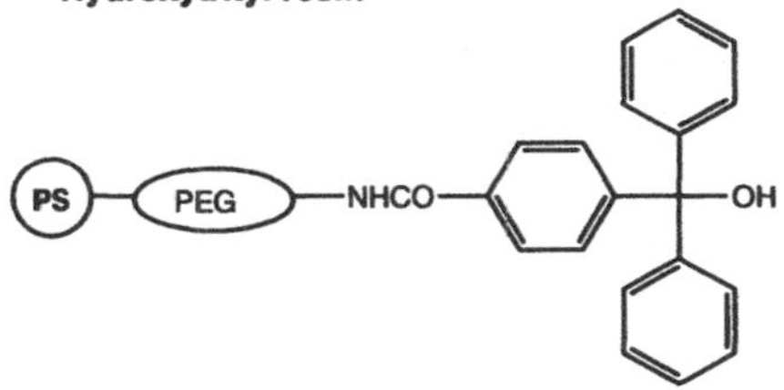

Aldehyde resin

Hydroxymethylphenyloxy resin

5-(4-Aminomethyl-3,5-dimethoxyphenoxy)pentanoylamino resin

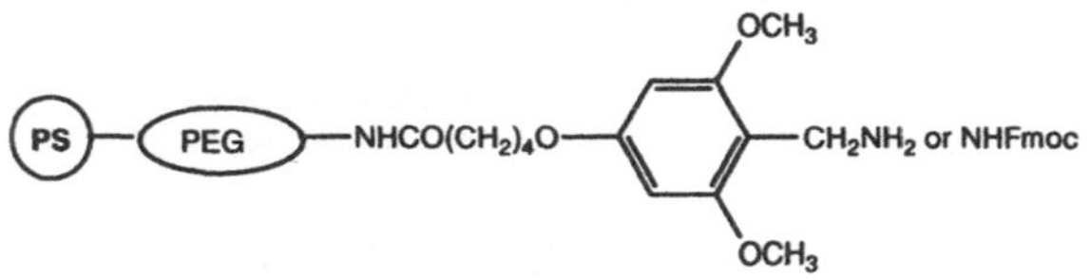

4-(2',4'-dimethoxyphenyl-N-Fmoc-aminomethyl)phenoxyacetylamino resin

41

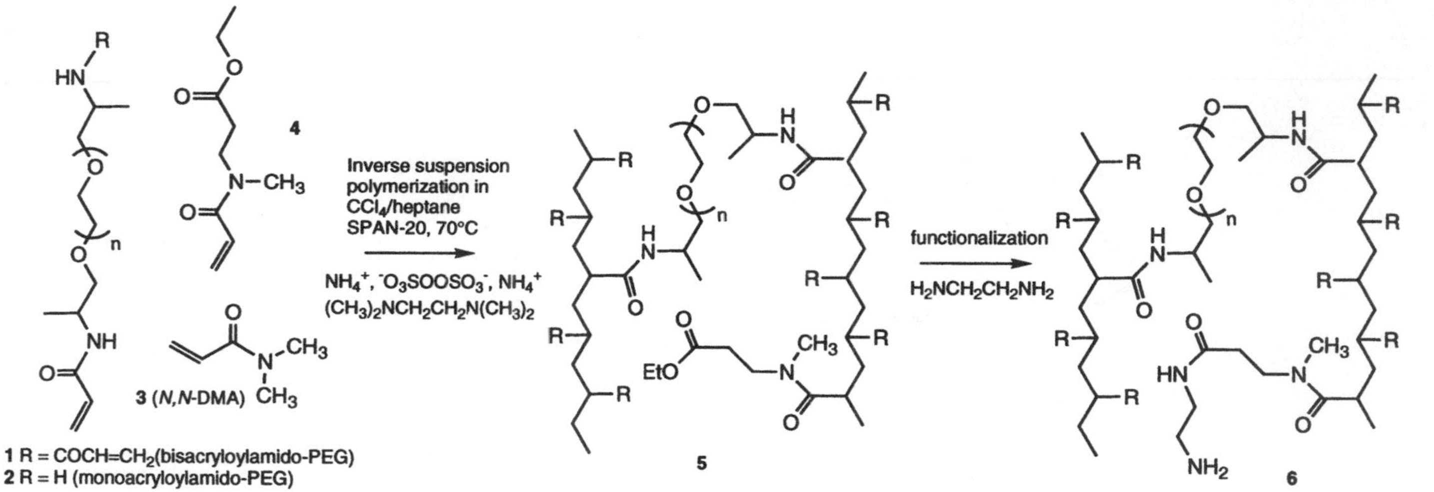

Figure 4 Preparation of PEGA resins.

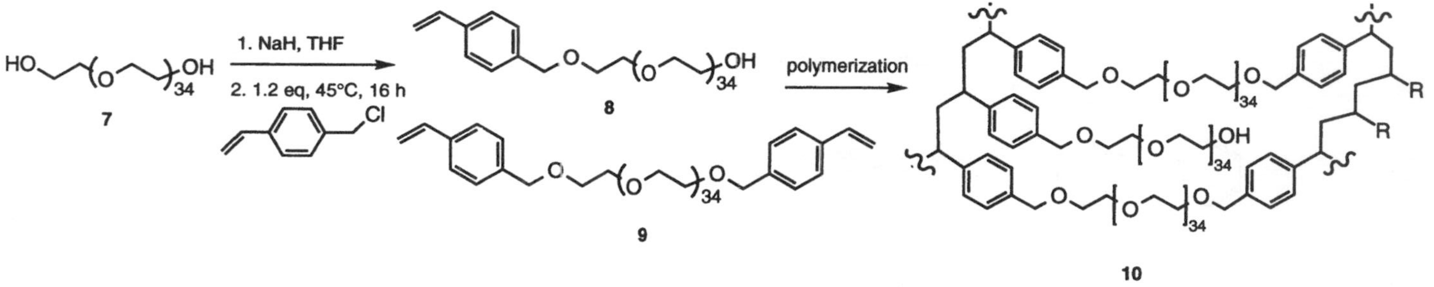

Figure 5 Preparation of POEPS resins.

ized in a beaded form by inverse suspension polymerization or bulk polymerization followed by swelling and granulation [30]. This resin contains ~10% polystyrene. The resin shows good solvation in water as well as in nonpolar media such as toluene. The hydroxyl group capacities of these supports can be easily varied from 0.1 to 0.6 mmol/g by adjusting the monomer composition.

a. Preparation of the POEPS Resin. The macromonomers **8** and **9** are synthesized by dissolving PEG$_{1500}$ (5 g, 6.6 mmol) in dry tetrahydrofuran (THF) (5 mL) and adding sodium hydride (55% in oil) (0.15 g, 3.9 mmol) with stirring. After 6 h, vinylbenzyl chloride (0.57 mL, 4 mmol) is added to the reaction mixture at 45°C, and the mixture is allowed to stir for an additional 16 h. The products are precipitated using diethyl ether, filtered off, and purified by dissolving in DCM and filtering to remove any insoluble components. The products are precipitated using cold diethyl ether, filtered, and dried under high vacuum to yield 4.8 g (95% of theoretical). The macromonomers **8** and **9** (2.5 g) are bulk homopolymerized in a Pyrex tube under argon at 100°C using a free radical initiator (organic peroxides, 0.10 g) for 12 h to afford the PEG cross-linked polymer **10**. It is washed with methanol and granulated through 1-mm sieves; further washed with DCM, methanol, and ether; and dried under high vacuum to yield 1.5 g (60%). The polymer is esterified by dissolving Fmoc-Gly (3 equiv.) and 1-(2-mesitylenesulfonyl)-3-nitro-1,2,4-triazole (MSNT) (3 equiv.) in dry DCM in the presence of *N*-methylimidazole (2.25 equiv.) and reacting with the resin. The hydroxyl group capacity of resin **10** is determined by measuring the UV absorbance of the dibenzofulvene–piperidine adduct formed by treatment of a weighed polymer sample with piperidine–DMF (1:4). Rink amide handle has been attached to support **10** by the MSNT esterification method.

3. Polyoxyethylene–Polyoxypropylene Resins (POEPOP)

Partial reaction of polyethylene glycol (**7**) with sodium hydride and epichlorohydrin yields a mixture of PEG mono- and bismethyloxirane (**12**) (Fig. 6). This mixture can be bulk polymerized at high temperatures in the presence of *t*BuOK to yield an inert polymer with only ether bonds and hydroxyl groups as the functional groups [30]. The hydroxyl groups can be alkylated or acylated for synthesis.

a. Preparation of the POEPOP Resin. The macromonomers **11** and **12** are synthesized by dissolving PEG$_{1500}$ (5 g, 6.6 mmol) in dry THF (5 mL) and adding sodium hydride (55% in oil) (0.15 g, 3.9 mmol) with stirring. After 6 h, epichlorohydrin (0.32 mL, 3.9 mmol) is added to the reaction mixture at 45°C and allowed to stir for another 16 h. The products are precipitated with diethyl ether, filtered off, and purified by dissolving in

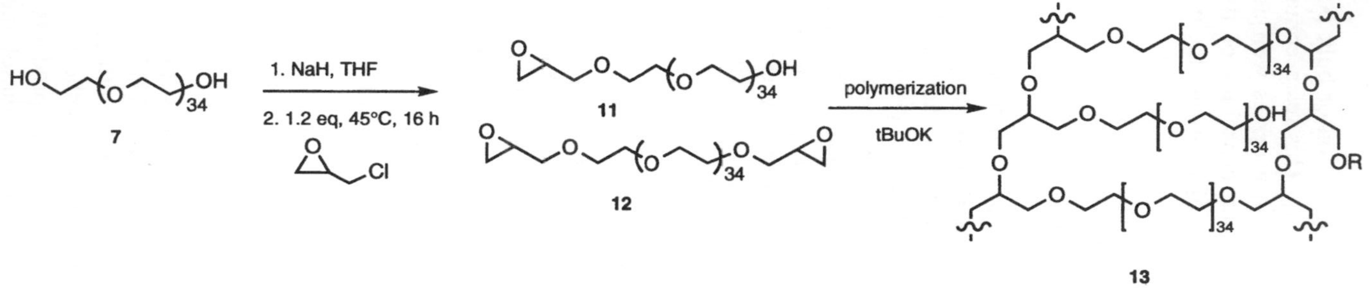

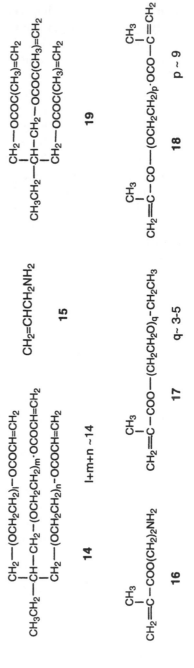

Figure 6 Synthesis of POEPOP resin.

Figure 7 Precursors used for the synthesis of CLEAR.

DCM and filtering to remove any insoluble components. The products are precipitated using cold diethyl ether, filtered, and dried under high vacuum to yield 4.5 g (90%). The anionic polymerization of macromonomers **11** and **12** (2.6 g) catalyzed by *t*BuOK (0.10 g) is carried out at 100°C under an argon atmosphere and the sticky point is reached after 30 min. The polymerization is then continued for 12 h. The reaction is quenched by the addition of methanol and the product **13** is granulated through 1-mm sieves. The polymer is stirred in 4 M HCl for 2 h to remove potassium salts and then washed with water and ethanol and dried under high vacuum to yield 2.5 g (96%). The hydroxyl group capacity of resin **13** is determined following esterification with Fmoc-Gly as described earlier. Rink amide handle is attached to support **10** by the MSNT esterification method.

4. Cross-Linked Ethoxylate Acrylate Resin (CLEAR)

CLEAR is a highly cross-linked support (>95% by weight of cross-linker) that exhibits excellent solvation properties and performance in batchwise and continuous-flow syntheses. This support is prepared by radical copolymerization of the branched cross-linker trimethoxypropane ethoxylate (14/3 EO/OH) triacrylate **14** with other monomers (**15–19**) [31] (Fig. 7).

Either bulk polymerization or suspension polymerization results in CLEAR with adequate chemical and mechanical stability. For SPPS, the initial functional group is substituted with an appropriate handle.

a. Bulk Polymerization. Allylamine (**15**) (0.51 g, 9.0 mmol), trimethoxypropane ethoxylate (14/3 EO/OH) triacrylate (**14**) (8.21 g, 9.0 mmol), and AIBN (0.1 g, 0.6 mmol) are placed in a screw-cap culture tube, dissolved in cyclohexanol (17.5 mL), and purged with a stream of nitrogen for 5 min. The tube is sealed and irradiated overnight (350 nm) in a Rayonet photochemical reactor. The resulting polymer is ground in a mortar and wet-sieved with water (total ~6 L) through 106- and 125-μm sieves. The 106–125-μm fraction is collected and washed on a sintered glass filter funnel with acidified water (~300 mL) and methanol (~150 mL). The fines are removed by repetitive (~10×) sedimentations (by gravity) and decantations using methanol (~50 mL each time). The remaining particles are dried in vacuo overnight to yield 2.6 g. Elemental analysis found C, 54.84; H, 8.01; N, 0.90, consistent with molar incorporation of **14:15** = 5:3 (approximately half of the **15** incorporated provides free amino groups). Note that the polymer is 96% by weight derived from **14**.

b. Suspension Polymerization. Spherical beads are prepared using a reactor and an overhead stirrer following the procedure described by Arshady and Ledwith [32]. An aqueous phase consisting of deionized water (120 mL; previously purged with a stream of argon for 5 min), 1% polyvinyl alcohol in water (6 mL), and ammonium laureate solution (5 mL of 1% lauric acid in water, adjusted to pH 10.3 with concentrated aqueous NH₄OH) and an organic phase (which had been purged with a stream of argon for 5 min) consisting of allylamine (**15**) (2.86 g, 50 mmol), trimethylolpropane ethoxylate (14/3 EO/OH) triacrylate (**14**) (10.94 g, 12 mmol), trimethylolpropane trimethacrylate (**19**) (1.02 g, 3.0 mmol), AIBN (0.5 g, 3.0 mmol), and toluene (17 mL) are stirred (400 rpm) under an argon atmosphere for 1 h at 70°C. The beads are collected on a sintered glass funnel and washed with water (~2 L) and methanol (~300 mL). The beads were sieved, and the major fraction (106–125 μm) is suspended in methanol and repetitively sedimented and decanted (~10×) to remove remaining smaller beads. The beads are dried in vacuo overnight to yield 2.38 g. Elemental analysis found C, 55.05; H, 8.11; N, 0.86.

c.   Derivatization of CLEAR [Including Addition of Norleucine (Nle) as an Internal Reference].   Fmoc-Nle-OH (0.64 g, 1.8 mmol), dissolved in DMF (3 mL), is added to the CLEAR resin (1 g). Coupling is initiated by the addition of 7-aza-1-hydroxybenzotriazole (HOAt) (0.23 g, 1.8 mmol) in DMF (3 mL) followed by *N,N'*-diisopropylcarbodiimide (DIPCDI) (0.25 g, 1.8 mmol) in DMF (3 mL). The mixtures are shaken at 25°C for 48 h and then filtered and washed with DMF (5 × 15 mL) and DCM (5 × 15 mL). At this point an acetylation step was carried out using acetic anhydride (1 g) in 16 mL DCM–pyridine (1:1) for 30 min, followed by washing with DCM (5 × 15 mL) and DMF (5 × 15 mL). The Fmoc groups are removed by treatment with piperidine–DMF (1:4) (5 + 15 min), followed by washings with DMF (5 × 15 mL) and DCM (5 × 15 mL). A solution of Fmoc-5-(4-aminomethyl-3,5-dimethoxyphenoxy)valeric   acid   (Fmoc-PAL-OH) (0.89 g, 1.8 mmol) in DMF (3 mL) is added, and the next steps (activation, coupling, acetylation) follow exactly the outline given previously for the introduction of Fmoc-Nle-OH.

IV.   CONTROLLED PORE GLASS AND OTHER SILICA-BASED SUPPORTS

A.   Controlled Pore Glass (CPG)

Controlled pore glass is a nonswelling, inorganic matrix composed of polar silanol groups mainly hydrophilic and nondeformable. The loadings on CPG are usually too low to be of practical use (60 μmol), and problems associated with synthesis of long peptides on the rigid surface have not been evaluated. This support has been used in oligonucleotide [129–132] and peptide [133,134] synthesis. CPG has been shown to prevent peptide chain aggregation during SPPS [134].

Inorganic solids derived from silicates with reactive organic groups on the surface are prepared by polymerization of the corresponding monomers (trichloro)-[3-(4-chloromethylphenyl)] propylsilane (a), (dichloro)-[4-(4-chloromethylphenyl)butyl] methylsilane (b), and (chloro)-[4-(4-chloromethylphenyl)butyl] dimethylsilane (c) on the surface (Fig. 8). The inorganic matrices with reactive chlorobenzyl groups on the surface can be used in SPS. Also, the 3-aminopropyltriethoxysilane has been used to derivatize CPG [135], where the beads are immersed in a 2% solution of this silane in acetone. Excess liquid is decanted off and the beads are allowed to stand at 45°C for 24 h. This direct polymerization produced beads containing 80–90 μmol NH_2/g support.

The most serious problems of CPG are the presence, in the crude product, of silica and polymeric siloxanes released during cleavage and deprotection of oligonucleotides. Furthermore, the silane coupling chemistry used to functionalize the inorganic surface of CPG beads is complex, leading to variation in substitution levels from preparation to preparation.

B. Silica Gels

Silica gel is a rigid matrix that does not swell in common organic solvents. Silica gels can be modified with aminopropylsilane to yield a satisfactory loading and synthetic performance for smaller peptides. The loading is performed [129] by refluxing (3-aminopropyl)triethoxysilane with silica gel in dry toluene for 3 h (Fig. 9). In order to obtain a carboxylic acid group used for oligonucleotide synthesis, succinic anhydride is reacted with the amino groups. Excess silanol groups are eliminated by treatment with trimethylsilylchloride. This support has been used in high-yield (60–70%) enzyme conversions, such as those with glycotransferases to afford glycopeptides [136].

1. Synthesis of Silica Gel Support

HPLC grade silica gel (2 g; Vydac TP-20, 100 m^2/g surface area, 200 Å pore size, 20 μm particle size) is exposed to an atmosphere of 15% relative humidity (saturated LiCl) for at least 24 h. The silica is then treated with 3-(triethoxysilyl)propylamine (2.3 g, 0.01 M in toluene) for 2 h at 20°C and 12 h refluxing under a Drierite drying tube [137]. Reaction is completed on a shaking apparatus, as magnetic stir bars pulverize the silica gel and must be avoided. Aminopropylsilica was isolated by centrifugation; washed successively (twice each) with toluene, methanol, and ether; and air dried. The carboxylic acid groups are introduced by agitation of the aminopropylsilica (2 g) and succinic anhydride (2.5 g, 0.025 M in water). The pH is controlled

a   b   c

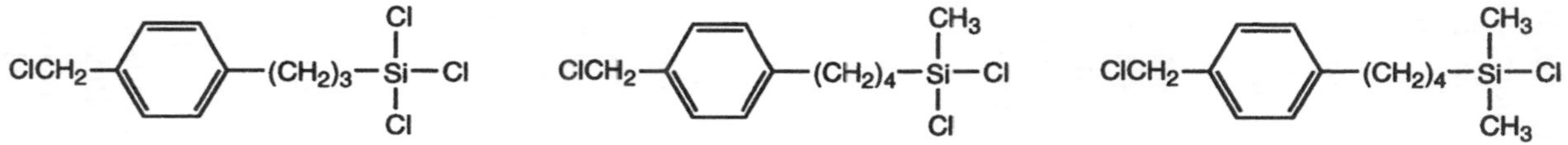

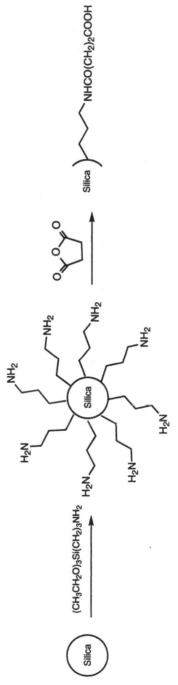

Figure 8 Monomers used for the synthesis of CPG.

Figure 9 Synthesis of silica gels.

(pH 2–6) by addition of 2 M NaOH. Completeness of the carboxylation reaction is qualitatively monitored by using a picrate sulfate test [138]. An aliquot of silica (approximately 2 mg) is treated with 0.5 mL of 0.1 M picrate sulfate in saturated sodium borate buffer (pH 10). The aminopropylsilica reacts within 10 min and turns a bright yellow, whereas the carboxylic derivative remains white. The succinic anhydride reaction is allowed to continue until the silica gel remains white by the picrate sulfate test. Usually the total reaction time is 1 h, and a second addition of succinic anhydride is required. After being washed successively (twice each) with water, 0.1 M trichloroacetic acid, water, methanol, and ether, the carboxylic derivative is air dried, dried in vacuo, and treated with trimethylsilyl chloride (1.25 mL, 0.01 M) in pyridine (7 mL) for 24 h at 25°C. The carboxylic derivative is then washed with methanol (four times) and ether. Analysis for the extent of carboxylation involves a two-step procedure. An accurately weighed aliquot is treated with DCC and p-nitrophenol in pyridine. After several washings with THF to remove unreacted p-nitrophenol, piperidine–pyridine (1:9) is added to the silica gel, and the amount of p-nitrophenol released is measured at $\lambda = 410$ nm by using 1.57×10^4 M^{-1} cm^{-1} as the extinction coefficient of p-nitrophenoxide. The incorporation of carboxylic acid is 200 μmol/g.

V. MEMBRANES

A. Cellulose

Beaded cellulose was one of the first carriers tested by Merrifield [1] for solid-phase synthesis but was found to be unsuitable. However, it has subsequently been shown [139] that peptides synthesized on cellulose paper support can be obtained in reasonable yield and purity. Cellulose fibers from filter paper [139–142] or cotton [143–145] are inexpensive sources of large-surface polymers and can be modified by O-acylation with protected amino acids. Derivatization of cellulose can be achieved by direct attachment to hydroxyls on the cotton [144] or paper [139,141] to form amino acid esters. Alternatively, amino groups can be obtained by cyanoethylation followed by borane reduction. Synthesis is performed on the surface of the large cellulose fibers. The large surface area of the rigid fibers provides a reasonable quantity of functional groups. The fibrous cellulose sheets or woven cotton pieces are easy to manipulate and are well suited for multiple combinatorial synthesis [146]. The rigid structure and low swelling properties of cellulose papers allow multiple coupling reactions to be carried out under low-pressure continuous-flow conditions. Cellulose-type supports cannot be used with strong acids, such as HF, owing to the cleavage of glycosidic bonds under such conditions. However, they are surprisingly stable to TFA for short time periods. Both cotton and paper need a mild acid pretreatment [139,147] that affords limited and irreversible swelling of the cellulose matrix. By using 10% or less TFA in DCM, the swelling process can be controlled.

1. Derivatization of the Cellulose Paper

One hundred paper disks are added to 30 mL of TFA–DCM (1:9) in a stoppered glass filter funnel. After 20 min the acid solution is filtered off and excess DMF is added. The disks are then washed twice with 2-propanol and DCM and dried in vacuo. The disks are carefully pressed successively into the reaction column and connected to the synthesizer. The column is flushed with pyridine until all air bubbles have disappeared. MSNT (2.5 mmol, 0.74 g) is dissolved in 4 mL pyridine and transferred with a gastight glass syringe to a flask containing the linker reagent 4-methoxytrityl-6[4'-(oxymethyl)phenoxyl]hexanoic acid (0.5 mmol, dried twice by coevaporation with 10 mL pyridine). N-Methylimidazole (2.5 mmol, 0.2 mL) is added and the solution transferred to the reaction column. The mixture is recirculated for 2 h, followed by flushing the column with pyridine for 10 min. If required, this treatment may be repeated. Ten milliliters of acetylation mixture (acetic anhydride–collidine–5% DMAP in acetonitrile, 2:3:15) is injected and recirculated for 30 min. After washing the column with pyridine for 10 min and DCM for 10 min, 3% dichloroacetic acid in DCM is passed through until the orange color dissipates (~10 min). The column is then flushed for 15 min with DCM. The MeOTr$^+$-containing efflux is collected and used to determine the yield of linker attachment. The disks are dried in the column by passing a stream of N_2 and then stored at $-20°C$.

B. Organic Membranes

Polymeric membranes are monolithic, continuously porous materials. Membranes can be produced from numerous organic polymers including polyalkanes (polyethylene and polypropylene) and their fluorinated derivatives [polyvinylidene fluoride and polytetrafluoroethylene (PTFE)]. Once formed, a membrane can be chemically functionalized by a number of methods including direct conversion of functional groups in the bulk polymer, coating of the surface with a preformed polymer, or "graft" copolymerization of reactive monomers onto the membrane surface.

Polyethylene fibers can be functionalized by chromium trioxide and ozone oxidation to generate carboxylic acid groups [148]. Surface properties may be modified by derivatization with a polar carbohydrate or PEG spacer for direct biomolecular solid-phase assays (see Fig. 10).

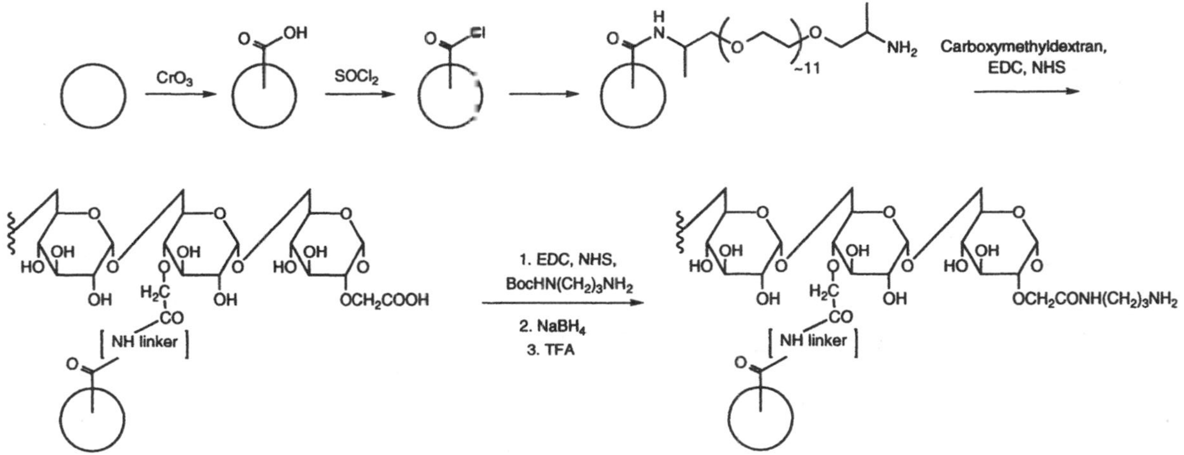

Figure 10 Derivatization of polyethylene fibers. EDC, *N*-ethyl-*N'*-(3-dimethylaminopropyl)carbodiimide; NHS, *N*-hydroxysuccinimide.

DNA synthesis has utilized PTFE membranes of 0.2 μm pore size, surface modified with a terpolymer coating to provide an amino derivatized support [149]. The terpolymer was formed by polymerization of N,N-dimethylacrylamide (1), methylene-bis-acrylamide (2), and aminopropyl-methacrylamide (3) (Fig. 11).

A hydroxypropylacrylate-coated polypropylene membrane has been used as a solid support for the stepwise synthesis of peptides [150]. An acid-labile linker was attached to the coated membrane (Fig. 12).

VI. MISCELLANEOUS

A. Polyacrylamide Bead Gels

Sheppard et al. [23] developed a cross-linked polyamide resin whose fundamental nature was very similar to that of peptide chains. It was thus anticipated that a single solvent might effectively solvate both the peptide and the carrier matrix. The polyacrylamide resins are well solvated in polar solvents such as DMF and have improved properties compared with PS for synthesis under polar conditions [151]. Sufficient solvation in aqueous buffers has allowed the development of solid-phase enzyme assays [152]. Polydimethylacrylamide obtained after copolymerization is further functionalized with ethylenediamine, whose primary amino groups provide appropriate linkage points for the growing peptide chain or linkage agents. The procedures for polymerization and functionalization of polyamide beads described in the following are applicable for a variety of linkage agents, including 2-(4-hydroxymethyl-3-methoxyphenoxy)acetic acid (HALLA, 3), 4-[(N-9-fluorenyl-methoxycarbonyl)-2,4-dimethoxyphenylaminomethyl]-phenoxyacetic acid (Rink linker, 4), and 4-(((hydroxysuccinyl)amino)-methyl)-4'-nitrobenzophenone oxime (Kaiser oxime linker 5) (see Fig. 13).

1. Synthesis of Copoly(Dimethylacrylamide-bisacryloylethylenediamine-acryloylsarcosine Methyl Ester)

Cellulose acetate butyrate (12.5 g) is completely dissolved in dichloroethane (300 mL) and placed in a cylindrical fluted polymerization vessel fitted with a stirrer and nitrogen inlet and maintained at 50 $\pm$ 1°C in a thermostatically controlled water bath. The solution is stirred at 450 $\pm$ 20 rpm (counter or stroboscope) and flushed with nitrogen for 10 min before adding the monomer mixture, which consists of dimethylacrylamide (15 g, 0.152 mol), acryloylsarcosine methyl ester (1.25 g, 7.96 mmol), and bisacryloylethylenediamine (1.75 g, 10.4 mmol), diluted with 150 mL cooled (5°C) DMF–water (1:2) and mixed well with ammonium persulfate (2.25 g).

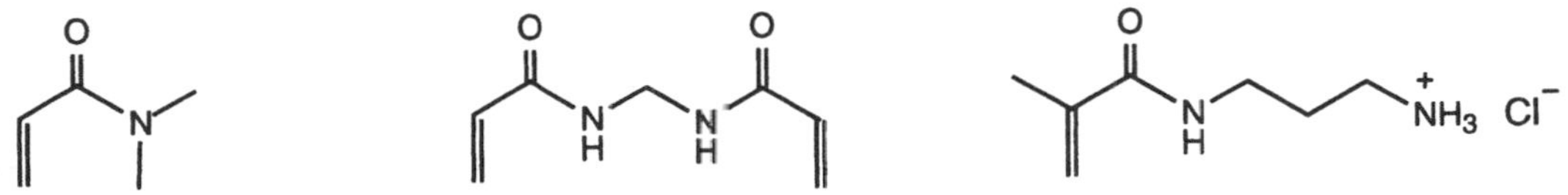

Figure 11 Precursors for synthesis of terpolymer.

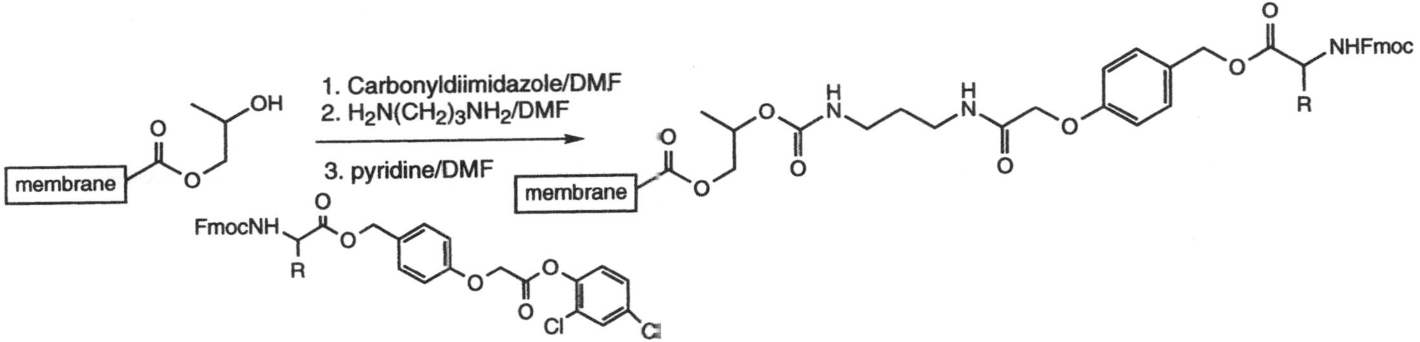

Figure 12 Attachment of the linker to hydroxypropylacrylate coated polypropylene.

53

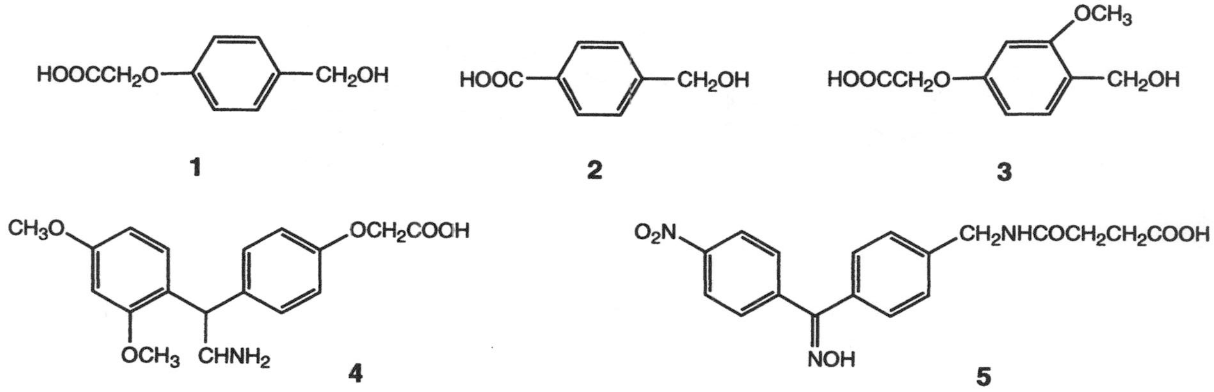

Figure 13 Linkers used in conjunction with polyacrylamido bead gels.

Polymerization is allowed to continue under a very slow stream of nitrogen for ~15 h. The mixture is cooled, diluted with acetone–water (1:1), stirred until a homogenous suspension is obtained, and filtered. The recovered polymer is washed and fine particles are removed by stirring and decantation using acetone–water (1:2) (3 × 1 L) and then acetone (3 or 4 × 500 mL). The polymer is washed with ether (2 × 500 mL), collected by filtration, and dried (P_2O_5) in vacuo to yield ~15 g of completely beaded resin. The average bead size is 50–100 μm, and the sarcosine content is found to be 0.35 mmol/g.

2. Functionalization of Polyamide Resins

a. Aminoethylamidation. Place polyamide resin (1 g) in a 50-mL conical flask, add 30 mL ethylenediamine, stir gently to mix the contents, and leave the mixture to stand overnight. Transfer to a solid-phase reactor, filter the ethylenediamine off, and wash the resin with DMF (10 × 10 mL), *N,N*-diisopropylethylamine (DIEA)–DMF (1:9, 3 × 10 mL), and DMF (5 × 10 mL). Perform a ninhydrin test on the filtrate from the last wash (1 drop) to confirm that all of the ethylenediamine has been removed from the resin. If the test is positive repeat the last five DMF washes and perform the ninhydrin test again. Repeat until the test is negative.

b. Coupling of Internal Reference Standard Using HBTU. Place Fmoc Nle (2.5 equiv.) and *N*-[(1*H*-benzotriazol-1-yl)(dimethylamino)methylene]-*N*-methylmethanaminium hexafluorophosphate *N*-oxide (HBTU) (2.38 equiv.) as dry powders in a beaker followed by DMF (~10 mL). Add DIEA (5 equiv.) and stir the mixture gently for 2–3 min. Transfer the solution to the resin and add DMF if necessary to allow the resin to be mobile under nitrogen agitation. Stir the mixture gently at intervals to ensure thorough mixing. After 1 h remove a sample of resin (4–5 mg) and place it in a vial. Wash the sample (by decantation or filtration) using DMF (2 × 1 mL), DCM (1 mL), and diethyl ether (2 × 1 mL). Submit this sample to a ninhydrin test. A negative test result indicates that the coupling is complete. The reaction can be left overnight if required. When the coupling is complete, draw off the reaction solution and wash the resin with DMF (10 × 10 mL).

c. Fmoc Removal. Add 10 mL piperidine–DMF (1:4) to the resin, mix well, and allow the mixture to stand for 10 min. Draw off the piperidine–DMF and repeat the treatment with fresh piperidine–DMF. Draw off the piperidine–DMF and wash the resin with DMF (10 × 10 mL).

d. Attachment of the Linkage Agent. Place the linkage agent (3 equiv.) and HOBt (6 equiv.) as dry powders in the reactor followed by the minimum amount of DMF required to make the resin mobile to nitrogen agitation. Add *N,N'*-diisopropylcarbodiimide (DIPCDI) (4 equiv.) and stir the mixture gently at intervals to ensure thorough mixing. After 1 h remove a sample of resin (4–5 mg) and place it in a vial. Wash the sample (by decantation or filtration) using DMF (2 × 1 mL), DCM (1 mL), and diethyl ether (2 × 1 mL). Submit the sample to a ninhydrin test. A negative test result indicates that the coupling is complete. If the reaction is not complete, add additional DIPCDI (1 equiv.), stir the mixture, and then allow the reaction to proceed overnight. When the coupling is complete, draw off the reaction solution and wash the resin with DMF (10 × 10 mL), DCM (3 × 10 mL), DCM–diethyl ether (1:1, 2 × 10 mL), and diethyl ether (3 × 10 mL). Pass nitrogen through the resin for 10 min to evaporate off residual diethyl ether. Cover with filter paper and leave the resin to air dry overnight.

B. Polyacrylamide-Supported Gel

Polystyrene and polydimethylacrylamide gels are not sufficiently stable for the mechanical wear of continuous-flow pressure. To alleviate this problem, composites have been constructed in which a rigid matrix supports the soft gel.

1. Kieselguhr-Polyamide

Kieselguhr, a porous, inert, inorganic, but highly heterogeneous material prepared as granules from natural sediments, has been used as a matrix for dimethylacrylamide polymerization to afford a resin that was stable to flow and had excellent properties for synthesis [153]. However, problems caused by aggregation were often observed during difficult peptide synthesis, and reaction rates were slow compared with the unsupported gel polymer [154]. The porous kieselguhr structure allows rapid diffusion and permeation of solvents and reagent to the polyamide gel, where peptide elongation takes place. The resin functionality is provided by sarcosine methyl ester (typically 0.12 mmol/g). Before use, the resin must be further derivatized by treatment with neat ethylenediamine to generate the corresponding aminomethyl resin. Introduction of an appropriate carboxylic acid–containing linker can then be carried out using standard methods of amide bond formation.

a.   Physically Supported Copoly(Dimethylarylamide-bisacryloylethylenediamine-acryloylsarcosine Methyl Ester).   N,N-Dimethylacrylamide (33.3 g, 336 mmol) and acryloylsarcosine methyl ester (2.83 g, 18 mmol) are added to a solution of bisacryloylethylenediamine (3.90 g, 23.3 mmol) in water–DMF (3.12:2, v/v; 137 mL). Ten percent aqueous ammonium persulfate (25 mL) is added and the mixture poured immediately onto the kieselguhr support (335–500 μm; 100 g) and stirred thoroughly with a glass rod. The polymerization mixture is placed into a desiccator and evacuated with a water pump vacuum for 2 min. It is stored in vacuo for 15 min, after which the vacuum is broken by the introduction of nitrogen and the mixture left for 2.5 h. After being washed thoroughly with water on a sintered glass filter, the kieselguhr-polymer is passed through a 700-μm sieve, which releases excess of surface polymer, and is washed on a sintered funnel by backflowing with water. The support is then washed by decantation four times before being filtered off, washed with acetone and ether, and dried under vacuum over P_2O_5 to yield 125.5 g. The sarcosine content is found to be 0.108 mequiv./g.

b. Functionalization of Methoxycarbonyl-Polymer Support. The kieselguhr-polymer (2.5 g, ~0.27 mequiv. sarcosine) is shaken gently in a sealed round-bottomed flask with an excess of ethylenediamine (sufficient to cover the resin) overnight. The ethylenediamine is removed and the resin washed with DMF by swirling and decantation. Most of the resin is transferred to a flow synthesizer glass column and washed with DMF until the effluent gives no coloration upon ninhydrin treatment (~30 min). The symmetrical anhydride (0.475 mmol) of Fmoc-Nle is coupled using DMF as solvent. After 40 min a ninhydrin test of the resin was negative. The resin is washed with DMF, the Fmoc groups removed with piperidine–DMF (1:4) for 10 min, and the resin washed with DMF. Addition of the reversible linkage agent followed. A solution of *p*-hydroxymethylphenoxyacetic acid 2,4,5-trichlorophenyl ester (0.184 g, 0.5 mmol) and HOBt (0.076 g, 0.5 mmol) in DMF (2 mL) is added to the resin. After 25 min the resin gave negative ninhynidrin and trinitrobenzenesulfonic acid tests.

2. Highly Cross-Linked Polystyrene-Polyamide (Polyhipe)

Improvement of flow properties and homogeneity is obtained by substituting the kieselguhr with highly cross-linked (50%) macroporous polystyrene granules as the rigid matrix [155–157]. The poly-N,N-dimethylacrylamide is grafted to the surface by introduction of aminomethyl groups on the polystyrene and acryloylation [156]. The resin functionality is provided by sarcosine methyl ester (typically 0.2–0.4 mmol). The macroporous polystyrene granules have a very high (~90%) pore volume, which leads to high reactivity and reagent permeability. Before use, the resin must be further derivatized to aminomethyl resin and a linker introduced by standard methods of amide bond formation (see functionalization of polyamide resin before this).

C. Multipin Supports

A mobile polymeric surface suitable for peptide synthesis can be radiation grafted to more rigid plastic (polypropylene or polyethylene) pins or small caps attachable to pins (Fig. 14) [158]. The gel supported on an array of rigid pins is then used for multiple synthesis of up to 96 compounds simultaneously. The mobile polymeric surfaces used are polyacrylic and polymethacrylic acid, polystyrene, and poly-2-hydroxyethylmethacrylate (HEMA). Once the pins are grafted, further derivatization is performed.

The preparation of derivatized polyethylene crowns is initiated with the preparation of detachable pins or "crowns" injection molded from granular high-density polyethylene to a dimension of approximately 5.4 mm³. The detachable crowns are immersed in the various monomer solutions, deaerated, and γ-irradiated in a ⁶⁰Co source. Subsequently, the ungrafted polymer is removed by extensive washing to provide a reproducible grafted crown weight. Incorporation of graft polymer is determined by weight average of ~100 crowns before and after grafting.

For both polyacrylic and polymethacrylic acid grafts, *t*-butyloxycarbonyl (Boc)–derivatized 1,6-hexadiamine is coupled and subsequently Boc deprotected to yield an amine handle [159]. Polystyrene grafts are aminomethylated in a manner similar to that of polystyrene resins [45]. Poly-(HEMA) grafts are derivatized by esterification of Fmoc-β-Ala (50 mM) with DCC and 4-dimethylaminopyridine (DMAP) (10 nM) in DMF–DCM (1:3, 0.15 mL per crown). The remaining hydroxyl groups are capped by acetylation with acetic anhydride–triethylamine–DMF (5:1:50, v/v/v), and subsequent Fmoc removal with piperidine–DMF (1:4) results in a free amine. A Rink amide handle has been used for functionalization (Fig. 15, **a** [160,161] and **b** [162]), as has an HMP handle (Fig. 15, **c** [162]). The initial loading obtained is typically 0.44 mmol/g graft polymer, 1.5 mol/cm² grafted surface, or 2 μmol/pin.

1. Preparation of Functionalized Polyethylene Crowns

HEMA functionalized crowns are esterified with Fmoc-β-Ala as follows. All equipment and solid reagents are dried in a desiccator over silica under vacuum for at least 3 h before use. Fmoc-β-Ala is activated with 1 equiv. of DIPCDI at a concentration of 0.05 M and 0.01 M DMAP in DMF–DCM (1:3). After activation for 3 min, the solution is added to HEMA crowns (0.15 mL/crown) and reacted for 25 min at 27°C in a constant-temperature bath. The crowns are washed with DMF and DCM, and any excess HEMA hydroxyl groups are capped by acetylation with acetic anhydride–DIEA–DMF (5:1:50), 0.15 mL/crown, for 90 min at room temperature. The average loading per crown of Fmoc-β-Ala is determined in triplicate using quantitative Fmoc analysis and is found to be typically 1.5 μmol, with a variation of ~0.1 μmol.

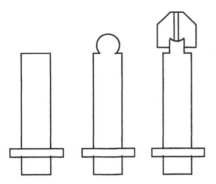

Figure 14   Multipin supports.

2.   Addition of HMP Handle

Fmoc-β-Ala functionalized crowns are treated with piperidine–DMF (1:4) for 20 min, washed with DMF for 2 min and three times with methanol for 2 min, and air dried in an acid-free fume hood for 30 min. A solution of hydroxymethylphenoxyacetic acid-O-pentafluorophenyl ester (0.05 M) and HOBt (0.06 M) in DMF is added to the crowns (0.15 mL/crown) and allowed to react at room temperature overnight. The crowns are filtered, washed with DMF and twice with methanol, and allowed to dry under suction. The presence of unreacted amino groups on the crowns is qualitatively analyzed by the trinitrobenzenesulfonic acid color test. A negative test indicates that the HMP handle coupled in high yield.

3.   Addition of the Rink Amide Handle

Fmoc-β-Ala functionalized crowns are treated with piperidine–DMF (1:4), washed, and dried as just described for the HMP handle. A solution of p-[(R,S)-α-[1-(9H-fluoren-9-yl)methoxyformamido]-2,4-dimethoxybenzyl]-phenoxyacetic acid (0.05 M) and HOBt (0.06 M) in DMF is added to the crowns (0.15 mL/crown) and allowed to react at room temperature overnight. The crowns are washed as for HMP handle addition, and the trinitrobenzenesulfonic acid test is used to analyze the coupling.

D.   ASPECTS

Powdered polyolefin materials, such as ASPECTS (augmented surface polyethylenes prepared by or used for chemical transformations), have been developed as solid supports [163]. In order to functionalize native ASPECT particles and to obtain a higher surface area, an oxidative treatment is per-

polyethylene pin—MA/DMA—NH(CH₂)₆NH—Gly—COCH₂O

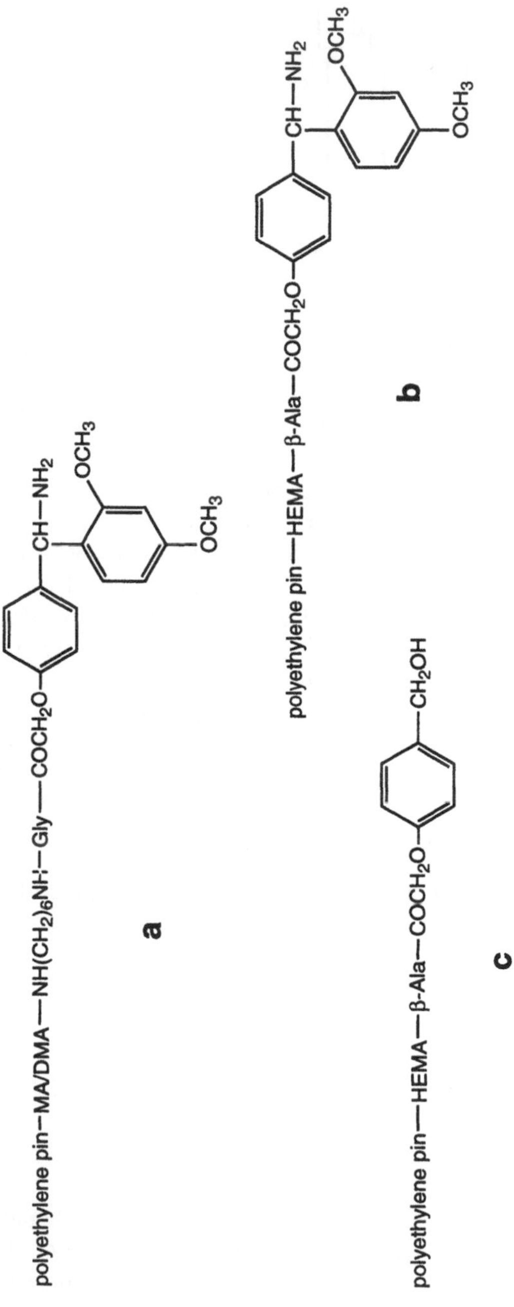

$$\text{polyethylene pin—MA/DMA—NH(CH}_2)_6\text{NH—Gly—COCH}_2\text{O}$$

CH—NH₂

OCH₃

OCH₃

a

polyethylene pin—HEMA—β-Ala—COCH₂O

CH—NH₂

OCH₃

OCH₃

b

polyethylene pin—HEMA—β-Ala—COCH₂O

CH₂OH

c

Figure 15 Handles attached to derivatized polyethylene pins.

formed to introduce groups such —COOH, —OH, or —C=O. The —COOH functionalization requires repeated CrO_3 oxidation or treatment with O_2–O_3 in the presence of a transition metal ion catalyst. This second treatment also introduces hydroxyl and carbonyl groups. Further derivatization is carried out by benzotriazolyl N-oxytris(dimethylamino)- phosphonium hexafluorophosphate (BOP)–mediated addition of 4,7,10- trioxa-1,13-tridecanediamine and $NaBH_4$ or $LiAlH_4$ reduction of extraneous carbonyl functionalities. ASPECTS are mechanically, chemically, and pressure stable and more hydrophilic than the native material. This support is well suited for SPOC because of its inertness to the conditions of numerous chemical reactions.

E. Chitin Support

Chitin is a β-1,4-linked polymer of N-acetyl-2-D-glucosamine found in the shells of crustaceans and the structural matter of insects and fungi. The average chain length of prepared chitin is 5200, corresponding to an average molecular weight of 1.0×10^6. Chitosan, a partially deacetylated form of chitin, contains amino groups that can be used as sites for peptide synthesis.

Peptide-chitin complexes resulting from peptide synthesis on chitin could prove useful for the induction of antipeptide antibodies. Chitin is stable to the conditions of Fmoc chemistry, including methods used to release peptides. The properties of chitin are similar to those of cellulose. Chitosan, unlike cellulose, does not have to be derivatized to create an amino group for attachment of the linker.

1. Chitosan-Linker Preparation [164]

Chitosan, derived from crab shell chitin, is ~80% deacetylated. It is dissolved in 1 M HOAc (5 g/L) and freeze dried to yield a white, soft material. The chitosan is washed with 0.9 M N-methylmorpholine (NMM) in DMF followed by DMF. The Rink linker (0.4 mmol) is dissolved in 6 mL of DMF containing N-[(1H-benzotriazol-1-yl)(dimethylamino)methylene]-N-methyl- methanaminium tetrafluoroborate N-oxide (TBTU) (0.3 M), HOBt (0.3 M), and NMM (0.4 M) and added to 150 mg (dry weight) of chitosan. The mixture is incubated at 45°C for 1 h, washed with DMF, and the chitosan capped with acetic anhydride–dry pyridine (1:1, v/v) for 1 h at 45°C. This procedure yields Fmoc-linker substituted chitin (Fig. 16). After drying in vacuo, the degree of substitution is determined by measuring the Fmoc released after treatment of a sample with piperidine–DMF (3:7) for 30 min at room temperature. Typically, chitosan substitution levels are 0.08–0.35 mmol/g.

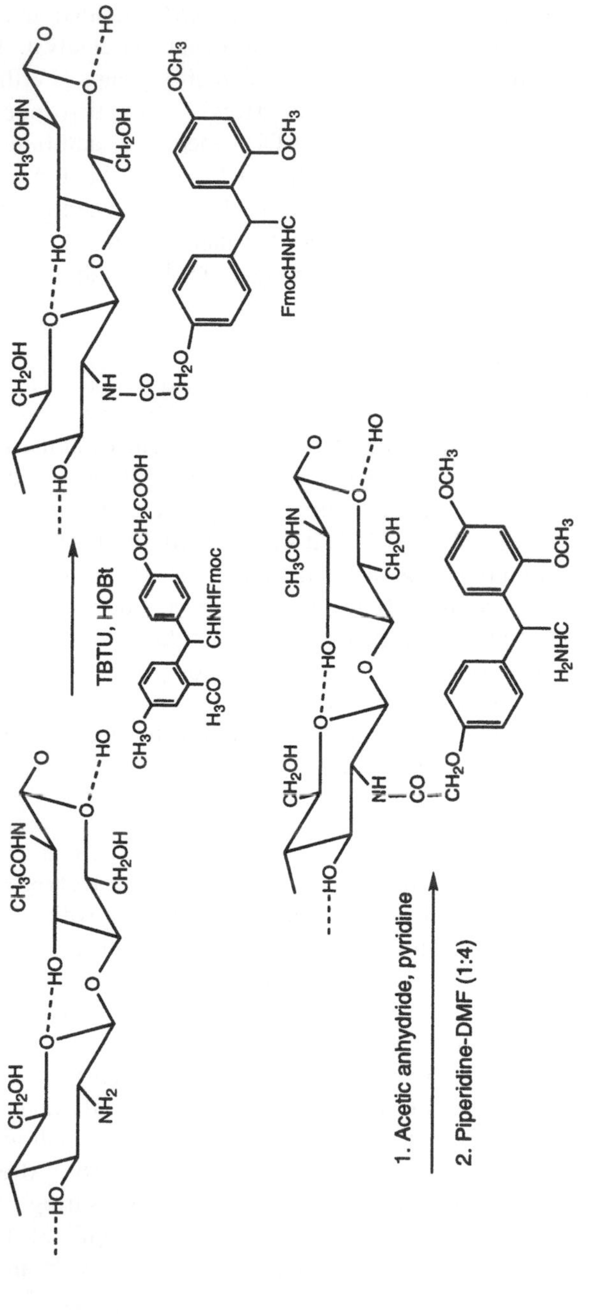

Figure 16 Preparation of linker-chitosan.

F. Sepharose and LH-20

Sepharose and LH-20 [165–167] have been used as supports for SPS. Both are synthetic dextran polymer beads composed of fibers in a porous network. The fragile nature and acid lability of the glycosidic bonds of the dextran are serious limitations to the use of these materials as solid supports. However, they are perfectly compatible with aqueous buffers and can be used in combined chemoenzymatic syntheses [167].

G. Beaded Cellulose

Beaded cellulose has been tested as a more hydrophilic support [168,169]. Although Merrifield [1] found beaded cellulose unsuitable for solid-phase synthesis, the Perloza beaded cellulose has shown promising results as a solid support. Perloza [168,169] is a beaded, non–cross-linked cellulose support that has adequate mechanical properties for the synthesis of small peptides by either batch or continuous-flow methods. It has good solvation properties in a wide variety of solvents including water, dioxane, DMF, dimethyl sulfoxide (DMSO), DCM, and THF. Perloza must be maintained in a solvent swollen state at all times, as upon drying it will not reswell to its original volume.

Perloza functionalization is performed by treatment with acrylonitrile followed by reduction with diborane to yield aminopropylcellulose. Substitution levels are typically 0.3–1.2 mmol/g. Bromoacetyl linker [168], 4-hydroxymethylphenoxyacetic (HMPA) acid linker [168,169], and p-[(R,S)-α-(9H-fluoren-9-yl)-methoxyformamido-2,4-dimethoxybenzyl]phenoxyacetic acid linker [168,169] have been introduced to functionalized Perloza. SPPS on Perloza requires modified Fmoc and Boc protocols [168,169], with Fmoc methodology being more suitable. The advantage of using a hydrophilic support is that the resin-bound peptide can be used for affinity chromatography or generation of antibodies.

1. Functionalization

a. Cyanoethylation. A weighed amount of water wet resin is washed twice with a 2% NaOH solution so that the second wash leaves the resin moist with the NaOH solution. The wet resin is transferred to a conical flask and a volume of dioxane in milliliters equal to the wet weight of the resin in grams is added. The slurry is mixed using a magnetic stirrer and then a calculated amount of acrylonitrile is added. The resulting mixture is stirred for 1 h at room temperature. The resin is filtered and washed with distilled water until the pH of the washings is neutral to pH paper. A sample of the resin is dried for determination of nitrogen content by elemental analysis.

b. Reduction of Cyanoethyl Group. Diborane is generated in a three-necked round-bottom flask using boron trifluoride etherate and sodium borohydride in a solution of THF–diglyme under oxygen-free nitrogen. The nitrogen is exhausted through a paraffin oil bubbler. The stochiometry is calculated to produce a four- to eightfold excess of diborance over cyano groups on the resin. Water-wet cyanoethyl Perloza is washed twice with dioxane, three times with THF, and three times with THF distilled from sodium wire. The flow of oxygen-free nitrogen to the diborane reactor is increased and the cyanoethyl resin is added slowly with swirling to the diborane solution to control the vigorous bubbling that results. The resin slurry is refluxed under oxygen-free nitrogen for 3 h, cooled, and treated cautiously with 95% ethanol to decompose any remaining diborane. The resin is filtered on a sintered glass funnel, washed twice with 1 M HCl, allowed to stand in 1 M HCl for 5–10 min, and then washed with distilled water until washings are neutral to pH paper. The resin is washed twice with 2% NaOH solution and allowed to stand 5–10 min after the second addition of NaOH solution. The resin is washed with distilled water until the washings are neutral to pH paper and stored swollen with distilled water. The amine substitution is determined by titration with picric acid [170].

c. Attachment of the Amide Linker. Amino Perloza (0.44 mmol/g, 0.5 mmol total amine) is washed twice with dioxane and five times with DMF. To the DMF-moist resin is added 0.75 mmol of the *p*-[(*R,S*)-α-(9*H*-fluoren-9-yl)-methoxyformamido-2,4-dimethoxybenzyl]phenoxyacetic acid linker (1.5 equiv.) and 1 mmol HOBt·H₂O (2 equiv.). DMF (5 mL) is added and the slurry mixed thoroughly to dissolve the reagents. DIPCDI (0.83 mmol, 1.66 equiv.) is added to the slurry and mixing is continued. After 9 h the residual amino groups are acetylated.

H. Protein Precipitates

Protein precipitates of bovine serum albumin (BSA) have been used for synthesis of small peptides with reasonable yields [171]. The use of protein supports provides a practical method for immunological studies in which the product can be used directly in immunizations.

VII. ANALYTICAL METHODS

A. Amino Acid Analysis

Quantitative amino acid analysis can be used to determine the loading of a resin. Following esterification of the first amino acid, the amino acid-resin is hydrolyzed with a 12 N HCl–propionic acid (1:1) solution for 90 min at

150°C [172]. Derivatization of the hydrolyzate with O-phthalaldehyde (OPA) [173], phenyl isocyanate (PTC chemistry) [174,175], or aminoquinolylhydroxysuccinimidyl carbamate (AQC) [176,177] allows amino acid detection and quantitation by fluorescence (OPA or AQC) or UV absorbance (PTC chemistry) and HPLC separation.

Procedure: Weigh accurately a sample of the amino acid-resin (~10 mg) into a glass hydrolysis tube having a Teflon-lined screw cap. Add 0.5 mL propionic acid and 0.5 mL 12 N HCl. Freeze the sample in an acetone–dry ice bath, then evacuate the tube on a water aspirator to remove air from the resin beads. Seal the tube using the screw cap, and allow the sample to thaw. Repeat the "freeze–evacuate–thaw" steps two more times. Briefly flush the tube with N_2 (do NOT bubble through the liquid; HCl will be lost, as the solution is supersaturated) and quickly screw on the cap. Place the tube in a preheated 130°C heating block. Hydrolyze single amino acids on Merrifield and hydroxymethyl resins for 2 h. Single amino acids on BHA and MBHA resins need heating for 16 h. Remove the tube from the block carefully and allow it to cool. Quantitatively transfer the contents of the hydrolysis tube to a 10-mL volumetric flask with distilled water. Remove resin particles by passing the hydrolysate and washings through a disposable pipet filled with a small amount of glass wool. This filtering device is washed with water prior to use. The resin may also be removed by filtering through Whatman No. 1 filter paper in a small Büchner funnel. Remove an appropriate aliquot and dilute with sample buffer for application to the amino acid analyzer.

B. Elemental Analysis

Elemental analysis is used for determination of the loading of different resins, such as amino-, bromo-, chloro-, and sulfo-resins, from the N, Br, Cl, and S contents, respectively. An accurate weight of resin is submitted for elemental analysis. Combustion of the sample provides the various elements present in the sample. The substitution level of the resin is calculated as follows:

X% of element A = X g A/100 g of resin

(X g A) × 10/(MW of A) = mmol A/g resin

C. Color Tests

1. Ninhydrin (Kaiser) Test

The qualitative ninhydrin, or Kaiser, test can be used for monitoring functionalization of an amino-resin by a linker. Alternatively, the quantitative ninhydrin test can be used to determine the loading of an amino resin.

a. Qualitative Ninhydrin Test
Reagents:

1. Cyanide. Dissolve 33 mg KCN in 50 mL water to create a 0.01 M KCN solution. Dilute 2 mL of the solution to 100 mL with pyridine.
2. Ninhydrin. Dissolve 2.5 g ninhydrin in 50 mL ethanol.
3. Phenol. Dissolve 80 g phenol in 20 mL *t*BuOH.

Procedure: A small resin sample is removed from the reaction vessel. The resin is washed through a fritted glass funnel by suction or in the test tube by decantion three times with 2 mL ethanol–HOAc (1:1) and three times with 2 mL ethanol. Two to three drops of reagents 1, 2, and 3 are added to the test tube containing the resin. The tube is placed in a preheated block at 100°C for 5 min. Observe the color of the beads by holding them against a white background. The test is negative and the coupling is complete if the solution is yellow with no coloration in the beads. The degree of coupling and the color of the beads and solution were determined by Kaiser et al. [178] for the attachment of Boc-Ala to Phe-resin (Table 4). A control test is recommended prior to coupling to provide a background color intensity.

b. Quantitative Ninhydrin Test
Reagents:

1. Cyanide reagent from above. Stir the solution with 4 g of Amberlite MB-3 mixed bed resin for 45 min to remove traces of ammonia, then filter.
2. Ninhydrin reagent from above.
3. Phenol reagent from above. Treat with MB-3 resin to remove ammonia, as above.

Table 4   Completeness of Coupling of Boc-Ala to Phe-Resin[a]

Percent reacted	Ninhydrin color reaction
76.0	Beads, dark blue; solution, dark blue
84.0	Beads, dark blue; solution, moderately blue
94.0	Beads, moderately blue; solution, slight blue
99.4	Beads, slightly blue; solution, trace of blue

[a]These results indicate that as little as 5 μmol/g peptide-resin free amino group can be detected by this method. If the resin contains no quaternary charged groups (e.g., PAM or BHA resins), the blue color will all be in the solution.

4. Tetraethylammonium chloride. Dissolve 8.28 g Et$_4$NCl in DCM to create 100 mL of a 0.5 M solution.
5. Water–ethanol (2:3).
6. Triethylamine–DCM (1:19).

Procedure: Wash the resin sample twice with reagent 6 and three times with DCM, then dry the sample in vacuo. Weigh 2–5 mg into a 10 × 75 mm test tube. Add 0.1 mL reagent 1, 0.04 mL reagent 2, and 0.05 mL reagent 3. Mix, heat at 100°C for 10 min, let cool, then chill in cold water. For Fmoc-resins premature removal of the Fmoc group (by pyridine) is minimized by adding 2–3 drops (20–40 μL) of glacial HOAc to each resin sample and heating the reaction mixture for 5 min instead of 10 min. Add 1 mL reagent 5. Mix and filter through a glass wool plug in a disposable pipet. Rinse twice with 0.2 mL of reagent 4. Dilute the combined filtrate and washes to 2.0 mL with reagent 5. Read the abosrbance at λ = 570 nm against a reagent blank. The concentration of free amino groups is calculated as follows:

$$\text{Free amino groups (mmol/g)} = \frac{A_{570} \times 10^6 \ \mu\text{mol/mol} \times 0.005 \ \text{L}}{15000 \ \text{M}^{-1} \ \text{cm}^{-1} \times 1 \ \text{cm} \times \text{mg of resin}}$$

Note that this test is not applicable for N-terminal Pro residues.

2. 2,4,6-Trinitrobenzenesulfonic Acid (TNBS) Test

TNBS provides a qualitative test for monitoring coupling reactions [179].

Reagents:

Ethanol–HOAc (1:1).
1% (w/v) picrylsulfonic acid (TNBS) in very pure DMF.

Procedure: Remove a resin sample and wash with ethanol–HOAc (3 × 2 mL) and ethanol (3 × 2 mL). Transfer the resin to a small centrifuge tube with 1 mL ethanol. Add 1 drop of the TNBS reagent, mix, and let stand at room temperature for 10 min. Centrifuge if necessary to collect the beads. Free amino groups result in a red–orange color of the beads. Completely negative beads have a steel-gray color. The TNBS test has approximately the same sensitivity as the Kaiser test.

D. Fmoc Monitoring

The Fmoc group, when present on the resin or introduced via a derivatized amino acid (usually Fmoc-Nle or Fmoc-Gly) esterified to the resin, can be quantitated to determine the resin loading. Removal of the Fmoc group by piperidine results in the formation of a dibenzofulvene-piperidine adduct (Fig. 17). Monitoring the UV absorbance of this adduct upon release from the resin allows calculation of the resin substitution level.

Procedure: Fmoc–amino acid–resin (4–8 mg) is shaken or stirred in piperidine–DMF (3:7) (0.5 mL) for 30 min, following which methanol (6.5 mL) is added and the resin is allowed to settle. The resultant fulvene-piperidine adduct has UV absorption maxima at $\lambda = 267$ nm ($\varepsilon = 17{,}500$ M^{-1} cm^{-1}), 290 nm ($\varepsilon = 5800$ M^{-1} cm^{-1}), and 301 nm ($\varepsilon = 7800$ M^{-1} cm^{-1}). For reference, a piperidine–DMF–methanol solution (0.3:0.7:39) is prepared. Spectrophotometric analysis is typically carried out at 301 nm, with comparison with a free Fmoc-amino acid (i.e., Fmoc-Ala) of known concentration treated under identical conditions. The substitution level is calculated from [180]

$$\text{Substitution level (mmol/g)} = \frac{A_{301} \times 106 \ \mu\text{mol/mol} \times 7 \ \text{mL}}{7800 \ M^{-1} \ cm^{-1} \times 1 \ \text{cm} \times \text{mg of resin}}$$

E. Picric Acid

Picric acid can be used for rapid measurement of the amine content of an insoluble support [170]. The amine-containing resin is reacted with picric acid, then treated with an excess of base, causing quantitative release of the picrate from the polymer into solution. The concentration of picrate in solution is determined spectrophotometrically and reflects the amine content of the resin.

Reagents:

1. Picric acid (trinitrophenol), 0.1 M in DCM (229.1 g/L).
2. DIEA–DCM (1:19).

Procedure: Swell the resin with DCM in a fritted glass funnel that has a Teflon stopcock in the stem. Neutralize with reagent 2 twice for 1 min each time. Wash well with DCM five times for 1 min each. Treat with reagent 1 twice for 1 min each. Wash well with DCM five times for 1 min each. Elute the picrate with reagent 2 twice for 1 min each and save the eluate. Dilute the eluted picrate with 95% ethanol to obtain a suitable absorbance. The final solution should not contain more than 20% DCM. Read the absorbance at $\lambda = 358$ nm. DIEA-picrate has an $\varepsilon_{358} = 14{,}500$ M^{-1} cm^{-1} (i.e., a 10^{-5} M solution has $A_{358} = 0.145$).

F. Modified Volhard Method

The modified Volhard method is used for determination of chloro-resin loading. The sample is acidified with HNO_3 and the chloride is precipitated with

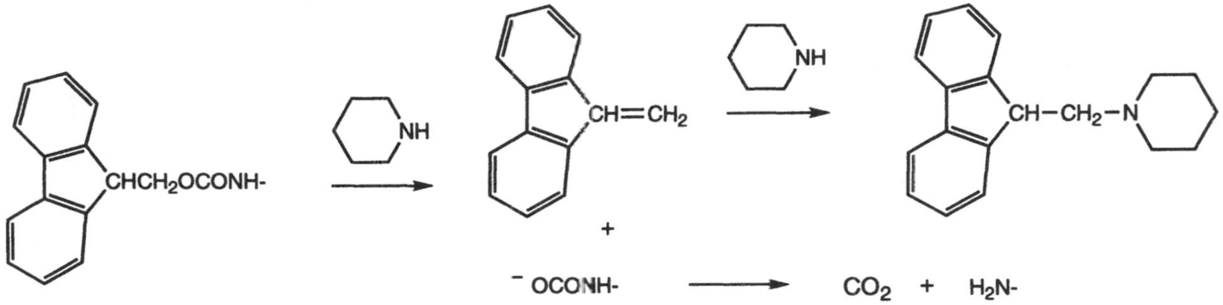

Figure 17 Formation of dibenzofulvene-piperidine adduct by treatment of an Fmoc-protected amine with piperidine.

a measured excess of standard $AgNO_3$ solution. The AgCl that is formed is coated with toluene and the excess $AgNO_3$ is back titrated with standard NH_4SCN solution, using ferric alum [$FeNH_4(SO_4)_2 \cdot 12H_2O$] as indicator. A red color, due to the formation of $Fe(SCN)_3$, indicates that an excess of SCN is present and that the end point has been reached.

Reagents:

1. Standard $AgNO_3$, 0.1 N in water (16.989 g/L).
2. Standard NH_4SCN, 0.1 N in water (7.612 g/L). Because the NH_4SCN must be standardized against $AgNO_3$ to ensure that 1.0 mL of NH_4SCN is equivalent to 1.0 mL of standard $AgNO_3$, prepare a slightly more concentrated solution of NH_4SCN and dilute it to 0.1 N after titration. The titration procedure is the same as described in the following, except that no sample is added. Calculate the amount of water that should be added to the NH_4SCN to make 10 mL of NH_4SCN exactly equivalent to 10 mL of $AgNO_3$. Repeat the procedure in the presence of a known concentration of NaCl (1 mL of 1 N NaCl).
3. NaCl, 1 N in water (58.45 g/L).
4. Saturated ferric alum, 124 g of $NH_4Fe(SO_4)_2 \cdot 12H_2O$ dissolved in 100 mL water.
5. HNO_3, 1 N in water.

Procedure: Pipet the sample or standard to be titrated into ~10 mL of water in a 250-mL Erlenmeyer flask protected from bright light. For titration of HCl-HOAc, a 1-mL sample should be used. To the flask add about 3 drops of reagent 4 and 1 mL of reagent 5. Add a magnetic stirring bar to the flask and place it on a magnetic stirrer. Slowly add 20 mL of reagent 1 with stirring. Stop the stirrer and let the mixture stand for 5 min. Add ~50 mL of water, followed by toluene, so that about a 1/4-inch layer of toluene is left on the water surface. Mix well with the stirrer. With the stirrer on, titrate with reagent 2. The first permanent tinge of red–brown indicates the end point.

G. Infrared Spectroscopy

Infrared spectroscopy is used as a monitoring method when the resin is functionalized with linkers that have chemical groups that absorb in the IR spectral region, such as carbonyls, esters, amides, or oximes. IR spectroscopy has also been used to monitor quantitative loading of a resin by comparing the difference in absorbance between two peaks, one from the starting material and the other one from the final product [45].

Procedure: Dry an aliquot of resin over P_2O_5 in vacuo and transfer a 4-mg sample to an agate mortar. Grind the beads thoroughly, add 100 mg of IR-quality KBr, and regrind. Press the pellet by standard methods. If the resin is not adequately dispersed in the KBr, the pellet may break upon removal of pressure. If this happens, regrind and repress the resin. Run a reference spectrum on the support alone to provide a background. The functional group bands can be identified by comparison with the reference.

ACKNOWLEDGMENTS

We gratefully acknowledge the support of the National Institutes of Health (CA77402, HL62427, and AR 01929). P.F. is a recipient of a grant for training of postdoctoral staff from Ministerio de Educación y Culture, Spain, and G.B.F. is a recipient of an NIH Research Career Development Award.

REFERENCES

1. RB Merrifield. J Am Chem Soc 85:2149–2154, 1963.
2. JB Backes, JA Ellman. J Am Chem Soc 116:11171–11172, 1994.
3. JB Backes, AA Virgilio, JA Ellman. J Am Chem Soc 118:3055–3056, 1996.
4. LM Gayo, MJ Suto. Tetrahedron Lett 38:211–214, 1997.
5. I Sucholeiki. Tetrahedron Lett 35:7307–7310, 1994.
6. MJ Plunkett, JA Ellman. J Org Chem 60:6006–6007, 1995.
7. B Chenera, JA Finkelstein, DF Veber. J Am Chem Soc 117:11999–12000, 1995.
8. K Ngu, DV Patel. Tetrahedron Lett 38:973–976, 1997.
9. W Heitz, KL Platt. Makromol Chem 127:113–140, 1969.
10. T Wieland, C Birr, F Flor. Justus Liebigs Ann Chem 727:130–137, 1969.
11. KC Pugh, EJ York, JM Stewart. Int J Peptide Protein Res 40:208–213, 1992.
12. GB Fields, CB Fields. J Am Chem Soc 113:4202–4207, 1991.
13. E Paetzold, G Oehme, H Pracejus. React Polym 14:75–80, 1991.
14. KW Pepper, D Reichenberg, D Hale. J Chem Soc 3129–3136, 1952.
15. HP Gregor, GK Hoeschele, J Potenza, AG Tsuk, R Feinland, M Shida, P Teyssié. J Am Chem Soc 87:5525–5534, 1965.
16. B Gutte, RB Merrifield. J Biol Chem 246:1922–1941, 1971.
17. J Rudinger, P Buetzer. In: Y Wolman, ed. Peptides 1974. New York: Wiley, 1975, pp 211–219.
18. JP Tam, Y-A Lu. J Am Chem Soc 117:12058–12063, 1995.
19. VK Sarin, SBH Kent, RB Merrifield. J Am Chem Soc 102:5463–5470, 1980.
20. L Andresson, M Lindqvist. In: CH Schneider, AN Eberle, eds. Peptides 1992, Proceedings of the Twenty-Second European Peptide Symposium. Leiden, The Netherlands: ESCOM, 1993, pp 265–266.

21. BF McGuinness, SD Britt, N Mu, D Whitney, N Afeyan. In: HLS Maia, ed. Peptides 1994, Proceedings of the Twenty-Third European Peptide Symposium. Leiden, The Netherlands: ESCOM, 1994, pp 277–278.
22. M Meldal. Methods Enzymol 289:83–104, 1997.
23. R Arshady, E Atherton, DLJ Clive, RS Sheppard. J Chem Soc Perkin Trans I 529–537, 1981.
24. P Kanda, RC Kennedy, JT Sparrow. Int J Peptide Protein Res 38:385–391, 1991.
25. W Rapp, L Zhang, R Häblish, E Bayer. In: G Jung, E Bayer, eds. Peptides 1988, Proceedings of the Twentieth European Peptide Symposium. Berlin: de Gruyter, 1989, pp 199–201.
26. DD Ho, AV Neumann, AS Perelson, W Chen, JM Leonard, M Markowitz. Nature 373:123–126, 1995.
27. S Zalipsky, JL Chang, F Albericio, G Barany. React Polym 22:243–258, 1994.
28. M Meldal. Tetrahedron Lett 33:3077–3080, 1992.
29. M Renil, M Meldal. Tetrahedron Lett 36:4647–4650, 1995.
30. M Renil, M Meldal. Tetrahedron Lett 37:6185–6188, 1996.
31. M Kempe, G Barany. J Am Chem Soc 118:7083–7093, 1996.
32. R Arshady, A Ledwith. React Polym 1:159–174, 1983.
33. S Wang, X Zhang. Polym Adv Technol 2:93, 1991.
34. F Svec, JMJ Frechet. Science 273:205–211, 1996.
35. R Arshady. J Chromatogr 586:199–219, 1991.
36. M Eggenweiler, N Clausen, H Fritz, L Zhang, E Bayer. In: HLS Maia, ed. Peptides 1994, Proceedings of the Twenty-Third European Peptide Symposium. Leiden, The Netherlands: ESCOM, 1995, pp 275–276.
37. J-M Guenet, M Klein. Macromol Chem Macromol Symp 39:85, 1996.
38. H Sillescu, R Brüssau. Chem Phys Lett 5:525, 1970.
39. F Müller-Plathe. Chem Phys Lett 252:419–424, 1996.
40. LM Dowling, GR Stark. Biochemistry 8:4728–4734, 1969.
41. JA Patterson. In: GR Stark, ed. Biochemical Aspects of Reactions in Solid Supports. New York: Academic Press, 1971, pp 189–213.
42. NM Weinshenker, C-M Shen. Tetrahedron Lett: 3281–3284, 1972.
43. JT Sparrow. J Org Chem 41:1350–1353, 1976.
44. AR Mitchell, SBH Kent, BW Erickson, RB Merrifield. Tetrahedron Lett: 3795–3798, 1976.
45. AR Mitchell, SBH Kent, M Engelhard, RB Merrifield. J Org Chem 43:2845–2852, 1978.
46. DH Rich, SK Gurwara. Tetrahedron Lett: 301–304, 1975.
47. DH Rich, SK Gurwara. J Am Chem Soc 97:1575–1579, 1975.
48. MS Bernatowicz. Tetrahedron Lett 30:4645–4648, 1989.
49. PG Pietta, GR Marshall. Chem Commun 650–651, 1970.
50. PG Pietta, O Brenna. J Org Chem 40:2995–2996, 1975.
51. CD Floyd, CN Lewis, SR Patel, M Whittaker. Tetrahedron Lett 37:8045–8048, 1996.
52. PG Pietta, PF Cavallo, K Takahashi, GR Marshall. J Org Chem 39:44–48, 1974.

53. J Rivier, P Brazeau, W Vale, R Guillemin. J Med Chem 18:123–126, 1975.
54. RC Orlowski, R Walter, D Winkler. J Org Chem 41:3701–3705, 1976.
55. PH Chapman, D Walker. J Chem Soc, Chem Commun:690–691, 1975.
56. GL Southard, GS Brooke, JM Pettee. Tetrahedron Lett:3505–3508, 1969.
57. GL Southard, GS Brooke, JM Pettee. Tetrahedron 27:2701–2703, 1971.
58. SS Wang. J Org Chem 40:1235–1239, 1975.
59. RL Letsinger, MJ Kornet. J Am Chem Soc 86:5163–5165, 1964.
60. AM Felix, RB Merrifield. J Am Chem Soc 92:1385–1391, 1970.
61. H Kinz, B Dombo. Angew Chem Int Ed Engl 27:711–713, 1987.
62. MJ Farral, JM Fréchet. J Org Chem 41:3877–3882, 1976.
63. JT Sparrow. Tetrahedron Lett:4637–4638, 1975.
64. H Yajima, H Kawatani, H Watanabe. Chem Pharm Bull 18:1333–1339, 1970.
65. RB Merrifield. Adv Enzymol Relat Areas Mol Biol 32:226–233, 1969.
66. MA Tilak. Tetrahedron Lett:6323–6326, 1968.
67. CH Li, D Yamashiro, L-F Tseng, HH Loh. J Med Chem 20:325–328, 1977.
68. SS Wang. J Org Chem 41:3258–3261, 1976.
69. T Mizoguchi, K Shigezane, N Takamura. Chem Pharm Bull 18:1465–1474, 1970.
70. RL Letsinger, V Mahadevan. J Am Chem Soc 88:5319–5324, 1966.
71. JS Panek, B Zhu. Tetrahedron Lett 37:8151–8154, 1996.
72. TM Fyles, CC Leznoff. Can J Chem 54:935–942, 1976.
73. F Camps, J Castells, MJ Ferrando, J Font. Tetrahedron Lett:1713–1714, 1971.
74. LT Scott, J Rebeck, L Ovsyanko, CL Sims. J Am Chem Soc 99:625–626, 1977.
75. JM Burlitch, RC Winterton. J Am Chem Soc 97:5605–5606, 1975.
76. RL Letsinger, MJ Kornet. J Am Chem Soc 85:3045–3046, 1963.
77. CR Harrison, P Hodge, J Kemp, GM Perry. Makromol Chem 176:267–274, 1975.
78. JM Fréchet, C Schuerch. J Am Chem Soc 93:492–496, 1971.
79. R Arshady, GW Kenner, A Ledwith. Makromol Chem 177:2911–2918, 1976.
80. RS Feinberg, RB Merrifield. Tetrahedron 28:5865–5871, 1972.
81. J Blake, CH Li. J Chem Soc, Chem Commun:504–505, 1976.
82. JJ Dahlmans. In: H Hanson, HD Jakubke, eds. Peptides 1972. Amsterdam: North-Holland, 1973, pp 171–172.
83. GE Martin, MB Shambhu, SR Shakhshir, GA Digenis. J Org Chem 43:4571–4574, 1978.
84. H Rink. Tetrahedron Lett 28:3787–3790, 1987.
85. SS Wang, RB Merrifield. J Am Chem Soc 91:6488–6491, 1969.
86. ET Wolters, GI Tesser, RJ Nivard. J Org Chem 39:3388–3392, 1974.
87. JD Glass, IL Schwartz, R Walter. J Am Chem Soc 94:6209–6211, 1972.
88. P Sieber. Tetrahedron Lett 28:2107–2110, 1987.
89. X Beebe, NE Schore, MJ Kurth. J Am Chem Soc 114:10061–10062, 1992.
90. MA Tilak, CS Holliden. Tetrahedron Lett:1297–1300, 1968.
91. R Sheppard, BJ Williams. Int J Peptide Protein Res 20:451–454, 1982.

92. A Flörsheimer, B Riniker. In: E Giralt, D Andreu, eds. Peptides 1990, Proceedings of the Twenty-First Europena Peptide Symposium. Leiden, The Netherlands: ESCOM, 1991, pp 131–133.
93. M Mergler, R Tanner, J Gosteli, P Grogg. Tetrahedron Lett 29:4005–4008, 1988.
94. R Arshady, GW Kenner, A Ledwith. J Polym Sci Polym Chem Ed 12:2017–2025, 1974.
95. GI Tesser, JT Buis, ET Wolters, EG Bothé-Helmes. Tetrahedron 32:1069–1072, 1976.
96. E Bayer, E Breitmaker, G Jung, W Parr. Hoppe-Seylers Z Physiol Chem 352:759–760, 1971.
97. BF Gisin, RB Merrifield. J Am Chem Soc 94:6165–6170, 1972.
98. BW Erickson, RB Merrifield. J Am Chem Soc 95:3757–3763, 1973.
99. SS Wang. J Am Chem Soc 95:1328–1333, 1973.
100. G Lu, S Mojsov, JP Tam, RB Merrifield. J Org Chem 46:3433–3436, 1981.
101. J Blake, CH Li. Int J Protein Res 3:185–189, 1971.
102. E Flanigan, GR Marshall. Tetrahedron Lett:2403–2406, 1970.
103. DL Marshall, IE Liener. J Org Chem 35:867–868, 1970.
104. RB Merrifield. Biochemistry 3:1385–1390, 1964.
105. RA Laursen. Eur J Biochem 20:89–102, 1971.
106. DH Rich, SK Gurwara. J Chem Soc Chem Commun:610–611, 1973.
107. WF DeGrado, ET Kaiser. J Org Chem 47:3258–3261, 1982.
108. JR Morphy, Z Rankovic, DC Rees. Tetrahedron Lett 37:3209–3212, 1996.
109. DM Dixit, CC Leznoff. J Chem Soc Chem Commun:798–799, 1977.
110. DM Dixit, CC Leznoff. Isr J Chem 17:248–252, 1978.
111. J Alsina, F Rabanal, E Giralt, F Albericio. Tetrahedron Lett 35:9633–9636, 1994.
112. GW Kenner, JR McDermott, RC Sheppard. J Chem Soc D:636–637, 1971.
113. LC Dorman, J Love. J Org Chem 34:158–165, 1969.
114. F Cramer, H Köster. Angew Chem Int Ed Engl 7:473–474, 1968.
115. JM Fréchet, LJ Nuyens. Can J Chem 54:926–934, 1976.
116. JM Fréchet, KE Haque. Tetrahedron Lett:3055–3056, 1975.
117. D Bellof, M Mutter. Chimia 39:317–320, 1985.
118. SJ Treague. Tetrahedron Lett 37:5751–5754, 1996.
119. JM Stewart, JD Young. Solid Phase Peptide Synthesis. 2nd ed. Rockford, IL: Pierce, 1984.
120. RH Berg, K Amdal, WB Pedersen, A Holm, JP Tam, RB Merrifield. J Am Chem Soc 111:8024–8026, 1989.
121. PHH Hermkens, HCJ Ottenheijm, D Rees. Tetrahedron 52:4525–4554, 1996.
122. PHH Hermkens, HCJ Ottenheijm, D Rees. Tetrahedron 53:5643–5678, 1997.
123. H Gehrhardt, M Mutter. Polym Bull 18:487–493, 1987.
124. M Mutter, R Uhmann E Bayer. Liebigs Ann Chem:901–915, 1975.
125. W Rapp, L Zhang, E Bayer. In: R Epton, ed. Innovation and Perspectives in Solid Phase Synthesis. Birmingham, UK: SPCC, 1990, pp 205–210.
126. S Zalipsky, F Albericio, G Barany. In: CM Deber, VJ Hruby, KD Kopple,

eds. Peptides 1985, Proceedings of the Ninth American Peptide Symposium. Rockford, IL: Pierce, 1986, pp 257–260.

127. FI Auzanneau, M Meldal, K Bock. J Peptide Sci 1:31–44, 1995.
128. M Meldal, FI Auzanneau, K Bock. In: R Epton, ed. Innovation and Perspectives in Solid Phase Synthesis. Birmingham, UK: Mayflower Worldwide, 1994, pp 259–266.
129. MD Matteucci, MH Caruthers. J Am Chem Soc 103:3185–3191, 1981.
130. GR Gough, MJ Brunden, PT Gilham. Tetrahedron Lett 22:4177–4180, 1981.
131. SP Adams, KS Kavka, EJ Wykes, SB Holder, GR Gallupi. J Am Chem Soc 105:661–663, 1983.
132. R Fathi, M Rudolph, RG Gentles, R Patel, EW MacMillan, MS Reitman, D Pelham, AF Cook. J Org Chem 61:5600–5609, 1996.
133. K Büttner, H Zahn, WH Fischer. In: GR Marshall, ed. Peptides: Chemistry and Biology. Leiden, The Netherlands: ESCOM, 1988, pp 210–211.
134. H Gausepohl, W Rapp, E Bayer, RW Frank. In: R Epton, ed. Innovation and Perspectives in Solid Phase Synthesis. Andover, UK: Intercept Ltd, 1992, pp 381–385.
135. PJ Robinson, P Dunnill, MD Lilly. Biochim Biophys Acta 242:659–661, 1971.
136. M Schuster, P Wang, JC Paulson, C-H Wong. J Am Chem Soc 116:1135–1136, 1994.
137. R Majors, M Hopper. J Chromatogr Sci 12:767–778, 1974.
138. DM Benjamin, JJ McCormak, DW Gump. Anal Chem 45:1531, 1973.
139. R Frank, R Döring. Tetrahedron 44:6031–6040, 1988.
140. F Albericio, J Bacardit, G Barany, JM Coull, M Egholm, E Giralt, GW Griffin, SA Kates, E Nicolas, NA Sole. In: HLS Maia, ed. Peptides 1994, Proceedings of the Twenty-Third European Peptide Symposium. Leiden, The Netherlands: ESCOM, 1995, pp 271–272.
141. B Blankemeyer-Menge, R Frank. In: E R., ed. Innovation and Perspectives in Solid Phase Synthesis. Birmingham, UK: SPCC, 1990, pp 465–472.
142. B Blankemeyer-Menge, M Nimtz, R Frank. Tetrahedron Lett 31:1701–1704, 1990.
143. J Eichler, M Bienert, NF Sepetov, P Stolba, V Krchnák, O Smékal, V Gut, M Lebl. In: R Epton, ed. Innovation and Perspectives in Solid Phase Synthesis. Birmingham, UK: SPCC, 1990, pp 337–343.
144. J Eichler, A Beinert, A Stierandova, M Lebl. Peptide Res 4:296–307, 1991.
145. M Rinnová, J Jekez, P Malon, M Lebl. Peptide Res 6:88–94, 1993.
146. R Frank. In: R Epton, ed. Innovation and Perspective in Solid Phase Synthesis. Birmingham, UK: Mayflower Worldwide, 1994, pp 509–512.
147. M Lebl, J Eicher. Peptide Res 2:297–300, 1989.
148. JA Buettner, D Hudson, CR Johnson, MJ Ross, K Shoemaker. In: R Epton, ed. Innovation and Perspective in Solid Phase Synthesis. Birmingham, UK: Mayflower Worldwide, 1994, pp 169–174.
149. R Fitzpatrick, P Goddard, R Stankowski, J Coull. In: R Epton, ed. Innovation and Perspectives in Solid Phase Synthesis. Birmingham, UK: Mayflower Worldwide, 1994, pp 157–162.

150. SB Daniels, MS Bernatowicz, JM Coull, H Köster. Tetrahedron Lett 30:4345–4348, 1989.
151. JT Sparrow, NG Kneib-Cordonier, P Kanda, NU Obeyesekere, JS McMurray. In: HLS Maia, ed. Peptides 1994, Proceedings of the Twenty-Third European Peptide Symposium. Leiden, The Netherlands: ESCOM, 1995, pp 281–282.
152. M Meldal, I Svendsen, K Breddam, FI Auzanneau. Proc Natl Acad Soc U S A 91:3314–3318, 1994.
153. E Atherton, E Brown, RC Sheppard. J Chem Soc Chem Commun 1151–1152, 1981.
154. M Meldal, RC Sheppard. In: D Theodoropoulos, ed. Peptides 1986, Proceedings of the Nineteenth European Peptide Symposium. Berlin: de Gruyter, 1987, pp 131–134.
155. DP Gregory, N Bhaskar, RC Sheppard, S Singleton. In: R Epton, ed. Innovation and Perspectives in Solid Phase Peptide Synthesis. Andover, UK: Intercept Limited, 1992, pp 391–396.
156. DC Sherrington. In: R Epton, ed. Innovation and Perspectives in Solid Phase Synthesis. Birmingham, UK: SPCC, 1990, pp 71–86.
157. PW Small, DC Sherrington. J Chem Soc Chem Commun:1589–1591, 1989.
158. AM Bray, NJ Maeji, HM Geysen. Tetrahedron Lett 31:5811–5814, 1990.
159. HM Geysen, SJ Rodda, TJ Mason, G Tibbick, PG Schoofs. J Immunol Methods 102:259–274, 1987.
160. AA Virgilio, JA Ellman. J Am Chem Soc 116:11580–11581, 1994.
161. AM Bray, DS Chiefari, RM Valeiro, NJ Maeji. Tetrahedron Lett 36:5081–5084, 1995.
162. RM Valeiro, AM Bray, NJ Maeji. Int J Peptide Protein Res 44:158–165, 1994.
163. RM Cook, D Hudson. In: PTP Kaumaya, RS Hodges, eds. Proceedings of the Fourteenth American Peptide Symposium. Kingswinford, UK: Mayflower Scientific, 1996, pp 39–41.
164. W Neugebauer, RE Williams, J-R Barbier, R Brzezinski. Int J Peptide Protein Res 47:269–275, 1996.
165. GP Vlasov, AY Bilibin, NN Skvortsova, U Kalejs, NY Kozhevnikova, G Aukone. In: HLS Maia, ed. Peptides 1994, Proceedings of the Twenty-Third European Peptide Symposium. Leiden, The Netherlands: ESCOM, 1995, pp 273–274.
166. A Orlowska, E Holodowicz, S Drabarek. Pol J Chem 55:2349–2354, 1981.
167. O Blixt, T Norberg. J Carbohydrate Chem 16:143–154, 1997.
168. DR Englebretsen, DRK Harding. Int J Peptide Protein Res 40:487–496, 1992.
169. DR Englebretsen, DRK Harding. Int J Peptide Protein Res 43:546–496, 1994.
170. BF Gisin. Anal Chim Acta 58:248–249, 1972.
171. PR Hansen, A Holm, G Houen. Int J Peptide Protein Res 41:237–245, 1993.
172. F Westfall, H Hesser. Anal Chem 61:610, 1974.
173. JR Benson, PE Hare. Proc Natl Acad Sci U S A 72:619–622, 1975.
174. SA Cohen, DJ Strydom. Anal Biochem 174:1–16, 1988.
175. I Molnar-Perl. J Chromatogr 661:45, 1994.
176. SA Cohen, DP Michaud. Anal Biochem 211:279–287, 1993.

177. DJ Strydom, SA Cohen. Anal Biochem 222:19–28, 1994.
178. E Kaiser, RL Colescott, CD Bossinger, PI Cook. Anal Biochem 34:595–598, 1970.
179. WS Hancock, JE Battersby. Anal Biochem 71:260–264, 1976.
180. GB Fields, Z Tian, G Barany. In: GA Grant, ed. Synthetic Peptides: A User's Guide. New York: Freeman, 1992, pp 77–183.

2
Strategy in Solid-Phase Peptide Synthesis

T. Scott Yokum and George Barany
University of Minnesota, Minneapolis, Minnesota

I. INTRODUCTION

Merrifield's invention of solid-phase peptide synthesis (SPPS) revolutionized the field of peptide chemistry [1]. Prior to the development of SPPS, peptides were synthesized via classical solution-phase methods, which are customarily quite tedious and time consuming, and require considerable expertise [2,3]. In solution-phase syntheses, purification and characterization are required after each step, and the solubility of the peptide worsens and becomes more unpredictable with increased chain length. With the advent of SPPS, particularly with aid of automated instruments, the time required for the synthesis of peptides has been reduced from weeks or months to hours or days.

Merrifield's basic concept was to covalently attach the first amino acid to an insoluble support and to elongate the peptide chain from this support-bound residue. Following incorporation of the desired number of monomers through a series of coupling and deprotection steps, the peptide was removed from the solid support. The enabling insights of solid-phase methodologies are that reactions may be driven to completion by use of excess reagents, and that the only purification required after each step is simply the washing away of excess reagents. Because the majority of the reactions of SPPS are repetitive and proceed in high yield, the process can be automated, and even the nonspecialist can produce peptides in good overall yields and purities. The significance of Merrifield's work was recognized by the award of the Nobel Prize in Chemistry in 1984 [4]. Solid-phase synthesis has expanded

79

from peptides to other biomolecules, including DNA [5,6], peptide nucleic acids (PNA) [7,8], and oligosaccharides [9,10], and the basic principles have been exploited to create the emerging field of combinatorial chemistry [11–14].

The two types of strategies for solid-phase synthesis of peptides are stepwise SPPS and convergent SPPS [15]. Stepwise SPPS proceeds by repetitive coupling and deprotection steps to introduce individual amino acid building blocks. Once the desired peptide length has been obtained, the peptide is cleaved from the solid support, and side-chain protecting groups are (usually) removed at the same time. Stepwise SPPS is most efficient for short- to medium-length peptides, with the routine upper limit being approximately 40 residues.

Alternatively, convergent approaches are often preferred when embarking on the synthesis of longer sequences [16]. Convergent approaches exploit efficient stepwise SPPS to create short segments, which are then purified and joined together further to form the target peptide. Convergent approaches can be subdivided into protected segment couplings and chemical ligations. In the former, segments that are fully protected, except for the termini to be coupled, are condensed via traditional methods involving carboxyl activation. In the latter, highly specific reacting groups on unprotected peptide fragments are used [16–18].

A number of variables, the relative importance of which depends on the specific circumstances of the synthesis, affect the success of any SPPS effort and must be considered during its planning and execution. Important choices that influence the overall outcome of a synthesis include the nature of the solid support, coupling chemistries, protection scheme, and the linkage for anchoring the peptide to the support. Variations in the protection scheme/anchoring strategy mirror the needs of the application planned, e.g., cyclic peptides, protected fragments, or modified C-terminal functionalities.

II. STEPWISE SPPS

Stepwise SPPS normally proceeds in the $C \rightarrow N$ direction. (SPPS in the $N \rightarrow C$ direction has been carried out in a few cases but suffers from inherent problems that limit its generality.) A general scheme of stepwise SPPS synthesis is shown in Fig. 1, and specific implementations are described in later chapters. The insoluble solid support must intrinsically bear an appropriate functional group or be capable of functionalization. The next step is to attach a handle (also called a linker), which is defined as a bifunctional spacer to connect the first amino acid to the solid support. The handle must be de-

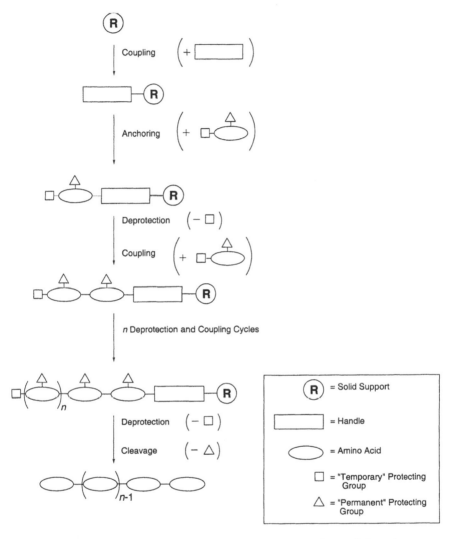

Figure 1 General scheme for SPPS. In practice, many variants of the scheme are used, as discussed in this and other chapters of this book.

signed to incorporate features of a cleanly removable, permanent protecting group (see later) for the *C*-terminal amino acid; thus, the free peptide with the desired *C*-terminal functionality is formed ultimately upon treatment with appropriate reagents. The other functional group of the handle must be such as to allow facile reaction with the functionalized solid support.

Once the handle has been attached to the support, an appropriately protected amino acid is bound covalently to the handle. Stepwise synthesis proceeds by selective removal of the group providing temporary protection to the first amino acid residue, followed by coupling of the second, appropriately protected, amino acid. Such successive deprotection–coupling cycles continue until the peptide sequence is elongated to the desired length. The temporary protecting group must be such that its removal is rapid, quantitative, and without effect on the anchoring linkage and permanent protecting groups; the latter groups are designed to prevent untoward side reactions during stepwise chain assembly. Subsequently, the protected peptide-resin is treated with a reagent cocktail that releases the peptide from the resin and (usually) removes the permanent protecting groups at the same time. This general scheme may be modified in a variety of different ways depending on the desired target.

A. Solid Supports (see Chapter 1)

A central idea of SPPS is that by anchoring a growing peptide chain to a solid support, reagents can be used in excess so as to drive reactions to completion, following which the excess reagents plus reaction coproducts are readily removed by simple filtration and washing. A myriad of supports for peptide synthesis have been developed, with varying levels of success and implementation [19]. Certain physical and chemical properties are required for a support to be useful for peptide synthesis. The support must be inert to all conditions used during synthesis, exhibit good mechanical stability, and be completely insoluble in all solvents used. In addition, the support must be obtained with functional groups, either direct or by a suitable functionalization reaction, so that the first amino acid, with a handle, may be bound to the support.

Solid supports for SPPS are usually not static materials as the term might imply, but rather many useful supports increase significantly in size, i.e., "swell," when solvated [20]. A swollen support is a gel in which reactions take place throughout the support (interior as well as surface areas) [19,21]. Good results for SPPS generally correlate with efficient swelling of the support in the solvents chosen for synthesis.

At present, two major families of supports are used for SPPS. Polystyrene, lightly cross-linked (e.g., 1%) with divinylbenzene, was used in most of Merrifield's seminal work and is still very popular for SPPS and solid-phase organic synthesis [1,22]. Polyethylene glycol (PEG) grafted onto polystyrene has found widespread application. The PEG chain can either be covalently attached to the polystyrene support (PEG-PS) [23–25] or ethylene oxide polymerized onto the polystyrene (Tentagel and ArgoGel) [26,27].

PEG-PS supports were developed to allow better solvation of the growing peptide chain in polar aprotic solvents commonly used for peptide synthesis. The PEG resins were also shown to have excellent mechanical stability, which allows their use in continuous-flow systems [19]. Other PEG-containing supports include poly(trimethylolpropane ethoxylate [14/3 EO/OH)triacrylate–co–allylamine] (CLEAR) [28] and poly(N,N-dimethylacrylamide–co–bisacrylamido polyethylene glycol–co–monoacrylamidopolyethylene glycol) (PEGA) [29,30], and have demonstrated promising properties.

B. Anchoring (see Chapter 5)

Solid supports ultimately require a functional group, e.g., an amino, hydroxyl, chloromethyl, or other functionality, that can be a starting point for SPPS. In some cases, a suitable functionalized monomer is included in the polymerization mix that gives the final support. Alternatively, the desired functional group can be introduced by a postpolymerization transformation [15]. The classic example of functionalization of a solid support is the chloromethylation of cross-linked polystyrene to produce chloromethyl-resin, also known as "Merrifield resin" [31]. This may then be modified further to introduce other functional groups such as aminomethyl or hydroxymethyl [15,32].

Attachment of the first amino acid to a functionalized solid support can be done according to two general strategies. The most straightforward approach is direct attachment, as was shown in Merrifield's original work, in which a *tert*-butyloxycarbonyl (Boc) amino acid salt was bound to the chloromethylated support by an O-alkylation reaction, resulting in a substituted benzyl ester linkage. The alternative and more controlled approach uses handles, which allow the amino acid to be anchored to the support in two distinct steps [33–35]. Depending on the precise structure of the handle, removal of the final peptide chain can be achieved concomitantly with removal of permanent protecting groups, or the protecting groups can be retained. In addition, a variety of C-termini besides carboxyl can be obtained, e.g., amides, hydrazides, esters, alcohols, aldehydes, and thioacids [35]. Additional options are available with side-chain and backbone anchoring [36–38]. These methods leave the C-terminus free for cyclization, segment condensation, or other transformations that introduce different functional end groups.

Handle approaches are subdivided further: in one method (Fig. 1) an appropriately protected handle is attached to the solid support first, followed by deprotection and coupling of the first protected amino acid [39]. A somewhat more laborious variation, the preformed handle approach, involves in-

itial coupling of the amino acid to the handle, intermediate purification, and
later coupling to the support [40,41]. The preformed handle method is often
preferred, because it offers a way to achieve almost quantitative loading
onto the resin, thereby providing precise substitution levels on the support.
Figures 2 and 3 show, respectively, the preformed handle strategy with

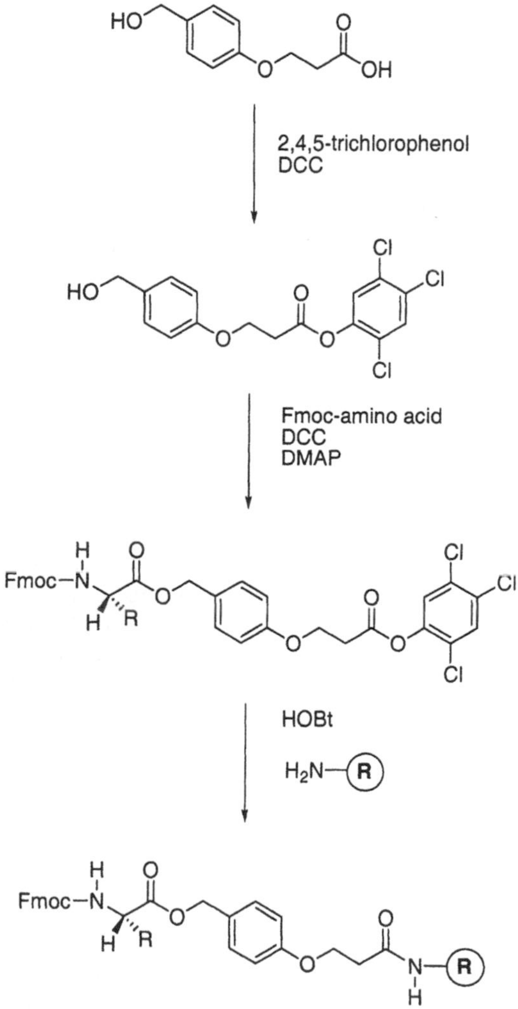

Figure 2 Preformed handle approach for the preparation of *C*-terminal peptide
acids with PAB anchoring and Fmoc chemistry.

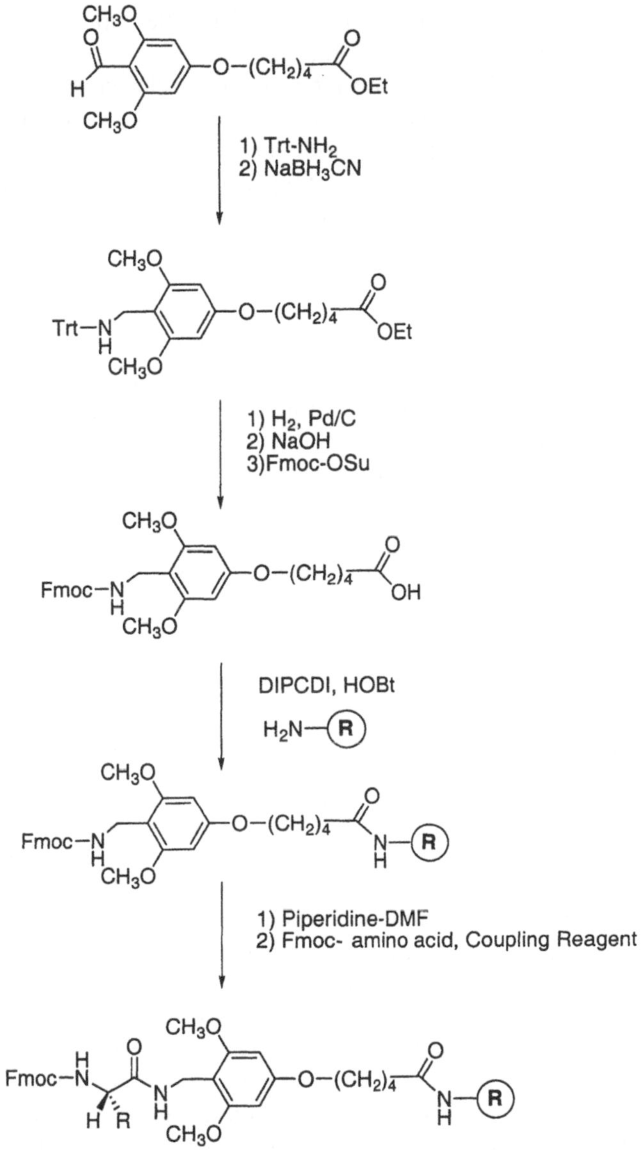

Figure 3 "Universal" handle approach for the preparation of C-terminal peptide amides with PAL anchoring and Fmoc chemistry.

p-alkoxybenzyl (PAB) esters [40] and the "universal" handle approach for tris(alkoxy)benzylamide (PAL) supports [39,42,43].

Multiple options are available for the mechanism of cleavage of the bond connecting the peptide to the handle. These include acid, base, fluoride ion, Pd^0, and light ($h\nu$) [15,34,35,44]. Acid-labile handles are prevalent and can be divided further into those cleaved by strong acid (e.g., HF–scavengers), moderate acid [e.g., trifluoroacetic acid (TFA)–scavengers], and mild acid (e.g., using very low percentages of TFA so that peptide segments retaining side-chain protection are obtained).

Cleavage of Merrifield's original *p*-alkylbenzyl ester linkage with HF gave a *C*-terminal peptide acid. Because the aforementioned linkage is not completely stable to repetitive TFA exposure [45], problems can arise; these can be solved by the use of *p*-(carbamoylmethyl)benzyl ester (PAM) linkage [45,46]. Relatedly, the 4-methylbenzhydrylamine (MBHA) [32,47] support is a popular choice for the generation of peptide amides via Boc chemistry. Correspondingly, peptide amides are made by 9-fluorenylmethyloxycarbonyl (Fmoc) chemistry usually in conjunction with PAL [39] or with 4-(2',4'-dimethoxyphenylaminomethyl)phenoxymethyl-resin (Rink amide resin) [48], both of which are cleaved by moderate acid. Fmoc synthesis of peptide acids is carried out with 4-alkoxybenzyl alcohol resin (Wang resin) [49] or PAB [40].

C. Protection Schemes

The choice of N^α-amino protecting group (temporary protection) dictates the choice of the handle and/or linkage to the resin as well as the array of side-chain protecting groups (permanent protection). Numerous N^α-protecting groups have been developed, but two have risen to the top and withstood the tests of time to become the standards in SPPS. These are the acid-labile Boc group and the base-labile Fmoc group (see Chapter 3).

The Boc–benzyl strategy, used first by Merrifield and substantially refined over the years, depends on graduated acid lability (Fig. 4). The N^α-Boc group is removed by TFA, usually 25–50% (v/v) in dichloromethane, or by strong mineral or Lewis acids in polar solvents. TFA treatment of a Boc-protected amine produces isobutylene, carbon dioxide, and a protonated amine, which must be neutralized before the next acylation may occur [50]. Permanent protection compatible with Boc generally involves substituted benzyl-type protecting groups (urethane, ester, ether). When the unsubstituted benzyl derivative is either too labile (premature loss during temporary deprotection) or too stable (incomplete removal during final cleavage), its properties can be fine-tuned by the introduction onto the aromatic ring of electron-withdrawing or electron-donating groups, respectively. In addition,

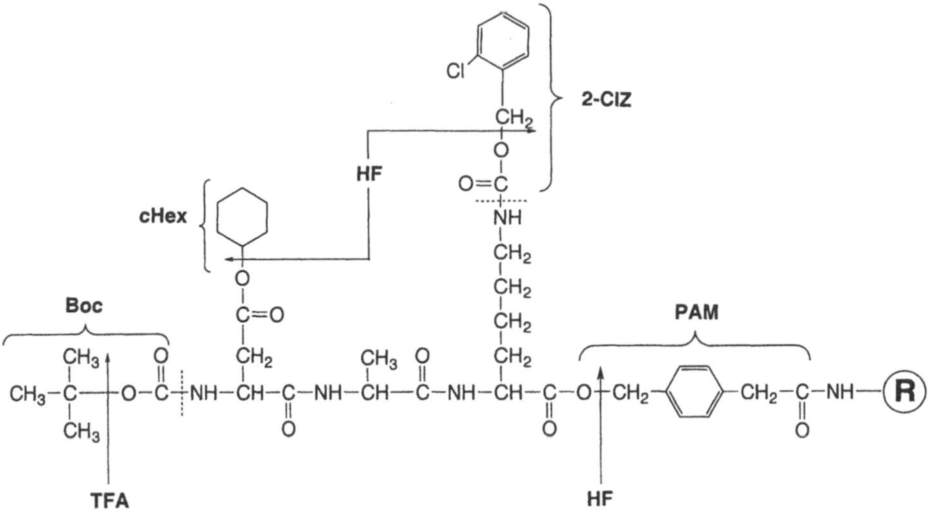

Figure 4 Representative protection scheme for Boc–benzyl strategy.

cyclohexyl (cHex) [51], p-toluenesulfonyl (Tos) [51], and other specialized protecting groups have sometimes found value. Removal of these families of side-chain protecting groups occurs with strong acids such as HF or trifluoromethanesulfonic acid (TFMSA), in the presence of suitable scavengers, usually anisole [52]. However, some protecting groups used in conjunction with Boc chemistry [e.g., acetamidomethyl (Acm), 2,4-dinitrophenyl (Dnp), and sometimes formyl (CHO) and 3-nitro-2-pyridinesulfenyl (Npys)] are stable to strong acids and therefore must be removed in a separate step, either on the resin or after cleavage [34].

The Boc scheme has proven to be successful for many peptide synthesis applications, but difficulties have been noted, particularly for fragile targets that do not survive the relatively harsh final acidic cleavage conditions. This provided an impetus for the development of milder methods. The Fmoc–t-butyl strategy has emerged as the preeminent example of an orthogonal protection scheme [53,54]. Independent classes of protecting groups, removed by different mechanisms so that they may be removed in any order in the presence of all other types of groups, are said to be orthogonal [55]. Orthogonal protection schemes are inherently milder because the selective deprotection is governed by alternative mechanisms of cleavage rather than by reaction rates.

The N^α-Fmoc group is removed by a variety of secondary amines, usually 20–50% piperidine in N,N-dimethylformamide (DMF) or by the stronger base 1,8-diazabicyclo[5.4.0]undec-7-ene (DBU) in the presence of a relatively low concentration of piperidine needed as a scavenger [2% DBU, 2% piperidine in DMF] [56]. Treatment of Fmoc-protected amines and peptides with base generates, via an $E1_cB$ mechanism, a highly reactive fulvene intermediate; this must be scavenged by a secondary amine such as piperidine [57,58].

Permanent protection compatible with Fmoc is often provided by t-butyl-type protecting groups (urethane, ester, ether). In addition, triphenylmethyl (Trt) [59,60], 2,2,5,7,8-pentamethylchroman-6-sulfonyl (Pmc) [61], 2,4,6-trimethoxybenzyl (Tmob) [62,63], and other specialized protecting groups have been used successfully (see Chapter 4). Removal of these groups, as well as the standard t-butyl-type groups, is normally achieved with high percentages of TFA in the presence of appropriate scavengers [64]. As with Boc-compatible protecting groups, some protecting groups used in conjunction with Fmoc chemistry [e.g., Acm and allyloxycarbonyl (Alloc)] are stable to cleavage conditions and must therefore be removed in a separate step, either on the resin or after cleavage [34].

Standard stepwise Fmoc chemistry with PAL or p-alkoxybenzyl anchoring is orthogonal in only two dimensions: Fmoc removal by base; side-chain and handle cleavage via acidolysis. Orthogonality may be extended into three or even four dimensions by using more specialized protecting groups and handles that are removed by Pd^0, light, thiols, metals, hydrazine, and other mechanisms [34,65]. An excellent example of a triply orthogonal protection scheme, that also illustrates the *B*ackbone *A*mide *L*inker (BAL) [38] approach, is shown in Fig. 5. The N^α-amino group is deprotected with base, the side chains are deprotected with acid, and the C-terminal ester is removed by Pd^0. Because orthogonal protection schemes allow the independent removal of protecting groups, synthesis of more complex molecules is possible.

D. Coupling Chemistries (see Chapter 6)

Exceptionally high yields for the repetitive amide bond formation reactions, or couplings, are a prerequisite for the success of stepwise SPPS. This entails appropriate activation of the C^α-carboxylic acid function, followed by displacement of the activating group by the incoming N^α-amino nucleophile. Fortunately, several good ways to achieve efficient coupling under mild conditions exist. There are two major categories of activation: preactivation and in situ activation. In the former, the C^α-carboxyl of a protected amino acid is activated selectively to give a reasonably stable derivative, and in a separate step is reacted with an amine. In the latter, an amine, carboxylic acid,

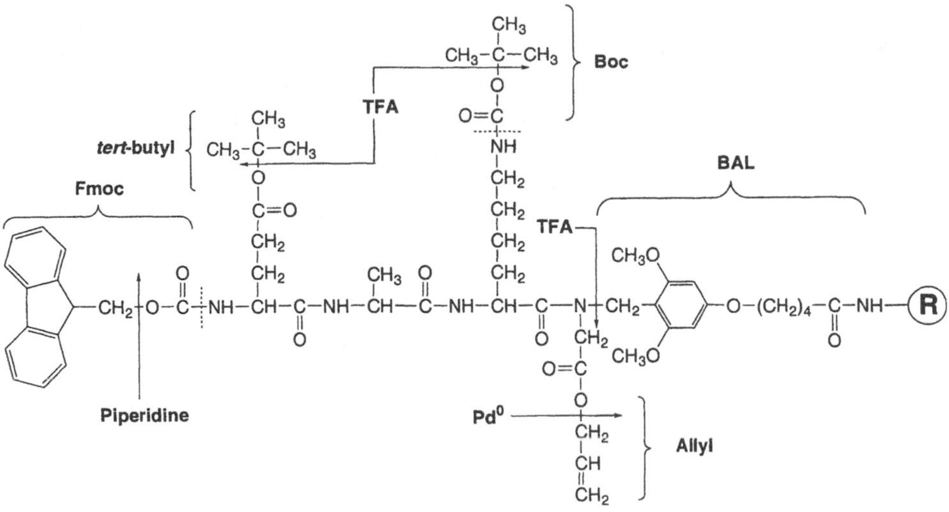

Figure 5 Representative triply orthogonal protection scheme for Fmoc–*t*-butyl strategy.

and condensing agent are mixed together simultaneously, and amide bond formation is in situ.

A variety of preactivated acyl derivatives has been developed for peptide bond formation. These include acyl halides [chlorides [66], fluorides [67,68]], acyl azides [69,70], active esters [e.g., pentafluorophenyl (OPfp) [71], *o*-nitrophenyl (ONp) [72], 3,4-dihydro-4-oxo-1,2,3-benzotriazin-3-yl (ODhbt) [73]], mixed anhydrides [3], and symmetrical anydrides [74,75].

Methods of amide bond formation involving the use of in situ coupling reagents are operationally simple and therefore very popular. The classic prototype reagent in this class is *N,N'*-dicyclohexylcarbodiimide (DCC) [76], which is still widely useful more than 40 years after this application was introduced. *N,N'*-Diisopropylcarbodiimide (DIC or DIPCDI) is even more convenient to use because of handling considerations and the improved solubility of the corresponding urea. Several side reactions are possible when carbodiimides are used; they are explained by overactivation of the C^α-carboxyl as an *O*-acylurea intermediate, which is believed to serve as the active coupling species. The majority of these side reactions may be circumvented by the addition of additives (1-hydroxybenzotriazole (HOBt) [77] or 1-hydroxy-7-azabenzotriazole (3-hydroxy-3*H*-1,2,3-triazolo-[4,5-*b*]pyridine) (HOAt) [78]), which react with the *O*-acylurea to form a less reactive but still quite effective acylating agent.

Several other in situ reagents have gained tremendous popularity. These include phosphonium salts as well as aminium salts (the original literature refers to the compounds as uronium salts but the correct structure was proved by X-ray crystallographic analysis [79]). The most widely used salts include N-[(1H-benzotriazol-1-yl)(dimethylamino)methylene]-N-methylmethanaminium hexafluorophosphate N-oxide (HBTU) [80,81], N-[(dimethylamino)-1H-1,2,3-triazolo-[4,5b]pyridin-1-yl-methylene]-N-methylmethanaminium hexafluorophosphate N-oxide (HATU) [82], 7-aza-benzotriazol-1-yl-N-oxy-tris(pyrrolidino)phosphonium hexafluorophosphate (PyAOP) [83], and benzotriazol-1-yl-N-oxy-tris(dimethylamino) phosphonium hexafluorophosphate (BOP) [84]. These reagents have the additional advantage that they contain of the effective additives as a leaving group, which is released upon reaction of the phosphonium or aminium salt with a carboxylate of an amino acid to form an active intermediate. The phosphonium and aminium salts are stable in the presence of a protected amino acid and react only with the corresponding carboxylate, so use of base is required with these reagents. Therefore, when using these reagents, additives per se to suppress side reactions are, strictly speaking, not required, although this is still often done in practice. A similar reagent used for the in situ generation of acid fluorides is 1,1,3,3-tetramethyl-2-fluoroformamidinium hexafluoro-phosphate (TFFH) [85].

E. Analysis

One of the major criticisms of SPPS during the years when the methodology was being developed was related to the fact that intermediates could not be isolated and characterized, and hence the progress of reactions could not be judged readily. To address this, many techniques have become available for the monitoring of coupling reactions and for analyses of peptides while they are still bound to the solid support. The extent of a coupling reaction, or confirmation that coupling is complete, may be determined by qualitative or quantitative reactions of unreacted primary amines with reagents such as ninhydrin [86], chloranil [87], picric acid [88,89], and bromophenol blue [90,91] (see Chapter 6). Coupling of Fmoc-amino acids can also be assessed spectrophotometrically by quantifying the fulvene–piperidine adduct produced upon Fmoc removal [92–94] (see Chapter 1). Many techniques used in the final characterization of the peptide may also be used to characterize resin-bound intermediates. These include amino acid analysis (see Chapter 19) and sequencing. Also, aliquots of the peptide-resin may be treated with the appropriate reagents to effect removal of free peptide from the support, following which traditional characterization methods [i.e., capillary zone electrophoresis (CZE), high-performance liquid chromatography (HPLC), and mass spectrometric techniques] may be applied (see Chapter 19). Finally, noninvasive techniques such as gel-phase nuclear magnetic resonance (NMR) can be very informative and will become increasingly important as their sensitivity and resolution improve.

III. CONVERGENT SYNTHESIS (see Chapter 9)

Despite significant advances in stepwise SPPS methodologies, the synthesis of longer peptides and proteins remains an arduous task. Incomplete couplings or deprotections, residue-specific side reactions, and other factors contribute to the formation of a variety of incorrect products that may nevertheless show characteristics similar to those of the target molecule [15,16,95–97]. Therefore, separation and identification of undesired products may be difficult or even impossible as the length of the target increases. Convergent approaches have been developed in an attempt to address complications associated with the stepwise synthesis of large peptides and proteins.

A. Solid-Phase Protected Segment Condensations

The classical approach to the synthesis of large peptides is called segment condensation, i.e., the convergent coupling of purified, smaller pieces that retain protection on the side-chains and on the termini not involved in the coupling [16,95,96]. A general scheme of this approach is shown in Fig. 6. The protected peptide fragment is assembled and removed from the solid support by a mechanism that leaves only the C-terminal carboxyl free. After purification, this protected segment is activated and coupled to the free amine of the segment(s) still on the solid support. Such cycles are continued until the target sequence is completed.

Whereas stepwise SPPS can often be carried out with two classes of protecting groups, three distinct cleavage mechanisms are required for convergent SPPS. The temporary (N^{α}-amino) protection and the permanent (side-chain) protection are the same as in stepwise, but a suitable semipermanent linkage to the solid support is required to allow generation of the protected segment with a free C-terminal carboxyl. The extreme C-terminal segment of the target may be synthesized on a permanent-type handle compatible with chemistry used in stepwise SPPS, and is typically not purified prior to segment condensation. Alternatively, this piece may be synthesized using a multidetachable handle or double-linker approach (Fig. 7) [98]. The latter approach is predicated on the idea that cleavage of a first handle provides a protected peptide anchored to a second handle. After purification,

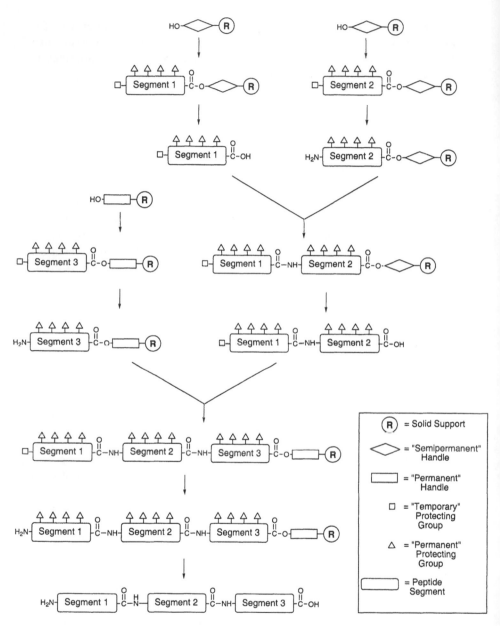

Figure 6 General scheme for solid-phase segment condensation approach.

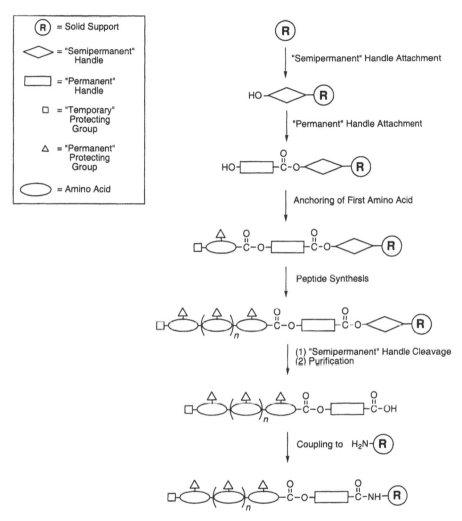

Figure 7 Multidetachable handle for synthesis, purification, and reattachment of protected peptide segments.

the second handle serves to reattach the peptide to the support, whereupon further manipulations can occur. This approach avoids difficulties associated with esterification of a C-terminal carboxylic acid onto a solid support.

Both Fmoc–t-butyl and Boc–benzyl chemistries have been used for the preparation of protected peptide segments. However, the milder Fmoc–t-butyl approach offers more alternatives for mild removal of the protected peptide fragment from the support [95,97]. Whatever the choice of chemistry, global protection of side-chains is preferred over minimal protection in order to lessen the occurrence of side reactions.

A number of handles have been used in both Fmoc and Boc schemes to generate protected peptides for subsequent use in segment condensations. The Fmoc–t-butyl approach frequently uses highly acid-labile handles for the attachment of the peptide to the support. The peptide is released with low percentages of TFA (or sometimes even acetic acid/hexafluoroisopropanol), which leaves the t-butyl–based side-chains intact. Triphenylmethyl (trityl) [99,100] and bis- and tris(alkoxy)benzyl alcohol-type [101–103] handles are the most popular choices for such strategies. Allyl-type handles (cleaved by Pd^0) [104–106] and o-nitrobenzyl handles (cleaved by photolysis) [41,107,108] are alternatives that provide a truly orthogonal system for the generation of protected peptide segments with the Fmoc–t-butyl strategy.

The traditional method for generation of protected peptide fragments in concert with the Boc–benzyl strategy is saponification or transesterification to cleave the PAM anchor. However, such methods are used less often today because of certain incompatibilities, as well as the development of more efficient techniques [95]. Allyl, o-nitrobenzyl, phenacyl (also cleaved by photolysis) [109,110], and fluorenylmethyl (cleaved by secondary amines) [111–113] handles all offer orthogonal mechanisms of release. The Kaiser oxime-resin is also popular for the production of protected segments; it is cleaved via aminolysis with an amino acid ester or transesterification with hydroxypiperidine (reduction of this ester provides the free carboxyl) [97,114].

A major advantage of convergent SPPS is that the intermediates (peptide segments) are purified prior to the segment coupling. Therefore, following successful assembly and cleavage of the protected segment, the daunting purification task must be addressed. The major obstacles encountered are related to solubility, but many "tricks" have been developed in attempts to overcome these. Thus, modification of the peptide backbone, the use of "solubilizing" protecting groups or residues [115–117], and the development of special solvent systems have all been tried [97]. After purification, the protected segment is normally characterized by amino acid analysis and mass spectrometry.

Solid-phase coupling of protected peptide segments to one another generally does not proceed with nearly the facility of stepwise SPPS with protected amino acid building blocks [95,97]. Coupling methods are nominally the same, but because of solubility considerations and overall limits on achievable concentrations, reaction times are often on the order of hours or even days. As reaction times increase, so does the likelihood of racemization, so special precautions must be taken to reduce this risk [41,118].

Monitoring the progress of solid-phase segment condensation is sometimes difficult because, as peptide length increases, the ninhydrin and related tests become less sensitive [16]. The progress or extent of coupling may be investigated by removing aliquots of the resin and performing amino acid analysis and/or solid-phase sequencing. Alternatively, the peptide can be cleaved from the resin at intermediate stages and the resultant free peptide characterized by traditional means.

B. Chemical Ligations

Chemical ligation strategies take advantage of the reactivities of chemoselective groups on unprotected peptides in aqueous solution in order to achieve condensation. Chemical ligations circumvent many of the problems such as poor solubility, slow reaction rates, and difficulty in characterization of intermediates that are associated with conventional segment couplings. These strategies are expected to gain popularity because they integrate well-developed processes in peptide chemistry, including efficient stepwise SPSS of small- to medium-length peptides; relatively easy purification of unprotected, water-soluble peptides; and the battery of methods for characterization of free peptides.

Two general methods of chemical ligations have been proposed; these are differentiated by the type of linkage formed between the peptide segments. The first class of ligations uses a chemoselective reaction between functional groups to form an unnatural linkage between fragments at the ligation site [16,119–122]. Second, the method of "native chemical ligation" [123] forms a natural peptide bond and a proteinogenic residue between the two peptide segments after rearrangement of an intermediate.

Examples of efficient ligation methods for creating artifically linked proteins or peptides include formation of thioesters by reaction of a thioacid and a bromoacetamide [119], formation of pseudoproline residues from an aldehyde and Cys or Thr [122,124], formation of an oxime from an aminooxy function and an aldehyde or ketone [120,121], and formation of a hydrazone from a hydrazide and an aldehyde [121,125,126]. Several of the preceding methods may be used in series because of different reactivities.

All native chemical ligations described to date use C-terminal thioesters or thioacids for reaction with an N-terminal nucleophile or electrophile, respectively. In the original method, an N-terminal Cys reacts with a C-terminal thioester to form a thioester intermediate that rearranges spontaneously to form a natural peptide bond with a Cys incorporated (Fig. 8) [123]. This method has been extended further to allow native bond formation without the exclusive use of N-terminal Cys. To accomplish this, an oxyethanethiol was attached to the amino function of the N-terminal residue,

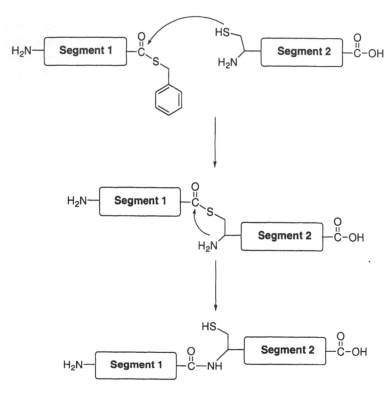

Figure 8   Mechanism of amide bond formation in native chemical ligation.

which was then reacted with a C-terminal thioester-containing fragment to form a tertiary amide [127]. The oxyethanethiol group was then removed to unmask the native backbone. Other methods of native ligations use thioacids as the nucleophiles, which react with β-bromoalanine [128] or derivatized Cys (containing a leaving group on the sulfur) [129]; these reactions proceed through similar intermediates, which then rearrange to form the native backbone.

        Chemical ligations require the synthesis of peptide fragments containing various C- and N-terminal functional groups, e.g., aldehydes, ketones, amino-oxy functions, and C-terminal thioesters and thioacids, which are not routine in traditional SPPS. Some of these groups present difficulties with regard to their synthesis and/or stability. Although functional groups may be introduced after synthesis onto a free amino group, e.g., the free N-terminus or a free N^{ε}-amino side-chain of Lys, formation of C-terminal thioesters and thioacids requires synthesis on specialized sulfur-containing supports. Peptides with thioester or thioacid functions are assembled using Boc chemistry because of extreme susceptibility of the thioester linkage to nucleophilic cleavage under standard Fmoc deprotection conditions. Peptide thioesters are generally obtained via the corresponding peptide thioacids, by reaction of the thioacid with either benzyl bromide [123] or a symmetrical disulfide such as 5,5'-dithiobis(2-nitrobenzoic acid) (DTNB) [123].

The preceding methods are designed to ligate two unprotected peptide segments. Convergent approaches bringing together more than two fragments use one type of ligation chemistry for the assembly of precursors and a different type of ligation chemistry to form the desired product [18,130]. Convergent sequential native ligations chemistries have also been described.

To date, only one example of ligations on the solid support has been reported. Muir and coworkers [131] achieved both intermolecular and intramolecular solid-phase native ligations using a 3-mercaptopropionic acid handle bound to amino-functionalized PEGA (HS-PEGA). The thioester linkage is stable to conditions of SPPS (Boc chemistry) and to side-chain deprotection conditions (HF) but is cleaved readily by the thiol of an *N*-terminal Cys residue. The Cys residue may be the *N*-terminal residue of the peptide on the support, in which case a cyclic peptide forms from intramolecular self-ligation, or may be on another incoming segment that reacts to form the intermolecular product.

IV. CONCLUSION

By necessity, the present chapter has provided only a brief and selective overview of historical perspectives and the current state of the art of stepwise and convergent SPPS. There can be little doubt about the enormous impact and power of solid-phase synthetic methodologies for studying important problems in peptide science.

REFERENCES

1. RB Merrifield. J Am Chem Soc 85:2149–2154, 1963.
2. T Wieland, M Bodanszky. The World of Peptides: A Brief History of Peptide Chemistry. Berlin: Springer-Verlag, 1991.
3. M Bodansky. Principles of Peptide Synthesis. 2nd ed. Berlin: Springer-Verlag, 1993.
4. RB Merrifield. Science 232:341–347, 1986. This is an updated version of Merrifield's Nobel Prize lecture, also published in several other journals.
5. MH Caruthers. Science 230:281–285, 1985 and references cited therein.
6. SL Beaucage, RP Iyer. Tetrahedron 48:2223–2311, 1992.

7. M Egholm, O Buchardt, PE Nielsen, RH Berg. J Am Chem Soc 114:1895–1897, 1992.
8. SA Thomson, JA Josey, R Cadilla, MD Gaul, CF Hassman, MJ Luzzio, AJ Pipe, KL Reed, DJ Ricca, RW Wiethe, SA Noble. Tetrahedron 52:6179–6194, 1995 and references cited therein.
9. JT Randolph, KF McClure, SJ Danishefsky. J Am Chem Soc 117:5712–5719, 1995.
10. KC Nicolaou, N Watanabe, J Li, J Pastor, N Winssinger. Angew Chem Int Ed Engl 37:1559–1561, 1998 and references cited therein.
11. EM Gordon, RW Barrett, WJ Dower, SPA Fodor, MA Gallop. J Med Chem 37:1385–1401, 1994.
12. LA Thompson, JA Ellman. Chem Rev 96:555–600, 1996.
13. JS Früchtel, G Jung. Angew Chem Int Ed Engl 35:17–42, 1996.
14. C Blackburn. Biopolymers (Peptide Sci) 47:311–351, 1998.
15. G Barany, RB Merrifield. In: E Gross, J Meienhofer, eds. The Peptides: Analysis, Synthesis, Biology, vol 2. New York: Academic Press, 1979, pp 1–284, and earlier reviews cited therein.
16. P Lloyd-Williams, F Albericio, E Giralt. Chemical Approaches to the Synthesis of Peptides and Proteins. Boca Raton, FL: CRC Press, 1997.
17. MA Walker. Angew Chem Int Ed Engl 36:1069–1071, 1997.
18. TW Muir, PE Dawson, SBH Kent. Methods Enzymol 289:266–298, 1997.
19. G Barany, M Kempe. In: CW Czarnik, SH DeWitt, eds. A Practical Guide to Combinatorial Chemistry. Washington, DC: American Chemical Society, 1997, pp 51–97.
20. RB Merrifield. Br Polym J 16:173–178, 1984.
21. J Vágner, G Barany, KS Lam, V Krchnák, NF Sepetov, JA Ostrem, P Strôp, M Lebl. Proc Natl Acad Sci U S A 93:8194–8199, 1996.
22. JS Kiely, TK Hayes, MC Griffith, Y Pei. In: CW Czarnik, SH DeWitt, eds. A Practical Guide to Combinatorial Chemistry. Washington, DC: American Chemical Society, 1997, pp 99–122.
23. G Barany, F Albericio, NA Solé, GW Griffin, SA Kates, D Hudson. In: CH Schneider, AN Eberle, eds. Peptides 1992: Proceedings of the Twenty-Second European Peptide Symposium. Leiden, The Netherlands: ESCOM Science Publishers, 1993, pp 267–268.
24. S Zalipsky, JL Chang, F Albericio, G Barany. React Polym 22:243–258, 1994 and references cited therein.
25. SA Kates, BF McGuinness, C Blackburn, GW Griffin, NA Solé, G Barany, F Albericio. Biopolymers (Peptide Sci) 47:365–380, 1998.
26. E Bayer, W Rapp. In: JM Harris, ed. Poly(Ethylene Glycol) Chemistry: Biotechnical and Biomedical Applications. New York: Plenum, 1992, pp 325–345.
27. JA Porco, TL Deegan, W Devonport, OW Gooding, K Heisler, JW Labadie, B Newcomb, C Nguyen, P van Eikeren, J Wong, P Wright. Mol Divers 2: 197–206, 1996.
28. M Kempe, G Barany. J Am Chem Soc 118:7083–7093, 1996.
29. M Meldal. Tetrahedron Lett 33:3077–3080, 1992.

30. F-I Auzanneau, M Meldal, KJ Bock. Peptide Sci 1:31–44, 1995.
31. RB Merrifield. Biochemistry 3:1385–1390, 1964.
32. JH Adams, RM Cook, D Hudson, V Jammalamadaka, MH Lyttle, MF Songster. J Org Chem 63:3706–3716, 1998.
33. G Barany, N Kneib-Cordonier, DG Mullen. Int J Peptide Protein Res 30:705–739, 1987.
34. GB Fields, Z Tian, G Barany. In: GA Grant, ed. Synthetic Peptides. A User's Guide. New York: WH Freeman, 1992, pp 77–183.
35. MF Songster, G Barany. Methods Enzymol 289:126–174, 1997.
36. F Albericio, R Van Abel, G Barany. Int J Peptide Protein Res 35:284–286, 1990 and references cited therein.
37. C Blackburn, SA Kates. Methods Enzymol 289:175–198, 1997.
38. KJ Jensen, J Alsina, MF Songster, J Vágner, F Albericio, G Barany. J Am Chem Soc 120:5441–5452, 1998.
39. F Albericio, N Kneib-Cordonier, S Biancalana, L Gera, RI Masada, D Hudson, G Barany. J Org Chem 55:3730–3743, 1990 and references cited therein.
40. F Albericio, G Barany. Int J Peptide Protein Res 26:92–97, 1985 and references cited therein.
41. N Kneib-Cordonier, F Albericio, G Barany. Int J Peptide Protein Res 35:527–538, 1990 and references cited therein.
42. F Albericio, G Barany. Int J Peptide Protein Res 30:206–216, 1987.
43. SK Sharma, MF Songster, TL Colpitts, P Hegyes, G Barany, FJ Castellino. J Org Chem 58:3696–3699, 1993.
44. C Blackburn, F Albericio, SA Kates. Drugs Future 22:1007–1035, 1997.
45. AR Mitchell, BW Erickson, MN Ryabtsev, RS Hodges, RB Merrifield. J Am Chem Soc 98:7357–7362, 1976.
46. AR Mitchell, SBH Kent, M Engelhard, RB Merrifield. J Org Chem 43:2845–2852, 1978 and references cited therein.
47. GR Matsueda, JA Stewart. Peptides 2:45–50, 1981.
48. H Rink. Tetrahedron Lett 28:3787–3790, 1987.
49. SS Wang. J Am Chem Soc 95:1328–1333, 1973.
50. P Alewood, D Alewood, L Miranda, S Love, W Meutermans, D Wilson. Methods Enzymol 289:14–29, 1997.
51. JP Tam, RB Merrifield. In: S Udenfriend, J Meienhofer, eds. The Peptides. Vol 9. New York: Academic Press, 1987, pp 185–248.
52. MW Pennington. In: MW Pennington, BM Dunn, eds. Peptide Synthesis Protocols. Totowa, NJ: Humana Press, 1994, pp 41–62.
53. LA Carpino, GY Han. J Org Chem 37:3404–3409, 1972.
54. E Atherton, H Fox, D Harkiss, CJ Logan, RC Sheppard, BJ Williams. J Chem Soc Chem Commun 537–539, 1978.
55. G Barany, RB Merrifield. J Am Chem Soc 99:7363–7365, 1977.
56. GB Fields, RL Noble. Int J Peptide Protein Res 35:161–214, 1990.
57. RAM O'Ferrall, SJ Slae. J Chem Soc (B) 260–268, 1970.
58. RAM O'Ferrall. J Chem Soc (B) 268–274, 1970.
59. K Barlos, D Papaioannou, D Theodoropoulos. J Org Chem 47:1324–1326, 1982.

60. I Photaki, J Taylor-Papadimitriou, C Sakarellos, P Mazarakis, L Zervas. J Chem Soc (C) 2683–2687, 1970.
61. R Ramage, J Green. Tetrahedron Lett 28:2287–2290, 1987.
62. F Weygand, W Steglich, J Bjarnason. Chem Ber 101:3642–3648, 1968.
63. MC Munson, C García-Echeverría, F Albericio, G Barany. J Org Chem 57: 3013–3018, 1992.
64. F Dick. In: MW Pennington, BM Dunn, eds. Peptide Synthesis Protocols. Totowa, NJ: Humana Press, 1994, pp 63–72.
65. BW Bycroft, WC Chan, SR Chhabra, ND Hone. J Chem Soc Chem Commun 778–779, 1993.
66. LA Carpino, BJ Cohen, KE Stephens Jr, SY Sadat-Aalaee, J-H Tien, DC Langridge. J Org Chem 51:3732–3734, 1986.
67. LA Carpino, D Sadat-Aalaee, HG Chao, RH DeSelms. J Am Chem Soc 112: 9651–9652, 1990.
68. LA Carpino, E-SME Mansour, D Sadat-Aalaee. J Org Chem 56:2611–2614, 1991.
69. J Honzl, J Rudinger. Collect Czech Chem Commun 26:2333–2344, 1961.
70. AM Felix, RB Merrifield. J Am Chem Soc 92:1385–1391, 1970.
71. K Kovacs, B Penke. In: H Hanson, HD Jakubke, eds. Peptides 1972. Amsterdam: North-Holland, 1973, pp 187–188.
72. M Bodansky, KW Funk. J Org Chem 38:1296–1300, 1973.
73. W König, R Geiger. Chem Ber 103:2034–2040, 1970.
74. CD Chang, AM Felix, MH Jimenez, J Meienhofer. Int J Peptide Protein Res 15:485–494, 1980.
75. RB Merrifield, LD Vizioli, HG Boman. Biochemistry 21:5020–5031, 1982.
76. JC Sheehan, GP Hess. J Am Chem Soc 77:1067–1068, 1955.
77. W König, R Geiger. Chem Ber 103:788–798, 1970.
78. LA Carpino. J Am Chem Soc 115:4397–4398, 1993.
79. I Abdelmoty, F Albericio, LA Carpino, BM Foxman, SA Kates. Lett Peptide Sci 1:57–67, 1994.
80. V Dourtoglou, J-C Ziegler, B Gross. Tetrahedron Lett 19:1269–1272, 1978.
81. R Knorr, A Trzeciak, W Bannwarth, D Gillessen. Tetrahedron Lett 30:1927–1930, 1989.
82. LA Carpino, A El-Faham, CA Minor, F Albericio. J Chem Soc Chem Commun 201–203, 1994.
83. F Albericio, M Cases, J Alsina, SA Triolo, LA Carpino, SA Kates. Tetrahedron Lett 38:4853–4856, 1997.
84. B Castro, JR Dormoy, G Evin, C Selve. Tetrahedron Lett 16:1219–1222, 1975.
85. LA Carpino, A El-Faham. J Am Chem Soc 117:5401–5402, 1995.
86. E Kaiser, RL Colescott, CD Bossinger, PI Cook. Anal Biochem 34:595–598, 1970.
87. T Christensen. Acta Chem Scand 33 B:763–766, 1979.
88. BF Gisin. Anal Chim Acta 58:248–249, 1972.
89. RS Hodges, RB Merrifield. Anal Biochem 65:241–272, 1975.

90. V Krchnák, J Vágner, P Safár, M Lebl. Collect Czech Chem Commun 53: 2542–2548, 1988.
91. M Flegel, RC Sheppard. J Chem Soc Chem Commun 536–538, 1990.
92. J Meienhofer, M Waki, EP Heimer, TJ Lambros, RC Makofske, C-D Chang. Int J Peptide Protein Res 13:35–42, 1979.
93. CD Chang, AM Felix, MH Jimenez, J Meienhofer. Int J Peptide Protein Res 15:485–494, 1980.
94. A Dryland, RC Sheppard. J Chem Soc Perkin Trans I 125–137, 1986.
95. H Benz. Synthesis 337–358, 1994.
96. P Lloyd-Williams, F Albericio, E Giralt. Tetrahedron 49:2279–2282, 1994.
97. F Albericio, P Lloyd-Williams, E Giralt. Methods Enzymol 289:313–336, 1997.
98. JP Tam, FS Tjoeng, RB Merrifield. J Am Chem Soc 102:6117–6127, 1980.
99. K Barlos, D Gatos, J Kallitsis, G Papaphotiu, P Sotiriu, Y Wenqing, W Schäfer. Tetrahedron Lett 30:3942–3946, 1989.
100. K Barlos, O Chatzi, D Gatos, G Stavropoulos. Int J Peptide Protein Res 37: 513–520, 1991.
101. A Flörsheimer, B Riniker. In: E Giralt, D Andreu, eds. Peptides 1990. Proceedings of the Twenty-First European Peptide Symposium. Leiden, The Netherlands: ESCOM, 1991, pp 131–133.
102. M Mergler, R Tanner, J Gosteli, P Grogg. Tetrahedron Lett 29:4005–4008, 1988.
103. F Albericio, G Barany. Tetrahedron Lett 32:1015–1018, 1991.
104. H Kunz, B Dombo. Angew Chem Int Ed Engl 27:711–713, 1988.
105. P Lloyd-Williams, G Jou, F Albericio, E Giralt. Tetrahedron Lett 32:4207–4210, 1991.
106. F Guibé, O Dangles, G Balavoine, A Loffet. Tetrahedron Lett 29:5871–5874, 1988.
107. DH Rich, SK Gurwara. J Am Chem Soc 97:1575–1579, 1975.
108. G Barany, F Albericio. J Am Chem Soc 107:4936–4942, 1985.
109. SS Wang. J Org Chem 41:3258–3261, 1976.
110. FS Tjoeng, GA Heaver. J Org Chem 48:355–359, 1983.
111. M Mutter, D Bellof. Helv Chim Acta 67:2009–2016, 1984.
112. YZ Liu, SH Ding, JY Chu, AM Felix. Int J Peptide Protein Res 35:95–98, 1990.
113. F Rabanal, E Giralt, F Albericio. Tetrahedron 51:1449–1458, 1995.
114. WF DeGrado, ET Kaiser. J Org Chem 47:3258–3261, 1982.
115. P Lloyd-Williams, F Albericio, M Gairí, E Giralt. In: R Epton, ed. Innovations and Perspectives in Solid-Phase Synthesis and Combinatorial Libraries. Kingswinford, England: Mayflower Scientific, 1997, pp 195–200.
116. T Haack, M Mutter. Tetrahedron Lett 33:1589–1592, 1992.
117. T Wöhr, M Mutter. Tetrahedron Lett 36:3847–3848, 1995.
118. L Carpino, A El-Faham, F Albericio. Tetrahedron Lett 35:2279–2282, 1994.
119. M Schölzner, SBH Kent. Science 256:221–225, 1992.
120. K Rose. J Am Chem Soc 116:30–33, 1994.
121. J Shao, JP Tam. J Am Chem Soc 117:3893–3899, 1995.

122. CF Lui, JP Tam. Proc Natl Acad Sci U S A 91:6584–6588, 1994.
123. PE Dawson, TW Muir, I Clark-Lewis, SBH Kent. Science 266:776–779, 1994.
124. C Liu, C Rao, JP Tam. J Am Chem Soc 118:307–312, 1996.
125. K Rose, LA Vilaseca, R Werlen, A Meunier, I Fisch, RLM Jones, RE Offord. Bioconjugate Chem 2:154–159, 1991.
126. K Rose, LA Vilaseca, RE Offord. Bioconjugate Chem 3:147–153, 1992.
127. LE Canne, SJ Bark, SBH Kent. J Am Chem Soc 118:5891–5896, 1996.
128. JP Tam, Y Lu, C Liu, J Shao. Proc Natl Acad Sci U S A 92:12485–12489, 1995.
129. C Liu, C Rao, JP Tam. Tetrahedron Lett 37:933–936, 1996.
130. LE Canne, AR Ferré-D'Amaré, SK Burley, SBH Kent. J Am Chem Soc 117:2998–3007, 1995.
131. JA Camarero, GJ Cotton, A Adeva, TW Muir. J Peptide Res 51:303–316, 1998.

3

Amino Acids: Alpha-Amino Protecting Groups

John D. Wade
Howard Florey Institute, University of Melbourne, Parkville, Victoria, Australia

I. INTRODUCTION

An essential requirement in chemical peptide synthesis is the appropriate selection and use of a temporary protecting group for the α-amino group of amino acids. As described in Chapter 2 of this volume, the experimental design of the synthesis and subsequently chosen orthogonal protection scheme largely dictate the choice of N^α-protection. Such protection is mandatory in solid phase synthesis (SPS) as it permits α-carboxy activation and subsequent selective condensation of the N^α-protected amino acid onto the free amine of the terminal amino acid of the growing peptide chain. The temporary protecting group must meet several strict requirements in order to allow successful SPS. The group must not affect the optical integrity of the amino acid derivative. It must withstand completely the conditions of SPS, cause no side reactions, and be rapidly, efficiently, and selectively cleaved under conditions that leave unaffected any amino acid side-chain ("permanent") protecting groups.

The first N^α-protecting group that largely satisfied the preceding requirements was the alkoxycarbonylamino (urethane) derivative benzyloxycarbonyl (Z, carbobenzoxy, carbobenzyloxy, **1** [all structures cited within the chapter can be found in the Appendix]), which was developed by Bergmann and Zervas [1] in 1932. Removal of the Z group was readily achieved by catalytic hydrogenation without damage to newly formed peptide bonds. It was also shown that the urethane structure imparted great resistance to racemization. The usefulness of the Z group was further enhanced when it was later shown to be cleaved by acidolysis, particularly by HBr in acetic acid [2] and by HF [3]. This finding led to efforts to develop alternative urethanes having greater acid lability. Foremost among these was the *tert*-butyloxy-carbonyl (Boc, **2**) group (see Section II.A). Its adoption and utilization by Merrifield as a key component of the SPS methodology [4] have contributed enormously to the success of modern peptide synthesis and research.

With further refinement and development of SPS over the past 30 years, a rich variety of alternative N^α-protecting groups have been invented. Particularly important are the base-labile urethanes removed mechanistically by β-elimination. These have enabled both a considerable relaxation of the conditions of and enhanced orthogonality of SPS (see Chapter 2). Other groups (not necessarily urethanes) were devised that could be removed by alternative methods such as photolysis, electrolysis, nucleophilic cleavage, reduction, and metal cleavage, thus enabling substantial broadening of the scope and application of SPS. This chapter reviews only the N^α-protecting groups of general interest and focuses on those of practical or emerging importance in SPS.

II. ACID-LABILE GROUPS

Since the inception of SPS, the great majority of syntheses have been based on the principle of graduated acid lability as developed by Merrifield [4] and preferentially utilized the Boc group for temporary N^α-protection. However, many other acid-labile derivatives have been developed and assessed for their suitability for use in SPS (for review, see Refs. 5 and 6). Most are urethanes and include, in order of increasing acid lability, *tert*-amyloxycarbonyl (Aoc, **3**) [7], α,α-dimethyl-3,5-dimethyloxybenzyloxycarbonyl (Ddz, **4**) [8], and 2-(4-biphenylyl)-isopropoxycarbonyl (Bpoc, **5**) [9]. In 75% aqueous acetic acid, the relative rates of N^α-deprotection are Boc:Ddz:Bpoc 1:1400:3000 [9]. The latter two groups allow considerably milder SPS to be undertaken, although the sensitivity of the Ddz and Bpoc groups to the free α-carboxyl groups requires these N^α-protected amino acids to be stored as their cyclohexylammonium (CHA) or dicyclohexylammonium (DCHA) salts. A number of nonurethane derivatives have also been used in SPS [5,6]. These include the triphenylmethyl (Trt, **6**) [10] and 2-nitrophenysulfenyl (Nps, **7**) amino acids [11]. The former is very stable to base but is exceptionally acid labile and is removed with aqueous acetic acid and also by tetrazole in trifluoroethanol. The Nps group, which is also resistant to base, has been used with success in SPS, although the corresponding derivative also requires special handling because of its lability to the free α-carboxyl of the amino acid derivative. Removal of Nps is most conveniently achieved with dilute solutions of HCl in chloroform−acetic acid [12]. It may also be removed by nucleophiles including 2-mercaptopyridine in acetic acid [13].

A. *tert*-Butyloxycarbonyl (Boc)

The development of the Boc group by Carpino [14] and the subsequent demonstration of its suitability for use in peptide synthesis as a result of its much greater acid lability then the Z group [15,16] marked a major turning point in chemical peptide synthesis. It was shown to be particularly suited for use in SPS [4,17]. The Boc group is stable to both bases and nucleophiles and is unaffected by catalytic hydrogenation. Boc-amino acids are generally obtained in crystalline form and are stable at room temperature for extended periods, although storage at 4°C is recommended. Deprotection is rapid with organic and inorganic acids and most commonly with solutions of trifluoroacetic acid (TFA) in dichloromethane (DCM), although many other acids have been tried and used in SPS. These include 4 N HCl in dioxane and 2 M methanesulfonic acid in dioxane [18].

1. Preparation

Early attempts to prepare Boc-amino acids with *tert*-butyl chlorocarbonate [19] met with only limited success because of the poor stability of this reagent. However, excellent yields of Boc-amino acids were obtained when the reaction was carried out using the corresponding azide [20], itself readily obtained from the Boc-hydrazide. This was long the method of choice for preparing Boc derivatives. However, reports of azide-caused laboratory explosions prompted the development of the highly active and stable di-*tert*-butyl dicarbonate (Boc-anhydride) [21] and 2-(*tert*-butoxycarbonyl-oxyimino)-2-phenylacetonitrile (Boc-ON) [22] as alternative reagents. Both are readily commercially available, safe, and can be stored in the refrigerator for extended periods. Their use in the preparation of Boc-amino acids in good yields is straightforward, although the greater cost of the anhydride makes it somewhat less popular. A list of Boc-amino acid derivatives commonly employed in SPS and their physical characteristics is given in Table 1.

a. 2-[tert-Butoxycarbonyl-oxyimino]-2-phenyacetonitrile (Boc-ON) [23,24]. Amino acid (5 mmol) is dissolved in THF−water (1:1, v/v) (15 mL) and the resulting solution brought to pH 9.5 by adding 10% aqueous sodium carbonate. The solution is cooled in an ice bath and Boc-ON (1.36 g, 5.5 mmol) in dioxane (5 mL) is added. After 15 min of stirring in an ice bath, the reaction is continued at 25°C and kept at pH 9.5 by adding further

Table 1 Boc-Amino Acid Derivatives

Compound	Molecular weight	M.p. (°C)	$[\alpha]_D$	Solvent[a] (conc.)
Boc-Ala-OH	189.2	80–82	−24.5	H (1)
Boc-Arg(Tos)-OH	428.5	96–97	−3.3	D (4)
Boc-Asn-OH	232.2	176–177	−7.0	H (1)
Boc-Asn(Xan)-OH	412.4	184	+9.0	M (1)
Boc-Asp(OChx)-OH	315.1	93–95	+5.9	H (2)
Boc-Cys(Acm)-OH	292.3	112	−35.5	W (1)
Boc-Cys(Fm)-OH	399.5	74–75	−10.1	D (1)
Boc-Cys(MeBzl)-OH	325.4	87	−37.4	M (1)
Boc-Cys(Npys)-OH	375.4	163–164	−8.5	M (1)
Boc-Gln-OH	246.3	116–118	−3.0	E (2)
Boc-Gln(Xan)-OH	426.5	148	−39.0	M (1)
Boc-Glu(OChx)-OH	329.4	133–136	−8.3	M (1)
Boc-Gly-OH	175.2	94–95	—	
Boc-His(Dnp)-OH	421.4	94	+55.3	H (1)
Boc-His(Bom)-OH	375.4	150–160	+7.5	M (1)
Boc-Ile-OH.1/2H$_2$O	240.2	66–68	+3.8	M (2)
Boc-Leu-OH.H$_2$O	249.3	78–81	−25.0	H (2)
Boc-Lys(2-ClZ)-OH	414.9	63–75	−10.5	D (2)
Boc-Met-OH	249.3	49	−20.0	M (1.3)
Boc-Phe-OH	265.3	84–86	+24.5	E (2)
Boc-Pro-OH	215.2	134–135	−60.2	H (2)
Boc-Ser(Bzl)-OH	295.3	61	+23.5	E (2)
Boc-Thr(Bzl)-OH	309.4	116	+15.8	M (1)
Boc-Trp(For)-OH	332.4	126	+21.5	E (2)
Boc-Tyr(2-BrZ)-OH	494.3	90–108	+21.0	M (2)
Boc-Val-OH	217.3	72–73	−5.8	H (1)

[a]Abbreviations for solvents: D, DMF; E, ethanol; H, acetic acid; M, methanol; W, water.
Source: Refs. 18 and 92 and Dr. D. Scanlon, Auspep, Melbourne, Australia.

sodium carbonate solution. Thin-layer chromatography (TLC) is used to monitor the presence of unreacted amino acid. Reaction is usually complete after 2 h. The mixture is washed with ether (2 × 50 mL) to remove excess Boc-ON, acidified with 1 N aqueous HCl to pH 2–3, and extracted with ethyl acetate (3 × 50 mL). The extracts are pooled, washed with water (2 × 25 mL) (see note 1 following), dried over anhydrous sodium or magnesium sulfate, and, after filtration, evaporated in vacuo. The product is recrystallized by dissolving in a few drops of ethyl acetate, adding hexane to incipient turbidity, and allowing to stand overnight at $-20°C$ (see note 2 following).

b. Di-tert-Butyl-dicarbonate [tert-Butyl Pyrocarbonate, Boc Anhydride, (Boc)$_2$O] [23,25]. A solution of the amino acid (10 mmol) in a mixture of dioxane (20 mL), water (10 mL), and 1 N sodium hydroxide (10 mL) is stirred and cooled in an ice bath. Boc-anhydride (2.4 g, 11 mmol) is added dropwise for 1 h and stirring is continued at room temperature for 30 min. The solution is concentrated in vacuo to about 10 to 15 mL, cooled in an ice-water bath, and extracted with ether (2 $\times$ 100 mL). The aqueous solution is acidified with 1 N aqueous HCl to pH 2–3 and workup continued as described in the preceding paragraph.

Notes:

1. Certain Boc-amino acids (serine and threonine) are fairly soluble in water and cannot be satisfactorily extracted into ethyl acetate. Saturate the aqueous phase with NaCl before the extraction [18].
2. If crystallization fails, the dicyclohexylammonium salt can be prepared by adding an equimolar amount of dicyclohexylamine to a solution of Boc-amino acid in absolute ethanol and diluting with ether [23].

2. Deprotection Schemes

a. Purification [26]

1. TFA is distilled cautiously with slow heating at atmospheric pressure, boiling point (b.p.) 72°C. Rapid heating gives anomalous boiling points. It is stored in amber bottles.
2. N,N'-Diisopropylethylamine (DIEA) is initially distilled from ninhydrin (1–2 g/L) and then redistilled under nitrogen from KOH (5–10 g/L), b.p. 125–127°C. DIEA is stored in amber bottles under nitrogen in a cool place.

b. Peptide Synthesis. If employing the in situ neutralization method of SPS [27,28] (Chapter 6), neat TFA (**caution:** toxic, corrosive) is employed for removal of the N^α-deprotection. In earlier SPS protocols, a solution of 25–50% TFA in DCM is employed (1 $\times$ 5 min, 1 $\times$ 15 min). Boc-group removal via the latter step must be followed by neutralization of the resulting TFA salt-bound N^α-amino group by treatment of the peptide-resin with 5–10% TEA or, preferably, DIEA in DMF or NMP (2 $\times$ 5 min).

3. Acidolysis-Mediated Side Reactions

No evidence has been found of N^α-*tert*-butylation following N^α-deprotection in SPS [29]. However, the free side chain of Trp residues can be significantly

N^{in}-*tert*-butylated [30] although in modern SPS this can be simply circumvented by used of suitable N^{in}-protection, usually by the formyl group (Chapter 4). *tert*-Butyl cations also alkylate the nucleophilic thioether of Met residues and, rarely, the hydroxyl of Tyr [5]. To prevent the former side reaction, it has been recommended that a cationic scavenger, 1,2-ethanedithiol, be added to the TFA solution [31]. Use of neat TFA after a DMF wash in the in situ neutralization method of SPS causes a substantial elevation of resin temperature, which, in the case of N^{α}-deprotection of terminal Gln residues, leads to substantial pyroglutamic acid formation with subsequent termination of the growing peptide chain [27]. This may be prevented by introducing an intermediate DCM flow wash immediately before and after the TFA washes.

B. 2-(4-Biphenylyl)-isopropoxycarbonyl (Bpoc)

There was much early interest in the use of Bpoc-amino acids in SPS and a number of successful applications have been described [32,33]. Like the Boc group, the Bpoc group is stable to bases and nucleophiles, but in contrast to the Boc group, it is cleaved by catalytic hydrogenation. However, the principal disadvantage of such derivatives was that most could be obtained only as oils and had to be isolated and, because of the problem of premature free α-carboxyl-mediated N^{α}-deprotection, stored as their CHA or DCHA salts. Such inconvenience clearly led to the Bpoc group being abandoned as a serious alternative to the Boc group. However, interest in the use of Bpoc-based SPS has been renewed by the availability of simpler methods for the preparation of Bpoc-amino acids (including those with novel side-chain protection) and their convenient O-pentafluorophenyl (O-Pfp) esters [34,35]. Although successful use of the O-Pfp derivatives in SPS has been reported [35], it remains to be seen whether the Bpoc group regains a place among SPS protocols.

1. Preparation

Bpoc-amino acid derivatives were originally prepared via their azides [5]. The inconvenience of such reagents, particularly their toxicity, led to the development of Bpoc-phenyl carbonate and the more reactive and stable Bpoc-p-methoxycarbonylphenyl carbonate as alternatives [5,36]. Unfortunately neither of these reagents is commercially available and they must be prepared immediately before their use in the preparation of Bpoc-amino acids. A list of Bpoc-amino acid derivatives and their physical characteristics is given in Table 2.

Table 2 Bpoc-Amino Acid Derivatives

Compound	Molecular weight	M.p. (°C)	$[\alpha]_D$	Solvent[a] (conc.)
Bpoc-Ala-OH	327.1	102–111	−39.4	D (1)
Bpoc-Ala-O.DCHA	508.1	167–168	+3.4	M (1)
Bpoc-Ala-OPfp	493.5	96–99	−47.0	T (1)
Bpoc-Arg(Tos)-OH	566.2	Amor.		
Bpoc-Asn-OH	370.1	176–178	−13.8	D (1)
Bpoc-Asn(Trt)-OH	612	134–135	−17.3	T (1)
Bpoc-Asn(Trt)-OPfp	778	128–130	−55.0	T (1)
Bpoc-Asp(OtBu)-OPfp	593	81–85	−25.5	T (1)
Bpoc-Cys(tSBu)-OPfp	613	89–91	−62.0	T (1)
Bpoc-Gln(Trt)-OH	626	102–104	+6.4	T (1)
Bpoc-Gln(Trt)-OPfp	792	148–150	−16.4	T (1)
Bpoc-Glu(OtBu)-O.DCHA	623.1	138	+12.9	M (1)
Bpoc-Glu(OtBu)-OPfp	607	25–27	−21.6	T (1)
Bpoc-Gly-O.DCHA	494.1	192–193	—	
Bpoc-Gly-OPfp	479	113–114	—	
Bpoc-His(Boc)-OH	493.2	134	+21.0	M (1)
Bpoc-His(Trt)-OH	668	104–105	+36.7	T (1)
Bpoc-His(Trt)-OPfp	801	82–84	−26.6	T (1)
Bpoc-Ile-OH	369.2	215–218	−2.0	M (1)
Bpoc-Ile-OPfp	535	71–74	−35.4	T (1)
Bpoc-Leu-OH	369.2	227–230	−12.0	M (1)
Bpoc-Leu-OPfp	535	95–98	−35.7	T (1)
Bpoc-Lys(Tfa)-OPfp	646	115–119	−16.8	T (1)
Bpoc-Met-O.DCHA	568.2	143–145	+14.0	M (1)
Bpoc-Met-OPfp	553	69–70	−31.0	T (1)
Bpoc-Phe-O.DCHA	584.2	116–119	+33.0	M (1)
Bpoc-Phe-OPfp	570	123–125	−34.4	T (1)
Bpoc-Pro-O.DCHA	538	173–175	−12.0	M (1)
Bpoc-Pro-OPfp	519	97–100	−36.3	T (1)
Bpoc-Ser(tBu)-O.DCHA	581.1	183	+21.0	M (1)
Bpoc-Ser(tBu)-OPfp	565	25–27	−28.0	M (1)
Bpoc-Thr(tBu)-O.CHA	513.1	192–194	+7.0	M (1)
Bpoc-Thr(tBu)-OPfp	579	68–72	−39.6	T (1)
Bpoc-Trp-OH	442.2	145	+36.8	M (1)
Bpoc-Trp-OPfp	608	40–42	−18.9	T (1)
Bpoc-Tyr(OAl)-OPfp	627	135–136	−18.1	T (1)
Bpoc-Val-OH	355.1	225	−14.9	M (1)
Bpoc-Val-OPfp	521	47–54	−33.9	T (1)

[a]Abbreviations for solvents: D, DMF; M, methanol; T, tetrahydrofuran.
Source: Refs. 18, 34, and 35.

a. 2-Phenoxycarbonyloxy-2-(p-biphenylyl)-propane (Bpoc-OPh) [34]. In a three-necked flask equipped with a mechanical stirrer, a solution of commercially available 2-(*p*-biphenylyl)-2-propanol (0.5 mmol) in 500 mL of DCM is stirred and cooled in an ice–salt bath during the addition of 0.61 mol of dry pyridine. After 10 min at −7°C, 0.56 mol of phenyl chloroformate in 150 mL of DCM is added dropwise over 1 h, and the resulting mixture is stirred overnight at 4°C. The mixture is filtered and the collected salt washed with DCM until colorless. The combined filtrates are poured over 600 mL of ice water, the layers are separated, and the organic phase is washed with 3 × 1 L water and 1 × 500 mL brine, then dried over anhydrous sodium sulfate and evaporated to yield after drying a white solid (80%), melting point (m.p.) 111–115°C. Recrystallization is from hot ethyl acetate. The product is moisture sensitive and must be stored in the freezer.

b. Bpoc-Amino Acid [34,35]. The amino acid zwitterion (162 mmol) is solubilized in Triton B (162 mmol of a 40% solution in methanol) and then concentrated on a high-vacuum rotary evaporator to remove any excess water and methanol. The white syrupy solid is dissolved in DMF (40 mL) and the solution evaporated to a syrup on a high-vacuum rotary evaporator. This step is repeated two to three times. The resulting heavy syrup is mixed with DMF (200 mL) and Bpoc-phenyl carbonate (162 mmol) and warmed to 55°C. The reaction mixture is stirred for 3 h and then the DMF is removed with a high-vacuum rotary evaporator. The pasty solid or oil is diluted with water (100 mL), sodium sulfate (2 g, helps to prevent emulsions), and 5% sodium bicarbonate (10 mL) and then overlayered with ether (200 mL). The layers are separated and the aqueous phase extracted twice more with ether (2 × 200 mL). The combined ether washes are back extracted with 1% aqueous sodium bicarbonate (20 mL) and the aqueous phases combined, cooled in a 0°C ice bath, and overlayered with ether. Dropwise addition of pH 3.5 citrate buffer to the biphasic mixture causes clouding in the aqueous layer that clears upon swirling. Addition is continued until pH 3.5 is reached in the mixture. The aqueous phase is extracted with ether (2 × 150 mL) and the ether phases are combined and washed with citrate buffer (1 × 100 mL), water (3 × 200 mL), and brine (1 × 100 mL); dried over anhydrous magnesium sulfate; and concentrated to an oily solid (yields 60–95%). Derivatives are converted immediately to dicyclohexylamine (DCHA) or cyclohexylamine (CHA) salts for storage (Section II.A.1) or to *O*-pentafluorophenyl esters (see the following).

c. Bpoc-Amino Acid-OPfp [35]. Bpoc-amino acid (150 mmol) is dissolved in ethyl acetate (400 mL) and cooled to −10°C in an acetone ice bath. Pentafluorophenol (150 mmol) and dicyclohexylcarbodiimide (DCC) (150 mmol) are added sequentially in one portion and the reaction is stirred for 6 h at $-10°C$ and then placed in a freezer overnight. The solution is filtered to remove dicyclohexylurea (DCU), and the ethyl acetate is removed in vacuo to a dry foam or solid. This is taken up in ether and allowed to sit overnight in a freezer at $-20°C$. Residual DCU is removed by filtration and the ether removed in vacuo. The resulting foam or solid is recrystallized from hexane containing a small quantity of a more polar cosolvent (ether, isopropanol).

2. Deprotection

a. Peptide Synthesis. The Bpoc group is conveniently removed by treatment for 15–30 min with a solution of 0.2–0.5% TFA in DCM. Higher concentrations of acid (up to 5%) have been employed but the risk of premature cleavage of the side chain *tert*-butyl protecting groups is substantially increased. The deprotection step must be followed by neutralization of the N^α-amine as described in Section II.A.2.

3. Acidolysis-Mediated Side Reactions

One study showed that the released Bpoc carbocation causes approximately 0.5% N^{in}-alkylation of Trp residues per typical 30-min cycle of deprotection with 0.5% TFA in DCM [34]. Met thioethers were comparatively unaffected. It was further shown that this level of N^{in} side reaction could be reduced, but not eliminated, by addition of thiophenol. Suitable N^{in}-protection for Trp during Bpoc SPS is now afforded by the allyl group (Chapter 4) [35].

III. BASE-LABILE GROUPS

A recognized weakness of the Boc-polystyrene SPS procedure lies in the use of temporary and permanent protecting groups of graded lability to the same reagent type, namely acid. There was growing awareness that repeated cycles of neat TFA-mediated N^α-deprotection over the course of a long synthesis may lead to modification and/or degradation of sensitive peptide sequences. The situation could be compounded by the use of the harsh reagent, liquid hydrogen fluoride, for final side-chain deprotection and simultaneous peptide-resin cleavage [37,38]. It was reasoned that this limitation could be minimized by the use of base-labile N^α-protecting groups, and subsequently two laboratories [39,40] independently explored the use in SPS of the 9-fluorenylmethoxycarbonyl (Fmoc, **8**) group of Carpino and Han [41]. It was found to be particularly well suited for SPS, being rapidly and efficiently removed by proton abstraction with secondary amines, conditions that do not affect acid-sensitive *tert*-butyl side-chain protecting groups. This, together with the use of *tert*-butyl–based side-chain protection and a TFA acid-labile peptide-resin linker (see Chapter 2), resulted in a considerably milder means of SPS. The use of Fmoc-based SPS has increased enormously to the point where it is probably now the method of choice for the chemical synthesis of peptides.

In contrast to the extensive development of new acid-labile N^{α}-protecting groups, comparatively little work has been carried out on developing alternative base-labile N^{α}-protecting groups. The 2-[4-(methylsulfonyl)phenylsulfonyl]ethyloxycarbonyl group (Mpc, **9**) [42] was shown to have higher solvent stability than the Fmoc group and to allow successful SPS [43]. Base removal of the 2,2-bis(4′-nitrophenyl)ethan-1-oxycarbonyl group (Bnpeoc, **10**) [44] is accompanied by a blue color that permits automated monitoring of the extent of deprotection. Two new base-labile groups have been developed and show promise as alternatives to the Fmoc group. The nonurethane N-4,4-dimethyl-2,6-dioxocyclohexylidenemethyl (Dcm, **11**) [45] is removed by treatment with either 1% hydrazine in DMF or 20% morpholine in the same solvent. Because of incompatibility of the Dcm amino acids with uronium activating agents, coupling is mediated via DIPCDI-HOBt. In contrast, organic bases remove the urethane derivative 2-(4-nitrophenysulfonyl)ethoxycarbonyl (Nsc, **12**) [46] via a β-elimination mechanism. This reaction proceeds three to eight times more slowly than that with the Fmoc group and is dependent on the base and solvent. On the other hand, it exhibits greater stability in neutral and weakly basic aprotic solutions. A further advantage of the Nsc group is that its cleavage by piperidine yields a yello color that provides a visual indication that N^{α}-deprotection has occurred. The liberated N-[2-(4-nitrophenylsulfonyl)ethyl]-piperidine has a long wave shoulder at 320–380 nm, which would allow selective on-line monitoring during SPS. Addition of the stronger base 1,8-diazabicyclo[5.4.0]undec-7-ene (DBU) (1%) to a solution of 20% piperidine in DMF was required to compensate for the slower rate of N^{α}-deprotection. Caution is obviously indicated when using this base mixture in the presence of aspartimide-sensitive sequences (see the following).

A. 9-Fluorenylmethoxycarbonyl (Fmoc)

The Fmoc group has several properties that make it nearly ideal for use in SPS. It displays exceptional acid stability; has a high ultraviolet absorption (λ_{max} 267 nm), which permits monitoring of acylation and deprotection reactions; and is fully compatible with *tert*-butyl–based side-chain protection [47]. Furthermore, Fmoc-amino acids are generally easy to prepare in crystalline form in high yield and are stable as the free acid when stored at 4°C. The Fmoc group is labile to organic bases, particularly secondary amines

Table 3   Half-Life of Fmoc group in Amines

Amine	Approximate half-life[a]
20% piperidine	6 s
5% piperazine	20 s
50% morpholine	1 min
10% *N,N*-dicyclohexylamine	35 min
10% *p*-dimethyaminopyridine	85 min
50% *N,N*-diisopropylethylamine	10.1 h

[a]Determined by dissolving Fmoc-Val-OH in the amine-DMF mixture and monitoring the amount of free valine generated by amino acid analysis.
Source: Ref. 47.

such as piperidine, but also to primary and tertiary amines. Rates of deprotection vary considerably among bases, but deprotection is fastest with unhindered secondary amines such as piperidine (Table 3) [47]. The liberated dibenzofulvene (DBF) intermediate is trapped by excess secondary amine to yield an inert adduct. A number of other bases have also been used for N^α-deprotection in SPS, including *cis*-2,6-dimethylpiperidine and hexamethyleneimine [48]. In addition, fluoride ion has been shown to effect smooth removal of the Fmoc group [49]. However, it does not quench the released DBF, so its use alone is recommended only in continuous flow SPS to prevent the DBF from reacting with the resulting resin-bound peptide terminal free N^α-amine. In batch systems, piperidine is added to the DBU solution to quench the liberated DBF.

1.   Preparation

The Fmoc group is normally introduced under weakly basic conditions via the Schotten–Baumann reaction using the readily commercially available chloroformate (also known as chloride) [46]. However, certain amino acids, particularly glycine, alanine, phenyalanine, and N^ε-trifluoroacetyl-lysine, may be contaminated with significant levels (2–20%) of their corresponding Fmoc-dipeptide and even tripeptide [50,51]. These impurities require removal by rigorous recrystallization with correspondingly reduced overall yield of product. For this reason, the chloroformate has been largely superseded by the less reactive hydroxysuccinimido ester [50,52]. This compound remains the reagent of choice although a variety of alternative protocols have been reported for the introduction of the Fmoc group free of the amino acid oligomer side reaction. These include the use of prior amino acid bis-trimethylsilylation followed by reaction with fluorenylmethyl chloroformate [51], fluorenylmethyl azidoformate [46,53], and [(fluorenylmethyl)-phenyl]dimethylsulfonium methyl sulfate (Fmoc-ODSP) in water [54]. A useful reagent is fluorenylmethyl pentafluorophenyl carbonate, which has the added advantage of allowing subsequent one-pot preparation of the efficient acylating agents O-pentafluorophenyl esters of the Fmoc-amino acids [55] (see the following and Chapter 6). Alternatively, these active esters may be prepared with N,N'-dicyclohexylcarbodiimide and pentafluorophenol [26]. A list of Fmoc-amino acid derivatives commonly employed in SPS is shown in Table 4.

a. Via Fmoc-Choloroformate (Fmoc-Cl) [26,52]. L-Amino acid (50 mmol) is dissolved in a mixture of dioxane (50 mL) and threefold molar excess of 10% sodium carbonate (133 mL) and stirred briskly at ice temperature. To this stirred solution, a 10% molar excess of crystalline Fmoc-Cl (55 mmol) dissolved in peroxide-free dioxane (50 mL) is added dropwise via a separatory funnel over a 15-min period. The ice bath is removed and the reaction mixture stirred for 1 h or longer (see note 1 following) until reaction is complete as determined by TLC on silica gel using C:M:A (85: 10:5, v/v) as a solvent system [ultraviolet (UV) and ninhydrin monitoring] (see Section V). The reaction mixture is poured into water (500 mL) and the clear solution extracted three times with ether (150 mL each) to remove excess chloroformate and acidified (**Caution:** effervescence) with aqueous citric acid or dilute hydrochloric acid to pH 3. The precipitated white solid is extracted into ethyl acetate (3 × 150 mL) and the combined extracts washed thoroughly with water (3 × 200 mL) until the pH is near neutral. The organic phase is dried over anhydrous sodium or magnesium sulfate, filtered, and evaporated. Crystallization is from ethyl acetate—petroleum ether.

Notes:

1. Occasionally, overnight reaction is required before complete reaction occurs.

b. Via Fmoc-Succinimide (Fmoc-OSu) [26,52]. The lower solubility of Fmoc-OSu in dioxane necessitates its dissolution in acetone before its use as described before. However, precipitation of either the reagent or the base is occasionally observed. Best restuls are obtained when the reagent is dissolved in dimethoxyethane [56]. Work-up is as described before.

c. Fmoc Pentafluorophenyl Carbonate (Fmoc-OPfp) [55]. Fluorenyl-methyl chloroformate (25.9 g, 0.1 mol) and pentafluorophenol (18.4 g, 0.1 mol) are dissolved in diethyl ether (200 mL) at 0°C. Triethylamine (14.0

Table 4 Fmoc-Amino Acid Derivatives

Compound	Molecular weight	M.p. (°C)	$[\alpha]_D$	Solvent[a] (conc.)
Fmoc-Ala-OH	311.3	143–144	−18.6	D (1)
Fmoc-Arg(Mtr)-OH	608.9	118–120	+7.9	M (0.5)
Fmoc-Arg(Pmc)-OH	662.8	70–120 (dec.)	−0.2	D (1)
Fmoc-Arg(Pbf)-OH	648.8	103–115	−5.3	D (1)
Fmoc-Asn-OH	354.4	185–186	−11.4	D (1)
Fmoc-Asn(Trt)-OH	596.7	209–220	−15.5	M (1)
Fmoc-Asp(OtBu)-OH	411.5	148–149	−20.3	D (1)
Fmoc-Cys(Acm)-OH	414.5	150–154	−27.5	E (1)
Fmoc-Cys(StBu)-OH	431.8	74–76	−84.6	E (1)
Fmoc-Cys(Trt)-OH	585.7	174–178	+20.6	D (1)
Fmoc-Gln-OH	368.4	220–225	−17.0	D (1)
Fmoc-Gln(Trt)-OH	610.7	110–125 (dec.)	+2.3	M (1)
Fmoc-Glu(OtBu)-OH	425.5	76–77	+0.9	D (1)
Fmoc-Gly-OH	297.3	173–176	—	
Fmoc-His(Boc)-OH	477.5	149–151	+15.2	D (1)
Fmoc-His(Trt)-OH	619.7	130–145 (dec.)	+85.0	C (5)
Fmoc-Ile-OH	353.4	143–145	−11.9	D (1)
Fmoc-Leu-OH	353.4	153–154	−24.0	D (1)
Fmoc-Lys(Boc)-OH	468.6	123–124	−11.7	D (1)
Fmoc-Met-OH	371.5	129–132	−28.3	D (1)
Fmoc-Phe-OH	387.4	178–179	−37.6	D (1)
Fmoc-Pro-OH	337.4	114–115	−33.2	D (1)
Fmoc-Ser(tBu)-OH	383.4	126–129	−1.5	D (1)
Fmoc-Thr(tBu)-OH	397.5	129–132	−4.5	D (1)
Fmoc-Trp-OH	426.5	165–166	−26.6	D (1)
Fmoc-Trp(Boc)-OH	526.6	97–105	−19.4	D (1)
Fmoc-Tyr(tBu)-OH	459.5	150–151	−27.6	D (1)
Fmoc-Val-OH	339.4	143–144	−16.1	D (1)

[a]Abbreviations for solvents: C, chloroform; D, DMF; E, ethanol; M, methanol.
Source: Refs. 26 and 47 and Dr. D. Scanlon, Auspep, Melbourne, Australia.

mL, 0.1 mol) is added dropwise and the mixture stirred for 2 h. The solution is washed with water (3 × 200 mL), dried over anhydrous sodium sulfate, filtered, and evaporated. The solid residue is crystallized from *n*-hexane (120 mL) to give the product (29.6 g, 73.6%). A second crop may be obtained from the mother liquor. Melting point 84–86°C; R_f (TLC, silica gel, ethyl acetate) 0.70.

2.   Deprotection

a.   Piperidine (**Caution:** harmful vapor, toxic).   *Purification* [26]: Synthe-sis grade reagent is readily available from reputable suppliers. Laboratory grade or old piperidine must be purified before use. It is distilled under nitrogen from potassium hydroxide pellets (10–20 g/L). The initial distillate, approximately 5% of the total, is discarded, and then the main fraction, b.p. 106°C, collected. Distilled piperidine is stored in amber bottles in a cool place. Stock solutions for Fmoc removal should be made up fresh before each peptide synthesis. Do not use if yellow in color.

Peptide synthesis: A solution of 20% in DMF (v/v) is most commonly employed for both batch and continuous-flow syntheses, although concen-trations as high as 50% in DMF (v/v) have also been reported [26,57]. For some difficult syntheses, DBU (see next) has been added to the piperidine solution to accelerate the rate of Fmoc group removal [58]. Solvents other than DMF may be used successfully. These include DMA and NMP or mixtures of these together with DCM.

b.   1,8-Diazabicyclo[5.4.0]undec-7-ene (DBU) (**Caution:** harmful vapor, toxic).   *Purification:* This is generally used without further purification.

Peptide synthesis: DBU is used in continuous-flow SPS as a 1–2% solution in DMF (v/v) [59]. In batch SPS, an additive is required to quench the liberated dibenzofulvene, usually piperidine to 2% (v/v) [60].

3.   Side Reactions

With increasing use of Fmoc SPS, a number of base-mediated side reactions have been identified and reported that require careful awareness [61]. Some of these are described elsewhere in this volume (Chapters 2 and 4). The principal base-mediated side reactions are diketopiperazine formation caused by cyclization, particularly during N^{α}-deprotection of the residue adjacent to C-terminal resin-linked proline [62], and aspartimide formation, particularly at Asp-X residues [63,64]. The former can now be prevented by use of the substantially sterically hindered 2-chlorotrityl linker [65]. The latter side reaction is more difficult to control and appears to be largely sequence de-pendent. Asp-Gly, -Ser, -Thr, -Asn, and -Gln pairs are most at risk of po-tential imide formation, although several other Asp-X combinations have also been observed to cyclize [66,67]. For one sensitive peptide sequence, use of piperazine for N^{α}-deprotection eliminated this side reaction [68]. However, for another peptide, this base was ineffective (J Wade, unpub-lished). Reduction of the basicity of the piperidine solution by addition of 0.1 M 1-hydroxybenzotriazole is beneficial [69,70]. Complete protection against aspartimide formation was afforded by the use of N-(2-hydroxy-4-methoxyl) protection of the aspartyl amide bond [71,72].

Cysteine-containing peptides are particularly prone to base-mediated side reaction. Acid-labile linker-bound C-terminal Cys undergoes significant racemization (approximately 0.5%) with each cycle of piperidine treatment [73]. This could be reduced to a modest level (0.1%) by use of 1% DBU in DMF [59]. However, both piperidine and DBU were observed to cause dehydration of C-terminal cysteine residues to a 3-(1-piperidinyl)alanine or dehydroalanine [74]. No measures could be found to prevent completely this partially side chain protection– and sequence-dependent problem. Recourse to Boc-SPS may be necessary.

IV. OTHER PROTECTING GROUPS

An astonishing variety of N^{α}-protecting groups has been developed over the past 30 years and shown to have properties that make them suitable for SPPS. However, for various reasons, such as complexity of preparation, inadequate shelf life, or difficulty with complete deprotection, none have achieved the popularity of Boc of Fmoc groups. It is beyond the scope of this chapter to list these alternative groups; instead, the reader is referred to comprehensive treatises for further details [5,6,75]. However, the following N^{α}-protecting groups have been chosen to illustrate the versatility and scope afforded to SPS.

A. Thiol Labile Groups

The dithiasuccinoyl (Dts, **13**) [76] is remarkably stable to acid, including 6 N HCl at 110°C, and is selectively removed by base-catalyzed reduction such as with 0.2 M mercaptoethanol in DCM. However, the cumbersome preparation of Dts-amino acids and the unavailability of a complete range of adequately side-chain protected derivatives have limited its use in SPS. The TFA acid-labile 2-nitropyridinyl (Nps, **7**) group is also cleaved by thiolysis [13].

B. Photolabile Groups

Little work has been undertaken on the development of photolabile N^{α}-protecting groups presumably because of the technical difficulties associated with using such derivatives. However, this may change following a recent ingenious application of the nitroveratryloxycarbonyl (Nvoc, **14**) group [77], which is photochemically cleaved at 320 nm. Nvoc derivatives were used to study molecular recognition by producing large libraries of peptides by light-directed, spatially addressable parallel SPS on a glass surface [78]. The very acid-labile α,α-dimethyl-3,5-dimethyloxybenzyloxycarbonyl (Ddz, **4**) is also removed by photolysis at wavelengths above 280 nm. This property makes the Ddz group also potentially very useful in SPS library screening procedures [79], although, as with the Nvoc group, special care is required to prevent premature cleavage or photolysis-mediated modification of sensitive peptides.

C. Other

Several N^{α}-protecting groups have been further chemically modified to make them amenable to post-SPS affininty or ion-exchange chromatography as an aid to peptide purification. These are typically used on the terminal N^{α}-amine and, after chromatography-mediated removal of accumulated impurities, are removed under controlled conditions. First was the 9-(2-sulfo)-fluorenylmethyloxycarbonyl group (Sulfmoc, **15**) [80], which provides the only anion in the peptide at a pH below 2. Therefore, the Sulfmoc-peptide can be bound to a weakly basic ion-exchange column under acidic conditions in which the underivatized peptides will not be retained. The very hydrophobic tetrabenzo[a,c,g,i]fluorenyl-17-methoxycarbonyl group (Tbfmoc, **16**) [81] allows retention of the labeled peptide on porous graphitized carbon or separation from nonlabeled peptide by simple reversed-phase HPLC (RP-HPLC). Its removal is achieved by treatment with aqueous pH 8.5 solution [82]. The Fmoc group has also been chemically modified to accommodate a biotin molecule, which in turn enables affinity purification of the labeled peptide via an immobilized monomeric avidin column [83].

V. ANALYTICAL METHODS

Despite the enormous advances made in simplifying, accelerating, and automating SPS, considerable expense and skilled operator management are required to ensure successful peptide synthesis. It is a basic tenet of SPS that all solvents and reagents used are of the highest purity. This is particularly the case with N^{α}-protected amino acid derivatives, which are invariably introduced to the growing peptide-resin in large excess. It is therefore mandatory that all derivatives, whether purchased from a commercial source or, in particular, prepared in the laboratory, be carefully subjected to quality control. Commercial suppliers usually rigorously undertake such analysis and provide comprehensive data sheets with each lot of synthetic compound. However, there is much anecdotal evidence of failed solid-phase peptide syntheses caused by the careless use of partially decomposed or even incorrectly labeled derivatives. Another source of derivative-mediated side reaction is decomposition of trace ethyl acetate used as recrystallization solvent to acetic acid. This is activated in SPS and causes N^α-acetylation and subsequent peptide chain termination [27,84]. This may be prevented by simple overnight freeze-drying of the amino acid derivative to remove solvent.

It is recommended that for an accurate assessment of the purity of amino acid derivatives, a minimum of three of the following analytical methods be used. At least one of these must be TLC or RP-HPLC.

A. Thin-Layer Chromatography (TLC)

This is the easiest of the chromatographic techniques to perform and requires only a simple apparatus. It is preferred over paper chromatography as it is more sensitive, requires less amino acid derivative for analysis, and is a more rapid technique [85]. Many texts describe the practice of TLC in detail [e.g., 86,87]. Numerous media are available for TLC, including silica gel, kieselguhr, alumina, cellulose, and polyamide. Plastic and aluminum plates precoated with such media are readily commercially available. Particularly useful are the precoated silica gel plates containing a fluorescent indicator. These enable ready postchromatography UV viewing of the separated substances before further identification by a chemical reagent spray.

TLC is performed in closed glass tanks, the walls of which should be lined with chromatography paper moistened with the developing solvent to ensure a saturated atmosphere. Plates may be cut to any size to fit within the dimensions of the tank. The amino acid derivative sample is dissolved in either methanol or ethyl acetate (10–50 mg/mL) and 1–5 μL carefully spotted onto the plate. After air drying, chromatography is performed using one of the many solvent systems currently available. The most commonly employed systems are listed in Table 5. The solvent system chosen for chromatography determines the extent of migration of the component(s) (R_f) on the plate and subsequent resolution. Those containing a higher proportion of nonpolar organic solvent result in the generally nonpolar N^α-protected derivatives migrating farther on the plate (high R_f), while amino acids lacking N^α-protection, being more polar, remain on or close to the plate's origin (low R_f). It is strongly recommended that TLC be performed using at least two different solvent systems.

After chromatography, the plate is thoroughly dried in air or by blowing warm air from a fan heater (fume hood!). After derivative visualization under UV light (if using silica gel plates containing a fluorescent indicator), the plates are sprayed with a general identification reagent (see Section V.B). These react with all amino acids, usually via the amino group. Before spraying, the Boc group is removed by placing the plate in a closed

Table 5 Most Commonly Employed Solvent Systems

Solvents	Composition (parts by volume)
Acetone–acetic acid	98:2
n-Butanol–acetic acid–water	3:1:1
n-Butanol–acetic acid–water–ethyl acetate	1:1:1:2
n-Butanol–acetic acid–water–pyridine	15:3:8:10
Chloroform–acetic acid	1:3
Chloroform–acetone	5:1
Chloroform–methanol–acetic acid	85:10:5
Chloroform–methanol–acetic acid	7:1:2
Ethyl acetate–acetic acid	18:1
Ethyl acetate–hexane	1:4
Ethyl acetate–hexane	3:2
Ethyl acetate–pyridine–acetic acid–water	80:20:5:10
Methanol–chloroform	1:4

glass chamber containing a beaker of fresh concentrated HCl. After a 15-min exposure to the fumes, the plates are heated for 10 min in a 105°C oven. Similarly, the Fmoc group is removed by 15 min exposure to diethylamine. No subsequent heating of the plate is required.

B. TLC Identification Sprays

1. Ninhydrin

The reaction of amino acids (and peptides) with ninhydrin has been known for nearly a century. Ninhydrin reacts rapidly with all primary amines to produce characteristic blue-colored spots.

> *Reagent:* 0.2–0.5% ninhydrin in acetone (w/v). This may be stored in an amber bottle at 4°C for up to 2 weeks.
> *Method:* The dried plate is sprayed with reagent and then heated at 105°C for 5 min. The chromatogram may be preserved by spraying with 0.5% $NiSO_4 \cdot 6H_2O$ in water. The colors (which fade to pink-red) will remain stable for many months if kept in the dark.

2. o-Phthalaldehyde

This is similar in sensitivity to ninhydrin, although some batches of reagent react poorly.

> *Reagent:* 0.5% o-phthalaldehyde in acetone (w/v).

Method: The plate is sprayed with reagent and then heated at 40°C for 5 min. Different amino acids give different colored spots. The spots are also fluorescent.

3. Fluorescamine (Fluram)

This reagent reacts with primary amines to produce a fluorescent that can be detected under UV light at 350, 366, or 390 nm. It is expensive but more sensitive than ninhydrin.

Reagent A: Triethylamine (10% in DCM)
Reagent B: Fluorescamine (20 mg in 100 mL dry acetone)
Method: The plate is lightly sprayed with reagent A and dried briefly. Reagent B is sprayed and the plate is again allowed to air dry briefly before repeating with reagent A. The fluorescence appears immediately and is stable for at least 24 h.

4. Isatin

This reacts with proline and hydroxyproline to produce intense blue spots on a yellow background.

Reagent: Freshly prepared 0.5% isatin in acetone.
Method: Chromatograms are sprayed with reagent and immediately heated at 105°C for 10 min.

5. β-Napthaquinone Sulfonic Acid (NQSA, Folin Reagent)

If chromatograms have been developed in acid solvents, it is more satisfactory to make sheets alkaline by spraying with 10% aqueous sodium carbonate solution and, when dry, applying reagent. Reaction is immediate and all spots appear blue (except red for proline) on a yellow–brown background.

Reagent: 0.3% NQSA in either water or 10% aqueous sodium carbonate.
Method: Color forms rapidly after spraying reagent onto the plate but does not remain for more than a few days.

6. Iodine

Detection of amines is based on the readily formed and intensely colored ion radicals.

Method: The dried TLC plate is placed in a closed glass chamber containing several iodine crystals for 10–15 min.

7. Other

Several other reagents are available including some specific for certain amino acid side chains [18,86,87]. These include the Pauly spray for tyrosine and histidine but they are rarely used for N^{α}-protected amino acid derivatives.

C. High-Performance Liquid Chromatography (HPLC)

The high sensitivity, resolving power, ease, and speed of the method make this an excellent means of critical assessment of purity of amino acid derivatives [88,89]. Typically, elution of derivatives is from a small-diameter (5– 10 μm, 300 Å pore size) reversed phase (RP) support using a linear gradient of acetonitrile in aqueous buffer. Gradients are usually from 30% B to 100% B over 30 min. The following buffer systems are widely used, and it is recommended that at least one be employed for the determination of derivative purity.

1. Buffer A: 0.1% aqueous TFA; buffer B: 0.1% TFA in CH_3CN
2. Buffer A: 0.2 N triethylamine phosphate, pH 6.3; buffer B: CH_3CN
3. Buffer A: 0.01 M ammonium acetate, pH 4.5; buffer B: 10% A in CH_3CN

Ultraviolet detection is at 210 nm or, additionally for Fmoc derivatives, 266 nm. Note that such detection will not show free amino acids and, unless these have aromatic side chains or UV-absorbing side-chain protecting groups, it will be necessary to use a conductivity detector.

D. Elemental Analysis, Optical Rotation, Melting Point

These techniques, by themselves, do not give a precise indication of product purity. However, they do provide strong evidence of product identity when compared with standards. All novel amino acid derivatives require these analyses to be carried out.

E. Other

Nuclear magnetic resonance spectroscopy provides the most conclusive evidence of both identity and purity [90] but few laboratories are equipped with such a resource and even fewer researchers with the experience to interpret the resulting data. Gas chromatography can be used to assess the chiral purity of derivatives, and mass spectrometry (MS) is a particularly sensitive and accurate measure of product purity. Use of electrospray ionization MS in the negative ion mode allows rapid analysis of low-molecular-weight compounds such as amino acid derivatives [91].

ACKNOWLEDGMENTS

The author gratefully acknowledges the fine proofreading of this chapter and helpful comments by M. Mathieu, A. Clippingdale, and Professor G. W. Tregear (Howard Florey Institute, Australia). Dr. D. Scanlon, Auspep (Melbourne, Australia), also kindly provided useful information on certain amino acid derivatives, and Dr. R. C. Sheppard (Cambridge, UK) was generous with his advice and help.

APPENDIX

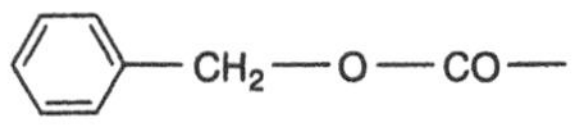

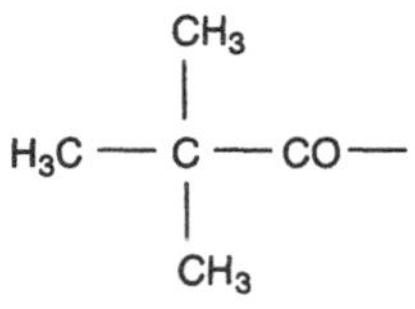

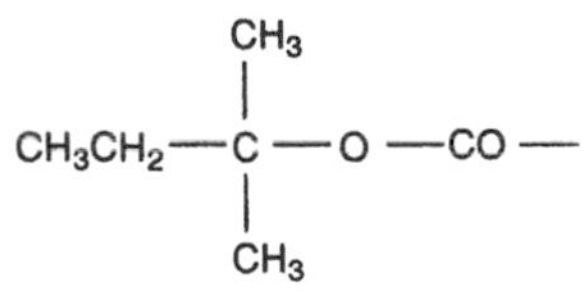

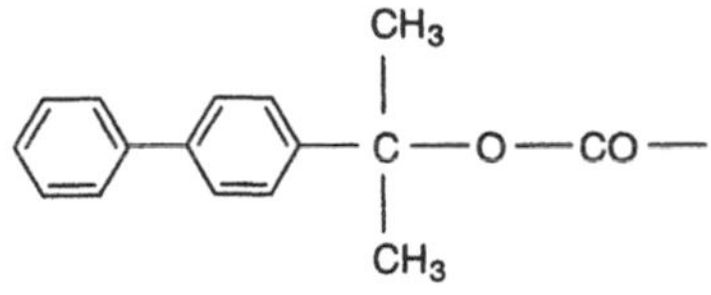

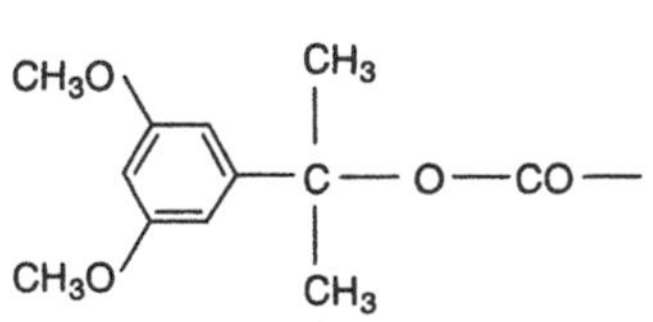

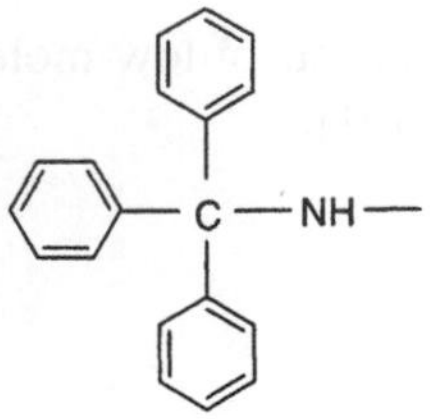

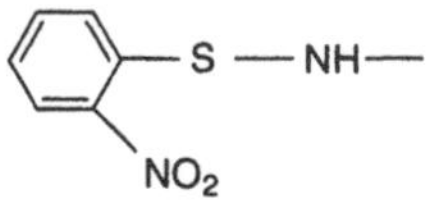

6

7

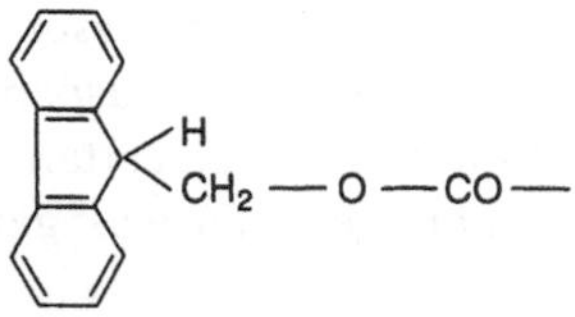

8

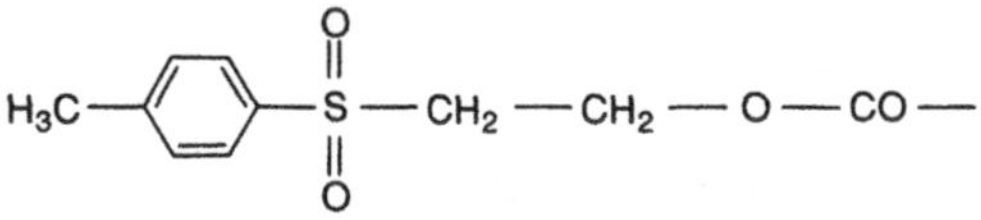

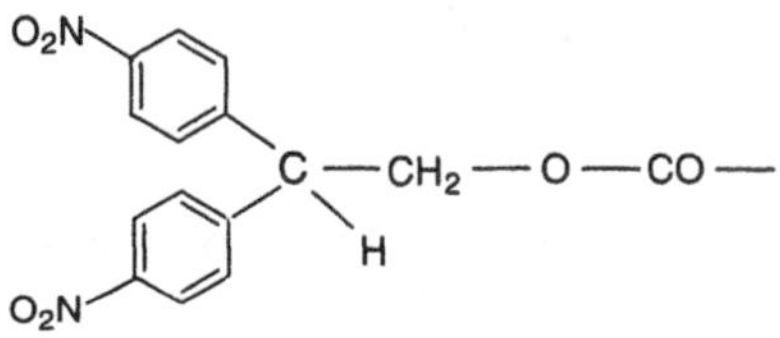

9

10

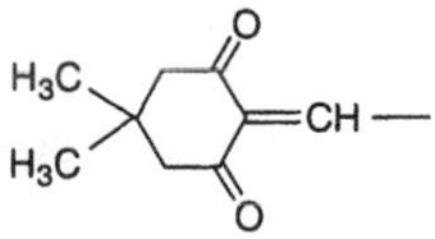

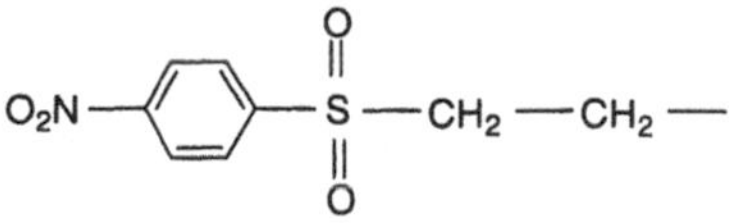

11

12

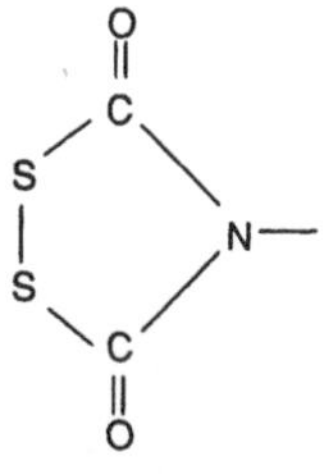

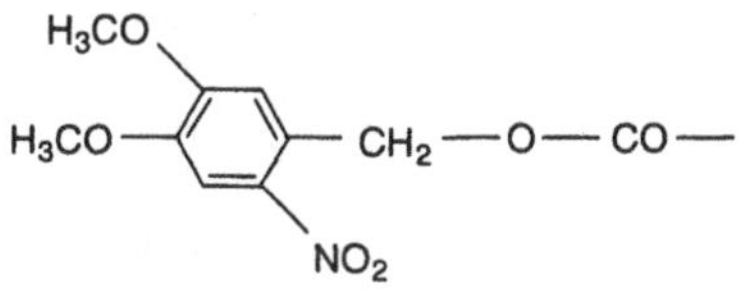

13

14

15   16

REFERENCES

1. M Bergmann, L Zervas. Ber Dtsch Chem Ges 65:1192–1201, 1932.
2. D Ben Ishai, A Berger. J Org Chem 17:1564–1570, 1952.
3. S Sakakibara, Y Shimonishi, Y Kishida, M Okada, N Inukai. Bull Chem Soc Jpn 38:1522–1525, 1967.
4. RB Merrifield. J Am Chem Soc 85:2148–2154, 1963.
5. G Barany, RB Merrifield. In: E Gross, J Meienhofer, eds. The Peptides. Analysis, Synthesis, Biology. Vol 2. New York: Academic Press, 1980, pp 1–284.
6. M Bodansky. Principles of Peptide Synthesis. 2nd ed. Berlin: Springer-Verlag, 1993.
7. S Sakakibara, M Shin, M Fujino, Y Shimonishi, S Inone, N Inukai. Bull Chem Soc Jpn 38:1522–1525, 1965.
8. C Birr, W Lochinger, C Stahnke, P Lang. Justus Liebigs Ann Chem 763:162–172, 1972.
9. P Sieber, B Iselin. Helv Chim Acta 57:2617–2621, 1968.
10. L Zervas, D Theodoropoulous. J Am Chem Soc 78:1359–1363, 1956.
11. L Zervas, D Borovas, E Grazis. J Am Chem Soc 85:3660–3666, 1963.
12. VA Najjar, RB Merrifield. Biochemistry 5:3765–3770, 1966.
13. A Tun-Kyi. Helv Chim Acta 61:1086–1090, 1978.
14. L Carpino. J Am Chem Soc 79:4427–4431, 1957.
15. FC McKay, NF Albertson. J Am Chem Soc 79:4686–4690, 1957.
16. GW Anderson, AC McGregor. J Am Chem Soc 79:6180–6183, 1957.
17. RB Merrifield. Adv Enzymol 32:221–296, 1969.
18. JM Stewart, JD Young. Solid Phase Peptide Synthesis. 2nd ed. Rockford, IL: Pierce Chemical Company, 1984.
19. L Carpino. Acc Chem Res 20:401–407, 1987.
20. E Schnabel. Justus Liebigs Ann Chem 702:188–196, 1967.
21. DS Tarbell, Y Yamamoto, BM Pope. Proc Natl Acad Sci U S A 69:730–732, 1972.
22. M Itoh, D Hagiwara, T Kamiya. Tetrahedron Lett 4393–4394, 1974.
23. M Bodanszky, A Bodanszky. The Practice of Peptide Synthesis. Berlin: Springer-Verlag, 1984.

24. F Albericio, E Nicolas, J Rizo, M Ruiz-Gayo, E Pedroso, E Giralt. Synthesis 119–122, 1990.
25. L Moroder, L Wackerle, E Wünsch. Z Physiol Chem 357:1651–1653, 1976.
26. E Atherton, RC Sheppard. Solid Phase Peptide Synthesis. A Practical Approach. Oxford: IRL Press, 1989.
27. M Schnölzer, P Alewood, A Jones, D Alewood, SBH Kent. Int J Peptide Protein Res 40:180–193, 1992.
28. P Alewood, D Alewood, L Miranda, S Love, W Meutermans, D Wilson. Methods Enzymol 289:14–29, 1997.
29. A Mitchell, RB Merrifield. J Org Chem 41:2015–2019, 1976.
30. H Ogawa, T Sasaki, H Irie, H Yajima. Chem Pharm Bull 26:3144–3149, 1978.
31. BF Lundt, NL Johansen, A Volund, J Markussen. Int J Peptide Protein Res 12: 258–268, 1978.
32. S Mosjov, RB Merrifield. Biochemistry 20:2950–2957, 1980.
33. Y Trudelle, F Heitz. Int J Peptide Protein Res 30:163–169, 1987.
34. DS Kemp, N Fotouhi, JG Boyd, RI Carey, C Ashton, J Hoare. Int J Peptide Protein Res 31:359–372, 1987.
35. RI Carey, LW Bordas, RA Slaughter, BC Meadows, JL Wadsworth, H Huang, JJ Smith, E Furusjö. J Peptide Res 49:570–581, 1997.
36. E Schnabel, G Schmidt, E Klauke. Justus Liebigs Ann Chem 743:69, 1971.
37. H Yajima, N Fuji. In: E Gross, J Meienhofer, eds. The Peptides. Analysis, Synthesis, Biology. London: Academic Press, 1983, pp 65–109.
38. J Tam, MW Reimen, RB Merrifield. Peptide Res 1:6–18, 1998.
39. E Atherton, H Fox, D Harkiss, CJ Logan, RC Sheppard, BJ Williams. J Chem Soc Chem Commun 537–539, 1978.
40. C-D Chang, J Meienhofer. Int J Peptide Protein Res 11:246–249, 1978.
41. LA Carpino, GY Han. J Org Chem 37:3404–3405, 1972.
42. CGJ Verhart, GI Tesser. Rec Trav Chem Pays-Bas 107:621–626, 1988.
43. WJG Schielen, HPHM Adams, W Nieuwenhuizen, GI Tesser. Int J Peptide Protein Res 37:341–346, 1991.
44. BB Bycroft, WC Chan, SR Chabra, PH Teesdale-Spittle, PM Hardy. J Chem Soc Chem Commun 776–777, 1993.
45. AN Sabirov, Y-D Kim, H-J Kim, VV Samukov. Protein Peptide Lett 4:307–312, 1997.
46. LA Carpino, GY Han. J Am Chem Soc 92:5748–5749, 1970.
47. E Atherton, RC Sheppard. In: S Udenfriend, J Meienhofer, eds. The Peptides. Analysis, Synthesis, Biology. Vol 9. New York: Academic Press, 1987, pp 1–38.
48. X Li, T Kawakami, S Aimoto. In: Y Shimonishi, ed. Peptide Chemistry. Proceedings of the First International Peptide Symposium. Amsterdam: Kluwer, 1999, pp 579–580.
49. M Ueki, M Amemiya. Tetrahedron Lett 28:6617–6620, 1987.
50. A Pacquet. Can J Chem 60:976–980, 1982.
51. DR Bolin, II Sytwu, F Humiec, J Meienhofer. Int J Peptide Protein Res 33: 353–359, 1989.
52. L Lapatsanis, G Milias, K Froussios, M Kolovos. Synthesis 671–673, 1983.

53. M Tessier, F Alberico, E Pedroso, A Grandas, R Eritja, E Giralt, C Granier, J van Rietschoten. Int J Peptide Protein Res 22:125–128, 1983.
54. I Azuse, M Tamura, K Kinomura, H Okai, K Kouge, F Hamatsu, T Koizumi. Bull Chem Soc Jpn 62:3103–3108, 1989.
55. I Schön, L Kisfaludy. Synthesis 303–305, 1986.
56. CG Fields, GB Fields, RL Noble, TA Cross. Int J Peptide Protein Res 33:298–303, 1989.
57. D Hudson. J Org Chem 53:617–624, 1988.
58. CG Fields, DJ Mickelson, SL Drake, JB McCarthy, GB Fields. J Biol Chem 268:14153–14160, 1993.
59. JD Wade, J Bedford, RC Sheppard, GW Tregear. Peptide Res 4:194–199, 1991.
60. M Dettin, S Pegoraro, P Rovero, S Bicciato, A Bagano, C di Bello. J Peptide Res 49:103–111, 1997.
61. JD Fontenot, JM Ball, MA Miller, CM David, RC Montelaro. Peptide Res 4: 19–25, 1991.
62. E Pedroso, A Grandas, X de las Heras, R Eritja, E Giralt. Tetrahedron Lett 27: 743–746, 1988.
63. E Nicolas, E Pedroso, E Giralt. Tetrahedron Lett 30:497–500, 1989.
64. I Schön, T Szirtes, A Rill, G Baloh, Z Vadász, J Seprödi, I Teplán, N Chino, K Yoshigawa Kumogaye, S Sakakibara. J Chem Soc Perkin Trans I 3213–3223, 1991.
65. R Steinauer, P White. In: R Epton, ed. Innovation and Perspectives in Solid Phase Synthesis. Birmingham, UK: Mayflower Worldwide, 1994, pp 689–690.
66. M Bodanszky, JZ Kwei. Int J Peptide Protein Res 12:69–74, 1978.
67. M Beyermann, E Krause, M Schmidt, H Wenschuh, M Brudel, M Bienert. In: R Epton, ed. Innovation and Perspectives in Solid Phase Synthesis. Proceedings of the Third International Symposium, Oxford, 1993. Birmingham, UK: Mayflower Worldwide, 1994, pp 245–250.
68. R Dölling, M Beyermann, J Haenel, F Kernchen, E Krause, P Franke, M Brudel, M Bienert. In: HLS Maia, ed. Peptide Chemistry. Proceedings of the 23rd European Peptide Symposium. Leiden: ESCOM, 1995, pp 244–245.
69. R Dölling, M Beyermann, J Haenel, F Kernchen, E Krause, P Franke, M Brudel, M Bienert. J Chem Soc Chem Commun 853–854, 1994.
70. JL Lauer, CG Fields, GB Fields. Lett Peptide Sci 1:197–205, 1994.
71. M Quibell, D Owen, LC Packman, T Johnson. J Chem Soc Chem Commun 2343–2345, 1994.
72. L Urge, L Otvos Jr. Lett Peptide Sci 1:207–212, 1994.
73. E Atherton, PM Hardy, DE Harris, BH Matthews. In: E Giralt, D Andreu, eds. Peptides 1990. Proceedings of the 21st European Peptide Symposium. Leiden: ESCOM, 1991, pp 243–244.
74. J Lukszo, D Patterson, F Albericio, SA Kates. Lett Peptide Sci 3:157–166, 1996.
75. J Jones. The Chemical Synthesis of Peptides. Oxford: Clarendon Press, 1994.
76. F Albericio, G Barany. Int J Peptide Protein Res 30:177–205, 1987.
77. A Patchornik, B Amit, RB Woodward. J Am Chem Soc 92:6333–6335, 1970.

78. SPA Fodor, JL Read, MC Pirrung, L Stryer, AT Lu, D Solas. Science 251:767–773, 1991.
79. J Vágner, V Krchňák, NF Sepetov, P Štrop, KS Lam, G Barany, M Lebl. In: R Epton, ed. Innovation and Perspectives in Solid Phase Synthesis. Birmingham, UK: Mayflower Worldwide, 1994, pp 347–352.
80. RB Merrifield, AE Bach. J Org Chem 43:4808–4816, 1978.
81. AR Brown, SL Irving, R Ramage. Tetrahedron Lett 34:7129–7132, 1993.
82. AR Brown, M Covington, RC Newton, R Ramage, P Welch. J Peptide Sci 2: 40–46, 1996.
83. HL Ball, G Bertolini, P Mascagni. J Peptide Sci 1:288–294, 1995.
84. L Caporale, R Nutt, J Levy, J Smith, B Ariston, C Bennett, G Aleber-Schonberg, S Pitzenberger, M Rosenblatt, R Hirschmann. J Org Chem 54:343–346, 1989.
85. I Smith, JWT Seakins. Chromatographic and Electrophoretic Techniques. 4th ed. Bath: William Heinemann, 1976.
86. B Fried, J Sherma. Thin-Layer Chromatography. Techniques and Applications. New York: Marcel Dekker, 1982.
87. N Grinsberg. Modern Thin-Layer Chromatography. New York: Marcel Dekker, 1990.
88. WS Hancock, JT Sparrow, HPLC Analysis of Biological Compounds. A Laboratory Guide. New York: Marcel Dekker, 1984.
89. WS Hancock, ed. Handbook of HPLC for the Separation of Amino Acids, Peptides and Proteins. Vol 1. Boca Raton, FL: CRC Press, 1984.
90. JKM Sanders, BK Hunter. Modern NMR Spectroscopy. Oxford: Oxford University Press, 1987.
91. JA McCloskey, ed. Methods Enzymol 193:1990.
92. GR Petit. Synthetic Peptides. Vol 3. New York: Academic Press, 1975.

4

Side-Chain Protecting Groups

Amanda L. Doherty-Kirby and Gilles A. Lajoie
University of Waterloo, Waterloo, Ontario, Canada

I. INTRODUCTION

Protection is warranted for a number of amino acids that have a reactive functionality on their side chain. Some amino acids (e.g., Asp, Glu, Cys, Arg, Lys, His, Ser, Thr) usually require protection; others (e.g., Asn, Gln, Met, Trp) may or may not depending on the sequence and length of the synthetic target, the type of chemistry, the coupling and cleavage methods, and the resin being used. Side reactions may occur during both synthesis and deprotection, but with a judicial choice of side-chain protecting groups, many known side reactions can be suppressed or avoided totally. Although this chapter focuses on the side-chain protection of amino acids for peptide synthesis, most of the protecting groups could be used for analogous functional groups found in the growing number of molecules amenable to solid-phase chemistry. Although the vast majority of the protected derivatives mentioned in this chapter are commercially available, references for their preparation are given. These citations could be of use for the preparation of uncommon or isotopically labeled amino acids as well as for other types of compounds. This chapter also describes the use of backbone amide protection, which has been shown to aid in a number of different syntheses and to also suppress some known side reactions.

Side-chain protecting groups that are called "semipermanent" are removed concomitantly with resin cleavage to give the unprotected peptide. "Temporary" protecting groups are present only for amino acid coupling and are removed with the next deprotection of the *N*-terminal protecting group. Protecting groups selectively removed without affecting other protecting groups provide orthogonality. Applications of this concept include the preparation of nonlinear peptide targets (e.g., lactam bridge formation, on-resin disulfide formation, multiple antigenic peptides) and enhancement of the solubility of protected peptide fragments for purification purposes. Most of the protecting groups described in the following are used for Boc or Fmoc solid-phase peptide synthesis (SPPS). However, the combination of α-amino and side-chain protection will undoubtedly be more flexible as new supports and linkers are continuously being developed. The final choice of protecting groups also depends on the peptide length, sequence, and synthetic scale as well as postsynthetic modification such as phosphorylation and formation of cyclic peptides, partially or fully protected fragments, etc.

II. ASPARTIC AND GLUTAMIC ACIDS

The β- and γ-carboxyls of Asp and Glu always require protection in classical SPPS or are activated and form a peptide bond with the free amine leading to the formation of a mixture of peptides. An exception to this practice is the use of a C-terminal Asp or Glu attached to the solid support via the side chain, in which case the α-carboxylate is protected. When the desired product is the isoglutamyl or isoaspartyl peptide, protection of the C^{α}-carboxyl is required. Benzyl (Bzl) type or t-butyl (tBu) groups have most commonly been used for the protection of side-chain carboxylates in SPPS based on t-butyloxycarbonyl (Boc) and 9-fluorenylmethoxycarbonyl (Fmoc) N-terminal protection, respectively.

Acid- or base-catalyzed aspartimide formation with subsequent reopening to form β-aspartyl peptides can be problematic in both Boc- and Fmoc-based SPPS (Scheme 1). The most frequently used protecting group in Boc chemistry is the benzyl group. However, Asp(OBzl) is prone to aspartimide formation, particularly in sensitive sequences such as Asp-Xxx where Xxx is Gly, Ser, Asn, or His [1]. This has led to the use of the more sterically hindered cyclohexyl (OcHx) and 2-adamantyl (O-2-Ada) protecting groups, which were found superior to the benzyl group but require harsher cleavage conditions [2–4]. Treatment of Asp(OcHx)-containing peptides with base is known to lead to significant imide formation. To overcome this problem, Karlström and Unden [5,6] introduced the 2,4-dimethyl-3-pentyl (ODmp) ester protecting group, which is not prone to base-catalyzed imide formation and provides an additional advantage as it is more readily cleaved using HF than the OcHx group.

The commonly used t-butyl group, which was initially believed to suppress base-catalyzed aspartimide formation during Fmoc SPPS, was also shown to be subject to this side reaction. Susceptible sequences are Asp-Xxx where Xxx is Gly, Ser, Thr, Cys, Asn, Asp, or Arg [7–9]. In addition,

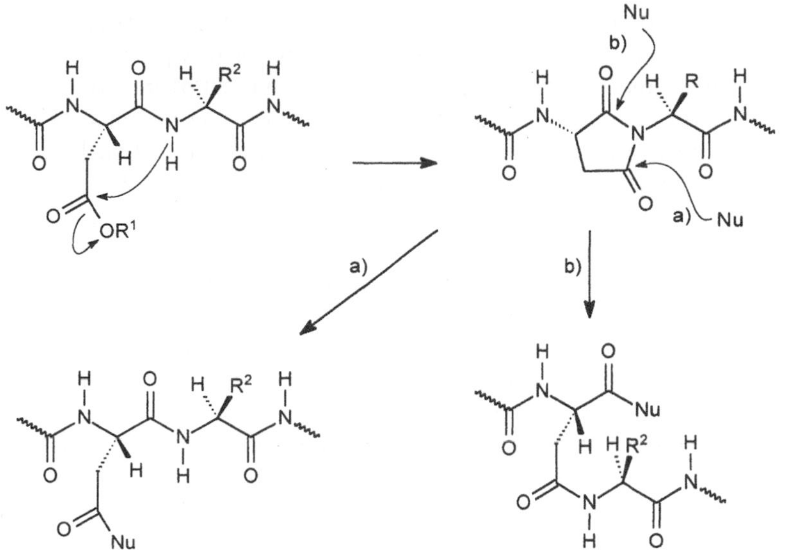

Scheme 1 Cyclization of Asp to form the aspartimide and subsequent reopening.

the formation of piperidide adducts has been noted when using typical conditions to remove the Fmoc group [20% piperidine in dimethylformamide (DMF) [7–9]]. Removal of the Fmoc group with 1,8-diazabicyclo[5.4.0]undec-7-ene (DBU) also promotes aspartimide formation in peptides containing Asp (OtBu) [8,10]. The orthogonal allyl (OAl) protecting group has been shown to be more prone to imide formation than Asp(OtBu) derivatives and should be used with caution [11]. Alternatively, 1-adamantyl (1-Ada) derivatives are available for Asp [3,4]. The 3-methylpent-3-yl (OMpe) ester group has been claimed to be superior in suppressing aspartimide formation to either the OtBu or the 1-Ada group [12]. Complete suppression of aspartimide formation can be accomplished through protection of the amide backbone with the 2-hydroxy-4-methoxybenzyl (Hmb) group [11,13,14].

Glu can also undergo cyclization and is normally protected in the same manner as Asp. Cleavage with HF in Boc SPPS can lead to dehydration and alkylation, which can be overcome with the low–high HF cleavage procedure developed by Tam et al. [15,16].

Asp and Glu are often incorporated for the formation of lactam-bridged peptides. In this case, a protecting group orthogonal to those normally used in SPPS must be used. The palladium-sensitive allyl (OAl) group is compatible with both Boc and Fmoc chemistries [17–19]. The side-chain fluorenylmethylcarbonyl esters of Asp and Glu, Boc-Asp(OFm) and Boc-Glu(OFm), are useful alternative protecting groups for this purpose in Boc SPPS [20]. For Fmoc SPPS, the 2-phenyl isopropyl (OPp) moiety, which can be selectively cleaved with dilute TFA (1–2%) in the presence of *t*Bu protecting groups, provides a viable alternative to allyl protection [21,22]. Also, the 4-{*N*-[1-(4,4'-dimethyl-2,6-dioxocyclohexylidene)-3-methylbutyl]aminobenzyl ester (ODmab) group has been reported as a quasi-orthogonal protecting group for the production of cyclic peptides for Fmoc SPPS [23]. The ODmab group should be particularly useful in conjunction with *N*-1-(4,4-dimethyl-2,6-dioxocyclohexylidene)ethyl (Dde) protection for Lys because both moieties are removed with dilute hydrazine in DMF, allowing facile on-resin lactam formation.

Simple ω-esters (Me, Et, allyl, benzyl) can easily be prepared by preferential acid catalysis of the ω-acid with the corresponding alcohol and the free Asp or Glu. For optimal yield, the alcohols must be dried and the amount of acid controlled by the addition of TMSCl or dry HCl [24–26] (Scheme 2). For the incorporation of other protecting groups, several methods have been used. Reaction of the N^α-Boc or N^α-Cbz protected benzyl ester with the appropriate alcohol in the presence of DCC and DMAP has been applied for Fm, 1-Ada, 2-Ada, and ODmab protection [3–6,27] (Scheme 3). After esterification, the N^α-Cbz and C^α-benzyl group is removed and either the Fmoc or Boc moiety is introduced. To obtain *t*Bu and Fm derivatives, formation of the N^α-Cbz protected 5-oxo-4-oxazolidinone followed by derivatization with *t*-butyl fluorocarbonate or fluorenylmethanol can be used (Scheme 4) [28,29]. Formation of the acid chloride followed by reaction with 3-methyl-3-pentanol and removal of the C^α-carboxyl protecting group yields Fmoc-Asp(OMpe)-OH from Fmoc-Asp(OH)-OBzl [12] (Scheme 5). Boron reagents such as triethylborane, tetrafluoroboric acid diethyletherate, and boron trifluoride diethyl etherate are also useful for the introduction of Fm and *t*Bu groups, presumably because of the simultaneous protection of the α-amino and α-carboxyl groups to liberate the side-chain

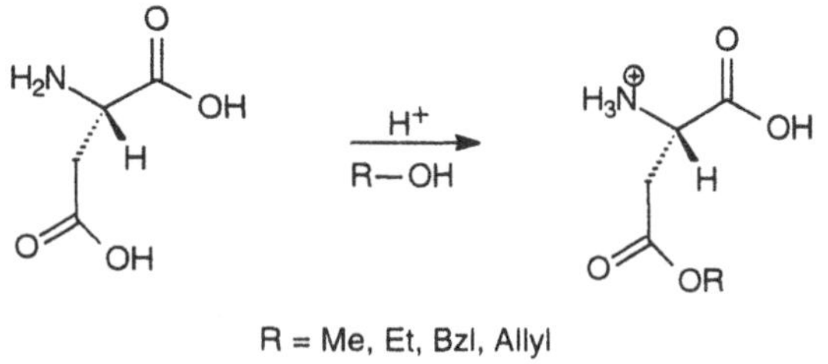

R = Me, Et, Bzl, Allyl

Scheme 2   Regioselective acid-catalyzed esterification of Asp.

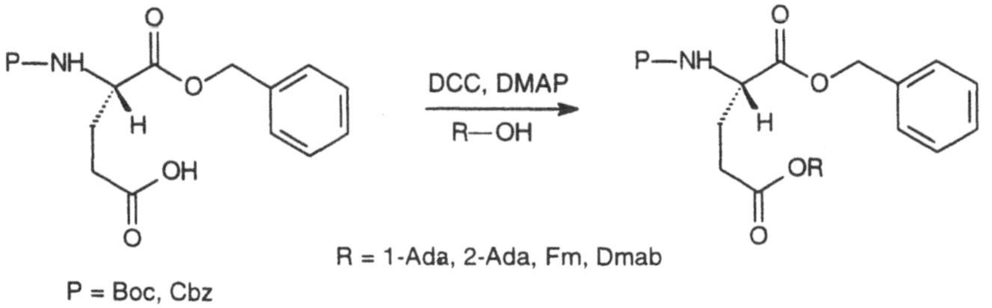

R = 1-Ada, 2-Ada, Fm, Dmab

P = Boc, Cbz

Scheme 3 General scheme for preparation of Ada, 2-Ada, Fm, and Dmab protected derivatives of Glu.

carboxyl function [30–32] (Scheme 6). The *t*Bu moiety may also be introduced on unprotected Asp or Glu using excess isobutene in the presence of 4-toluenesulfonic acid or using *t*-butyl 2,2,2-trichloroacetimidate [33,34] (Scheme 7).

III. ASPARAGINE AND GLUTAMINE

Asn and Gln can be incorporated into peptides without side-chain protection, but the unprotected derivatives have low solubility and often couple slowly. The carboxamide side chain can be dehydrated during activation to produce a cyano-containing peptide (Scheme 8), which is difficult to separate from the desired target peptide. In Boc SPPS, dehydration tends not to be a significant problem because it can apparently be reversed by treatment with HF followed by an aqueous workup [35]. Mosjov et al. [35] reported that TFA treatment is not sufficiently acidic to reverse this side reaction. However, the use of preformed active esters such as OPfp generally minimizes dehydration [36]. Gln at the *N*-terminal position can undergo cyclization to form a pyroglutamyl residue and result in the formation of a truncated product. This reaction, particularly relevant in Boc chemistry, is catalyzed by weak acids such as HOBt, the incoming Boc amino acid, and TFA during Boc deprotection but can be prevented using adequate side-chain protection [37] (Scheme 9).

Side-chain protection of Asn and Gln eliminates dehydration and pyroglutamate formation. In addition, such protection may inhibit hydrogen bonding, which can lead to the formation of secondary structures causing incomplete deprotection or reducing coupling efficiency [37]. The 9-xanthenyl (Xan) [38] and 4,4'-dimethoxybenzhydryl (Mbh) [39] protecting groups are used for Boc SPPS, and 2,4,6-trimethoxybenzyl (Tmob) [40] and

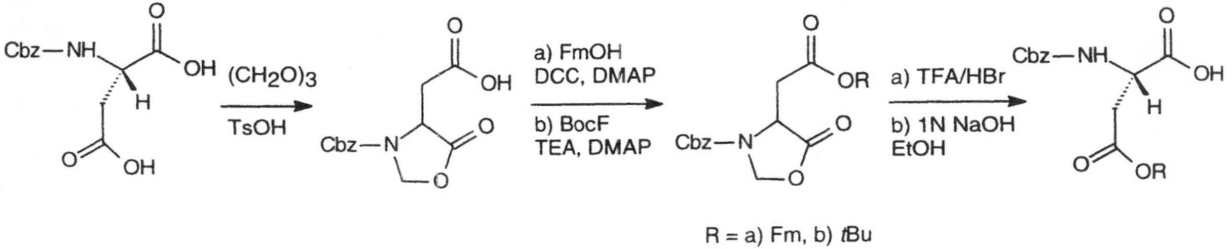

Scheme 4 Preparation of Fm and *t*Bu derivatives of Asp.

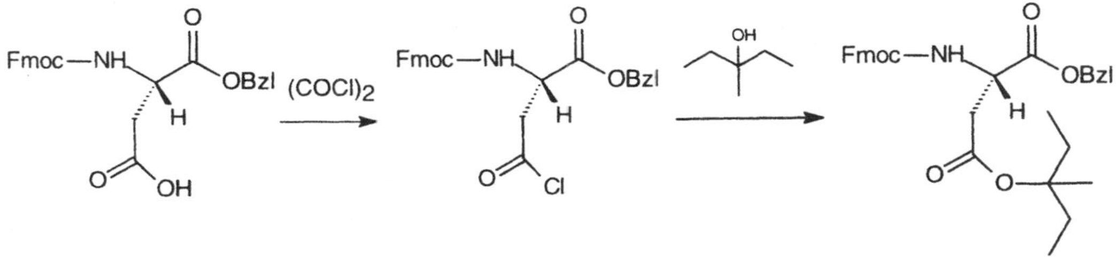

Scheme 5 Preparation of Mpe-protected Asp.

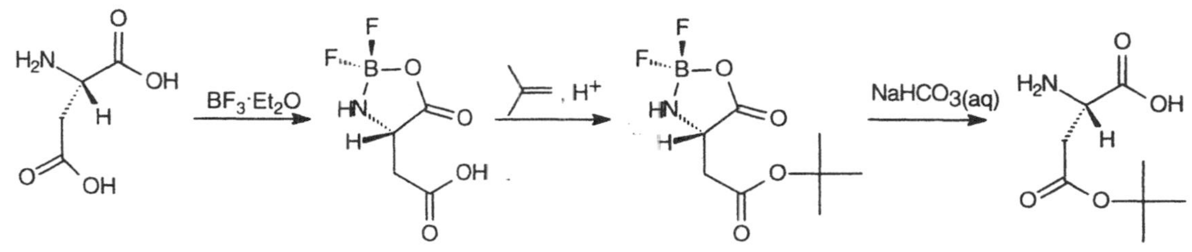

Scheme 6 Protection of Asp using boron reagents.

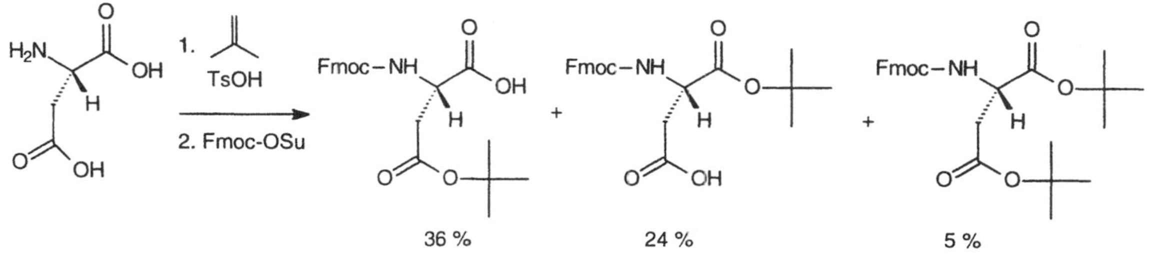

Scheme 7 Synthesis of Fmoc-Asp(OtBu) using isobutene and 4-toluenesulfonic acid.

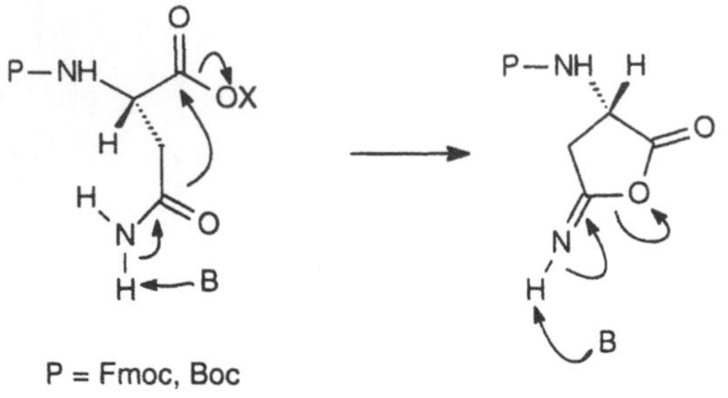

P = Fmoc, Boc

Scheme 8   Dehydration of the carboxamide of Asn during activation.

triphenylmethyl, commonly referred to as trityl (Trt) [41], protecting groups are used for Fmoc SPPS. These protecting groups minimize dehydration but generate stable carbonium ions, which can, in some cases, alkylate Trp [42]. Both Mbh protection and Xan protection are partially labile to TFA and can be liberated during removal of the Boc group as the synthesis progresses [43]. Removal of Trt with TFA tends to be incomplete when this function is positioned at the *N*-terminus, presumably because of its proximity to the free α-amino group [44]. This can be overcome with the use of the 4-methyltrityl (Mtt) group, which is more acid labile than trityl [45] and is efficiently deprotected when Asn is at the *N*-terminus [44]. The Xan protecting group is apparently applicable to Fmoc SPPS along with its 2-methoxy counterpart (2-Moxan) owing to their TFA sensitivity. Xan and 2-Moxan are more readily cleaved from an *N*-terminal Asn residue than Trt. In addition, 2-Moxan suppresses all Trp alkylation whereas Xan is significantly better than Trt for avoiding this undesired reaction [46].

The introduction of the primary amide-protecting group is straightforward and generally involves acid-catalyzed reaction of the corresponding

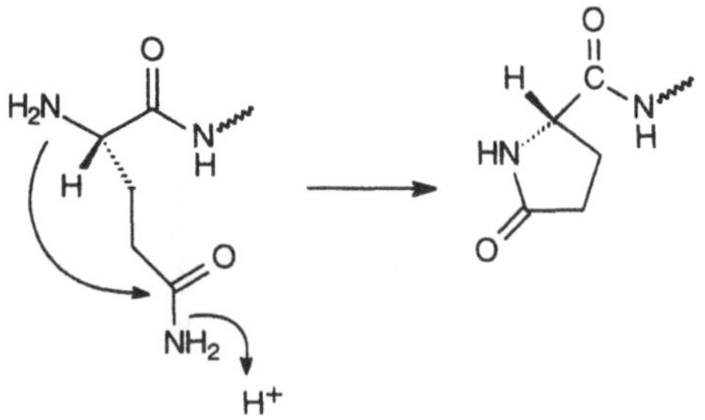

Scheme 9   Cyclization of *N*-terminal Gln.

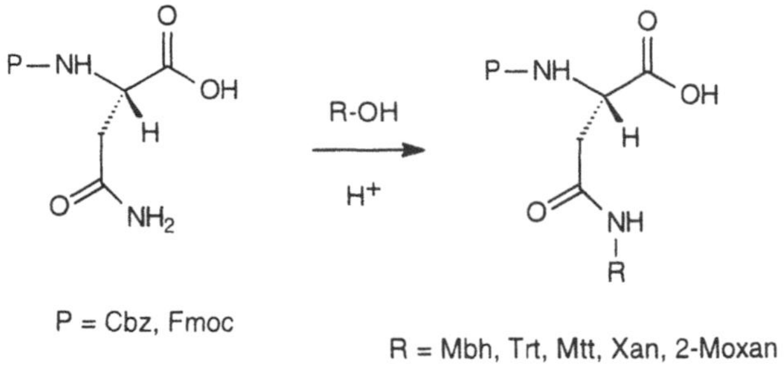

P = Cbz, Fmoc

R = Mbh, Trt, Mtt, Xan, 2-Moxan

Scheme 10 Synthesis of carboxamide-protected Asn derivatives.

alcohol with the carboxamide function. For both Fmoc and Boc derivatives, this is typically performed using the N^α-Cbz protected amino acid with subsequent hydrogenolysis of the Cbz group followed by introduction of N^α-Boc or N^α-Fmoc protection [38,39,41,45]. Xanthyl protection of Fmoc-Asn/Gln-OH can be achieved using the corresponding xanthydrol and acid [46] (Scheme 10). The Tmob group is added to Asp or Glu with trimethoxybenzylamine [40] (Scheme 11).

An alternative strategy for constructing peptides containing Asn or Gln at the C-terminal is attachment of the side-chain carboxylate of Asp or Glu, respectively, to a resin that will produce an amide upon cleavage. This is advantageous because of the slow, problematic reaction of these amino acids with the solid support [47,48] and has been applied in both Boc [49] and Fmoc [50,51] chemistry (Scheme 12).

IV. SERINE, THREONINE, AND HYDROXYPROLINE

Hydroxyl-containing amino acids may be incorporated into peptides without side-chain protection although the common practice is to use them as tBu ether derivatives in Fmoc SPPS or benzyl ether derivatives in Boc SPPS. Coupling of unprotected serine can lead to acylation of the hydroxyl group followed by O-to-N migration of the acyl group after deprotection (Scheme 13). Because of the less nucleophilic nature of the secondary alcohol, this side reaction is not often encountered with Thr and a number of syntheses have been carried out successfully using unprotected Thr [52]. A free hydroxyl group for serine may be desirable as in the case of phosphorylation, but this is preferentially achieved with the use of an orthogonal protecting group. Arzeno et al. [53] have shown that the tBu group can be used as a temporary group in Boc SPPS in a synthesis of luteinizing hormone–releasing hormone (LH-RH) analogues containing minimal side-chain protec-

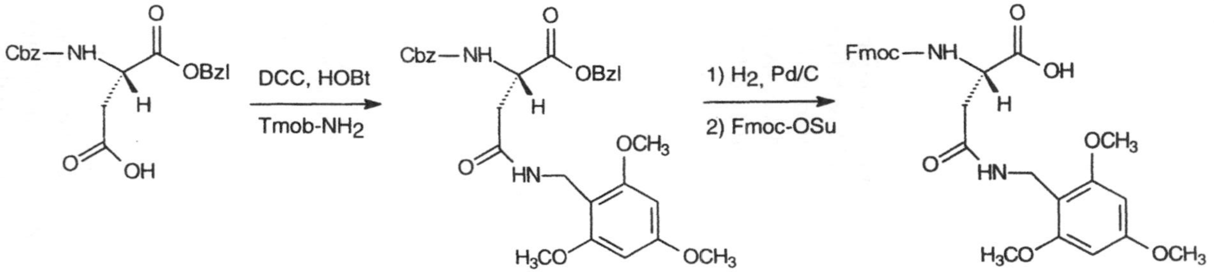

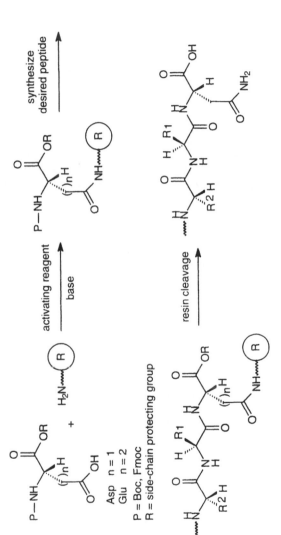

Scheme 11 Preparation of Fmoc-Asn(Tmob)-OH.

Scheme 12 Alternative synthesis of *C*-terminal Asn/Gln-containing peptides.

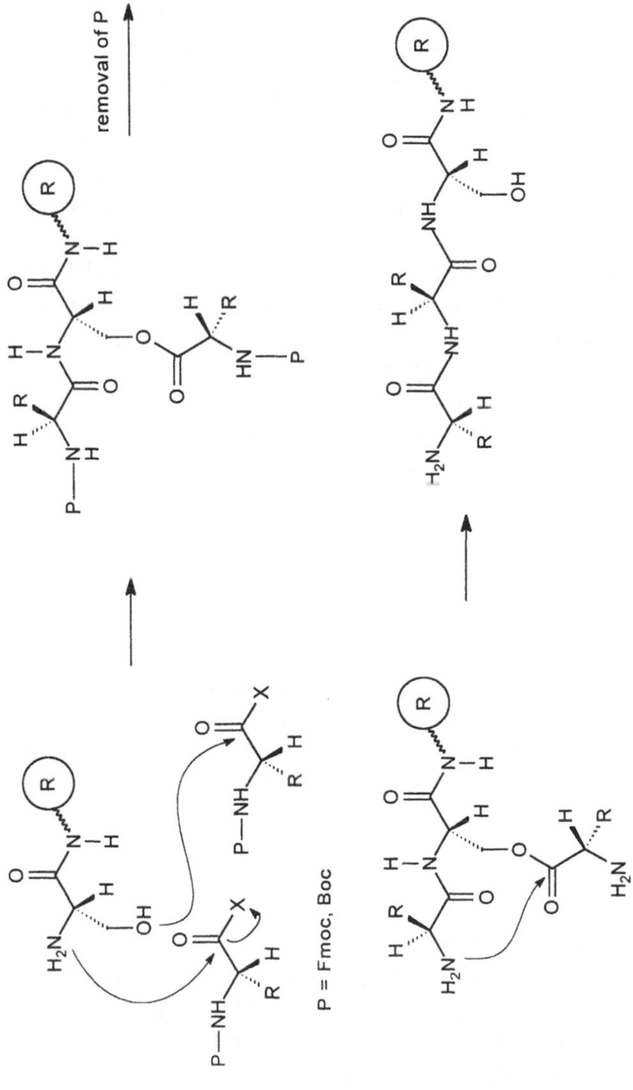

Scheme 13 Acylation of the side chain of Ser as a side reaction in SPPS.

tion. Such a scheme might be more practical on an industrial scale because large amounts of HF would not be required [53].

The Trt group may also be used for the protection of hydroxyl groups in Fmoc SPPS. Trt-protected Ser or Thr can be selectively deprotected on the solid support with dilute (1%) TFA [54]. Protection of all side chains with Trt (i.e., global protection) has been shown to produce a cleaner peptide than global *t*Bu protection [55]. Another acid-labile protecting group of interest is the *t*-butyldimethylsilyl (TBDMS) group, which can be removed from Ser and Thr in the presence of *t*Bu using mild acidic conditions (3:1:1 AcOH–THF–H_2O) [56]. These types of orthogonal protecting groups are advantageous for postsynthetic modification such as phosphorylation or glycosylation.

Many methods have been used to synthesize *t*Bu and benzyl derivatives of Thr and Ser. These include reaction of benzyl bromide with Ser under basic conditions, acid-catalyzed addition of benzyl alcohol to Thr, and acid-catalyzed addition of isobutylene for *t*Bu protection [57–61]. Benzyl and *t*Bu protection can also be achieved via formation of 2,2-difluoro-1,3,2-oxazaborolidin-5-ones by reaction of the sodium salt of Thr or the lithium salt of Ser with BF_3. Treatment with isobutylene or benzyl 2,2,2-trichloroacetimidate followed by destruction of the 2,2-difluoro-1,3,2-oxazaborolidin-5-ones generates the desired side-chain protected amino acid [62] (Scheme 14). The Trt and TBDMS groups are obtained via their respective chlorides in the presence of base [54,56] (Scheme 15).

Hydroxyproline (Hyp) can be incorporated into peptides without side-chain protection. The *t*Bu side-chain protecting group has been used for Hyp in Fmoc SPPS [63]. The 3-nitro-2-pyridinesulfenyl (Npys) group for Hyp, Ser, and Thr is compatible with N^α-Boc and Fmoc protection but has not been widely utilized for synthesis on the solid support. Preparation is via Npys-Cl in the presence of base [64] (Scheme 16).

V. CYSTEINE

The sulfhydryl functional group of Cys is an excellent nucleophile and is easily acylated or alkylated; hence it must be protected during SPPS. Oxidation to form the disulfide may also occur if Cys is left unprotected. Many of the suggested protecting groups are compatible with both Boc and Fmoc SPPS as they are not base or acid labile.

Boc-compatible protecting groups for Cys are the 4-methylbenzyl (Meb), 4-methoxybenzyl (Mob), 3-nitro-2-pyridinesulfenyl (Npys), *S*-[(*N*'-methyl-*N*'-phenylcarbamoyl)sulfenyl] (Snm), 9-fluorenylmethyl (Fm), and *S*-2-[2,4-dinitrophenyl]ethyl (Dnpe) groups. These protecting groups are re-

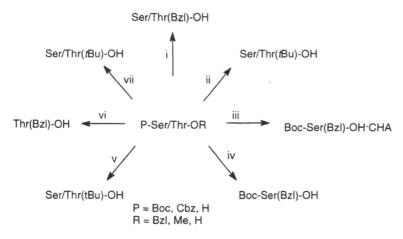

Scheme 14 Preparation of *t*Bu and Bzl derivatives of Ser and Thr. (i) 1) Cl₃CC(NH)OBzl, BF₃, 2) 1 M NaOH; (ii) 1) isobutylene, BF₃, H₃PO₄, 2) 1 M NaOH; (iii) 1) Na, NH₃, 2) a. BzlBr, b. HCl, c. CHA; (iv) 1) Na, 2) BzlBr; (v) 1) Bzl(NO₂)Br, 2) isobutylene, H⁺, 3) saponification, 4. H₂, Pd/C; (vi) 1) BzlOH, toluene, Δ, 2) saponification; (vii) 1) isobutylene, TsOH, 2) saponification.

moved with strong acid (Meb, Mob), through thiolysis (Npys, Snm), or with base (Fm, Dnpe) [30,65–72]. For Boc SPPS, the pyridyl (Pyr) protecting group has been shown to be useful when orthogonal protection of unprotected segments is required [73]. The ethylcarbamoyl (Ec) moiety is useful for protection of peptides containing a single Cys in the synthesis of immunogen conjugates. In addition, this group is stable to acidolytic cleavage as well as conditions for purification and storage; thus, the undesired oxidized peptide is not obtained. This group can be readily removed under basic conditions (NaOH) or with nucleophiles (hydrazine, NH₃) before conjugation [74,75].

Protecting groups that are Fmoc compatible include trityl (Trt), 2,4,6-trimethoxybenzyl (Tmob), monomethoxytrityl (Mmt), 9-phenylxanthen-9-yl (pixyl), 9*H*-xanthen-9-yl (xanthyl), and its corresponding 2-methoxy derivative (2-Moxan) [76–81]. The Trt protecting group is most commonly used in Fmoc SPPS because of its lability to TFA. However, *S*-detritylation of Cys is reversible, but this reversibility can be significantly reduced with the use of trialkylsilane scavengers [82]. Triisopropylsilane is recognized as more effective than triethylsilane as the latter can lead to reduction of Trp [82]. Fmoc-Cys(Trt) has been shown to undergo racemization under base-mediated activation conditions [83,84]. The Tmob group has been advocated

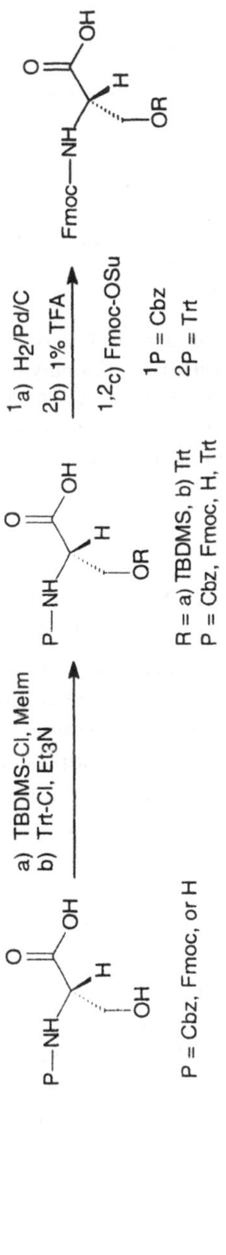

Scheme 15 Preparation of Ser(TBDMS) and Ser(Trt) derivatives.

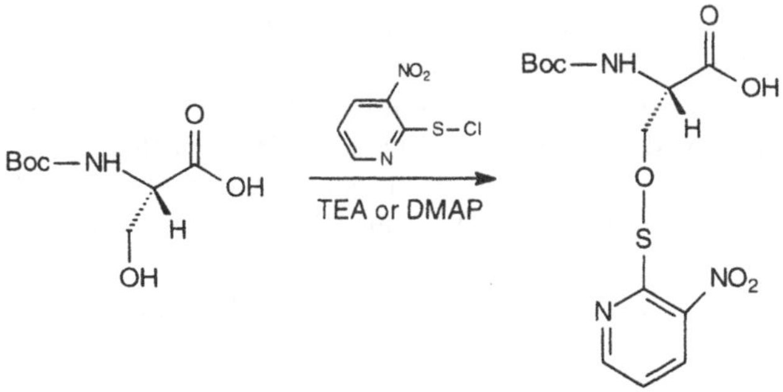

Scheme 16 Incorporation of the Npys protecting group for hydroxy amino acids.

as an alternative to Trt, but Tmob cations are known to alkylate Trp [44,78]. The problem of reversible deprotection was also encountered with the pixyl group [80]. The Xan, 2-Moxan, and Mmt derivatives of Cys show considerably more acid lability than Trt and can be selectively deprotected in the presence of *t*Bu protecting groups [79,81]. These protecting groups are cleaved with acid to generate the free thiol or can be cleaved with concomitant oxidation with I_2 or Tl(tfa)$_3$ to give a disulfide. Npys cannot normally be used for Fmoc SPPS because of its base lability but has been used for the *N*-terminal residue in Fmoc SPPS when incorporated as the Boc-Cys(Npys)-OH derivative. Npys protection can be performed on the solid support using Npys-Cl to obtain an internal protected Cys in Fmoc SPPS [85].

The *t*-butyl (*t*Bu), *t*-butylsulfenyl (S*t*Bu), acetamidomethyl (Acm), trimethylacetamidomethyl (Tacm), and phenylacetamidomethyl (Phacm) groups are compatible with both Boc and Fmoc SPPS [86–90]. The Acm group and variants are typically removed with mercury acetate or by concomitant oxidation of Cys to the disulfide with I_2 or Tl(tfa)$_3$. Some acid lability of Acm has been observed. This was found to be dependent on the type of scavengers present with triisopropylsilane in the absence of water consistently giving the highest yields of desired product [91]. It has been reported that Acm groups can be transferred to Ser, Thr, and Gln residues and can alkylate Tyr [92–94]. As with many other side reactions, this can be suppressed with the use of suitable scavengers including glycerol and glutamine [92–94]. Although allyl-based protecting groups are usually compatible with Boc and Fmoc SPPS, Cys(Aloc) derivatives are suitable only for Boc chemistry because of their base lability [18]. The allyloxycarbonylaminomethyl (Alocam) protecting group has been suggested for use with

Fmoc SPPS [95]. The Aloc or Alocam protecting groups can be removed by palladium-catalyzed hydrastannolysis [18,95].

As previously mentioned, Cys is susceptible to a number of side reactions that should be considered before a synthesis. Peptides that have a *C*-terminal Cys can undergo elimination of the sulfhydryl protected side chain in both Boc and Fmoc SPPS [96,97]. In the latter case 3-(1-piperidinyl)alanine is known to form [97]. This has been noted for Cys(Bzl) in Boc SPPS [96] as well as Cys(Acm) and Cys(Trt) in Fmoc SPPS and is dependent on the solid support, the linker, and the side-chain protecting group [97]. In a comparison of Cys(Acm) and Cys(Trt), elimination occurred more readily with Acm protection than with Trt protection [97]. A side reaction at the *C*-terminal residue has also been observed for Cys(*St*Bu) and is presumably the formation of the piperidinyl-alanine adduct [98,99]. Esterification of Cys to a solid support and standard preactivation and Fmoc deprotection protocols used in Fmoc SPPS tend to result in racemization [83,84,100,101].

Derivatization of Cys is easily accomplished because of the highly nucleophilic nature of the thiol side chain. Use of the desired alkyl halide or tosylate under a variety of conditions (acidic, basic) or the relevant alcohol under acidic conditions generally gives good yields of the desired derivative [30,65,67–69,72,75,76,78–81,86–89] (Scheme 17). Protection with the Snm group can be accomplished by displacing the Acm group with (chlorocarbonyl) sulfenyl chloride followed by addition of *N*-methylaniline [70] (Scheme 18). Benzyl-type protecting groups can also be introduced via reduction of the thiazolidine formed by the reaction of the suitable aldehyde with unprotected Cys in TFA [102] (Scheme 19).

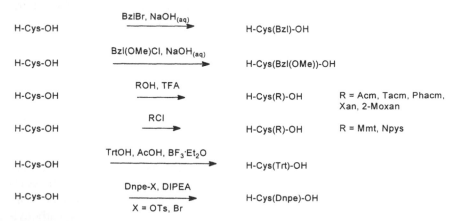

Scheme 17   Protection of the side chain of Cys.

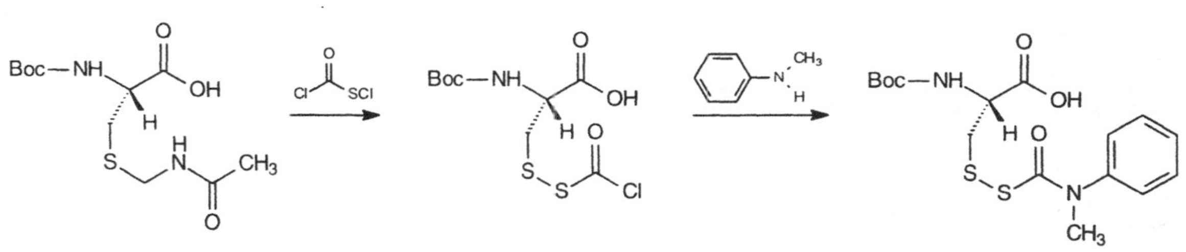

Scheme 18 Preparation of Boc-Cys(Snm)-OH.

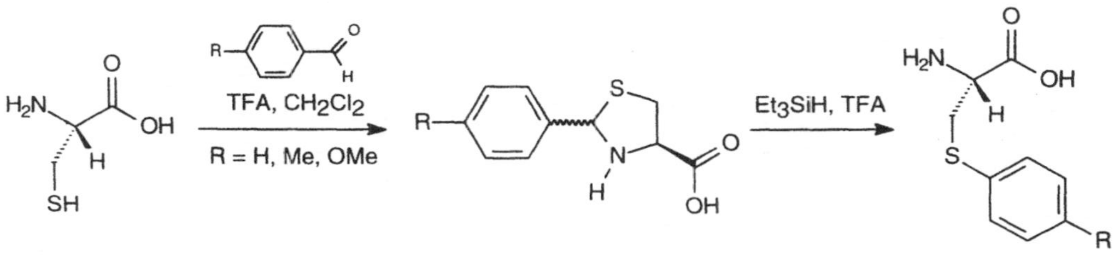

Scheme 19 Introduction of benzyl-type protecting groups via reductive *S*-alkylation of thiazolidones.

VI. METHIONINE

Methionine is generally coupled in an unprotected form in Fmoc SPPS but can undergo acid-catalyzed oxidation upon TFA cleavage if the proper scavengers are not used. Met can undergo oxidation to the sulfoxide or alkylation to generate sulfonium salts, which can in turn lead to the formation of side products such as the homoserine lactone under the strong acid conditions typically used for resin cleavage in Boc SPPS [103] (Scheme 20). To prevent these side reactions, Met can be introduced as its sulfoxide derivative, Met(O), which can be prepared by treatment of Met with H_2O_2 [104]. Reduction of Met(O) occurs during deprotection with 20 to 25% HF in dimethyl sulfide as in the low–high HF cleavage procedure developed by Tam et al. [16].

Alternatively, a number of different procedures can be used to reduce the sulfoxide to the sulfide after cleavage. These methods include treatment with N-methylmercaptoacetamide in 10% aqueous acetic acid, sulfur trioxide (SO_3) and 1,2-ethanedithiol (EDT, 5 eq) in 20% piperidine–DMF (Scheme 21), and titanium(IV) chloride (3 eq.)–sodium iodide (6 eq.) in methanol–acetonitrile–DMF (5:5:4) [105–107]. The first method requires long reaction times, which can be detrimental to disulfide bonds. The use of SO_3–EDT requires protection of side-chain hydroxyl groups to prevent sulfonation of these functionalities. TiCl$_4$-mediated sulfoxide reduction can lead to reduction of disulfide bonds and oxidation of Trp residues if extended reaction times (>4 min) are used. Trimethylsilylbromide (TMSBr, 0.1 M) and EDT (0.2 M) can be added to the deprotection cocktail in the last few minutes of cleavage if water is not added as a scavenger and anhydrous TFA is used. The incorporation of TMSBr into the cleavage mixture appears to be suitable for Trp-containing peptides [108]. The use of ammonium iodide (7 to 20 eq.) in TFA at 0°C is effective in reducing Met(O)-containing peptides that have a disulfide bond. The addition of dimethylsulfide increases the reaction rate and does not affect disulfide bonds, while unprotected Cys-containing peptides can be oxidized in the presence of NH$_4$I–DMS [109,110].

VII. TYROSINE

Unprotected Tyr can be acylated during SPPS because of the nucleophilic nature of the phenolate anion, which is generated under basic conditions. The phenol ring is also prone to alkylation at the position *ortho* to the hydroxyl group. These side reactions can be suppressed by the use of suitable protecting groups.

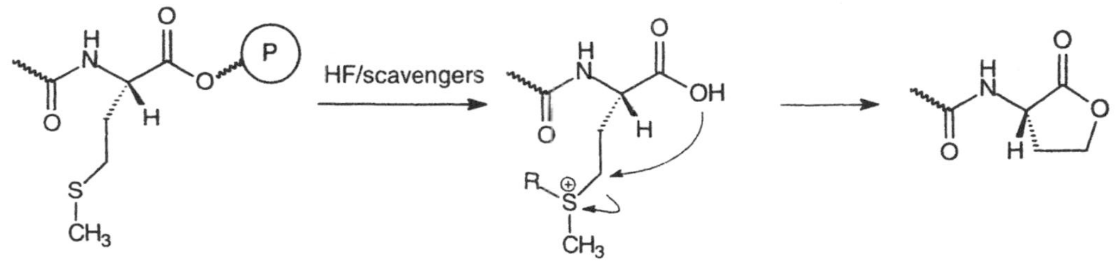

Scheme 20 Formation of the homoserine lactone as a side reaction in SPPS.

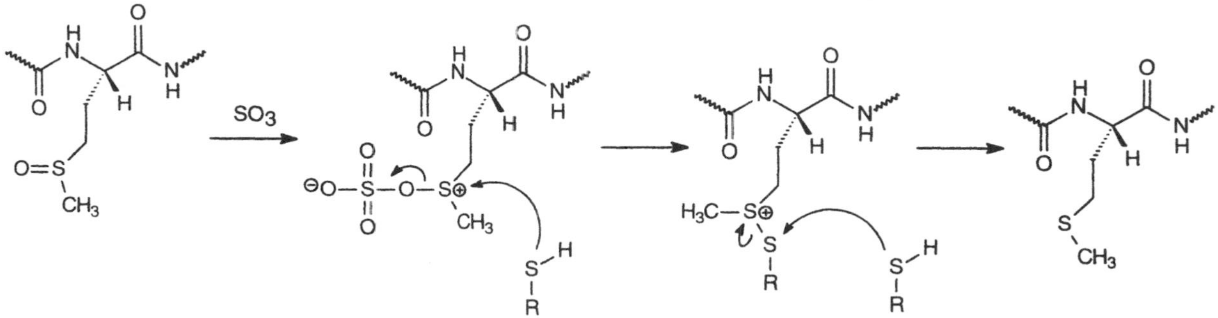

Scheme 21 Proposed mechanism for the reduction of sulfoxide by sulfur trioxide.

The benzyl ether of tyrosine has traditionally been used for Boc SPPS but can lead to benzylation of the tyrosine aromatic ring in strong acid [66]. The 2,6-dichlorobenzyl, cyclohexyl, and 2-bromobenzyloxycarbonyl ethers have been reported to increase acid stability and suppress alkylation [111,112]. The 4-methylsulfinylbenzyloxycarbonyl (Msz) protecting group for Tyr has been shown to be the most stable when compared with Bzl, Br-Z, Cl_2Bzl, and 4-methylsulfinylbenzyl (Msob) [113]. The Msz group can be removed via reductive thiolysis using TFA and tetrachlorosilane [113]. 2,4-Dimethylpent-3-yloxycarbonyl (Doc) protection has been proposed for Tyr when nucleophile or base-labile groups are present. Doc shows much greater stability to piperidine than the commonly used Br-Z group, although there is a slightly higher sensitivity to TFA [114]. The 2,4-dinitrophenol (Dnp) protecting group is practical for use in Boc SPPS as it is stable in TFA. It can be removed by thiolysis, which can be performed while the peptide is still attached to a solid support and may prove useful for on-resin modification of tyrosine. However, the Dnp group is labile to both piperidine and DBU, rendering it incompatible with Fmoc SPPS [115].

In Fmoc chemistry, the *t*Bu group is commonly used for the protection of Tyr. The TBDMS group has also been suggested for Fmoc SPPS, but, unlike the TBDMS ethers of serine and threonine, the TBDMS ether of tyrosine is less acid labile than the *t*Bu ether [56]. However, selective deprotection can be accomplished by treatment with tetrabutylammonium fluoride (TBAF) [56]. The trimethylsilyl (TMSE) and dimethylphenylsilyl (DMPSE) groups have been shown to be compatible with Bpoc-based synthesis, but their use in Fmoc SPPS has not been reported [116].

Tyr derivatives are prepared after protection of the amino and carboxyl groups either as a copper chelate or with more classical protecting groups. Protection of the phenol can be accomplished with the corresponding alkyl halide (e.g., BrZ, Doc) or with isobutylene (*t*Bu) [61,111–114] (Scheme 22).

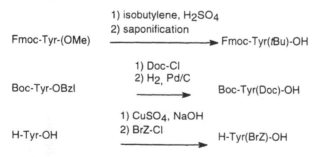

Scheme 22   Methods for the side-chain protection of Tyr.

VIII. HISTIDINE

Histidine is highly prone to racemization during activation and coupling. The τ-nitrogen of the imidazole ring is farther from the α-carbon than the π-nitrogen and more sterically accessible. The π-nitrogen is sufficiently basic to abstract the hydrogen atom from the α-carbon. The acylation of His can also be a serious problem, leading to branching or truncation if His is incorporated unprotected or with a temporary protecting group that is removed before synthesis is complete. Imidazole protecting groups for His either block the π-nitrogen or are attached to the τ-nitrogen, reducing the basicity of the π-nitrogen through inductive effects. Protection of the τ-nitrogen with the tosyl (Tos) group is commonly used in Boc SPPS, but this group is slightly labile to HOBt. Acylation of the unprotected imidazole by incoming amino acids or acetylating agents can be problematic. Migration of the acyl group to the unprotected N-terminus leads to chain termination or incorporation of additional amino acids [117,118]. The N^τ-Dnp group for His protection is useful in SPPS utilizing Boc chemistry but is stable to HF and must be removed in a separate deprotection step with thiophenol in DMF [119]. Gesquière et al. [120] have shown that His(Dnp) protection is not compatible with Lys(Fmoc). Such an orthogonal scheme results in Dnp-modified peptides upon cleavage of Fmoc with base, because of migration of the Dnp group from His to Lys. Okada et al. [121,122] have introduced the N^τ-2-adamantyloxycarbonyl (2-Adoc) group, which is stable to TFA but removed readily with HF. Like other τ-nitrogen protecting groups, Adoc diminishes but does not eliminate racemization [121,122]. N^τ-2,4-dimethylpent-3-yloxycarbonyl (Doc) has been described as a useful His-protecting group in Boc SPPS. Doc is stable to TFA, resistant to nucleophiles preventing N^{im} to N^α acyl transfer, and, although cleaved with 2% hydrazine in DMF, more stable to piperidine than Dnp [123].

The benzyloxymethyl (Bom) group has been reported as a π-nitrogen blocking group and gives racemization-free coupling of His [124–126]. Deprotection in HF or TFSMA also results in the formation of free formaldehyde, which can methylate or formylate other amino acid residues [127]. The use of His(Bom) with an N-terminal cysteine residue can lead to thiazolidine or thioproline formation because of the presence of formaldehyde (Scheme 23). Suppression of this side reaction is possible with formaldehyde scavengers such as cysteine derivatives or resorcinol [128,129]. N^π-2-Adamantyloxymethylhistidine [His(2-Adom)] has been shown to be stable to conditions required for Boc SPPS and labile to HF and TFSMA. The Boc-His(2-Adom) is reported to be highly soluble in dichloromethane and DMF and can be coupled without racemization [130,131].

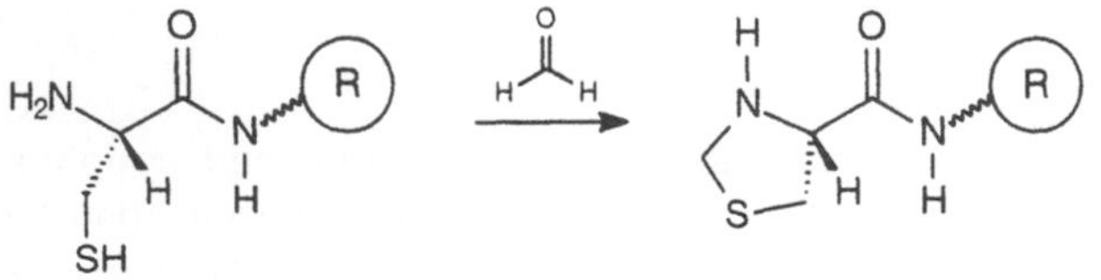

Scheme 23 Thiazolidine formation from N-terminal Cys.

The usual protecting group for His in Fmoc SPPS is the Trt group [132,133]. It blocks the less hindered τ-nitrogen and in general suppresses racemization, although it is not completely prevented. Esterification of Fmoc-His(Trt)-OH to p-alkoxybenzylalcohol resin occurs with substantial racemization, which can be decreased significantly if N^α-Trt-His(Trt)-OH is used. The N^α-trityl group can be selectively removed with dilute acid, leaving the N^τ-Trt protection intact [133]. Another approach involves changing the solid support to 2-chlorotrityl resin, which can be esterified without racemization [134]. The 4-methoxytrityl (monomethoxytrityl, Mmt) and methyltrityl (Mtt) groups have been described for His protection. Like the trityl group, these derivatives are resistant to nucleophiles. However, Mmt and Mtt are significantly more acid labile than Trt and can be selectively cleaved in the presence of tBu groups, whereas Trt cleavage requires harsher acidic conditions that result in partial loss of tBu groups [135].

The *tert*-butoxymethyl (Bum) protecting group protects the π-nitrogen of the imidazole ring and prevents epimerization in Fmoc SPPS [136]. Okada et al. [137,138] have suggested the N^π-1-adamantyloxymethylhistidine (N^π-1-Adom) derivative for Fmoc chemistry. This derivative is more soluble than Fmoc-His(Bum)-OH and can be obtained in better synthetic yields [137,138].

The synthesis of τ-nitrogen protected derivatives is quite straightforward because of the more reactive nature of the τ-nitrogen. Cbz-His can be reacted with the appropriate alkyl halide. Alternatively, free His is reacted with dichlorodimethylsilane followed by addition of the required alkyl or aryl chloride and methanolysis to give the desired τ-protected His [121–123,132,133,135] (Scheme 24). Protection of the π-nitrogen first requires modification at the τ-nitrogen, typically by acylation with Boc or acetyl (Ac) moieties on N^α-Boc or N^α-Cbz-His-OMe. Introduction of the π-protecting group and removal of the N^τ-Boc or N^τ-Ac groups is followed by hydrolysis of the methyl ester (Scheme 25). Removal of the N^α-protecting group and derivatization with Fmoc-OSu generates the Fmoc protected derivatives. The yields of N^π-protected derivatives tend to be low and commercially available derivatives are relatively expensive [124,125,130,131,136–138].

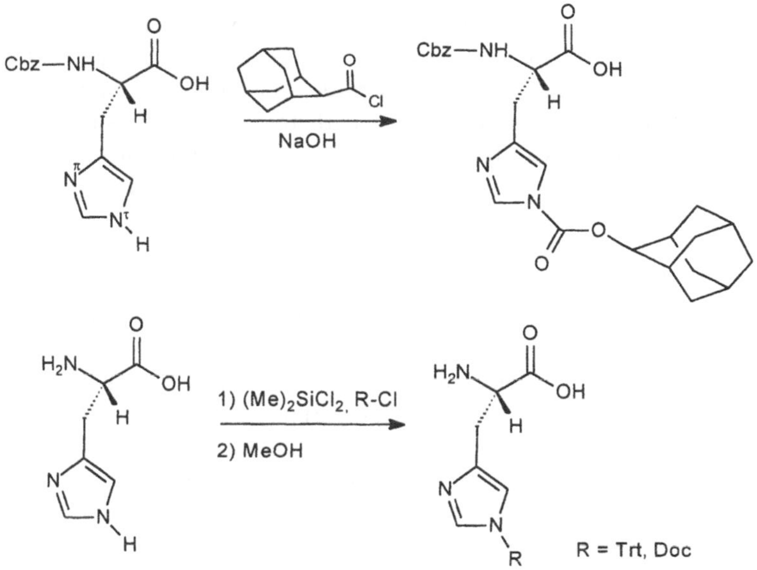

Scheme 24 Preparation of N^τ-protected derivatives of His.

IX. TRYPTOPHAN

In typical conditions for SPPS, Trp can undergo both oxidation and alkylation of the indole ring. In Boc SPPS, the formyl (For) protecting group, which can be removed with base or acid treatment with the appropriate thiol scavenger, is typically used [139]. Other indole protecting groups compatible with Boc SPPS have been described: the N^{in}-diphenylphosphinothioyl (Ppt) group, which can be removed with methanesulfonic acid or tetra-n-butyl ammonium fluoride (TBAF) [140]; the N^{in}-2,4-dimethylpent-3-yloxycarbonyl (Doc) group, which is cleaved with HF or TFMSA containing the appropriate scavengers [141]; and the cyclohexyloxycarbonyl (Hoc) group, which can be readily cleaved with HF [142]. The Doc and Hoc groups have the advantage that they are stable to base, allowing an additional degree of orthogonality in the synthesis [141,142].

In Fmoc-based SPPS, Trp is most often incorporated with no indole protection, but this can be problematic for sequences containing Arg protected by the 4-methoxy-2,3,6-trimethylbenzenesulfonyl (Mtr), 2,2,5,7,8-pentamethylchroman-6-sulfonyl (Pmc), or 2,2,4,6,7-pentamethyldihydrobenzofuran-5-sulfonyl (Pbf) protecting groups, which can readily alkylate the indole ring of Trp [143,144]. This is sequence dependent and difficult to predict, although it appears that the worst-case scenario is Arg-Xxx-Trp based on a study using Arg(Pmc) [144]. The use of t-butoxycarbonyl (Boc) protected Trp suppresses this alkylation [145–147]. It has been speculated that during normal TFA cleavage of the peptide, only the tBu group is lost and the indole ring remains protected as a carbamic acid. Removal of this group is then concomitant with lyophilization in acidic aqueous media [148]. The allyloxycarbonyl (Aloc) protection of Trp also eliminates oxidation side products and suppresses alkylation. This group is cleaved using palladium catalyst and is compatible with Boc SPPS. Although the indole-Aloc is unstable in 20% piperidine in DMF, the Aloc protecting group is stable to DBU and can be utilized for Fmoc SPPS provided DBU is used for Fmoc deprotection [149].

Preparation of indole-protected derivatives is usually straightforward, involving reaction of N^{α}-Z or Boc protected carboxyl-tBu, Bzl, or phenacyl Trp ester with di-t-butyl dicarbonate or the appropriate chloroformate or in the presence of a tertiary base (Scheme 26). Subsequent removal of the carboxyl and/or amino groups followed by N-terminal derivatization yields the Boc or Fmoc derivatives [141,142,147–149]. The formyl group is introduced with excess formic acid [150].

X. LYSINE AND ORNITHINE

Lysine and ornithine each have a primary amino group, which can be readily acylated or alkylated if not protected. The free amine may be desired at a later stage of the synthesis and thus requires an orthogonal protection scheme. In Boc SPPS, the benzyloxycarbonyl (Cbz, Z) group has been used for protection of Lys but is somewhat labile to TFA, which has led to the adoption of the 2-chlorobenzyloxycarbonyl (2-Cl-Z) group [151]. The allyloxycarbonyl (Aloc) group, which is cleaved using palladium catalyst in the presence of a nucleophile, is compatible with both Fmoc and Boc chemistry [18,19,152]. Alternative derivatives for temporary protection in Boc SPPS are Boc-Lys(Fmoc) and the trifluoroacetylated Boc-Lys(tfa), which are removed under basic conditions [30,153]. In addition, the 3-nitro-2-pyridinesulphenyl (Npys) protecting group, which is cleaved with 2-mercaptopyridine-N-oxide or triphenylphosphine, can be used [154].

In Fmoc chemistry, Lys and Orn are conveniently blocked with the Boc protecting group [155]. For more temporary protection, the 1-(4,4'dimethyl-2,6-dioxocyclohexylidene)ethyl (Dde) protecting group is useful in the preparation of branched peptides, multiple antigenic peptides (MAPs), cyclic peptides, template-assembled synthetic proteins (TASPs), and templates for combinatorial chemistry [156–161]. Dde is quasi-orthogonal because the use of dilute hydrazine to cleave Dde partially removes the N^{α}-

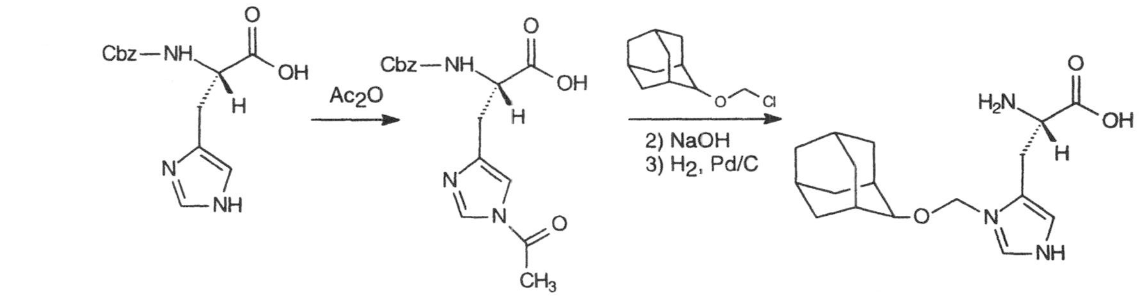

Scheme 25 Preparation of N^π-protected derivatives of His. Synthesis of His(2-Adom)-OH.

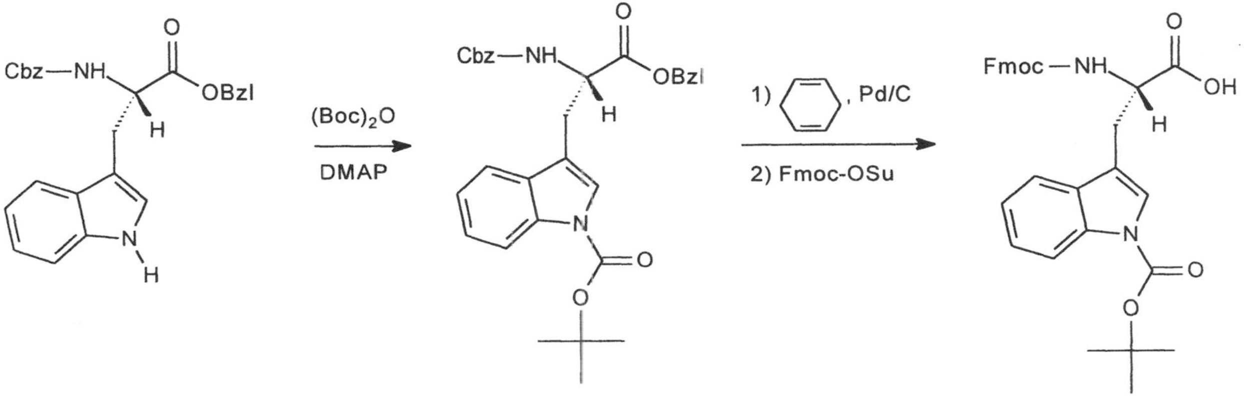

Scheme 26 Synthesis of Fmoc-Trp(Boc)-OH.

Fmoc protecting group. In addition, Dde is partially labile to piperidine and can migrate to an unprotected Lys [162]. Aloc and Dde protecting groups are not totally compatible because of partial reduction of the Aloc group by dilute hydrazine [161]. This side reaction can be prevented by the addition of allyl alcohol as a scavenger when Dde groups are deprotected [163]. Variants of Dde that have bulkier alkyl [R = CH_2CH_3, $CH_2CH_2CH_3$, $CH_2CH(CH_3)_2$] or benzyl groups at the exocyclic alkene position (R = CH_3 for Dde) are all completely stable to piperidine, and the 1-(4,4'-dimethyl-2,6-dioxocyclohexylidene)methylbutyl derivative [R = $CH_2CH(CH_3)_2$], has been shown to be less susceptible to migration [164]. Fmoc-Lys(Mtt) and Fmoc-Orn(Mtt) can be incorporated when temporary protection is desired. Mtt has been selectively removed in the presence of tBu groups using AcOH–TFE–DCE (1:2:7) [165]. These conditions failed to remove the Mtt group when hydrophilic resins (e.g., TentaGel, cellulose) were used. To overcome this problem, the monomethoxytrityl (Mmt) and dimethoxytrityl (Dmt) protecting groups have been examined. The Mmt group could be readily cleaved using weak acid for a test peptide prepared on TentaGel resin while the Mtt group was not removed [166].

Selective protection of the ω-NH_2 of Lys (and Orn) can be accomplished by preparing copper chelates of the carboxyl and N^α-amino group. Side-chain derivatization is followed by removal of the chelate with thioacetamide and protection of the N^α-group [18,19,151,167] (Scheme 27). Trityl-type protecting groups can be incorporated on the ω-NH_2 via treatment with TMSCl followed by reaction with the desired chloride derivative. In this case, the N^α-amino group also becomes protected but can be preferentially deblocked using mild conditions that do not affect the N^ϵ-amino group [165,166] (Scheme 28). Alternatively, side-chain protection of Lys has been performed on N^α-protected derivatives (e.g., Dde, Npys) [156,168] (Scheme 29).

XI. ARGININE

The side-chain of Arg is nucleophilic and, during SPPS, can undergo acylation followed by intramolecular decomposition to produce ornithine, and it is also susceptible to δ-lactam formation (Scheme 30). Arg is typically protected only at the ω-nitrogen, although it has been suggested that all three side-chain nitrogens should be protected in order to eliminate these side reactions completely. Several types of protecting groups have been described for Arg, including nitro, urethane, arylsulfonyl, alkyl, and simple protonation.

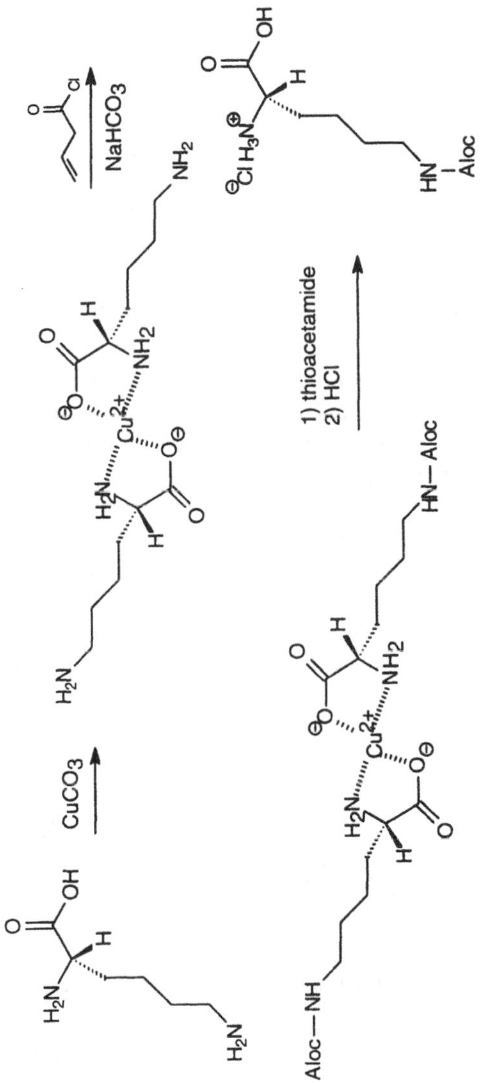

Scheme 27 Preparation of HCl Lys(Aloc)-OH.

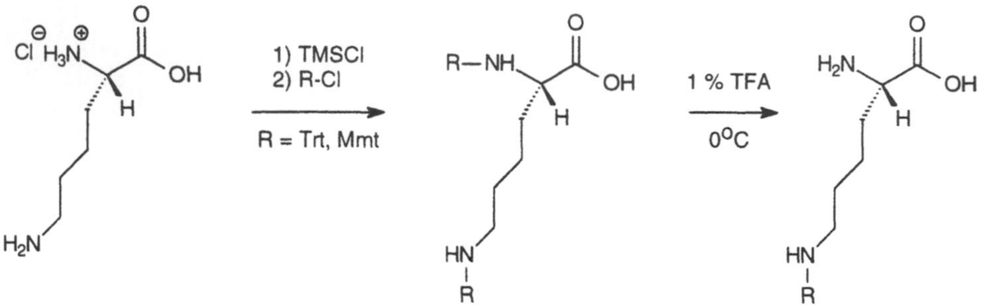

Scheme 28   Protection of Lys with trityl-based protecting groups.

The nitro group has been employed for both Boc and Fmoc SPPS but undergoes lactam formation and can transfer the nitro group to free amines, thus preventing further incorporation of amino acid residues. The nitro group can be removed by HF or by catalytic hydrogenolysis. Long reaction times may be detrimental to sensitive sequences, and with hydrogenolysis, stable, partially reduced products may be obtained [169–171].

The 4-toluenesulfonyl (tosyl, Tos) [172] and mesitylene-2-sulfonyl (Mts) [173,174] groups are compatible with Boc SPPS. The tosyl group is generally removed by HF or Na/NH$_3$, and can also be cleaved using TFSMA–TFA–thioanisole [175]. The more acid-labile Mts group is readily cleaved with TFSMA–TFA–thioanisole. The Tos group had been shown to lead to substantial amounts of δ-lactam formation. Protection of Arg as the

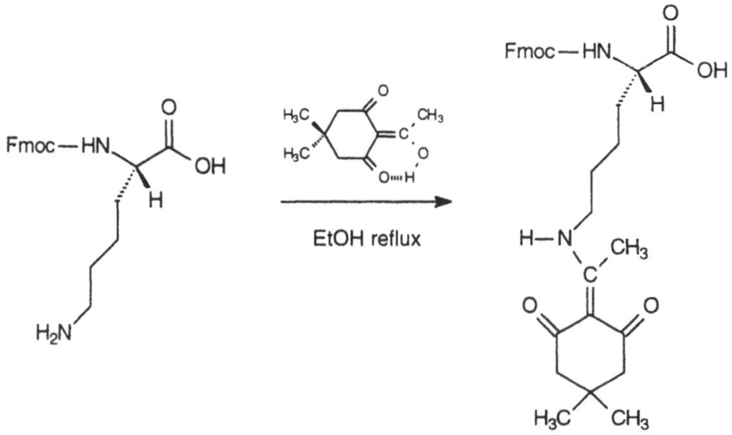

Scheme 29   Synthesis of Fmoc-Lys(Dde)-OH.

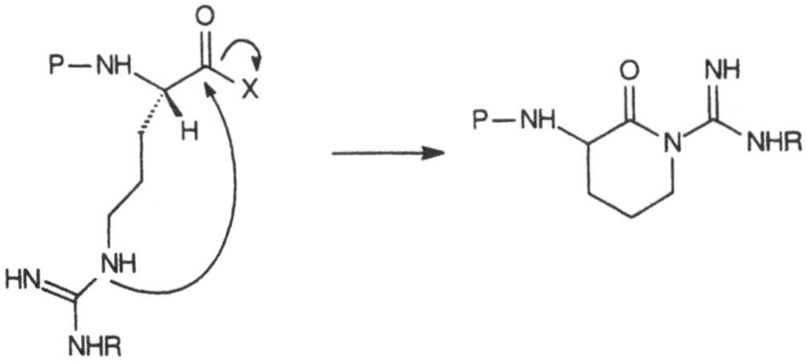

Scheme 30 γ-Lactam formation from Arg.

bis-Aloc derivative has been determined to be compatible with Boc SPPS, but treatment with piperidine results in formation of the mono-Aloc compound [19].

The ω-Boc and δ,ω-bis-adamantyloxycarbonyl (Adoc) urethane derivatives of Arg have been used for Fmoc SPPS but appear to be inadequate for protection of the guanidino functionality and show substantial ornithine formation [176]. A ω,ω'-bis-Boc derivative has been reported to eliminate ornithine production, although lactam formation is not completely prevented [177].

More commonly, the arylsulfonyl 4-methoxy-2,3,6-trimethylsulfonyl (Mtr), 2,2,5,7,8-pentamethylchroman-6-sulfonyl (Pmc), or 2,2,4,6,7-pentamethyldihydrobenzofuran-5-sulfonyl (Pbf) derivatives are used for Fmoc SPPS [178–182]. The 9-anthracenesulfonyl (Ans) moiety has been reported but has not been widely used [183]. Mtr is difficult to remove using standard TFA deprotection cocktails but can be readily removed in 1 M TMSBr in TFA. Pmc was originally shown to be comparable to tBu groups in terms of acid lability but later was found to be acid stable with sequences containing multiple Arg residues. However, cleavage is quantitative with the use of TMSBr [184]. Pbf is slightly more acid labile than Pmc and causes less Trp alkylation than either Pmc or Mtr [146]. The arylsulfonyl-based groups can also lead to sulfonation of Arg, Ser, and Thr residues [185,186]. With regard to alkylation of Trp residues, this side reaction can be suppressed substantially by judicious use of scavengers, although there are cases in which total suppression is not possible [145]. As previously mentioned, the use of Trp(Boc) derivatives has been shown to reduce this problem considerably [146].

The 10,11-dihydro-5H-dibenzo[a,d]cyclohepten-5-yl (5-dibenzosuberyl, Sub), 5H-dibenzo[a,d]-cyclohepten-5-yl (5-dibenzosuberenyl, Suben), and 2-methoxy-10,11-dihydro-5H-dibenzo[a,d]-cyclohepten-5-yl (2-methoxy-5-dibenzosuberyl, Me-Sub) protecting groups were introduced by Noda and Kiffe [187] as an alternative for Arg protection. These residues showed significantly more acid lability than the arysulfonyl derivatives, with complete deprotection of a peptide containing four Arg moieties in 1 h when 50% TFA was used. The alkylation of Trp was reported to be suppressed by using low temperature (6°C) for deprotection [187]. Slightly higher production of ornithine was observed for these derivatives when compared with the Pmc protecting group [185]. The trityl group has also been investigated as a protecting group for Arg, but because of low solubility of the Fmoc derivative has not been widely used [188].

The synthetic strategy for the preparation of Arg derivatives depends on the desired protecting group. The NO_2 group is added to the ω'-nitrogen with NH_4NO_3–H_2SO_4 [189] (Scheme 31). The Boc group can be introduced at the ω-nitrogen with Boc-N_3 [190]. The δ- and ω-nitrogens can be simultaneously protected by Cbz using trimethylsilyl chloride and benzyloxychloroformate [191]. The ω,ω' bis Boc derivative can be obtained by reaction of ornithine with N,N'-bis(t-butoxycarbonyl)-S-methylurea [192]. The Adoc group is introduced with adamantylfluoroformate [193]. Arylsulfonyl derivatives are obtained by reaction of N^α-Z or N^α-Boc-Arg with the required sulfonyl chlorides [178–182]. For the preparation of Fmoc derivatives, reaction at the ω-nitrogen is performed with the N^α-Z derivative. The Z group is then removed and replaced by Fmoc (Scheme 32). Suberyl protection is achieved by reaction of the corresponding chlorides with Fmoc-Arg-OH using slightly acidic conditions [187].

Incorporation of a suitably protected ornithine followed by derivatization with guanylating agents such as 1-guanylpyrazole hydrochloride to form arginine is an alternative method of incorporating Arg residues in a peptide [194,195]. This has been applied to both Boc and Fmoc SPPS. How-

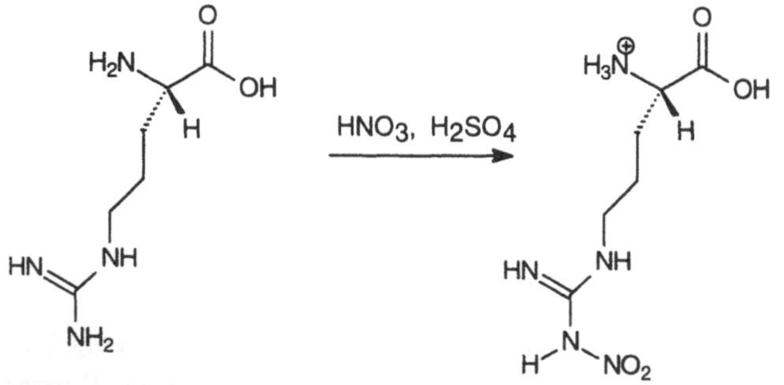

Scheme 31   Preparation of Arg(NO_2).

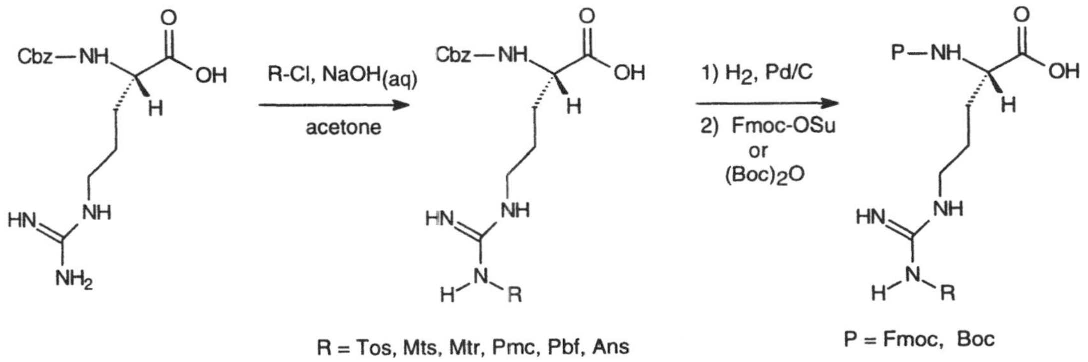

Scheme 32 Synthesis of sulfonyl derivatives of Arg.

ever, with the latter strategy, the N-terminal amino acid must be added as the N^α-Boc derivative as guanylation requires basic conditions, which can remove the Fmoc moiety [194,195].

XII.   BACKBONE AMIDE PROTECTION

The peptide backbone is capable of forming β-sheets or other secondary structures through intramolecular or intermolecular hydrogen bonding. This aggregation results in shrinkage of the peptide resin, leading to a decrease in solvation and resin penetration, and additional steric interactions at the growing end of the peptide chain. Typically, decreased coupling and deprotection rates for these difficult sequences are observed [196]. One way to overcome aggregation problems is through masking of the peptide bond with a protecting group. Early experiments with methyl and phenylthiomethyl backbone protection illustrated the feasibility of this approach [197,198].

The 2-hydroxy-4-methoxybenzyl (Hmb) group has been shown to be useful as a backbone amide protecting group in a number of different syntheses. Sheppard et al. [199] first demonstrated the use of Hmb in the synthesis of the acyl carrier protein (ACP) 65–74 decapeptide, a well-known difficult sequence. This protecting group is TFA labile and fully compatible with Fmoc SPPS. Since its introduction, Hmb backbone protection has been used for various applications such as the stepwise synthesis of several peptides with difficult sequences [199–205] and to eliminate base-mediated aspartimide formation in Asp(OtBu)-containing peptides [13,14]. It is particularly attractive for the synthesis of Asn N-linked glycopeptides (with no aspartimide formation) [11], phosphopeptides [206], and small proteins via protected fragments [207,208].

Hmb backbone protection also increases the solubility of protected peptide fragments, which allows purification by standard chromatographic methods and subsequent enhancement of coupling rates in convergent SPPS [209]. Increased solubility has also been observed for some side-chain deprotected, Hmb backbone protected peptides [210]. The normally TFA-labile Hmb protecting group can be reversibly acetylated at the 2-hydroxy group before side-chain deprotection (note: the amino terminus must be Boc protected unless N-terminal acetylation is desired). Acetylation of the Hmb group increases TFA stability and the crude, backbone protected, side-chain deprotected peptide can be purified more readily because of increased solubility [210]. Removal of the Hmb-acetyl group with 20% piperidine or 5% hydrazine in DMF followed by cleavage of the Hmb group with TFA yields the fully deprotected pure peptide [210].

Hmb derivatives of amino acids are formed through a Schiff base intermediate by the reaction of the desired amino acid with 2-hydroxy-4-methoxybenzaldehyde followed by reduction with $NaBH_4$ [211,212] (Scheme 33). Initially, Hmb protection was performed with the N,O-bis-Fmoc derivative. Two groups have shown that incorporation as the mono-Fmoc amino acid derivative is also possible, although, depending on the nature of the protected amino acid, coupling can be slow or incomplete [205,212]. If the mono-Fmoc derivatives are used, the activated species is the 4,5-dihydro-8-methoxy-1,4-benzoxazepin-2(3H)-one [212]. Acylation of the secondary amine occurs through a base-catalyzed acylation of the 2-hydroxyl group followed by intramolecular acyl transfer [198]. Several "rules" have been formulated for the use of Hmb backbone protection into growing peptides: (1) backbone protection needs to be introduced only about every sixth residue because of long-range effects; (2) in a sequence of Fmoc-Yyy-(Hmb)Xxx, Yyy can be any residue if Xxx is Gly, but may not be Ile, Thr, or Val if Xxx is not Gly [200]; (3) N^α-Fmoc-N^{in}-Boc protected Trp should be used in conjunction with Hmb protection because the cation generated from the cleavage of Hmb alkylates Trp and cannot be completely suppressed using scavengers [199].

In a study examining the effect of Hmb protection of the C-terminal amino acid on fragment condensation, the formed 4,5-dihydro-8-methoxy-1,4-benzoxazepin-2(3H)-one species between the hydroxyl group and the activated group resulted in slow, incomplete couplings [213]. This has led to the development of the 2-hydroxy-4-methoxy-5-nitrobenzyl protecting group, which is better than Hmb in terms of coupling efficiency and suppressing epimerization at the C-terminal. Unfortunately, because of its TFA stability, study of this protecting group was abandoned. A safety-catch amide-bond protecting group, 6-hydroxy-5-methyl-1,3-benzoxathiolyl, has been developed that can be reduced from the sulfoxide to the sulfide using ammonium iodide and dimethyl sulfide to increase TFA lability (Scheme 34). This group has been shown to enhance coupling efficiency compared with Hmb derivatives and significantly decrease epimerization compared with

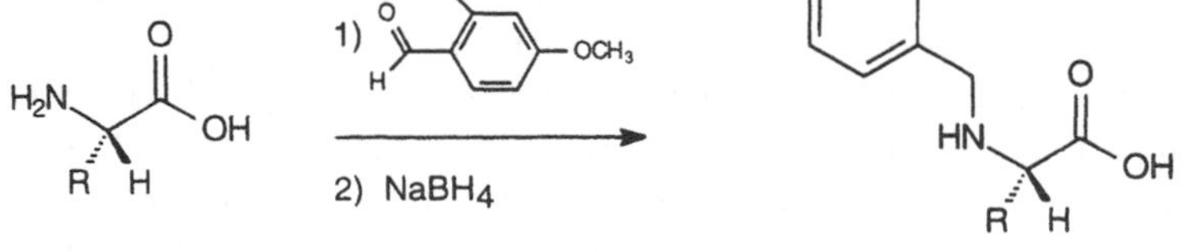

Scheme 33 Preparation of Hmb-protected amino acids.

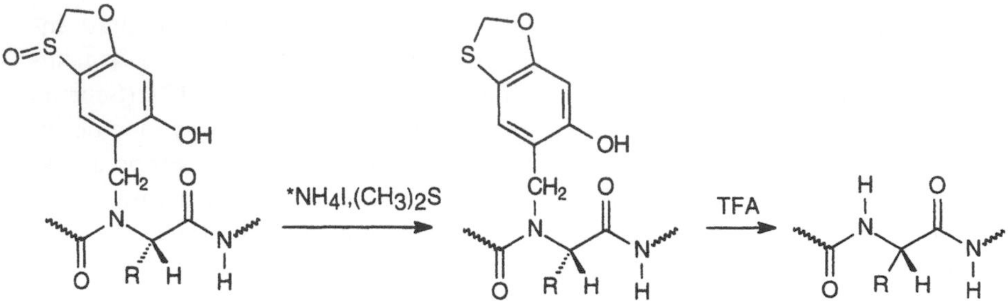

Scheme 34 Removal of the 6-hydroxy-5-methyl-1,3-benzoxathiolyl backbone protecting group. *One-pot cleavage, i.e., NH_4I/dimethyl sulfide added to TFA.

non–backbone-protected *C*-terminal amino acids. The role of *C*-terminal backbone protection in reducing epimerization in segment coupling is being investigated further to determine its generality [213].

The 2-hydroxybenzyl (Hbz) amide protecting group has been evaluated for use in Boc SPPS because it is stable to TFA but labile to TFSMA. To date, ACP (65–74) is the only reported synthesis using Hbz protection [214]. A comparison of sequences prepared using optimized Boc chemistry and Fmoc chemistry with Hmb backbone protection showed no significant difference between these two methods [205].

Because of the cost of preparing Hmb derivatives of amino acids, Bayer et al. [215] introduced the 2,4,6-trimethoxybenzyl (Tmob) backbone protecting group on Gly residues for use with Fmoc SPPS. Similarly to Hmb, the Tmob protecting group increased solvation of the resin-bound peptide and the solubility of protected fragments and has potential utility in convergent SPPS [215].

XIII. ANALYTICAL METHODS

The purity of protected amino acids is especially important for the synthesis of longer peptides. Standard techniques such as melting point determination, nuclear magnetic resonance (NMR) spectroscopy, mass spectrometry, and optical rotation are effective means of characterization. The optical purity can also be evaluated by high-performance liquid chromatography (HPLC) after derivatization with Marfey's reagent [216,217]. The "advanced Marfey method" refers to analysis by mass spectrometry after derivatization with Marfey's reagent [218–221]. Purification of side-chain protected amino acids by recrystallization is usually sufficient.

XIV. SELECTED PROCEDURES

A. Aspartic and Glutamic Acid

1. L-Aspartic Acid β-Allylester Hydrochloride [24]

L-Aspartic acid (4.0 g, 3.0 mmol) is suspended in dry allyl alcohol (150 mL) under N_2. Chlorotrimethylsilane (9.5 mL, 7.5 mmol) is added dropwise with a pressure-equalizing funnel and the resulting solution is stirred at room temperature (RT) for 18 h. Diethyl ether (1 L) is added to give a white precipitate. The precipitate is collected via filtration, washed with diethyl ether, and dried under vacuum to give the desired product in high yield (5.82 g, 93%); m.p. 183–185°C; $[\alpha]_D^{20}$ +19.0° (c = 1, MeOH).

2. L-Glutamic Acid γ-Allylester Hydrochloride [24]

L-Glutamic acid γ-allylester hydrochloride is prepared similarly to L-aspartic acid β-allylester hydrochloride except that chlorotrimethylsilane is added more slowly (dropwise over 2 h). The reaction is stopped upon appearance of the diallyl ester on TLC: yield 77%; m.p. 130–132°C; $[\alpha]_D^{20}$ +22.5° (c = 1, MeOH).

3. N-(9-Fluorenylmethoxycarbonyl)aspartic Acid β-tert-Butyl Ester [33]

A suspension of aspartic acid (3.9 g, 29 mmol) and 4-toluenesulfonic acid (11.1 g, 58 mmol) in 1,4-dioxane (140 mL) is stirred under isobutylene gas (5 psi) at RT for 72 h. After addition of Na_2CO_3(aq) (10% w/v, 175 mL), Fmoc-succinimide (9.9 g, 29 mmol) in dioxane (50 mL) is added dropwise while the amino acid solution is stirred at 0°C. The mixture is stirred overnight at RT, poured into ice water (300 mL), and extracted with diethyl ether (3 × 300 mL). The aqueous phase is cooled to 0°C, acidified to pH 5.5 with 1 N HCl, and extracted with ethyl acetate (3 × 300 mL). The extract is washed with brine, dried ($MgSO_4$), and evaporated to yield a white solid residue corresponding to a mixture of α- and β-mono-t-butyl esters. The solid is dissolved in CH_2Cl_2–petroleum ether (1:1) and kept at 0°C overnight. The resulting white crystals are filtered and washed with petroleum ether to give Fmoc-Asp(OtBu)-OH (5.53 g, 35%) as the only product; m.p. 149–150°C; $[\alpha]_D^{25}$ −21.3° (c = 1, DMF).

B. Asparagine: L-Asn(Trt)-OH.0.5 H₂O [41]

Asparagine (13.2 g, 100 mmol), trityl alcohol (52 g, 200 mmol), acetic anhydride (18.9 mL, 200 mmol), and concentrated H_2SO_4 (6.1 mL, 115 mmol) are suspended in glacial acetic acid (300 mL) and heated (60°C) for 75 min. The resulting solution is added slowly to cold water (600 mL), adjusted to pH 6 with 10 N NaOH, and maintained at 0°C for 1 h. The crystals obtained are filtered, washed thoroughly with water, washed with toluene, and dried to give 28.7 g of product (75% yield); m.p. >240°C dec.; $[\alpha]_D^{20}$ −3.7° (c = 1, 1 N NaOH).

C. Serine and Threonine

1.   O-tert-Butyl-L-serine Methyl Ester 4-Toluenesulfonate [61]

Serine methyl ester hydrochloride (10.2 g, 65.7 mmol), 4-toluenesulfonic acid (25.0 g, 131 mmol), and CH_2Cl_2 (500 mL) are stirred under isobutylene gas (5 psi) for 72 h. After degassing, the solution is evaporated to one third of the original volume, and Et_2O (2.5 L) is added. The mixture is chilled to give a white crystalline product in excellent yield (20.2 g, 89%); m.p. 141– 142°C; $[\alpha]_D^{20}$ +13.0 (c = 1, DMF).

2.   O-tert-Butyl-L-threonine Methyl Ester 4-Toluenesulfonate [61]

Threonine methyl ester hydrochloride (2.00 g, 11.8 mmol) and 4-toluene- sulfonic acid (11.2 g, 58.9 mmol) in CH_2Cl_2 (100 mL) are stirred under isobutylene gas (5 psi) for 72 h. The solution is degassed, evaporated to one third of the original volume, washed with cold saturated $NaHCO_3$, dried (Na_2SO_4), and evaporated to 5 mL. Petroleum ether (30–60°C, 150 mL) is added and the product is precipitated as a white solid (2.85 g, 67%); m.p. 130–131°C; $[\alpha]_D^{20}$ +5.5° (c = 1, DMF).

3.   N-(9-Fluorenylmethoxycarbonyl)-O-tert-butyl-L-threonine [61]

TsOH.Thr(tBu)-OCH$_3$ (0.58 g, 1.61 mmol) and NaOH (0.129 g, 3.22 mmol) in H_2O (10 mL) are stirred for 2 h at 0°C and then neutralized with con- centrated HCl. Na_2CO_3 is added to 10% (w/v) and Fmoc-ONSu (0.57 g, 1.69 mmol) in 1,4-dioxane (25 mL) is added dropwise. After stirring for 24 h with warming to RT, the mixture is washed with Et_2O (3 × 25 mL) and the aqueous phase is chilled to 0°C, acidified to pH 2 with concentrated HCl, and extracted with Et_2O (3 × 25 mL). The Et_2O extract is dried (Na_2SO_4), evaporated to an oil, and crystallized from CH_3NO_2 to give white crystals (0.58 g, 91%); m.p. 131–132°C; $[\alpha]_D^{20}$ +15.3° (c = 1.0, EtOAc). This pro- cedure may also be used to prepare Fmoc-Ser from TsOH.Ser(tBu)-OCH$_3$; 95% yield, m.p. 129–130.5°C; $[\alpha]_D^{20}$ +25.9 (c = 1, EtOAc).

D. Tyrosine: *N*-(9-Fluorenylmethoxycarbonyl)-*O-tert*-butyl-L-tyrosine [61]

1. *N*-(9-Fluorenylmethoxycarbonyl)-L-tyrosine Methyl Ester

Fmoc-succinimide (37.4 g, 110 mmol) in 1,4-dioxane (160 mL) is added dropwise to tyrosine methyl ester hydrochloride in a mixture of 10% Na_2CO_3 (170 mL) and 1,4-dioxane (80 mL) at 0°C. The reaction mixture is stirred for 20 h while gradually warming to RT. The solution is poured into ice water (1.3 L) and extracted with diethyl ether (3 × 400 mL). The organic extract is washed with brine (500 mL), dried (Na_2SO_4), and evaporated to an oil. The oil is crystallized from EtOAc–hexane in high yields (35.0 g, 97%); m.p. 122–125°C; $[\alpha]_D^{20}$ −17.0 ($c = 1$, DMF).

2. *N*-(9-Fluorenylmethoxycarbonyl)-*O-tert*-butyl-L-tyrosine Methyl Ester

Fmoc-Tyr-OMe, (5.00 g, 12.0 mmol) and concentrated H_2SO_4 (0.33 mL, 6.0 mmol) in CH_2Cl_2 (100 mL) are stirred under isobutylene gas (5 psi) for 6 h at RT. The solution is washed with cold $NaHCO_3$ (2 × 100 mL) and brine (100 mL), dried (Na_2SO_4), and rotovapped to dryness. The residue is dissolved in 1:1 MeOH–CCl_4 (400 mL), washed with water (300 mL), and extracted with a 1:1 mixture of MeOH and water (2 × 200 mL). The organic layer is dried (Na_2SO_4) and evaporated to produce a white solid. The solid is recrystallized from CH_2Cl_2–hexane in excellent yields (4.70 g, 83%); m.p. 90–92°C; $[\alpha]_D^{20}$ −22.1 ($c = 1$, DMF).

3. *N*-(9-Fluorenylmethoxycarbonyl)-*O-tert*-butyl-L-tyrosine

A mixture of Fmoc-Tyr(*t*Bu)-OMe (2.00 g, 4.22 mmol) in CH_3CN (250 mL) and 3% Na_2CO_3 (375 mL) is stirred at RT for 15 h. The reaction mixture is washed with hexane (3 × 500 mL), acidified to pH 3–4 with 2 N HCl, and extracted with $CHCl_3$ (2 × 600 mL). The $CHCl_3$ extract is washed with brine (500 mL), dried (Na_2SO_4), and evaporated to an oil. The oil is crystallized from EtOAc–hexane to give 1.43 g (74%) of desired product; m.p. 150–151°C; $[\alpha]_D^{23-25}$ −28.0 ($c = 1$, DMF).

E. Histidine: *N*^{im}-Tritylhistidine [132]

Dichlorodimethylsilane (1.21 mL, 10 mmol) is added to a suspension of histidine (1.55 g, 10 mmol) in CH_2Cl_2 (15 mL) and the reaction mixture is refluxed for 4 h. Triethylamine (2.79 mL, 20 mmol) is added and reflux is continued for 15 min. Additional triethylamine (1.39 mL, 10 mmol) is added, followed by a solution of Trt-Cl (2.79 g, 10 mmol) in CH_2Cl_2 (10 mL), while the solution is stirred at RT. The reaction mixture is stirred at RT for an additional 2 h, excess MeOH is added, and the solvent is removed in vacuo. Water is added to the residue and the pH is adjusted to pH 8–8.5 using triethylamine. The slurry obtained is washed with $CHCl_3$ and filtered. The solid is washed with water and then with diethyl ether to give 3.85 g (97%) of product; m.p. 220–222°C; $[\alpha]_D^{25}$ −2.1° ($c = 1$, THF-H_2O (1:1)).

F. Lysine: N^ϵ-Allyloxycarbonyl-L-lysine Hydrochloride [167]

A solution of lysine hydrochloride (50.0 g, 280 mmol) and basic copper (II) carbonate (63.5 g, 290 mmol) is refluxed in water for 30 min. Solids formed during reflux are removed via hot filtration. The filtrate is cooled to 0°C and adjusted to pH 9 using solid $Na_2CO_3 \cdot H_2O$ (approximately 5 g used), and allylchloroformate (42 mL, 410 mmol) is added dropwise over 1 h. For the addition, the temperature is maintained at 0°C and the reaction is kept at pH 9 by the addition of solid $Na_2CO_3 \cdot H_2O$ (70 g total). The reaction mixture is stirred for 12 h while warming to RT. The blue solid obtained is collected by filtration (quantitative yield) and suspended in H_2O (1 L), and thioacetamide (42.1 g, 560 mmol) is added. The suspension is stirred at 50°C for 3 h, over which time the solid slowly dissolves. The solution is adjusted to pH 2 with 2 N HCl and boiled for 5 min. The precipitate (CuS) is removed by filtration. The filtrate is concentrated to approximately 300 mL, at which time a white precipitate, HCl·Lys(Aloc), forms. This precipitate is collected via filtration in quantitative yield (74.7 g, 280 mmol); m.p. 228–231°C, dec.; $[\alpha]_D^{20}$ +9.2 [$c = 1$, 8% K_2CO_3 (aq)].

G. Arginine: N^ω-2,2,5,7,8-Pentamethylchroman-6-sulfonyl-L-arginine [181]

1. 2,2,5,6,8-Pentamethylchroman

2,3,5-Trimethylphenol (200 g, 1.47 mol) and isoprene (147 mL, 1.47 mol) are stirred in the presence of zinc chloride (23.5 g, 0.17 mol) in glacial acetic acid (180 mL) for 12 h at 23°C. The solution is refluxed on an oil bath (150°C) for 7 h, during which the solution turns black. The reaction is cooled to RT, H_2O (1 L) is added, and the black oil is separated. The aqueous solution is extracted with 40–60°C petroleum ether (3 × 800 mL). The combined organic extracts and oil are washed with Claisen's alkali (KOH–H_2O–MeOH, 2.45:1.75:7) (3 × 700 mL), water (3 × 1 L), and brine (2 × 800 mL). The solution is dried ($CaCl_2$) and the solvent is removed under reduced pressure. The residue is distilled under vacuum (0.3 mm Hg) to give a pale yellow liquid that solidifies on cooling to give 131.6 g (44%) of the desired product; b.p. 99–108°C (0.3 mm Hg); m.p. 32–38°C.

2. 2,2,5,7,8-Pentamethylchroman-6-sulfonyl Chloride, Pmc-Cl

2,2,5,7,8-Pentamethylchroman (51.7 g, 0.25 mol) is dissolved in dry $CHCl_3$ (1 L) and cooled to $-5°C$. Chlorosulphonic acid (70 mL, 1.05 mol) in dry $CHCl_3$ (800 mL) is added while the mixture is kept at $-5°C$. After addition, the mixture is stirred for 15 min at $-5°C$ and for a further 1 h after removal of the cooling bath. The dark brown solution is poured over crushed ice. The organic layer is separated; washed with 5% Na_2CO_3 (1.5 L), saturated $NaHCO_3$ (1.5 L), H_2O (1.5 L), and brine (1.5 L); and dried ($MgSO_4$). The solution is stirred with activated charcoal to decolorize, filtered through kieselguhr, and evaporated. The residue is crystallized from 40–60°C petroleum ether (40.5 g, 53%); m.p. 79–82°C.

3. N^{α}-Benzyloxycarbonyl-N^{ω}-(2,2,5,7,8-pentamethylchroman-6-sulfonyl)-arginine Cyclohexylamine Salt, Z-Arg(Pmc)-OH.CHA

Z-Arg-OH (34.79 g, 113 mmol) is dissolved in 3.2 M NaOH (aq) (146 mL) and acetone (400 mL) and cooled to 0°C. A solution of Pmc-Cl (54.92 g, 181 mmol) in acetone (250 mL) is added and the mixture is stirred at 0°C for 2 h and then at RT for a further 2 h. The reaction mixture is acidified to pH 6.5 with a saturated citric acid solution and the acetone removed under reduced pressure. The remaining solution is adjusted to pH 3 with a saturated citric acid solution, diluted with H_2O (500 mL), and extracted with EtOAc (3 × 500 mL). The combined extracts are filtered to remove insoluble by-product ($PmcO^-Na^+$). The EtOAc extract is washed with H_2O (2 × 700 mL) and brine (2 × 700 mL) and dried ($MgSO_4$). The solution is concentrated in vacuo to a volume of approximately 500 mL, cooled in an ice-water bath, and cyclohexylamine (12.9 mL, 113 mmol) is added. Anhydrous ether is added to give a thick white gum that solidifies on standing at 4°C overnight. The solid is recrystallized from MeOH–ether to give a white crystalline material (76.05 g, 59%); m.p. 156°C; $[\alpha]_D^{25}$ +6.6° ($c = 1$, MeOH).

4. N^{ω}-(2,2,5,7,8-Pentamethylchroman-6-sulfonyl)-arginine, H-Arg(Pmc)-OH

Z-Arg(Pmc)-OH.CHA (33.57 g, 49.8 mmol) is converted to the free acid to give a foam, which is then taken up in MeOH (250 mL). Palladium (10%) on charcoal (3.05 g) is added under N_2 and hydrogenolysis is continued overnight. Catalyst is removed by filtration through kieselguhr, the solvent is evaporated, and ether is added to give Arg(Pmc)-OH as a white powder (18.67 g, 85%); m.p. 95°C, then 145°C; $[\alpha]_D^{25}$ −4.2° (c = 1, MeOH).

5. N^α-Fluorenylmethoxy-N^ω-(2,2,5,7,8-pentamethylchroman-6-sulfonyl)-arginine, Fmoc-Arg(Pmc)-OH

H-Arg(Pmc)-OH (2.49 g, 5.65 mmol) is dissolved in 6% sodium carbonate (aq) (21 mL) and the solution is cooled to 0°C. A solution of Fmoc-OSu (1.92 g, 5.66 mmol) in DMF (10 mL) is added dropwise, the ice bath is removed, and the reaction is stirred for 1 h. The solution is diluted with H_2O (100 mL), washed with ether (2 × 50 mL), and acidified with a saturated citric acid solution (30 mL). The solution is extracted with ethyl acetate (3 × 100 mL), and the organic extract is washed with H_2O (2×) and brine, and dried (MgSO$_4$). The solution is evaporated in vacuo and hexane is used to precipitate the product (3.33 g, 89%); m.p. 80–93°C; $[A]_D^{25}$ +3.6° (c = 1, CHCl$_3$).

ADDITIONAL SOURCE FOR RESEARCH

The Peptide Synthesis Database, located on the World Wide Web at http:// ChemLibrary. BRI.NRC.CA/home.html, can be searched for amino acid derivatives. It contains structures, approximate prices, and links to many commercial suppliers.

REFERENCES

1. JP Tam, MW Riemen, RB Merrifield. Peptide Res 1:6–18, 1988 and references therein.
2. JP Tam, TW Wong, MW Riemen, FS Tjoeng, RB Merrifield. Tetrahedron Lett 42:4033–4036, 1979.
3. Y Okada, S Iguchi, K Kawasaki. J Chem Soc Chem Commun 1532–1534, 1987.
4. Y Okada, S Iguchi. J Chem Soc Perkin Trans I 2129–2136, 1988.
5. AH Karlström, AE Undén. Tetrahedron Lett 36:3909–3912, 1995.
6. A Karlström, A Undén. Int J Peptide Protein Res 48:305–311, 1996.
7. R Dolling, M Beyerman, J Haenel, F Kernchen, E Krause, P Franke, M Brundel, M Biernart. J Chem Soc Chem Commun 853–854, 1994.
8. JL Lauer, CG Fields, GB Fields. Lett Peptide Sci 1:197–205, 1994.
9. Y Yang, WV Sweeney, K Schneider, S Thornqvist, BT Chait, JP Tam. Tetrahedron Lett 35:9689–9692, 1994.
10. EA Kitas, R Knorr, A Trzeciak, W Bannworth. Helv Chim Acta 1314–1327, 1991.

11. J Offer, M Quibell, T Johnson. J Chem Soc Perkin Trans I 175–182, 1996.
12. A Karlström, A Undén. Tetrahedron Lett 37:4243–4246, 1996.
13. M Quibell, D Owen, LC Packman, T Johnson. J Chem Soc Chem Commun 2343–2344, 1994.
14. LC Packman. Tetrahedron Lett 36:7523–7526, 1995.
15. RS Feinberg, RB Merrifield. J Am Chem Soc 97:3485–3496, 1975.
16. JP Tam, WF Heath, RB Merrifield. J Am Chem Soc 105:6442–6455, 1983.
17. H Kunz, H Waldmann, C Unverzagt. Int J Peptide Protein Res 26:493–497, 1985.
18. A Loffet, HX Zhang. Int J Peptide Protein Res 42:346–351, 1993.
19. MH Lyttle, D Hudson. In: JA Smith, JE Rivier, eds. Peptides Chemistry and Biology, Proceedings of the 12th American Peptide Symposium. Leiden: ES-COM, 1992, pp 583–584.
20. AM Felix, CT Wang, EP Heimer, A Fournier. Int J Peptide Protein Res 31: 231–238, 1988.
21. F Dick, U Fritschi, G Haas, O Hässler, R Nyfeler, E Rapp. In: R Ramage, R Epton, eds. Peptides 1996. Leiden: ESCOM, 1998, pp 339–340.
22. C Yue, J Thierry, P Potier. Tetrahedron Lett 34:323–326, 1993.
23. WC Chan, BW Bycroft, DJ Evans, PD White. J Chem Soc Chem Commun 2209–2210, 1995.
24. PJ Belshaw, S Mzengeza, GA Lajoie. Synth Commun 20:3157–3160, 1990.
25. JE Baldwin, M North, A Flinn, MG Moloney. J Chem Soc Chem Commun 828–829, 1988.
26. D Coleman. J Chem Soc 2294–2295, 1951.
27. DR Bolin, CT Wang, AM Felix. Org Prep Proc Int 21:67–74, 1989.
28. A Loffet, N Galeotti, P Jouin, B Castro. Tetrahedron Lett 30:6859–6860, 1989.
29. F Al-Obedi, DG Sanderson, VJ Hruby. Int J Peptide Protein Res 35:215–218, 1990.
30. F Albericio, E Nicolas, J Rizo, M Ruiz-Gayo, E Pedroso, E Giralt. Synthesis 119–122, 1990.
31. PJ Belshaw, JG Adamson, GA Lajoie. Synth Commun 22:1001–1005, 1992.
32. J Wang, Y Okada, Z Wang, Y Wang, W Li. Chem Pharm Bull 44:2189–2191, 1996.
33. G Lajoie, A Crivici, JG Adamson. Synthesis 571–572, 1990.
34. A Armstrong, I Brackenridge, RFW Jackson, JM Kirk. Tetrahedron Lett 29: 2483–2486, 1988.
35. S Mosjov, AR Mitchell, RB Merrifield. J Org Chem 45:555–560, 1980.
36. H Gausepohl, M Kraft, RW Frank. Int J Peptide Protein Res 34:287–294, 1989.
37. RD Dimarchi, JP Tam, SBH Kent, RB Merrifield. Int J Peptide Protein Res 19:88–93, 1982.
38. Y Shimonishi, S Sakabara, S Akabori. Bull Chem Soc Jpn 35:1966–1970, 1962.
39. W Konig, R Geiger. Chem Ber 103:2041–2051, 1970.
40. D Hudson. US Patent 5935536, 1990 (Chem Abs 110:213367b).

41. P Sieber, B Riniker. Tetrahedron Lett 32:739–742, 1991.
42. D Shah, A Schneider, S Babler, R Gandhi, E van Noord, E Chess. Peptide Res 5:241–244, 1992.
43. G Barany, RB Merrifield. In E Gross, J Meinhofer, eds. The Peptides. Vol 2, New York: Academic Press, 1979, pp 199–208.
44. M Friede, S Denery, J Neimark, S Kieffer, H Gausepohl, JP Briand. Peptide Res 5:145–148, 1992.
45. B Sax, F Dick, R Tanner, J Gosteli. Peptide Res 5:245–246, 1992.
46. Y Han, NA Solé, J Tejbrant, G Barany. Peptide Res 9:166–173, 1996.
47. CR Wu, JD Wade, GW Traeger. Int J Peptide Protein Res 31:47–57, 1988.
48. A Grandas, X Jorba, E Giralt, E Pedroso. Int J Peptide Protein Res 33:386–390, 1989.
49. SS Wang, R Makofske, A Bach, RB Merrifield. Int J Peptide Protein Res 15:1–4, 1980.
50. F Albericio, R van Abel, G Barany. Int J Peptide Protein Res 35:284–286, 1990.
51. G Breipohl, J Knolle, W Stuber. Int J Peptide Protein Res 35:281–283, 1990.
52. PM Fischer, KV Retson, MI Tyler, MEH Howden. Int J Peptide Protein Res 38:491–493, 1991.
53. HB Arzeno, W Bingenheimer, R Blanchette, DJ Morgans, J Robinson III. Int J Peptide Protein Res 41:342–346, 1993.
54. K Barlos, D Gatos, S Koutsogianni, W Schäfer, G Stravropoulous, Y Yenging. Tetrahedron Lett 32:471–474, 1991.
55. K Barlos, D Gatos, S Koutsogianni. J Peptide Res 51:194–200, 1998.
56. PM Fischer. Tetrahedron Lett 33:7605–7608, 1992.
57. CD Chang, M Waki, M Ahmad, J Meienhofer, EO Lundell, JD Haug. Int J Peptide Protein Res 15:59–66, 1980.
58. T Mizoguchi, G Levin, DW Woolley, JM Stewart. J Org Chem 33:903–904, 1968.
59. VJ Hruby, KW Ehler. J Org Chem 35:1690, 1970.
60. H Sugano, M Miyoshi. J Org Chem 41:2352–2354, 1976.
61. JG Adamson, MA Blaskowitch, H Groenvelt, GA Lajoie. J Org Chem 56:3447–3449, 1991.
62. J Wang, Y Okada, W Li, T Yokoi, J Zhu. J Chem Soc Perkin Trans I 621–624, 1997.
63. CG Fields, CM Lovdahl, AJ Miles, VL Matthias Hagen, GB Fields. Biopolymers 33:1695–1707, 1993.
64. O Rosen, S Rubinraut, M Fridkin. Int J Peptide Protein Res 35:545–549, 1990.
65. BW Erickson, RB Merrifield. J Am Chem Soc 95:3750–3756, 1973.
66. S Akabori, S Sakakibara, Y Shimonishi, Y Nobuhara. Bull Chem Soc Jpn 37:433–434, 1964.
67. MS Bernatowicz, R Matsueda, GR Matsueda. Int J Peptide Protein Res 28:107–112, 1986.
68. R Matseuda, R Walter. Int J Peptide Protein Res 16:392–401, 1980.

69. F Albericio, D Andreu, E Giralt, C Navalpotro, E Pedroso, B Ponsati, M Ruiz-Gayo. Int J Peptide Protein Res 34:124–128, 1989.
70. AL Schroll. J Org Chem 54:244–247, 1989.
71. M Bodansky, MA Bednarek. Int J Peptide Protein Res 20:434–437, 1982.
72. M Royo, C Garcia-Echeverria, E Giralt, R Eritja, F Albericio. Tetrahedron Lett 33:2391–2394, 1992.
73. H Huang, RI Carey. J Peptide Res 51:290–296, 1998.
74. J Blake, BA Woodworth, L Litzi-Davis, WL Cosand. Int J Peptide Protein Res 40:62–65, 1992.
75. S Guttman. Helv Chim Acta 49:83–96, 1966.
76. I Photaki, J Taylor-Papadimitriou, C Sakarelios, P Mazarakis, L Zervas. J Chem Soc 2683–2687, 1970.
77. SN McCurdy. Peptide Res 2:147–152, 1989.
78. MC Munson, C Garcia-Echeverria, F Albericio, G Barany. J Org Chem 57: 3013–3018, 1992.
79. K Barlos, D Gatos, O Hatzi, N Koch, S Koutsogianni. Int J Peptide Protein Res 47:148–153, 1996.
80. H Echner, W Voelter. In: R Epton, ed. Innovation and Perpsectives in Solid Phase Synthesis. Andover: Intercept, 1992, pp 371–375.
81. Y Han, G Barany. J Org Chem 62:3841–3848, 1997.
82. DA Pearson, M Blanchette, ML Baker, CA Guindon. Tetrahedron Lett 30: 2739–2742, 1989.
83. T Kaiser, GJ Nicholson, HJ Kohlbau, W Voelter. Tetrahedron Lett 37:1187–1190, 1996.
84. Y Han, F Albericio, G Barany. J Org Chem 62:4307–4312, 1997.
85. RG Simmonds, DE Tupper, JR Harris. Int J Peptide Protein Res 43:363–366, 1994.
86. E Atherton, M Pinori, RC Sheppard. J Chem Soc Perkin Trans I 2057–2064, 1985.
87. U Weber, P Hartter. Hoppe-Seylers Z Physiol Chem 351:1384–1388, 1970.
88. DF Veber, JD Milkowski, SL Varga, RG Denkewalter, R Hirschmann. J Am Chem Soc 94:5456–5461, 1972.
89. Y Kiso, M Yoshida, T Kimura, Y Fuijiwara, M Shimokura. Tetrahedron Lett 30:1979–1982, 1989.
90. M Royo, J Alsina, E Giralt, U Slomcyznska, F Albericio. J Chem Soc Perkin Trans I 1095–1102, 1995.
91. PR Singh, M Rajopadhye, SL Clark, NE Williams. Tetrahedron Lett 37:4117–4120, 1996.
92. H Lamthanh, C Roumestand, C Deprun, A Menez. Int J Peptide Protein Res 41:85–95, 1993.
93. H Lamthanh, H Virelizier, D Frayssinhes. Peptide Res 8:316–320, 1995.
94. M Engebretsen, E Agner, J Sandosham, PM Fischer. J Peptide Res 49:341–346, 1997.
95. AM Kimbonguila, A Merzouk, F Guibé, A Loffet. Tetrahedron Lett 35:9035–9038, 1994.
96. EA Hallinan. Int J Peptide Protein Res 38:601–602, 1991.

97. J Lukszo, D Patterson, F Albericio, SA Kates. Lett Peptide Sci 3:157–166, 1996.
98. ER Eritja, JP Zieheler-Martin, PA Walker, TD Lee, K Legesse, F Albericio, BE Kaplan. Tetrahedron 43:2675–2680, 1987.
99. BH Rietman, RFR Peters, GI Tesser. Rec Trav Chim Pays-Bas 114:1–5, 1995.
100. E Atherton, PM Hardy, DE Harris, BH Matthews. In: E Giralt, D Andreu, eds. Peptides 1990. Leiden: ESCOM, 1991, pp 243–244.
101. Y Fujiwara, K Akaji, Y Kiso. Chem Pharm Bull 42:724–726, 1994.
102. LS Richter, JC Marsters, TR Gadek. Tetrahedron Lett 35:1631–1634, 1994.
103. M Gairi, P Lloyd-Williams, F Albericio, E Giralt. Tetrahedron Lett 35:175–178, 1994 and references therein.
104. B Iselin. Helv Chim Acta 44:61–78, 1961.
105. RA Houghten, CH Li. Anal Biochem 98:36–46, 1979.
106. S Futaki, T Yagami, T Taike, T Akita, K Kitagawa. J Chem Soc Perkin Trans I 653–658, 1990.
107. MW Pennington, ME Byrnes. Peptide Res 8:39–43, 1995.
108. W Beck, G Jung. Lett Peptide Sci 1:31–37, 1994.
109. H Yajima, N Fujii, S Funakoshi, T Watanabe, E Murayama, A Otaka. Tetrahedron 44:805–819, 1988.
110. E Nicolás, M Vilaseca, E Giralt. Tetrahedron 51:5701–5710, 1995.
111. D Yamashiro, CH Li. J Am Chem Soc 95:1310–1315, 1973.
112. D Yamashiro, CH Li. J Org Chem 38:591–592, 1973.
113. Y Kiso, S Tanaka, T Kimura, H Itoh, K Akaji. Chem Pharm Bull 39:3097–3099, 1991.
114. K Rosenthal, A Karlström, A Undén. Tetrahedron Lett 38:1075–1078, 1997.
115. R Philosof-Oppenheimer, I Pecht, M Fridkin. Int J Protein Peptide Res 45:116–121, 1995.
116. N Fotouhi, DS Kemp. Int J Peptide Protein Res 41:153–161, 1993.
117. T Ishiguro, C Eguchi. Chem Pharm Bull 37:506–508, 1989.
118. M Kusunoki, S Nakagawa, K Seo, T Hamana, T Fukuda. Int J Peptide Protein Res 36:381–386, 1990.
119. S Shatiel. Biochem Biophys Res Commun 29:178–183, 1967.
120. JC Gesquière, J Najib, T Latailler, P Maes, A Tartar. Tetrahedron Lett 34:1921–1924, 1993.
121. Y Nishiyama, N Shintomi, Y Kondo, Y Okada. J Chem Soc Chem Commun 2515–2516, 1994.
122. Y Nishiyama, N Shintomi, Y Kondo, T Izumi, Y Okada. J Chem Soc Perkin Trans I 2309–2313, 1995.
123. A Karlström, A Undén. J Chem Soc Chem Commun 959–960, 1996.
124. T Brown, JH Jones. J Chem Soc Chem Commun 648–649, 1981.
125. T Brown, JH Jones, JD Richards. J Chem Soc Perkin Trans I 1553–1561, 1982.
126. R Pipkorn, B Ekberg. Int J Peptide Protein Res 27:583–588, 1986.
127. MA Mitchell, TA Runge, WR Mathews, AK Ichhpurani, NK Harn, PJ Dobrowolski, FM Eckenrode. Int J Peptide Protein Res 36:350–355, 1990.

128. KY Kumagaye, T Inui, K Nakajima, T Kimura, S Sakakibara. Peptide Res 4: 84–87, 1991.

129. JC Gesquiere, J Najib, E Diesis, D Barbry, A Tartar. In: JA Smith, JE Rivier, eds. Peptides, Chemistry and Biology, Proceedings of the 12th American Peptide Symposium. Leiden: ESCOM, 1992, pp 641–642.

130. Y Okada, J Wang, T Yamamoto, Y Mu. J Chem Soc Perkin Trans I 753–754, 1996.

131. Y Okada, J Wang, T Yamamoto, T Yokoi, Y Mu. Chem Pharm Bull 45:452–456, 1997.

132. K Barlos, D Papaioannou, D Theodoropoulos. J Org Chem 47:1324–1326, 1982.

133. P Sieber, B Riniker. Tetrahedron Lett 28:6031–6034, 1987.

134. K Barlos, O Chatzi, D Gatos, G Stravropoulos. Int J Peptide Protein Res 37: 513–520, 1991.

135. K Barlos, O Chatzi, D Gatos, G Stavropoulos, T Tsegenidis. Tetrahedron Lett 32:475–478, 1991.

136. R Colombo, F Colombo, JH Jones. J Chem Soc Chem Commun 292–293, 1984.

137. Y Okada, J Wang, T Yamamoto, Y Mu. Chem Pharm Bull 44:871–873, 1996.

138. Y Okada, J Wang, T Yamamoto, Y Mu, T Yokoi. J Chem Soc Perkin Trans I 2139–2143, 1996.

139. GR Matseuda. Int J Peptide Protein Res 20:26–34, 1982.

140. Y Kiso, T Kimura, M Shimokura, T Narukami. J Chem Soc Chem Commun 287–289, 1988.

141. A Karlström, A Undén. J Chem Soc Chem Commun 1471–1472, 1996.

142. Y Nishiuchi, H Nishio, T Inui, T Kimura, S Sakakibara. Tetrahedron Lett 37: 7529–7532, 1996.

143. P Sieber. Tetrahedron Lett 28:1637–1640, 1987.

144. A Stierandova, N Sepetov, GV Nikiforovich, M Lebl. Int J Peptide Protein Res 41:31–38, 1994.

145. H Choi, JV Aldrich. Int J Peptide Protein Res 42:58–63, 1993.

146. CG Fields, GB Fields. Tetrahedron Lett 34:6661–6664, 1993.

147. P White. In: JA Smith, JE Rivier, eds. Peptides, Chemistry and Biology, Proceedings of the 12th American Peptide Symposium. Leiden: ESCOM, 1992, pp 537–538.

148. H Franzen, L Grehn, U Ragnarsson. J Chem Soc Chem Commun 1699–1700, 1984.

149. T Vorherr, A Trzeciak, W Bannwarth. Int J Peptide Protein Res 48:553–558, 1996.

150. M Ohno, S Tsukamoto, N Izumiya. Bull Chem Soc Jpn 45:2852–2855, 1972.

151. BW Erickson, RB Merrifield. J Am Chem Soc 95:3757–3763, 1973.

152. O Dangles, F Guibe, G Balavoine, S Lavielle, A Marquet. J Org Chem 52: 4984–4993, 1987.

153. EE Schallenberg, M Calvin. J Am Chem Soc 77:2779–2783, 1955.

154. S Rajagopalan, TJ Heck, T Iwamoto, JM Tomich. Int J Peptide Protein Res 45:173–179, 1995.

155. R Schwyzer, W Rittel. Helv Chim Acta 44:159–169, 1961.
156. BW Bycroft, WC Chan, SR Chhabra, ND Hone. J Chem Soc Chem Commun 778–779, 1993.
157. GB Bloomberg, D Askin, AR Gargaro, MJA Tanner. Tetrahedron Lett 34: 4709–4712, 1993.
158. P Dumy, IM Eggleston, S Cerviani, U Sila, X Sun, M Mutter. Tetrahedron Lett 36:1255–1258, 1995.
159. D Lelièvre, D Daguet, A Brack. Tetrahedron Lett 36:9317–9320, 1995.
160. WC Chan, BW Bycroft, DJ Evans, PD White. In: RS Hodges, JA Smith, eds. Peptides, Chemistry, Structure and Biology, Proceedings of the 13th American Peptide Symposium. Leiden: ESCOM, 1994, pp 727–728.
161. J Eichler, AW Lucka, RA Houghten. Peptide Res 7:300–307, 1994.
162. K Augustyns, W Kraas, G Jung. J Peptide Res 51:127–133, 1998.
163. B Rohwedder, Y Mutti, P Dumy, M Mutter. Tetrahedron Lett 39:1175–1178, 1998.
164. SR Chhabra, B Hothi, DJ Evans, PD White, BW Bycroft, WC Chan. Tetrahedron Lett 39:1603–1606, 1998.
165. A Aletras, K Barlos, D Gatos, S Koutsogianni, P Mamos. Int J Peptide Protein Res 45:488–496, 1995.
166. S Matysiak, T Böldicke, W Tegge, R Frank. Tetrahedron Lett 39:1733–1734, 1998.
167. A Crivici, G Lajoie. Synth Commun 23:49–53, 1993.
168. KC Pugh, L Gera, JM Stewart. Int J Peptide Protein Res 42:159–164, 1993.
169. B Rzeszotarska, E Masiukiewicz. Org Prep Proc Int 20:427–464, 1988.
170. J Leonard. J Org Chem 32:250–251, 1967.
171. A Turan, A Pathy, S Bajusz. Acta Chim Acad Sci Hung 85:327–332, 1975.
172. J Ramachandran, CH Li. J Org Chem 27:4006–4009, 1962.
173. H Yajima, K Akaji, K Mitani, N Fujii, S Funakoshi, H Adachi, M Oishi, Y Akazawa. Int J Peptide Protein Res 14:169–176, 1979.
174. H Yajima, M Takeyama, J Kanaki, K Mitani. J Chem Soc Chem Commun 482–483, 1978.
175. Y Kiso, M Satomi, K Ukawa, T Akita. J Chem Soc Commun 1063–1064, 1980.
176. H Rink, P Sieber, F Raschdorf. Tetrahedron Lett 25:621–624, 1984.
177. AS Verdini, P Lucietto, G Fossat, C Giordan. In: JA Smith, JE Rivier, eds. Peptides, Chemistry and Biology, Proceedings of the 12th American Peptide Symposium. Leiden: ESCOM, 1992, pp 562–563.
178. E Atherton, RC Sheppard, JD Wade. J Chem Soc Chem Commun 1060–1062, 1983.
179. R Ramage, J Green. Tetrahedron Lett 28:2287–2290, 1987.
180. J Green, OM Ogunjobi, R Ramage, ASJ Stewart. Tetrahedron Lett 34:4341–4344, 1988.
181. R Ramage, J Green. Tetrahedron 32:6353–6370, 1991.
182. LA Carpino, H Shroff, SA Triolo, ESME Mansour, H Wenschuh, F Albericio. Tetrahedron Lett 34:7829–7832, 1993.
183. HB Arzeno, DS Kemp. Synthesis 32–36, 1988.

184. PM Fischer, KV Retson, MI Tyler, MEH Howden. Int J Peptide Protein Res 40:19–24, 1992.
185. AG Beck-Sickinger, G Schnorrenberg, J Metzger, G Jung. Int J Peptide Protein Res 38:25–31, 1991.
186. E Jaeger, HA Remmer, G Jung, J Metzger, W Oberthür, KP Rücknagel, W Schäfer, J Sonnenbichler, I Zetl. Biol Chem Hoppe-Seyler 374:349–362, 1993.
187. M Noda, M Kiffe. J Peptide Res 50:329–335, 1997.
188. GB Fields, RL Noble. Int J Peptide Protein Res 35:161–214, 1990.
189. K Hofman, WD Peckham, A Rheiner. J Am Chem Soc 78:238–242, 1956.
190. FC Gronvald, NL Johansen. In: K Brunfeldt, ed. Peptides 1980. Copenhagen: Scriptor, 1981, pp 111–115.
191. M Jetten, AM Peters, JWFM van Nispen, HCJ Ottenheijm. Tetrahedron Lett 32:6025–6028, 1991.
192. AS Verdini, P Lucietto, G Fossati, C Giordani. Tetrahedron Lett 33:6541–6542, 1992.
193. R Presentini, G Antoni. Int J Peptide Protein Res 27:123–126, 1986.
194. MS Bernatowicz, Y Wu, GR Matseuda. J Org Chem 57:2497–2502, 1992.
195. MS Bernatowicz, GR Matsueda. In: RS Hodges, JA Smith, eds. Peptides, Chemistry, Structure and Biology, Proceedings of the 13th American Peptide Symposium. Leiden: ESCOM, 1994, pp 107–109.
196. SBH Kent, D Alewood, P Alewood, M Baca, A Tones, M Schnolzer. In: R Epton, ed. Innovations and Perspectives in Solid Phase Synthesis. Andover: Intercept Limited, 1992, pp 1–22.
197. RCdeL Milton, SCF Milton, PA Adams. J Am Chem Soc 112:6039–6046, 1990.
198. R Bartl, KD Klöppel, R Frank. In: JA Smith, JE Rivier, eds. Peptides: Chemistry and Biology, Proceedings of the 12th American Peptide Symposium, Leiden: ESCOM, 1992, pp 505–506.
199. T Johnson, M Quibell, D Owen, RC Sheppard. J Chem Soc Chem Commun 369–372, 1993.
200. C Hyde, T Johnson, D Owen, M Quibell, RC Sheppard. Int J Peptide Protein Res 43:431–440, 1994.
201. M Quibell, WG Turnell, T Johnson. J Org Chem 59:1745–1750, 1994.
202. M Quibell, WG Turnell, T Johnson. J Chem Soc Perkin Trans I 2019–2024, 1995.
203. RG Simmonds. Int J Peptide Protein Res 47:36–41, 1996.
204. LC Packman, M Quibell, T Johnson. Peptide Res 7:125–131, 1994.
205. W Zeng, PO Regamy, K Rose, Y Wang, E Bayer. J Peptide Res 49:273–279, 1997.
206. T Johnson, LC Packman, CB Hyde, D Owen, M Quibell. J Chem Soc Perkin Trans I 719–728, 1996.
207. M Quibell, LC Packman, T Johnson. J Am Chem Soc 117:11656–11668, 1995.
208. M Quibell, LC Packman, T Johnson. J Chem Soc Perkin Trans I 1227–1234, 1996.

209. M Quibell, T Johnson. In: HLS Maia, ed. Peptides 1994, Proceedings of the 23rd European Peptide Symposium. Leiden: ESCOM, 1995, pp 173–174.
210. M Quibell, WG Turnell, T Johnson. Tetrahedon Lett 35:2237–2238, 1994.
211. T Johnson, M Quibell, RC Sheppard. J Peptide Sci 1:11–25, 1995.
212. E Nicolàs, M Pujades, J Bacardit, E Giralt, F Albericio. Tetrahedron Lett 38: 2317–2320, 1997.
213. J Offer, T Johnson, M Quibell. Tetrahedron Lett 38:9047–9050, 1997.
214. T Johnson, M Quibell. Tetrahedron Lett 35:463–466, 1994.
215. N Clausen, C Goldhammer, K Jauch, E Bayer. In: PTP Kaumaya, R Hodges, eds. Peptides: Chemistry, Structure and Biology, Proceedings of the 14th American Peptide Symposium. Birmingham, UK: Mayflower Scientific, 1996, pp 71–72.
216. JG Adamson, T Hoang, A Crivici, GA Lajoie. Anal Biochem 202:210–214, 1992.
217. P Marfey. Carlsberg Res Commun 49:591–596, 1984.
218. K Fujii, I Yoshitomo, H Oka, M Suzuki, K Harada. Anal Chem 69:5146–5151, 1997.
219. DR Goodlett, AA Perlette, PA Savage, KA Kowalski, TK Mukherjee, K Tarit, JW Tolan, N Corkum, G Goldstein, JB Crowther. J Chromatogr A 707:233–244, 1995.
220. K Harada, K Fujii, T Mayumi, Y Hibino, M Suzuki. Tetrahedron Lett 36: 1515–1518, 1995.
221. T Kishnamurthy. J Am Soc Mass Spectrom 5:724–730, 1994.

Appendix A Side-Chain Protecting Groups

Name, structure	Derivatives	Stability	Deprotection	Synthesis references
Acetamidomethyl (Acm) 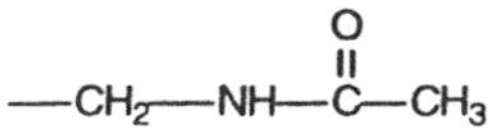 $-CH_2-NH-\overset{\overset{\displaystyle O}{\displaystyle \|}}{C}-CH_3$	Cys(Acm)	Piperidine, acids	Hg(II) (SH), I$_2$ (S—S), Tl(tfa)$_3$ (S—S)	88
1-Adamantyl (1-Ada) 	Asp/Glu(O-1-Ada)	Base	TFA	3,4
2-Adamantyl (2-Ada) 	Asp/Glu(O-2-Ada)	TFA, piperidine	HF, TFMSA	3,4
1-Adamantyloxycarbonyl (Adoc) 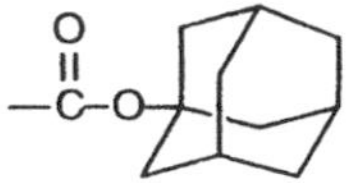	Arg(Adoc$_2$)	Base	TFA	193
2-Adamantyloxycarbonyl (2-Adoc) 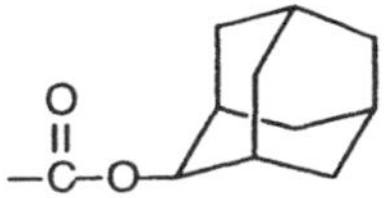	His(2-Adoc)	TFA, tertiary amines, HOBt	HF, TFMSA	121,122

177

Appendix A Continued

Name, structure	Derivatives	Stability	Deprotection	Synthesis references
1-Adamantyloxymethyl (1-Adom)	His(1-Adom)	Piperidine	TFA	137,138
2-Adamantyloxymethyl (2-Adom)	His(2-Adom)	TFA,piperidine	HF, TFMSA	130,131
Allyl (All, Al)	Asp/Glu(OAl)	Acids, base	Pd(0)/Bu$_3$SnH	18,19,24
—CH$_2$—CH=CH$_2$	Tyr(Al)	TFA, piperidine	Pd(0)/Bu$_3$SnH	18
Allyloxycarbonyl (Alloc, Aloc)	Cys(Aloc)	TFA	Pd(0)/Bu$_3$SnH, piperidine	18,19
	Trp(Aloc)	TFA, DBU	Pd(0)	149
	Lys/Orn(Aloc)	Acids, base	Pd(0)/Bu$_3$SnH	18,19
O	Arg(Aloc$_2$)	TFA	Pd(0)/Bu$_3$SnH, piperidine [Arg(Aloc) formed]	18
‖	His(Aloc)	TFA	Pd(0)/Bu$_3$SnH, piperidine	18
—C—O—CH$_2$—CH=CH$_2$	Ser/Thr(Aloc)	TFA, piperidine	Pd(0)/Bu$_3$SnH	18,19

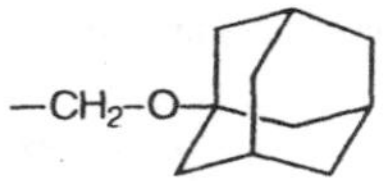

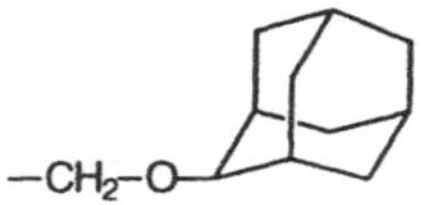

Protecting group	Amino acid			Ref.
Allyloxycarbonylaminomethyl (Allocam, Alocam) —CH$_2$—NH—C(=O)—O—CH$_2$—CH=CH$_2$ 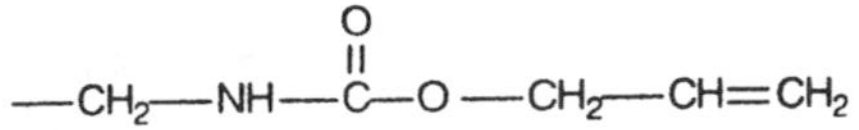	Cys(Alocam)	Piperidine	Pd(0)/Bu$_3$SnH	95
9-Anthracenesulfonyl (Ans) 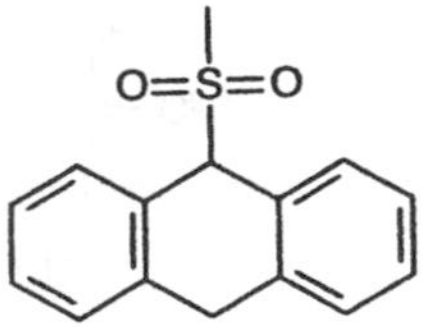	Arg(Ans)	Piperidine, dilute TFA	TFA, Ru/1-benzyl-1,4-dihydronicotinamide, aluminum	183
Benzyl (Bzl) —CH$_2$—	Asp/Glu(OBzl) Ser/Thr/Tyr(Bzl) Cys(Bzl)	TFA TFA, base Mild acid, base	HF, TFMSA, HBr/AcOH HF, TFMSA, TFSOTf HF, Na/NH$_3$	58–60,62 102
Benzyloxymethyl (Bom) —CH$_2$—O—CH$_2$— 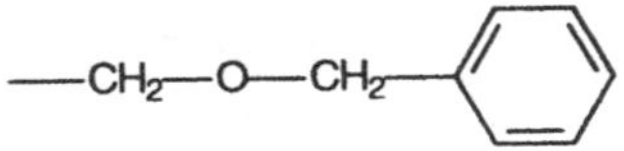	His(Bom)	TFA	HF/scavengers, TFMSA, TMSOTf, HBr/AcOH	124,125
2-Bromobenzyloxycarbonyl (2-Br-Z) 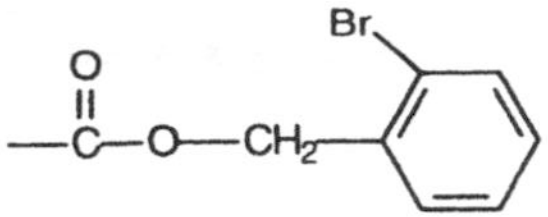	Tyr(2-Br-Z)	TFA	HF, TFMSA, TMSOTf, HBr/AcOH	112

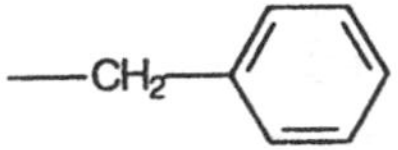

Appendix A Continued

Name, structure	Derivatives	Stability	Deprotection	Synthesis references
t-Butoxycarbonyl (Boc)	Lys/Orn(Boc)	Base	TFA	155
	Trp(Boc)	Piperidine	TFA/lyophilization	147,148
	Arg(Boc)	Piperidine	TFA	190, 192
	Arg(Boc$_2$)			
t-Butoxymethyl (Bum) 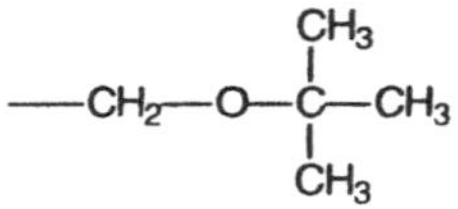	His(Bum)	Base, hydrogenolysis	TFA, dry HCl or HBr	136
t-Butyl	Asp/Glu(O*t*Bu)	Base	TFA	25,26,28,32–34
	Ser/Thr/Tyr/Hyp(*t*Bu)	Base	TFA	57,61,62
	Cys(*t*Bu)	TFA, base	HF, Hg(II), TFMSA	86
t-Butyldimethylsilyl (TBDMS)	Ser/Thr(TBDMS)	Piperidine	Dilute TFA	56
	Tyr(TBDMS)	Piperidine	Anhydrous TFA, TBAF	56

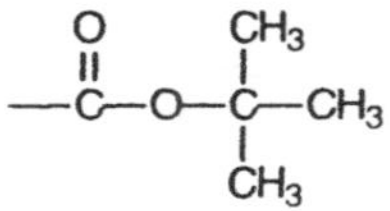

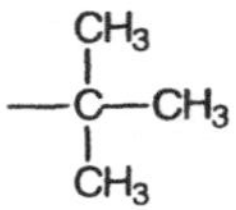

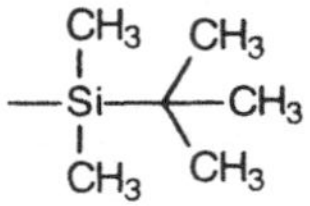

Protecting group	Amino acid	Reagents		Ref.
t-Butylsulfenyl (S-tBu) $-S-\overset{\overset{\displaystyle CH_3}{\vert}}{\underset{\underset{\displaystyle CH_3}{\vert}}{C}}-CH_3$ 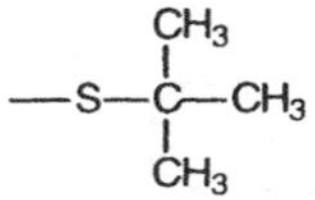	Cys(S-tBu)	Piperidine, TFA, hydrazine	Thiols, phosphines	86
2-Chlorobenzyloxycarbonyl (2-Cl-Z) 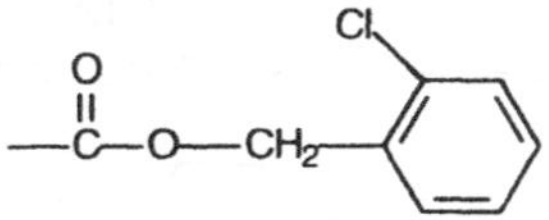	Lys(2-Cl-Z)	TFA	HF, TFMSA, TMSOTf, HBr/AcOH	151
Cyclohexyl (cHx) 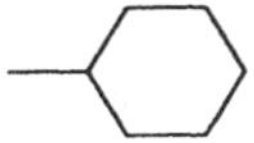	Asp/Glu(OcHx)	TFA	HF, TMSOTf	2
Cyclohexyloxycarbonyl (Hoc) 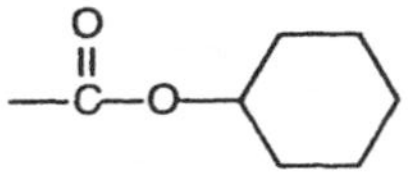	Trp(Hoc)	TFA, base	HF	142
5H-Dibenzo[a,d]cyclohepten-5-yl (5-dibenzosuberenyl, Suben) 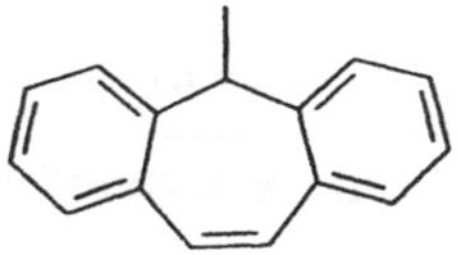	Arg(Suben)	Base, nucleophiles	TFA	187

181

Appendix A Continued

Name, structure	Derivatives	Stability	Deprotection	Synthesis references
2,6-Dichlorobenzyl (2,6-Cl₂Bzl) 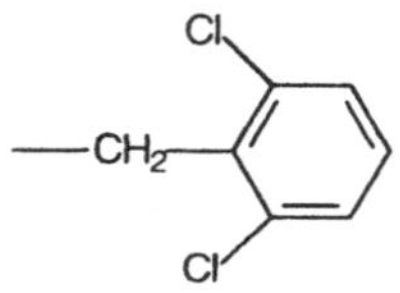	Tyr(2,6-Cl₂Bzl)	TFA	HF, TMSOTf	111
10,11-Dihydro-5H-dibenzo[a,d]cyclohepten-5-yl (5-dibenzosuberyl, Sub) 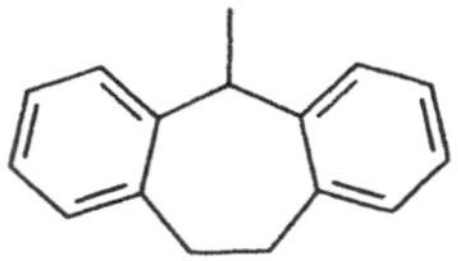 	Arg(Sub)	Base, nucleophiles	TFA	187
4,4'-Dimethoxybenzhydryl (Mbh) 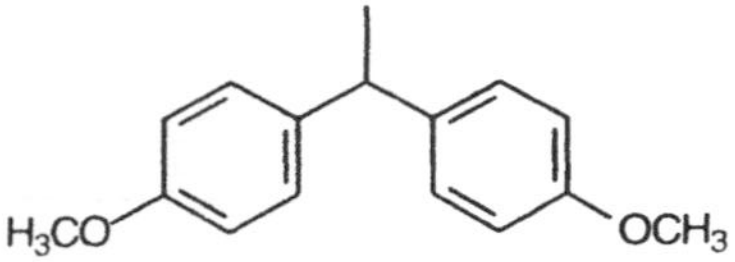 	Asn/Gln(Mbh)	Base	TFA/anisole	39
N-1-(4,4'-Dimethyl-2,6-dioxo-cyclohexylidene)-3-ethyl (Dde)	Lys/Orn(Dde)	Piperidine, TFA	Dilute hydrazine	156

Structure	Name			Ref
	Lys/Orn(Dde)	Piperidine, TFA	Dilute hydrazine	164
	N-1-(4,4'-Dimethyl-2,6-dioxocyclohexylidene)-3-methylbutyl (Ddiv)			
	Lys(Dmt) Dimethoxytrityl (Dmt)	Piperidine	Dilute acid	166
	Asp/Glu(ODmab) 4{N-[1,(4,4'-dimethyl-2,6-dioxocyclohexylidene)-3-methylbutyl]aminobenzyl (Dmab)	Piperidine	Dilute hydrazine	23

183

Appendix A Continued

Name, structure	Derivatives	Stability	Deprotection	Synthesis references
2,4-Dimethyl-3-pentyl (Dmp) 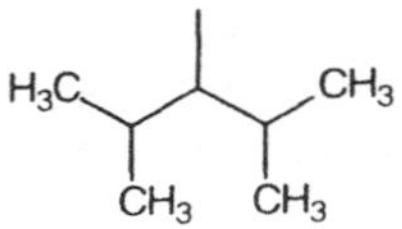	Asp/Glu(ODmp)	TFA, base	HF	5,6
2,4-Dimethylpent-3-yloxycarbonyl (Doc) 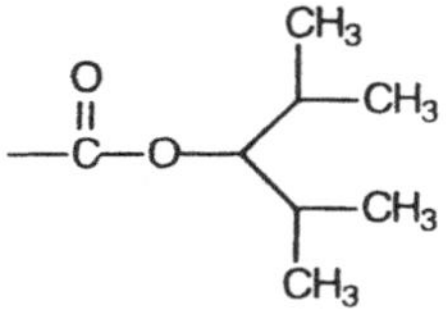	Tyr(Doc) His(Doc) Trp(Doc)	Piperidine, TFA TFA, nucleophiles TFA, nucleophiles	HF HF HF, TFMSA	114 123 141
Dimethylphenylsilylethyl (DMPSE) 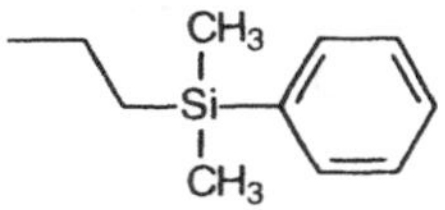	Tyr(DMPSE)	Trialkylamines, HOBt, trialkylphosphines, nucleophiles	Neat TFA	116
2,4-Dinitrophenyl (Dnp) 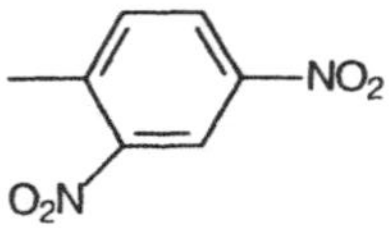	His(Dnp)	Acids	Thiophenol	119

2-(2,4-Dinitrophenyl)ethyl (Dnpe) 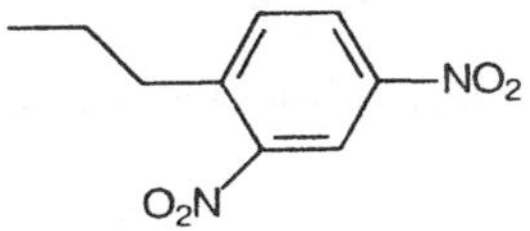	Cys(Dnpe)	TFA, I$_2$, Tl(tfa)$_3$	Piperidine	72
Diphenylphosphinothioyl (Ppt) 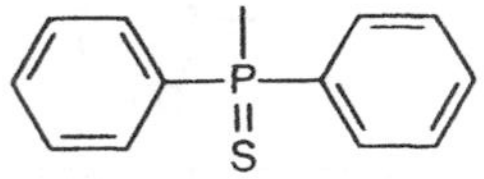	Trp(Ppt)	TFA	Methanesulfonic acid, TBAF	140
Ethylcarbamoyl (Ec) 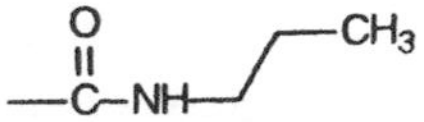	Cys(Ec)	Acids	Aqueous NaOH, hydrazine, NH$_3$	75
9-Fluorenylmethoxy (Fmoc) 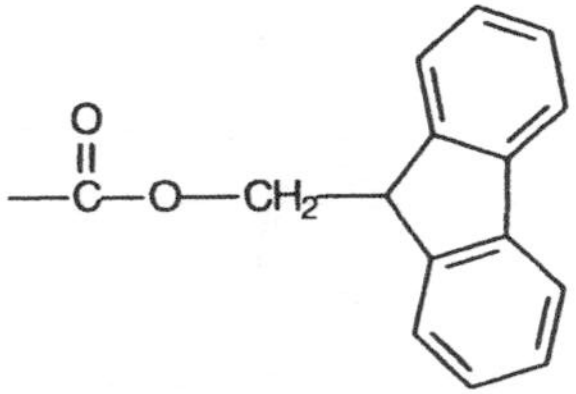	Lys(Fmoc)	TFA	Piperidine, TBAF	30

Appendix A Continued

Name, structure	Derivatives	Stability	Deprotection	Synthesis references
9-Fluorenylmethyl (Fm) 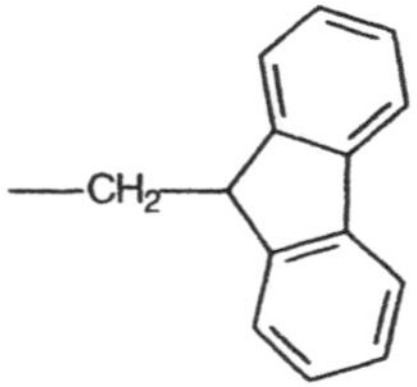	Asp/Glu(OFm) Cys(Fm)	TFA, HF TFA, HF	Piperidine, TBAF Piperidine, TBAF	27,29–31 30,71
Formyl (For) 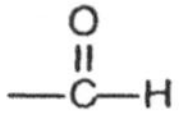	Trp(For)	TFA, high (90%) HF	TFMSA, piperidine, low HF/scavengers	150
Mesitylene-2-sulfonyl (Mts) 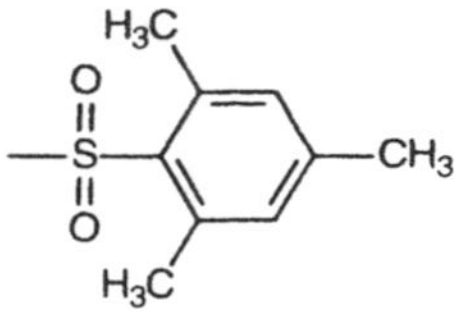	Arg(Mts)	TFA	HF, TFMSA, TFSOTf, HBr/AcOH	173,174
2-Methoxybenzyl (Mob) 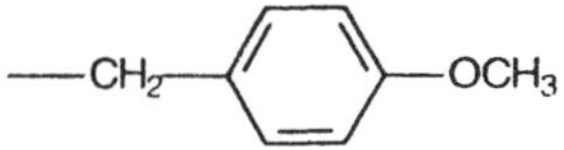	Cys(Mob)	TFA, I$_2$, base	HF, Hg(II) (S—H), Tl(tfa)$_3$ (S—S) TFMSA	66,102

2-Methoxy-10,11-dihydro-5H-dibenzo[a,d]cyclohepten-5-yl (2-methoxy-5-dibenzosuberyl, Me-Sub) 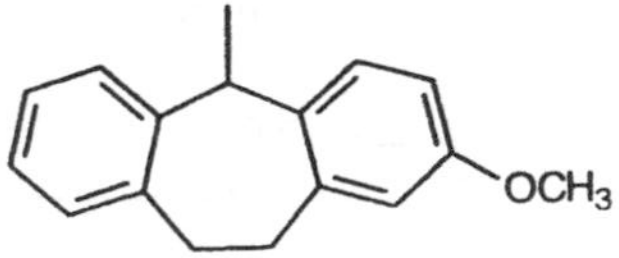	Arg(Me-Sub)	Base, nucleophiles	TFA	187
4-Methoxy-2,3,6-trimethylbenzenesulfonyl (Mtr) 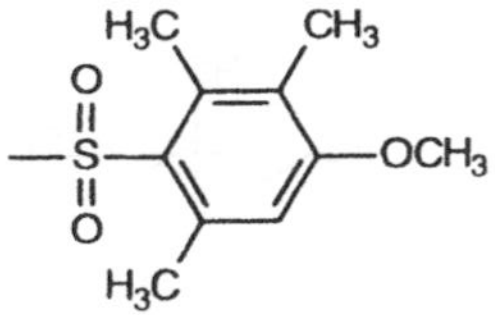	Arg(Mtr)	Piperidine	TMSBr, TFA (slow)	178
2-Methoxy-9H-xanthan-9-yl (2-Moxan)	Asn/Gln(2-Moxan)	Piperidine	TFA	46
	Cys(2-Moxan)	Piperidine	TFA	81

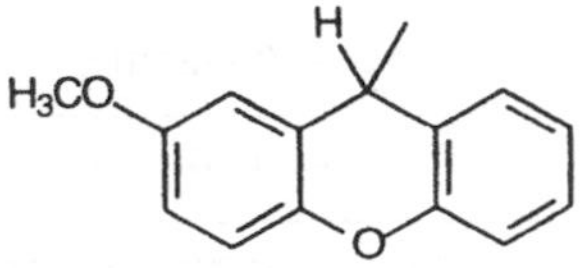

187

Appendix A Continued

Name, structure	Derivatives	Stability	Deprotection	Synthesis references
4-Methylbenzyl (Meb) 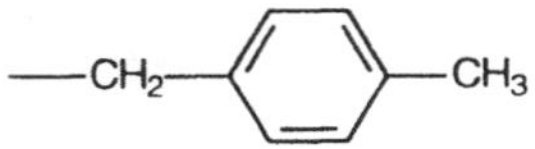	Cys(Meb)	TFA	HF, HBr/AcOH, Tl(tfa)$_3$	102
3-Methylpent-3-yl (Mpe) 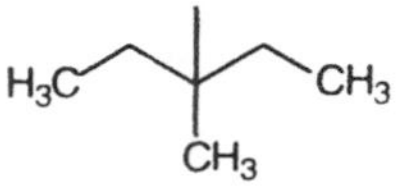	Asp/Glu(OMpe)	Piperidine	TFA	12
S-[(N'-methyl-N'-phenylcarbamoyl)sulfenyl] (Snm) 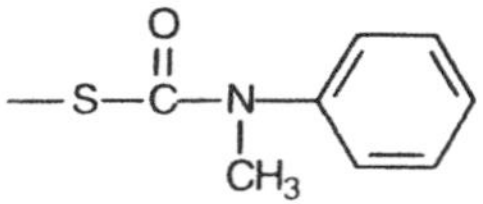	Cys(Snm)	HF, TFA	Thiols	70
4-Methylsulfinylbenzyl (Msob) 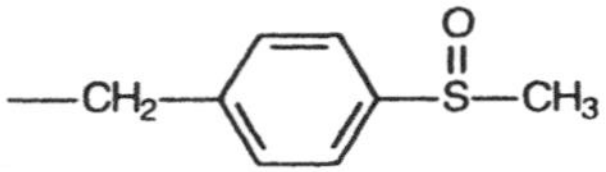	Ser/Thr/Tyr(Msob)	TFA-anisole, methanesulfonic acid, DIPEA	SiCl$_4$–TFA–scavengers	113

Name	Structure	Amino acid	Reagent	Cleavage	Ref
4-Methylsulfinylbenzyloxycarbonyl (Msz) 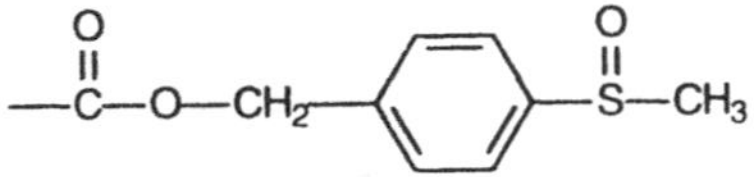		Tyr(Msz)	TFA-anisole, methanesulfonic acid, DIPEA	SiCl$_4$–TFA–scavengers	113
4-Methyltrityl (Mtt) 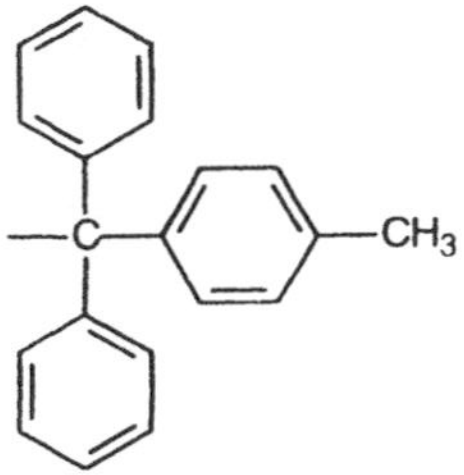		Asn/Gln(Mtt)	HOBt, AcOH	TFA	45
		His(Mtt)	Base	Dilute TFA	135
		Lys(Mtt)	Piperidine	Dilute TFA, AcOH–TFE–CH$_2$Cl$_2$	165
Monomethoxytrityl (4-methoxytrityl, Mmt) 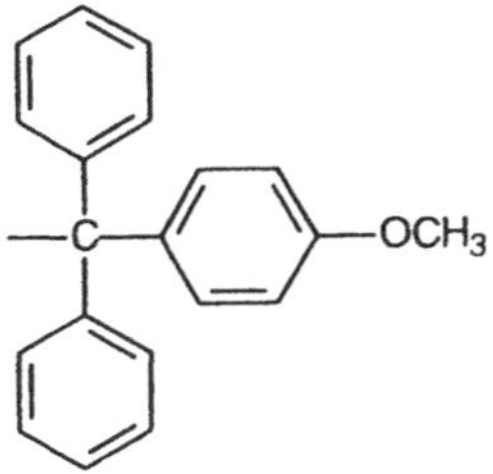		Cys(Mmt)	Piperidine	Dilute TFA	79
		His(Mmt)	Base	Dilute TFA	135
		Lys(Mmt)	Piperidine	Dilute TFA	166

Appendix A Continued

Name, structure	Derivatives	Stability	Deprotection	Synthesis references
Nitro (NO₂)	Arg(NO₂)	TFA	HF–anisole	189
3-Nitro-2-pyridinesulfenyl (Npys)	Cys(Npys)	Acids	Thiols, phosphines	67,68
	Ser/Thr/Hyp(Npys)	Acids	Thiols	64
	Lys(Npys)	Acids	Thiols	154,168
2,2,5,7,8-Pentamethylchroman-6-sulfonyl (Pmc)	Arg(Pmc)	Piperidine	TFA, TMSBr	179–181

2,2,4,6,7-Pentamethyldihydrobenzofuran-5-sulfonyl (Pbf) 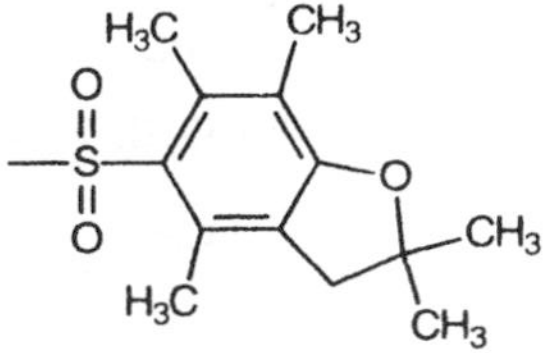	Arg(Pbf)	Piperidine	TFA	182
Phenylacetamidomethyl (Phacm) 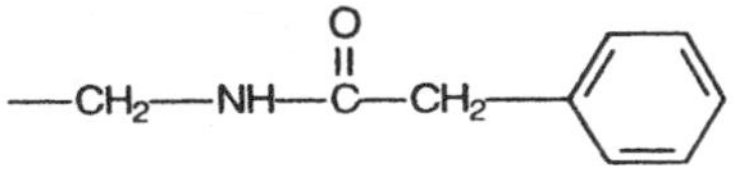	Cys(Phacm)	HF, base	Hg(II) (SH), penicillin amidohydrolase (SH), Tl(tfa)$_3$ (S—S), I$_2$ (S—S)	90
2-Phenyl isopropyl (Pp) 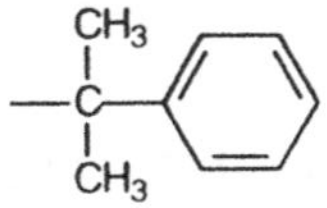	Asp/Glu(OPp)	Piperidine	Dilute TFA	22
9-Phenylxanthen-9-yl (pixyl, Pix) 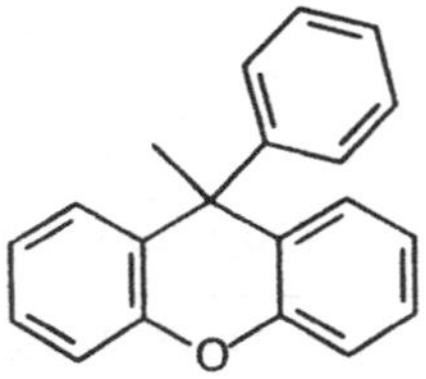	Cys(Pix)	Base	Dilute TFA (SH), Hg(II) (SH), I$_2$ (S—S), Tl(tfa)$_3$ (S—S)	80

191

Appendix A Continued

Name, structure	Derivatives	Stability	Deprotection	Synthesis references
S-Pyridyl (Pyr) 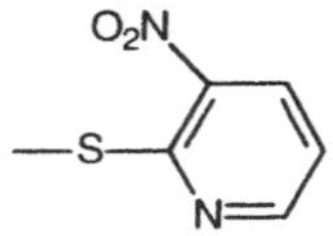 [structure]	Cys(S-Pyr)	Acids	Thiols	73
Sulfoxide (on thioether) 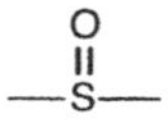 [structure]	Met(O)	TFA, base	Low HF, thiols, SO$_3$–EDT, TiCl$_4$, TMSBr, NH$_4$I	104
4-Toluenesulfonyl (Tos) 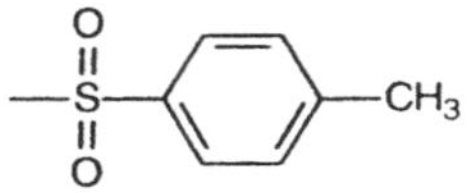 [structure]	Arg(Tos)	TFA	HF	172
Trifluoroacetyl (tfa) 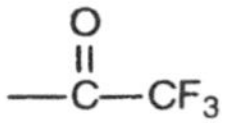 [structure]	Lys(tfa)	Acid, weak bases	NaOH	153
2,4,6-Trimethoxybenzyl (Tmob) 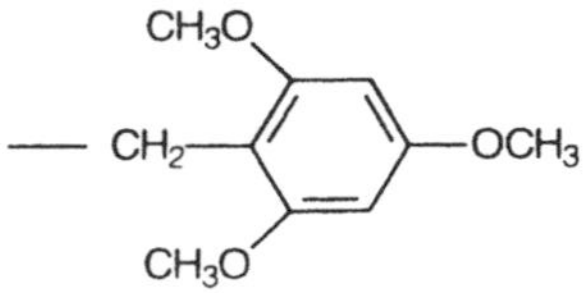 [structure]	Asn/Gln(Tmob) Cys(Tmob)	Piperidine Base, nucleophiles	TFA TFA (SH), I$_2$ (S—S), Tl(tfa)$_2$ (S—S)	40 78

		Acid, base		
Trimethylacetamidomethyl (Tacm) 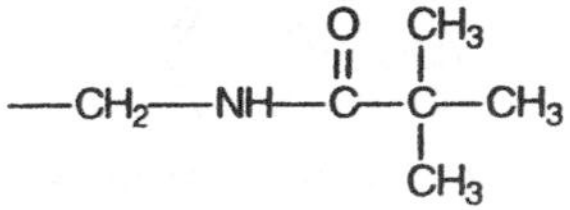 	Cys(Tacm)	Acid, base	Hg(II) (SH), I$_2$ (S—S)	89
Trimethylsilylethyl (TMSE) 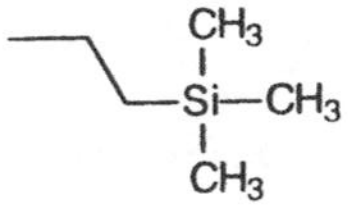 	Tyr(TMSE)	Trialkylamines, HOBt, trialkylphosphene, nucleophiles	Neat TFA	116
Triphenylmethyl (trityl, Trt) 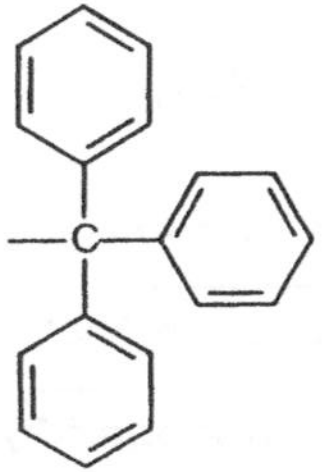 	Asn/Gln(Trt)	Piperidine	TFA	41
	Cys(Trt)	Piperidine	TFA (SH), Hg(II) (SH) I$_2$ (S—S)	76
	His(Trt)	Piperidine	TFA	132,133
	Ser/Thr(Trt)	Piperidine	Dilute TFA	54
9H-Xanthen-9-yl (Xan) 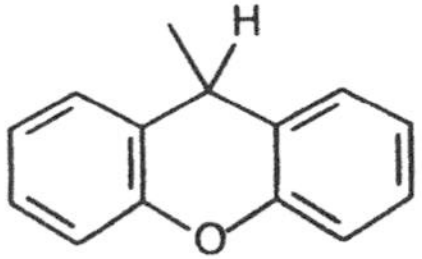 	Cys(Xan)	Piperidine, HOBt	TFA	81
	Asn/Gln(Xan)	Piperidine, HOBt	TFA	38,46

193

Appendix B Amide Backbone Protecting Groups

Name, structure	Stability	Deprotection	Synthesis references
2-Acetoxy-4-methoxybenzyl (AcHmb) 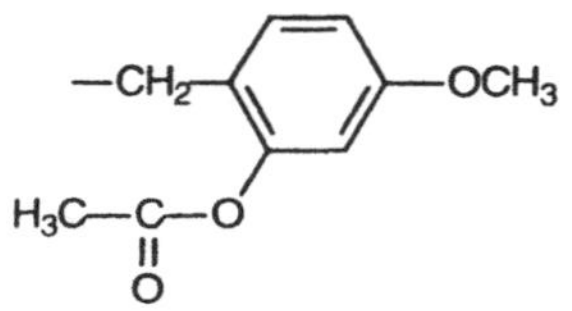	TFA	(1) Piperidine (removes Ac) (2) TFA	210
2-Hydroxybenzyl (Hbz) 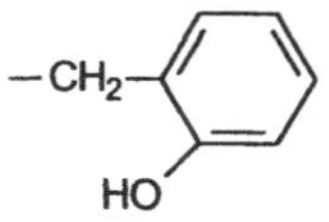	TFA	HF	214
2-Hydroxy-4-methoxybenzyl (Hmb) 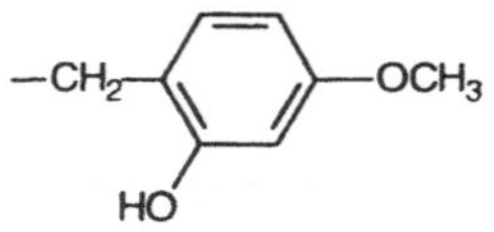	Piperidine	TFA	211,212

6-Hydroxy-5-methyl-1,3-benzoxathiolyl 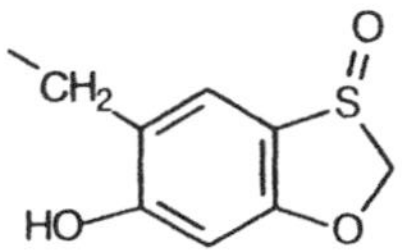	Piperidine, TFA	(1) Reduction (2) TFA	213
2,4,6-trimethoxybenzyl (Tmob) 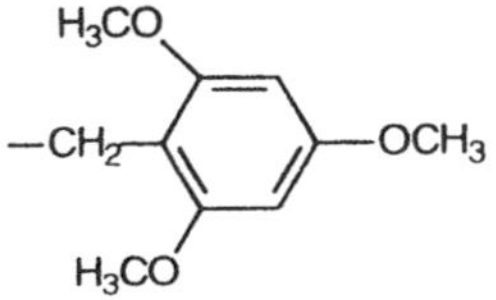	Piperidine	TFA	215

195

5

Solid Supports for the Synthesis of Peptides and Small Molecules

Christopher Blackburn
Millennium Pharmaceuticals, Cambridge, Massachusetts

Parallel synthesis methods have evolved in order to keep pace with developments in the search for therapeutic lead compounds by high-throughput screening. Thus, synthesis techniques for generating compound mixtures or arrays of single compounds have become an integral part of combinatorial chemistry [1–6] and its application to drug discovery. Solid-phase synthesis [7–16] is a key component of high-throughput synthesis because compounds anchored to a solid support can be treated with large excesses of reactants to drive reactions to completion. The resin-bound product can then be purified by a simple solvent wash to remove excess reactants, other reagents, and any high-boiling solvents. Moreover, there are several examples [7,8] of cyclization reactions that have been shown to be more efficient when conducted on a polymeric solid support than in solution, where oligomerizations compete readily. This is generally attributed to site isolation [17], a feature that also permits the solid support to act as a protecting group for one of two identical functional groups in the same molecule. Solid-phase synthesis is also readily automated, and a number of synthesis instruments are commercially available [16,18]. The creation of defined libraries from a single aliquot of resin by the mix and split method [19–22] further improves throughput if sensitive and/or on-bead screening methods for mixtures are available and appropriate deconvolution methods can be used.

Solid-phase synthesis was first applied to the assembly of peptides [23–27] and subsequently extended to other biopolymers such as oligonucleotides [28–30] and oligosaccharides [31–34]. Solid-phase synthesis is particularly valuable for the construction of linear oligomers because their construction proceeds by a repetitive sequence of high-yield reactions involving coupling and deprotection steps. Recently, organic chemists have applied solid-phase methods to a much wider range of organic transformations, primarily in the search for new therapeutic agents. This review focuses on the most common *insoluble* solid supports used for the synthesis and release of small organic molecules. Solid-phase synthesis in which the subsequent biological screening is conducted with the compounds remaining anchored to the support, resin scavenge techniques [35], and syntheses conducted on soluble polyethylene glycol polymers [36] are not covered here.

The most commonly used base resins for solid-phase synthesis are polystyrene–divinylbenzene copolymers (with ~1–2% cross-linking) and derived amphiphilic resins with polyethylene glycol (PEG) chains grafted [37–42] (see also Chapter 1). The open channels of the former support permit rapid diffusion of dissolved compounds to reaction sites, but reaction rates in polar protic solvents are often low because of poor swelling. In contrast, amphiphilic polystyrene–polyethylene glycol copolymers (PEG-PS) swell equally well in protic and aprotic solvents. The ArgoPore resins [43,44] are highly cross-linked polystyrenes designed to exhibit little shrinkage or swelling in any solvent, thereby permitting rapid diffusion of reactants to functional group sites under all conditions. PEG derivatives of these resins have also been reported. Improvements in synthesis efficiency arising from the incorporation of the PEG spacer are thought to be due to a general increase in the swelling of the resin in all the solvents encountered during the construction of a sequence rather than to the spacer effect that distances the reaction site from the cross-linked regions of the solid support [45]. Although there are also examples of improved synthetic efficiency in nonpeptide reactions when using aqueous-based solvents, PEG spacers are not always advantageous because derived resins have lower loadings and are reputed to be less stable mechanically and to treatment with trifluoroacetic acid (TFA).

As solid-phase synthesis has developed, new functionalized supports have been designed to extend the range of organic transformations that can be conducted on the polymer support without affecting the anchoring linkage. In the case of peptides, new linker designs allow final products to be cleaved under mild acidic conditions or under neutral or basic conditions. These permutations have been further expanded as solid-phase synthesis has been extended to a wider range of compound classes. In some cases this has involved the use of preformed handles that consist of the first building block prederivatized with the linker [46]. After construction of the target, cleavage releases the final product into solution while the linker remains attached to the resin. In this review resins and linkers are classified according to the functional group on the solid support to which the first building block is anchored. The chemistry of these supports is illustrated by a small but representative number of experimental procedures for compound classes synthesized on the solid phase generally for biological evaluation.

I. BENZYLIC HALIDE AND RELATED RESINS

Merrifield's chloromethyl-PS **1a** (Table 1) was designed for anchoring [23–27] protected amino acids by nucleophilic displacement of chlorine [47] to give resin-bound benzyl esters. Following a sequence of amino acid coupling reactions, the peptide is typically cleaved from the resin by treatment with HF or trifluoromethanesulfonic acid (TFMSA) with the resin presumably forming a transient benzyl cation. Resin **1a** has been applied to several nonpeptide syntheses. For example, the synthesis of tetrahydroisoquinolines commenced with the anchoring of *tert*-butoxycarbonyl (Boc)-protected dimethoxyphenylalanine [48] and treatment of the resin with TFA to remove the protecting group according to Scheme 1. A second carboxylic acid was then coupled and the product was subjected to Bischler–Napierelski cyclization on resin followed by reduction of the dihydroisoquinoline. Cleavage was then effected using HF.

Procedure [48]. The cesium salt of (L)-3,4-dimethoxy-*N*-Boc-phenylalanine (1.5 equiv.) in *N,N*-dimethylformamide (DMF) solution was added to chloromethyl-PS and heated at 50°C for 16 h. The Boc group was removed by treatment of the resin with TFA–CH$_2$Cl$_2$ (2:3) for 60 s and then repeating the deprotection for 10 min. The resin was then washed with CH$_2$Cl$_2$, CH$_2$Cl$_2$–MeOH (1:1), and DMF. The resin-bound amine was acylated by treatment with the appropriate carboxylic acid (4 equiv.), *N*-[(1*H*-benzotriazol-1-yl)(dimethylamino)methylene]-*N*-methylmethanaminium hexafluorophosphate *N*-oxide (HBTU, 4 equiv.), and *N,N*-diisopropylethylamine (DIEA, 5 equiv.) in DMF (0.5 M in the first two reactants) for 10 min. After washing with solvent, the resin was treated with freshly distilled POCl$_3$ (30 equiv.) in toluene and heated at 80°C for 8 h. Reduction to the tetrahydroisoquinoline was accomplished by suspending the resin (2 g) in MeOH (10 mL) and adding NaBH$_3$CN (5 equiv.). A few drops of methanolic HCl were added intermittently to maintain evolution of hydrogen gas. The reduction was repeated a second time and by-products removed by washing the resin with solvent. To cleave products, the resin was suspended in *p*-cresol and HF was distilled into the reaction vessel at 0°C and the mixture stirred for 1 h. The filtrate was evaporated under vacuum and the residue treated with a small volume of ether to precipitate the product. Clean products with the expected molecular ions were observed by liquid chromatography–mass spectrometry (LC-MS) for a series of eight carboxylic acids.

Table 1 Solid Supports and Linkers with Reactive Halogens

1a

1b-1e

1f-1h

1i

1j

1k

1l

1m

1n

1o-1q

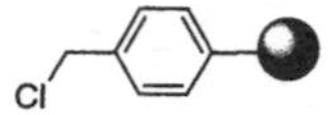

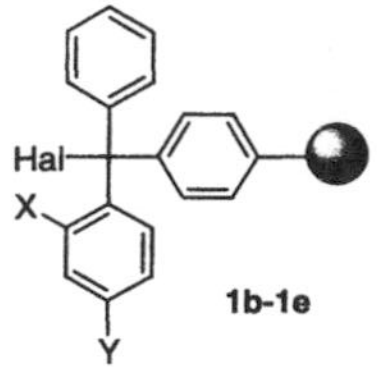

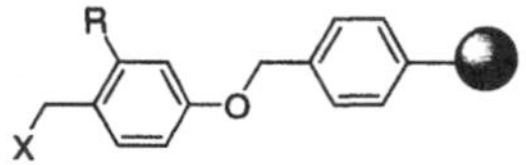

Name and identifier	Anchoring conditions	Cleavage conditions[a]	References
Merrifield's chloromethyl-PS **1a**	RCOOH, $CsCO_3$	HF	23–27,47
			48
		ROH, RO^- [RCOOR]	52–54
		ROH, H^+ [RCOOR]	55
		NH_3 [$RCONH_2$]	56,57

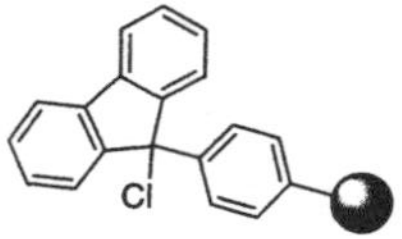

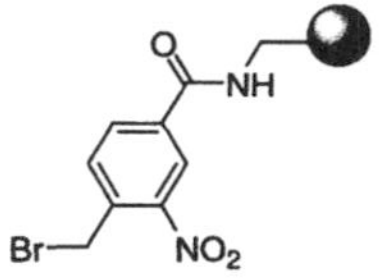

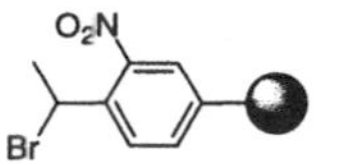

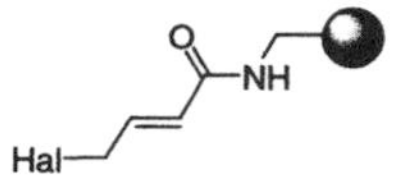

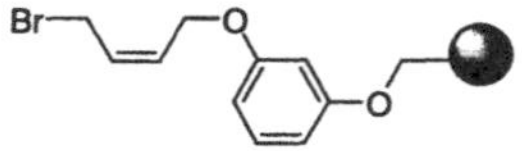

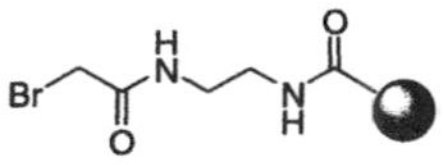

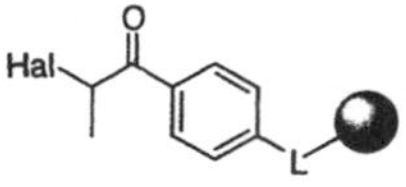

Resin	Loading	Cleavage [product]	Ref
	ROH	N_2H_4 [$RCONHNH_2$]	58
	RNH_2	1. DMAE, 2. H_2O	51
		OH^-	46,49
		DIBAL [RCH_2OH]	60
		RNH_2 [lactam]	61
		$SnCl_4$	62
		$ClCOCH_2CH_2Cl$	63
Trityl halide resins			
1b X = Y = H	$RCOOH$	TFA–CH_2Cl_2 (50:1)	65–68
1c X = Cl	$RCOOH$, DIEA	HOAc–CH_2Cl_2 or TFA–CH_2Cl_2	69
	ROH, pyridine	HCOOH [$ROCHO$]	70–72
	$ArOH$, pyridine, THF, 50°C	TFA–CH_2Cl_2 (1:99)	76
1d X = H, Y = Me	RNH_2, DMF, or CH_2Cl_2	10–25% TFA/CH_2Cl_2	64,74,75
1e X = F, Y = H			
CONHCH$_2$-spacer to base resin	$RCOOH$, DIEA	TFA–CH_2Cl_2 (0.1:100)	68
1f R = H, X = Br Bromo-Wang	RNH_2	1. $ArSO_2Cl$, 2. 95% TFA	77
	R_2NH	R'COCl [R_2NCOR']	78
1g R = OMe, X = Br	RNH_2	[$ArSO_2NHR$]	77
1h R = H, X = Cl. Chloro-Wang	ROH, NaH	1. $ArSO_2Cl$, 2.5% TFA [$ArSO_2NHR$]	80
	$ArOH$	DDQ	76,81
		TFA–H_2O (19:1)	
1i PhFl acetic acid	$RCOOH$, NMM, 80°C	20% TFA/CH_2Cl_2–MeOH	82,83
	$ArOH$, NMM, 80°C	20% TFA/CH_2Cl_2–MeOH	
	$ArNH_2$, NMM, 80°C	20% TFA/CH_2Cl_2–MeOH	
1j o-Nitrobenzyl resin	$RCOOH$, DIEA, heat	hv 250 nm MeOH	84–87

Table 1 Continued

Name and identifier	Anchoring conditions	Cleavage conditions[a]	References
1k α-Methyl-o-nitrobenzyl resin	RNH_2	1. Acylate, 2. $h\nu$ 350 nm MeOH [R'CONHR]	88
1l Hycram	RCOOH, DIEA	$h\nu$ 320–350 nm EtOH–CH_2Cl_2 (1:1)	89
1m Allylic halide	$RCOOCs$	$Pd(PPh_3)_4$, morpholine–THF	90–92
	RNH_2	1. Boc-2-amino-4-pentenoic acid 2. Grubb's catalyst [cycloalkene]	93
1n Glycolamides	$RCOOCs$	aq. NaOH	94,95
		MeOH, base [RCOOMe]	96
		NH_3 [$RCONH_2$]	
1o L = $CH_2CONHCH_2$	RCOOH	$h\nu$ 250 nm MeOH	97
1p L = $OCH_2CONHCH_2$		$BnMe_3NOH$	98,99
1q L = no spacer		KCN, DMF	100
		N_2H_4, Et_3N, MeOH [$RCONHNH_2$]	102
			97,103
		OH^-, dioxane	101

[a]The general structure of the product formed as a result of the cleavage process is shown in brackets except where cleavage regenerates the functional group that was used in the initial anchoring process.

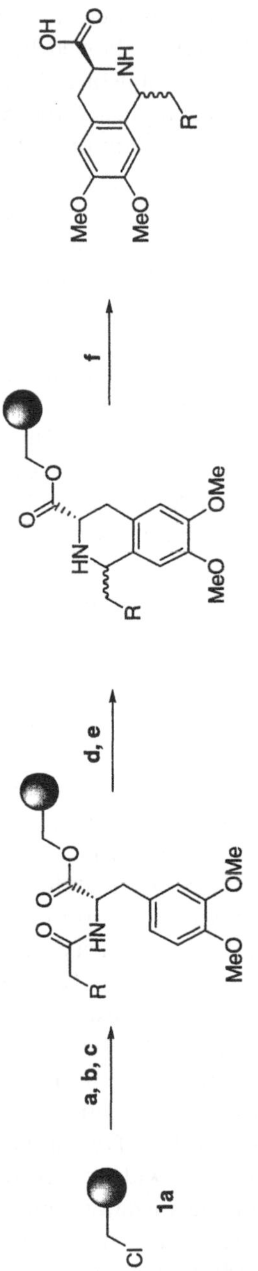

Scheme 1 (a) Boc-AA-OCs, DMF, 50°C; (b) TFA; (c) RCH₂COOH, HBTU, DIEA; (d) POCl₃, PhMe, 80°C; (e) NaBH₃CN; (f) HF.

Base-induced cleavages of peptides from resin **1a** have also been re-ported [49,50], as well as a two-step route to the acid involving reaction with dimethylaminoethanol (DMAE) followed by hydrolysis of the DMAE ester in solution by treatment with aqueous DMF [51]. Cleavage of ester anchoring groups may also be accomplished by transesterification under ei-ther basic [52–54] or acidic [55] conditions. Protected peptide segments have been cleaved by ammonolysis [56,57] or hydrazinolysis [58,59]. An-choring ester linkages may also be cleaved reductively using diisobutyl-aluminum hydride, resulting in release of alcohols from the support [60]. Intramolecular nucleophilic attack at the ester anchoring group by an amine has been shown to be an effective cleavage mechanism, giving rise to 1,4-benzodiazepin-2-ones [61] (Table 1).

Ether and thioether anchoring groups can be formed by displacement of chloride ion from chloromethyl-PS **1a** by appropriate nucleophiles, but these anchors are not readily cleaved directly. However, it has been reported that cleavage of an ether anchoring group to release an alcohol into solution can be effected using $SnCl_4$ [62]. Amines can also be anchored to resin **1a**, but their release under acidic conditions is not easily achieved without prior acylation to convert the molecule into a better leaving group. Thus, second-ary amines have been shown to displace the chlorine from resin **1a** to give solid-supported benzylamines, which were unaffected by treatment with strong acids or bases. Cleavage can be achieved, however, by reaction with α-chloroethyl chloroformate to give the free amine or with methyl chloro-formate to give the corresponding carbamate with regeneration of chloro-methyl-PS [63].

The trityl linkers were introduced to permit anchoring of carboxylic acids and other nucleophiles to a solid support and to effect cleavage reac-tions under very mild acidic conditions [64–67]. Various trityl resins, such as **1b-1e** (Table 1), have been developed that differ in the substitution pattern of the aromatic ring substituents in order to modify the cleavage properties by their influence on the stability of the trityl cation. For carboxylic acids, amines, and phenols, the chlorotrityl resin **1c** affords a more stable anchor [65–67] than does resin **1b**. Similarly, resin **1e**, which contains both fluoro and carbonyl ring substituents, proved to be very stable toward nucleophiles and was fully compatible with piperidine N^{α}-Fmoc (9-fluorenylmethoxycar-bonyl) deprotections used in a model peptide synthesis. Cleavage of acids from **1e** could be effected using dilute TFA in dichloromethane [68].

Chlorotrityl resin **1c** has been widely used for anchoring the carboxyl group of amino acids without racemization and, in the case of C-terminal proline sequences, without interference from cyclizations to give diketopi-perazines. Cleavage of the carboxylic acid from the support is usually ac-complished by treatment with dilute TFA or acetic acid (HOAc) in CH_2Cl_2.

In one example [69], acrylic acid was anchored to support **1c** in the presence of base and the product allowed to react with sulfonamides and aldehydes. The resulting three-component condensation product was cleaved from the support using TFA–CH$_2$Cl$_2$ (1:19) (Scheme 2).

Procedure [69]. 2-Chlorotritylchloride-PS [1% divinylbenzene (DVB)] **1c** (50 mg, loading 1.3 mmol/g) was suspended in DMF–CH$_2$Cl$_2$ (1:1) (0.6 mL) at 0°C and treated with acrylic acid (18 μL, 4 equiv.) and triethylamine (55 μL, 6 equiv.) for 4 h. The resin was filtered, washed, resuspended in dioxane (0.6 mL), and treated with 4-trifluoromethylbenzaldehyde (176 μL, 16 equiv.), 4-methoxyphenylsulfonamide (236 mg, 20 equiv.), and 1,4-dia-zabicyclo[2.2.2]octane (DABCO, 14 mg, 1.6 equiv.) for 16 h at 70°C. After filtration, the resin was washed with solvent and treated with CH$_2$Cl$_2$–TFA (19:1) for 30 min to cleave the product. The cleavage solution was evaporated and the product lyophilized from tBuOH–H$_2$O (4:1) and characterized by ^{1}H and ^{13}C nuclear magnetic resonance (NMR) spectroscopy as well as by mass spectrometry. Product purity was 78% as determined by high-performance liquid chromatography (HPLC).

A useful application of trityl linkers involves anchoring one function of a symmetrical diol [70–73] or diamine [74,75], enabling subsequent transformations to be conducted on the second functional group before acid-induced cleavage. Trityl resins can be used for anchoring phenols, such as 4-hydroxybenzaldehyde, in the presence of pyridine. The product was transformed via an oxime into isoxazolines, which were cleaved from the resin using TFA–CH$_2$Cl$_2$ (1:99) [76].

A benzyl halide solid support **1f** (Table 1) has been reported [77] that affords anchoring groups which are less labile than trityl anchors **1b-1e** but more labile than anchors to chloromethyl-PS (**1a**). Resin **1f** has been used to prepare amides and sulfonamides by reaction with an amine followed by acylation or sulfonylation, respectively, and cleavage using 95% TFA. When the same reaction sequence was conducted on an analogous resin **1g** prepared from Tentagel AC (acid-cleavable linker), the products could be cleaved under milder conditions (5% TFA) because of the greater stability

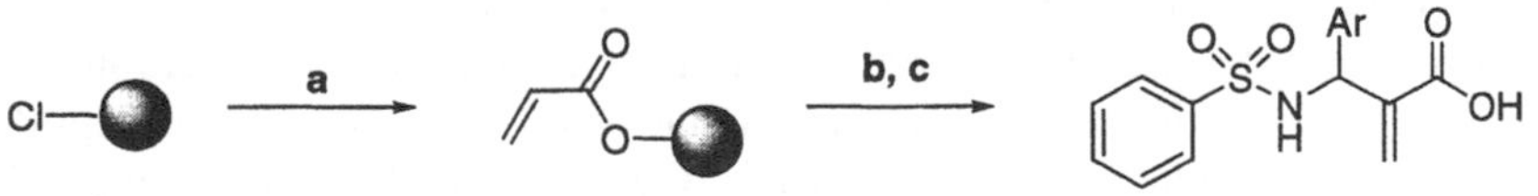

1c

Scheme 2 (a) Acrylic acid, Et$_3$N; (b) PhSO$_2$NH$_2$, ArCHO, DABCO; (c) TFA–CH$_2$Cl$_2$ (1:19).

of the resin cation conferred by an additional ring methoxy group. Reaction of **1f** with *N*-benzylpiperazine gives a resin-bound tertiary amine that reacts with acid chlorides to release the corresponding tertiary amide into solution owing to activation of the anchor toward cleavage by the 4-alkoxy group of the resin [78].

A related chloro derivative [79] of Wang resin **1h** has been used to anchor alcohols by reaction with the corresponding alkoxide ion. Treatment with TFA led to release of alcohols from the ether anchor but the corresponding esters were also formed. Oxidative cleavage using 2,3-dichloro-5,6-dicyanobenzoquinone followed by resin scavenge of excess quinone was the preferred method [80]. Phenols have been anchored similarly and ultimately cleaved using TFA–H_2O (19:1) [81].

A new linker and a derived solid support based on the chlorofluorenyl residue **1i** have been reported [82,83] that are analogous to the trityl supports in that carboxylic acids, amines, and phenols can be anchored. The resulting linkages proved to be more stable than those to the 2-chlorotrityl resins but could be cleaved by treatment with TFA. In the case of a carboxylic acid (Scheme 3), cleavage was achieved using 20% TFA in CH_2Cl_2–MeOH (9:1).

Procedure [82,83]. 9-Phenylfluoren-9-yl-PS (PhFl) modified aminomethylpolystyrene **1i** was suspended in DMF and treated with 4-bromobenzoic acid (10 equiv.) and *N*-methylmorpholine (10 equiv.) at 80°C for 4 days. The reaction mixture was quenched by addition of MeOH and the resin was then washed with DMF, MeOH, and CH_2Cl_2. To the resin-bound bromide in 1,2-dimethoxyethane (DME) was added phenylboronic acid (2 equiv.), Na_2CO_3 (2.5 equiv. from a 2 M aqueous solution), and tetrakis(triphenylphosphine)palladium (0.005 equiv.). The mixture was heated at 80°C for 16 h and then washed successively with DME–H_2O, sodium diethyldithiocarbamate–DIEA in DMF (20 mmol), DMF, MeOH, and CH_2Cl_2. 4-Phenylbenzoic acid was cleaved from the support by treatment with 20% TFA in CH_2Cl_2–MeOH (9:1) for 2 h. The product was isolated in 67% yield after chromatography on kieselgel.

Benzylic halide supports with electron-withdrawing nitro groups at the *ortho* position such as **1j** have been used for carboxyl anchoring of protected amino acids. The resulting esters are photolabile with cleavage taking place by a rearrangement involving oxygen transfer to the benzylic position followed by elimination to give a resin-bound *ortho*-nitrosoaldehyde. Peptide syntheses have been accomplished using Boc amino acids with benzyl (Bzl) and tosyl (Tos) side-chain protecting groups; following chain elongation, peptides were cleaved photochemically by irradiation at 350 nm in the absence of oxygen [84–87]. Amines have also been anchored to this resin and

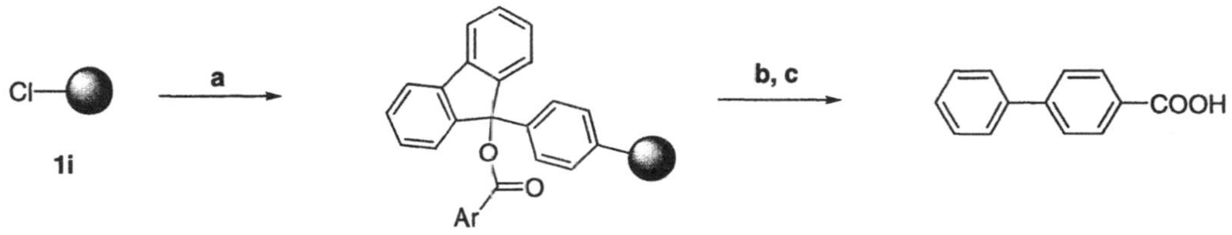

Scheme 3 (a) 4-Bromobenzoic acid, *N*-methylmorpholine; (b) PhB(OH)$_2$, Pd(PPh$_3$)$_4$, DME; (c) TFA, CH$_2$Cl$_2$, MeOH.

peptides constructed from the secondary amine before photochemical cleavage [88]. Support **1k**, which carries a methyl group at the benzylic position, is not susceptible to secondary photoreactions such as those of **1j**, in which the initial resin nitroso-aldehyde undergoes disproportionation to give a resin azobenzene derivative that can act as a light filter, retarding the cleavage process [89]. Related hydroxyl resins are discussed in Section V.

The allyl halide Hycram resins **1l** were developed for anchoring the carboxyl group of amino acids to give supports that are compatible with the use of both acid- and base-labile protecting groups because cleavages are effected using Pd(0) reagents [90–92]. The related allylic halide resin **1m** has been shown to react with amines [93]; subsequent coupling of Boc-allylglycine gave a resin-bound diene that, upon treatment with Grubb's catalyst, underwent intramolecular metathesis resulting in cleavage of a cycloalkene from the resin.

Glycoamidic ester anchoring groups [94,95] were developed for use in peptide synthesis on polyacrylic acid-base resins (supports **1n**) with the first amino acid anchored via the Cs salt. During peptide elongation the ester anchor is stable toward acid-induced deprotections and is also compatible with Fmoc strategies. Cleavage can be achieved by treatment with strong nucleophiles; thus, products can be released as acids, methyl esters, or amides by base hydrolysis, reaction with methanol in the presence of base, or reaction with ammonia, respectively [96].

Photolabile α-halophenacyl resins (**1o-1q**) have been used to form an acid-stable anchor to the carboxyl group of amino acids prior to peptide elongation [97]. Cleavage of the final products was achieved by irradiation with light of wavelength 350 nm. Related linkers that differ in the mode of attachment to the resin have been reported [98–100]. Carboxylic acids can also be released by treatment of these supports with base [101] or cyanide ion [102] and peptide hydrazides have been released by treatment with hydrazine [97,103].

II. CARBONATES AND ACTIVATED VINYL SUPPORTS

Hydroxyl resins have been converted to chloroformates or active carbonates to provide solid supports that can be used for reversible anchoring of amines as carbamates. Thus, nitrophenylcarbonate **2a** (Table 2) has been used to anchor symmetrical diamines and, after further reactions, treatment with TFA released products as trifluoroacetamides [104]. A chloroformate derivative of Tentagel-S **2b** has been used to anchor the amino group of α-amino acids and the products used to synthesize dipeptides that could be cleaved by TFA [105]. Succinimidylcarbonate resins **2c** have been prepared by treatment of

PAC/HMPP supports (see Section V) with disuccinimidyl carbonate (DSC) and used to anchor amino acids as carbamates [106]. After further chain elongations using the Fmoc strategy, lactamization reactions were conducted on resin and the resulting cyclic peptides cleaved by treatment with TFA in the presence of cation scavengers.

Procedure [106]. HMPP-Ala-MBHA-resin (where Ala was used an internal amino acid) (100 mg, 0.47 mmol/g) preswollen in DMF (0.5 mL) was treated with DSC (120 mg, 10 equiv.) and DMAP (5.7 mg, 1 equiv.) under Ar and shaken for 2 h. A solution of the TFA salt of Fmoc-Lys-OAl (245 mg, 10 equiv.) and DIEA (0.16 mL, 20 equiv.) in DMF (0.5 mL) was then added under Ar and the mixture shaken for 5 h. Chain assembly in the C $\rightarrow$ N direction was carried out by the standard Fmoc/tBu method. Fmoc removal was carried out by treatment of the resin with piperidine–DMF (1:4, 3 $\times$ 1 min, 2 $\times$ 5 min) followed by DMF washes (5 $\times$ 0.5 min). Fmoc–amino acid (10 equiv.) couplings were conducted in the presence of HOAt (10 equiv.) and DIPCDI (10 equiv.) in DMF for 0.5 h and were followed by DMF washes (5 $\times$ 0.5 min). The Tyr side chain was protected by tBu. After assembly of the target sequence, Fmoc-Val-Phe-Sar-Tyr(tBu)-DTrp-Lys(COO-HMPP-Ala-MBMA)-OAl, the C-terminal ally ester was deprotected by treatment of the resin (20 mg, 0.29 mmol/g) with Pd(Ph$_3$)$_4$ (24 mg, 0.029 mmol, 5 equiv.) in dimethyl sulfoxide (DMSO)–THF–0.5 N aqueous HCl–morpholine (2:2:1:0.1, 1.02 mL) under Ar for 2.5 h. The resin was then washed successively with THF (3 $\times$ 2 min), DMF (3 $\times$ 2 min), CH$_2$Cl$_2$ (3 $\times$ 2 min), DIEA–CH$_2$Cl$_2$ (1:19, 3 $\times$ 2 min), CH$_2$Cl$_2$ (3 $\times$ 2 min), diethyldithiocarbamic acid, sodium salt (0.03 M in DMF, 3 $\times$ 15 min), DMF (5 $\times$ 2 min), CH$_2$Cl$_2$ (3 $\times$ 2 min), DMF (3 $\times$ 1 min). The N-terminal Fmoc protecting group was removed by treatment with piperidine–DMF (1:4, 3 $\times$ 1 min, 2 $\times$ 5 min, 2 $\times$ 1 min) followed by washing with DMF (5 $\times$ 0.5 min). The resin-bound peptide was cyclized by addition of a solution of BOP (13 mg, 5 equiv.) and HOBt (4.4 mg, 5 equiv.) in DMF (0.3 mL) followed by DIEA (9.8 μL, 10 equiv.). After 2 h, the resin was washed with DMF (5 $\times$ 0.5 min) and the cyclic peptide cleaved from the support with concomitant removal of the tBu side-chain protecting groups from the Tyr residue using TFA–thioanisole–1,2-ehtanedithiol–anisole (90:5:3:2) at 25°C for 2 h. The crude product was obtained by precipitation using ether, characterized by fast atom bombardment–mass spectrometry (FAB-MS), and shown to be >70% pure by HPLC analysis.

Attempts to construct peptides by first anchoring the hydroxyl group of serine to the PAC analogue of **2c** were less successful because the anchor was labile toward treatment with piperidine. However, the succinimidyl carbonate derivative of hydroxymethyl-PS could be used to anchor serine and

Table 2 Carbonate, Activated Vinyl, and Related Supports

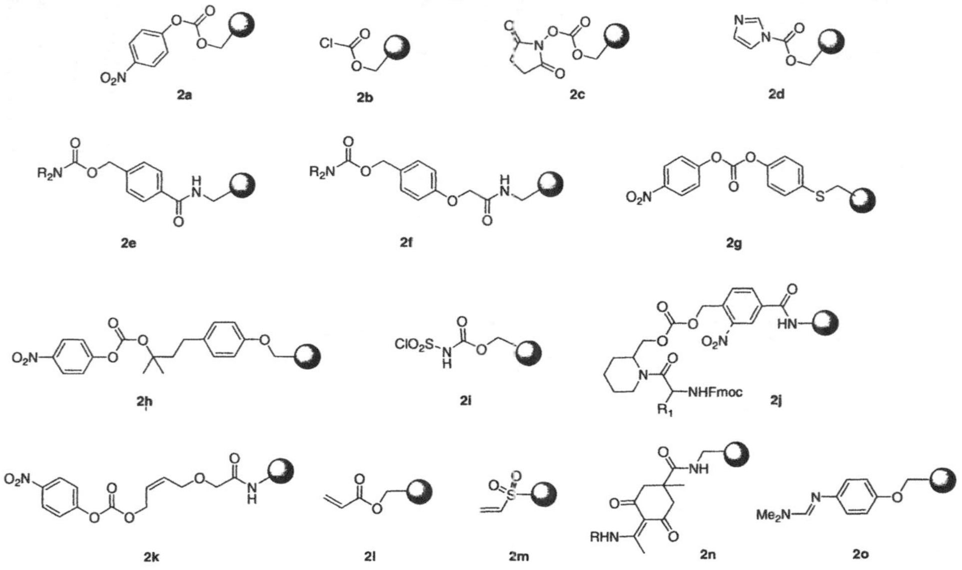

Name and identifier	Anchoring conditions	Cleavage conditions[a]	References
2a Base resin hydroxymethyl-PS or Wang	RNH_2	$TFA–CH_2Cl_2$ [$RNHCOCF_3$]	104
	$ArNH_2$, HOBt, DIEA	Et_3N, MeOH, heat [cyclization]	110,111
	RNH_2, HOBt, DIEA	$TFA–CH_2Cl_2$ (1:1)	109
	RNH_2	1. RSO_2Cl, 2. MeO^- [RSO_2NHR]	112

Support	Anchoring group	Cleavage	Ref.
PAC-PS base resin	$ArCH(NH)NH_2$, dioxane, Et_3N	TFA–CH_2Cl_2–H_2O	113,114
2b Tentagel chloroformate hydroxymethyl-PS base resin	RNH_2	TFA	105
2c PAC base resin	$ArNH_2$, dioxane	HF–anisole	108
2c Hydroxymethyl-PS base resin	RNH_2,DMF	TFA–scavengers	106
	ROH	HF	106
2d Wang imidazolide carbamate	RNH_2	TFA–CH_2Cl_2 (1:1) or $Pd(OAc)_2$, DMF, H_2O	105 107
2e HMB carbamate	Preformed handle	TFMSA–TFA	115
2f PAC carbamate	Preformed handle	TFA–$HSCH_2CH_2SH$–PhSMe	115–117
2g Thiophenoxycarbonyl support	R_2NH	1. H_2O_2, 2. R_2NH [ureas]	118
2h DHPP carbonate	RNH_2	TFA–CH_2Cl_2 (1:5)	119
2i Chlorosulfonyl carbamate	RNH_2	$R'NH_2$ [$RNHSO_2NHCONHR'$]	120
2j Photolabile carbonate	Preformed handle	$h\nu$ 365 nm, H_2O	121
2k Allyl carbonate	RNH_2	$(Ph_3P)_2PdCl_2$, Bu_3SnH	122
2l REM	R_2NH	1. R'I, 2. DIEA [$R_2R'N$]	123–125
2m Vinyl sulfonate	R_2NH	1. R'I, 2. DIEA [$R_2R'N$]	126,127
2n ADCC	Preformed handle	N_2H_4–DMF (1:50)	128
2o Formamidine	R^1R^2NH	N_2H_4	129

aThe general structure of the product formed as a result of the cleavage process is shown in brackets except where cleavage regenerates the functional group that was used in the initial anchoring process.

tyrosine, and this carbonate proved to be stable to the conditions used for peptide assembly by the Boc/Bzl strategy. The final cyclic peptide was cleaved by treatment with HF [106]. The imidazole carbamate derivative of Wang resin **2d** can also be used to anchor peptides through the N-terminus, with cleavage being effected by either acidolysis or hydrogenolysis in the presence of palladium acetate [105,107]. A chloroformate resin prepared from high-load hydroxymethyl-PS has been used to anchor p-phenylenedi-amine. Peptides were then assembled from the other amino group using Boc protection schemes and TFA deprotection, which did not affect the anchoring group. Final products were cleaved using anhydrous HF [108].

An active carbonate resin **2a** has also been prepared from Wang resin and used to anchor a multifunctional benzylamine derivative for a combinatorial application. Final products were cleaved using TFA–CH$_2$Cl$_2$ (1:1) [109]. Carbonate resins have proved valuable for construction of heterocycles by a cyclization-induced cleavage that involves intramolecular attack of an amide nitrogen at the anchor in the presence of base [110,111]. Another example of base-induced cleavage of resin-bound carbamates is provided by the observation that after sulfonylation, treatment with sodium methoxide released sulfonamides [112]. Support **2a** has also been found to react with 3-hydroxybenzamidine, leading to anchoring via the amidine function. The hydroxyl group was further functionalized on resin and the amidine products cleaved using TFA–CH$_2$Cl$_2$ [113]. A related approach to amidine therapeutic candidates using a preformed handle has also been reported [114]. The influence of the structure of the base resin on the acid lability of carbamates is illustrated by a synthesis of polyamine derivatives using preformed handles [115,116]. The hydroxymethylbenzamide anchor (**2e**) required TFMSA–TFA (1:1) mixtures for cleavage, whereas carbamates derived from the PAC-type anchor **2f** could be cleaved using TFA and cation scavengers [117].

The synthesis of ureas has been accomplished on a carbonate resin [118] by first anchoring an amine to 4-nitrophenylcarbonate resin **2g**. To render the anchor more susceptible to nucleophilic attack, the thioether was oxidized to the sulfone before reaction with a second amine that released ureas into solution. An active carbonate support **2h** derived from a tertiary alcohol resin related to the 4-(1′,1′-dimethyl-1′-hydroxypropyl)phenoxy-acetyl (DHPP) support (Section V) has been used for N-anchoring of amino acids. The resulting carbamate anchor is analogous to the Boc group and thus final products can be cleaved from the support using dilute TFA [119]. Reaction of Wang resin with chlorosulfonylisocyanate (CSI) has been reported to give a chlorosulfonyl carbamate support **2i** that can be used to anchor primary and secondary amines. Subsequent reaction with a second amine at elevated temperatures resulted in the release of substituted aminosulfonylureas by nucleophilic displacement at the carbamate, although a competing solvolysis to give primary sulfonamides was also observed [120].

Photolabile carbonate supports **2j** have been constructed by attachment of a preformed handle consisting of the reaction scaffold and an *ortho*-nitrobenzyl alcohol photolabile linker to PEG-PS [121]. Peptide sequences were constructed from the scaffold, and final products were liberated in a form suitable for direct biological screening by irradiation at 365 nm in an aqueous medium.

An allyl carbonate linker **2k** has been used [122] to synthesize pseudoargipinine III on a 4-methylbenzhydrylamine base resin (to which alanine had been attached as a spacer and internal standard). The nitrophenylallyl carbonate was prepared from the allyl alcohol and used to anchor monoprotected diamines. During further synthetic steps the anchoring group was shown to be stable toward concentrated TFA solutions and piperidine but could be efficiently cleaved by Pd-catalyzed allyl transfer.

A traceless linker for anchoring secondary amines has been developed [123,124] and is referred to as the REM linker **2l** (*re*generated after cleavage, functionalized by *M*ichael addition). Tertiary amines can be prepared on this support by conversion to a resin-bound quaternary ammonium salt followed by cleavage by base-induced Hofmann elimination [125]. The analogous vinyl sulfone resin **2m** can also be used to anchor secondary amines and is stable to a greater range of reactants [126,127].

The 4-acetyl-3,5-dioxo-1-methylcyclohexane carboxylic acid (ADCC) linker **2n** [128] was developed to provide a support for anchoring primary amines via a preformed handle method. The anchor is stable toward acids and bases including piperidine and 1,8-diazabicyclo[5.4.0]undec-7-ene (DBU) and is thus compatible with Fmoc chemistry. The amine can be released by rapid hydrazinolysis on treatment with 2% hydrazine in DMF. A formamidine linker **2o** has been obtained from Merrifield resin by reaction with the appropriate phenol. Secondary amines can be anchored to the support by an amidine exchange reaction and after further transformations cleaved by treatment with hydrazine [129].

III. ALDEHYDE RESINS

Formylpolystyrene **3a** (Table 3) [130] has been used in reactions with nitromethane to give a resin-bound nitrophenylethanol that could be dehydrated and subjected to cycloaddition reactions. The resulting adducts released tetrahydrofuran derivatives into solution by iodolactonization [131,132]. The 4-formylphenoxymethyl-PS support **3b** has been applied to the solid-phase synthesis of polyketides but cleavage conditions were not reported [133].

Table 3 Aldehyde Functionalized Supports

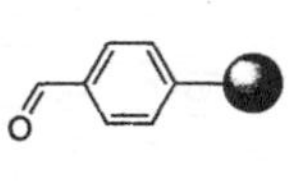

3a

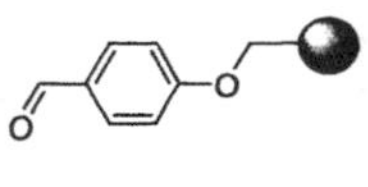

3b

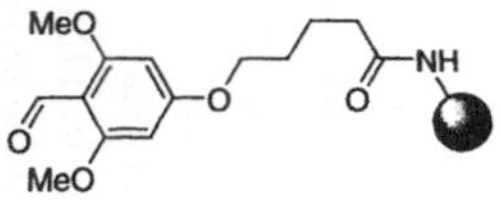

3c

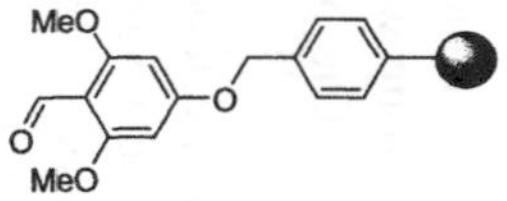

3d

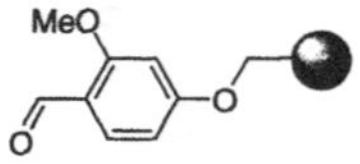

3e

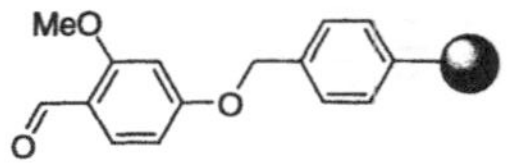

3f

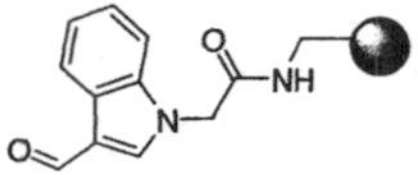

3g

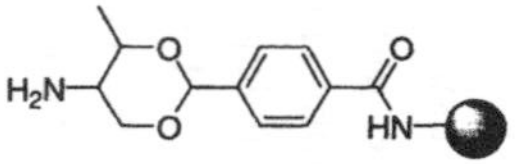

3h

Name and identifier	Coupling conditions	Cleavage conditions [product][a]	References
3a 4-Formyl-PS	MeNO₂, Et₃N	ICl [2-(cyanomethyl-5-(iodomethyl)tetrahydrofuran]	130–132
3b	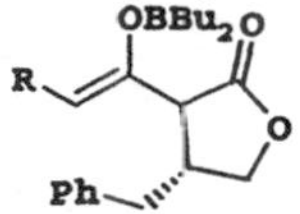	None	133
3c BAL	RNH₂ ArNH₂, NaBH(OAc)₃, HOAc–DMF (1:99)	1. Acylate, 2. HF/TFMSA [R′CONHR] TFA–H₂O–M3₂S (90:5:5)	134–136 137
3d	RNH₂, NaBH(OAc)₃, HOAc–DMF (1:99)	TFA–H₂O–Me₂S (90:5:5) [lactams]	138 139
3e Argogel MB-CHO resin	RNH₂, TMOF, NaBH₄ RNH₂, NaBH(OAc)₃, HOAc, DMF	TFA HOAc, heat	141 142
3f AMEBA (SASRIN base resin) MALDRE	RNH₂, NaBH(OAc)₃	1. ArSO₂Cl, 2. TFA–CH₂Cl₂ (1:19) [RNHSO₂Ar]	143 140
3g Indole resin	RNH₂, Me₄NBH(OAc)₃	1. Acylate, 2. TFA–CH₂Cl₂ (1:20)	144
3h 4-Carboxybenzamide resin	Preformed handle	TFA, cation scavengers	145

[a]The general structure of the product formed as a result of the cleavage process is shown in brackets except where cleavage regenerates the functional group that was used in the initial anchoring process.

The most widely used aldehyde-functionalized resins contain additional electron-donating groups to facilitate the cleavage of reaction products, which are typically amides or amines. Thus, linkers with aldehyde attachment points have been used for peptide synthesis on PEG-PS supports commencing with the anchoring of an amino acid by reductive amination. Chain extension leads to a peptide with a backbone anchoring linkage (hence BAL linker, **3c**) as distinct from the traditional C-terminal anchoring strategy [134–136]. This linker has been attached to amine-derivatized crowns and used to anchor a purine derivative through an arylamine function. After further transformations on resin, the arylamine could be cleaved under acidic conditions without need for prior acylation [137]. A related linker has also been attached to chloromethyl-PS to give resin **3d**, which has been used in several nonpeptide applications such as the synthesis of 1,4-benzodiazepine-2,5-diones [138,139].

Supports **3e** and **3f** derived from a related linker with only one methoxy group *ortho* to the aldehyde have been reported [140] to have a more reactive aldehyde group than **3c** or **3d** due to lower steric hindrance. Resin **3e** was used to synthesize aminothiazoles, which were cleaved under acidic conditions [141]. Thus, amines were anchored reductively to **3e** and converted to resin-bound thioureas by reaction with Fmoc-isothiocyanate followed by deprotection. Cyclization by reaction with α-bromomketones gave the desired heterocycles, which were cleaved from the resin by treatment with TFA–H_2O (19:1) (Scheme 4).

Procedure [141]. Argogel-MB-CHO **3e** (366 mg, loading 0.41 mmol/g) was suspended in trimethylorthoformate (5 mL) containing 10 equiv. of the appropriate primary amine in a sealed tube. The reaction mixture was heated to 70°C for 2 h and filtered. The resin was then subjected to the same reaction conditions a second time before washing with trimethylorthoformate (5 mL) and anhydrous methanol (3 × 5 mL). The resin was resuspended in anhydrous methanol, treated with sodium borohydride (133 mg, 20 equiv.) for 8 h at ambient temperature, and then washed with methanol (3 × 5 mL), methanol–H_2O (1:1, 3 × 5 mL), DMF–H_2O (1:1, 3 × 5 mL), DMF (3 × 5 mL) and CH_2Cl_2 (3 × 5 mL). A 0.2 M solution of fluorenylmethoxycarbonylisothiocyanate in CH_2Cl_2 was added to the resin and allowed to react for 20 min before washing with CH_2Cl_2 (2 × 5 mL) and DMF (3 × 5 mL). The Fmoc protecting group was removed by treatment with piperidine–DMF (1:4, 3 × 5 mL for 2.5 min each) followed by washing with DMF (3 × 5 mL) and dioxane (3 × 5 mL). The resin-bound thiourea was treated with a 0.2 M solution of the appropriate α-bromoketone in dioxane (5 mL). After 1 h at ambient temperature, the resin was washed with dioxane (3 × 5 mL) and the bromoketone reaction and wash steps were repeated twice. The resin was washed with CH_2Cl_2 (5 × 5 mL), dried under nitrogen, and

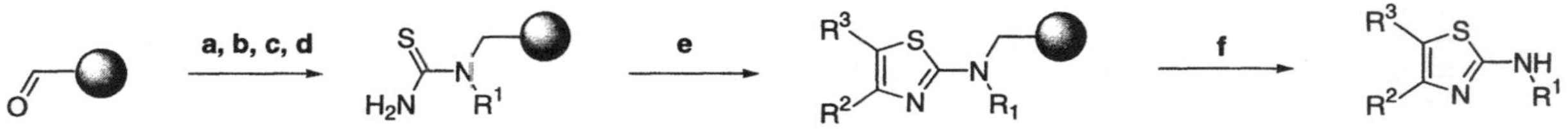

Scheme 4 (a) R^1NH_2, $(MeO)_3CH$; (b) $NaBH_4$; (c) FmocNCS; (d) piperidine; (e) $R^2COCH(R^3)Br$, dioxane; (f) TFA–H_2O (19:1).

the product cleaved by treatment with $TFA-H_2O$ (19:1, 5 mL) for 2 h. The cleavage solution was drained and the resin washed with TFA (2×2.5 mL); the combined filtrates were evaporated under reduced pressure to give crude products, which were eluted over a short column of basic alumina in methanol to give the aminothiazole derivatives, which were characterized by ^1H NMR spectroscopy and mass spectrometry.

A synthesis of imidazoles has also been reported using support **3e** in which the anchoring nitrogen atom becomes incorporated in the imidazole ring. The imidazole ring nitrogen did not function as a leaving group in the presence of 95% TFA, but cleavage could be effected by treatment with acetic acid at elevated temperatures [142].

Other aldehyde resins such as the acid-sensitive methoxybenzyl resin (AMEBA) (**3f**), obtained by oxidation of SASRIN using SO_3-pyridine, and the related 2-methoxy-4-benzyloxy-PS aldehyde resin (MALDRE) [108] have been reported. Aliphatic or aromatic amines have been anchored reductively to **3f** and the resulting secondary amine was shown to be stable toward concentrated TFA solutions. However, following conversion to amides, ureas, or sulfonamides, products could be cleaved using $TFA-CH_2Cl_2$ (1:19), although amides derived from aromatic amines required more concentrated acidic conditions to effect cleavage [143].

3-Formylindole-1-acetic acid has been coupled to aminomethyl-PS to give a new aldehyde linker **3g** to which amines could be anchored by reductive amination [144]. Conversion to amides, sulfonamides, and ureas could be accomplished on the solid phase and the products could be cleaved using dilute TFA in CH_2Cl_2 with the indole ring providing a driving force for the cleavage by resonance stabilization of the resin-bound cation. Interestingly, aromatic amines could be cleaved directly from this resin without the need for prior acylation, although this was not the case with benzylamines. The acetal derived from Fmoc-threoninol and 4-carboxybenzaldehyde has been attached to Rink amide (RAM) base resin. The resulting support **3h** was used to assemble model peptides using Fmoc amino acids and HBTU/HOBt/DIEA-induced couplings. Cleavage of the final C-terminal alcohol peptide was conducted using 90% TFA in the presence of scavengers [145].

IV. LINKERS FOR ALCOHOLS AND PHENOLS. DIHYDROPYRAN, CARBOXYLATE, AND SULFONATE RESINS AND THEIR DERIVATIVES

In order to permit the coupling and release of alcohols under mild acidic conditions, a dihydropyran linker **4a** has been introduced (Table 4) [146].

Table 4 Dihydropyran, Carboxylic, and Sulfonic Acid Supports

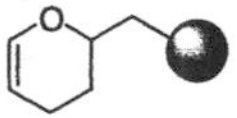

Name and identifier	Coupling conditions	Cleavage conditions[a]	References
4a DHP linker	ROH, TsOH, CH$_2$Cl$_2$	TFA–H$_2$O (20:1)	146–148
	ROH, PPTS, dichloroethane	PPTS–BuOH–dichloroethane, 60°C	149
	ROH, PPTS, dichloroethane, heat	HCl–Et$_2$O–THF	150
	Tetrazole, TFA	HCl–MeOH (1:33)	151
	2,6-Dichloropurine, CSA	TFA–CH$_2$Cl$_2$ (1:8)	152
4b	ArOH	NaOMe, THF–MeOH (4:1)	154,155
	ROH, DMAP, Et$_3$N	Na$_2$CO$_3$, MeOH–THF	156
	Isoquinoline, TMSCN	Base	157
4c	ROH, DIPCDI, HOBt	NaOMe, MeOH	158
4d	ROH, Et$_3$N, CH$_2$Cl$_2$	KI, 2-butanone, 65°C	159
	ArOH, Et$_3$N, CH$_2$Cl$_2$	Pd(OAc)$_2$, Et$_3$N, HCOOH, heat	160
	(NH$_2$)$_2$CHNAr, KOH, H$_2$O–dioxane	HF, anisole	161
Dowex base resin	ROH, Et$_3$N, CH$_2$Cl$_2$	R'NH$_2$, 60°C [RNRH']	162,163

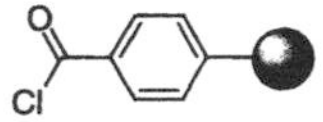

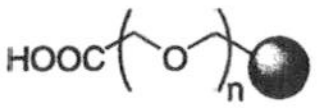

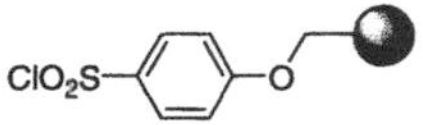

[a]The general structure of the product formed as a result of the cleavage process is shown in brackets except where cleavage regenerates the functional group that was used in the initial anchoring process.

Alcohols can be anchored to **4a** in the presence of acid to give a tetrahy-dropyran anchoring function. Following a synthesis, products can be cleaved by transacetalization or by hydrolysis using TFA–H_2O [147,148]. For example, 2-pyrrolidinemethanols have been synthesized [149] by first anchoring a carbamate derivative of 4-hydroxyproline methyl ester to resin **4a** in the presence of pyridine p-toluenesulfonate (PPTS) (Scheme 5). Subsequent reaction with a Grignard reagent followed by base-induced deprotection of the carbamate gave the resin-bound 2-pyrrolidinemethanol, which was acylated and then reduced using Red-Al. Cleavage of final products was accomplished using 3 equiv. of PPTS in butanol–dichloroethane (1:1).

Procedure [149]. Dihydropyran (DHP)-resin **4a** (6.0 g) in dichloro-ethane (70 mL) was treated with N-ethoxycarbonyl-*trans*-4-hydroxy-L-pro-line methyl ester (21 mmol) and PPTS (6.7 mmol) and the suspension heated to 80°C for 48 h, filtered, washed with CH_2Cl_2 (8 × 50 mL), and dried. The resin-bound methyl ester (1.0 g, 0.43 mmol) in tetrahydrofuran (THF) (25 mL) was treated with phenylmagnesium bromide in THF (50 equiv.) at 0°C and the mixture allowed to warm to ambient temperature and stirred for 12 h. The resin was filtered and washed with THF (5 × 30 mL), DMF–H_2O (1:1) (5 × 30 mL), DMF (5 × 30 mL), and CH_2Cl_2 (5 × 30 mL). Hydrolysis of the carbamate was effected by treatment with KOH (1.5 g) in 1:2 BuOH–1,4-dioxane (50 mL) under reflux overnight. The resin was filtered and washed with H_2O (5 × 30 mL), 1:1 DMF–H_2O (5 × 30 mL), DMF (5 × 30 mL), and CH_2Cl_2 (5 × 30 mL). The resin-bound secondary amine was resuspended in pyridine (40 mL) and treated with acetic anhydride (4.3 mmol, 10 equiv.) with stirring at ambient temperature for 3 h. The resin was filtered, washed with CH_2Cl_2 (8 × 30 mL) and toluene (5 × 30 mL) and dried under vacuum. The support was resuspended in toluene (40 mL) and treated with 65 wt% Red-Al in toluene (28 mmol) at ambient temperature overnight. After filtration, the resin was washed with toluene (5 × 30 mL) and CH_2Cl_2 (5 × 30 mL) and product was cleaved by treatment with PPTS (1.2 mmol) in 1-butanol (1-BuOH)–1,2-dichloroethane (1:1) (20 mL) at 60°C overnight. The cleavage solution was drained and evaporated and the crude product purified by chromatography, affording a 75% yield.

Orthogonally protected steroidal triols have been anchored to resin **4a** under similar conditions, subjected to further elaboration, and cleaved from the resin using HCl–ether [150]. Tetrazoles [151] and purines [152] have also been anchored via their nitrogen atoms to a related linker, functional-ized, and then cleaved using 3% HCl in MeOH.

Carboxylated-PS [153] has found most use in the form of the acid chloride derivative **4b** [154]. An example of a combinatorial synthesis on this support [155] commenced with anchoring a substituted phenol for fur-

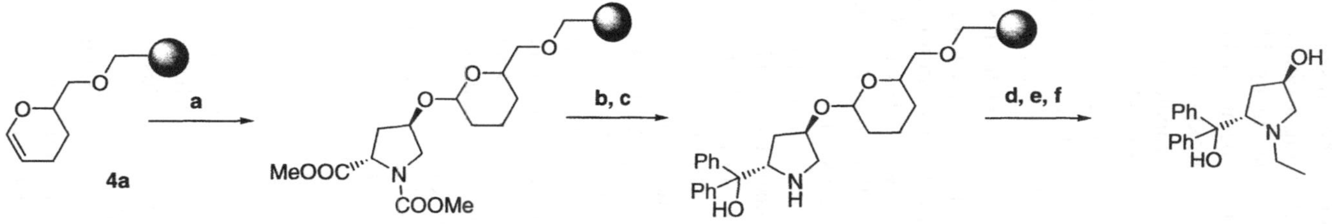

Scheme 5 (a) *N*-Ethoxycarbonyl-*trans*-4-hydroxy-L-proline methyl ester, PPTS, CH₂Cl₂; (b) PhMgBr, THF; (c) KOH; (d) Ac₂O; (e) Red-Al; (f) PPTS.

ther transformations. Products were released from the resin by nucleophilic attack at the anchoring group by methoxide ion. Alcohols have also been anchored to **4b** and ultimately cleaved by base hydrolysis [156]. Resin-bound acid chloride **4b** has been shown to form Reissert complexes with isoquinolines in the presence of trimethylsilylcyanide [157]. After cycloaddition reactions on resin, treatment with base led to cleavage of the anchoring group with elimination of cyanide ion to regenerate the isoquinoline ring. Anchoring a bromonitroaryl–allylic alcohol to carboxy-PS **4c** in the presence of N,N'-diisopropylcarbodiimide (DIPCDI) followed by reduction gave a substrate that was subjected to radical cyclization in the presence of tributyltin hydride prior to cleavage with sodium methoxide [158].

Resin-bound sulfonic acids have begun to receive attention and can be activated for anchoring nucleophiles by conversion to sulfonyl halides. Thus, support **4d** prepared from Merrifield chloromethyl resin has been used for oligosaccharide synthesis with products released by iodide ion displacement of the hydroxymethyl group [159]. Conversion of cross-linked benzenesulfonic acid ion-exchange resin (Dowex) to the sulfonyl halide gave a support for anchoring phenols. This acid-stable sulfonate anchor is traceless because palladium-induced cleavage releases products with a hydrogen atom at the former anchoring position provided that the ring contains electron-withdrawing groups [160]. The sulfonyl halide support **4d** can also be used to anchor the side chain of guanidine, giving an anchor that is stable toward base and TFA but can be cleaved using HF [161]. Reaction of **4d** with alcohols affords resin-bound sulfonates that can be used to alkylate primary and secondary amines [162,163].

V. HYDROXYLATED RESINS

A. Hydroxyalkylpolystyrenes

The hydroxyl-containing resins (Table 5), together with their amine counterparts (Section VII), are the most widely used solid supports for organic synthesis. Their availability stems from developments in peptide chemistry that led to a series of supports for anchoring and releasing carboxylic acids over a range of conditions. Carboxylic acids anchored to hydroxymethyl-PS **5a** carry the same ester function as that obtained by reaction of chloromethyl-PS **1a** with a carboxylate salt and thus the same cleavage conditions can be used. Alternative cleavage methods have been devised for use in nonpeptide synthesis including base hydrolysis [164]. Analogues of the natural product lavendustin A [165] have been synthesized beginning with the esterification of resin **5a** by reaction with Fmoc–aminobenzoyl chloride. Deprotection, reductive alkylation, and further reaction with 2-methoxyben-

Table 5 Hydroxylic Resins and Linkers

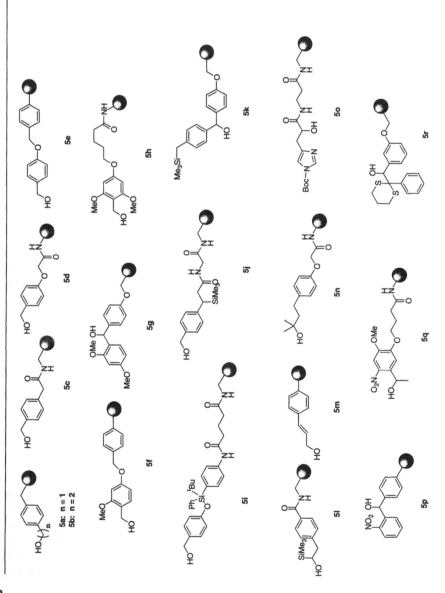

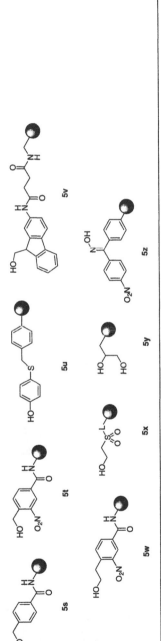

Name and identifier	Coupling conditions	Cleavage conditions[a]	References
5a Hydroxymethyl-PS	RCOOH, diimide	Me₄NOH, THF	164
	ArCOCl	BBr₃	165
	RCOCl, DIEA, CH₂Cl₂	TMSCl, MeOH [RCOOMe]	167
	RCOOtBu, toluene, 100°C	DIBAL [RCH₂OH]	168
	RCOOH, diimide	Toluene 100°C [pyrazolone]	169
5b Hydroxyethyl-PS		R'NH₂, AlMe₃ [RCONHR']	170,171
		RNH₂ [diketopiperazine]	172
5c PAM	RCOOH, CDI, CH₂Cl₂	HBr in HOAc–TFA (1:1)	173,174
		Me₂NCH₂CH₂NH₂ heat [RCONHCH₂CH₂NMe₂]	175
5d PAC	RCOOH, diimide	TFA–CH₂Cl₂	46,176
	ArOH, preformed handle	TFA–CH₂Cl₂	177–180
53 Wang	RCOOH, DIPCDI, DMAP	TFA–CH₂Cl₂–H₂O (16:3:1)	181,183–188
	RCOOH, PPh₃, DEAD	TFA–CH₂Cl₂	189
	RCOOH, pyridine, 2,6-dichlorobenzoyl chloride	TFA–CH₂Cl₂ (1:1)	190

223

Table 5 Continued

Name and identifier	Coupling conditions	Cleavage conditions[a]	References
	Succinic anhydride, pyridine vinylamide, HCl–THF	TFA–CH$_2$Cl$_2$ (1:5)	191
	Meldrum's acid, TFH, heat	TFA–CH$_2$Cl$_2$ (1:5)	192
		TFA–H$_2$O (19:1)	193
	RCOOH, diimide	HCl–toluene, heat [pyrimidinediones or hydantoins]	74,75,194
	Preformed handle	NaOtBu, THF [benzodiazepine-2,5-diones]; Bu$_4$N$^+$OH$^-$ [tetramic acids]	195
		MeOH, Et$_3$N, KCN [RCOOMe]	196
	RCOCl, Et$_3$N	NaOMe, THF–MeOH (4:1) [RCOOMe] R'NH$_2$, AlCl$_3$, CH$_2$Cl$_2$ [RCONHR']	197
		1. NMO, 2. TFA–CH$_2$Cl$_2$–H$_2$O (13:6:1) [ArOPO(OH)$_2$]	198
	RCOOH, DIPCDI, DMAP	TFA–CH$_2$Cl$_2$ (1:9) [RPO(OR')OH]	199
	Phosphoramidites, tetrazole, THF 2-chloro-4H-1,3,2-benzodioxaphosphorin-4-one	TFA–CH$_2$Cl$_2$ (1:1)	200
			201
	ArOH, DEAD, PPh$_3$	TFA–CH$_2$Cl$_2$ (1:1)	202,203
	1. CCl$_3$CN, DBU, 2. ROH, BF$_3$		204,205
5f SASRIN	Diketene, DMAP	TFA–CH$_2$Cl$_2$ (1:50) [ArCOOH]	206,209,211
		TFA–CH$_2$Cl$_2$ (1:9)	210
5g Rink	RCOOH, DCC, DMAP, NMM	TFA–CH$_2$Cl$_2$ (1:99)	207
5h HAL	RCOOH, DIPCDI, DMAP	0.1% TFA/CH$_2$Cl$_2$	208
5i Pbs	Preformed handle	TBAF, DIEA, PhSH, DMF	213
5j	Preformed handle	TBAF, DMF	214,215
5k	RCOOH, DIPCDI, DMAP	TBAF, DMF	216
5l SAC	Preformed handle	TFA–PhSMe–PhOH	217
5m Cinnamyl alcohol	RBr	O$_3$	31–34
	ArSO$_2$NHR, DEAD, PPh$_3$	Grubbs's [cycloalkene]	218

5n DHPP	RCOCl, pyridine, DMAP	TFA–H$_2$O (19:1)	219,220
5o	MSNT/MeIm	1. TFA, 2. Phosphate pH 7.5	221
5p NBH	RCOOH, DCC	$h\nu$ 365 nm, MeOH	226
5l α-Methyl-6-nitroveratryl alcohol	RCOOH, DIPCDI, pyridine, HOBt, THF	$h\nu$ 365 nm 2-PrOH	227,228 229,230
5r	RCOOH, DIPCDI, DIEA, HOBt	1. HgCl$_2$, THF–H$_2$O, 2. $h\nu$ 350 nm	231 232
5s HMBA	RCOOH, diimide	Et$_3$N, MeOH [RCOOMe] R'NH$_2$ [RCONHR']	233,235,236 15
5t nitroHMB	RCOOH, MSNT	0.1 M NaOH	234
	RCOOH, DIPCDI, DMAP	TBAF, H$_2$O	237
5u Phenol-sulfide linker	RCOOH, DIPDCI, DMAP	DIEA, MeOH [RCOOMe] R'NH$_2$ [RCONHR']	
		R'NH$_2$ [RCONHR'] [lactam]	238,239 240
5v HMFS MBHA base resin	RCOOH, DCC, DMAP	Morpholine–DMF (1:4)	243,244
52 NPE MBHA base resin	Preformed handle	0.1 M DBU, dioxane	247
5x Hydroxyethylsulfones			
L = CH$_2$	RCOOH, DCC	Dioxane–MeOH–4 M NaOH (30:9:1)	248,249
L = p-C$_6$H$_4$CO	RCOOH, DCC	0.1 M NaOH	250
L = —Ch$_2$CONH—	Preformed handle	0.1 M NaOH	251
5y Diol	RCHO, TsOH, Na$_2$SO$_4$, toluene, heat	3 M HCl–dioxane (1:1)	252–254
5z Oxime	RCOOH, DCC	R'NH$_2$ [RCONHR']	255,256
	RNCO	R$_2$NH, toluene, heat [ureas]	257
	RCOOH, DIPCDI	NH$_3$, EtOH [RCONH$_2$]	258
	RCOOH, DIPCDI	TBSONH$_2$, DCE, 90°C [RCONHOTBS]	259

[a]The general structure of the product formed as a result of the cleavage process is shown in brackets except where cleavage regenerates the functional group that was used in the initial anchoring process.

zyl bromide (Scheme 6) were followed by cleavage of the product and removal of methyl ether protecting groups in a single step using BBr₃.

Procedure [165].   Hydroxymethyl-PS resin **5a** (250 mg, 0.26 mmol) was washed with *N,N*-dimethylacetamide (DMA) (2 × 2 mL) and THF (2 × 2 mL). Pyridine (1 mL) was added, followed by a solution of 5-[(flu-orenylethoxycarbonyl)-amino]-2-methoxybenzoyl chloride (0.82 mmol, freshly prepared from the carboxylic acid) in THF (3.2 mL). After shaking for 45 min, the reaction mixture was treated with 4-dimethylaminopyridine (DMAP, 0.16 mmol) and the mixture was shaken for a further 2 h. The resin was washed with DMA (6 × 1 mL) and CH_2Cl_2 (4 × 1 mL). The substitution level was found to be 0.73 mmol/g by deprotection of the Fmoc group using piperidine–DMF (1:4) and quantification of the piperidine–dibenzfulvene adduct by ultraviolet (UV) spectroscopy [166]. The bulk of the resin was then deprotected by treatment with piperidine–DMA (1:4) (2 × 2 mL for 7 min) and washed with DMA (5 × 2 mL) followed by HOAc–DMA (1:99) (5 × 2 mL). The resin-bound amine was treated with 2-methoxy-benzaldehyde (3.2 equiv.) followed by HOAc–DMA (1:99) (2 mL) and the mixture shaken for 5 min. NaBH₃CN (18 equiv.) was then added over a 2-h period and the mixture was shaken overnight. The resin was filtered and washed with HOAc–DMA (1:99) (5 × 2 mL) followed by MeOH (4 × 1 mL). The resin-bound secondary amine was washed with DMSO (4 × 2 mL), and a solution of 2-methoxybenzyl bromide (5.2 equiv.) in DMSO (2.5 mL) was added. The mixture was shaken for 30 min, DUB (1.5 equiv.) was added, and shaking was continued for a further 2.5 h. The resin was washed with DMSO (5 × 1 mL) and CH_2Cl_2 (6 × 2 mL) and dried over NaOH. The final product was cleaved by cooling the resin to −78°C under nitrogen and adding a 1.0 M solution of BBr₃ in CH_2Cl_2 (10 mL). After agitating at −78°C for 30 min and ambient temperature for 3.5 h, the reaction mixture was recooled to −78°C and quenched by addition of MeOH (5 mL). After 15 min at ambient temperature, the mixture was filtered and the filtrate evaporated under high vacuum. The lavendustin A analogue was purified by HPLC on a C_8 column, eluting with an H_2O–acetonitrile (MeCN) gradient.

Ester linkages to hydroxymethyl-PS (**5a**) have also been cleaved by transesterification [167] in the presence of trimethylsilyl chloride (TMSCl) or by reduction [168] to give alcohols. Heterocycles have been synthesized using support **5a** with a cyclization step in toluene serving to cleave the product from the resin, leaving acyclic impurities bound to the support [169]. Esters derived from hydroxyethyl-PS **5b** do not afford benzylic cations when treated with acid and are thus not cleaved as easily. Nonetheless, resin **5b** has been applied to syntheses in which the final cleavage of the ester anchor is effected by an intermolecular reaction with an amine in the presence of

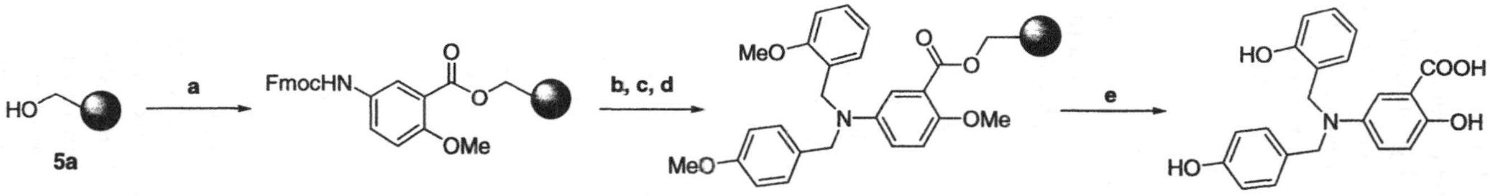

Scheme 6 (a) ArCOCl, pyridine, THF; (b) piperidine; (c) ArCHO, NaBH$_3$CN; (d) ArCH$_2$Br; (e) BBr$_3$, CH$_2$Cl$_2$.

a Lewis acid to afford amides [170,171] or by an intramolecular attack at the ester link by an amine to release heterocycles into solution [172].

The support 5c obtained by attaching the 4-(hydroxymethyl)phenylacetic acid linker to aminomethyl-PS is commonly known as *p*-hydroxymethylphenyl acetic acid (PAM) resin [173,174]. The electron-withdrawing effect of the carboxamidomethyl group increases the stability of derived esters toward TFA–CH_2Cl_2 (1:1) 100-fold compared with the corresponding anchors to resins 1a or 5a. As a result, losses of peptide during acid-induced deprotection steps are insignificant when using resin 5c. Peptides can be cleaved from support 5c using liquid HF at 0°C, which also leads to the removal of most side-chain protecting groups. Aminolysis by treatment of ester linkages to this resin with dimethylaminopropylamine has been used to cleave pyrrole-containing polyamides from an amino acid spacer [175].

B. Alkoxybenzyl Alcohol Supports

Linkers and resins derived from alkoxy-substitutde benzyl alcohols (Table 5) were developed initially to enable peptides prepared by the Fmoc/*t*butyl (*t*Bu) method to be cleaved without the use of HF, which is required for Merrifield resin 1a and PAM resin 5c. Their design is based upon the principle that cleavages can be effected under conditions of lower acid strength when the transient resin-bound carbocation is stabilized by the presence of electron-donating groups. Cleavage of the ester anchoring groups to these supports can also be carried out by nucleophilic attack by hydroxide, alkoxide, or amine nucleophiles. Solid supports 5d based on the peptide acid linker derived from phenoxyacetic acid (PAC) have been widely used for anchoring peptide acids with products cleaved with TFA [46,176]. Support 5d has also been applied to the anchoring of phenols using the preformed handle approach, as in the synthesis of phenolic benzodiazepines [177], quinazoline-2,4-diones [178], indoles [179], and benzimidazoles [180]. After the synthesis, treatment with TFA causes cleavage of the benzylic ether anchor, releasing the phenol from the support.

Wang resin, 5e, is one of the most widely used supports for anchoring carboxylic acids for further transformations [181,182]. Because resin 5e is structurally related to solid support 5d based on the PAC linker (Table 5), carboxylic acids can be anchored and cleaved under comparable conditions, although esters to the Wang resin are slightly more labile under acidic conditions. Thus, couplings mediated by diimide [183–188], couplings under Mitsunobu conditions [189], and activation in the presence of 2,6-dichlorobenzoyl chloride [190] have been reported. Reaction of 5e with succinic anhydride gave a resin-bound carboxylic acid that was shown to undergo the Ugi four-component condensation (4CC) reaction before cleavage with

TFA–CH$_2$Cl$_2$ [191]. Ugi 4CC products prepared in solution from the "interconvertible isonitrile" may be captured on a solid support by nucleophilic attack of the benzylic alcohol of resin **5e** on the cyclic dipole that is formed in the presence of acid; carboxylic acids can then be released from the resin by treatment with TFA [192].

A malonic acid derivative of Wang resin has been prepared by reaction with Meldrum's acid and the product converted to a malonamide via the acid chloride. Condensation with an aldehyde under Knoevenagel conditions gave methylene malonic acid-amides that were cleaved from the resin with TFA–H$_2$O [193] (Scheme 7).

Procedure [193]. Wang resin was converted to the resin-bound malonic acid by reaction with Meldrum's acid in refluxing THF. Conversion to the malonamic ester was achieved by sequential treatment with oxalyl chloride and benzylamine. Resin-bound *N*-benzylmalonamide (0.33 g, 0.276 mmol) was suspended in toluene (3.5 mL) and treated with piperidine (4 μL, 0.28 mmol), acetic acid (4 μL), and benzaldehyde (84 μL, 0.83 mmol). The mixture was heated at 85°C for 24 h with stirring. After filtration, the resin was washed successively with toluene, methanol, CH$_2$Cl$_2$, and ether (three aliquots of each). Product was cleaved by treatment with TFA–H$_2$O (19:1) for 1 h followed by washing with TFA–CH$_2$Cl$_2$ (1:1). The combined filtrates were evaporated under reduced pressure to afford 43 mg (89%) of product at >90% purity as estimated by ^{1}H NMR analysis.

A cyclization–cleavage process has been used [61] to prepare hydantoins using support **5e**. Reaction of an isocyanate with a resin-bound amino acid was followed by acid-induced cyclization and cleavage. 5,6-Dihydropyrimidine-2,4-diones were also obtained by an acid-induced cyclization–cleavage of appropriate precursors bound to **5e** [194]. Examples of base-induced cyclization–cleavages of functionalized amino acids anchored to Wang resin include the synthesis of 1,4-benzodiazepine-2,5-diones [195] and tetramic acids [196]. Base-induced cleavage affording methyl esters has also been carried out [197] by treatment with methanol in the presence of triethylamine and cyanide ion. The synthesis of tetrahydro-1,4-benzodiazepine-2-ones on support **5e** has been reported and involves a transesterification–cleavage process [198]. As noted earlier for the hydroxymethyl-PS resins, cleavage of an ester anchoring group to the Wang resin can also be carried out by reaction with amines [199]. Wang resin can be used to anchor phosphates and phosphonates. Thus, reaction with phophoramidites provided, following oxidation, a phosphate anchor that could be cleaved using TFA–CH$_2$Cl$_2$–H$_2$O [200]. Phosphonates have also been generated on resin and cleaved under similar conditions [201].

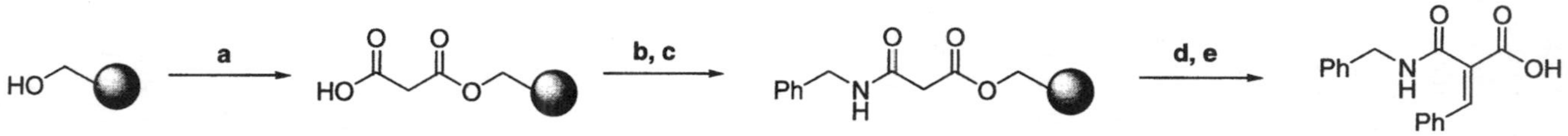

Scheme 7   (a) Meldrum's acid, THF; (b) $(COCl)_2$; (c) $PhCH_2NH_2$; (d) PhCHO, piperidine; (e) TFA.

The Wang resin can also be used for anchoring of alcohols and phenols as ethers under Mitsunobu conditions [202,203]; the ethers, unlike ethers derived from the Merrifield resin, can readily be cleaved using TFA–CH$_2$Cl$_2$ (1:1). Treatment with TFA–THF (1:2) enabled selective cleavage of an enol ether with the phenol remaining anchored to the support. Alcohols and phenols can also be anchored as ethers by reaction with the trichloroacetimidate derivatives of Wang **5e** and a related PEG resin [204,205] with products cleaved under similar conditions.

The greater combined electron donor strength of the aryl substituents in the SASRIN (super-acid-sensitive resin) **5f** [206] and Rink acid [207] resin (**5g**), and the hyper-acid-sensitive resin HAL (**5h**) [208] compared with Wang and PAC supports enables carboxylic acids to be cleaved using lower acid concentrations (typically 1–10% TFA in CH$_2$Cl$_2$) than used for the Wang (**5e**) and PAC (**5d**) supports [209,210]. Acetoacetic acid has been anchored to SASRIN **5f** by reaction of the resin with diketene and used to synthesize substituted pyridines, which were cleaved by treatment with TFA–CH$_2$Cl$_2$ (1:33) [211]. The same products and related heterocycles [212] synthesized on Wang resin required treatment with 95% TFA to effect cleavage. Cleavage of peptides from Rink ester resin can be effected using HOAc–CH$_2$Cl$_2$ (1:9), leaving Boc or tBu side-chain protecting groups intact [207].

The silyl ether–derived N-(3 or 4)[[4-(hydroxymethyl)phenoxy]-t-butylphenylsilyl]phenyl pentanedioic acid monoamide (Pbs) linker was developed to provide base-stable preformed handles for amino acids that can be attached to amine resins to give **5i** and used to assemble peptides. Treatment with 1 equivalent of tetrabutylammonium fluoride (TBAF) in DMF [213] results in a 1,6 elimination with release of the desired peptide and a quinone methide into solution. Side reactions of the quinone methide by-product were prevented by trapping it with thiophenol. Anchors to the Pbs resin **5i** have been shown to be compatible with Fmoc/tBu and with Dts (dithiasuccinoyl, thiol-cleavable protecting groups) methods but incompatible with the acidic conditions required for the Boc/Bzl strategy. Other hydroxylic supports with silyl groups appropriately positioned for quinonemethide formation such as **5j** [214,214] (Table 5) have been reported to which carboxylic acids could be anchored by a preformed handle approach. Product cleavage from these supports using fluoride ion results in the quinonemethide remaining anchored to the support. Resin **5j** has been used for the synthesis and cleavage of a fully protected peptide using the Fmoc/tBu strategy [214,215]. Support **5k** is readily synthesized from 4-formyl-PS and carboxylic acids can be anchored in the presence of DIPCDI without the need to prepare a preformed handle [216]. Ester anchoring groups proved to be relatively acid stable but could be cleaved by treatment with fluoride ion in

DMF. The silyl acid support (SAC) **5l** [217] has been used for peptide synthesis with cleavage being effected with either TFA or fluoride ion. It was shown that when using support **5l** diketopiperazine formation and tryptophan alkylation are suppressed compared with construction of the same sequence of PAC resin.

The cinnamyl alcohol resin **5m** was first used [31–34] for anchoring glycosyl bromide, which was then converted to a disaccharide that was cleaved by ozonolysis. This support was used to construct a resin-bound diene commencing with Mitsunobu anchoring of a sulfonamide, which underwent ring closure–cleavage by olefin metathesis on treatment with Grubb's catalyst [218]. Although this reaction was conducted on a hydroxyl resin, this cleavage method may find more general use.

C. Tertiary Alcohol Linkers

Peptides synthesized on supports derived from benzylic alcohols are prone to premature cyclization–cleavage by diketopiperazine formation when the peptide contains C-terminal proline. To encumber the intramolecular attack of the amine terminus at the anchoring ester linkage, the sterically hindered tertiary alcohol 4-(1',1'-dimethyl-1'-hydroxypropyl)phenoxyacetic acid (DHPP) (**5n**) was developed to which proline can be anchored via the Fmoc acid chloride [219]. An application of the DHPP linker involved the synthesis of enalapril and its diastereomer, which were cleaved from the resin using TFA (Scheme 8). On-resin base hydrolysis was also carried out prior to cleavage to give enalaprilat (the active metabolite of enalapril) together with a diastereoisomer [220].

Procedure [220]. High-load PEG-PS (2.5 g, 0.6 mmol/g) was suspended in a solution of 4-(1',1'-dimethyl-1'-hydroxypropyl)phenoxyacetic acid (1.43 g, 6 mmol) in DMF (12 mL) and treated with DIPCDI (0.75 g, 6 mmol) and HOBt (0.82 g, 6 mmol). After 16 h at ambient temperature, a ninhydrin test of an aliquot of reaction beads indicated complete acylation of the resin amino groups. The DHPP-derivatized resin (**5n**) was washed successively with DMF, MeOH, and CH_2Cl_2. Resin **5n** (1.6 g, 1 mmol) was swelled in CH_2Cl_2–pyridine (4:1) (20 mL) and treated with Fmoc-Pro-Cl (1.8 g, 5 mmol) for 20 h at ambient temperature. The resin was washed successively with CH_2Cl_2, MeOH, and CH_2Cl_2 and then deprotected by treatment with piperidine–DMF (1:4), (2 × 7 min). The loading was 0.5 mmol/g, determined by UV spectroscopic determination of the released piperidine–dibenzfulvene adduct [166]. Fmoc-Ala-OH (0.62 g, 2 mmol) was dissolved in DMF (2 mL) and treated with tetrafluoroformamidinium hexafluorophosphate (TFFH, 0.53 g, 2 mmol) and DIEA (0.7 mL, 4 mmol); after allowing the acid fluoride to form for 10 min, the mixture was added

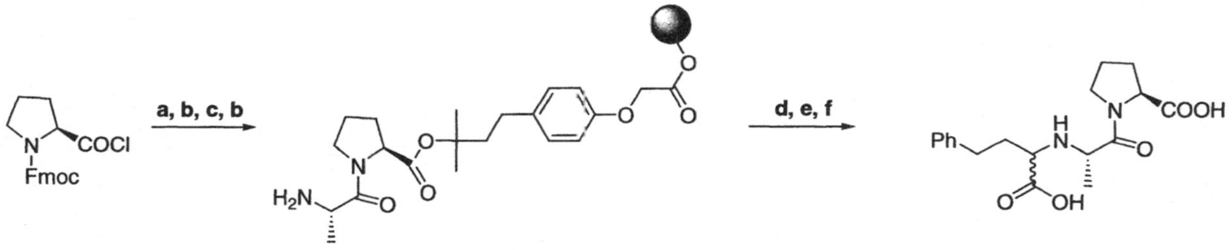

Scheme 8 (a) Resin **5n**, pyridine; (b) piperidine, DMF; (c) Fmoc-Ala-OH, TFFH, DIEA; (d) ethyl-2-oxo-4-phenylbutyrate, NaBH₃CN; (e) NaOH; (f) TFA–H₂O.

to the resin-bound proline (1.0 g, 0.5 mmol) and allowed to couple for 30 min. The resin was washed and deprotected and the loading (0.45 mmol/g) determined as described previously. The Ala-coupled resin (325 mg, 0.15 mmol) was swelled in HOAc–DMF (1:99) (3.5 mL) and treated with ethyl-2-oxo-4-phenylbutyrate (0.82 g, 3 mmol, 20 equiv.). After 5 min, NaBH$_3$CN (378 mg, 6 mmol, 40 equiv.) was added and the mixture shaken at ambient temperature for 16 h. The resin was filtered and washed successively with DMF, MeOH, and CH$_2$Cl$_2$. The ethyl ester group of resin-bound enalapril was hydrolyzed by treatment of the resin with 1 N NaOH for 20 h. After washing the support successively with H$_2$O, MeOH, and CH$_2$Cl$_2$, the product was cleaved by treatment with TFA–H$_2$O (95:5) and characterized by electrospray ionization–mass spectrometry (ESI-MS).

A safety-catch hydroxyl linker **5o** derived from imidazole has been reported. Peptides have been constructed from the hydroxyl group using Fmoc amino acids as the building blocks and piperidine for deprotections. At the completion of the synthesis, treatment of the support with TFA leads to removal of the Boc group but the peptide remains anchored to the resin when the imidazole ring is pronated. Subsequent treatment with phosphate buffer converts the imidazole to the free base, which catalyzes peptide cleavage by intramolecular attack at the ester anchoring group [221].

D. Photocleavable Hydroxyl Linkers

Photochemical cleavage mechanisms are attractive because reaction conditions are usually orthogonal to the protecting groups used in the preceding synthetic steps [222] and because compounds can be released from the support at neutral pH for direct biological screening [223,224] uncontaminated by any reagents or to release coding tags while leaving biological test molecules bound to the support [225]. The 2'-nitrobenzhydryl (NBH) resin **5p** has been used for model peptide syntheses using Boc amino acids and the products cleaved photochemically. The anchoring ester group has been shown to be less acid sensitive than anchors to the related benzyl halide supports discussed earlier [226].

To address the problems of slow cleavage kinetics and methionine oxidation associated with the use of photolabile support **1j** (Table 1), a linker based α-methyl-6-nitroveratryl alcohol [227,228] was designed. The additional ring alkoxy groups and the α-methyl group of support **5q** facilitate cleavage and lead to fewer side products that result from the photoproduct, a resin-bound nitroso ketone, compared with **1j** and other conventional photolabile linkers. Support **5q** has been used for the synthesis of sultams [229], which were cleaved from the resin by irradiation in 2-propanol at 365 nm.

An application of a related resin involves the synthesis of oligonucleotides via photolabile carbonate and carbamate anchors [230]. Photocleavable alkoxylbenzyl safety-catch solid supports **5r** (Table 5) have been prepared [231,232]. Carboxylic acids were esterified by reaction with the secondary alcohol function of **5r** and alcohols and amines could be attached via carbonate and carbamate derivatives, respectively. Unlike many previously reported photolabile linkers, these anchors were stable to irradiation with sunlight. In each case, product release was effected by treatment with mercuric chloride to remove the safety catch followed by irradiation with light of wavelength 350 nm, which leaves a benzofuran by-product tethered to the resin.

E. Base-Labile Supports

The hydroxymethylbenzoic acid linker (HMB) was designed for the construction of peptides from amino acids with acid-labile side-chain protecting groups. As a consequence of the carbonyl group in the *para* position, the ester anchor is activated to nucleophilic attack but is stable toward treatment with acid [15,233,234]. Thus, the HMB linker attached to Tentagel (resin **5s**, Table 5), could be used in the acid-promoted solid-phase Fischer synthesis of indoles [235] and spiroindoles [236]. Treatment with triethylamine in MeOH released methyl esters from the anchoring site. Peptides have been synthesized using a support derived from the HMB linker and the PEGA base resin with cleavage being effected using aqueous base [234]. The α-(4-hydroxymethyl-3-nitrobenzamido)benzyl-PS resin **5t** [237] has been used to anchor Fmoc-amino acids in the presence of DIPCDI and DMAP. The anchor was stable to brief treatments with piperidine to remove the protecting groups and thus short peptides could be constructed. High-yielding cleavages were accomplished using aqueous TBAF, DIEA or KCN in MeOH, or primary amines to give free acids, esters, or amides, respectively.

The 4-hydroxythiophenol solid support or phenol-sulfide linker **5u** [238] can be used for anchoring carboxylic acids in the presence of carbodiimides to give an acid-stable ester linkage that can be activated toward nucleophilic attack by amines by oxidation to the sulfone. However, although the acid stability has been confirmed, it has been reported that prior oxidation of the anchor is unnecessary for release of amides by reaction with amines by either an intermolecular [239] or intramolecular pathway [240].

Supports based on hydroxymethylfluorene have been used for the anchoring of Boc-amino acids for peptide synthesis with final cleavage by a base (typically piperidine)–induced β-elimination process [241,242]. Side reactions were encountered, however, involving premature cleavage during peptide elongation due to intramolecular attack by the amine terminus.

The N[9-hydroxymethyl-2-fluorenyl]succinamic acid (HMFS) resin **5v** [243,244] attached via a different spacer to the base resin was therefore developed and leads to improved performance in peptide syntheses with Boc-amino acids (using TFA-induced N^α-deprotection). The final sequence was cleaved most effectively by use of morpholine–DMF (1:4). Other resins that cleave by β-elimination are the nitrophenylethanols **5w** but these supports have thus far been applied mainly for oligonucleotide syntheses [245,246]. Peptides have also been assembled on resin **5u** using Boc protecting groups and shown to be stable to DIEA–CH_2Cl_2 (1:9) but cleaved by treatment with DBU [247].

The hydroxyethylsulfones **5x** derived from Merrifield's resin or from polydimethylacrylamide with an interposed linker have been used to synthesize several model peptides by use of Boc-protected amino acids. The anchor was stable toward acid used in deprotections but, following the assembly, could be cleaved by a β-elimination induced by hydroxide ion. The base-induced cleavage of the same sequence was shown to be faster from **5x** than from support **1a**, resulting in fewer transesterification side reactions [248,249]. A related sulfone has also been applied to oligonucleotide syntheses [250]. The corresponding support **5x**, derived from aminomethyl-PS, has been used in peptide synthesis and is compatible with both the 5 M HCl–dioxane used to remove Boc protecting groups and piperidine used in Fmoc deprotections but can be cleaved using hydroxide ion [251].

F.   Diols

Reaction of chloromethyl-PS with the alkoxide of Solketal followed by hydrolysis of the acetonide affords a diol linker **5y** [252,253] to which aldehydes can be anchored in the presence of acid and a dehydrating agent. An application of this resin involved anchored bromobenzoic acids that were subjected to Suzuki couplings and products cleaved by heating in 3 M HCl–dioxane (1:1) [254].

G.   Oxime Resins

The oxime resin **5z** [255,256], which forms an acid-stable anchor for the carboxyl group of amino acids, has been used for some time for the synthesis of C-terminal modified peptides and homodetic cyclic peptides. The key step in lactam formation is intramolecular nucleophilic attack of the N-terminus at the anchoring group. The resin also reacts with isocyanates and the products undergo nucleophilic attack by amines resulting in the release of ureas [257]. Resin-bound Boc-tryptophan has been used in the acid-catalyzed Pictet–Spengler synthesis with final products cleaved as primary amides by treatment with ammonia [258]. Cleavage of anchors by nucleophilic displacement using O-t-butyldimethylsilylhydroxylamine [259] followed by removal of the t-butyldimethylsilyl (TBS) protecting group in solution with TFA has been used to synthesize hydroxamic acids.

VI. THIOLS

The thiol support **6a** (Table 6) [260,261] has been prepared from an amine-functionalized polyacrylamide base resin and used to anchor thiols, such as cystamine, oxidatively through a disulfide bond. Peptides could then be assembled from the amino group and the resulting cysteamide peptides cleaved by thiolysis using dithiothreitol (DTT) or by reduction using the phosphine tris(2-carboxyethyl)phosphine (TCEP). The anchor was shown to be stable to piperidine used in Fmoc deprotections and to TFA used for side-chain deprotections. A related support has also been used for the construction of β-turn mimetics involving anchoring a thiol by disulfide exchange and a final reductive cleavage process followed by cyclization in solution [262].

Thiomethyl-PS **6b**, prepared from Merrifield's resin **1a** by reaction with thioacetate followed by reduction, can be acylated to give thioesters [263]. The resin-bound thioesters have been converted to silyl enol ethers, which were shown to form aldol products that could be released from the resin by three methods [264]. Thus, reduction with lithium borohydride or diisobutylaluminum hydride (DIBAL) gave diols and aldehydes, respectively; alternatively, base hydrolysis afforded carboxylic acids. Resin **6b** thereby extends the range of functional groups available compared with cleavage of related molecules from an ester anchor.

Tentagel-thiol resin **6c** was shown to displace the halogen of a 2-chloropyrimidine to give a safety-catch thioether that was stable to subsequent synthetic transformations [265]. Activation of the safety catch was accomplished by peracid oxidation to a sulfone, which underwent nucleophilic cleavage when treated with amines. This synthesis is analogous to the reaction of resin **1a** with thiourea to give a resin-bound thiouronium salt [266], which was used to synthesize substituted pyrimidines according to Scheme 9. The cleavage process again involved peracid oxidation to give a sulfone followed by nucleophilic displacement of the anchor by reaction with an amine. The resin-bound thiouronium salt can also be used to synthesize substituted guanidines. The two nitrogen atoms are protected as Boc derivatives and, after alkylation, guanidines are released via nucleophilic cleavage by amines [267].

Procedure [266].   A mixture of high-load Merrifield resin (50 g, 0.17 mol) and thiourea (64.7 g, 0.85 mol) in dioxane−ethanol (EtOH, 4:1) (350

Table 6 Thiol Functionalized Supports

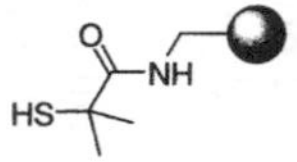

6a 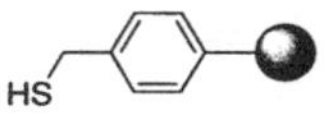   6b 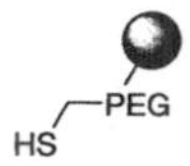   6c 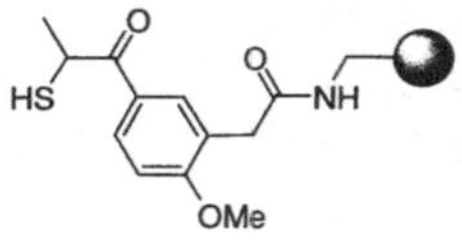   6d

Name and identifier	Coupling conditions	Cleavage conditions [product][a]	References
6a Disulfide linker			
Polyacrylamide base resin	$S_2(CH_2CH_2NH_2)_2$, H_2O_2	DTT or TCEP	260,261
Aminomethyl-PS base resin	$BtSS(CH_2)_nOMs$	TCEP	262
6b Thiomethyl-PS	$RCOCl$, Et_3N	$LiBH_4$, Et_2O [RCH_2OH]	263
		DIBAL, CH_2Cl_2 [RCHO]	264
		1 N NaOH–dioxane (1:4)	
6c Tentagel thil	2-Chloropyrimidine	1. RCO_3H, 2. RNH_2 [2-substituted aminopyrimidine]	265
6d Photolabile support, Tentagel base resin	$ArCH_2Br$, DIEA, DMF	hv 350 nm MeCN	268,269

[a]The general structure of the product formed as a result of the cleavage process is shown in brackets except where cleavage regenerates the functional group that was used in the initial anchoring process.

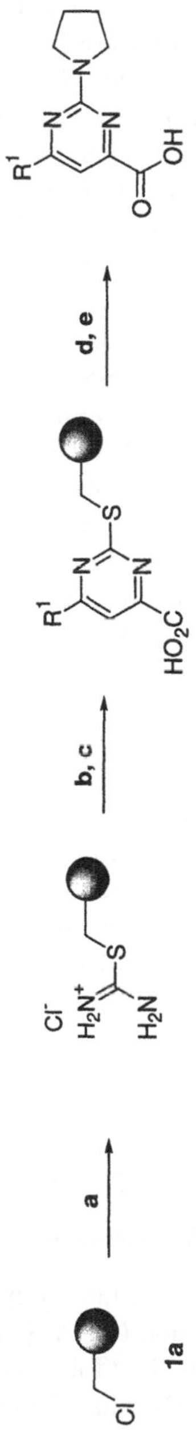

Scheme 9 (a) Thiourea, dioxane, EtOH; (b) $R^1COC \equiv CCOO^tBu$; (c) TFA; (d) mCPBA; (e) pyrrolidine.

mL) was shaken at 85°C for 15 h. The resin was filtered and washed with ethanol at 70°C (4 × 150 mL) and with dioxane (2 × 150 mL) and pentane (2 × 150 mL) at ambient temperature to give the thiouronium salt, which was dried at 60°C under high vacuum. The resin-bound thiouronium salt (4.0 g, 10.56 mmol) was swollen in DMF (20 mL), drained, and resuspended in a solution of the appropriate acetylenic ketone (1.2 equiv.) in DMF (30 mL). DIEA (1.5 equiv.) in dioxane (10 mL) was then added dropwise to the reaction mixture via a syringe pump over 24 h. The reaction mixture was vortexed for 24 h and washed successively with DMF (3 × 50 mL), 2-propanol (i-PrOH) (3 × 50 mL), dioxane (3 × 50 mL), i-PrOH (3 × 50 mL), CH_2Cl_2 (3 × 50 mL), and pentane (3 × 50 mL) and dried under vacuum. The polymer-bound t-butyl ester was swollen in CH_2Cl_2, drained, and treated with TFA–CH_2Cl_2 (1:1) (30 mL) for 15 h. The resin was filtered and washed successively with CH_2Cl_2 (4 × 50 mL), CH_2Cl_2–Et_3N (4:1) (4 × 50 mL), DMF (2 × 50 mL), i-PrOH (2 × 50 mL), and dioxane–2 N HCl to pH 2 followed by DMF (2 × 50 mL), i-PrOH (2 × 50 mL), and pentane (2 × 50 mL). The resin-bound pyrimidinecarboxylic acid was swollen in CH_2Cl_2, drained, and treated with CH_2Cl_2 (3 mL/mmol) followed by 3-chloroperoxybenzoic acid (3 equiv.). The mixture was vortexed at ambient temperature for 15 h and washed successively with the following solvents using 10 mL per mmol resin: CH_2Cl_2 (3×), i-PrOH (3×), pentane (3×), CH_2Cl_2 (3×), and dioxane (2×). Anhydrous dioxane (3 mL/mmol resin) was added followed by pyrrolidine (1.5 equiv.). The mixture was vortexed for 6 h at ambient temperature, filtered, and the filtrate evaporated to give the corresponding 2-(pyrrolodin-2-yl) pyrimidine-3-carboxamide.

A photolabile thiol support **6d** has been reported that can be used for anchoring reactive halides such as benzylic bromides [268,269]. After synthetic steps, photolysis at 350 nm in MeCN releases products with a hydrogen atom at the former point of attachment to the linker (which is defined as a traceless linker).

VII. LINKERS WITH AMINO GROUPS

A. 4-Methylbenzhydrylamine Resins

The first supports bearing primary amino groups including the aminomethylpolystyrene [270], benzhydrylamine (BHA) [271], and p-methylbenzhydrylamine (MBHA) [272,273] resins **7a** (Table 7) were developed initially for the synthesis of peptide amides. Because the carbocations derived from resins **7a** are not especially stabilized, cleavage requires the use of a strong acid such as TFMSA or treatment with HF and usually requires an amide leaving group [274,275]. However, an application has been reported in which amines were cleaved directly from resin using HF–anisole [276,277].

B. Polyalkoxyaminobenzyl and Alkoxydiphenylamino Resins

The polyalkoxyaminobenzyl and alkoxydiphenylamino resins (Table 7) have been used for some time to prepare peptide amides and can be cleaved under milder acidic conditions than those used for the BHA and related resins. The most widely used linkers for peptide amide synthesis are 5-(4-aminomethyl-3,5-dimethoxyphenoxy)valeric acid, known as the peptide amide linker (PAL) **7b** [278] and the amine analogue of the Rink alcohol support known as the Rink amide support (RAM) **7c** [207] together with the related Knorr linker **7d** [279]. The linkers were designed on the same principles as noted earlier for the corresponding hydroxyl resins. Cleavage rates of Fmoc-Val-OH from a series of related peptide amide supports using TFA–phenol (19: 1) showed that the electron-rich aromatic rings of the Knorr and PAL supports result in optimal cleavage rates, with the latter being slightly faster [279]. The synthesis of amino acid coenzyme chimeras by DIPCDI/HOBt-mediated couplings using the PAL support with products cleaved from the resin with TFA–thioanisole–1,2-ethanedithiol–anisole (90:5:3:2) has been reported [280,281]. Solid supports derived from the PAL linker such as **7b** have been used for nonpeptide applications such as a synthesis of dihydro-pyridines [282] in which the heterocyclic nitrogen originates from the resin anchoring site. The initial condensation reaction with methylacetoacetate (Scheme 10) was reported to be faster on the PAL support **7b** than on the more sterically hindered RAM resin **7c**. The resulting tethered enamino ester was condensed with a benzylidene acetoacetic ester and then subjected to acid-induced cleavage to give the dihydropyridine. It was suggested that cyclization occurred in solution after cleavage, in which case the enamino ester acts as the leaving group in the cleavage step. Cleavage of final products required higher TFA concentrations, TFA–THF (19:1) for the PAL support **7b** compared with TFA–CH$_2$Cl$_2$ (1:32) for RAM **7c**. This observation contrasts that noted before and presumably reflects decreased lability arising from the presence in the Knorr linker of the carbonyl group that is absent from the RAM support.

The properties of the p-benzyloxybenzylamine (BOBA) resin **7e** have been reported. Support **7e** forms imines that undergo aldol reactions with silyl enol ethers and the resulting γ-ketoamines can be cleaved oxidatively using 2,3-dichloro-5,6-dicyano-1,4-benzoquinone (DDQ). In contrast, treatment of the support with triflic acid (TfOH) or trimethylsilyltriflate (TMSOTf) resulted in cleavage at the benzylic ether group of **7e** affording

Table 7 Amine-Functionalized Resins and Linkers

7a

7b

7c

7d

7e

7f

7g

7h

7i

7j

7k

7l

7m

7n

7o

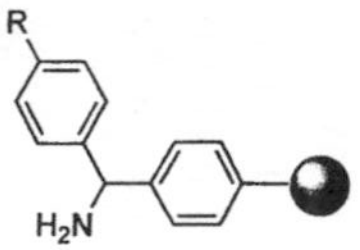

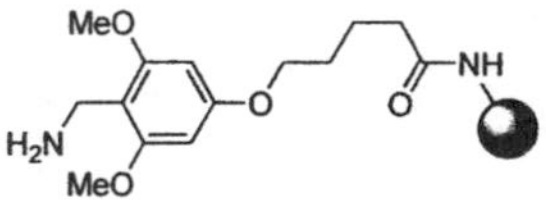

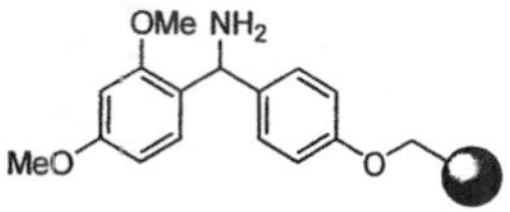

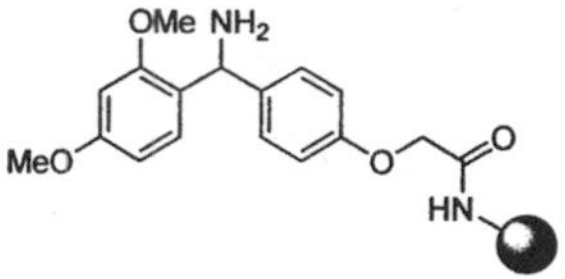

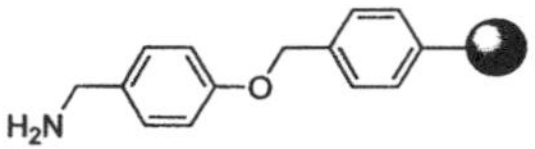

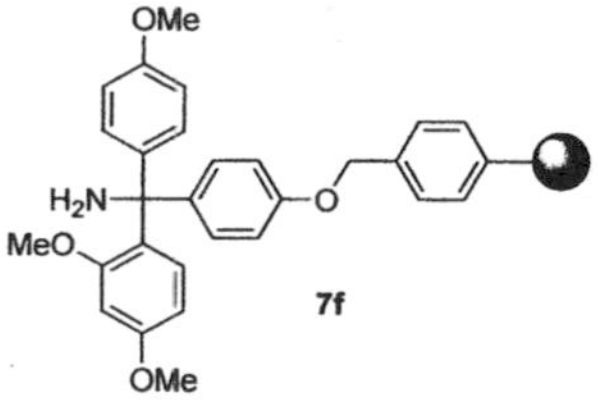

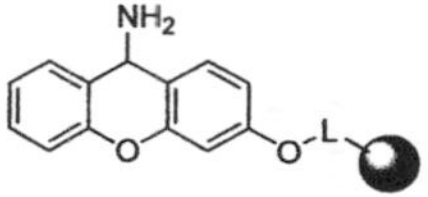

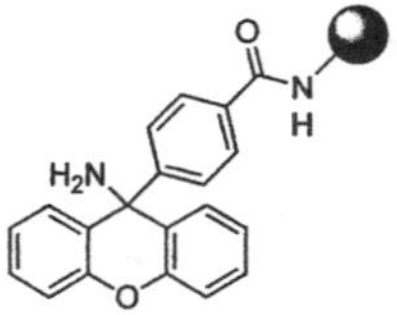

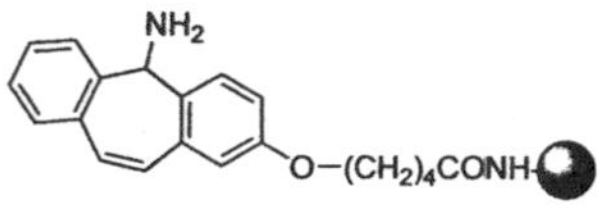

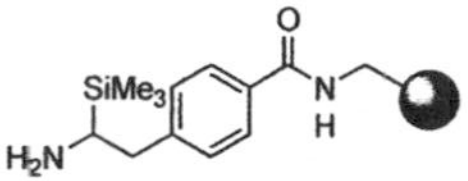

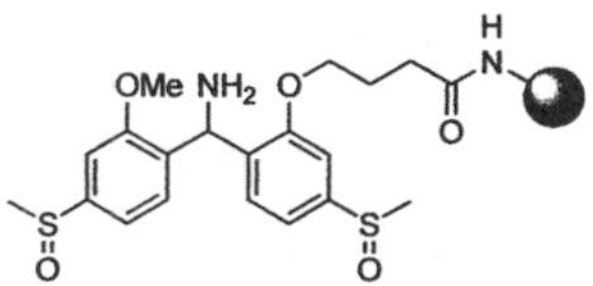

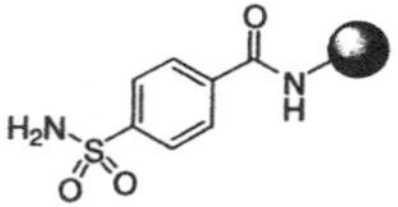

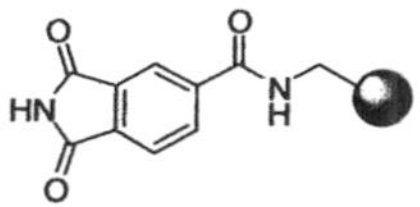

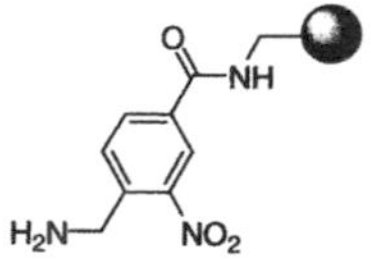

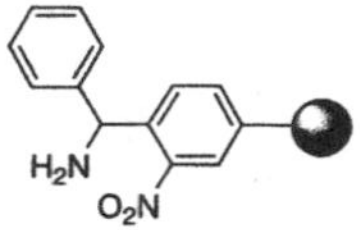

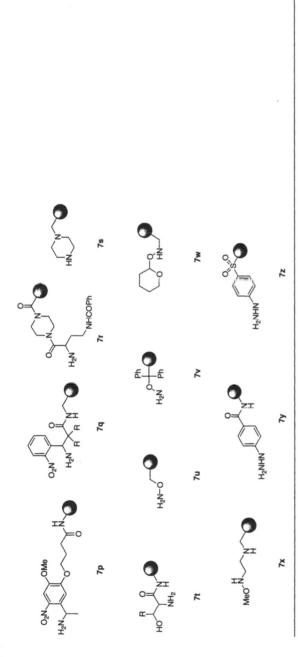

Name and identifier	Coupling conditions	Cleavage conditions[a]	References
7a BHA (R = H)	RCOOH, DCC	HF, TFMSA [RCONH₂]	270–273
MBHA (R = Me)	RCOOH	1. Acylate, 2. HF–anisole [RCONH₂]	274,275
		HF–anisole [RNH₂]	276
7b PAL	RCOOH	TFA–H₂O (19:1)	278
PEG-PS base resin		TFA–PhSMe–1,2-ethanethiol–PhOMe (90:5:3:2)	280,281
	MeCOCH₂COOMe, molecular sieves, CH₂Cl₂	TFA–THF (19:1)	282
7c RAM	RCOOH, DIPCDI, DMF	TFA–H₂O (19:1)	286–295

Table 7 Continued

Name and identifier	Coupling conditions	Cleavage conditions[a]	References
	RCOOH, HATU	TFA–H_2O (19:1)	296,300,303
	RCOOH, HBTU	TFA–CH_2Cl_2 (1:5)	297,302
	RCHO, NaBH₃CN	1. Acylate, 2. TFA–CH_2Cl_2 (1:19) [RCH₂NCOR']	284
	ROCOOAr, HOBt, DIEA, NMP	1. TFA–CH_2Cl_2 (1:9), 2. TFA–H_2O (9:1) [ROCONH₂]	304
	RNHCOOAr, DIEA, CH_2Cl_2	1. TFA–CH_2Cl_2 (1:9), 2. TFA–H_2O (9:1) [RNHCONH₂]	305
	RSO₂Cl, pyridine, CH_2Cl_2	TFA–CH_2Cl_2 (1:5) [RSO₂NH₂]	306
	Ugi reaction	TFA–CH_2Cl_2 [RCONH₂]	307–309
	1. TFA–CH_2Cl_2 (7:3), 2. ArNH₂, CH_2Cl_2	TFA–CH_2Cl_2 (7:3)	310
7d Knorr linker	RCOOH	TFA [RCONH₂]	279
	RCHO, NaBH(OAc)₃, HOAc–DMF (1:99)	TFA–CH_2Cl_2 (1:1) [R₃N]	285
7e BOBA	1. PhCHO, HOAc, DMF, 2. Me₂CH=C(OMe)OSiMe₃	DDQ, PhH [MeOCOCMe₂CH(R)NH₂]	271
7f Trityl anchor	RCOOH, DCC, PSA	TFA–CH_2Cl_2 (1:99)	311
7g Sieber; L = CH₂	RCOOH, DCC, HOBt	TFA–1,2-dichloroethane (1:50)	312
XAL; L = (CH₂)₄CONH—	RCOOH, DIPCDI, HOAt	TFA–CH_2Cl_2 (1:19)	313

Linker	Reagent	Cleavage conditions [product]	Ref.
7h	RCHO–NaBH$_4$, HOAc–DMF (1:99)	1. Acylate, 2. TFA–iPr$_3$SiH, 1,2-ethanethiol–H$_2$O (0, 1:4.5:4.5) [RCH$_2$NCOR']	314
7i CHA CHE	1. RCHO, HOAc–DMF (1:99) 2. (Me$_3$SiO)$_2$PH	TFA–CH$_2$Cl$_2$ (1:1) [RCH(NH$_2$)PHOOH]	315
7j SAL	RCOF, DIEA	TFA–CH$_2$Cl$_2$ (1:9)	316
	PyBOP, HOBt, NMM	TFA–PhOH–CH$_2$Cl$_2$ (10:5:85)	317
7k SCAL	RCOOH, BOP, DIEA, DMF	TFA–1,2-ethanethiol–PhOH–PhSMe (90:5:3:2)	318
7l Kenner's safety catch	RCOOH	TFA–Me$_3$SiBr–PhOMe	319
	RCOOC$_6$F$_5$, DMAP	1. CH$_2$N$_2$ 2. OH$^-$ [RCOOH] or R'NH$_2$ [RCONHR']	320 321
7m	ROH, PPh$_3$, DEAD	N$_2$H$_4$ [RNH$_2$]	322
7n	RCOOH	NH$_3$, hv [RCONH$_2$]	323
7o NBHA	RCOOH	hv 250 nm DMF–MeOH (1:1) [RCONH$_2$]	324
7p o-Nitrobenzyl photolabile linker	RCOOH, DIPCDI, DMF	hv 365 nm, DMSO–pH 7.4 PBS (1:19)	227,228 325
7q R = H	RCOOH	hv, THF [RCONH$_2$]	326,327
7q R = Me		hv 365 nm MeOH, H$_2$O [RCONH$_2$]	328

Table 7 Continued

Name and identifier	Coupling conditions	Cleavage conditions[a]	References
7r	RCOOH, diimide, HOBt	40 mM NaOH–iPrOH–H$_2$O (7:3)	329,330
7s Piperazine linker	1. HC≡CCH$_2$PPh$_3$Br, CH$_2$Cl$_2$ 2. R'CHO	TFA–CH$_2$Cl$_2$ (1:33) [R'CH=CHCOMe]	331 332
7t Serine and threonine supports	RCHO, MeOH–DIEA, 60°C	HOAc, H$_2$O, 60°C	333
7u Hydroxylamine linker Wang base resin	RCOOH	TFA–CH$_2$Cl$_2$–iPr$_3$SiH [RCONHOH]	334
7v Hydroxylamine linker Trityl base resin	1. Succinic anhydride, THF, 60°C, 2. RNH$_2$, DCC	HCOOH–THF (1:3) [RNHCOCH$_2$CH$_2$CONHOH]	335,336
7w O-THP hydroxylamine linker	RCOOH, DIPCDI, DMF	TFA–CH$_2$Cl$_2$ (3:2) [RCONHOH]	337
7x Weinreb amide linker	RCOOH, BOP, DIEA	LiAlH$_4$ [RCHO] RMgBr [RCOR']	338,339 340
7y Carboxyphenylhydrazine support	Preformed handle	1. NBS, 2. H$_2$O 1. NBS, 2. NH$_3$ [RCONH$_2$]	341
7z	RCOOH, DIC, HOBt	CuSO$_4$, HOAc, H$_2$O, DMF	342

[a]The general structure of the product formed as a result of the cleavage process is shown in brackets except where cleavage regenerates the functional group that was used in the initial anchoring process.

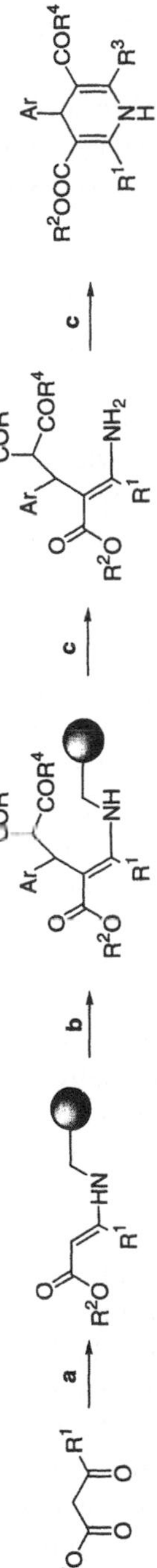

Scheme 10 (a) Resin **7b**; (b) $R^3COCH_2COR^4$, ArCHO, pyridine; (c) TFA–THF (19:1).

the corresponding phenol [283]. A related resin was reported previously [229].

Reductive alkylation of RAM resin **7c** using aldehydes or ketones provides a route to resin-bound secondary amines that is an alternative to the reductive amination of aldehyde resins discussed in Section III. After acylation, secondary amides can be cleaved from the support using TFA–CH_2Cl_2 (1:19) [284]. Reductive alkylation of the Knorr resin by aldehydes has also been reported. The products were further alkylated and the resulting tertiary amines could be cleaved directly from the support using TFA–CH_2Cl_2 (1:1) [285].

Supports **7c** and **7d** [279] have been widely used for numerous other nonpeptide syntheses commencing with the anchoring of carboxylic acids, further elaboration, and final cleavage using TFA to give primary amides. For example, the RAM support **7c** has found widespread use for the synthesis of linear [286,287], cyclized [288], and saccharide-containing [291] peptoids beginning with carbodiimide-mediated coupling of α-bromoacetic acid to resin **7c**. The final oligomers are typically cleaved from the resin using TFA–H_2O (19:1). Numerous small-molecule potential therapeutic compounds have been synthesized on RAM resin using an amide anchoring point. The initial carboxylic acid building block can be anchored to resin **7** using various coupling agents, including carbodiimides [292–295], N-[(dimethylamino)-1H-1,2,3-triazolo[4,5-b]pyridin-ylmethylene]-N-methylmethanaminium hexafluorophosphate N-oxide (HATU) [296], or HBTU [297], and final products are typically cleaved rapidly from the resin using TFA in the presence of H_2O or CH_2Cl_2 even though the anchoring amide function to RAM resin is stable to dilute aqueous acids such as HCl [296]. Reactions conducted on RAM resin usually commence with removal of the Fmoc protection group from the commercially available resin prior to the initial coupling step. A separate deprotection step has been shown to be unnecessary because treatment of the resin with a carboxylic acid, EDC, HOBt, and an excess of DIEA affords the resin-bound amide directly [298].

We have developed a Lewis acid–catalyzed three-component condensation [299] between a 2-aminoazine, an aldehyde, and an isonitrile that affords 3-aminoimidazo[1,2-a]pyridines and related compounds. This reaction was adapted to the solid phase [300] by anchoring any one of the three reactants to RAM resin **7c**. Thus, 6-aminonicotinic and 3-formylbenzoic acids could be anchored to support **7c** in the presence of HATU and subjected to the 3CC reaction. Alternatively, a resin-bound isonitrile was obtained by coupling γ-aminobutyric acid (GABA) to **7c**, followed by formylation of the amino group and dehydration (Scheme 11). After the 3CC reaction, cleavage was effected using TFA–CH_2Cl_2 (1:1).

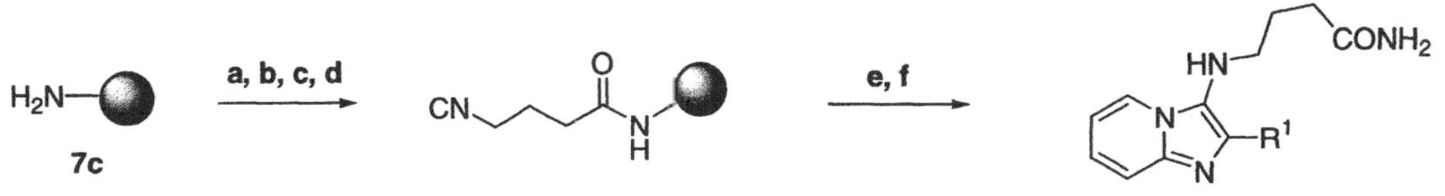

Scheme 11 (a) Fmoc-GABA-OH, HATU, DIEA; (b) piperidine, DMF; (c) trichlorophenylformate, DMF; (d) CCl$_4$, PPh$_3$, DIEA; (e) 2-aminopyridine, R^{1}CHO, Sc(OTf)$_3$; (f) TFA–CH$_2$Cl$_2$.

Procedure [300].   A solution of Fmoc–4 aminobutanoic acid (750 mg, 2.3 mmol, 2.7 equiv.) in DMF (6 mL) was treated with HATU (875 mg, 2.3 mmol, 2.7 equiv.) and DIEA (802 μL, 4.6 mmol, 5.4 equiv.). After 10 min at ambient temperature the activated acid was added to Rink amide resin (1.7 g, 0.85 mmol) that had been preswollen in CH_2Cl_2. The coupling reaction was allowed to proceed for 30 min until the resin gave a negative ninhydrin test [301] before draining and washing the resin with DMF (3 $\times$ 10 mL) and CH_2Cl_2 (3 $\times$ 10 mL). The loading was determined to be 0.50 mmol/g by UV analysis of piperidine-dibenzfulvene adduct [166] that was released on treatment of an aliquot of the resin with piperidine-DMF. Resin-bound Fmoc–4-aminobutanoic acid (1.54 g, 4.7 mmol) was deprotected by treatment with piperidine–DMF (1:4) (2 $\times$ 7 min), drained, washed with DMF (3 $\times$ 10 mL) and CH_2Cl_2 (3 $\times$ 10 mL), and then formylated by reaction with 2,4,5-trichlorophenylformate (1.54 g, 6.8 mmol) in DMF (8 mL) at ambient temperature until the resin gave a negative ninhydrin test ($\sim$2 h). The product was filtered and washed with DMF (3 $\times$ 10 mL), MeOH (3 $\times$ 10 mL), and CH_2Cl_2 (3 $\times$ 10 mL). The formamide resin (800 mg, 0.4 mmol) swollen in CH_2Cl_2 (6 mL) was cooled to $-20°C$ and treated with triphenylphosphine (524 mg, 2 mmol), triethylamine (290 μL, 2 mmol), and carbon tetrachloride (193 μL, 2 mmol). The reaction mixture was allowed to warm to 4°C and was gently agitated for a further 120 h. The resin was drained and washed with CH_2Cl_2 (6 $\times$ 10 mL) [IR (CH_2Cl_2 gel) ν_{max} 2152 cm^{-1}].

A solution of 2-aminopyridine (141 mg, 1.5 mmol) in CH_2Cl_2–MeOH (3:1) (3 mL) was treated with p-anisaldehyde (204 mg, 1.5 mmol) and $Sc(OTf)_3$ (37 mg, 0.08 mmol). After allowing the imine to form for 15 min, the resin-bound isonitrile (150 mg, 0.06 mmol) was added and the reaction mixture was agitated at ambient temperature for 80 h. The resin was filtered; washed with CH_2Cl_2 (3 $\times$ 3 mL), MeOH (3 $\times$ 3 mL), and CH_2Cl_2 (3 $\times$ 3 mL); and dried under vacuum. The product was cleaved from the resin by treatment with TFA–CH_2Cl_2 (1:1) (2 $\times$ 2 mL) for 1 h. The cleavage solution was evaporated and a stock solution of the product was prepared in MeOH and analyzed by reverse-phase LC-MS on a C_{18} column eluting with an H_2O–MeCN gradient. The purity (94%) was assessed from the relative peak area of the desired product with the UV detector set at 254 nm, and the yield (40%) was calculated by comparison of the peak area with that of a known amount of a related compound assuming an ε_{max} value of 21,400 M^{-1} cm^{-1} at 254 nm [299].

The Suzuki reaction has been applied successfully to the solid phase with the RAM linker. For example, iodobenzoic acids have been coupled to RAM–Tentagel resin and shown to undergo Suzuki couplings (Scheme 12)

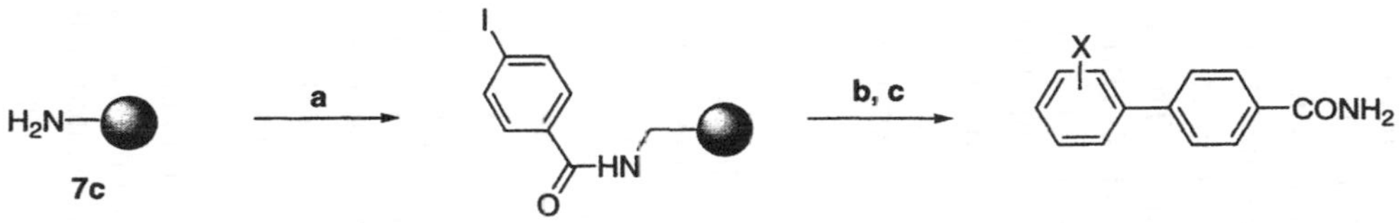

Scheme 12 (a) 4-Iodobenzoic acid, HBTU, DIEA, DMF; (b) ArB(OH)$_2$, microwave irradiation, 45 W, 3.8 min; (c) TFA–H$_2$O.

that could be accelerated by microwave irradiation. The resulting biaryl carboxamides were cleaved using TFA–H$_2$O (99:1) [266].

Procedure [302]. RAM–Tentagel resin was coupled to 4-iodobenzoic acid (4 equiv.) in the presence of HBTU (4 equiv.) and DIEA (4 equiv.) in DMF solution for 2 h and residual amino groups were then capped by acylation. The resin-bound iodo compound (100 mg, 0.023 mmol) was suspended in 2 M Na$_2$CO$_3$ (0.1 mL), H$_2$O (0.3 mL), EtOH (0.19 mL), and DME (0.75 mL) and treated with the appropriate arylboronic acid (0.2 mmol) and Pd(PPh$_3$)$_4$ (1.2 mg, 1 μmol) in a sealed Pyrex tube. The reaction mixture was subjected to microwave irradiation at 2450 MHz for 3.8 min (45 W) and then cooled to room temperature. The resin was washed successively with H$_2$O, DME, DMF, saturated KCN–DMSO, MeOH, H$_2$O, MeOH, and CH$_2$Cl$_2$ (2 × 3 mL of each). The resin was next treated with TFA–H$_2$O (99:1) for 1 h, filtered, and washed with TFA and CH$_2$Cl$_2$. The combined filtrates were evaporated to give the biaryl compound, which was shown to be >95% pure by gas chromatography–mass spectrometry (GC-MS) and was characterized by proton nuclear magnetic resonance spectroscopy (^{1}H NMR) and mass spectrometry.

Substituted tetrahydroquinoxalin-2-ones have been prepared on the RAM support **7c** [303] by first coupling 4-fluoro-3-nitrobenzoic acid to the support in the presence of HATU using a procedure comparable to that described earlier followed by displacement of fluorine by an α-aminoester. Reduction of the nitro group useing tin (II) chloride lead to ring closure (Scheme 13) and further alkylation was followed by TFA-induced cleavage.

Procedure [303]. The resin-bound fluoro compound (1 g, ~0.5 mmol) in DMF (5 mL) was treated with the appropriate amino ester (10 equiv.) and DIEA (20 equiv.). The suspensions were mixed for 3 days at ambient temperature and filtered. The resin was washed successively with DMF, MeOH, CH$_2$Cl$_2$, and ether and dried under vacuum. The product (0.9 g, ~0.45 mmol) in DMF (5 mL) was treated with SnCl$_2$·H$_2$O (1.78 g, 9.4 mmol) and the suspension mixed at ambient temperature for 24 h. The resin was filtered and washed as described previously. The product resin (160 mg, ~0.8 mmol) in acetone (2 mL) was treated with the appropriate alkyl halide (2 mmol, 25 equiv.) and K$_2$CO$_3$ (276 mg, 2 mmol). The reaction mixture was heated at 55°C for 24 h. After cooling, the suspension was filtered and the resin washed with H$_2$O, DMF, CH$_2$Cl$_2$, and ether and dried under vacuum. The product was cleaved by treatment with TFA–H$_2$O (19:1) (2 mL). The cleavage suspension was allowed to warm to room temperature and mixed for a further 50 min. The suspension was filtered and the resin washed with MeOH (3 × 2 mL). The combined filtrates were evaporated under

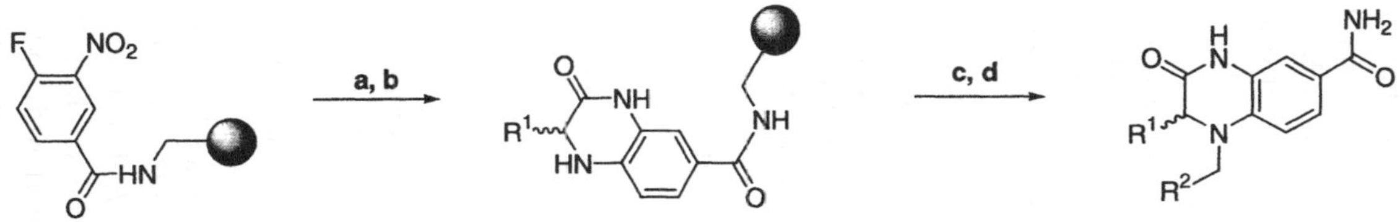

Scheme 13 (a) $H_2NCH(R^1)COOR$, DIEA; (b) $SnCl_2$; (c) R^2CH_2Br, K_2CO_3; (d) TFA–H_2O (99:1).

vacuum to give the crude product, which was analyzed by LC-MS. Purification was carried out by flash chromatography.

Carbamate and urea anchoring groups to the RAM support have also been reported with cleavage being conducted under acidic conditions comparable to those used to release amides. Reaction of RAM resin **7c** with Fmoc−amino-*p*-nitrophenyl carbonates affords a carbamate anchor [304]. Pentameric carbamates were then assembled and cleaved from the support using TFA. Coupling an azidonitrophenylcarbamate building block to the resin **7c** leads to a urea link [305]. After azide reduction, a second building block can be coupled that can be further elaborated to give oligoureas, which were cleaved from the resin using TFA. Reaction of RAM resin with sulfonyl halides leads to a sulfonamide link that can be cleaved using TFA [306].

The four-component Ugi reaction has received renewed attention because of the large number of diverse substituents that can be arrayed over a series of scaffolds. Although the four-component (4CC) acylaminoamide synthesis proceeds readily for many aliphatic and aromatic amines, ammonia does not readily afford a 4CC product. A solution to this problem involves use of the RAM resin **7c** as an ammonia equivalent [307−309] with products being cleaved from the resin using TFA.

It has been reported that the resin-bound carbocation obtained by treatment of RAM resin **7c** with acid can be isolated and subsequently allowed to react with amine, alcohol, or carboxylic acid nucleophiles to give the corresponding anchor [310].

C. Miscellaneous Amine Supports

As an extension of the trityl halide supports discussed earlier, tritylamine supports such as **7f** have been developed in order to synthesize and cleave protected peptide amides under correspondingly mild acidic conditions [311]. The first amino acid can be anchored to the support in the presence of *N,N'*-dicyclohexylcarbodiimide (DCC) and the chain assembled using N^α-Fmoc−protected amino acids. The peptide fully protected by *t*Bu side chains could be cleaved using TFA–CH$_2$Cl$_2$ (1:99). The Sieber [312] and related xanthenyl amide (XAL) linkers **7g** [313] are xanthone analogues of BHA. The cation stabilizing effect of the xanthone ring system enables amides to be released from the solid support with low (1−5%) TFA concentrations. These mild cleavage conditions have been used advantageously in the synthesis of tryptophan-containing peptides. The amino group can also be subjected to reductive alkylation [314]. Imines formed on this support have been converted to phosphinic acids, which could be cleaved from the resin by TFA treatment with the amino group serving as the leaving group [315]. A related resin **7h** with an additional aryl ring has also been prepared and used to anchor amino acids via the corresponding acyl fluoride. Cleavages were carried out using TFA–CH_2Cl_2 [316].

The closely related resins based on the handles 5-{[(*R,S*)-5-[9-fluorenylmethoxycarbonylamino]-10,11-dihydrodibenzo[*a,d*]cyclohepten-2-yl]oxy}valeric acid (CHE) **7i** and CHA (the corresponding cycloheptan-2-yl in which the double bond is reduced to an alkane) [317] have been shown to form amides that are more acid labile than those derived from PAL- or RAM-based resins and can thus be used for the preparation of side-chain protected peptide amides. The CHA resin proved to be the better choice for the synthesis of peptides containing acid-labile residues. A silyl amide linker (SAL) **7j** analogous to the hydroxyl resin SAC discussed earlier was designed for the cleavage of amides under acidic conditions by a β-elimination process. This cleavage mechanism would be expected to form a resin-bound styrene rather than a stable carbocation and therefore reduce irreversible back-alkylation side reactions. Support **7j** proved useful for the synthesis of peptide amides with a C-terminal Trp residue, which are particularly prone to back-alkylation, but cation scavengers were still needed in the cleavage solution [318].

Safety-catch linkers containing an amino group such as safety-catch amide linker (SCAL) **7k** have been used for anchoring carboxylic acids with the resulting amide anchor being stable toward acid treatment unless the sulfoxides are reduced to electron-donating thioethers. Thus, cleavage can be effected by reductive acidolysis [319]. The amino group of a sulfonamide resin **7l** has been used to attach carboxylic acids according to the Kenner safety-catch approach [320]. The resulting anchoring group to resin **7l** is stable to basic conditions because of formation of an acylsulfonamide anion but upon alkylation becomes susceptible to nucleophilic attack by hydroxide ion or amines [321]. A phthalimide functionalized resin **7m** has been reported to which primary alcohols could be anchored under Mitsunobu conditions. Subsequent treatment with hydrazine led to cleavage of the corresponding primary amine [322].

As discussed in Section IV, photolysis is a mild cleavage method that is attractive for combinatorial drug discovery because products can be released into neutral aqueous solution suitable for direct screening. Amine analogues of the hydroxyl resins described earlier have been reported. Thus, (aminomethyl)-*o*-nitrobenzyl supports **7n** have been reported on which short peptide amides could be synthesized with cleavage occurring upon irradiation at 350 nm in the presence of ammonia [323]. 2-Nitrobenzhydryl (NBHA) supports **7o** have also been used for peptide amide assembly using Boc-protected amino acids. The anchor was unaffected by 4 N HCl–dioxane used in the deprotection steps, and final peptide amides were cleaved by irradiation at 350 nm [324]. In the case of resin **7o**, the resin photoproduct is a nitrosoketone, which, unlike nitrosoaldehydes, does not rearrange to an azobenzene derivative and is therefore less of a light filter. Use of the α-methyl-6-nitroveratrylamine linker has provided resin **7p**, which gives improved synthetic routes to amides [227,228,325] because the alkoxy ring substituents and benzylic methyl group facilitate the photocleavage process. The 3-amino(2-nitrophenyl)propionyl (ANP) photolabile linker **7q** [326] and linker **7n** have been used to synthesize oligosaccharides containing a glycolamide spacer that were cleaved by irradiation in THF [327]. Substitution of both benzylic hydrogens by methyl groups gives a support for anchoring carboxylic acids that shows improved stability toward acids, bases, and amine–Lewis acid mixtures but can be photocleaved under identical conditions [328].

A piperazine-based resin **7r** functionalized by diaminopropionic acid to which acids can be anchored at the amino group has been reported. Cleavage is effected under basic conditions that lead to formation of an isocyanate, which then cyclizes onto an adjacent amino group on the peptide chain. Peptide acids or amides can then be released by hydrolysis or ammonolysis, respectively, of the resulting acylurea [329,330]. The piperazine-functionalized resin **7s** has been shown to react with propargyltriphenylphosphonium bromide to give a phosphonium salt that undergoes a conjugate Wittig reaction with aldehydes. The resulting enamine was treated with dilute TFA in CH_2Cl_2 to release an α,β-unsaturated ketone into solution [331,332] or subjected to cycloaddition reactions prior to cleavage. Amine-functionalized supports based on serine or threonine **7t** that permit stable but reversible anchoring of aldehydes as oxazolidines have been reported [333]. The cleavage conditions were variable, depending on the structure of the aldehyde.

D. Hydroxylamine Supports

Several approaches to hydroxylamine linkers attached to the solid phase through the oxygen atom have been investigated because of interest in hydroxamic acids as metalloenzyme inhibitors. Mesylated Wang resin has been converted to a hydroxylamine **7u** via an N-hydroxyphthalimide and used to assemble peptide sequences by Fmoc procedures and peptoids using the submomomer approach [334]. In each case, cleavage to give hydroxamic acids was accomplished using TFA–CH_2Cl_2 in the presence of carbocation scavengers. A related support **7v**, prepared by reaction of a trityl halide resin with N-hydroxyphthalimide [335] or N-Fmoc-hydroxylamine [336], afforded hydroxamic acids under milder cleavage conditions. Support **7w**, which retains protection of the hydroxamate oxygen as a tetrahydropyran (THP) ether and in which the nitrogen is anchored to the resin throughout the synthesis, has been obtained by reductive amination of the BAL linker using *O*-THP-protected hydroxylamine [337]. The preformed handle was then attached to Tentagel and used to synthesize a model compound of therapeutic interest. Treatment with dilute TFA in CH_2Cl_2 selectively removed the THP protecting group and higher concentrations of TFA cleaved the hydroxamic acid from the resin.

Methoxyamine solid supports **7x** can be used to anchor carboxylic acids in the presence of BOP/DIEA as Weinreb amides [338,339]. The resultant anchor is compatible with peptide assembly using Boc or Fmoc strategies, and treatment of the final sequence with $LiAlH_4$ in THF led to release of peptide aldehydes. A related support with RAM as the base resin has also been used to prepare resin-bound Weinreb amides, which afforded ketones on cleavage induced by reaction with Grignard reagents [340].

Using a preformed handle to the first amino acid, carboxyphenylhydrazine support **7y** has been applied to peptide syntheses with Boc protection schemes. Cleavage of final products was accomplished by oxidation of the hydrazide to an azo compound from which the carboxylic acid is readily liberated by hydrolysis, although direct ammonolysis is also effective in giving the corresponding peptide amide [341]. A related resin **7z** has also been used for peptide synthesis, with Boc amino acids coupled in the presence of diimide and final products again cleaved oxidatively in the presence of Cu(II) ions [342].

VIII. SOLID SUPPORTS WITH SILICON AND PHOSPHORUS AS THE ATTACHMENT POINT

Resins **8a** containing the aryldialkylsilyl chloride function (Table 8) have been used for some time [343] for anchoring the hydroxyl groups of diols and carbohydrates as silyl ethers [344,345], which can subsequently be cleaved by treatment with tetrabutylammonium fluoride (TBAF). The diethylsilane-substituted resin **8b** also forms silyl ether anchoring groups when treated with alcohols or phenols or with carbonyl compounds in the presence of Wilkinson's catalyst. These ethers are more stable than those derived from **8a** [346] but cleavage to give alcohols could be effected by HF–pyridine followed by scavenging the excess HF with methoxytrimethylsilane to give volatile by-products. Biaryls prepared by Suzuki couplings on a silane support **8c** may be cleaved by electrophilic ipso substitution, resulting in the incorporation of a halogen atom at the original anchorng site [347].

The presence of an invariant functional group derived from the anchoring and cleavage process (typically an amide or carboxylic acid) in each

Table 8 Miscellaneous Supports Based on Silicon and Phosphorus and Isonitrile Attachment Points

8a

8b

8c

8d

8e

8f

8g

8h

8i

8j

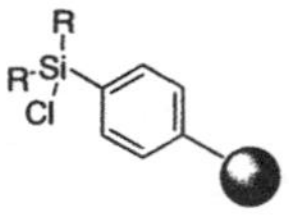

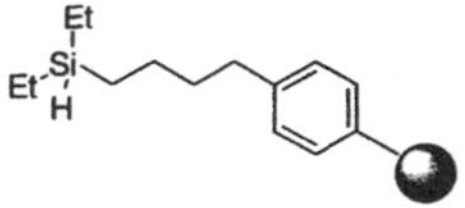

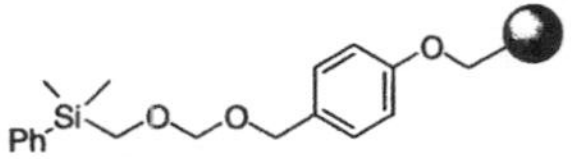

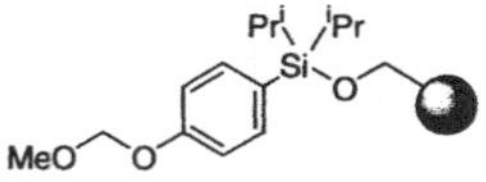

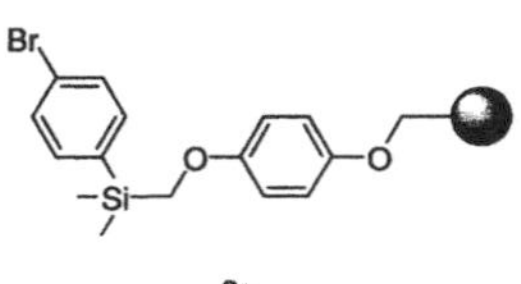

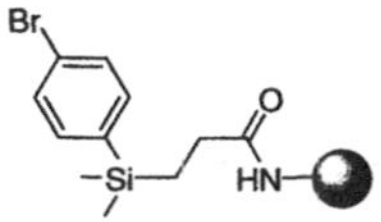

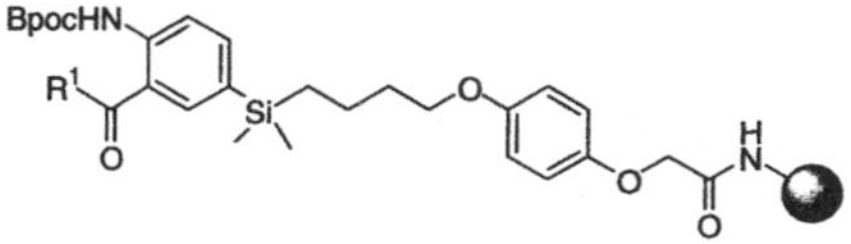

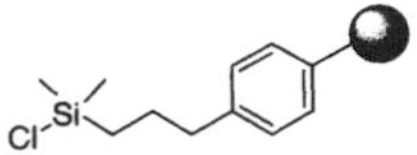

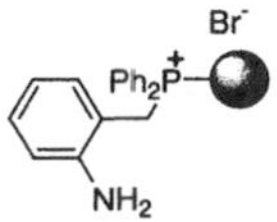

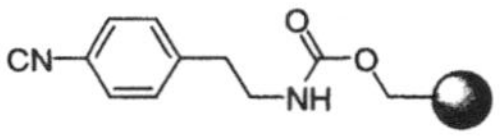

Name and identifier	Coupling conditions	Cleavage conditions [product]	References
8a	ROH, DIEA, DMAP, CH₂Cl₂	TBAF–HOAc, THF 40°C	343–345
8b Trialkylsilane support	ROH, NMP, RhCl(PPh₃)₃ 60°C	HF–pyridine–THF	346
8c Wang base resin	R^2R^3CO, NMP, RhCl(PPh₃)₃; Preformed handle	HF–pyridine–THF [R^2R^3CHOH]; ICl, CH₂Cl₂ [ArCl]; Br₂, pyridine, CH₂Cl₂ [ArBr]	347
8d Traceless silyl ether	Preformed handle	TBAF, DMF, 65°C [traceless release of benzofurans]	348
8e Arylsilane linker aminomethyl-PS base resin	Preformed handle	CsF, DMF–H₂O; TFA [traceless release of biaryls]	349
8f BHA base resin	Preformed handle	TFA–CH₂Cl₂ (1:1) [traceless release of biaryls]	350
8f BHA base resin	Preformed handle	TFA vapor [traceless release of benzylamine derivatives]	351
8g	Preformed handle	HF [traceless release of benzodiazepinones]	352
8h	ArLi, THF	TBAF, THF [traceless release of pyridine-based tricyclics]	353
8i Phosphonium traceless support	1. ArCOCl, pyridine, CH₂Cl₂	RCHO, NaOMe [alkenes]	354 355
8j Isonitrile safety catch	Ugi 4CC	1. Boc₂O, 2. NaOMe, THF, MeOH	356

member of a screening library is a limitation of the use of conventional linkers and resins. Silyl linkers have proved valuable as traceless linkers wherein a C—H bond appears in the product at the point of cleavage. For example, silyl ether linker **8d** prepared by coupling a preformed handle, MOM-protected 4-(chlorodiisopropylsilyl)phenol, to hydroxymethyl-PS has been used to construct benzofurans with final products cleaved from the resin using TBAF in DMF [348].

Related resins **8e** [349] and **8f** [350] have been used in Suzuki coupling reactions; cleavage was possible under conditions of lower acid concentration for **8f**, attributable to anchimerically assisted ipso protonation by the amide carbonyl. Similar observations have been made concerning the conditions required to cleave this silane anchoring group from a linker that had been attached to benzhydrylamine resin [351].

The solid-phase synthesis of 1,4-benzodiazepines has been conducted using a traceless silicon anchor **8g** that was prepared via a preformed handle. Cleavage from the resin was effected with HF [352] because the anchoring group was inert to concentrated aqueous TFA due to the highly electron-deficient character of the benzodiazepine ring.

The traceless anchors based on silicon just described are limited by the need to synthesize a preformed handle consisting of the first building block and the silicon linker. Moreover, traceless release from conventional aryldialkylsilanes is complicated by their low hydrolytic stability and loss of silicon from the resin in the cleavage process. The silyl chloride resin **8h** can be used to anchor aryl rings by reaction with aryllithium reagents. The anchor is stable to concentrated TFA and bases, but treatment with TBAF cleaves the final products [353].

Polymer-bound triphenylphenylphosphine [354] has been converted to a phosphonium salt **8i** by reaction with 2-nitrobenzyl bromide. The nitro group was then reduced to an amine and acylated. The resulting product formed a resin-bound ylide in the presence of base, which underwent a Wittig reaction with aldehydes that released alkenes into solution with the phosphine oxide by-product remaining bound to the solid support [355].

Isonitrile solid support **8j** functions as a safety-catch anchor for acyl-aminoamides formed by a Ugi 4CC reaction [356]. Boc derivatization of the secondary amide anchoring site causes activation toward nucleophilic attack, and thus products could be cleaved by reaction with alkoxide ion.

IX. CONCLUSIONS

Interest in solid-phase techniques that began with peptide synthesis followed by extension to other natural oligomers has been revived by numerous applications in small-molecule synthesis. It has been suggested that the rapid

growth of solid-phase synthesis applications will slow as the quality of solution-phase parallel synthesis libraries improves with developments in scavenger resin techniques (see Chapter 10) that allow compounds to be purified
in parallel. Because solution-phase reactions are easier to monitor and therefore easier to optimize, a trend away from solid-phase synthesis has been
predicted. However, the number of reports of new bond-forming reactions
successfully applied to the solid phase continues to grow, as does the number
of new linkers. Thus, most compound classes will eventually be accessible
by solid-phase synthesis. At this stage of development, however, rigorous
optimizations of reaction conditions are required before the efficiency of a
small-molecule synthesis even approaches that of a biopolymer synthesis.
Developments in traceless linkers that permit more ready access to compound series devoid of a polar invariant functional group can be expected
to continue as these will enable wider structure–activity correlations to be
investigated. Finally, improvements in the design of reaction blocks and
automation will improve the efficiency of reaction optimization and the
throughput of library synthesis on the solid phase, making it more attractive
for parallel synthesis of single compounds.

ABBREVIATIONS

AA, amino acid; AC, acid-cleavable linker [4-(hydroxymethyl)-3-methoxy-
phenoxy]acetic acid; ADCC, 4-acetyl-3,5-dioxo-1-methylcyclohexane carboxylic acid; AMEBA, acid-sensitive methoxybenzyl resin; ANP, 3-
amino(2-nitrophenyl)propionyl; BAL, backbone amide linker; 9-BBN,
9-borabicyclo[3.3.1]nonane; BHA, benzhydrylamine; Boc, *tert*-butoxycar-
bonyl; BOP, (benzotriazol-1-yloxy)tris(dimethylamino)phosphonium hexafluorophosphate; BOPA, *p*-benzyloxybenzylamine; Bpoc, [[2-(4-bi-
phenyl)isopropyl]oxy]carbonyl; *t*Bu, *t*-butyl; 1-BuOH, 1-butanol; Bt,
2-benzothiazolyl; Bzl, benzyl; CC, multicomponent condensation; CDI, carbonyldiimidazole; CHA, 5-{[(*R,S*)-5-[(9-fluorenylmethoxycarbonylamino]-
10,11-dihydrodibenzo[*a,d*]cyclohepten-2-yl]oxy}valeric acid; CHE, 5-
{[(*R,S*)-5-[(9-fluorenylmethoxycarbonylamino]-10,11-dihydrodibenzo[*a,d*]-
cyclohepten-2-yl]oxy}valeric acid; *m*CPBA, *m*-chloroperoxybenzoic acid;
CSA, camphorsulfonic acid; CSI, chlorosulfonylisocyanate; DABCO, 1,4-
diazabicyclo[2.2.2]octane; DBU, 1,8-diazabicyclo[5.4.0]undec-7-ene; DCC,
N,N,-dicylcohexylcarbodiimide; DDQ, 2,3-dichloro-5,6-dicyano-1,4-benzo-
quinone; DEAD, diethyl azodicarboxylate; DHP, dihyropyran; DHPP, 4-
(1′,1′-dimethyl-1′-hydroxypropyl)phenoxyacetyl; DIBAL, diisobutylaluminum hydride; DIEA, *N,N*-diisopropylethylamine; DIPCDI,
N,N-diisopropylcarbodiimide; DMA, *N,N*-dimethylacetamide; DMAE, di-

methylaminoethanol; DMAP, 4-*N*,*N*-dimethylaminopyridine; DME, 1,2-di-methoxyethane; DMF, *N*,*N*-dimethylformamide; DMS, dimethyl sulfide; DMSO, dimethyl sulfoxide; Dts, dithiasuccinoyl; DTT, dithiothreitol; DVB, divinylbenzene; EDC, ethyl-3-[3-(dimethylamino)propyl]carbodiimide; ESI-MS, electrospray ionization mass spectrometry; EtOH, ethanol; Fmoc, 9-fluorenylmethoxycarbonyl; GABA, γ-aminobutyric acid; GC-MS, gas chromatography–mass spectrometry; HATU, *N*-[(dimethylamino)-1*H*-1,2,3-triazolo[4,5-*b*]pyridin-ylmethylene]-*N*-methylmethanaminium hexafluoro-phosphate *N*-oxide; HBTU *N*-[(1*H*-benzotriazol-1-yl)(dimethylamino)-methylene]-*N*-methylmethanaminium hexafluorophosphate *N*-oxide; HMFS, *N*-[9-hydroxymethyl)-2-fluorenyl]succinamic acid; HMB, 4-hydroxymethyl-benzoic acid; HMPB, 4-(4-hydroxymethyl-3-methoxyphenoxy)butyric acid; HMPP, 3-(4-hydroxymethylphenoxy)propionic acid; HOAc, acetic acid; HOAt, 1-hydroxyazabenzotriazole; HOBt, 1-hydroxybenzotriazole; HPLC, high-performance liquid chromatography; LC-MS, liquid chromatography–mass spectrometry; LDA, lithium diisopropylamide; MBHA, *p*-methylbenz-hydrylamine; MeCN, acetonitrile; MeIm, 1-methylimidazole; MeOH, methanol; MOM, methoxymethyl; MSNT, 2,4,6-mesitylenesulfonyl-3-nitro-1,2,4-triazole; Nb, 4-amino or hydroxymethyl-3-nitrobenzoic acid; NBHA, 2-nitrobenzhydrylamine; NCS, *N*-chlorosuccinimide; NMM, *N*-methylmor-pholine; NPE, 3-nitro-4-(2-hydroxyethyl)benzoic acid; NMR, nuclear mag-netic resonance spectroscopy; PAC, peptide acid linker phenoxyacetic acid; PAL, peptide amide linker; 5-(4-aminomethyl-3,5-dimethoxyphen-oxy)valeric acid; PAM, *p*-hydroxymethylphenyl acetic acid; Pbs, *N*-(3 or 4)[[4-(hydroxymethyl)phenoxy]-*t*-butylphenylsilyl]phenyl pentanedioic acid monoamide handle; PEGA, 2-acrylamidoprop-1-yl polyethyleneglycol cross-linked dimethyl acrylamide; PEG-PS, polyethylene glycol graft polystyrene; POE-PS polyoxyethylene–polystyrene; PPTS, pyridinium *p*-toluenesulfon-ate; *i*PrOH, 2-propanol; PS, polystyrene; PS-DVB, polystyrene–divinylbenzene; RAM, Rink amide resin; REM, regenerated after cleavage, functionalized by Michael addition; SASRIN, super-acid-sensitive resin; SAC, silyl acid support; SAL, silyl amide linker; SCAL, safety-catch amide linker; TBAF, tetrabutylammonium fluoride; TBS, *t*-butyldimethylsilyl; TBTU *N*-[(dimethylamino)-1*H*-1,2,3-triazolo[4,5-*b*]pyridin-ylmethylene]-*N*-methylmethanaminium tetrafluoroborate *N*-oxide; Teoc, 2-(trimethylsi-lyl)ethoxycarbonyl; TCEP, tris(2-carboxyethyl)phosphine; TFA, trifluoroace-tic acid; TFFH, tetrafluoroformamidinium hexafluorophosphate; TFMSA, trifluoromethanesulfonic acid; TfOH, triflic acid; THF, tetrahydrofuran; THP, tetrahydropyran; TMG, tetramethylguanidine; TMOF, trimethylorthofor-mate; TMS, trimethylsilyl; TMSCl, trimethylsilyl chloride; TMSOTf, tri-methylsilyltriflate; Tos, tosyl; TsOH, *p*-toluenesulfonic acid; XAL, xanthenyl amide linker.

REFERENCES

1. SR Wilson, AW Czarnik, eds. Combinatorial Chemistry Synthesis and Application. New York: Wiley, 1997.
2. G Lowe. Chem Soc Rev 24:309–317, 1995.
3. BA Bunin. The Combinatorial Index. San Diego: Academic Press, 1998.
4. AW Czarnik, SH DeWitt, eds. A Practical Guide to Combinatorial Chemistry. Washington, DC: American Chemical Society, 1997.
5. LA Thompson, JA Ellman. Chem Rev 96:555–600, 1996.
6. NK Terret, M Gardern, DW Gordon, RJ Kobylecki, J Steele. Tetrahedron 51: 8135–8173, 1995.
7. CC Leznoff. Chem Soc Rev 3:65–85, 1974.
8. CC Leznoff. Acc Chem Res 11:327–333, 1978.
9. JI Crowley, H Rapoport. Acc Chem Res 9:135–144, 1976.
10. JS Früchtel, G Jung. Angew Chem Int Ed Engl 35:17–42, 1996.
11. NK Terret, F Balkenhohl. Angew Chem Int Ed Engl 35:2288–2337, 1996.
12. RA Houghten, A Nefzu. Chem Rev 97:449–472, 1997.
13. C Blackburn, F Albericio, SA Kates. Drugs Future 22:1007–1025, 1997.
14. Novabiochem Combinatorial Chemistry Catalog. San Diego: Calbiochem-Novabiochem International, 1999.
15. E Atherton, RC Sheppard. Solid-Phase Peptide Synthesis: A Practical Approach. Oxford: IRL Press, 1989.
16. RHH Hermkens, HCJ Ottenheijm, DC Rees. Tetrahedron 53:5643–5678, 1997.
17. S Mazur, P Jayalalekshmy. J Am Chem Soc 101:677–681, 1979.
18. WA Warr. J Chem Inf Comput Sci 37:134–140, 1997.
19. A Furka, F Sebestyen, M Asgedom, G Dibo. Int J Peptide Protein Res 37: 487–493, 1991.
20. KS Lam, SE Salmon, EM Hersch, VJ Hruby, WM Kazmierski, RJ Knapp. Nature 354:82–84, 1991.
21. RA Houghten, C Pinilla, SE Blondelle, JR Appel, CT Dooley, JH Cuervo. Nature 354:84–86, 1991.
22. WC Stille. Acc Chem Res 29:155–163, 1996.
23. RB Merrifield. J Am Chem Soc 85:2149–2152, 1963.
24. B Gutte, RB Merrifield. J Biol Chem 246:1922–1941, 1971.
25. G Barany, RB Merrifield. In: E Gross, J Meienhofer, eds. The Peptides, Vol 2. New York: Academic Press, 1979, pp 1–284.
26. RB Merrifield. In: B Gutte, ed. Peptides: Synthesis, Structures and Applications. San Diego: Academic Press, 1995, pp 93–169.
27. GB Fields, Z Tian, G Barany. In: GA Grant, ed. Synthetic Peptides: A User's Guide. New York: WH Freeman, 1992, pp 77–183.
28. RL Letsinger, MJ Kornet. J Am Chem Soc 85:3045–3046, 1963.
29. SL Beaucage, RP Iyer. Tetrahedron 48:2223–2311, 1992.
30. SL Beaucage, RP Iyer. Tetrahedron 48:10441–10488, 1993.
31. JM Frechet, C Schuerch. J Am Chem Soc 93:492–496, 1971.

32. SJ Danishevsky, KF McClure, JT Randolph, RB Ruggeri. Science 260:1307–1309, 1993.
33. JT Randolph, SJ Danishevsky. Angew Chem Int Ed Engl 33:1470–1472, 1994.
34. S Hanessian, T Ogawa, Y Guindon, J Kamennof, R Roy. Carbohydr Res 38: C15–18, 1974.
35. DL Flynn, RV Devraj, JJ Parlow. Curr Opin Drug Discov Dev 1:41–50, 1998.
36. DJ Gravert, KD Janda. Chem Rev 97:489–509, 1997.
37. W Rapp. In: G Jung, ed. Peptides and Nonpeptide Libraries—A Handbook. VCH, Berlin, 1996, pp 425–464.
38. E Bayer. Angew Chem Int Ed Engl 30:113–129, 1991.
39. W Rapp, E Bayer. In: R Epton, ed. Innovation and Perspectives in Solid Phase Synthesis: Peptides, Polypeptides and Oligonucleotides. Andover: Intercept Limited, 1992, pp 259–266.
40. F Albericio, J Bacardit, G Barany, JM Coull, M Egholm, E Giralt, GW Griffin, SA Kates, E Nicolás, NA Solé. In: HLS Maia, ed. Peptides 1994: Proceedings of the Twenty-Third European Peptide Symposium, Escom, Leiden, 1995, pp 271–272.
41. G Barany, F Albericio, NA Solé, GW Griffin, SA Kates, D Hudson. In: CH Schneider, AN Eberle, eds. Peptides 1992: Proceedings of the Twenty-Second European Peptide Symposium, Escom, Leiden, 1993, pp 267–268.
42. S Zalipsky, JL Chang, F Albericio, G Barany. React Polym 22:243–258, 1994.
43. O Gooding, PDJ Hoeprich, JW Labadie, JA Porco, P van Eikeren, P Wright. In: IM Chaiken, KD Janda, eds. Molecular Diversity and Combinatorial Chemistry Libraries and Drug Discovery. Washington, DC: American Chemical Society, 1996, pp 199–226.
44. EE Swayze. Tetrahedron Lett 38:8465–8468, 1997.
45. JH Adams, RM Cook, D Hudson, V Jammalamadaka, MH Lyttle, MF Songster. J Org Chem 63:3706 3716, 1998.
46. RC Sheppard, BJ Williams. Int J Peptide Protein Res 20:451–454, 1982.
47. BF Gisin. Helv Chim Acta 56:1476–1482, 1973.
48. WDF Meutermans, PF Alewood. Tetrahedron Lett 36:7709–7712, 1995.
49. GR Marshall, RB Merrifield. Biochemistry 4:2394–2401, 1965.
50. MK Answer, AF Spatola. Tetrahedron Lett 33:3121–3124, 1992.
51. MA Barton, RU Lemieux, JY Savoiw. J Am Chem Soc 95:4501–4506, 1973.
52. R Frennette, RW Friesen. Tetrahedron Lett 35:9177–9180, 1994.
53. DR Tortolani, SA Biller. Tetrahedron Lett 37:5687–5690, 1996.
54. S Chamoin, S Houldsworth, V Snieckus. Tetrahedron Lett 39:4175–4178, 1998.
55. S Marquis, M Arlt. Tetrahedron Lett 37:5491–5494, 1996.
56. M Bodansky, JT Sheehan. Chem Ind (Lond) 1423–1424, 1964.
57. J Blake, CH Li. Int J Peptide Protein Res 3:185–189, 1971.
58. W Kessler, B Iselin. Helv Chim Acta 49:1330–1344, 1966.
59. KD Kaufmann, S Bauschke. Z Chem 20:145–146, 1980.
60. MJ Kurth, LAA Randall, C Chen, C Melander, RB Miller. J Org Chem 59: 5862–5864, 1994.

61. SH DeWitt, JS Kiely, CJ Stankovic, MC Schroeder, DMR Cody, MR Pavia. Proc Natl Acad Sci U S A 90:6909–6913, 1993.

62. D Stones, DJ Miller, MW Beaton, TJ Rutherford, D Gani. Tetrahedron Lett 39:4875–4878, 1998.

63. P Conti, D Demont, J Cals, HCJ Ottenheijm, D Leysen. Tetrahedron Lett 38: 2915–2918, 1997.

64. K Barlos, D Gatos, J Kallitsis, D Papaioannou, P Sotiriou. Liebigs Ann Chem 1079–1081, 1988.

65. K Barlos, D Gatos, J Kallitsis, G Papaphotiu, P Sotiriu, Y Wenqing, W Schäfer. Tetrahedron Lett 30:3943–3946, 1989.

66. K Barlos, D Gatos, J Kallitsis, G Papaphotiu, P Sotiriu, Y Wenqing, W Schäfer. Tetrahedron Lett 30:3947–3951, 1989.

67. K Barlos, O Chatzi, D Gatos, W Schäfer. Int J Peptide Protein Res 37:513–520, 1991.

68. CC Zikos, NG Ferderigos. Tetrahedron Lett 35:1767–1768, 1994.

69. H Richter, G Jung. Tetrahedron Lett 39:2729–2732, 1998.

70. TM Fyles, CC Leznoff, J Weatherston. Can J Chem 56:1031–1041, 1978.

71. JMJ Frechet, L Nuyens. Can J Chem 54:926–934, 1976.

72. C Chen, LAA Randall, RB Miller, AD Jones, MJ Kurth. J Am Chem Soc 116:2661–2662, 1994.

73. C Chen, LA Ahlberg, RB Miller, AD Jones, MJ Kurth. Tetrahedron 53:6595–6609, 1997.

74. BJ Egner, N Cardno, M Bradley. J Chem Soc Chem Commun 2163–2164, 1995.

75. MA Youngman, SL Dax. Tetrahedron Lett 38:6347–6350, 1997.

76. BB Shankar, DY Yang, S Girton, AK Ganguly. Tetrahedron Lett 39:2447–2450, 1998.

77. K Ngu, DV Patel. Tetrahedron Lett 38:973–976, 1997.

78. MW Miller, SF Vice, SW McCombie. Tetrahedron Lett 39:3429–3432, 1998.

79. B Raju, TP Kogan. Tetrahedron Lett 38:4965–4968, 1997.

80. TL Deegan, OW Gooding, S Baudart, JA Porco. Tetrahedron Lett 38:4973–4976, 1997.

81. DA Heerding, DT Takata, C Kwon, WF Huffman, J Samanen. Tetrahedron Lett 39:6815–6918, 1998.

82. KH Bleicher, JR Wareing. Tetrahedron Lett 39:4587–4590, 1998.

83. KH Bleicher, JR Wareing. Tetrahedron Lett 39:4591–4594, 1998.

84. DH Rich, SH Gurwara. J Am Chem Soc 97:1575–1579, 1975.

85. E Giralt, F Albericio, D Andreu, R Eritja, P Martin, E Pedroso. An Quim 77C:120–125, 1981.

86. E Giralt, C Celma, MD Ludevid, E Pedroso, F Albericio. Int J Peptide Protein Res 29:647–656, 1987.

87. C Celma, F Albericio, E Pedroso, E Giralt. Peptide Res 5:62–71, 1992.

88. A Ajayaghosh, VNR Pillai. J Org Chem 55:2826–2829, 1990.

89. A Ajayaghosh, VNR Pillai. Tetrahedron 44:6661–6666, 1988.

90. H Kunz, B Dombo. Angew Chem Int Ed Engl 27:711–713, 1988.

91. F Guibe, D Dangles, G Balavoine, A Loffet. Tetrahedron Lett 30:2641–2644, 1989.
92. F Albericio, P Lloyd-Williams, G Jou, E Giralt. Tetrahedron Lett 32:4207, 1991.
93. JH van Maarseveen, JAJ Hartog, V Engelen, E Finner, G Visser, CG Kruse. Tetrahedron Lett 37:8249–8252, 1996.
94. F Baleux, J Daunis, R Jacquier. Tetrahedron Lett 25:5893, 1984.
95. F Baleux, B Calas, J Mery. Int J Peptide Protein Res 28:22–28, 1986.
96. B Calas, J Mery, J Parello, A Cave. Tetrahedron 41:5331–5339, 1985.
97. SS Wang. J Org Chem 41:3258–3261, 1976.
98. FS Tjoeng, GA Heaver. J Org Chem 48:355–359, 1983.
99. D Bellof, M Mutter. Chimia 39:317–320, 1985.
100. NA Abraham, G Fazal, JM Ferland, S Rakhit, J Gauthier. Tetrahedron Lett 32:577–580, 1991.
101. C Birr, M Wengert-Muller. Angew Chem Int Ed Engl 18:147–148, 1979.
102. JP Tam, WF Cunningham-Rundles, BW Erickson, RB Merrifield. Tetrahedron Lett 46:4001–4004, 1977.
103. M Tomishige, K Shigezane, T Takamura. Chem Pharm Bull 18:1465–1474, 1970.
104. DM Dixit, CC Leznoff. Isr J Chem 17:248–252, 1978.
105. JR Hauske, P Dorff. Tetrahedron Lett 36:1589–1592, 1995.
106. J Alsina, F Rabanal, C Chiva, E Giralt, F Albericio. Tetrahedron 54:10125–10152, 1998.
107. DP Rotella. J Am Chem Soc 118:12246–12247, 1996.
108. DJ Burdick, ME Struble, JP Burnier. Tetrahedron Lett 34:2589–2592, 1993.
109. SW Kim, CY Hong, K Lee, EJ Lee, JS Koh. Bioorg Med Chem Lett 8:735–738, 1998.
110. BA Dressman, LA Spangle, SW Kaldor. Tetrahedron Lett 37:9337–9340, 1996.
111. L Gouilleux, JA Fehrentz, F Winternitz, J Martinez. Tetrahedron Lett 37:7031–7034, 1996.
112. B Raju, TP Kogan. Tetrahedron Lett 38:3377–3380, 1997.
113. R Mohan, W Yan, BO Buckman, A Liang, L Trinh, MM Morrisey. Bioorg Med Chem Lett 8:1877–1882, 1998.
114. P Roussel, M Bradley, I Matthews, P Kane. Tetrahedron Lett 38:4861–4864, 1997.
115. IR Marsh, H Smith, M Bradley. J Chem Soc Chem Commun 941–942, 1996.
116. P Page, S Burrage, L Baldock, M Bradley. Bioorg Med Chem Lett 8:1751–1756, 1998.
117. G Breipohl, J Knolle, R Geiger. Tetrahedron Lett 28:5647–5650, 1987.
118. BA Dressman, U Singh, SW Kaldor. Tetrahedron Lett 39:3631–3634, 1998.
119. R Leger, R Yen MW She, VJ Lee, SJ Hecker. Tetrahedron Lett 39:4171–4174, 1998.
120. LJ Fitzpatrick, RA Rivero. Tetrahedron Lett 38:7479–7482, 1997.
121. JJ Baldwin, JJ Burbaum, I Henderson, MHJ Ohlmeyer. J Am Chem Soc 117:5588–5589, 1995.

122. K Kaljuste, A Uden. Tetrahedron Lett 37:3031–3034, 1996.
123. JR Morphy, Z Rankovic, DC Rees. Tetrahedron Lett 37:3209–3212, 1996.
124. AR Brown, DC Rees, Z Rankovic, JR Morphy. J Am Chem Soc 119:3288–3295, 1997.
125. O Xiaohu, RW Armstrong, MM Murphy. J Org Chem 63:1027–1032, 1998.
126. FEK Kroll, R Morphy, D Rees, D Gani. Tetrahedron Lett 38:8573, 1997.
127. P Heinonen, H Lonnberg. Tetrahedron Lett 38:8569–8572, 1997.
128. W Bannwarth, J Huebscher, R Barner. Bioorg Med Chem Lett 6:1525–1528, 1996.
129. PS Furth, MS Reitman, AF Cook. Tetrahedron Lett 38:5403–5406, 1997.
130. JT Ayres, CK Mann. J Polym Sci Polym Lett Ed 3:505–508, 1965.
131. X Beebe, NE Schore, MJ Kurth. J Am Chem Soc 114:10061–10062, 1992.
132. X Beebe, NE Schore, MJ Kurth. J Org Chem 60:4196–4203, 1995.
133. M Reggelin, V Brenig. Tetrahedron Lett 37:6851–6852, 1996.
134. KJ Jensen, MF Songster, J Vágner, J Alsina, F Albericio, G Barany. In: PTP Kaumaya, RS Hodges, eds. Peptides—Chemistry and Biology: Proceedings of the Fourteenth American Peptide Symposium. Birmingham, UK: Mayflower Worldwide, 1996, pp 30–32.
135. C Holmes. Presented at the 213th National Meeting of the American Chemical Society, San Francisco, April 1997, paper ORGN 383.
136. KJ Jensen, J Alsina, MF Songster, J Vágner, F Albericio, G Barany. J Am Chem Soc 120:5441–5452, 1998.
137. NS Gray, S Kwon, P Schultz. Tetrahedron Lett 38:1161–1164, 1997.
138. CG Boojamra, K Burow, JA Ellman. J Org Chem 60:5742–5743, 1995.
139. CG Boojamra, K Burow, JA Ellman. J Org Chem 62:1240–1256, 1997.
140. D Sarantakis, JJ Bicksler. Tetrahedron Lett 38:7325–7328, 1997.
141. PC Kearney, M Fernandez, JA Flygare. J Org Chem 63:196–197, 1998.
142. MT Bilodeau, AM Cunningham. J Org Chem 63:2800–2801, 1998.
143. AM Fivush, TM Willson. Tetrahedron Lett 38:7151–7154, 1997.
144. KJ Estepp, CE Neipp, LM Stephens Stramiello, MD Adam, MP Allen, S Robinson, EJ Roskamp. J Org Chem 63:5300–5301, 1998.
145. Y-T Wu, H-P Hsieh, C-Y Wu, H-M Yu, S-T Chen, K-T Wang. Tetrahedron Lett 39:1783–1784, 1998.
146. LA Thompson, JA Ellman. Tetrahedron Lett 35:9333–9336, 1994.
147. EK Kick, JA Ellman. J Med Chem 48:1427–1430, 1995.
148. JS Koh, JA Ellman. J Org Chem 61:4494–4495, 1996.
149. G Liu, JA Ellman. J Org Chem 60:7712–7713, 1995.
150. G Wess, K Bock, H Kleine, M Kurz, W Guba, H Hemmerle, E Lopez-Calle, KH Baringhaus, H Glombik, A Enhsen, W Kramer. Angew Chem Int Ed Engl 35:2222–2224, 1996.
151. S Yoo, J Seo, K Yi, Y Gong. Tetrahedron Lett 38:1203–1206, 1997.
152. DA Nugiel, LAM Cornelius, JW Corbett. J Org Chem 62:201–203, 1997.
153. TM Fyles, CC Leznoff. Can J Chem 54:935–942, 1976.
154. MJ Farrell, MJ Frechet. J Org Chem 41:3877–3882, 1976.
155. HV Meyers, GJ Dilley, TL Durgin, TS Powers, NA Winssinger, H Zhu, MR Pavia. Mol Divers 1:13–20, 1995.

156. JS Panek, B Zhu. Tetrahedron Lett 37:8151–8154, 1996.
157. BA Lorsbach, RB Miller, MJ Kurth. J Org Chem 61:8716–8717, 1996.
158. A Routledge, C Abell, S Balasubramanian. Synlett 61–62, 1997.
159. JA Hunt, WR Rousch. J Am Chem Soc 118:9998–9999, 1996.
160. S Jin, DP Holub, DJ Wustrow. Tetrahedron Lett 39:3651–3654, 1998.
161. HZ Zhong, MN Greco, BE Maryanoff. J Org Chem 62:9326–9330, 1997.
162. JK Rueter, SO Nortey, EW Baxter, GC Leo, AB Reitz. Tetrahedron Lett 39: 975–978, 1998.
163. EW Baxter, JK Rueter, SO Nortey, AB Reitz. Tetrahedron Lett 39:979–982, 1998.
164. JM Goldwasser, CC Leznoff. Can J Chem 56:1562–1568, 1978.
165. J Green. J Org Chem 60:4287–4290, 1995.
166. J Meinhofer, M Waki, EP Heimer, TJ Labros, RC Makofske, CD Change. Int J Peptide Protein Res 13:35–42, 1973.
167. C Sylvain, A Wagner, C Miosowski. Tetrahedron Lett 38:1042–1046, 1997.
168. LF Tietz, A Steinmetz. Angew Chem Int Ed Engl 35:651–652, 1996.
169. LF Tietz, A Steinmetz. Synlett 667–668, 1996.
170. T Ruhland, H Kunzer. Tetrahedron Lett 36:2757–2760, 1996.
171. O Prien, K Rolfing, M Thiel, H Kunzer. Synlett 325–326, 1997.
172. A van Loevezijn, JH van Marseveen, K Stegman, GM Visser, G-J Koomen. Tetrahedron Lett 39:4737–4740, 1998.
173. AR Mitchell, BW Erickson, MN Ryabtsev, RS Hodges, RB Merrifield. J Am Chem Soc 98:7357–7362, 1976.
174. AR Mitchell, SBH Kent, M Engelhard, RB Merrifield. J Org Chem 43:2845–2852, 1978.
175. ED Baird, PB Dervan. J Am Chem Soc 118:6141–6146, 1996.
176. F Albericio, G Barany. Int J Peptide Protein Res 26:92–97, 1985.
177. MJ Plunkett, JA Ellman. J Am Chem Soc 117:3306–3307, 1995.
178. R Buckman, R Mohan. Tetrahedron Lett 37:4439–4442, 1996.
179. W Yun, R Mohan. Tetrahedron Lett 37:7189–7192, 1996.
180. GB Phillips, GP Wei. Tetrahedron Lett 37:4887–4890, 1996.
181. SS Wang. J Am Chem Soc 95:1328–1333, 1973.
182. G Lu, RB Merrifield. J Org Chem 46:3433–3436, 1981.
183. F Wang, JR Hauske. Tetrahedron Lett 38:8651–8654, 1997.
184. KL Yu, MS Deshpande, DM Vyas. Tetrahedron Lett 35:8919–8922, 1994.
185. JW Guiles, SJ Johnson, WV Murray. J Org Chem 61:5169–5171, 1996.
186. MA Blaskovich, M Kahn. J Org Chem 63:1119–1125, 1998.
187. P Wipf, A Cunningham. Tetrahedron Lett 36:7819–7822, 1995.
188. JP Mayer, GS Lewis, C McGee, D Bankaitis-Davis. Tetrahedron Lett 39: 6655–6658, 1998.
189. A Nouvet, F Lamaty, R Lazaro. Tetrahedron Lett 39:3469–3472, 1998.
190. GL Bolton, JC Hodges, JR Rubin. Tetrahedron 53:6611–6614, 1997.
191. AM Strocker, TA Keating, PA Tempest, RW Armstrong. Tetrahedron Lett 37: 1149–1152, 1996.
192. TA Keating, RW Armstrong. J Am Chem Soc 118:2574–2583, 1996.

193. BC Hamper, SA Kolodziej, AM Scates. Tetrahedron Lett 39:2047–2050, 1998.
194. SA Kolodziej, BC Hamper. Tetrahedron Lett 37:5277–5280, 1996.
195. JP Mayer, J Zhang, K Bjergarde, DM Lenz, JJ Gaudino. Tetrahedron Lett 37: 8081–8084, 1996.
196. BA Kulkarni, A Ganesan. Tetrahedron Lett 39:4369–4372, 1998.
197. GJ Kuster, HW Scheeren. Tetrahedron Lett 39:3613–2616, 1998.
198. G Bhalay, P Blaney, VH Palmer, AD Baxter. Tetrahedron Lett 38:8375–8378, 1997.
199. DR Barn, JR Morphy, DC Rees. Tetrahedron Lett 37:3213–3216, 1996.
200. CA Metcalf, CB Vu, R Sundaramoorthi, VA Jacobsen, EA Laborde, J Green, Y Green, KJ Macek, TJ Merry, SG Pradeepan, M Uesugi, VM Varkhedkar, DA Holt. Tetrahedron Lett 39:3435–3438, 1998.
201. X Cao, AMM Mjalli. Tetrahedron Lett 37:6073–6076, 1996.
202. Y Wang, SR Wilson. Tetrahedron Lett 38:4021–4024, 1997.
203. C Chen, B Munoz. Tetrahedron Lett 39:6781–6784, 1998.
204. S Hannessian, F Xie. Tetrahedron Lett 39:733–736, 1998.
205. S Hannessian, F Xie. Tetrahedron Lett 36:737–740, 1998.
206. M Mergler, R Tanner, J Gosteli, P Grogg. Tetrahedron Lett 29:4005–4008, 1998.
207. H Rink. Tetrahedron Lett 28:3787–3790, 1987.
208. F Albericio, G Barany. Tetrahedron Lett 32:1015–1018, 1991.
209. SM Hutchins, KT Chapman. Tetrahedron Lett 35:4055–4058, 1994.
210. MM Murphy, JR Schullck, EM Gordon, MA Gallop. J Am Chem Soc 117: 7029–7030, 1995.
211. MF Gordeev, DV Patel, J Wu, EM Gordon. Tetrahedron Lett 37:4643–4646, 1996.
212. MF Gordeev, HC Hui, EM Gordon, DV Patel. Tetrahedron Lett 38:1729–1732, 1997.
213. DG Mullen, G Barany. J Org Chem 53:5240, 1988.
214. R Ramage, CA Barron, S Bielecki, DW Thomas. Tetrahedron Lett 28:4105–4108, 1987.
215. R Ramage, CA Barron, S Bielecki, R Holden, DW Thomas. Tetrahedron 48: 499–514, 1992.
216. A Routledge, HT Stock, SL Flitsch, NJ Turner. Tetrahedron Lett 38:8287–8290, 1997.
217. H Chao, MS Bernatowicz, PD Reiss, CE Klimas, GR Matsueda. J Am Chem Soc 116:1746, 1994.
218. AD Piscopico, JF Miller, K Koch. Tetrahedron Lett 38:7143–7146, 1997.
219. K Akaji, Y Kiso, LC Carpino. J Chem Soc Chem Commun 584–586, 1990.
220. C Blackburn, A Pingali, T Kehoe, LW Herman, H Wang, SA Kates. Bioorg Med Chem Lett 7:823–826, 1997.
221. S Hoffman, R Frank. Tetrahedron Lett 35:7763–7766, 1994.
222. P Lloyd-Williams, F Albericio, E Giralt. Tetrahedron 49:11065–11133, 1993.
223. AJ You, RJ Jackman, GM Whitesides, SL Schreiber. Chem Biol 4:969–975, 1997.

224. A Borchardt, SD Liberles, SR Biggar, GR Crabtree, SL Schreiber. Chem Biol 4:961–968, 1997.
225. MJH Ohlmeyer, RN Swanson, LW Dillard, JC Reader, G Asouline, R Kobayashi, M Wigler, WC Still. Proc Natl Acad Sci U S A 90:10922, 1993.
226. A Ajayaghosh, VNR Pillai. J Org Chem 52:5714–5717, 1987.
227. CP Holmes. J Org Chem 62:2370–2380, 1997.
228. CP Holmes, DG Jones, BT Frederick, L-C Dong. In: PTP Kaumaya, RS Hodges, eds. Peptides: Chemistry, Structure and Biology. Proceedings of the 14th American Peptide Symposium. Birmingham, UK: Mayflower Scientific, 1995, pp 44–45.
229. MF Gordeev, EM Gordon, DV Patel. J Org Chem 62:8177–8181, 1997.
230. DL McMinn, R Hirsch, MC Greenberg. Tetrahedron Lett 39:4155–4158, 1998.
231. RR Rock, SI Chan. J Org Chem 61:1526–1529, 1996.
232. A Routledge, C Abel, S Balasubramanian. Tetrahedron Lett 38:1227–1230, 1997.
233. E Atherton, CJ Logan, RC Sheppard. J Chem Soc Perkin Trans I 538–546, 1981.
234. M Renil, M Meldal. Tetrahedron Lett 36:4647–4650, 1995.
235. SM Hutchins, KT Chapman. Tetrahedron Lett 37:4869–4872, 1996.
236. Y Cheng, KT Chapman. Tetrahedron Lett 38:1497–1500, 1997.
237. E Nicolas, J Clemente, M Perello, F Albericio, E Pedroso, E Giralt. Tetrahedron Lett 33:2183–2186, 1992.
238. DL Marshal, IE Liener. J Org Chem 35:867–868, 1970.
239. JG Breitenbucher, CR Johnson, M Haight, JC Phelan. Tetrahedron Lett 39:1295–1298, 1998.
240. PP Fantauzzi, KM Yager. Tetrahedron Lett 39:1291–1294, 1998.
241. M Mutter, D Bellhof. Helv Chim Acta 67:2009–2016, 1984.
242. YZ Liu, SH Ding, JY Chu, AM Felix. Int J Peptide Protein Res 35:95–98, 1990.
243. F Rabanal, E Giralt, F Albericio. Tetrahedron Lett 33:1775–1778, 1992.
244. F Rabanal, E Giralt, F Albericio. Tetrahedron 51:1449–1458, 1995.
245. R Eritja, J Robles, JA Avino, F Albericio, E Pedroso. Tetrahedron 40:4171–4182, 1992.
246. R Eritja, J Robles, D Fernandez-Forner, F Albericio, E Giralt, E Pedroso. Tetrahedron Lett 32:1511–1514, 1991.
247. F Albericio, R Eritja, E Giralt. Tetrahedron Lett 32:1515–1518, 1991.
248. GI Tesser, JTWRAM Buis, ETM Wolters, EGAM Bothe-Helmes. Tetrahedron 32:1069–1072, 1976.
249. JTWRAM Buis, GI Tesser, RJF Nivard. Tetrahedron 32:2321–2325, 1976.
250. R Shwyzer, E Felder, P Failli. Helv Chim Acta 67:1316–1327, 1984.
251. SB Katti, PK Mistra, W Haq, KB Mathur. J Chem Soc Chem Commun 843–844, 1992.
252. CC Leznoff, JY Wong. Can J Chem 51:3756, 1973.
253. CC Leznoff, W Sywanyk. J Org Chem 42:3203, 1997.

254. S Chamoin, S Houldsworth, CG Kruse, WI Bakker, V Snieckus. Tetrahedron Lett 39:4179–4182, 1998.
255. WF DeGrado, ET Kaiser. J Org Chem 45:1295–1300, 1980.
256. WF DeGrado, ET Kaiser. J Org Chem 47:3258–3261, 1982.
257. MA Scialdone. Tetrahedron Lett 37:8141–8144, 1996.
258. R Mohan, YL Chou, MM Morrissey. Tetrahedron Lett 37:3963–3966, 1996.
259. A Golebiowski, SR Klopfenstein. Tetrahedron Lett 39:3397–3400, 1998.
260. J Mery, J Brugidou, J Derancourt. J Peptide Res 5:233–240, 1992.
261. J Mery, C Granier, M Juin, J Brugidou. Int J Peptide Protein Res 42:44–52, 1993.
262. AA Virgilio, SC Schurer, JA Ellman. Tetrahedron Lett 37:6961–6964, 1996.
263. S Kobayashi, I Hachiya, S Suzuki, M Moriwaki. Tetrahedron Lett 36:2809–2812, 1996.
264. S Kobayashi, I Hachiya, M Yasuda. Tetrahedron Lett 36:5569–5572, 1996.
265. M Suto. Tetrahedron Lett 38:211–214, 1997.
266. D Obrecht, C Abrecht, A Grieder, JM Villalgordo. Helv Chim Acta 80:65–72, 1997.
267. DS Doss, OB Wallace. Tetrahedron Lett 39:5701–5704, 1998.
268. I Sucholeiki. Tetrahedron Lett 35:7307–7310, 1994.
269. FW Forman, I Sucholeiki. J Org Chem 60:523–528, 1995.
270. AR Mitchell, SBH Kent, M Engelhard, RB Merrifield. J Org Chem 48:2845–2852, 1978.
271. PG Pietta, GR Marshall. J Chem Soc Chem Commun 650–651, 1970.
272. GR Matsueda, JM Stewart. Peptides 2:45–50, 1981.
273. ST Gaehde, GR Matsueda. Int J Peptide Protein Res 18:451–458, 1981.
274. Y Pei, RA Houghten, JS Keily. Tetrahedron Lett 38:3349–3352, 1997.
275. A Nefzi, M Giulianotti, RA Houghten. Tetrahedron Lett 39:3671–3674, 1998.
276. A Nefzi, JM Ostrech, JP Meyer, RA Houghten. Tetrahedron Lett 38:931–934, 1997.
277. AR Katritzky, L Xie, G Zhang, M Griffith, K Watson, JS Keily. Tetrahedron Lett 38:7011–7014, 1997.
278. F Albericio, N Kneib-Cordonier, S Biancalana, L Gera, RI Masada, D Hudson, G Barany. J Org Chem 55:3730–3743, 1990.
279. MS Bernatowicz, SB Daniels, H Koster. Tetrahedron Lett 30:4645–4648, 1989.
280. B Imperiali, RS Roy. J Org Chem 60:1891–1894, 1995.
281. RS Roy, B Imperiali. Tetrahedron Lett 37:2129–2132, 1996.
282. MF Gordeev, DV Patel, EM Gordon. J Org Chem 61:924–928, 1996.
283. S Kobayashi, Y Aoki. Tetrahedron Lett 39:7345–7348, 1998.
284. EG Brown, JM Nuss. Tetrahedron Lett 38:8457–8460, 1997.
285. AV Purandare, MA Poss. Tetrahedron Lett 39:935–938, 1998.
286. SM Miller, RJ Simon, S Ng, RN Zuckermann, JM Kerr, WH Moos. Bioorg Med Chem Lett 4:2657–2660, 1994.
287. RN Zuckermann, JM Kerr, SBH Kent, WH Moos. J Am Chem Soc 114:10646–10647, 1992.
288. Y Pei, WH Moos. Tetrahedron Lett 35:5825–5828, 1994.

289. SM Hutchins, KT Chapman. Tetrahedron Lett 37:4865–4868, 1996.
290. DA Goff, RN Zuckermann. Tetrahedron Lett 37:6247–6250, 1996.
291. H Yuasa, Y Kamata, S Kurono, H Hashimoto. Bioorg Med Chem Lett 8: 2139–2144, 1998.
292. AA Virgilio, JA Ellman. J Am Chem Soc 226:11580–11581, 1994.
293. A Marzinik, ER Felder. J Org Chem 63:723–727, 1998.
294. JP Mayer, GS Lewis, C McGee, D Bankaitis-Davis. Tetrahedron Lett 39: 6655–6658, 1998.
295. DH Drewry, SW Gerritz, JA Linn. Tetrahedron Lett 38:3377–3380, 1997.
296. RD Wilson, SP Watson, SA Richards. Tetrahedron Lett 39:2827–2830, 1998.
297. AMM Mjalli, S Sarshar, TJ Baiga. Tetrahedron Lett 37:2943–2946, 1996.
298. MA Lago, TT Nguyen, P Bhatnagar. Tetrahedron Lett 39:3885–3888, 1998.
299. C Blackburn, B Guan, P Fleming, K Shiosaki. Tetrahedron Lett 39:3635–3638, 1998.
300. C Blackburn. Tetrahedron Lett 39:5469–5472, 1998.
301. E Kaiser, RL Colescott, CD Bossinger, PI Cook. Anal Biochem 34:595–598, 1970.
302. M Larhed, G Lindeberg, A Hallberg. Tetrahedron Lett 37:8219–8222, 1996.
303. J Lee, WV Murray, RA Rivero. J Org Chem 62:3874–3879, 1997.
304. SJ Pakoff, TE Wilson, CY Cho, PG Shultz. Tetrahedron Lett 37:5653–5656, 1996.
305. JM Kim, Y Bi, SJ Paikoff, PG Shultz. Tetrahedron Lett 37:5305–5308, 1996.
306. KA Beaver, AC Siegmund, KL Spear. Tetrahedron Lett 37:1145–1148, 1996.
307. X Cao, EJ Moran, D Siev, A Lio, C Ohasi, AMM Mjalli. Bioorg Med Chem Lett 5:2953–2956, 1995.
308. PA Tempest, SD Brown, RW Armstrong. Angew Chem Int Ed Engl 35:640–642, 1996.
309. SW Kim, SM Bauer, RW Armstrong. Tetrahedron Lett 39:6993–6996, 1998.
310. RA Tommassi, PG Nantermet, MJ Shapiro, J Chin, WK-D Brill, K Ang. Tetrahedron Lett 39:5477–5480, 1998.
311. M Meisenbach, W Voelter. Chem Lett 1265–1266, 1997.
312. PA Sieber. Tetrahedron Lett 28:2107–2110, 1987.
313. Y Han, SL Bontems, P Hegyes, MC Munson, CA Minor, SA Kates, F Albericio, G Barany. J Org Chem 61:6326–6339, 1996.
314. WC Chan, SL Mellor. J Chem Soc Chem Commun 1475–1477, 1995.
315. EA Boyd, WC Chan, VM Loh Jr. Tetrahedron Lett 37:1647–1650, 1996.
316. B Henkel, W Zeng, E Bayer. Tetrahedron Lett 38:3511–3512, 1997.
317. M Noda, M Yamaguchi, E Ando, K Takeda, K Nokihara. J Org Chem 59: 7968–7975, 1994.
318. H-G Chao, MS Bernatowicz, GR Matseuda. J Org Chem 58:2640–2644, 1993.
319. M Patek, M Lebl. Tetrahedron Lett 32:3891–3894, 1991.
320. GW Kenner, JR McDermott, RC Sheppard. J Chem Soc Chem Commun 636–637, 1971.
321. BJ Backes, JA Ellman. J Am Chem Soc 116:11171–11172, 1994.
322. AM Aronov, MH Gelb. Tetrahedron Lett 39:4947–4950, 1998.

323. RP Hammer, F Albericio, L Gera, G Barany. Int J Peptide Protein Res 36: 31–45, 1990.
324. A Ajayaghosh, VNR Pillai. Tetrahedron Lett 36:777–780, 1995.
325. CP Holmes, DG Jones. J Org Chem 60:2318–2319, 1995.
326. BB Brown, DS Wagner, HM Geysen. Mol Divers 1:4, 1995.
327. R Rodebaugh, S Joshi, B Fraser-Reid, HM Geysen. J Org Chem 62:5660–5661, 1997.
328. SM Sternson, SL Schreiber. Tetrahedron Lett 39:7451–7455, 1998.
329. R Sola, P Sauguer, M-L David, R Pascal. J Chem Soc Chem Commun 1786–1788, 1993.
330. J Sola, R Mery, R Pascal. Tetrahedron Lett 37:9195–9198, 1996.
331. NW Hird, K Irie, K Nagai. Tetrahedron Lett 38:7111–7114, 1997.
332. M Crawshaw, NW Hird, K Irie, K Nagai. Tetrahedron Lett 38:7115–7118, 1997.
333. NJ Ede, AM Bray. Tetrahedron Lett 38:7119–7122, 1997.
334. LS Richter, MC Desai. Tetrahedron Lett 38:321–324, 1997.
335. U Bauer, W Ho, AMP Koskinen. Tetrahedron Lett 38:7233–7236, 1997.
336. L Mellor, C McGuire, WC Chan. Tetrahedron Lett 38:3311–3314, 1997.
337. K Ngu, DV Patel. J Org Chem 62:7088–7089, 1997.
338. JA Fehrentz, M Paris, A Heitz, J Velek, C-F Liu, F Winternitz, J Martinez. Tetrahedron Lett 36:7871–7874, 1995.
339. JA Fehrentz, M Paris, A Hietz, J Velek, C-F Liu, F Winternitz, J Martinez. J Org Chem 62:6792–6796, 1997.
340. TQ Dinh, RW Armstrong. Tetrahedron Lett 37:1161 1164, 1996.
341. T Weiland, J Lewalter, C Birr. Justus Liebigs Ann Chem 740:31–74, 1970.
342. AN Semenov, KY Gordeev. Int J Peptide Protein Res 45:303–304, 1995.
343. TH Chan, WQ Huang. J Chem Soc Chem Commun 909–910, 1985.
344. JT Randolph, KF McClure, SJ Danishevsky. J Am Chem Soc 117:5712–5719, 1995.
345. C Zheng, PH Seeberger, SJ Danishevsky. J Org Chem 63:1126–1130, 1998.
346. Y Hu, JA Porco. Tetrahedron Lett 39:2711–2714, 1998.
347. Y Han, SD Walker, RN Young. Tetrahedron Lett 37:2703–2706, 1996.
348. TL Boehm, HDH Showalter. J Org Chem 61:6498–6499, 1996.
349. B Chenera, JA Finkelstein, DF Veber. J Am Chem Soc 117:11999–12000, 1995.
350. ND Hone, SG Davies, NJ Devereux, SL Taylor, AD Baxter. Tetrahedron Lett 39:897–900, 1998.
351. KA Newlander, BC Chenera, DF Veber, NCF Yim, ML Moore. J Org Chem 62:6726–6732, 1997.
352. MJ Plunkett, JA Ellman. J Org Chem 60:6006–6007, 1995.
353. FX Woolard, J Paetsch, JA Ellman. J Org Chem 62:6102–6103, 1997.
354. M Bernard, WT Ford. J Org Chem 48:326–332, 1983.
355. I Hughes. Tetrahedron Lett 37:7595–7598, 1996.
356. C Hulme, J Peng, G Morton, JM Salvino, T Herpin, R Labaudiniere. Tetrahedron Lett 39:7227–7230, 1998.

6

Coupling Methods: Solid-Phase Formation of Amide and Ester Bonds

Fernando Albericio
University of Barcelona, Barcelona, Spain

Steven A. Kates
Consensus Pharmaceuticals, Inc., Medford, Massachusetts

I. INTRODUCTION

The formation of amides (Fig. 1) is a key step in the construction (synthesis) of natural organic compounds as well as in *de novo* design [1]. The amide bond is present in peptides (peptide bond), peptoids, oligocarbamates, oligoamides, and β-lactams as well as many compounds containing rigid scaffolds such as diketopiperazines, benzodiazepines, and hydantoins. Furthermore, compounds containing amide bonds are excellent precursors for the preparation of heterocyclic compounds [2] (see Chapter 15). Thus, amide bonds play an important role in the synthesis of a majority of the compounds of biological interest.

The controlled formation of the amide bond requires activation of the carboxylic group prior to reaction with the amino component. Another important feature present in many organic compounds is an ester bond, which is typically constructed via activation of a carboxylic acid followed by reaction with an alcohol function. In this chapter coupling methods primarily based on the activation of a carboxylic acid leading to the formation of amides or esters on a solid phase are discussed. Other methods based on the formation of esters via the displacement of halides or activated alcohol moi-

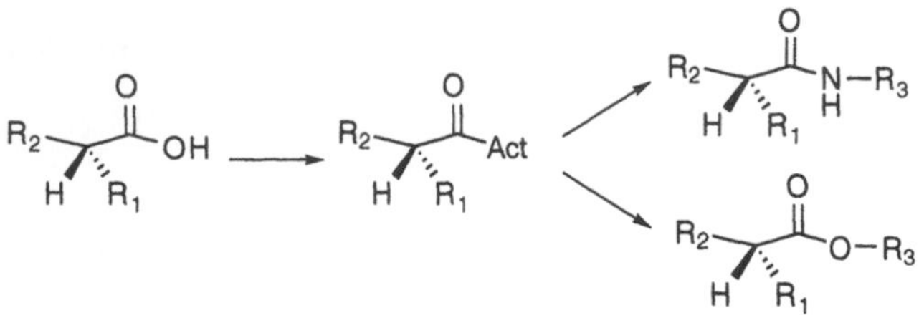

Figure 1   Formation of amides and esters.

eties by the carboxylic–carboxylate function are outside the scope of this chapter.

Coupling techniques have evolved rapidly [3–5], primarily from the field of solid-phase peptide synthesis because of the need to prepare long sequences and/or incorporate noncoded hindered amino acids [6–9]. In addition, with advances in combinatorial chemistry, in which a large number of compounds are prepared simultaneously or in parallel via the solid phase, coupling methods have attracted considerable interest [10]. Thus, coupling techniques should be efficient and reliable because they will typically be applied repetitively for a broad range of substrates [11]. Configurational integrity must also be maintained when the carboxylic component contains a chiral center in the α-position. The combination of yield and absence of racemization is often difficult to achieve because the best methods usually involve conversion of the acid to a derivative bearing a good leaving group. Such leaving groups tend to increase the acidity of the α-proton, which favors enolization and, if applicable, formation of a 5(4H)-oxazolone (oxazolonium ion in the case of N-alkyl amino acids), which are the two main mechanisms of racemization (Fig. 2) [12].

The two main classes of coupling techniques are those that require in situ activation of the carboxylic acid and those that depend on an activated species that has previously been prepared, isolated, purified, and characterized. As discussed in the following, the precise experimental conditions, e.g., organic solvent, preactivation step (when applicable), presence of base, temperature, time, and concentration, are often critical and should be considered when a difficult coupling is attempted. Furthermore, some of the coupling protocols most currently implemented are suggested by the vendors of automatic synthesizers and are often dictated by the technical and engineering characteristics of the instruments. Therefore, coupling reactions should be rationally evaluated before they are implemented in manual synthesis.

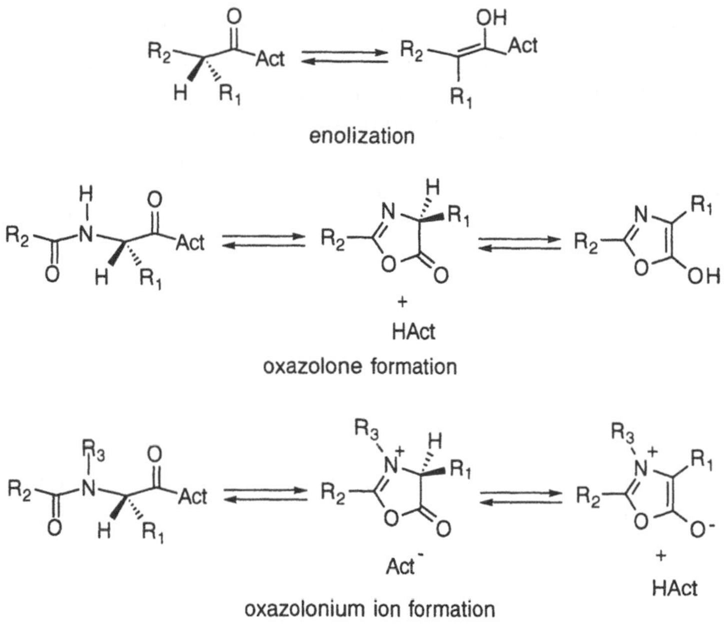

enolization

+

HAct

oxazolone formation

Act⁻

+

HAct

oxazolonium ion formation

Figure 2 Mechanisms of racemization.

II. COUPLING REAGENTS

There has been an evolution in the development of new coupling reagents and their application to solid-phase methodology. The formerly predominant carbodiimide reagents are being replaced by onium (phosphonium and aminium) salts. As discussed later, depending on the conditions, the activation step with onium salts occurs faster than with carbodiimides and usually leads to less perishable intermediates.

A. Carbodiimides

Carbodiimides were the coupling reagents most commonly used in solid-phase synthesis (SPS) until about 1985 (Fig. 3) [13–15]. For conducting SPS involving an intermediate treatment with trifluoroacetic acid (TFA) [e.g., solid-phase peptide synthesis (SPPS) by the *tert*-butoxycarbonyl (Boc)/

R–N=C=N–R

Figure 3 General structure of carbodiimides.

benzyl (Bzl) strategy], the N,N'-dicyclohexyl derivative (1) (DCC) is the most widely used carbodiimide, as the by-product N,N'-dicyclohexylurea (DCU) is eliminated from the reaction vessel owing to its solubility during treatment with this acid. The optimal carbodiimide for syntheses conducted with N,N'-dimethylformamide (DMF) as the main solvent [e.g., SPPS by the 9-fluorenylmethoxycarbonyl (Fmoc)/*tert*-butyl (*t*Bu) strategy] is N,N'-diisopropylcarbodiimide (2) (DIPCDI) because of the high solubility of the derived urea. It is important to note that DCU can very easily plug the frits of a reaction vessel or a column in the case of batch or continuous-flow syntheses. Substituting DIPCDI for DCC does not affect the activation rates [16]. Finally, 1-ethyl-3-(3'-dimethylaminopropyl)carbodiimide hydrochloride (3) [EDC, WSD (water-soluble carbodiimide)], whose urea is soluble in aqueous solvent mixtures, is mainly used for syntheses carried out in solution.

The mechanism of carbodiimide-mediated activation is complex, strongly dependent on the solvent, and still not totally understood (Fig. 4) [14,17–21]. The first step is a proton transfer, followed by addition of the carboxylic acid to form the O-acylisourea (4). This is a very reactive intermediate that attacks the amino component to give the corresponding amide. The O-acylisourea can undergo a rearrangement to give the N-acylurea (5), which is not as reactive as the corresponding O-derivative, or attack other carboxylic acid functions to give the symmetrical anhydride (6). If the carboxylic acid is an N-carboxamide or carbamate α-amino acid, the O-acylisourea can sustain an intramolecular cyclization to give a 5(4H)-oxazolone (7). Formation of the oxazolone occurs more readily because the carbonyl group of amides is more nucleophilic than that of urethanes [22]. Both symmetrical anhydrides and 5(4H)-oxazolone are also acylating reagents. The O-acylisourea can be trapped by a nucleophile present in the medium, predominantly hydroxylamine derivatives (R''R'NOH), to give a less reactive but more stable species (8).

The first step of the mechanism, the formation of the ion-pair intermediate, explains why the activation of a carboxylic acid is much slower in the presence of a base. Thus, DCC reacts 30-fold more slowly with acetic acid (HOAc) in the presence of triethylamine (9) (TEA) [18]. This lower activation rate in the presence of a base [N,N'-diisopropylethylamine (10) (DIEA)] was corroborated when Fmoc-amino acids were activated with DIPCDI in the presence of 1-hydroxybenzotriazole (11) (HOBt) [23] as a trapping agent [24]. However, once the activated species is formed, addition of a base accelerates the coupling reaction [24,25]. When the activation of carboxylic acids with carbodiimides is carried out in a solvent of low dielectric constant such as $CHCl_3$ or DCM, the formation of the O-acylisourea occurs instantaneously [26,27]. In the absence of a nucleophile or a base,

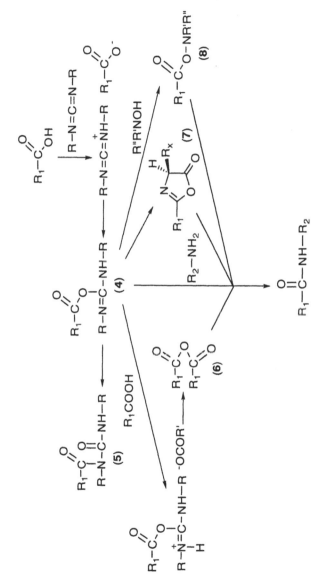

Figure 4 Mechanisms of peptide bond formation through carbodiimide activation.

the *O*-acylisourea can be stable for many hours [26]. However, if the activation is carried out in a more polar solvent such as DMF, no immediate reaction can be detected, and a complex mixture of starting amino acid derivative, symmetrical anhydride, *N*-acylurea, and urea is formed [27]. Thus, activation may be slower in a polar as opposed to a nonpolar solvent. Alternatively, an equilibrium between the carboxylic acid and the *O*-acylisourea may develop. In this case, even in the absence of a second equivalent of a carboxylic acid, the formation of the symmetrical anhydride occurs because the reaction of the *O*-acylisourea with the unreacted carboxylic acid is much faster than the addition of the acid to the carbodiimide. Furthermore, rearrangement of the *O*-acylisourea to the *N*-acylisourea occurs rapidly when DMF is incorporated as the solvent, which is a serious limitation of this process [27]. Formation of 2-alkoxy-5(4*H*)-oxazolone from the corresponding alkoxycarbonylamino acids [e.g., Boc, 9-fluorenylmethoxycarbonyl (Fmoc) derivatives] during carbodiimide activation occurs only in the presence of tertiary base salts. Under these conditions, racemization of 2-alkoxy-5(4*H*)-oxazolone derivatives does not appear to be problematic [12,27]. However, the alkylcarbonylamino acids (e.g., acetyl or peptides) are more prone to form the 5(4*H*)-oxazolone and thus more sensitive to racemization [12]. Addition of $CuCl_2$ has been shown to suppress the racemization of the 5(4*H*)-oxazolone intermediate [28].

When the activation step is performed in the presence of an equivalent of a hydroxylamine derivative [mainly HOBt [23] or 7-aza-1-hydroxybenzotriazole (**12**) (HOAt) [29]], the corresponding active esters are cleanly obtained. The main advantage of using these additives as trapping agents of the *O*-acylisourea is to increase the concentration of an active species when DMF is used as a solvent and to reduce loss of configuration at the carboxylic residue when applicable. The use of 1-oxo-2-hydroxydihydrobenzotriazine (**13**) (HODhbt or HOOBt) [30] as a potential additive is recommended less. Although the ODhbt esters are very reactive, formation of this derivative is accompanied by a side product, 3-(2-azido-benzyloxy)-4-oxo-3,4-dihydro-1,2,3-benzotriazine (**14**) (Fig. 5), which can react with the amino group to terminate chain growth [30].

HOAt has been described as superior to HOBt as an additive for both solution and solid-phase synthesis. HOAt enhances coupling rates and reduces the risk of racemization [31–33], possibly because it incorporates into the HOBt structure a nitrogen atom strategically placed at position 7 of the aromatic system. Incorporation of a nitrogen atom in the benzene ring has two consequences. First, the electron-withdrawing influence of a nitrogen atom (regardless of its position) effects stabilization of the leaving group, leading to greater reactivity. Second, placement specifically at position 7 provides a classic neighboring group effect that can both increase the reac-

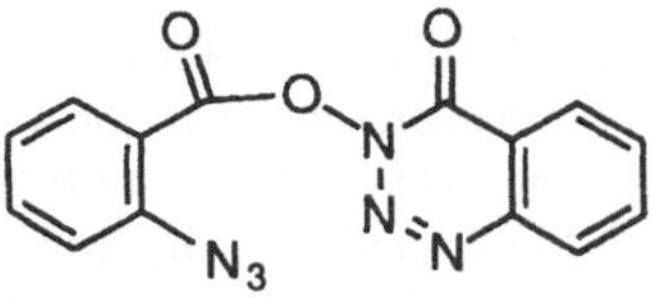

(1 4)

Figure 5 3-(2-Azido-benzyloxy)-4-oxo-3,4-dihydro-1,2,3-benzotriazine (**14**).

tivity and reduce loss of configuration (Fig. 6) [33]. The corresponding 4-isomer (**15**), although more acidic than HOAt and HOBt but lacking the ability to participate in such a neighboring group effect, has no influence on the extent of stereomutation during the segment coupling reaction relative to HOAt [34]. DIPCDI–HOAt coupling is also very favorable for the solid-phase preparation of peptides containing consecutive *N*-methylamino acids [35,36].

Adding a tertiary amine (DIEA) to a HOXt ester coupling has been described to be as efficient as the use of onium salts, which are the most powerful coupling reagents [24].

Two 1-hydroxytriazole derivatives have been described. Both present less steric hindrance than the benzotriazole derivatives and do not absorb in the ultraviolet (UV) at ~302 nm, allowing real-time monitoring of each coupling cycle. Ethyl 1-hydroxytriazole-4-carboxylate (**16**) (HOCt) [37] and 5-chloro-1-hydroxytriazole (**17**) [38] have been reported to be as effective as HOBt and HOAt, respectively, in providing fast coupling, although with less suppression of racemization.

The highly reactive preformed symmetrical anhydrides [39] are also very useful for both increasing the coupling rates, specifically with hindered nucleophiles [40], and reducing racemization [41,42].

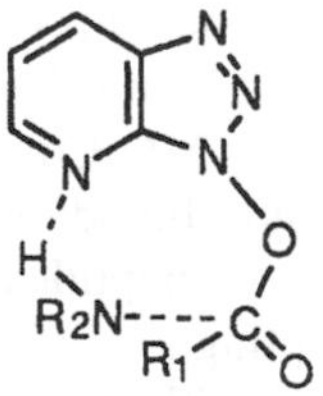

Figure 6 Assisted basic catalysis during the coupling of At esters.

The use of 4-dimethylaminopyridine (DMAP) (**18**) as a catalyst for enhancement of carbodiimide-based peptide couplings has been proposed [43] but the potential for racemization is increased. This side reaction is more significant when the method is applied to the acylation of hydroxy-methyl-based resin [44]. If the amount of DMAP is reduced to 10% with respect to protected amino acids and carbodiimide and it is added to the resin when the active species is already formed, the racemization is reduced substantially. In most cases this procedure is a convenient method for in-corporating the first amino acid in the resin [44]. Carbodiimides in the pres-ence of DMAP have also been applied for the solid-phase formation of depside and depsipeptide bonds with low racemization [45,46].

Additives such as HOBt or DMAP can be used while attached to a polymer. Thus, the polymeric *N*-benzyl-1-hydroxybenzotriazole-6-sulfona-mide (**19**) [47] and the polymeric 1-hydroxybenzotriazole (**20**) [48] have been shown to be highly efficient for the solution synthesis of amides. The efficiency of **19** could be attributed to its high acidity, conferred by the sulfonyl moiety. The procedure for amide construction involves the forma-tion of an activated ester on the derivatized polymer followed, in a second step, by treatment with an amine to generate the amide in solution. This HOBt-supported polymer has also been applied for the preparation of *N*-hydroxysuccinimide esters, useful for the modification of proteins [49]. Poly-meric DMAP is a less basic compound and generally gives very low race-mization [50].

1. General Procedures for Coupling Reaction Using Carbodiimides

Couplings described in this section, as well as throughout this chapter, can be carried out either via automation or manually. Manual synthesis can be performed in a polypropylene syringe fitted with a polyethylene disk and a stopcock with occasional stirring or in a mechanically shaken silanized screw-cap reaction vessel with a Teflon-lined cap, a sintered glass frit, and a stopcock [21]. All procedures presented will be limited to protocols for manual synthesis, because those for automatic synthesizers are either dic-tated by the manufacturer or rapidly adapted from the manual procedure. The concentration of the activated species should be maintained at a maxi-mum (0.6–1 M based on the solubility of the reagents). NMP or other convenient solvents as discussed earlier can be used rather than DMF. Cou-pling times are dependent on the nature of the substrates and should be investigated for each case. For solid-phase standard assembly of peptides, typically a 15–60 min cycle time provides a quantitative coupling.

Carbodiimide-mediated couplings are usually carried out with preac-tivation of the protected amino acid at either 25°C or 4°C using dichloromethane (DCM) as a solvent. For Fmoc-amino acids that are not totally soluble in DCM, mixtures with DMF may be used [40]. The use of DCM may be optimal for the solid-phase acylation of isolated nucleophiles [40], but for linear assembly, where interchain aggregation may occur, the use of a more polar solvent to inhibit the tendency to form secondary structure is recommended. In these cases if DCM is used as an initial solvent, a more effective procedure would involve filtration of the urea by-product and DCM evaporation followed by addition of DMF as the coupling medium [51–53]. For large-scale synthesis, preactivation at 4°C is advised because of the exothermic nature of the reaction.

Carbodiimides as well as other coupling reagents are acute skin irritants for susceptible individuals and should be treated with care. Thus, manipulation in a well-ventilated hood, using glasses, gloves, and if possible, a face mask, is recommended. DCC, which has a low melting point, can be handled as a liquid by gentle warming of the reagent container [54]. HOBt normally crystallizes with one molecule of water. Use of the hydrated form is perfectly satisfactory, but if the anhydrous material is required the dehydration should be carried out with care. Heating of HOBt or HOAt above 180°C can cause rapid exothermic decomposition [55]. The use of N-hydroxytetrazoles as trapping reagents should be precluded because of their tendency to explode [38].

Boc-Amino Acids via Symmetric Anhydrides [56]. Boc-amino acid (8 equiv.) is dissolved in DCM at 4°C; then DCC (4 equiv.) is added and the mixture is stored for 5–15 min at 4°C. The solution is filtered and, optionally, DCM is evaporated and the solution diluted with DMF and added to the resin containing the free amino group. After 30 min the resin can be filtered and washed with DMF. For DMAP-catalyzed couplings, 0.1 equiv. of DMAP is added to the resin after the anhydride [44].

Fmoc-Amino Acids via Symmetric Anhydrides [40]. A procedure similar to the preceding one is used except that the Fmoc-amino acid is dissolved in DCM containing the minimum amount of DMF required for dissolution. Also, DIPCDI is substituted for DCC and, after preactivation, the solution is not filtered because the N,N-diisopropylurea does not precipitate. This method can be used for all protected amino acids, except Boc- and Fmoc-Asn/Gln, side-chain protected Boc- and Fmoc-Arg, and Boc-His(Dnp). For these derivatives, the use of HOXt esters is recommended.

Boc- and Fmoc-Amino Acids via HOXt Esters (X = B, A) [31]. A procedure similar to that just described is used except that only 4 equiv. of protected amino acid is dissolved in DCM or DCM–DMF followed by the addition of HOXt (4 equiv.). Following treatment of the resin with the active species, DIEA (4 equiv.) can be added optionally [24].

B. Phosphonium and Aminium or Uronium Salts

Independent work of Kenner [57], Castro [58], Hruby [59], and Yamada [60] has shown the potential of applying derivatives of trisdimethylaminophosphonium salts for the activation of carboxylic acids and subsequent preparation of amides and peptides (Fig. 7). The first phosphonium salt–based reagents commercially available were μ-oxo-bis-[tris(dimethylamino)-phosphonium]-bis-tetrafluoroborate (**22**) ("Bates reagent") [61] and benzotriazol-1-yl-N-oxy-tris(dimethylamino)phosphonium hexafluorophosphate (**23**) (BOP) [62]. Later, Coste et al. [63] described the pyrrolidino derivative of BOP, benzotriazol-1-yl-N-oxy-tris(pyrrolidino)phosphonium hexafluorophosphate (**24**) (PyBOP), which does not form carcinogenic hexamethylphosphoric triamide (HMPA) as a by-product [64]. Furthermore, phosphonium salts derived from HOAt, such as (7-azabenzotriazol-1-yloxy)-tris(dimethylamino)-phosphonium hexafluorophosphate (**25**) (AOP) and (7-azabenzotriazol-1-yloxy)-tris(pyrrolidino)phosphonium hexafluorophosphate (**26**) (PyAOP), have also been prepared and are generally more efficient than BOP and PyBOP [31,65,66].

The precise nature of the mechanism involving the use of phosphonium reagents is still not known. Since the introduction of these derivatives, a controversy has arisen regarding the possible intermediacy of an acyloxyphosphonium salt and its lifetime. Several authors [57,59–61] proposed that in the absence of a nucleophile incorporated in the reagent, such as a moiety of HOBt in BOP, the active species is the acyloxyphosphonium salt. Castro and Dormoy [67] postulated that this salt is very reactive and even at low temperatures reacts immediately with carboxylate ions present in the medium to give the symmetrical anhydride. This pathway is supported by kinetic studies carried out by Hudson [16]. Kim and Patel [68] reported that this intermediate should exist at $-20°C$ when BOP was used as a coupling reagent. However, Coste and Campagne [69] suggested that this species is

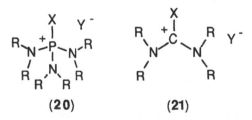

(20)   **(21)**

Figure 7 General structure of phosphonium (**20**) and aminium or uronium salts (**21**).

very unstable and even at low temperature undergoes conversion to an active ester.

The preparation and implementation in peptide synthesis of analogues of phosphonium salts such as HBTU (**27**) and HATU (**28**) bearing a positive carbon atom in place of the phosphonium residue have also been reported (Fig. 7). Although they were initially assigned a uronium-type structure [29,31,70,71], presumably by analogy to the corresponding phosphonium salts, it has been determined by X-ray analysis that *N*-[(1*H*-benzotriazol-1-yl)(dimethylamino)methylene]-*N*-methylmethanaminium hexafluorophosphate *N*-oxide (HBTU), *N*-[(dimethylamino)-1*H*-1,2,3-triazolo[4,5-*b*]pyridino-1-ylmethylene]-*N*-methylmethanaminium hexafluorophosphate *N*-oxide (HATU), and 1-(1-pyrrolidinyl-1*H*-1,2,3-triazolo[4,5-*b*]pyridin-1-ylmethylene)pyrrolidinium hexafluorophosphate *N*-oxide (**29**) (HAPyU) crystallize as aminium salts (guanidinium *N*-oxides) [72,73]. Nuclear magnetic resonance (NMR) studies in the case of HAPyU showed that the same structure is found in solution [73].* TBTU (**30**) and TATU (**31**), the tetrafluoroborate salts related to HBTU and HATU, have also been synthesized and incorporated in SPPS. The counterion (hexafluorophosphate or tetrafluoroborate) does not appear to affect reactivity [74]. In addition, the tetrafluoroborate derivatives are more soluble in DMF than their corresponding hexafluorophosphate salts.

Aminium salts can react with the *N*-terminal amino component, leading to a guanidino derivative, a process that terminates peptide chain elongation [75–78]. This side reaction is not prevalent during the solid-phase coupling of single amino acids, because the activation step often occurs before the addition of the amino component. Typically, carboxylic acid activation is fast and the aminium salt is rapidly consumed before exposure to a resin containing an amino terminus. However, during the much slower activation of hindered amino acids, protected peptide segments, or carboxylic acids before cyclization, the aminium salt may react with the amino component.

A study has correlated the reactivities of different onium salts based on HOXt [66]. The nature of the different reagents, phosphonium versus aminium or uronium, *N*-alkyl substituents (R), and OAt versus OBt influences were investigated. HXTU, HAPyU, PyXOP, XOP, the piperidino derivative, *O*-(7-azabenzotriazol-1-yl)-1,1,3,3-bis(pentamethylene)uronium hexafluorophosphate (**32**) (HAPipU), both dihydroimidazole derivatives, *O*-(7-azabenzotriazol-1-yl)-1,3-dimethyl-1,3-dimethyleneuronium hexaflu-

---

*For compounds for which X-ray data have not yet been obtained, generic structural representations are shown. For these reagents, traditional nomenclature, such as uronium salts, has arbitrarily been retained.

orophosphate (HAMDU) (**33**), *O*-(benzotriazol-1-yl)-1,3-dimethyl-1,3-dimethyleneuronium hexafluorophosphate (**34**) (HBMDU, BOI [79]), the trihydropyrimidine derivative, *O*-(7-azabenzotriazol-1-yl)-1,3-dimethyl-1,3-trimethyleneuronium hexafluorophosphate (**35**) (HAMTU), and the uronium salt derived from HODhbt, *O*-(3,4-dihydro-4-oxo-1,2,3-benzotriazin-3-yl)1,1,3,3-tetramethyluronium hexafluorophosphate (**36**) (HDTU) [71,80] were examined. Model experiments were designed to evaluate the stability of these reagents in solution and determine the ease of activation of a hindered carboxylic acid [Fmoc-diethylglycine (Fmoc-Deg-OH)] and coupling with an amino component (fluorenylmethyl ester of phenylalanine hydrochloride) as well as the amount of undesired side reactions.

The following conclusions were reported [66]:

1. Aza derivatives are more reactive in both steps, activation and coupling.
2. Phosphonium salts are more unstable than aminium or uronium derivatives in the absence of a base. In the presence of a base, the phosphonium reagents are slightly more stable, indicating that aminium or uronium salts are more reactive in coupling conditions.
3. The carbon skeleton structure has a determinate role in the efficiency of the reagent for the activation step. Thus, the pyrrolidino (HAPyU) derivative is much more reactive than the piperidino (HAPipU), which is more reactive than tetramethyl (HATU). The same phenomenon occurs with the phosphonium salts (PyXOP is preferred to XOP).
4. All the coupling reagents, except HAMDU, HBMDU, and HAMTU, can be incorporated efficiently in peptide synthesis.
5. For the optimal activation of a hindered amino acid, such as Fmoc-Deg-OH, the presence of DIEA rather than a weaker base such as 2,4,6-trimethylpyridine (**37**) (TMP, collidine) is recommended.
6. All coupling reagents including the most reactive HAMDU require the presence of a base for activation.
7. HDTU is very efficient for activating the carboxylic group, but the corresponding active ester is less reactive than both the OAt and OBt esters. Furthermore, the use of this reagent can lead to the formation of side products, possibly a triazine derivative. These results have been corroborated by the parallel synthesis of ACP (65–74) carried out with HATU and HDTU [80]. The appearance of extra peaks in the high-performance liquid chromatographic (HPLC) trace of the synthesis carried out with HDTU is indicative of side products.

8. When using the preceding reagents in automatic solid-phase peptide instruments, where the coupling reagent is stored in solution under an N_2 atmosphere, one can safely continue the synthesis for a time that exceeds that required to assemble a long sequence (2–3 weeks). When using peptide synthesizers in which the coupling reagent solutions are stored in an open vessel, the use of freshly prepared solutions of the least reactive HOAt-based reagents, such as HATU or HAPipU, is advised.

9. The formation of guanidino side products is not critical for the stepwise solid-phase synthesis of standard peptides, in which a preactivation of reactive amino acids is conducted. However, if activation in situ is carried out or activation of hindered amino acids, protected peptides, or a carboxylic moiety before a cyclization step is performed, the use of phosphonium salts may be more efficient. For slow activation reactions, the addition of more coupling reagent (e.g., PyAOP) during the course of the reaction is advisable because the coupling derivative is hydrolyzed after a few minutes.

HOAt-based onium salts have been shown to be very favorable for the coupling of hindered amino acids as demonstrated by excellent syntheses of peptides containing N-methylamino acids or dialkylglycine (Aib or Deg) [31,81]. Examples include [MeLeu^1]cyclosporin A (Cs A) [34,65] and alamethicinamide [82]. In the latter case, results obtained with HATU were only slightly inferior to those acquired via isolated acid fluorides, which represent the reagents of choice for such highly hindered amino acids [83]. For the acylation of an N-trialkoxybenzylamino acid, the optimal conditions for the reaction were to conduct it in DCM using HATU without preactivation [40]. In this example, the use of DMF as a solvent gave inferior results. In addition, the use of nonpolar solvents such as DCM reduces the racemization during the coupling [32]. These reagents have also been advantageous for the coupling of protected peptides [84] in a convergent approach [85] and for depside [86] and depsipeptide [87] bond formation. A Boc/Bzl strategy was described that employs HATU/DMSO in situ neutralization as the coupling method and incorporates a protected amino acid residue every 5 min to produce peptides of good quality [88]. This rapid coupling chemistry was successfully demonstrated by synthesizing several small to medium-sized peptides, including the "difficult" C-terminal sequence of human immunodeficiency virus type 1 (HIV-1) protease (residues 81–99); fragment 65–74 of the acyl carrier protein; conotoxin PnIA(A10L), a potent neuronal nicotinic receptor antagonist; and the proinflammatory chemotactic protein CP10, an 88-residue protein, by means of native chemical ligation.

Coupling the third amino acid to a peptide chain assembled on hydroxymethylbenzyl resins requires special attention. These resins are prone to lead to substantial amounts of diketopiperazine formation during N-deprotection of the second amino acid in the sequence [89,90]. This intramolecular cyclization is favored by the presence of Pro or Gly in the first two positions of the sequence. For Boc chemistry, the formation of DKP can be diminished by using coupling methods in which neutralization of the dipeptide-resin is carried out in situ during acylation of the third amino acid rather than by neutralization in a separate wash step by a tertiary amine. A convenient protocol involves removing of the Boc group with TFA–DCM and carrying out coupling of the next amino acid via PyBOP or PyAOP in the presence of DIEA [91]. For the Fmoc/tBu strategy, the second amino acid has to be introduced using N-trityl (Trt) protection, as the trityl group can be selectively removed with a very dilute acid solution (0.2–1% TFA in DCM) in the presence of tBu-based protecting groups [92].

Related coupling reagents based on hydroxy derivatives other than HXBt, such as 2-[2-oxo-1(2H)-pyridyl]-1,1,3,3-tetramethyluronium tetrafluoroborate (**38**) (TPTU), [93], 2-[2-oxo-1(2H)-pyridyl]-1,1,3,3-bis-(pentamethylene)uronium tetrafluoroborate (**39**) (TOPPipU) [94], 2-succinimido-1,1,3,3-tetramethyluronium tetrafluoroborate (**40**) (TSTU) [71,95], 2-phtalimido-1,1,3,3-tetramethyluronium tetrafluoroborate (**41**) (TPhTU) [75], 2-mercaptopyridine-1,1,3,3-tetramethyluronium tetrafluoroborate (**42**) (TMTU) [75], 2-(*endo*-5-norbornene-2,3-dicarboximido)-1,1,3,3-tetramethyluronium tetrafluoroborate (**43**) (TNTU) [71,74], 2-[6-(trifluoromethyl)benzotriazol-1-yl]-1,1,3,3-tetramethyluronium hexafluorophosphate (**44**) (CF$_3$-HBTU) [96], 2-pentafluorophenyl-1,1,3,3-tetramethyluronium hexafluorophosphate (**45**) (PfTU) [97], and bis(tetramethylene)pentafluorophenoxyformamidinium hexafluorophosphate (**46**) [(PfPyU) [97], (HPy-OPfp) [98,99]], [(6-nitrobenzotriazol-1-yl)oxy]tris(pyrrolidino)phosphonium hexafluorophosphate (**47**) (PyNOP) [100], [(6-trifluorobenzotriazol-1-yl)oxy]tris(pyrrolidino)phosphonium hexafluorophosphate (**48**) [(CF$_3$-PyBOP) [96], (PyFOP) [100]], [(3,4-dihydro-4-oxo-1,2,3-benzotriazin-3-yl)oxy]-tris(pyrrolidino)phosphonium hexafluorophosphate (**49**) (PyDOP) [100], [(pentafluorophenyl)oxy]tris(pyrrolidino) phosphonium hexafluorophosphate (**50**), [(PyPOP) [100], (PfOP) [97], (PyPfpOP) [98]], (pyridyl-2-thio)tris(pyrrolidino) phosphonium hexafluorophosphate (**51**) (PyTOP) [100], and 1,3-dimethyl-2-(1-pyrrolidinyl)-2-(3H-1,2,3-triazolo(4,5-b)-pyridin-3-yloxy)-1,3,2-diazaphospholidinium hexafluorophosphate (AOMP) [98] (**52**), have also been used successfully in both solid-phase and solution modes.

Coupling reagents in which the HOXt and HOPfp moieties were substituted by their corresponding thio derivatives, *S*-(7-azabenzotriazol-yl)-1,1,3,3-tetramethylthiouronium hexafluorophosphate (**53**) (HATTU) [98,101], *S*-(7-azabenzotriazol-yl)-1,1,3,3-bis(tetramethylene)thiouronium hexafluorophosphate (**54**) (HAPyTU) [98], and *S*-(pentafluorophenyl)-1,1,3,3-bis(tetramethylene)thiouronium hexafluorophosphate (**55**) (HPySPfp) [98] were investigated. The results with these derivatives demonstrated (1) that the thio derivatives afford more epimerization without an increase in yield and (2) the superiority of the HOAt-based reagents. In addition, it was shown that the use of HOPfp-based uronium salts can be improved by addition of 1 equiv. of HOAt [98,99].

Among the most reactive of the common coupling reagents are the preformed amino acid fluorides (see the following) [102]. Rather than obtaining an isolated acid fluoride, it is more convenient to prepare these efficient reagents by in situ generation via the aminium reagent tetramethylfluoroformamidinium hexafluorophosphate (**56**) (TFFH) [103]. A potential problem with carbamate-protected halides is conversion to the corresponding 5(4*H*)-oxazolone [102]. TFFH treatment to Fmoc-dimethylglycine (Fmoc-Aib-OH) in the presence of various solvents and bases was studied as a model [104]. Optimal conditions were obtained with 2 equiv. of DIEA in DMF. With DCM, large amounts of 5(4*H*)-oxazolone accompanied the formation of the acid fluoride. For this solvent, increasing the concentration of base gave less acid fluoride, and among several pyridine bases that were examined [pyridine (**57**), TMP, 2,6-di-*t*-butyl-4-methylpyridine (**58**), 2,6-di-*t*-butyl-4-(dimethylamino)pyridine [DB(DMAP)] (**59**), and 2,3,5,6-tetramethylpyridine (**60**)] [105,106], pyridine itself was the most effective although not as efficient as DIEA in DMF.

TFFH proved fully effective for the two proteinogenic amino acids (Arg and His) that cannot be converted to shelf-stable Fmoc-protected amino acid fluorides [104]. The acid fluoride of Fmoc-Arg(Pbf)-OH was generated using 2 equiv. of DIEA in DMF in less than 2 min, and although cyclization to the corresponding lactam occurred slowly, significant amounts of acid fluoride remained after 60 min. In this case TMP led to formation of the 5(4*H*)-oxazolone within 2 min but conversion of the 5(4*H*)-oxazolone to the acid fluoride required an additional 15 min. In DCM, cyclization to lactam occurred readily (30 min) regardless of the base used. TMP proved to be the most efficient activator base in the case of Fmoc-Asn(Trt)-OH [104]. For this residue, better results are obtained if 1 equiv. of HOAt is present during the coupling process [103].

The pyrrolidine and dihydroimidazole derivatives of TFFH, bis-(tetramethylene)fluoroformamidium hexafluorophosphate (**61**) (BTFFH) and

1,3-dimethyl-2-fluoro-4,5-dihydro-1*H*-imidazolium   hexafluorophosphate
(**62**) (DFIH), have also been prepared and tested in SPPS [107,108]. Like
the OAt analogues, both are more reactive than the parent TFFH, but this
increased reactivity does not lead to cleaner reaction products.

The chloro and bromo derivatives of phosphonium and uronium salts
have not been extensively used in solid-phase strategy. Thus, chloro- and
bromo-tris(pyrrolidino)-phosphonium hexafluorophosphate [PyCloP (**63**) and
PyBroP (**64**)] [109] and bromotris(dimethylamino)-phosphonium hexafluorophosphate (**65**) (BroP) [110], which have been used with success in solution for the synthesis of peptides containing *N*-methylamino acids, have
not been found to be effective in the solid-phase mode. The active species
detected for PyCloP, PyBroP, and BroP in the absence of HOBt are the
symmetrical anhydride, the 5(4*H*)-oxazolone, and, for Boc-amino acids, the
N-carboxyanhydride [99]. Furthermore, 2-chloro-1,3-dimethyl-4,5-dihydro-
1*H*-imidazolium hexafluorophosphate (**66**) (CIP, DCIH) [107,108,111] has
been found effective only in the presence of 1 equiv. of HOAt.

As previously mentioned, the use of these various onium salts requires
careful attention to the tertiary base and the preactivation time. Although
automated instrumentation may have limitations with regard to protocol editing, the proper preactivation time should be selected to ensure more effective coupling. In manual syntheses, this step can easily be monitored with
a stopwatch. For onium salts incorporating HOAt, the activation of proteogenic amino acids gives the corresponding OAt esters almost instantaneously. Therefore, the preactivation time should be kept to a minimum to
prevent the activated species experiencing several side reactions including
racemization and formation of δ-lactam (Arg), cyano derivatives (Asn or
Gln), or α-aminocrotonic acid (Thr) (see later). Similar considerations
should be applied to coupling reagents that incorporate HOBt. For TFFH,
longer preactivation times may be required depending on the specific amino
acid and activating base.

It has been reported that serine [112,113] and, primarily, cysteine
[41,42,114] derivatives can undergo substantial levels of racemization specifically during incorporation into a peptide chain by application of these
reagents. For example, using standard protocols involving a 5-min preactivation, for BOP, PyAOP, HBTU, and HATU, the level of racemization in a
model Cys-containing peptide was in the range 5–33% [42]. However, these
levels were generally reduced by a factor of 6 or 7 by avoiding the preactivation step [42]. Additional strategies to minimize racemization included
changing to a hindered and/or weaker base such as TMP or DB(DMAP),
twofold reduction of the amount of base, and modification of the solvent
from neat DMF to the less polar DCM–DMF (1:1) [115].

1. General Procedures for Coupling Reaction Using Onium Salts

Boc- and Fmoc-Amino Acids via Phosphonium and Aminium Salts.   Protected amino acid (4 equiv.) and the phosphonium or aminium salt (4 equiv.) are dissolved in DMF or NMP, the mixture is added to the resin bearing the free amino group, and finally the base (4–8 equiv.) is added. After 30 min the resin is filtered and washed with DMF or NMP. For Fmoc-Asn(Trt) coupling, HOXt (4 equiv.) should be added. Some researchers prefer to carry out a preactivation step involving the protected amino acid, the coupling reagent, the base, and HOXt for 10 min. This protocol is preferred in the case of aminium salt couplings, where chain termination via formation of the guanidino species can intervene. Alternatively, it is possible to reverse the addition of base and the other reagents in order to avoid contact of the aminium salt with the free amino group, because the activation step is faster than formation of the chain-terminating guanidino derivative.

Boc-Amino Acids via Phosphonium and Aminium Salts with In Situ Neutralization.   The previously described procedure was used except that 8–12 equiv. of base was added during preactivation. When this procedure is performed for sequences susceptible to DKP formation, the use of 8 equiv. of base is preferred.

C. Other Coupling Reagents

Other phosphorus reagents (Fig. 8) have also been successfully applied for the in situ activation of carboxylic acids and their posterior reaction with amine derivatives in solid-phase strategy.

Derivatives of phosphinic acids have been used by Ramage and co-workers [116,117] in both solution and solid-phase approaches. The mechanism of the coupling was postulated to involve a carboxylic–phosphinic mixed anhydride [116]. An advantage to these mixed anhydride intermediates in comparison with the biscarboxylic derivatives is regioselectivity. Bis-

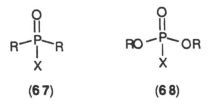

(6 7)   **(6 8)**

Figure 8   General structures of phosphinic (**67**) and phosphoric (**68**) acid coupling reagents.

carboxylic mixed anhydrides are governed by electronic and steric factors [118], and carboxylic–phosphinic derivatives are dependent on the nature of the nucleophile. Thus, aminolysis and alcoholysis occur at the carboxylic and phosphinic moieties, respectively (Fig. 9) [116,119,120]. Among all the derivatives studied, diphenylphosphinic chloride (69) (Dpp-Cl) and 1-oxo-1-chlorophospholane (70) (Cpt-Cl) are considered to be the optimal choices for routine use in peptide synthesis [120,121]. Furthermore, some "stand-alone" compounds, such as pentafluorophenyl diphenylphosphinate (71) (FDPP), have been implemented in the solid phase [122].

Derivatives of phosphoric acid, diphenylphosphoryl azide (72) (DPPA) [123], diethylphosphoryl cyanide (73) (DEPC) [124], bis(2-oxo-3-oxazolidinyl)phosphorodiamidic chloride (74) (BOP-Cl) [125,126], N-diethoxyphosphoryl benzoxazolone (75) (DEPBO) [127], N-(2-oxo-1,3,2-dioxaphosphorinanyl)-benzoxazolone (76) (DOPBO) [127], 3-[O-(2-oxo-1,3,2-dioxaphosphorinanyl)-oxy]-1,2,3-benzotriazin-4(3H)-one (77) (DOPBT) [129], 3-(diethoxyphosphoryloxy)-1,2,3-benzotriazin-4(3H)-one (78) (ENDPP) [127,128], and benzotriazol-1-yl-diethylphosphate (80) (BDP) [129] have performed well in solution but are totally unsuitable for solid-phase synthesis. DPPA has shown utility for cyclizations [130] and BOP-Cl for the acylation of N-methyl amino acid derivatives [131,132] and esterifications [126]. The effectiveness of BOP-Cl for these couplings is attributed to intramolecular base catalysis by the oxazolidinone carbonyl of the mixed anhydride active species (Fig. 10) [132].

A comparison of the dimethyl derivatives of phosphinic chloride (Me$_2$POCl) and phosphochloridate [(MeO)$_2$POCl] indicated that the latter was less reactive toward oxygen nucleophiles than dimethylphosphinic chloride, which suggests that phosphinic chloride derivatives should react with carboxylate anions to form mixed anhydrides more rapidly than phosphorochloridates [119]. Fast formation of mixed anhydrides is an important consideration in coupling reactions. A further advantage of the use of phosphinic carboxylic mixed anhydrides is the elimination of the potential substitution

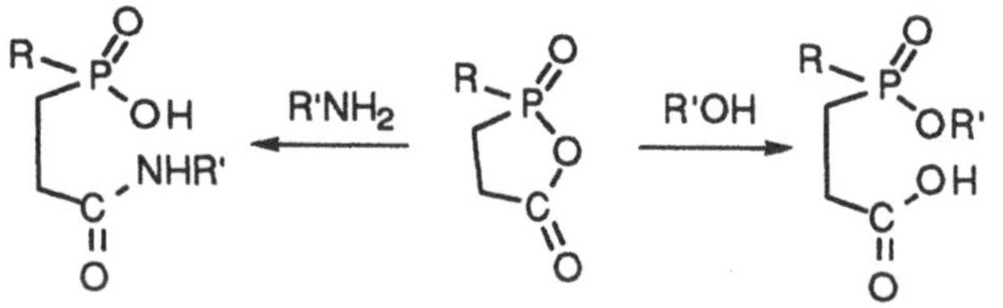

Figure 9 Regiospecific attack of carboxylic-phosphinic anhydrides depending on the nature of the nucleophile.

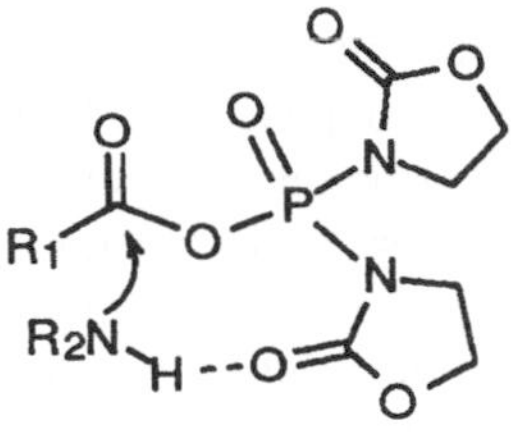

Figure 10   Intramolecular base catalysis during activation with BOP-Cl.

of the OR groups in mixed anhydride intermediate derivatives of phosphoric acids [119].

The oxidation–reduction method, developed initially by Mukaiyama et al. [133] and related to the previously described organophosphorus methods, has permitted a variety of important solid-phase applications. The mechanism of the activation is complex and involves the oxidation of the triaryl/alkyl-phosphine to the oxide as well as reduction of the disulfide to the mercapto derivative. However, different active species, such as **81** (Fig. 11), the 2-pyridyl thioester, or even the symmetrical anhydride, have been postulated to form. For the intermediate **81**, the peptide bond formation may proceed through a cyclic transition state. The method has been used for conventional stepwise synthesis [134], acylation of the first protected amino acid to a hydroxymethyl resin, and to achieve segment condensation on a solid support in the opposite direction (N → C) [135,136]. Lastly, it has been used for efficient grafting of a polyethylene glycol (molecular weight 2000) derivative to an aminomethyl resin to prepare PEG-PS resins [137].

Some sulfonates of strongly acidic N-hydroxy compounds, such as HOBt-substituted derivatives and HODhbt, are excellent coupling reagents for amide bond formation in solution [138] but of limited applicability in solid-phase mode. The main drawback of this method is the formation of the corresponding sulfonylamide [138]. The arenesulfonyl-1,2,4-triazoles,

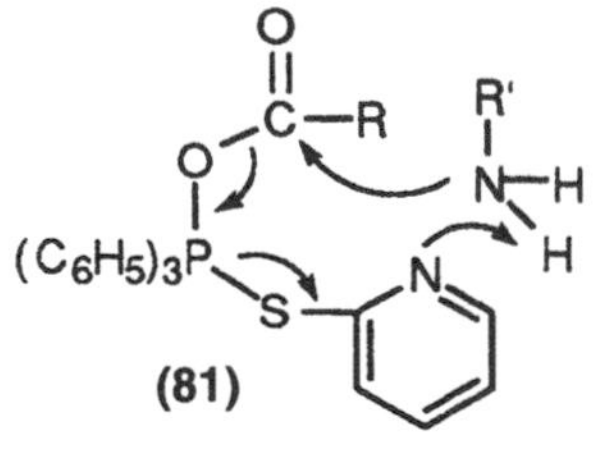

Figure 11   Active species and cyclic intermediate involved in the oxidation–reduction coupling method.

which were originally used as coupling reagents in phosphotriester oligo-nucleotide synthesis [139] have been applied to the solid phase for both elongation of the peptidic chain [140] and incorporation of the first amino acid in a hydroxymethyl resin [141]. In the former, the use of 1-(mesityle-nesulfonyl)-3-nitro-1,2,4-triazole (**82**) (MSNT) in the presence of *N*-meth-ylimidazole (**83**) (NMI) gave some sulfonylamide accompanied by racemi-zation of the carboxylic component [140]. The use of the most hindered sulfonyl derivative (1-(2,4,6-triisopropylbenzenesulfonyl)-3-nitro-1,2,4-tria-zole (**84**) (TPSNT) in the presence of 4-morpholine pyridine-1-oxide (**85**) (MPO) circumvents both problems [140]. However, the use of MSNT and NMI as opposed to carbodiimide-based methods is one of the best ways to acylate unreactive hydroxy resins [141].

1,3,5-Triazines (**86**) have also been used as coupling reagents in so-lution and in solid-phase mode (Fig. 12). 2-Chloro-4,6-dimethoxy-1,3,5-tri-azine (**87**) (CDMT) has been found useful for the coupling of hindered amino acids [142–144] and cephalosporin derivatives [145]. Serine can be incorporated without protection of its side-chain hydroxy group, and because of the weakly basic properties of the triazine ring, the side products and excess coupling reagent are easily removed by washing with dilute acids [143]. The related derivative 2,4,6-tris(pentafluorophenyloxy)-1,3,5-triazine (**88**) (TPfT) [97] was shown to be less reactive than Pfp phosphonium and uronium salt derivatives [97]. An advantage of TPfT is that this reagent is soluble even in very nonpolar solvents [97]. This property is useful if re-actions have to be carried out in solvent systems with low polarity. For CDMT- and TPfT-based couplings, the active species are the 2-acyloxy-4,6-dimethoxy-1,3,5-triazine [143], and the Pfp ester [97], respectively.

1. General Procedures for Coupling Reaction Using the Oxidation–Reduction Method [137]

The carboxylic acid derivative (4 equiv.) and 2,2'-dipyridyl disulfide (4 equiv.) are dissolved in DCM, the mixture is added to the resin bearing the free amino group, and finally tri-*n*-butylphosphine (4 equiv.) is added. Fol-

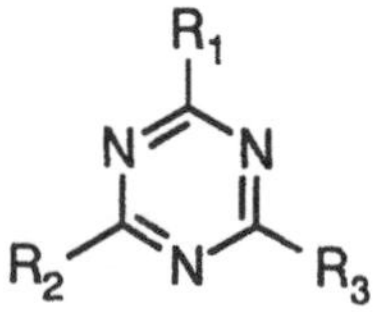

Figure 12   General structure of substituted 1,3,5-triazines (**86**).

lowing completion of the reaction, the solution is filtered off and the support extensively washed with DCM, DMF, and DCM.

2. General Procedures for Acylation of Hydroxymethyl Resins Using MSNTNMI [144]

The support material is carefully washed with DMF and dry DCM (distilled from P_2O_5). In a septum-stoppered flask, the dry amino acid derivative (freeze-dried from dioxane) (2 equiv.) and NMI (1.5 equiv.) are dissolved in dry DCM and, if required, with a few drops of THF to give a 0.1 M solution. This solution is transferred via syringe to another stoppered flask containing MSNT (2 equiv.) and the mixture is immediately added to the support. Upon reaction completion (30 min to 1 h), the solution is filtered off and the support extensively washed with DCM, DMF, and DCM.

III. ACTIVE SPECIES

The major difficulty in the preparation of an amide bond via isolated intermediates is the need to prepare or obtain an activated derivative. To apply this technique to the assembly of peptides, the full range of L- and nonproteinogenic amino acids prepared from the corresponding carboxylic acid is required. A disadvantage of activated derivatives is that the starting materials implemented in a coupling reagent–mediated synthesis are simpler because carboxylic acid functions are used. However, the use of activated species has some important advantages: (1) high reactivity for some derivatives, (2) simplicity of the manipulation (in some cases additives or base are not required), and (3) simplicity of the by-products released (e.g., carbon dioxide and fluoride ion). The active species most often handled are active esters, urethane-protected α-amino acid N-carboxyanhydrides (**89**) (UNCAs), acid fluorides (**90**), and acid chlorides (**91**) (Fig. 13). Symmetrical anhydrides can also be prepared analytically pure [146,147]. However, these reagents are

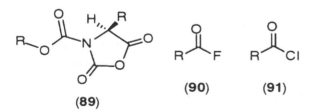

(89)   (90)   (91)

Figure 13 Structure of UNCA (**89**), acid fluorides (**90**), and acid chlorides (**91**).

preferably preformed in situ in DCM immediately before incorporation into a peptide chain (see earlier).

A. Active Esters

Active esters commonly used in solution (e.g., o-nitrophenyl and N-hydroxy-succinimide) [148] have limited applicability in the solid phase, primarily because of their low reactivity in comparison with carbodiimide-based methods. Thus, coupling reactions typically require a reaction time of 8–18 h [21]. More reactive esters, such as those derived from pentafluorophenol (HOPfp) [149] and HODhbt [30], have been predominantly implemented in automated continuous-flow Fmoc/tBu synthesis [152]. Both derivatives can be prepared by treatment of the protected amino acid and the hydroxy derivative with DCC [30,151]. HOPfp esters can also be synthesized from pentafluorophenyl trifluoroacetate [152]. During the preparation of HODhbt esters, the triazine side product can be removed. Therefore, the use of HODhbt preformed esters is safe and free of undesired side reactions, in contrast to employing HODhbt as an additive [30] or in a stand-alone derivative [80]. HOBt and HOAt are less stable than their corresponding HOPfp and HODhbt esters and are not generally isolated. HOPfp esters react moderately slowly, although the addition of HOAt accelerates their reaction [31,99]. Because HOPf esters are not extremely reactive derivatives, fewer or less extensive side reactions occur. As a result, the use of HOPfp allows incorporation of Asn and Gln without protection of the side-chain function [153]. Furthermore, Fmoc derivatives of Tyr, Ser, and Thr have been incorporated as preformed active esters without side-chain protection [154,155]. Because of their solubility, esters derived from 1-(4′-nitrophenyl)-pyrazolin-5-one (**92**) (HOHpp) [156] have been used successfully for the solid-phase incorporation of poorly soluble biomolecules such as biotin and fluorescein [157]. Finally, 2,3-dihydro-2,5-diphenyl-4-hydroxy-3-oxo-thiophen-1,1-dioxide (**93**) (TD) esters [158] incorporated the first amino acid in hydroxymethyl resins with very minimal racemization [159].

1. General Procedures for Coupling Reactions Using Preformed Active Esters [31,42,160,161]

The preformed ester of the amino acid (4 equiv.) and HOXt (4 equiv.) are dissolved in DMF and, after 5 min, the solution is added to the resin bearing the free amino group. After 30 min the resin is filtered and washed with DMF. For HODhbt ester coupling, the addition of an extra equivalent of HOXt may be avoided. For the coupling of cysteine derivatives, DCM–DMF is used instead of DMF to minimize the level of racemization.

2. General Procedures for Acylation of Hydroxymethyl Resins Using TDO Esters

A suspension of the resin in the minimal volume of DMF is shaken gently with a TDO ester of the amino acid [162] (3 equiv. with respect to hydroxyl function) and DIEA (3 equiv.) for 2 h at 25°C. The red solution is filtered and the resin washed thoroughly with DMF.

B. Urethane-Protected α-Amino Acid N-Carboxyanhydrides (UNCAs)

The main advantage of UNCAs [163] over classical NCAs or Leuchs' anhydrides [164] is their greater stability and lack of sensitivity to polymerization. Upon coupling, carbon dioxide is released, which is a distinct advantage compared with other methods [163,165]. UNCAs are conveniently prepared by initial phosgene treatment of bis-trimethylsilyl amino acids followed by reaction with Fmoc-Cl or BocON in the presence of N-methylmorpholine (94) (NMM) or pyridine, respectively [166]. These derivatives are crystalline solids that are stable in the absence of water for extended periods of time at low temperatures. UNCAs are very soluble in most organic solvents and are most reactive in DMF. Their reactivity is comparable to that of BOP- and HBTU-style reagents [167] and is particularly suitable for SPS of sterically hindered peptides [168]. Furthermore, hydroxyl-containing resins can be efficiently esterified with high levels of substitution and low racemization [169]. Excellent incorporation of His, a residue prone to racemize under traditional conditions, was observed [170]. Finally, UNCAs have been used for the preparation of β-keto esters and other derivatives by reaction with the appropriate organometallic reagents [171,172].

1. General Procedures for Coupling Reactions Using UNCAs [166]

The UNCA (4 equiv.) is dissolved in DMF and the solution is added to the resin bearing the free amino group. The addition of catalytic amounts of base (up to 1 equiv.) is optional, although recommended if it is performed as a second coupling of a double coupling protocol. After 30 min the resin is filtered and washed with DMF.

2. General Procedures for Acylation of Hydroxy-Containing Resins Using UNCAs

The UNCA (4 equiv.) dissolved in toluene and NMM (4 equiv.) is added to the hydroxy-containing resins and allowed to react for 15 and 90 min for the Wang-type and Rink resins, respectively.

C. Acid Halides

Although the first application of acid chlorides in peptide synthesis was reported in the 1930s, there are contrasting reports on the relative reactivity of the different halides and the mechanism of their action. It appears that for simple acid halides, the variance in relative order of reactivity depends on the nature of the nucleophile [102]. Thus, the fluorides are less reactive than the chlorides toward neutral oxygen nucleophiles, but the anion formation reverses the reactivity. With amines, benzoyl chloride is again more reactive than the corresponding fluoride [173]. Furthermore, there is a tendency for urethane-protected amino acid halides to convert to 5(4H)-oxazolones, which are less reactive intermediates. Thus, the benefits of employing amino acid halides sometimes may be negated [174,175].

D. Acid Chlorides

The practical use of amino acid chlorides is restricted to the amino acid derivatives that do not contain acid-labile protecting groups such as Boc or tBu [102]. Typically, only bifunctional Fmoc-amino acids fulfill this requirement. These derivatives can be prepared by treatment of the Fmoc-amino acid with thionyl chloride and can be stored indefinitely in a dry atmosphere [176]. Fmoc-derivatives of Tyr, Thr, and Ser with tBu-protected side-chains can be synthesized, but loss of the tBu group was observed on storage [175]. Coupling of these derivatives requires the presence of a hydrogen chloride acceptor. In the presence of a tertiary amine, such as DIEA or NMM, the corresponding 5(4H)-oxazolone is formed. If a hindered base such as 2,6-di-$tert$-butylpyridine is used as a hydrogen chloride scavenger, only a small amount of 5(4H)-oxazolone is formed [175].

Acid chlorides can be coupled in the presence of a 1:1 mixture of an amine [175] and HOBt or the potassium salts of HOBt [177]. In both cases, the corresponding OBt ester is formed initially. Fmoc-amino acid chlorides have been successfully applied in solid-phase syntheses [175], as well as for the incorporation of the first amino acid on hindered hydroxyl resins [178,179].

The utility of acid chlorides as effective activating species has been reestablished, in conjunction with other protecting groups for the α-amino function. For example, acid chlorides of derivatives containing hetero-arenesulfonamide N^{α}-protecting groups [180,181] as well as azido acids [182] coupled efficiently without significant racemization. In the latter, the azido function acts as a precursor of the amine.

E. Acid Fluorides

Advantages of fluorides relative to the chloro derivatives are their total compatibility with *t*Bu-based protecting groups, greater stability even in the presence of moisture, and lack of conversion to 5(4*H*)-oxazolone in the presence of tertiary amines [102]. Acid fluorides can easily be prepared by treatment of an amino acid with cyanuric fluoride in the presence of pyridine [183–185], (diethylamino)sulfur trifluoride (DAST) in the absence of base [186], or TFFH in the presence of DIEA [34]. These derivatives are very soluble in organic solvents (>1 M in DMF), thus facilitating acylation through a high concentration effect. Acid fluorides are stable when stored in DMF for extended periods of time (3 days or longer) and therefore suitable for use in multiple peptide synthesizers [187].

Acid fluorides have been successfully used for the SPPS of peptides containing difficult sequences, such as the (1–42) β-amyloid peptide [188], and most important for the synthesis of peptides containing very hindered units such as those incorporating two or more consecutive Aib residues (up to four) or *N*-substituted amino acids [189]. Thus, the incorporation of amino acid fluorides allowed the first solid-phase syntheses of petabiols, naturally occurring peptides containing up to 60% Aib residues [83]. A difference between the chlorides and fluorides is the ability of the latter to effect acylation in the total absence of base, thereby reducing the risk of racemization [190].

The effectiveness of these derivatives can be enhanced in the presence of a silylating agent, such as bis(trimethylsilyl)acetamide (BSA), as demonstrated by the coupling of Fmoc-MeAib-F to H-Aib-OMe [191]. BSA may act both as a base and as an *N*-silylating reagent enhancing the nucleophilicity of the amino function [191]. A carboxylic acid fluoride also proved to be most effective for acylating an α-(α-hydroxyisovaleryl)propionyl unit present in didemnins, a class of important biologically active cyclodepsipeptides [192]. Finally, Fmoc-amino acid fluorides have been successfully applied for the incorporation of the first amino acid on hydroxymethyl resins [193,194].

1. General Procedures for Coupling Reactions Using Protected Amino Acid Chlorides [175,176]

Fmoc-amino acid chloride (4 equiv.) and either the base (4 equiv.) or a mixture of HOXt base (1:1, 4 equiv.) are dissolved in DMF and added to the resin bearing the free amino group. After 30 min the resin is filtered and washed with DMF.

2. General Procedures for Coupling Reactions Using Protected Amino Acid Fluorides [83,190]

The Fmoc-amino acid fluoride (4 equiv.) is dissolved in DMF and added to the resin bearing the free amino group. The addition of base (4 equiv.) is optional. After 30 min the resin is filtered and washed with DMF.

3. General Procedures for Acylation of Hydroxymethyl Resins Using Protected Amino Acid Fluorides [193]

DCM–pyridine (6:4) is added to a mixture of Fmoc-amino acid fluoride (3–5 equiv.) and the resin. After 2 h, the solution is removed and the resin washed with DCM. Alternatively, Fmoc-amino acid fluoride (3 equiv.) dissolved in either toluene or THF is added to the resin, followed by DMAP (2 equiv.). The mixture is stirred for 30 min, and then the resin is filtered and washed with toluene or THF and DCM [194].

IV. SIDE REACTIONS

It is conventionally accepted that during stepwise SPPS, activation and coupling of N^α-urethane–protected amino acids proceeds without loss of stereomutation at substantial levels. However, this wisdom is erroneous for the latter when in situ activating reagents in the presence of a base are used. Thus, independent studies by various research groups have shown that both Cys [41,42,114] and Ser [112,113] Fmoc derivatives suffer racemization upon activation with onium salt reagents in the presence of a base such as DIEA or NMM. For Cys, loss of configuration was observed regardless of the side-chain protecting group [acetamidomethyl (Acm), Trt, 2,4,6-trimethoxybenzyl (Tmob), and 9H-xanthen-9-yl (Xan) were examined in conjunction with the Fmoc-based method] [42,115]. The stereomutation is more significant in DMF than in mixtures of DMF and DCM. For phosphonium or aminium coupling reagents, the loss of configuration increases notably with preactivation time, whereas for DIPCDI-based coupling it is reduced after 5 min of preactivation. Substitution of TMP or even more hindered bases, such as octahydroacridine (OHA), and 2,6-di-*tert*-butyl-4-(dimethylamino)pyridine [DB(DMAP)], for DIEA or NMM leads to cleaner products. In addition, it is preferable to add only 1 as opposed to 2 equivalents of base. Optimized conditions include (1) DIPCDI HOXt (X = A or B) in DMF–DCM (1:1) with 5 min of preactivation, (2) symmetric anhydride prepared in DCM–DMF (9:1) and coupled in DMF–DCM (1:1), (3) Pfp esters in either DMF or DMF–DCM (1:1), and (4) phosphonium and aminium salts in the presence of HOXt with 1 equivalent of TMP and no preactivation [41]. For automated instrumentation, implementing optimized conditions 1, 2, and 3 as well as use of the less reactive aminium salt, HBTU, in conjunction with HOBt and TMP with minimum preactivation time is recommended [195]. For Ser, the use of HATU in the presence of HOXt and TMP leads to an acceptable level of racemization (<0.5%) [112].

Aminium reagents should be used with caution, as such salts can react with the amino component leading to a guanidino derivative, a process that terminates the peptide chain [66,75–78]. This side reaction is not critical during the coupling of single protected amino acids, because activation is fast and the aminium salt is rapidly consumed or hydrolyzed before exposure to a resin containing an amino terminus. However, during a much slower activation process, the aminium salt may react with the amino component. This side reaction is more dramatic when the most reactive reagents such as HAPyU, HATU, and TFFH are used [66,196]. Model experiments carried out with phosphonium salts (PyAOP and PyBOP) have not shown any product related to the reaction of the phosphonium salts with the amino component [66].

Activation of α-hydroxycarboxylic acids with TBTU leads to the formation of N,N-dimethylcarbamoyl derivatives [197]. The coupling of this class of carboxylic acids can be safely accomplished with phosphonium salts.

Pyrrolidide derivatives have been observed as unwanted by-products in slow reactions of activated carboxylates with the nucleophilic amines from the reactions mediated by phosphonium salt coupling reagents (PyAOP, PyBOP, PyBroP) [198]. This side reaction, which leads to a product with a mass 53 daltons greater than that of the starting carboxylic acid component, has been attributed to the presence of small amounts (0.1–0.5%, w/w) of pyrrolidine as a contaminant of commercial phosphonium salts. Pyrrolidine formation does not occur when the reagents are crystallized immediately before use in coupling reactions. This side reaction is insignificant during conventional N → C SPPS, in which the carboxylic acid is activated in solution. However, this undesired product may form when the carboxylic acid to be activated is supported on a solid matrix (e.g., solid-phase preparation of cyclic peptides or solid-phase elongation in the C → N direction).

The use of HODhbt as an additive provokes the formation of 3-(2-azidobenzoloxy)-4-oxo-3,4-dihydro-1,2,3-benzotriazine, which can react with the amino group to terminate chain growth [30].

Asn and Gln should be coupled with their side-chains protected or in the form of an active ester; otherwise the carboxamide group is converted to the corresponding cyano derivative [153]. Likewise, Thr(Bzl) can undergo a β-elimination yielding α-aminocrotonic acid [199].

During the activation step, Arg can cyclize, giving the corresponding δ-lactam. Although this reaction is more pronounced when unprotected Arg is used, it is often encountered with ω-tosyl-based protection of Arg [21,200]. For reasons still not clear, the coupling of Fmoc-Asn(Trt)-OH via phosphonium or aminium reagents is subject to some deficiencies. Thus, the incorporation of both of these residues is more efficient if an equivalent of HOAt is present [31,34]. These two side reactions do not provoke any undesirable modification of the peptide; however, lowering the concentration of the activated species favors the formation of deletion peptides that do not contain Arg or Asn in the final product. For Fmoc chemistry, if urethane-type side-chain protecting groups are used for Arg, acylation of the guanidino residue occurs along with subsequent partial conversion of Arg to Orn [201].

Intramolecular rearrangement of symmetrical anhydrides gives bis-protected dipeptide derivatives, which can again be activated, leading to the incorporation of an extra residue. This side reaction is most favored when relatively unhindered amino acids such as Gly and Ala are incorporated [202].

An important side reaction can occur during the synthesis of His-containing peptides using the Boc/Bzl strategy if the imidazole ring of His has been protected with a Tos residue and HOBt is used after the coupling step. Detosylation can occur via HOBt [203], and the resulting free imidazole moiety can be acetylated or otherwise acylated. During the neutralization step, an N^{im}-N transfer of the acetyl or acyl moiety can occur, yielding a terminated peptide (acetylation) or insertion of an extra residue (acylation) [204–206]. To avoid these side reactions, it is recommended that the imidazole residue be protected by the Dnp group [207]. Removal of several Dnp protecting groups from peptides containing multiple His residues is not straightforward, and the preparation of these sequences using Boc/Bzl has not been optimized (C. Carreño, D. Andreu, personal communication).

V. EXPERIMENTAL CONSIDERATIONS: SOLVENTS, TEMPERATURE, MICROWAVE, AND ULTRASOUNDS TECHNIQUES

As previously discussed, typically the coupling protocols commonly used are derived from those established for automated instrumentation (DMF as solvent, several minutes of preactivation time, room temperature, conventional stirring methods such as vortex, N_2 bubbling, or continuous-flow). Although these conditions are effective for most acylations, they can fail for the coupling of steric and/or electronically low reactive building blocks in

SPPS or SPOS. In these cases, altering experimental conditions should be investigated.

In SPPS, incomplete acylations are often observed during the coupling of hindered amino acids, such as those having β-branched side-chains [208,209], or α-dialkyl and N-alkyl substitutents [31,81]. This effect is increased when several of these amino acids follow each other in the sequence. If a ninhydrin test indicates incomplete coupling, incorporation of the residue into the sequence can be improved by recoupling. In such cases, the use of a different coupling method and/or solvent is advisable for the second coupling. A more intractable problem is internal aggregation of peptide chains within the peptide-resin matrix [210,211], resulting in visible shrinking of the peptide resin in a batch reactor or continuous-flow column. These aggregations can be weakened by increasing the temperature and/or using appropriate solvents and/or additives. Furthermore, the use of resins containing functionalized sites evenly distributed and with low loadings (0.1–0.2 mmol/ g) can diminish the aggregation problem [211,212].

Performing an acylation at elevated temperature can have a dual purpose [213–218]. Aggregations caused by hydrogen bonding may become weaker [214,218] and coupling times and/or excesses or reagents can be reduced [216,217]. The latter is valid only for couplings with higher activation energies [215]. For coupling reactions with low activation energy, the reaction rate may be increased by methods other than simply increasing the temperature. The routine use of high temperatures (50–80°C) for peptides containing Asn, Gln, and Glu should be avoided because dehydration is possible for Asn and Gln and pGlu can be formed from Glu [214,216]. The use of high temperatures is also convenient for enhancing the rate of the deprotection step [216,217].

The use of polar solvents such as DMF, N-methylpyrrolidone (NMP), dimethyl sulfoxide (DMSO) [208,210,219–222] (the latter can cause oxidation of methionine to the sulfoxide [223]), or even mixtures of 2,2,2-trifluoroethanol (TFE) [224] and 1,1,1,3,3,3-hexafluoro-2-propanol (HFIP) [225] in DCM can disrupt aggregation. TFE is preferred to HFIP, as the latter can consume the activated species to give the corresponding ester [226].

Another type of aggregation can occur in regions containing apolar side-chain protecting groups [227]. In this case the use of polar mixtures during coupling is not sufficient to overcome the problem. In the Fmoc/tBu strategy the lack of polar side-chain protecting groups interferes with proper solvation of the peptide resin. In this case, solvent mixtures containing both a polar and a nonpolar component, such as THF–NMP (7:13) or TFE–DCM (1:4), can be useful [221].

The use of denaturants [228,229] such as urea or Triton X-100 or chaotropic agents [230,231] such as LiCl, LiBr, KSCN, LiClO$_4$, and NaClO$_4$ can increase resin swelling and improve coupling yields by disrupting β-sheets. Perchlorate salts do not oxidize Trp and KSCN does not modify free amino groups on the peptide resin, although the latter can catalyze rearrangement of the active O-acylisourea to the inactive N-acylurea. Preincubation of the peptide resin with 0.8 M KSCN and preactivation of the protected amino acid via either DCC or DCC/HOXt was found to be particularly effective [230]. In some cases, it has been reported that although the use of these chaotropic salts inhibits interchain aggregation and assists in the dissolution of protected peptides, the efficiency of coupling is lower than that with unmodified solvents [232].

On the basis of the results just described, Bayer and coworkers [229] proposed a solvent system, referred to as "magic mixture" [2 N ethylene carbonate in DMF–DCM–NMP (1:1:1) containing 1% Triton X-100 at 50°C], that was effective for the synthesis of peptides containing difficult sequences. This system incorporates DMF, NMP, and ethylene carbonate to break the α-helical and β-sheet structures and DCM and the nonionic detergent to mediate in the aggregation of the hydrophobic peptide chains.

Ultrasonic wave [233] and microwave [234] irradiation have been also applied to enhance reactions rates in SPPS. Although the initial results with both techniques were promising, no further developments have been described recently.

VI. MONITORING

The sensitivity of these tests decreases with the length of the peptide attached to the resin. Thus, for medium to small peptides (up to 15 amino acid residues) the accuracy is sufficient, but for longer sequences the results are often misleading.

Procedure for the Ninhydrin Test [235,236].   The most convenient method for rapid monitoring of the coupling process is the ninhydrin test [235]. Ninhydrin reacts with primary amines to give the dye known as Ruhemann's purple.

Two solutions are required and can be prepared as follows. (a) Mix 40 g of phenol with 10 mL of absolute ethanol. Warm until dissolved. Stir with 4 g of Amberlite mixed-bed resin MB-3 for 45 min and filter. In a separate vessel, dissolve 65 mg of KCN in 100 mL of water. Dilute 2 mL of this solution to 100 mL with pyridine (freshly distilled from ninhydrin). Stir with

4 g of Amberlite mixed-bed resin MB-3 for 45 min and filter. The solutions are mixed. (b) Dissolve 2.5 g of ninhydrin in 50 mL of absolute ethanol. Store in the dark under nitrogen. Test procedure: (1) transfer a sample containing about 1–3 mg of resin to a test tube; (2) add 9 drops of solution (a) and 3 drops of solution (b); (3) Mix well and place the tube in a heating block preadjusted to 110°C for 3 min; (4) place the tube in cold water. If free amino groups are present, a purple color is formed in the solution or on the resin beads. The sensitivity for detection is in the range of 1 μmol/g resin (99.5% coupling for resins having a functionalization level of 0.2–0.5 mmol/g).

Procedure for the Chloranil Test [237]. For Pro and other secondary amines, the chloranil test can be used [237]. Chloranil (2,3,5,6-tetrachloro-1,4-benzoquinone) reacts with secondary and primary amines in the presence of acetone or acetaldehyde, respectively, to give a green–blue benzoquinone derivative.

(1) Transfer a sample containing about 1 mg of resin to a test tube; (2) add 200 μL of acetone for secondary amines or acetaldehyde for primary amines; (3) add 50 μL of a saturated solution of chloranil in toluene; (4) shake the test tube occasionally for 5 min. If free amino groups are present, a green or blue color is formed on the resin beads. The sensitivity for detection of Pro is in the range 2–5 μmol/g resin (97–99% coupling for resins have a level of functionalization of 0.2–0.5 mmol/g) and for primary amines 5–8 μmol/g resin when approximately 1 mg of peptide resin is assayed.

Procedure for the Bromophenol Blue Test [238]. The coupling end point can also be monitored by a noninvasive test based on the use of the acid–base indicator bromophenol blue [238]. In the presence of free amino groups a deep blue color, which turns to greenish yellow in the absence of free amino groups, occurs. This monitoring can be performed continuously, with bromophenol blue simply being added to the acylating agent, or discontinuously, with a sample of the peptide resin being removed from the reactor and treated with the reagent.

Add 3 drops of a 1% solution of bromophenol blue in N,N-dimethylacetamide to the reactor containing the acylating reagents, causing the suspension to turn dark blue. After the suspension turns greenish yellow, the next step of the synthesis is carried out. Alternatively, (1) transfer a sample containing about 1–3 mg of resin to a test tube; (2) add 3 drops of a 1% solution of bromophenol blue in N,N-dimethylacetamide; (3) shake the test tube occasionally for 1 min. If free amino groups are present, a blue–green color is formed on the resin beads. The sensitivity is in the range of 3 mol/g resin (99.5% of coupling for resins having a level of functionalization of 0.7 mmol/g).

VII. CAPPING

Some protocols recommend a capping reaction for unreacted chains following each coupling. The capping step is usually carried out by acetylation
with Ac$_2$O-DIEA (1:1, 30 equiv.) in DMF. In some automated instruments
Ac$_2$O is replaced by a 0.3 M solution of N-acetylimidazole in DMF [21].

ACKNOWLEDGMENT

Work in the University of Barcelona is supported by funds from CICYT
(PB96-1490) and Generalitat de Catalunya (Grup Consolidat 1997SGR 430
and Centre de Referència en Biotecnologia).

APPENDIX

Table 1 Coupling Reagents

No.	Name	Abbreviation	Structure
1	N,N'-Dicyclohexylcarbodiimide	DCC	
2	N,N'-Diisopropylcarbodiimide	DIPCDI	
3	1-Ethyl-3-(3'-Dimethyl-aminopropyl)carbodiimide hydrochloride	ECC, WSC	
22	μ-Oxo-bis-[tris(dimethylamino)-phosphonium]-bis-tetrafluoroborate	Bates reagent	

Table 1 Continued

No.	Name	Abbreviation	Structure
23	Benzotriazol-1-yl-*N*-oxy-tris(dimethylamino)phosphonium hexafluorophosphate	BOP	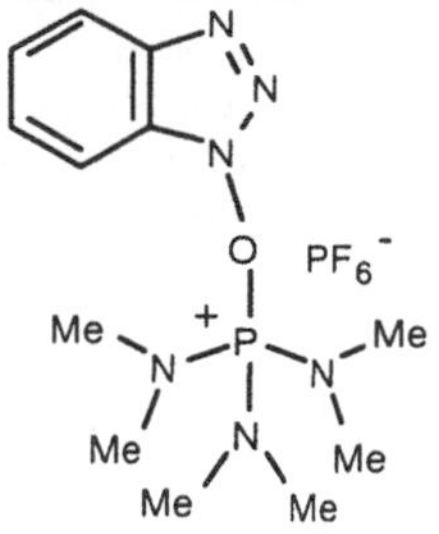
24	Benzotriazol-1-yl-*N*-oxy-tris(pyrrolidino)phosphonium hexafluorophosphate	PyBOP	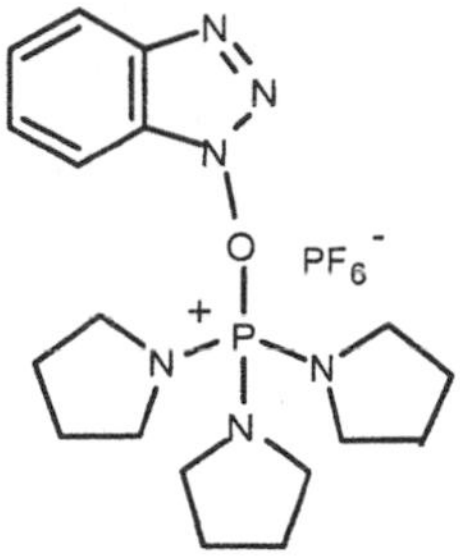
25	(7-Azabenzotriazol-1-yloxy)-tris(dimethylamino)phosphonium hexafluorophosphate	AOP	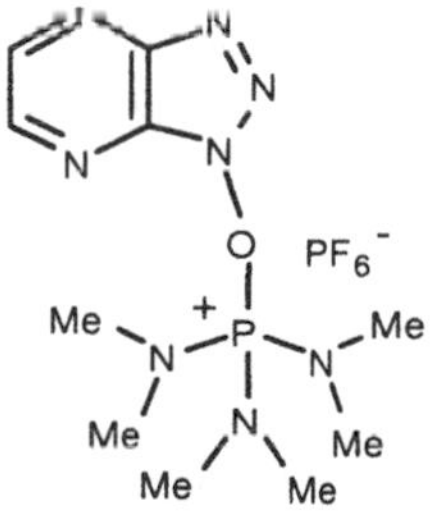
26	(7-Azabenzotriazol-1-yloxy)-tris(pyrrolidino)phosphonium hexafluorophosphate	PyAOP	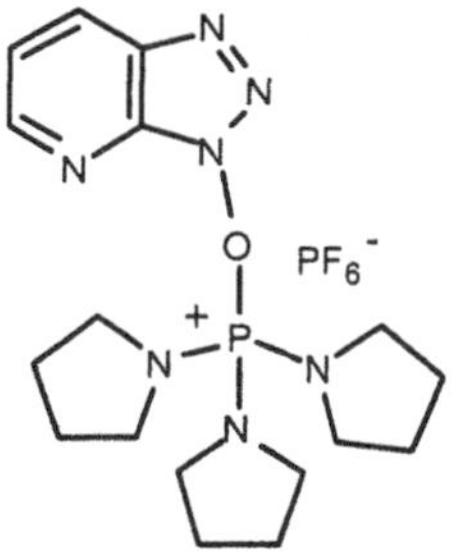

Table 1 Continued

No.	Name	Abbreviation	Structure
27	N-[(1H-Benzotriazol-1-yl)(diamethylamino)methylene]-N-methylmethanaminium hexafluorophosphate N-oxide	HBTU	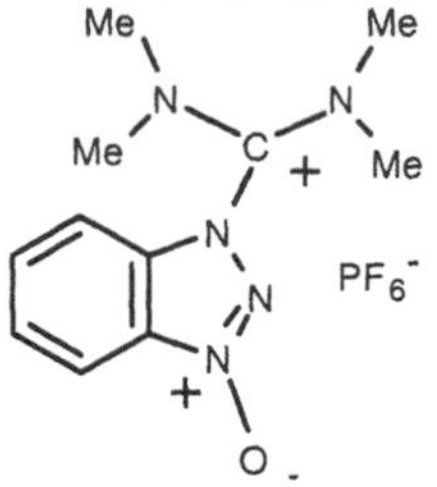
28	N-[(Dimethylamino)-1H-1,2,3-triazolo[4,5-b]pyridino-1-ylmethylene]-N-methylmethanaminium hexafluorophosphate N-oxide	HATU	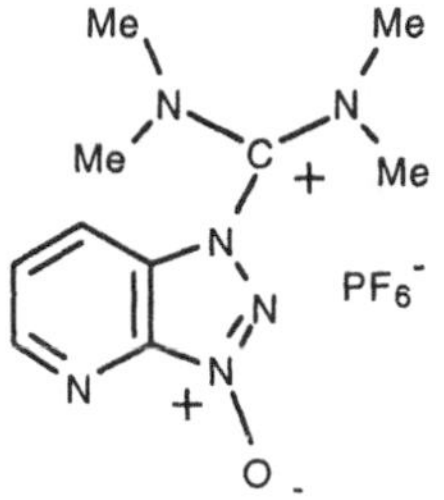
29	1-(1-Pyrrolidinyl-1H-1,2,3-triazolo[4,5-b]pyridin-1-ylmethylene)pyrrolidinium hexafluorophosphate N-oxide	HAPyU	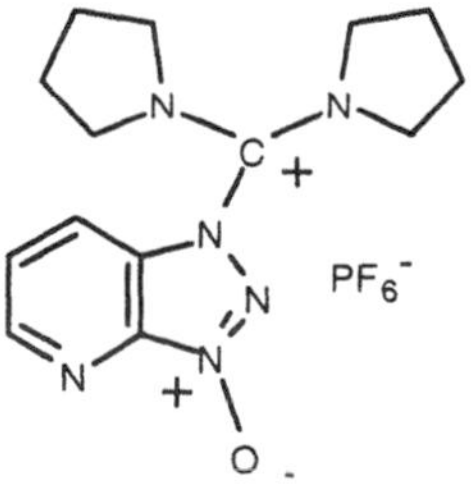
30	N-[(1H-Benzotriazol-1-yl)(dimethylamino)methylene]-N-methylmethanaminium tetrafluoroborate N-oxide	TBTU	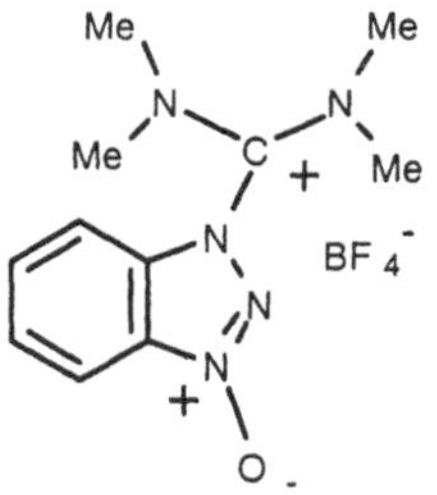

Table 1 Continued

No.	Name	Abbreviation	Structure
31	N-[(Dimethylamino)-1H-1,2,3-triazolo[4,5-b]pyridino-1-ylmethylene]-N-methylmethanaminium tetrafluoroborate N-oxide	TATU	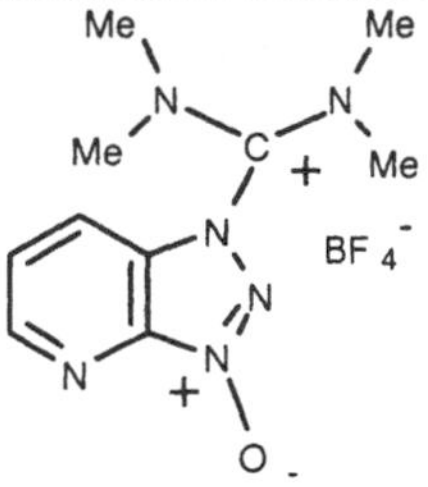
32	O-(7-Azabenzotriazol-1-yl)-1,1,3,3-bis(pentamethylene)-uronium hexafluoro-phosphate	HAPipU	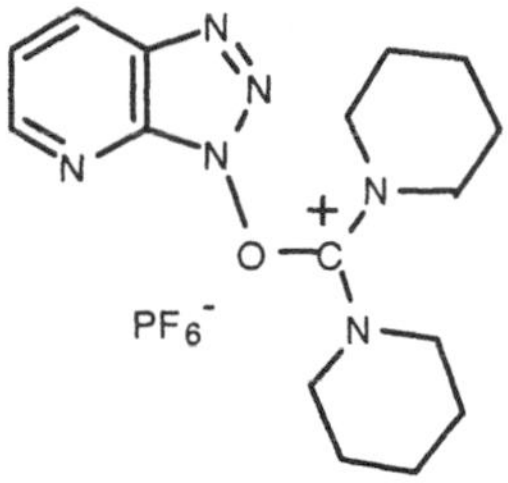
33	O-(7-Azabenzotriazol-1-yl)-1,3-dimethyl-1,3-dimethyleneuronium hexafluorophosphate	HAMDU	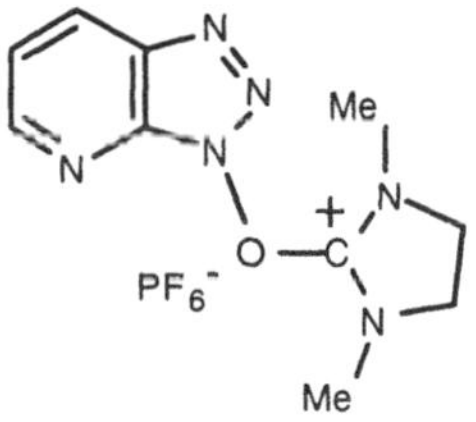
34	O-(Benzotriazol-1-yl)-1,3-dimethyl-1,3-dimethyleneuronium hexafluorophosphate	HBMDU, BOI	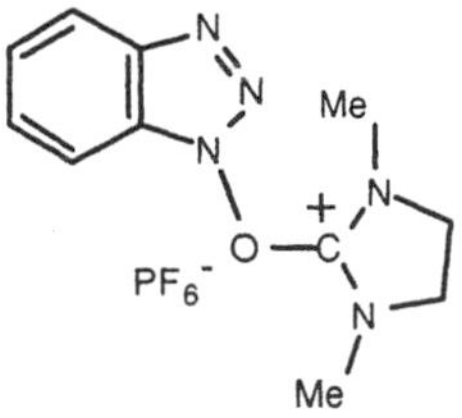
35	O-(7-Azabenzotriazol-1-yl)-1,3-dimethyl-1,3-trimethyleneuronium hexafluorophosphate	HAMTU	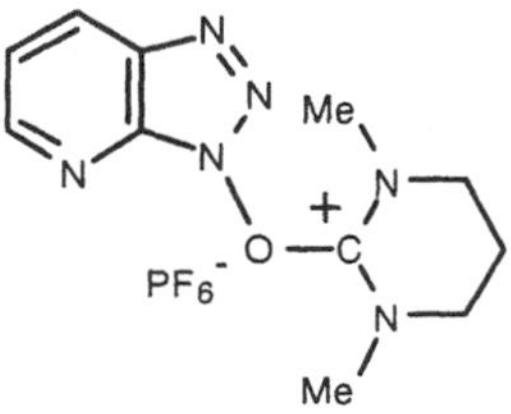

Table 1 Continued

No.	Name	Abbreviation	Structure
36	*O*-(3,4-Dihydro-4-oxo-1,2,3-benzotriazin-3-yl)1,1,3,3-tetramethyluronium hexafluorophosphate	HDTU	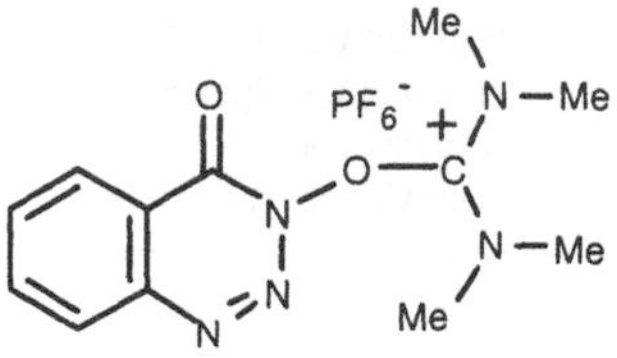
38	2-[2-Oxo-1(2*H*)-pyridyl]-1,1,3,3-tetramethyluronium tetrafluoroborate	TPTU	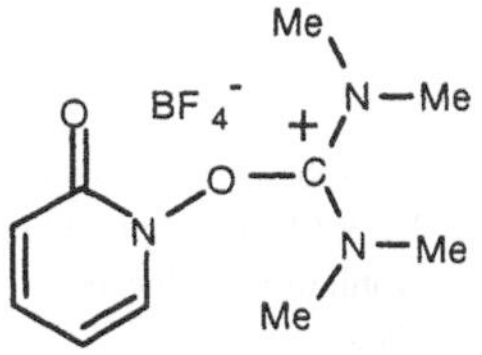
39	2-[2-Oxo-1(2*H*)-pyridyl]-1,1,3,3-bis(pentamethylene)uronium tetrafluoroborate	TOPPipU	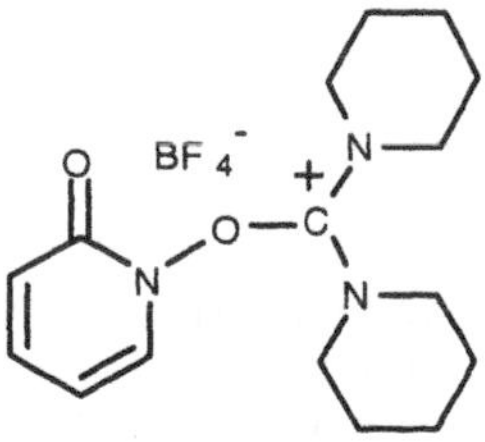
40	2-Succinimido-1,1,3,3-tetramethyluronium tetrafluoroborate	TSTU	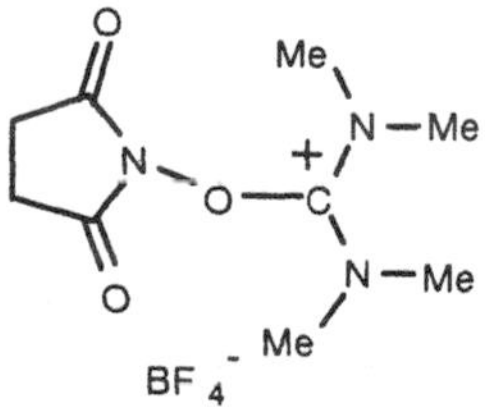
41	2-Phthalimido-1,1,3,3-tetramethyluronium tetrafluoroborate	TPhTU	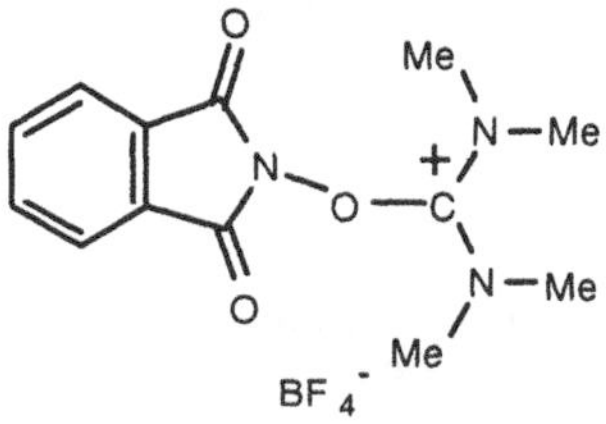
42	2-Mercaptopyridine-1,1,3,3-tetramethyluronium tetrafluoroborate	TMTU	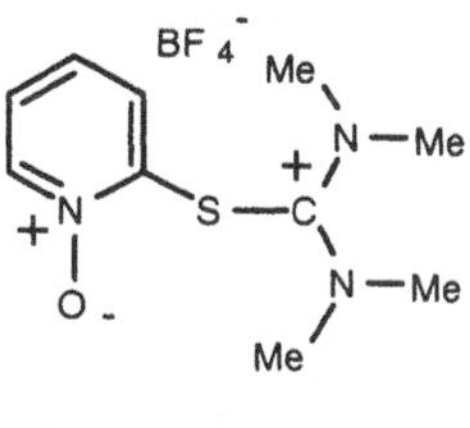

Coupling Methods

Table 1 Continued

No.	Name	Abbreviation	Structure
43	2-(*endo*-6-Norbornene-2,3-dicarboximido)-1,1,3,3-tetramethyluronium tetrafluoroborate	TNTU	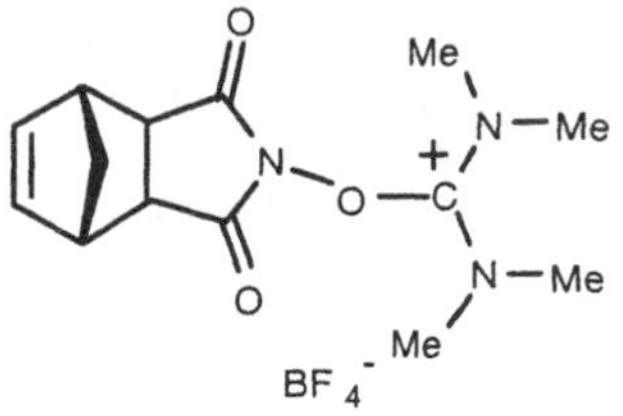
44	2-[6(Trifluoromethyl)benzotriazol-1-yl]-1,1,3,3-tetramethyluronium hexafluorophosphate	CF₃-HBTU	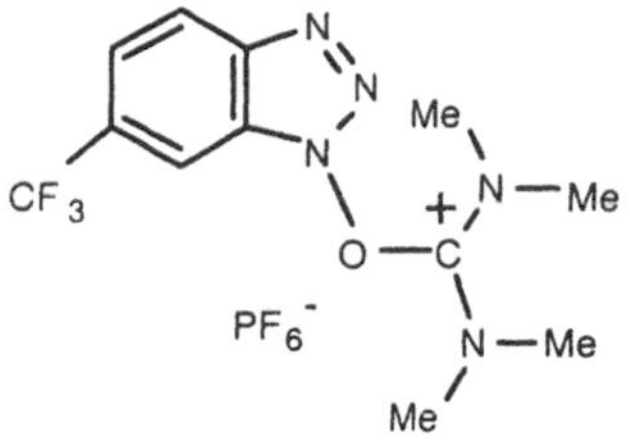
45	2-Pentafluorophenyl-1,1,3,3-tetramethyluronium hexafluorophosphate	PfTU	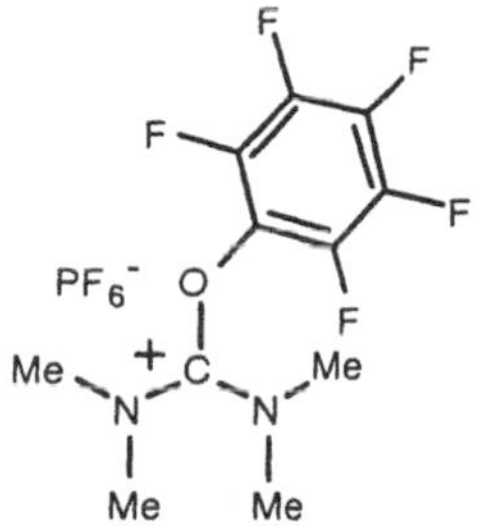
46	Bis(tetramethylene)-pentafluorophenoxy-formamidinium hexafluorophosphate	PfPyU, HPyOPfp	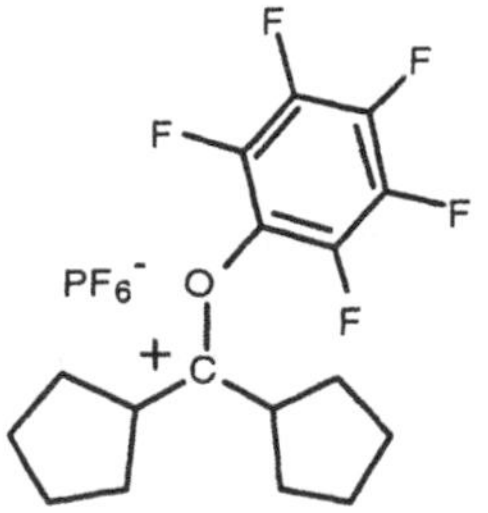

Table 1 Continued

No.	Name	Abbreviation	Structure
47	[(6-Nitrobenzotriazol-1-yl)-oxy]tris(pyrrolidino)-phosphonium hexafluorophosphate	PyNOP	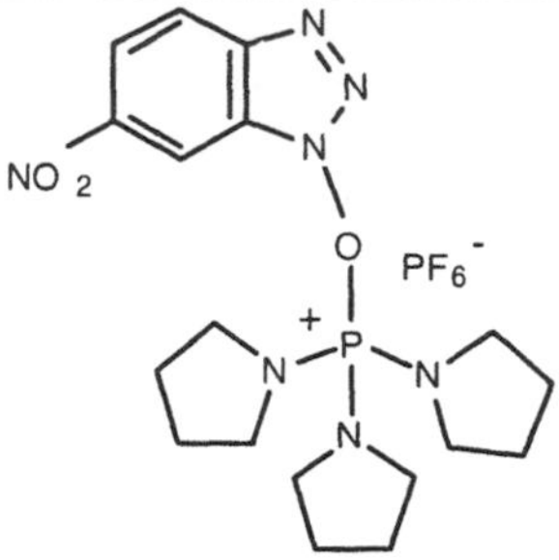
48	[(6-Trifluorobenzotriazol-1-yl)oxy]tris(pyrrolidino)phosphonium hexafluorophosphate	CF₃-PyBOP, PyFOP	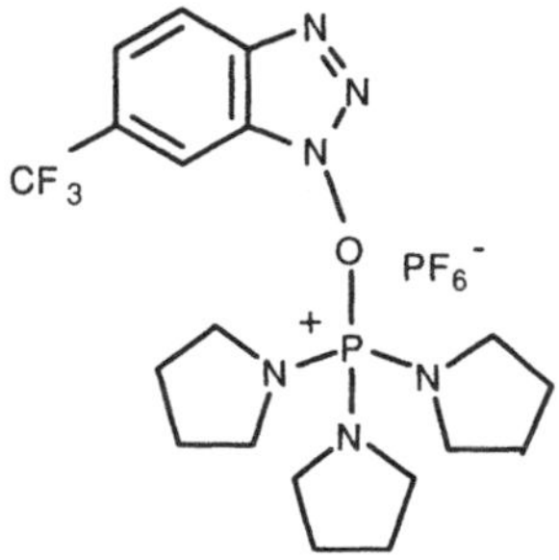
49	[(3,4-Dihydro-4-oxo-1,2,3-benzotriazin-3-yl)oxy]tris-(pyrrolidino)phosphonium hexafluorophosphate	PyDOP	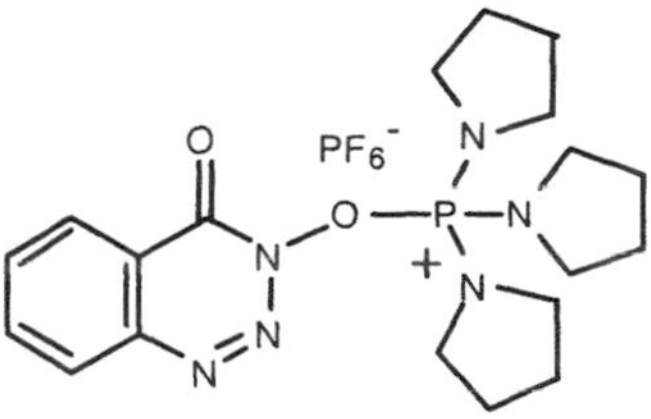
50	(Pentafluorophenyl)oxy]-tris(pyrrolidino)phos-phonium hexafluorophosphate	PyPOP, PfOP, PyPfpOP	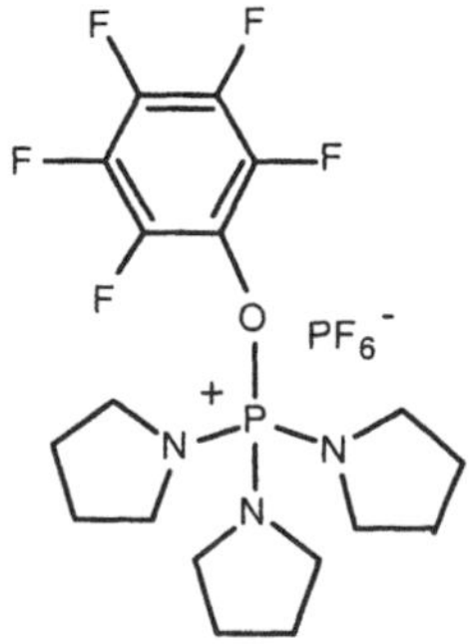

Table 1 Continued

No.	Name	Abbreviation	Structure
51	(Pyridyl-2-thio)tris(pyrrolidino)-phosphonium hexafluorophosphate	PyTOP	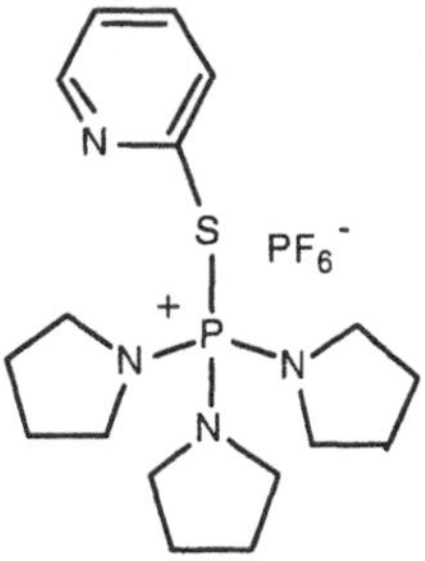
52	1,3-Dimethyl-2-(1-pyrrolidinyl)-2-(3*H*-1,2,3-triazolo(4,5-b)pyridin-3-yloxy)-1,3,2-diazaphospholidinium hexafluorophosphate	AOMP	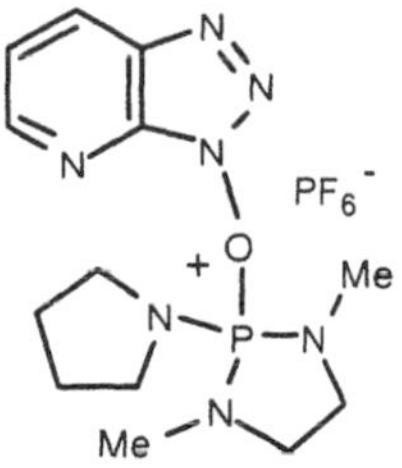
53	*S*-(7-Azabenzotriazol-yl)-1,1,3,3-tetramethylthiouronium hexafluorophosphate	HATTU	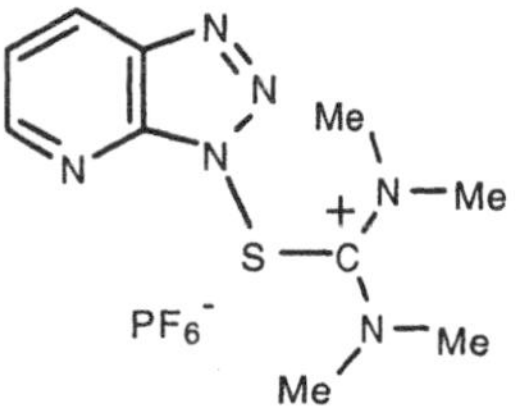
54	*S*-(7-Azabenzotriazol-yl)-1,1,3,3-bis(tetramethylene)thiouronium hexafluorophosphate	HAPyTU	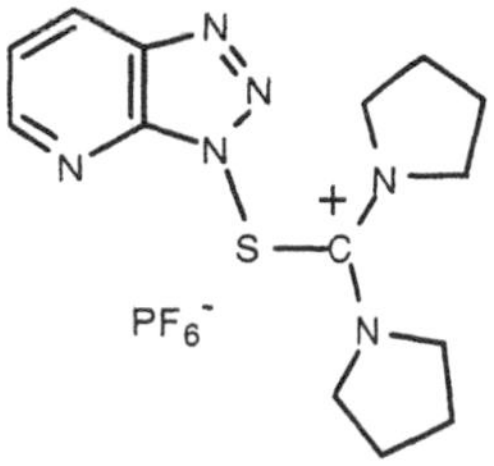

Table 1 Continued

No.	Name	Abbreviation	Structure
55	S-(Pentafluorophenyl)-1,1,3,3-bis(tetramethylene)thiouronium hexafluorophosphate	HPySPfp	
56	Tetramethylfluoroformamidinium hexafluorophosphate	TFFH	
61	Bis(tetramethylene)-fluoroformamidium hexafluorophosphate	BTFFH	
62	1,3-Dimethyl-2-fluoro-4,5-dihydro-1H-imidazolium hexafluorophosphate	DFIH	
63	Chlorotris(pyrrolidino)-phosphonium hexafluorophosphate	PyCloP	
64	Bromotris(pyrrolidino)-phosphonium hexafluorophosphate	PyBroP	

Table 1 Continued

No.	Name	Abbreviation	Structure
65	Bromotris(dimethylamino)- phosphonium hexafluorophosphate	BroP	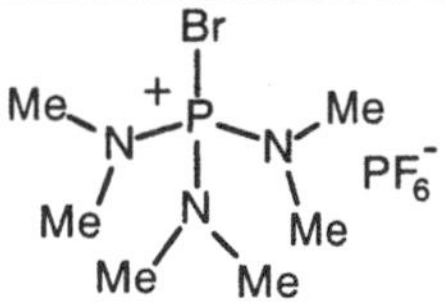
66	2-Chloro-1,3-dimethyl-4,5-dihydro- 1*H*-imidazolium hexafluorophosphate	CIP, DCIH	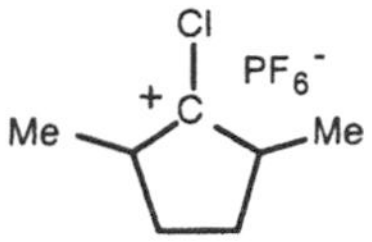
67	Diphenylphosphinic chloride	Dpp-Cl	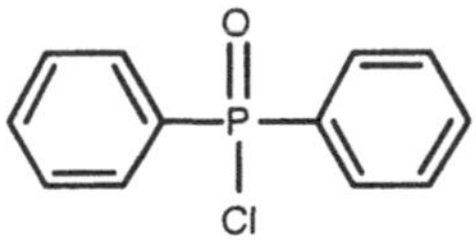
70	1-Oxo-1-chlorophospholane	Cpt-Cl	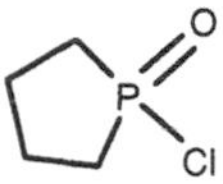
71	Pentafluorophenyl diphenylphosphinate	FDPP	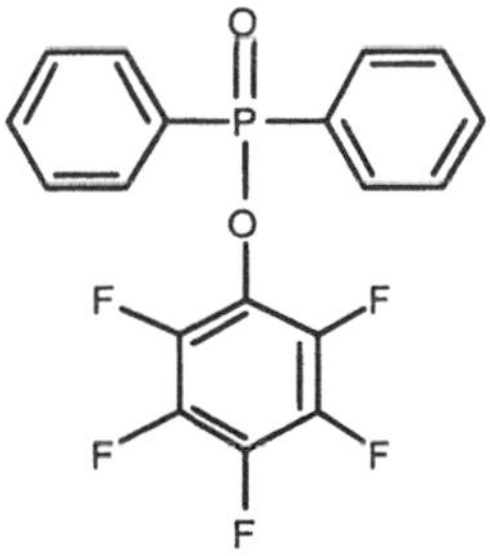
72	Diphenylphosphoryl azide	DPPA	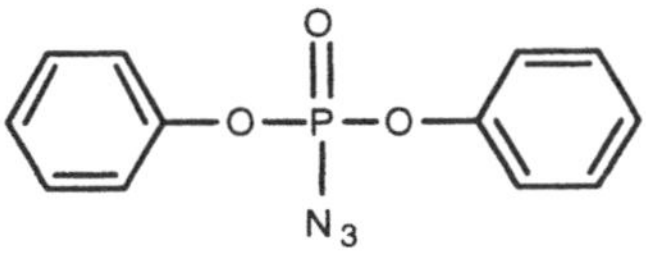
73	Diethylphosphoryl cyanide	DEPC	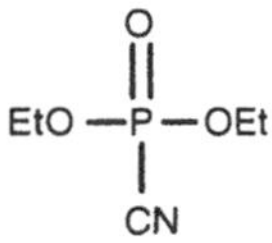

Table 1   Continued

No.	Name	Abbreviation	Structure
74	Bis(2-oxo-3-oxazolidinyl)phosphorodiamidic chloride	BOP-Cl	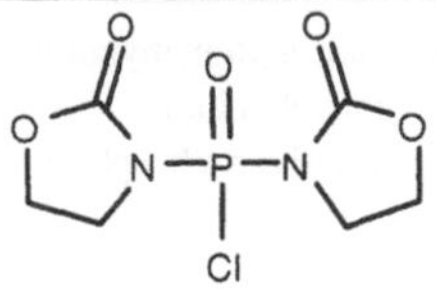
75	*N*-Diethoxyphosphoryl benzoxazolone	DEPBO	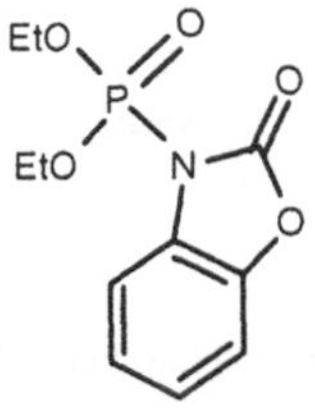
76	*N*-(2-Oxo-1,3,2-dioxaphosphorinanyl)benzoxazolone	DOPBO	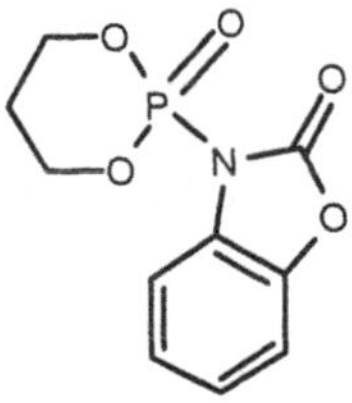
77	3-[*O*-(2-Oxo-1,3,2-dioxaphosphorinanyl)-oxy]-1,2,3-benzotriazin-4(3*H*)-one	DOPBT	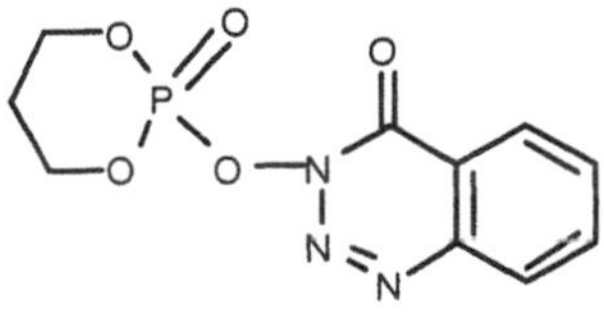
78	3-(Diethoxyphosphoryloxy)-1,2,3-benzotriazin-4(3*H*)-one	DEPBT	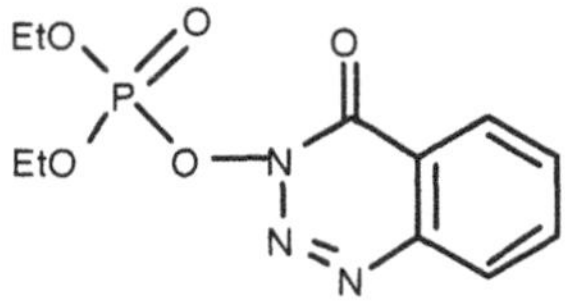
79	1,4-Epoxy-5-norbonene-2,3-dicarboximidodiphenyl-phosphate	ENDPP	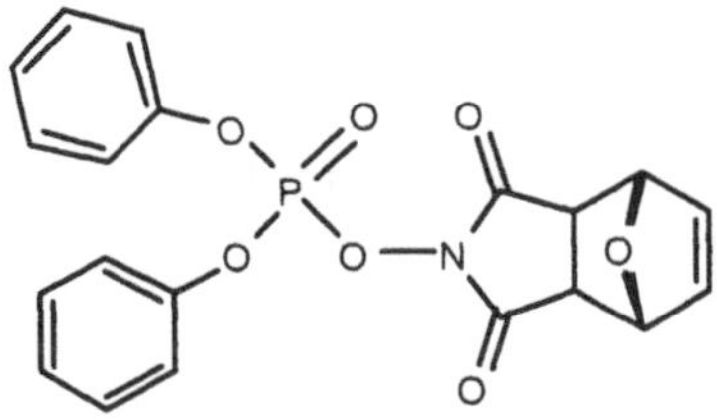

Table 1 Continued

No.	Name	Abbreviation	Structure
80	Benzotriazol-1-yl-diethylphosphate	BDP	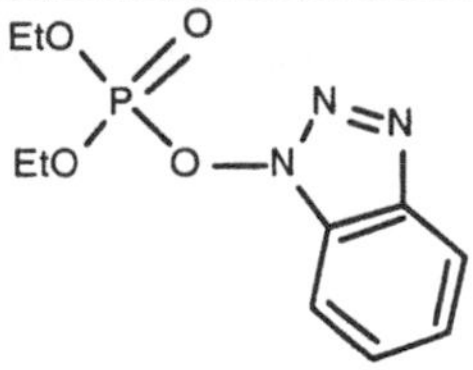
82	1-(Mesitylenesulfonyl)-3-nitro-1,2,4-triazole	MSNT	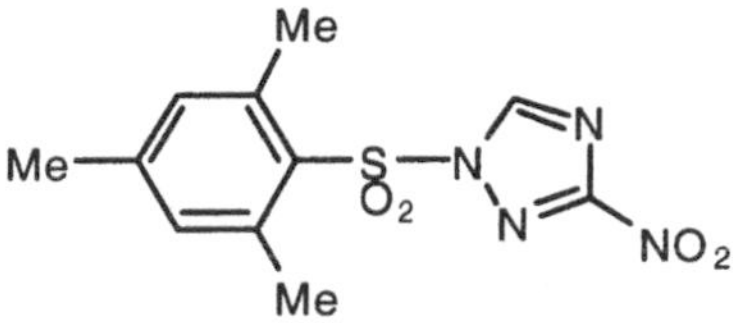
84	1-(2,4,6-Triisopropylbenzenesulfonyl)-3-nitro-1,2,4-triazole	TPSNT	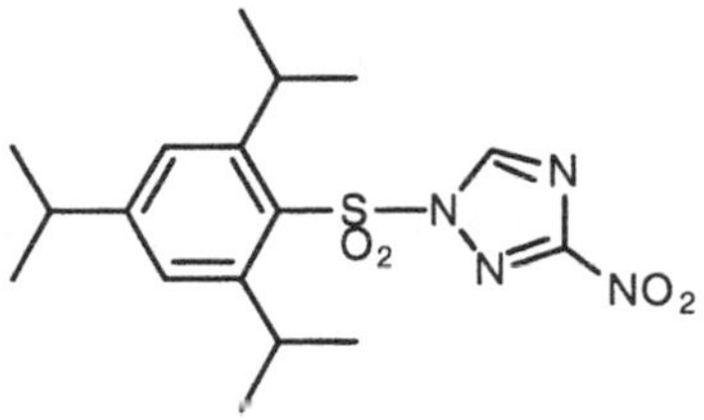
87	2-Chloro-4,6-dimethoxy-1,3,5-triazine	CDMT	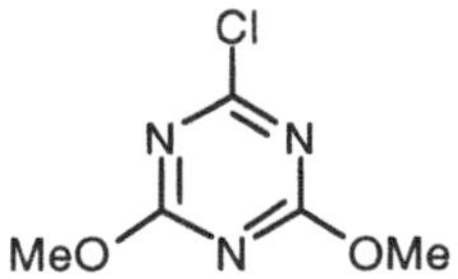
88	2,4,6-Tris(pentafluorophenyloxy)-1,3,5-triazine	TPft	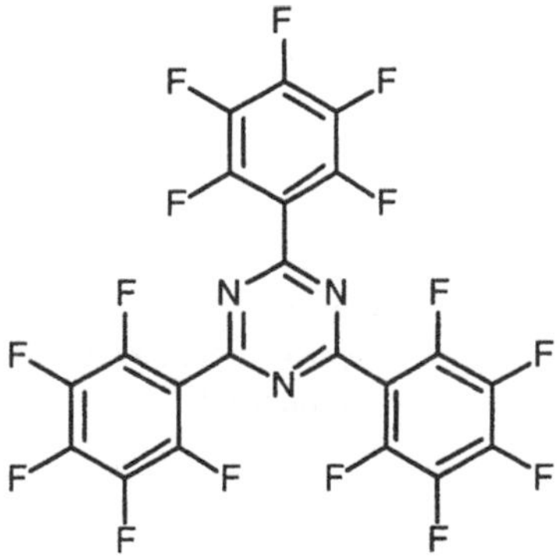

Table 2 Hydroxycompounds Used as Additives and/or Leaving Groups of Active Esters

No.	Name	Abbreviation	Structure
11	1-Hydroxybenzotriazole	HOBt	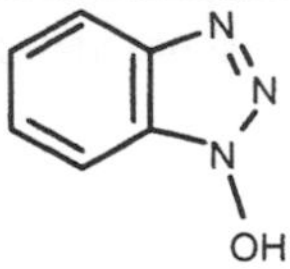
12	7-Aza-1-hydroxybenzotriazole	HOAt, 7-HOAt	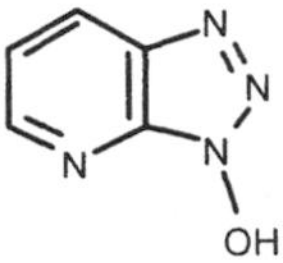
13	1-Oxo-2-hydroxydihydro-benzotriazine	HODhbt, HOOBt	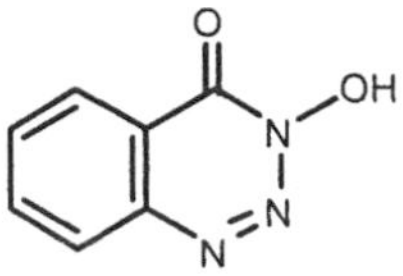
15	4-Aza-1-hydroxybenzotriazole	4-HOAt	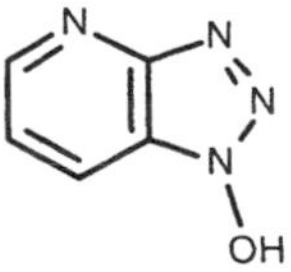
16	1-Hydroxytriazole-4-carboxylate	HOCt	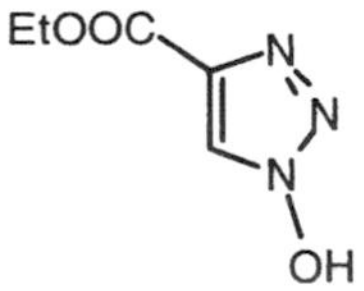
17	5-Chloro-1-hydroxytriazole		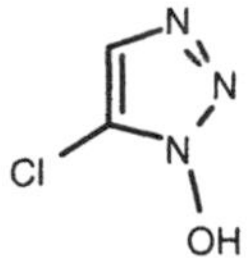
19	Polymeric *N*-benzyl-1-hydroxybenzotriazole-6-sulfonamide		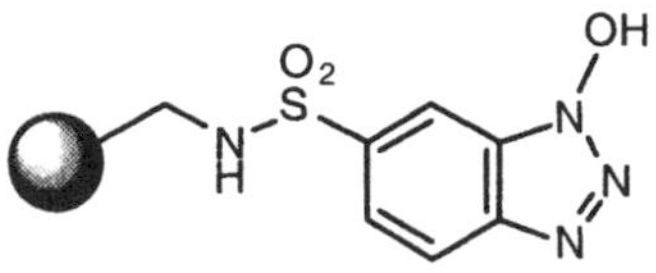

Table 2 Continued

No.	Name	Abbreviation	Structure
20	Polymeric 1-hydroxybenzotriazole	P-HOBt	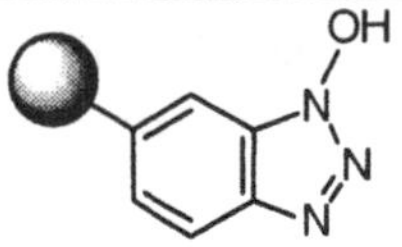
92	1-(4′-Nitrophenyl)-pyrazolin-5-one	HOHpp	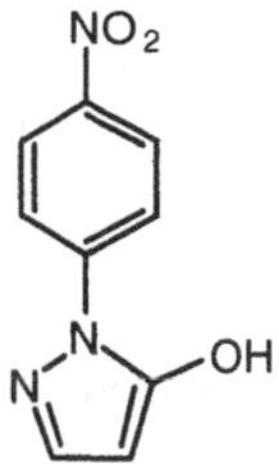
93	2,3-Dihydro-2,5-diphenyl-4-hydroxy-3-oxo-thiophen-1,1-dioxide	TDO	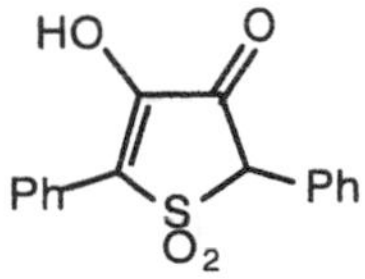

Table 3 Bases

No.	Name	Abbreviation	Structure
9	Triethylamine	TEA	
10	*N,N'*-Diisopropylethylamine	DIEA	
18	4-Dimethylaminopyridine	DMAP	
37	2,4,6-Trimethylpyridine	TMP, collidine	
57	Pyridine		
58	2,6-Di-*t*-butyl-4-methylpyridine		
59	2,6-Di-*t*-butyl-4-(dimethylamino)pyridine	DB(DMAP)	

Table 3 Continued

No.	Name	Abbreviation	Structure
60	2,3,5,6-Tetramethylpyridine		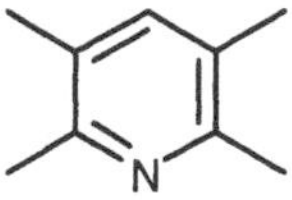
83	*N*-Methylimidazole	NMI	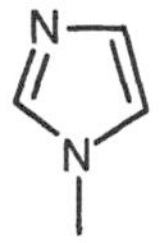
85	4-Morpholine pyridine-1-oxide	MPO	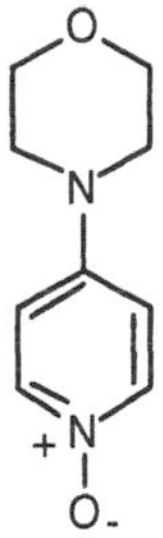
94	*N*-Methylmorpholine	NMM	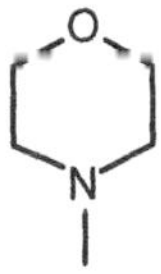

REFERENCES

1. JM Humphrey, AR Chamberlin. Chem Rev 97:2243–2266, 1997.
2. FJ Sardina, H Rapoport. Chem Rev 96:1825–1872, 1996.
3. E Gross, J Meinhofer, eds. The Peptides: Analysis, Synthesis, Biology, Vol 1. New York: Academic Press, 1979.
4. Y Kiso, H Yajima. In: B Gutte, ed. Peptides. Synthesis, Structures, and Applications. San Diego: Academic Press, 1995, pp 39–91.
5. F Albericio, LA Carpino. Methods Enzymol 289:104–126, 1997.
6. RB Merrifield. Angew Chem Int Ed Engl 24:799–810, 1985.
7. GB Fields, Z Tian, G Barany. In: GA Grant, ed. Synthetic Peptides: A User's Guide. New York: WH Freeman, 1992, pp 77–183.
8. GB Fields, ed. Methods in Enzymology, Vol 289, Solid-Phase Peptide Synthesis. Orlando, FL: Academic Press, 1997.

9. P Lloyd-Williams, F Albericio, E Giralt. Chemical Approaches to the Synthesis of Peptides and Proteins. Boca Raton, FL: CRC Press, 1997.
10. D Obrecht, JM Villalgordo, eds. Solid-Supported Combinatorial and Parallel Synthesis of Small-Molecular-Weight Compounds Libraries. Oxford: Pergamon, 1998.
11. LW Herman, G Tarr, SA Kates. Mol Divers 2:147–155, 1997.
12. DS Kemp. In: E Gross, J Meinhofer, eds. The Peptides: Analysis, Synthesis, Biology, Vol 1. New York: Academic Press, 1979, pp 315–383.
13. JC Sheehan, GP Hess. J Am Chem Soc 77:1067–1068, 1955.
14. DH Rich, J Singh. In: E Gross, J Meinhofer, eds. The Peptides: Analysis, Synthesis, Biology, Vol 1. New York: Academic Press, 1979, pp 241–261.
15. A Williams, IT Ibrahim. Chem Rev 81:589–636, 1981.
16. D Hudson. J Org Chem 53:617–624, 1988.
17. DF DeTar, R Silverstein. J Am Chem Soc 88:1013–1019, 1966.
18. DF DeTar, R Silverstein. J Am Chem Soc 88:1020–1023, 1966.
19. A Arendt, AM Kolodziejczyk. Tetrahedron Lett 3867–3868, 1978.
20. JJ Jones. In: E Gross, J Meinhofer, eds. The Peptides: Analysis, Synthesis, Biology, Vol 1. New York: Academic Press, 1979, pp 65–104.
21. G Barany, RB Merrifield. In: E Gross, J Meinhofer, eds. The Peptides: Analysis, Synthesis, Biology, Vol 2. New York: Academic Press, 1979, pp 1–284.
22. FL Scott, RE Glick, S Winstein. Experientia 13:183–185, 1957.
23. W König, R Geiger. Chem Ber 103:788–798, 1970.
24. M Beyermann, P Henklein, A Klose, R Sohr, M Bienert. Int J Peptide Protein Res 37:252–256, 1991.
25. CD Chang, AM Felix, MH Jimenez, J Meinhofer. Int J Peptide Protein Res 15:485–494, 1980.
26. HS Bates, JH Jones, MJ Witty. J Chem Soc Chem Commun 773–774, 1980.
27. HS Bates, JH Jones, WI Ramage, MJ Witty. In: K Brunfeldt, ed. Peptides 1980. Proceedings of the 16th European Peptide Symposium. Copenhagen: Scriptor, 1981, pp 185–190.
28. T Miyazawa, T Otomatsu, Y Fukui, T Yamada, S Kuwata. J Chem Soc Chem Commun 419–420, 1988.
29. LA Carpino. J Am Chem Soc 115:4397–4398, 1993.
30. W König, R Geiger. Chem Ber 103:2034–2040, 1970.
31. LA Carpino, A El-Fahan, CA Minor, F Albericio. J Chem Soc Chem Commun 201–203, 1994.
32. LA Carpino, A El-Fahan, F Albericio. Tetrahedron Lett 35:2279–2282, 1994.
33. YP Xu, MJ Miller. J Org Chem 63:4314–4322, 1998.
34. SA Kates, SA Triolo, GW Griffin, LW Herman, G Tarr, NA Solé, E Diekmann, A El-Faham, D Ionescu, F Albericio, LA Carpino. In: R Epton, ed. Innovation and Perspectives in Solid Phase Synthesis & Combinatorial Chemical Libraries. Kingswinford, England: Mayflower Scientific, 1997, pp 41–50.
35. YM Angell, C García-Echeverría, DH Rich. Tetrahedron Lett 35:5981–5984, 1994.
36. YM Angell, TL Thomas, GR Flentke, DH Rich. J Am Chem Soc 117:7279–7280, 1995.

37. L Jiang, A Davison, G Tennant, R Ramage. Tetrahedron 54:14233–14254, 1998.
38. JC Spetzler, M Meldal, J Felding, P Vedso, M Begtrup. J Chem Soc Perkin Trans I 1727–1732, 1998.
39. RB Merrifield, LD Vizioli, HG Boman. Biochemistry 21:5020–5031, 1982.
40. KJ Jensen, J Alsina, MF Songster, J Vágner, F Albericio, G Barany. J Am Chem Soc 120:5441–5452, 1998.
41. E Kaiser, GJ Nicholson, HJ Kohlbau, W Voelter. Tetrahedron Lett 37:1187–1190, 1996.
42. Y Han, F Albericio, G Barany. J Org Chem 62:4307–4312, 1997.
43. SS Wang, JP Tam, BSH Wang, RB Merrifield. Int J Peptide Protein Res 18:459–467, 1981.
44. E Atherton, NL Benoiton, E Brown, RC Sheppard, BJ Williams. J Chem Soc Chem Commun 336–337, 1981.
45. JS Davies, J Howe, M Lebreton. J Chem Soc Perkin Trans II 2335–2339, 1995.
46. O Kuisle, E Quiñoa, R Riguera. Tetrahedron Lett 40:1203–1206, 1999.
47. IE Pop, BP Déprez, AL Tartar. J Org Chem 62:2594–2603, 1997.
48. K Dendrinos, J Jeong, W Huang, AG Kalivretenos. J Chem Soc Chem Commun 499–450, 1998.
49. KG Dendrinos, AG Kalivretenos. Tetrahedron Lett 39:1321–1324, 1998.
50. FC Frontin, F Guendouz, R Jacquier, J Verducci. Bull Soc Chim Fr 129:463–467, 1992.
51. SBH Kent, RB Merrifield. In: K Brunfeldt, ed. Peptides 1980. Proceedings of the 16th European Peptide Symposium. Copenhagen: Scriptor, 1981, 328–333.
52. DH Live, SBH Kent. In: VJ Hruby, DH Rich, eds. Peptides—Structure & Function. Rockford, IL: Pierce Chemical Company, 1983, pp 65–68.
53. SBH Kent. In: CM Deber, VJ Hruby, KD Kopple. Peptides—Structure & Function. Rockford, IL: Pierce Chemical Company, 1985, pp 407–414.
54. JS Albert, AD Hamilton. In: LA Paquette, ed. Encyclopedia of Reagents for Organic Synthesis, Vol 3. Chichester, UK: Wiley, 1995, pp 1751–1754.
55. SA Kates, F Albericio, LA Carpino. In: LA Paquette, ed. Encyclopedia of Reagents for Organic Synthesis, Vol 4. Chichester, UK: Wiley, 1995, pp 2784–2785.
56. GB Fields, KM Otteson, CG Fields, RL Noble. In: R Epton, ed. Innovation and Perspectives in Solid Phase Synthesis. Birmingham, UK: SPCC (UK), 1991, pp 241–260.
57. G Gawne, GW Kenner, RC Sheppard. J Am Chem Soc 91:5670–5671, 1969.
58. B Castro, JR Dormoy. Bull Soc Chim Fr 3034–3036, 1971.
59. LE Barstov, VJ Hruby. J Org Chem 36:1305–1306, 1971.
60. S Yamada, Y Takeuchi. Tetrahedron Lett 3595–3598, 1971.
61. AJ Bates, IJ Galpin, A Hallett, D Hudson, GW Kenner, R Ramage, RC Sheppard. Helv Chim Acta 58:688–696, 1975.
62. B Castro, JR Dormoy, G Evin, C Selve. Tetrahedron Lett 1219–1222, 1975.
63. J Coste, D Le-Nguyen, B Castro. Tetrahedron Lett 31:205–208, 1990.

64. RR Dykstra. In: LA Paquette, ed. Encyclopedia of Reagents for Organic Synthesis, Vol 4. Chichester, UK: Wiley, 1995, pp 2784–2785.
65. F Albericio, M Cases, J Alsina, SA Triolo, LA Carpino, SA Kates. Tetrahedron Lett 38:4853–4856, 1997.
66. F Albericio, JM Bofill, A El-Faham, SA Kates. J Org Chem 63:9678–9683, 1998.
67. B Castro, JR Dormoy. Tetrahedron Lett 4747–4750, 1972.
68. MH Kim, DV Patel. Tetrahedron Lett 35:5603–5606, 1994.
69. J Coste, JM Campagne. Tetrahedron Lett 36:4253–4256, 1995.
70. V Dourtoglou, JC Ziegler, B Gross. Tetrahedron Lett 1269–1272, 1978.
71. R Knorr, A Trzeciak, W Bannwarth, D Gillessen. Tetrahedron Lett 30:1927–1930, 1989.
72. I Abdelmoty, F Albericio, LA Carpino, BM Foxman, SA Kates. Lett Peptide Sci 1:57–67, 1994.
73. P Henklein, B Costisella, V Wray, T Domke, LA Carpino, A El-Faham, SA Kates, A Abdelmoty, BM Foxman. In: R Ramage, R Epton, eds. Peptides 1996. Proceedings of the 24th European Peptide Symposium. Kingswinford, England: Mayflower Scientific, 1998, pp 465–466.
74. MA Bailén, R Chinchilla, D Dodsworth, C Nájera, JM Soriano, M Yus. Peptides 1998. Proceedings of the 25th European Peptide Symposium, (S Bajusz, F Hudecz, eds.) Akadémiai Kiadó, Budapest, 1999, pp 172–173.
75. H Gausepohl, U Pieles, RW Frank. In JA Smith, JE Rivier, eds. Peptides—Chemistry and Biology: Proceedings of the 12th American Peptide Symposium. Leiden: ESCOM, Science, 1992, 523–524.
76. SC Story, JV Aldrich. Int J Peptide Protein Res 43:292–296, 1994.
77. S Arttamangkul, B Arbogast, D Barofsky, JV Aldrich. Lett Peptide Sci 3:357–370, 1996.
78. D Delforge, M Dieu, E Delaive, M Art, B Gillon, B Devreese, M Raes, J Van Beeumen, J Remacle. Lett Peptide Sci 3:89–97, 1996.
79. Y Kiso, Y Fujiwara, T Kimura, A Nishitani, K Akaji. Int J Peptide Protein Res 40:308–314, 1992.
80. LA Carpino, A El-Faham, F Albericio. J Org Chem 60:3561–3564, 1995.
81. LA Carpino, A El-Faham, GA Truran, CA Minor, SA Kates, GW Griffin, H Shroff, SA Triolo, F Albericio. In: R Epton, ed. Innovation and Perspectives in Solid-Phase Peptide Synthesis: Biological & Biomedical Applications. Birmingham, England: Mayflower Worldwide, 1994, pp 95–104.
82. F Albericio, I Abdelmoty, JM Bofill, LA Carpino, A El-Faham, BM Foxman, M Gairí, GW Griffin, SA Kates, P Lloyd-Williams, LM Scarmoutzos, H Shroff, SA Triolo, H Wenschuh. In: HLS Maia, ed. Peptides 1994: Proceedings of the Twenty-Third European Peptide Symposium. Leiden: ESCOM, Science, 1995, pp 23–25.
83. H Wenschuh, M Beyermann, H Haber, JK Seydel, E Krause, M Bienert, LA Carpino, A El-Faham, F Albericio. J Org Chem 60:405–410, 1995.
84. I Dalcol, F Rabanal, MD Ludevid, F Albericio, E Giralt. J Org Chem 60: 7575–7581, 1995.

85. F Albericio, P Lloyd-Williams, E Giralt. Methods Enzymol 289:313–336, 1997.
86. JS Davies, J Howe, J Jayatilake, T Riley. Lett Peptide Sci 4:441–445, 1997.
87. M Miyashita, T Nakamori, T Murai, H Miyagawa, M Akamatsu, T Ueno. Biosci Biotech Biochem 62:1799–1801, 1998.
88. LP Miranda, PF Alewood. Proc Natl Acad Sci U S A 96:1181–1186, 1999.
89. BF Gisin, RB Merrifield. J Am Chem Soc 94:3102–3106, 1972.
90. E Giralt, R Eritja, E Pedroso. Tetrahedron Lett 22:3779–3782, 1981.
91. M Gairí, P Lloyd-Williams, F Albericio, E Giralt. Tetrahedron Lett 31:7363–7366, 1990.
92. J Alsina, E Giralt, F Albericio. Tetrahedron Lett 37:4195–4198, 1996.
93. R Knorr, A Trzeciak, W Bannwarth, D Gillessen. In: E Giralt, D Andreu, eds. Proceedings of the Twenty-First European Peptide Symposium. Leiden: ESCOM, Science, 1991, pp 62–64.
94. P Henklein, M Beyermann, M Bienert, R Knorr. In: E Giralt, D Andreu, eds. Proceedings of the Twenty-First European Peptide Symposium. Leiden: ESCOM, Science, 1991, pp 67–68.
95. W Bannwarth, R Knorr. Tetrahedron Lett 32:1157–1160, 1991.
96. JCHM Wijkmans, JAW Kruijtzer, GA van der Marel, JH van Boom, W Bloemhoff. Recl Trav Chim Pays Bas 113:394–397, 1994.
97. J Habermann, H Kunz. J Prakt Chem 340:233–239, 1998.
98. J Klose, P Henklein, A El-Faham, LA Carpino, M Bienert. Peptides 1998. Proceedings of the 25th European Peptide Symposium, (S Bajusz, F Hudecz, eds.) Akadémiai Kiadó, Budapest, 1999, pp 204–205.
99. J Klose, P Henklein, A El-Faham, LA Carpino, M Bienert. Tetrahedron Lett 40:2045–2048, 1999.
100. T Hoeg-Jensen, CE Olsen, A Holm. J Org Chem 59:1257–1263, 1994.
101. H Echner, W Voelter. In: R Epton, ed. Innovation and Perspectives in Solid-Phase Peptide Synthesis and Combinatorial Libraries: Peptides, proteins, and nucleic acids. Small molecule organic chemical diversity. Birmingham, UK: Mayflower Scientific, 1999, pp 279–282.
102. LA Carpino, M Beyermann, H Wenschuh, M Bienert. Acc Chem Res 29:268–274, 1996.
103. LA Carpino, A El-Faham. J Am Chem Soc 117:5401–5402, 1995.
104. SA Triolo, D Ionescu, H Wenschuh, NA Solé, A El-Faham, LA Carpino, SA Kates. In: R Ramage, R Epton, eds. Peptides 1996. Proceedings of the 24th European Peptide Symposium. Kingswinford, England: Mayflower Scientific, 1998, pp 839–840.
105. LA Carpino, A El-Faham. J Org Chem 59:685–698, 1994.
106. LA Carpino, D Ionescu, A El-Faham. J Org Chem 61:2460–2465, 1996.
107. A El-Faham. Chem Lett 671–672, 1998.
108. A El-Faham. Org Prep Proced Int 30:477–481, 1998.
109. J Coste, E Frérot, P Jouin. J Org Chem 59:2437–2446, 1994.
110. J Coste, MN Dufour, A Pantaloni, B Castro. Tetrahedron Lett 31:669–672, 1990.
111. K Akajii, N Kuriyama, Y Kiso. J Org Chem 61:3350–3357, 1996.

112. A di Fenza, M Tancredi, C Galoppini, P Rovero. Tetrahedron Lett 39:8529–8532, 1998.
113. K Witte, O Seitz, CH Wong. J Am Chem Soc 120:1979–1989, 1998.
114. HJ Musiol, F Siedler, D Quarzago, L Moroder. Biopolym (Peptide Sci) 34: 1553–1562, 1994.
115. YM Angell, Y Han, F Albericio, G Barany. In: JP Tam, PTP Kaumaya, eds. Peptides—Chemistry and Biology: Proceedings of the Fifteenth American Peptide Symposium. Dordrecht: Kluwer-ESCOM, Academic Publishers, 1999, pp 339–340.
116. AG Jackson, GW Kenner, GA Moore, R Ramage, WD Thorpe. Tetrahedron Lett 3627–3630, 1976.
117. R Ramage, CP Ashton, D Hopton, MJ Parrott. Tetrahedron Lett 25:4825–4828, 1984.
118. J Meinhofer. In: E Gross, J Meinhofer, eds. The Peptides: Analysis, Synthesis, Biology, Vol 1. New York: Academic Press, 1979, pp 263–314.
119. R Ramage, D Hopton, MJ Parrott, RS Richardson, GW Kenner, GA Moore. J Chem Soc Perkin Trans I 461–470, 1985.
120. R Ramage, B Atrash, D Hopton, MJ Parrott. J Chem Soc Perkin Trans I 1617–1622, 1985.
121. C Poulos, CP Ashton, J Green, OM Ogunjobi, R Ramage, T Tsegenidis. Int J Peptide Protein Res 40:315–321, 1992.
122. S Che, J Xu. Tetrahedron Lett 32:6711–6714, 1991.
123. T Shiori, K Ninomiya, S Yamada. J Am Chem Soc 94:6203–6205, 1972.
124. S Yamada, Y Kasai, T Shiori. Tetrahedron Lett 1595–1598, 1973.
125. J Cabré-Castellví, AL Palomo-Coll. Tetrahedron Lett 21:4179–4182, 1980.
126. J Diago-Msesguer, AL Palomo-Coll, JR Fernandez-Lizarbe, A Zugaza-Bilbao. Synthesis 547–551, 1980.
127. CX Fan, XL Hao, YH Ye. Synth Comm 26:1455–1460, 1996.
128. C Griehl, J Weigt, H Jeschkeit, Z Palacz. In: CH Schneider, AN Eberle, eds. Proceedings of the Twenty-Second European Peptide Symposium. Leiden: ESCOM, Science, 1993, pp 459–460.
129. S Kim, H Chang, YK Ko. Bull Koren Chem Soc 8:471–475, 1987.
130. KL McLaren. J Org Chem 60:6082–6084, 1995.
131. RD Tung, MK Dhaon, DH Rich. J Org Chem 51:3350–3354, 1986.
132. C van der Auwera, MJO Anteunis. Int J Peptide Protein Res 29:574–588, 1987.
133. T Mukaiyama, R Matsueda, M Ueki. In: E Gross, J Meinhofer, eds. The Peptides: Analysis, Synthesis, Biology, Vol 2. New York: Academic Press, 1979, pp 383–416.
134. T Mukaiyama, R Matsueda, M Suzuki. 1901–1904, 1970.
135. R Matsueda, H Maruyama, E Kitazawa, H Takahagi, T Mukaiyama. J Am Chem Soc 97:2573–2575, 1975.
136. H Maruyama, R Matsueda, E Kitasaura, H Takahagi, T Mukaiyama. Bull Chem Soc Jpn 49:2259–2267, 1976.
137. S Zalipsky, JL Chang, F Albericio, G Barany. Reactive Polym 22:243–258, 1994.

138. M Itoh, H Nojima, J Notani, D Hagiwara, K Takai. Bull Chem Soc Jpn 51: 3320–3329, 1978.
139. N Katagiri, K Itakura, SA Narang. J Chem Soc Chem Commun 325–326, 1974.
140. X Jorba, F Albericio, A Grandas, W Bannwarth, E Giralt. Tetrahedron Lett 31:1915–1918, 1990.
141. B Blankemeyer-Menge, M Nimtz, R Frank. Tetrahedron Lett 31:1701–1704, 1990.
142. ZJ Kaminski. Tetrahedron Lett 26:2901–2904, 1985.
143. ZJ Kaminski. Synthesis 917–920, 1987.
144. ZJ Kaminski. Int J Peptide Protein Res 43:312–319, 1994.
145. HW Lee, TW Kang, KH Cha, EN Kim, NH Choi, JW Kim, CI Hong. Synth Commun 28:1339–1349, 1998.
146. T Wieland, F Flor, C Birr. Justus Liebigs Ann Chem 1595–1600, 1973.
147. FM Chen, K Kuroda, LL Benoiton. Synthesis 928–929, 1978.
148. M Bodanszky. In: E Gross, J Meinhofer, eds. The Peptides: Analysis, Synthesis, Biology, Vol 1. New York: Academic Press, 1979, pp 105–196.
149. J Kovacs, L Kisfaludy, MQ Ceprini. J Am Chem Soc 89:183–184, 1967.
150. E Atherton, RC Sheppard. In: Solid Phase Peptide Synthesis. A Practical Approach. Oxford: IRL Press, 1989.
151. L Kisfaludy, I Schön. Synthesis 325–327, 1983.
152. M Green, J Berman. Tetrahedron Lett 31:5851–5854, 1990.
153. H Gausepohl, M Kraft, RW Frank. Int J Peptide Protein Res 34:287–294, 1989.
154. L Otvos, I Elekes, VMY Lee. Int J Peptide Protein Res 34:129–133, 1989.
155. CG Fields, GB Fields, RL Noble, TA Cross. 33:298–303, 1989.
156. D Hudson. Peptide Res 3:51–55, 1990.
157. J Kremsky, M Pluskal, S Casey, H Perry-O'Keefe, SA Kates, ND Sinha. Tetrahedron Lett 37:4313–4316, 1996.
158. O Hollitzer, S Seewald, W Steglich. Angew Chem Int Ed Engl 15:444–445, 1976.
159. R Kirstgen, RC Sheppard, W Steglich. J Chem Soc Chem Comm 1870–1871, 1987.
160. E Atherton, LR Cameron, RC Sheppard. Tetrahedron 44:843–857, 1988.
161. E Atherton, JL Holder, M Meldal, RC Sheppard, RM Valerio. J Chem Soc Perkin Trans I 2887–2894, 1988.
162. R Kirstgen, A Olbrich, H Rehwinkel, W Steglich. Justus Liebigs Ann Chem 437–440, 1988.
163. WD Fuller, MP Cohen, M Shabankareh, RK Blair, M Goodman. J Am Chem Soc 112:7414–7416, 1990.
164. H Leuchs. Ber Dtsch Chem Ges 39:857–861, 1906.
165. JA Fehrentz, C Genu-Dellac, H Amblard, F Winternitz, A Loffet, J Martinez. J Peptide Sci 1:124–131, 1995.
166. WD Fuller, M Goodman, F Naider, YF Zhu. Peptide Sci 40:183–205, 1996.
167. CB Xue, F Naider. J Org Chem 58:350–355, 1993.

168. JR Spencer, VV Antonenko, NGJ Delaet, M Goodman. Int J Peptide Protein Res 40:282–293, 1992.
169. PA Swain, BL Anderson, WD Fuller, F Naider, M Goodman. React Polym 22:155–163, 1994.
170. YF Zhu, RK Blair, WD Fuller. Tetrahedron Lett 35:4673–4676, 1994.
171. C Pothion, JA Fehrentz, P Chevallet, A Loffet, J Martinez. Lett Peptide Sci 4:241–244, 1997.
172. M Paris, JA Fehrentz, A Heitz, J Martinez. Tetrahedron Lett 39:1569–1572, 1998.
173. ML Bender, JM Jones. J Org Chem 27:3771, 1962.
174. NL Benoiton. Biopolym (Peptide Sci) 40:245–254, 1996.
175. LA Carpino, HG Chao, M Beyermann, M Biennert. J Org Chem 56:2635–2642, 1991.
176. LA Carpino, BJ Cohen, KE Stephens, SY Sadat-Aalee, JH Tien, DC Langridge. J Org Chem 51:3732–3734, 1986.
177. VVS Babu, HN Gopi. Tetrahedron Lett 39:1049–1050, 1998.
178. K Akaji, Y Kiso, LA Carpino. J Chem Soc Chem Comm 584–586, 1990.
179. K Akaji, H Tanaka, H Itoh, J Imai, Y Fujiwara, T Kimura, Y Kiso. Chem Pharm Bull 38:3471–3472, 1990.
180. E Vedejs, SZ Lin, A Klapars, JB Wang. J Am Chem Soc 118:9796–9797, 1996.
181. LA Carpino, D Ionescu, A El-Faham, P Henklein, H Wenschuh, M Bienert, M Beyermann. Tetrahedron Lett 39:241–244, 1998.
182. M Meldal, MA Juliano, AM Jansson. Tetrahedron Lett 38:2531–2534, 1997.
183. GA Olah, M Nojima, I Kerekes. Synthesis 487–488, 1973.
184. JN Bertho, A Loffet, C Pinel, F Reuther, G Sennyey. Tetrahedron Lett 32:1303–1306, 1991.
185. LA Carpino, EME Mansour, D Sadat-Aalee. J Org Chem 56:2611–2614, 1991.
186. C Kaduk, H Wenschuh, HC Beyermann, K Forner, LA Carpino, M Bienert. Lett Peptide Sci 2:285–288, 1995.
187. H Wenschuh, M Beyermann, S Rothemund, LA Carpino, M Bienert. Tetrahedron Lett 36:1247–1250, 1995.
188. SCF Milton, RCdL Milton, SA Kates, C Glabe. In: Techniques in Protein Chemistry VIII. New York: Academic Press, 1997, pp 865–873.
189. H Wenschuh, M Beyermann, E Krause, M Brudel, R Winter, M Schümann, LA Carpino, M Bienert. J Org Chem 59:3275–3280, 1994.
190. H Wenshcuh, M Beyermann, A El-Faham, S Ghassemi, LA Carpino, M Bienert. J Chem Soc Chem Commun 669–670, 1995.
191. H Wenschuh, M Beyermann, R Winter, M Beinert, D Ionescu, LA Carpino. Tetrahedron Lett 37:5483–5486, 1996.
192. SC Mayer, MM Jouillé. Synth Commun 24:2367–2377, 1994.
193. J Green, K Bradley. Tetrahedron 49:4141–4146, 1993.
194. D Granitza, M Beyermann, H Wenschuh, H Haber, LA Carpino, GA Truran, M Bienert. J Chem Soc Chem Commun 2223–2224, 1995.
195. YM Angell, J Alsina, Y Han, F Albericio, G Barany. J Peptide Res, submitted.

196. J Alsina, T Scott Yokum, F Albericio, G Barany. J Org Chem 64, 1999, in press.

197. D Bourgin, F Dick, M Schwaller. In: R Ramage, R Epton, eds. Peptides 1996. Proceedings of the 24th European Peptide Symposium. Kingswinford, England: Mayflower Scientific, 1998, pp 279–280.

198. J Alsina, F Albericio, G Barany, SA Kates. Lett Peptide Sci 6:243–245, 1999.

199. S Sakakibara. Biopolym (Peptide Sci) 37:17–28, 1995.

200. MHS Cezari, L Juliano. Peptide Res 9:88–91, 1996.

201. H Rink, P Sieber, F Raschdorf. Tetrahedron Lett 25:621–624, 1984.

202. RB Merrifield, AR Mitchell, JE Clarke. J Org Chem 39:660–668, 1974.

203. T Fujii, S Sakakibara. Bull Chem Soc Jpn 47:3146–3551, 1974.

204. T Ishiguro, C Eguchi. Chem Pharm Bull 37:506–508, 1989.

205. M Kusunoki, S Nakagawa, K Seo, T Hamara, T Fukuda. Int J Peptide Protein Res 36:381–386, 1990.

206. C Celma, F Albericio, E Pedroso, E Giralt. Peptide Res 5:62–71, 1992.

207. F Chillemi, RB Merrifield. Biochemistry 8:4344–4346, 1969.

208. SM Meister, SBH Kent. In: VJ Hruby, DH Rich, eds. Peptides. Structure and Function. Proceedings of the 8th American Peptide Symposium. Rockford, IL: Pierce Chemical Company, 1983, pp 103–106.

209. RCdL Milton, SCF Milton, PA Adams. J Am Chem Soc 112:6039–6046, 1990.

210. DH Live, SBH Kent. In: VJ Hruby, DH Rich, eds. Peptides. Structure and Function. Proceedings of the 8th American Peptide Symposium. Rockford, IL: Pierce Chemical Company, 1983, pp 65–68.

211. JP Tam, YA Lu. J Am Chem Soc 117:12058–12063, 1995.

212. KC Pugh, EJ York, JM Stewart. Int J Peptide Protein Res 40:208–213, 1992.

213. JP Tam. Int J Peptide Protein Res 29:421–431, 1987.

214. DH Lloyd, GM Petrie, RL Noble, JP Tam. In: JE Rivier, GR Marshall, eds. Peptides. Structure and Function. Proceedings of the Eleventh American Peptide Symposium. Leiden: ESCOM, 1990, pp 909–910.

215. S Wang, GL Foutch. Chem Eng 46:2373–2376, 1991.

216. AK Rabinovich, JE Rivier. In: RS Hodges, JA Smith, eds. Peptides. Structure and Function. Proceedings of the Thirteenth American Peptide Symposium. Leiden: ESCOM, 1994, pp 71–73.

217. SA Kates, NA Solé, M Beyermann, G Barany, F Albericio. Peptide Res 9: 106–113, 1996.

218. BE Kaplan, LJ Hefta, RC Blake III, KM Swiderek, JE Shively. J Peptide Res 52:249–260, 1998.

219. SBH Kent. In: CM Deber, VJ Hruby, KD Kopple, eds. Peptides. Structure and Function. Proceedings of the 9th American Peptide Symposium. Rockford, IL: Pierce Chemical Company, 1985, pp 407–414.

220. SBH Kent. Annu Rev Biochem 57:957–989, 1988.

221. GB Fields, CG Fields. J Am Chem Soc 113:4202–4207, 1991.

222. EM Cilli, E Oliveira, R Marchetto, CR Nakaie. J Org Chem 61:8992–9000, 1996.

223. SA Kates, SB Daniels, F Albericio. Anal Biochem 212:303–310, 1993.

224. D Yamashiro, J Blake, CH Li. Tetrahedron Lett 1469–1472, 1976.

225. SCF Milton, RCdL Milton. Int J Peptide Protein Res 36:193–196, 1990.

226. H Kuroda, YN Chen, T Kimura, S Sakakibara. Int J Peptide Protein Res 40: 294–299, 1992.

227. E Atherton, V Woolley, RC Sheppard. J Chem Soc Chem Commun 970–971, 1980.

228. FC Westall, AB Robinson. J Org Chem 35:2842–2844, 1970.

229. L Zhang, C Goldammer, B Henkel, F Zühl, G Panhaus, G Jung, E Bayer. In: R Epton, ed. Innovation and Perspectives in Solid Phase Synthesis. Peptides, Proteins and Nucleic Acids. Biological and Biomedical Applications. Birmingham, UK: Mayflower Worldwide, 1994, pp 711–716.

230. JM Stewart, WA Klis. In: R Epton, ed. Innovation and Perspective in Solid-Phase Synthesis and Related Technologies. Peptides, Polypeptides and Oligonucleotides. Macro-Organic REagents and Catalysts. Birmingham, UK: SPCC Ltd., 1990, pp 1–9.

231. A Thaler, D Seebach, F Cardinaux. Helv Chim Acta 74:628–643, 1991.

232. JC Hendrix, KJ Halverson, JT Jarrett, PT Lansbury. J Org Chem 55:4517–4518, 1990.

233. S Takahashi, Y Shimonishi. Chem Lett 51–56, 1974.

234. KT Wang, ST Chen, SH Chiou. In: Techniques in Protein Chemistry II. New York: Academic Press, 1991, pp 241–247.

235. E Kaiser, RL Colescott, CD Bossinger, PI Cook. Anal Biochem 34:595–598, 1970.

236. VK Sarin, SBH Kent, JP Tam, RB Merrifield. Anal Biochem 117:147–157, 1981.

237. T Christensen. Acta Chem Scand 33B:763–766, 1979.

238. V Krchnák, J Vágner, P Safar, M Lebl. Collect Czech Chem Commun 53: 2542–2548, 1988.

7
Homodetic Cyclic Peptides

Paolo Rovero
Università di Salerno, Fisciano, Italy

I. INTRODUCTION

A number of naturally occurring cyclic peptides, such as antibiotics and toxins, have been described in the past 50 years. They can be divided into two general classes: homodetic and heterodetic. In the first case, ring formation occurs only through usual peptide (amide) linkages connecting amino and carboxyl functions, whereas heterodetic cyclic peptides include any other linkages such as lactone, ether, thioether, and, most commonly, disulfide bridge linkages. Investigation of the biological properties of these molecules showed that cyclic structures may exhibit improved metabolic stability and increased potency, receptor selectivity, and bioavailability. Moreover, the constrained geometry of a cyclic peptide allows conformational investigations.

Cyclic peptides have been prepared entirely in solution or by assembling the linear precursor following a solid-phase approach, cleaving the peptide from the resin, and cyclizing in solution. Despite efforts to develop practical and convenient methods, solution-phase methodologies suffer from several drawbacks, such as cyclodimerizations and cyclo-oligomerization side reactions, which may occur even at high dilutions.

It has been shown that cyclization can be conveniently performed while the peptide remains anchored to the solid support, thus taking advantage of the *pseudodilution* phenomenon, which favors intramolecular reactions over intermolecular side reactions. This chapter surveys the literature on the syntheses of cyclic peptides in which all steps have been carried out in the solid-phase mode. Other reviews of this subject have appeared [1,2].

Homodetic cyclic peptides have been divided into five general sub-classes according to the cyclization topology (Scheme 1). In head-to-tail cyclic peptides, the ring is closed between the amino and carboxyl groups of *N*- and *C*-termini. Side chain–to–side chain cyclization connects side-chain functional groups, such as the ω-amino function of Lys or Orn and the ω-carboxyl function of Asp or Glu. If one of these side chains is linked to either an *N*- or *C*-terminus (e.g., Lys/Orn to a *C*-terminal carboxyl group or Asp/Glu to an *N*-terminal amino function) a side chain–to–end cycle is obtained. In the fourth class (branched cycles), suitable lactam bridges (either side chain to side chain or side chain to end) connect two chains of a branched peptide. Finally, a special class of backbone-to-backbone cyclic peptides is obtained when amide nitrogens of the peptide backbone are connected via a bridge consisting of alkyl groups and an amide bond.

II. HEAD TO TAIL

The first reported solid-phase synthesis of head-to-tail cyclic peptides was based on the intramolecular aminolysis of resin-bound *o*-nitrophenyl esters. The cyclization proceeds concurrently to cleave the peptide from the resin, after deprotection and neutralization of the *N*-terminal residue (Scheme 2A). Accordingly, Fridkin et al. [3] reported the preparation of several simple, unhindered cyclopeptides, such as *cyclo*(Ala-Gly-Ala-Ala). Similarly, Flanigan and Marshall [4] obtained activation of the resin-bound peptide ester, after elongation of the peptide chain, by oxidation of the 4-(methylthio)phenyl (MTP) linker to a sulfonyl ester. Subsequent deblocking of the *N*-terminal residue and intramolecular condensation yielded the desired cyclic peptide. However, this method was found not to be suitable for the synthesis of longer and more hindered cyclic peptides [5].

The strategy based on the intramolecular aminolysis of an activated linker was subsequently extended to use of the *p*-nitrobenzophenone oxime resin developed by DeGrado and Kaiser [6]. Ösapay et al. [7] reported the preparation of the cyclodecapeptide tyrocidine A, an antibiotic [*cyclo*(DPhe-Pro-Phe-DPhe-Asn-Gln-Tyr-Val-Orn-Leu)], by *t*-butyloxycarbonyl–benzyl (Boc-Bzl) chemistry. Boc-Leu-OH was linked to the oxime resin at a low substitution level (0.11 mmol/g) in order to minimize interchain reactions during cyclization (pseudodilution) and the peptide chain was assembled using the BOP coupling procedure. After deprotection of the *N*-terminal Boc-DPhe-OH residue (TFA–CH_2Cl_2, 1:3) and neutralization [*N,N*-diisopropylethylamine (DIEA), 1.5 equiv.], the free amino group cleaved the peptide from the resin by intrachain aminolysis in CH_2Cl_2 (24 h, room temperature) with a yield of 55%. Subsequent side-chain deprotection and final semipre-

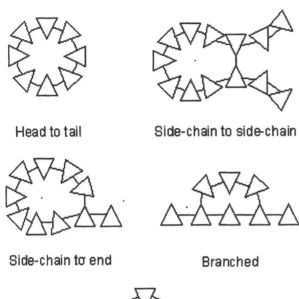

Head to tail   Side-chain to side-chain

Side-chain to end   Branched

Backbone to backbone

Scheme 1   Topological subclasses of homodetic cyclic peptides.

parative high-performance liquid chromatographic (HPLC) purification gave the desired product with an overall yield of 30%.

The same approach was described also by Nishino et al. [8] for the preparation of *cyclo*(Arg-Gly-Asp-Phg) (Phg, phenylglycine), a cyclic tetrapeptide with inhibitory activity toward cell adhesion. Following removal of the *N*-terminal Boc group, the peptide-resin was treated with CH_3COOH–triethylamine (TEA) (2 equiv. each) in *N,N*-dimethylformamide (DMF) at room temperature to effect cyclization of the peptide over a 24-h period (50–70% crude yield). In this case the substitution level of the oxime resin (0.5 mmol/g) was not reduced, contrary to the suggestion by Ösapay in order to achieve pseudodilution conditions. This study showed that the extent of oligomerization depended significantly on the choice of the *C*-terminal residue (up to 40% dimer formation when Gly was in the *C*-terminal position). In a follow-up of this study [9], the on-resin cyclodimerization reaction was optimized for the preparation of a more potent analogue of the previously

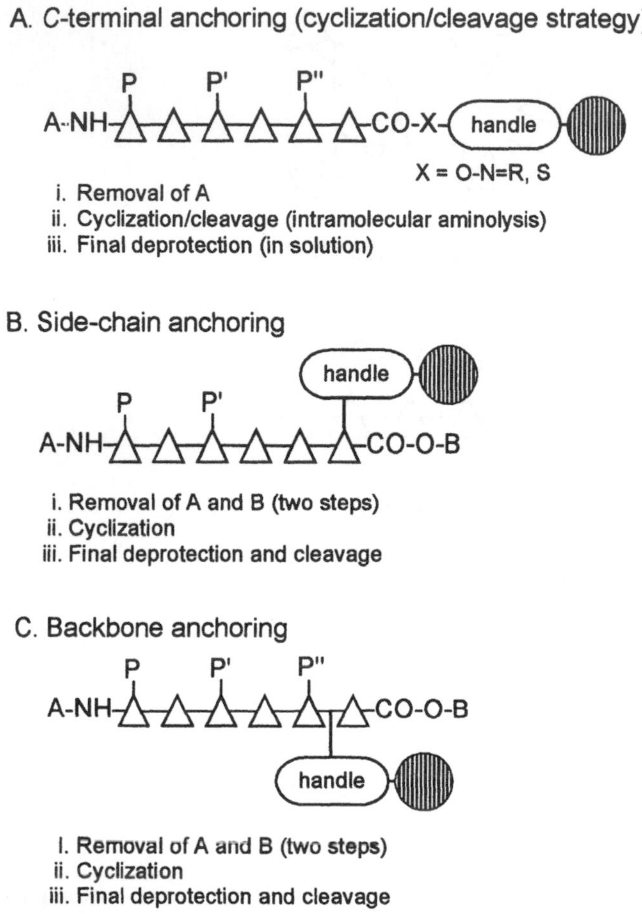

A. C-terminal anchoring (cyclization/cleavage strategy)
P P' P''
A-NH
CO-X
handle
X = O-N=R, S
i. Removal of A
ii. Cyclization/cleavage (intramolecular aminolysis)
iii. Final deprotection (in solution)
B. Side-chain anchoring
handle
P P'
A-NH
CO-O-B
i. Removal of A and B (two steps)
ii. Cyclization
iii. Final deprotection and cleavage
C. Backbone anchoring
P P' P''
A-NH
CO-O-B
handle
I. Removal of A and B (two steps)
ii. Cyclization
iii. Final deprotection and cleavage

Scheme 2 General strategies for the synthesis of head-to-tail cyclic peptides.

described cyclotetrapeptide: *cyclo*(Arg-Sar-Asp-Phg)$_2$ (Sar, sarcosine). The overall yield of the two-step cyclodimerization reaction, based on a first intermolecular aminolysis, yielding a resin-linked linear octapeptide, which subsequently undergoes an intramolecular aminolysis to give the cyclodimer, depends on the sequence of the tetrapeptide, as previously observed, and, most important, the substitution level of the resin. Accordingly, Arg(Tos)-Sar-Asp(O*c*Hx)-Phg-resin (0.49 mmol/g) gave 88% cyclodimer, whereas Phg-Arg(Tos)-Sar-Asp(O*c*Hx)-resin (0.49 mmol/g) gave only 59% cyclodimer. This observation may reflect the conformational preferences of the protected tetrapeptides on the oxime resin. However, when the latter peptide-

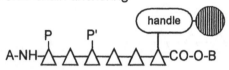

resin was prepared at a lower substitution level, the cyclodimerization yields decreased: 40% at 0.38 mmol/g, 27% at 0.23 mmol/g, and only 21% at 0.11 mmol/g. Another interesting observation is that longer sequences do not undergo cyclodimerization. Accordingly, the preformed resin-linked octapeptide upon cleavage gives only the cyclomonomer, independent of the sequence. Preparation of [DPya$^{4,4'}$]gramicidin S [cyclo(Val-Orn-Leu-DPya-Pro)$_2$, Pya, pyrenylalanine] by the intramolecular aminolysis of the preformed linear decapeptide was also reported [10]. Oxime resin (0.5 mmol/g) was used to obtain the desired cyclic peptide in 45% overall yield with only 8% dimerization product.

A similar approach, which does not suffer from oligomerization-related problems, was proposed by Richter et al. [11]. S-Trityl–protected thioglycolic acid was linked to a p-methylbenzhydrylamine (MBHA)-resin using a Phe residue as a spacer. After removal of the trityl group [trifluoroacetic acid (TFA)–CH$_2$Cl$_2$, 1:1 and triethylsilane, 5 equiv.], the first amino acid was attached to the thiol [N,N-diisopropylcarbodiimide (DIPCDI)–CH$_2$Cl$_2$ and N-methylmorpholine (NMM) or DIEA, 3 equiv.] through a thioester linkage. The linear peptide was assembled using standard Boc/Bzl chemistry, using the in situ neutralization protocol. Cyclization and cleavage by intramolecular aminolysis was achieved by treatment with DIEA (3 equiv.) and 4-N,N-dimethylaminopyridine (DMAP) (0.1 equiv.) in N,N-dimethylacetamide (DMA) for 2–7 days at 25°C, followed by side-chain deprotection in HF. Two model peptides, cyclo(DTyr-Lys-Gly-Ile-Trp-Gly) and cyclo(DPro-Arg-Ser-Ile-Trp-Gly), were prepared by the same protocol also on oxime resin with comparable yields but with less pure material than that obtained using the thioester linkage method. However, attempts to cyclize Nishino's model peptide (Asn-Phg-Arg-Gly, see earlier) using a thioester linkage failed to give the desired cyclic tetrapeptide.

Another method based on thioester resin linkage was proposed by Camarero et al. [12] for the preparation of large, Cys-containing head-to-tail cyclic peptides. These authors exploited the "native chemical ligation" approach [13], which uses a chemoselective reaction to link two unprotected peptide segments. Accordingly, in aqueous buffer at about neutral pH, a peptide having a Cys residue at its N-terminus will react with a second peptide containing an α-thioester moiety with resultant formation of an amide bond at the ligation junction. This method has been extended to intramolecular cyclization of a fully unprotected peptide that is still attached to the solid support via a thioester linkage. A 3-mercaptopropionamide–polyethylene glycol–poly-(N,N-dimethylacrylamide) copolymer support (HS-PEGA) was developed to which the first Boc-protected amino acid can be coupled through a thioester linkage using standard methods [1-hydroxybenzotriazole (HOBt) or HOSu active esters; substitution level 0.2 mmol/

g]. Next, the linear precursor was constructed using Boc/Bzl chemistry and an in situ neutralization–HBTU activation protocol. Side-chain deprotection was achieved with HF without affecting the thioester resin linkage. Finally, solid-phase cyclization–cleavage was achieved by treatment with 1 mM EDTA, 0.1 M sodium phosphate buffer, pH 7.5, for 1–2 h at room temperature. A model 15-mer peptide was prepared in 40% yield with only small amounts of oligomeric material being observed. The synthesis of a 47-mer cyclopeptide was also reported.

Small cyclopeptides based on the Arg-Gly-Asp (RGD) sequence have been prepared using the oxime resin approach by DeGrado's group [14]. The ring was closed by a semirigid nonpeptide linker in order to lock the intervening peptide backbone into a single conformer or a family of related conformers. Thus, cyclo(Gly-Arg-Gly-Asp-Mamb) [Mamb, m-(aminomethyl)benzoic acid] was prepared via esterificaion of Boc-Mamb-OH to oxime resin using HBTU/NMM. The remaining amino acids were sequentially coupled (Boc/Bzl chemistry, HBTU/NMM activation) and, after removal of the last protecting group, the peptide was cyclized by treatment with DIEA–CH$_2$Cl$_2$ (9:1) followed by CH$_3$COOH–DMF at 50°C. The crude product obtained showed a single major peak on HPLC and was deprotected with anhydrous HF, yielding the final product in 11% overall yield after HPLC purification.

Ösapay's method was used for the preparation of N-methylcyclodepsipeptides. Lee [15] reported the elegant synthesis of two analogues of the antiparasitic cyclodepsipeptide PF1022A based on the cyclization–cleavage method. PF1022A consists of eight residues; four N-methyl-L-leucines (MeLeu), two 3-phenyl-D-lactates (PheLac), and two D-lactates (Lac) in a 24-membered ring with alternating amide and ester bonds; in the two reported synthetic analogues one MeLeu residue was replaced by either Pro or pipecolic acid (Pip). The linear resin-bound octadepsipeptides MeLeu-PhLac-MeLeu-Lac-MeLeu-PhLac-Pip(or Pro)-Lac-O resin were obtained using a fragment condensation strategy; cyclization was achieved by heating in refluxing ethyl acetate for 2 days [yield 55% after preparative thin-layer chromatographic (TLC) purification].

A different approach to the solid-phase preparation of head-to-tail cyclic peptides that has found wide application is based on the principle of anchoring the peptide to the resin through a side chain. After elongation of the peptide chain, selective deprotection of the N- and C-termini enables cyclization on the resin, taking advantage of the pseudodilution effect, followed by a separate step of side-chain deprotection and peptide cleavage (Scheme 2B). This strategy was first applied by Isied et al. [16] and anchored the imidazole ring of Boc-His-OH to a polystyrene resin using 1,5-difluoro-2,4-dinitrobenzene to obtain a resin-linked N^{im}-2,4-Dnp histidine derivative

(0.2 mmol/g). H-Gly-OBzl was added at the free carboxyl function of His, using DCC/HOBt activation, and subsequently the dipeptide was further elongated at the N-terminus with four standard couplings until the resin-bound linear hexapeptide Boc-[His(Dnp)-Gly]$_2$-His(Dnp-Resin)-Gly-OBzl was obtained. Treatment of the peptide-resin with an HBr–CH$_3$COOH–TFA–CH$_2$Cl$_2$ mixture (4:6:5:5, 2 × 30 min) deprotected the N- and C-termini without affecting either the Dnp linker or His protecting groups. After washings (CH$_2$Cl$_2$) and neutralization (pyridine), cyclization was achieved by two treatments with 3 equiv. of N-(ethoxy-carbonyl)-2-ethoxy-1,2-dihydroquinoline (EDDQ) in pyridine–toluene (1:1, 24 h at room temperature). The peptide was removed from the resin by using deoxygenated 1.5 M thiophenol in DMF (yield 85%) and purified by HPLC with an overall yield of 42%.

An extension of this method uses the ω-carboxyl functions of Asp or Glu for the attachment to the solid support, as first proposed by Rovero et al. [17]. Depending on the nature of the linker connecting the side chain to the resin, cyclic peptides containing either Asp-Glu or Asn-Gln can be obtained after the final cleavage. Accordingly, the synthesis of a cyclic tachykinin antagonist, *cyclo*(Asp-Tyr-DTrp-Val-DTrp-DTrp-Val), was achieved by anchoring the side chain of Boc-Asp(OH)-OFm to a phenylacetamidomethyl (PAM) resin (0.35 mmol/g). After six cycles of peptide chain elongation, using Boc/Bzl chemistry and BOP/DIEA activation, the C- and N-termini of the resin-bound linear precursor were deprotected by treatment with TFA–CH$_2$Cl$_2$, (1:1, 5 + 15 min) and piperidine–DMF (1:4, 3 + 7 min), respectively. Cyclization was achieved using BOP (3 equiv.) and DIEA (6 equiv.) in DMF (3 h at room temperature, twice). The peptide was removed from the resin by using HF–anisole–Me$_2$S (20:2:1, crude yield 72%, HPLC purity 65%) and purified by semipreparative HLPC (Scheme 3).

This protocol was used by Taylor et al. [18] for the preparation of several cycloheptapeptides of general structure *cyclo*[DAla-Leu-DAsp(or DisoAsp)-Ala-Cys(or βAla)-DGLu(or isoGlu)-Asp], as models of the naturally occurring protein phosphatase inhibitor microcystin-LR. The solid-phase cyclization (HBTU/DIEA activation) was generally found to be more successful than solution-phase cyclizations, based on results obtained with several linear precursors and activation strategies, including HBTU, BOP, and HATU.

The same method can be used for the construction of Asn- or Gln-containing head-to-tail cyclic peptides, using a benzhydrylamine instead of a benzyl alcohol linker, as shown by Tromelin et al. [19]. The synthesis of an 18-mer head-to-tail cyclic peptide designed to mimic a loop involved in the curaremimetic action of a snake toxin protein, *cyclo*(Asn-Tyr-Lys-Lys-Val-Trp-Arg-Asp-His-Arg-Gly-Thr-Ile-Ile-Glu-Arg-Gly-Pro), was prepared.

Protections: Boc (α-NH$_2$)/OFm (α-COOH)/Bzl (side-chains)
Handles: PAM (X = O; AA$_5$ = Asp, Glu)
             MBHA (X = NH; AA$_5$ = Asn, Gln)

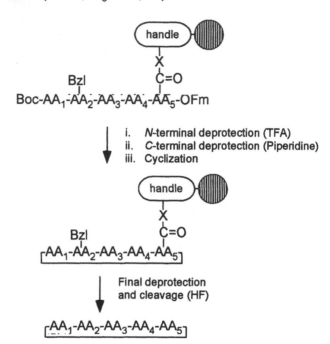

Scheme 3   Synthesis of head-to-tail cyclic peptides according to Rovero et al. [17] and Tromelin et al. [19].

Boc-Asp(OH)-OFm was attached to a methylbenzhydrylamine (MBHA)-resin, maintaining a low substitution level (0.4 mmol/g) in order to respect the pseudodilution principle; unreacted amino groups were capped by acetylation. After elongation of the peptide chain by standard methods on an automatic synthesizer using the Boc/Bzl strategy and deprotection of the *C*- and *N*-termini with TFA and piperidine, respectively, cyclization was obtained by treatment with BOP (3 equiv.) and DIEA (6 equiv.) in NMP (2 h at room temperature, twice). HF cleavage yielded the desired cyclic product with an overall 10% yield after HPLC purification (Scheme 3).

The protection scheme based on the combination Boc (α-NH$_2$)/OFm (α-COOH)/Bzl (side chains) groups was found to suffer from low yields and substantial epimerization in the incorporation of the initial Asp or Glu residues to the polymer through a hydroxy- or bromomethylphenylacetic acid handle. The chemistry of this step was studied by Valero et al. [20] and was found to be unsatisfactory for the conventional esterification procedure (amino acid–DIPCDI–DMAP 4:4:0.1 or 0.5 equiv., DMF, 3 × 1 h, room temperature) as well as several other anchoring protocols. The most effective method was the nucleophilic substitution of bromomethylphenylacetic handles by the Boc-Asp(OH)-OFm cesium or zinc carboxylate salt [amino acid (6 equiv.), CsHCO₃ pH 7, DMF, overnight, 50°C; quantitative yield, 2–3% epimerization]. An N,N-dicyclohexylcarbodiimide (DCC)-mediated coupling of the bromomethylphenylacetic acid handle to MBHA resin is strongly suggested (BrCH₂-C₆H₄-CH₂COOH/DCC 4:4 equiv., DMF, overnight, room temperature), because methods based on HOBt-derived reagents were found to be unsuccessful. Epimerization problems at the level of the resin-linked Asp residue have also been observed by Burgess et al. [21] during the synthesis of $cyclo$(Arg-Gly-Asp)₂ using McMurray's protocol (see later).

A first alternative to the Boc/OFm/Bzl strategy for the synthesis of cyclic peptides by side-chain anchoring was proposed by McMurray [22]. In order to use the Fmoc/tBu chemistry for the peptide chain elongation, the α-carboxylate of the resin-anchored Glu was anchored as a 2,4-dimethoxy-benzyl (Dmb) ester, obtaining Fmoc-Glu(OResin)-ODmb as starting material. The Dmb group could be removed by treatment with TFA–CH₂Cl₂ (1:99) without affecting tBu and 2,2,5,7,8-pentamethylchroman-6-sulfonyl (Pmc) side-chain protecting groups employed in the Fmoc strategy. However, this condition removes the Trt protection from Cys and His. The method is illustrated by the synthesis of $cyclo$(Ala-Ala-Arg-DPhe-Pro-Glu-Asp-Asn-Tyr-Glu), a rearranged sequence derived from the autophosphorylation site of the tyrosine kinase pp60[c-src]. Fmoc-Glu(OH)-ODmb was coupled to a 4-hydroxymethylphenoxyacetylaminomethyl resin (0.45 mmol/g) using DIPCDI/DMAP chemistry and the linear precursor was assembled using a standard protocol (BOP/HOBt/NMM activation). Subsequently, treatment with TFA–CH₂Cl₂, 1:99 (6 × 5 min), followed by piperidine–DMF (1:4) exposed the C- and N-termini, respectively. It is important to remove the Dmb group before the Fmoc function in order to obtain a cleaner crude product. On-resin cyclization was achieved with either (1) DIPCDI/HOBt (3:3 equiv., DMF, 16 h, room temperature) or (2) BOP/HOBt/NMM (3 equiv. each, DMF, 1 h, room temperature). In both cases, deprotection and cleavage were achieved by overnight treatment with TFA–phenol (19:1) and the cyclic peptide was separated from oligomeric products by gel filtration (Sephadex G-25). Yields were (1) 32% (22% after additional chromatographic purification) and (2) 35% (26% after additional chromatographic purification), and the weight ratio of oligomeric by-products to monomeric peptide was (1) 1.2 and (2) 1.3 (Scheme 4). It was concluded that the BOP-mediated cyclization, although faster, had no advantages in

Protections: Fmoc (α-NH₂)/ODmb (α-COOH)/*t*Bu (side-chains)

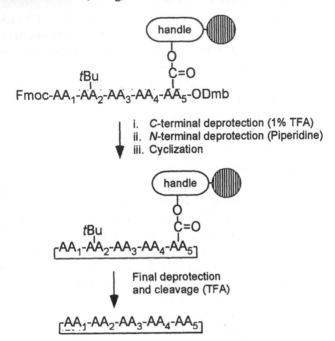

Scheme 4   Synthesis of head-to-tail cyclic peptides according to McMurray [22].

terms of yield or purity. In addition, oligomerization and racemization were resin dependent but solvent independent with DIPCDI [23].

McMurray's cyclization protocol, with the addition of a third level of protection, was applied by Brugghe et al. [24] in order to prepare cyclic peptides with an *S*-acetylmercaptoacetate (SAMA)–derivatized Lys residue for conjugation purposes. Accordingly, the model peptide *cyclo*[Thr-Asn-Asn-Asn-Leu-Lys(SAMA)-Thr-Lys-Asp] and several other analogues were prepared on an Fmoc-Asp(OR)-ODmb resin, according to the McMurray's protocol, including an N^ε-1-(4-4-dimethyl-2,6-dioxocyclohexylidene)ethyl (Dde)–protected Lys residue. After cyclization, the Dde group was selectively removed with hydrazine hydrate–DMA (1:99) and the free amino function was modified by reaction with pentafluorophenyl SAMA. After deprotection and cleavage (TFA–anisole–H₂O, 95:2.5:2.5), the SAMA-cyclopeptide can be specifically conjugated to bromoacetyl-peptides or proteins.

The first three-dimensional orthogonal protection scheme, fully compatible with the Fmoc/*t*Bu chemistry, was developed independently by Bannawarth's and Albericio's groups, combining Fmoc (α-NH₂)/allyl (α-

COOH)/tBu (side chains) protections. Trzeciak and Bannwarth [25] described the synthesis of the two cyclopeptides *cyclo*(Lys-Arg-Ser-Lys-Gly-Asx), where Asx = Asp or Asn. Fmoc-Asp(OH)-OAl was coupled to the resin using either an amide or a hydroxy linker and TBTU as a condensing reagent (substitution level not reported) and the linear peptides were elongated by standard Fmoc/tBu chemistry (TBTU activation). Subsequently, the allyl ester was cleaved with Pd(PPh₃)₄ and *N*-methylaniline in THF–DMSO–0.5 M HCl (2:2:1), the *N*-terminus was deprotected with piperidine–DMF (1:4), and the peptide was cyclized by TBTU/DIEA (1.5 equiv. each, 3 h) activation. Cleavage and deprotection with TFA–H₂O (9:1) yielded the expected cyclic products (Scheme 5).

Similarly, Kates et al. [26] reported four strategies for the preparation of the cyclic decapeptide described by McMurray (see earlier), using four different residues as the starting point: (1) Glu⁶ and (2) Glu¹⁰ [ω-carboxyl

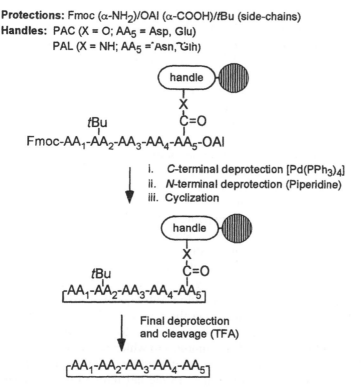

Protections: Fmoc (α-NH₂)/OAl (α-COOH)/tBu (side-chains)
Handles: PAC (X = O; AA₅ = Asp, Glu)
          PAL (X = NH; AA₅ = Asn, Gln)

Scheme 5 Synthesis of head-to-tail cyclic peptides according to Trzeciak and Bannwarth [25] and Kates et al. [26].

linked to p-alkoxybenzyl alcohol (PAC) polyethylene glycol–polystyrene (PEG-PS) support: Fmoc-Glu(OPAC-PEG-PS)-OAl]; (3) Asp[7] [Fmoc-Asp(OPAC-PEG-PS)-OAl]; (4) Asn[8] [β-carboxyl of an Asp residue linked to PEG-PS through a 5-(4-(9-fluorenylmethyl-oxycarbonyl)aminomethyl-3,5-dimethoxyphenoxy)valeric acid (PAL) handle: Fmoc-Asp(OPAL-PEG-PS)-OAl]. Note that in the latter case, the amide handle yields the required Asn residue upon cleavage. Syntheses of the linear precursors were carried out under continuous-flow conditions, using PyBOP/HOBt/DIEA (5 equiv. each) activation, followed by deprotection of the allyl esters with $Pd(PPh_3)_4$ in THF–DMSO–0.5 M HCl–morpholine (2:2:1:0.1) for 2 h at 25°C, then Fmoc removal and BOP/HOBt/DIEA (5:5:10 equiv.) mediated cyclization for 2–5 h. Cleavage from the resin and final deprotection were achieved with TFA–thioanisole–1,2-ethanedithiol–anisole (90:5:3:2) for 1 h at 25°C. All four strategies gave the desired peptide; the best yield and purity occurred with Asn[8] strategy (71% monomeric cyclic product in the crude, as evaluated by analytical HPLC), and the Glu[10] strategy gave the highest level of by-products.

In an extension of this work, the allyl deblocking procedure was automated. The standard method for allyl removal uses a suspension of palladium catalyst, and this reaction is not feasible on a peptide synthesizer because of problems related to the handling of insoluble materials. To avoid this problem, Kate et al. [27] optimized the procedure, performing the allyl removal with $Pd(PPh_3)_4$ (1 equiv.) in $CHCl_3$–CH_3COOH–NMM (37:2:1) for 2 h at 25°C, followed by washings with a solution of 0.5% DIEA and 0.5% sodium diethyldithiocarbamate in DMF. A detailed protocol for allyl deblocking and on-resin cyclization using an automated continuous-flow synthesizer was reported. As an example, Rovero's cyclic tachykinin antagonist (see earlier) prepared on the synthesizer gave results comparable to those obtained with manual cyclization.

The automated protocol was exploited by Delforge et al. [28] for the preparation of a "tailed" cyclic peptide, cyclo[DVal-Arg-Gly-Asp-Glu(εAHx-Cys-NH$_2$)], which contains the Arg-Gly-Asp adhesion motif recognized by cellular integrins. The tail of the peptide consists of a Cys residue linked to the β-carboxyl of Glu through a six-carbon spacer arm (6-aminohexanoic acid) and was designed to allow subsequent grafting of the cyclic peptide to polymers or proteins via the Cys thiol function. The automated synthesis was performed under continuous-flow conditions, starting from Fmoc-PAL-PEG-PS resin (0.16 mmol/g) to which was esterified Fmoc-Cys(Trt)-OH; the peptide tail was then completed by coupling Fmoc-εAhx-OH, and Fmoc-Glu(OH)-OAl was linked to the tail through its side chain. Next, the linear pentapeptide was constructed using a standard protocol and the allyl group was removed following Kates' procedure, with the addition of a final washing step with DIEA–CH$_2$Cl$_2$ (2:23) for 12 min, for complete removal of reagents and side products resulting from allyl deprotection and palladium wash. This further washing was found to be beneficial for the subsequent cyclization step. The cyclization was achieved on column, using a fourfold excess of TBTU (2 × 3 h) or HATU (2 × 1 h); after cleavage (TFA–thioanisole–1,2-ethanedithiol–anisole, 90:5:3:2, 2 h at room temperature) the expected tailed cyclic peptide was found to be the major product, with a minor side product corresponding to the dimeric form of the peptide; the yield was 85% (crude), 40% (purified). The cyclization performed with HATU activation gave slightly better results in terms of purity.

Another interesting extension of the "side-chain anchoring" approach to the synthesis of head-to-tail cyclic peptides, based on exploitation of the allyl chemistry, is the possibility of anchoring to the resin the side chain of Lys, introduced by Alsina et al. [29]. These authors used hydroxymethyl-phenoxypropionic acid to prepare an active carbonate resin, SuO-CO-OCH$_2$-C$_6$H$_4$-O(CH$_2$)$_2$-CO-R, to which Fmoc-Lys-OAl was attached by dissolution of the amino derivative as a salt (10 equiv.) in DMF followed by addition to the resin in the presence of DIEA (20 equiv.) for 4 h at 25°C. Next, the sequence of the model peptide Val-Phe-Sar-Tyr-DTrp-Lys was constructed using a standard Fmoc/tBu protocol (HOAt/DIPCDI activation) and the allyl protecting group was removed with Pd(PPh$_3$)$_4$ in THF–DMSO–0.5 M HCl–morpholine (2.2.1.0.1) for 150 min at 25°C. After Fmoc removal, BOP/HOAt/DIEA (5:5:10 equiv.)–mediated cyclization was carried out for 2 h. Cleavage from the resin and final deprotection, achieved with TFA–thio-anisole–1,2-ethanedithiol–anisole (90:4:3:2) for 2 h at 25°C, gave the expected cyclopeptide (yield 61%, HPLC purity 71%). The same group reported an extension of the active carbonate resin for the anchoring of the hydroxyl functions of alcohol or phenols [30]. Thus, Boc-Ser-OAl and Boc-Tyr-OAl were side-chain linked to active carbonates of 4-hydroxymethyl-PS-resin and 4-hydroxymethyl-Nbb-PS-resin, according to the previously reported methods. Two model peptides, *cyclo*[Leu-Phe-Gly-Gly-Tyr(or Ser)], were prepared on the latter resin and two others, *cyclo*[Glu-Ala-Ala-Arg-D-Phe-Pro-Glu-Asp-Asn-Tyr(or Ser)], on the former one, using standard Boc/Bzl chemistry. After the usual *C*- and *N*-terminus deprotections, cyclization was achieved with PyAOP/HOAt/DIEA (5:5:10) in DMF for 2 h at 25°C. Cleavage and final deprotection from the 4-hydroxymethyl-PS-resin were carried out with HF, and the protected peptides were released from the Nbb-resin by photolysis at 350 nm.

A much wider generalization of the side-chain anchoring principle was reported by Jensen et al. [31]. The growing peptide is linked to the resin through a backbone amide nitrogen using an appropriate handle called a backbone amide linker (BAL) (Scheme 2C). This approach is compatible with three-dimensional orthogonal schemes for solid-phase synthesis of cyclic peptides, as exemplified by the preparation of McMurray's decapeptide (see earlier). Accordingly, the fully protected resin-bound precursor Fmoc-Arg(Pmc)-DPhe-Pro-Glu(OtBu)-Asp(OtBu)-Asn(Trt)-Tyr(tBu)-Glu(OtBu)-Ala-(BAL-Ile-PEG-PS)Ala-OAl was deprotected at N- and C-termini using standard methods and subsequently cyclized (PyAOP/HOAt/DIEA activation). Acidolytic cleavage (TFA–Et$_3$SiH–H$_2$O, 92:5:3, for 3 h at 25°C) gave the desired cyclic peptide in 85% yield; however, 12% of this product was found to be C-terminal epimerized.

The preparation of head-to-tail cyclic tetra- and hexapeptides has been achieved by Sabatino et al. [32] on a solid support after side-chain attachment of an Fmoc-His-OAl residue to the resin via a trityl spacer.

A synthetic strategy based on the attachment to the resin of Lys/Orn side chains, as reported before and on Kates' on-resin cyclization method was exploited by Andreu et al. [33] for the preparation of the antibiotic gramicidin S, $cyclo$(Val-Orn-Leu-DPhe-Pro)$_2$, the corresponding Lys$^{2,2'}$ analogue, and a further derivative containing a dipeptide surrogate of type II' β-turns, the 2-amino-3-oxohexahydroindolizino[8,7-b]indole-5-carboxylate system (IBTM).

The availability of these new methods for solid-phase cyclization prompted several groups to exploit them for the synthesis of cyclic peptide libraries [34]. This target is particularly attractive for new drug lead discovery in view of the reduced degree of conformational freedom and improved chemical and biochemical stability of cyclic peptides. Thus, Rovero's protocol was employed by Spatola et al. [35,36] for the preparation of several cyclic peptide mixtures of various ring sizes (five, six, and seven residues) and, more recently [37], of cyclic pseudopeptide libraries containing one Ψ[CH$_2$NH] peptide bond surrogate (for example, $cyclo$[DPheΨ-(CH$_2$NH)Aaa-Arg-Gly-Asp], where Aaa is a proteogenic amino acid). Similarly, Mihara et al. [38] reported the preparation of a mixture of cyclic pentapeptides, $cyclo$(Arg-Gly-Asp-Xaa-Aca) (Xaa, 10 amino acids; Aca, ε-aminocaproic acid), by the cyclization–cleavage approach from oxime resin, previously described by the same group. Sanders et al. [39] used Kates' method to prepare a cyclopeptide library where three fixed residues are presented at many different distances and orientations to one another by merit of intercalating spacer molecules. Other examples of cyclic peptide libraries are described in the next section.

The versatility of the side-chain anchoring approach in the synthesis of head-to-tail cyclic peptides is shown by a synthesis of cyclic phosphorylated glycopeptides. In order to obtain a constrained peptidic inhibitor of mannose 6-phosphate receptor (MPR), Franzyk et al. [40] synthesized three cyclic hexapeptides α-glycosylated with two phosphorylated $\alpha(1\rightarrow2)$-

linked mannose disaccharides. Syntheses were performed starting from Fmoc-Asp-OAl, attached to the Rink-amide linker on the resin (0.24 mmol/g) via the β-carboxylic function. Chain elongation was achieved by addition of Fmoc-protected pentafluorophenyl esters (OPfp), including the two suitably protected phosphorylated mannosyl derivatives of Fmoc-Thr-OPfp, or by TBTU/DIEA activation. After allyl and Fmoc removal (following Kates' method), cyclization was achieved using two treatments with TBTU–DIEA (1:2 equiv.) in DMF for 4 h. The authors emphasize the necessity of adding the tertiary amine to the mixture containing free amino and carboxyl groups *before* TBTU is added in order to avoid having the free amino group react with TBTU before the carboxylic acid is converted into the acylating HOBt ester, a side reaction also observed by others (see later). The cyclic peptides were cleaved from the resin and deprotected with 95% TFA (41–57% cleavage yield after HPLC purification; 21–35% overall yield).

III. SIDE CHAIN TO SIDE CHAIN

Side chain–to–side chain cyclic peptides, based on the formation of a lactam bridge between the carboxylic acid function on the side chain of a residue and the amino function on the side chain of a second one, can be prepared using an on-resin cyclization strategy. This approach incorporates the pseudodilution phenomenon provided that the two residues to be linked can be selectively deprotected without affecting either all the other protections or the anchorage of the peptide to the resin (Scheme 6A). The first method that followed this strategy was developed by Schiller et al. [41] for the preparation of cyclic opioid peptide analogues such as $cyclo^{2,5}$[Tyr-Lys(or DLys)-Gly-Phe-Glu-NH$_2$], $cyclo^{2,4}$(Tyr-DLys-Phe-Glu-NH$_2$) and $cyclo^{2,4}$(Tyr-DGlu-Phe-Lys-NH$_2$). Linear peptides were assembled on BHA resin (0.4 mmol/g) using N^{α}-Fmoc amino acids with Bzl side-chain protecting groups and using Boc and tBu to protect the carboxy and amino functions, respectively, that were to be linked in the cycle. These were deprotected with TFA–CH$_2$Cl$_2$ (1:9), 2 + 10 min, and cyclization was achieved by adding DCC/HOBt (2.5 equiv. each) in DMF every 48 h for 5–10 days. Free cyclopeptides were obtained after HF cleavage (Scheme 7). In one case [$cyclo^{2,4}$(Tyr-DOrn-Gly-Glu-NH$_2$)] the cyclization was very slow (10 days) and HPLC analysis of the crude product revealed the presence of a second major component (55%) that was identified by fast atom bombardment (FAB)-MS as the side-chain linked antiparallel dimer. In a related study [42] it was shown that the substitution level of the BHA resin (0.4, 0.61, and 1.0 mmol/g) did not significantly affect the monomer/dimer ratio in the case of this peptide. However, a lower substitution level was not tested. Other syntheses of the identical

A. *C*-terminal anchoring

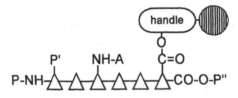

i.   Removal of B' and B"
ii. Cyclization
iii. Final deprotection and cleavage

B. Side-chain anchoring (cyclization/cleavage strategy)

i.   Removal of A
ii. Cyclization/cleavage (intramolecular aminolysis)
iii. Final deprotection (in solution) *or*
     removal of P" for fragment condensation

Scheme 6   General strategies for the synthesis of side chain–to–side chain cyclic peptides.

class of peptides have been reported by the same authors [43,44] who described a generalization of the method [45] for the preparation of more demanding and larger cyclic peptides, such as the following cyclic analogues of dynorphin A(1–13) (DynA, Tyr-Gly-Gly-Phe-Leu-Arg-Arg-Ile-Arg-Pro-Lys-Leu-Lys): *cyclo*(Orn^5, Asp^{10}) and *cyclo*(Orn^5, Asp^{13}). The linear sequences were assembled on MBHA resin (0.38 mmol/g) using Boc/Bzl chemistry for the *C*-terminal portions and switching to Fmoc/Bzl chemistry for the sections spanning the eventual lactam bridge. The Boc and *t*Bu side-chain protecting groups of Orn and Asp, respectively, were then removed by the standard method and, after neutralization, side-chain cyclization was achieved by adding DCC/HOBt (5 equiv. each) in DMF at room temperature every 48 h for 4–8 days. Next, the N^α-Fmoc group was removed, and Boc/Bzl chemistry was used again to complete the *N*-terminal section of the peptide. HF cleavage gave the desired cyclic monomer as the predominant product (>76%), showing that monomer formation if favored over cyclodimerization if the side chains to be cyclized are separated by two or more

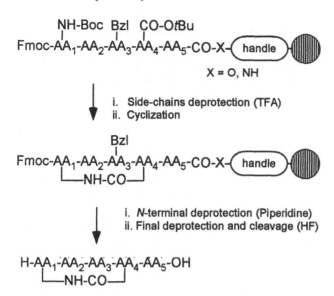

Scheme 7 Synthesis of side chain–to–side chain cyclic peptides according to Schiller et al. [41].

residues, whereas with only one residue in between dimer formation is generally predominant.

Essentially the same protocol was employed by Neugebauer et al. [46,47] for the preparation of three cyclic lactam analogues of the *C*-terminal portion of human parathryoid hormone [hPTH(20–34), Arg-Val-Glu-Trp-Leu-Arg-Lys-Lys-Leu-Gln-Asp-Val-His-Asn-Phe-NH$_2$]: *cyclo*(Lys^{26}, Asp^{30}), *cyclo*(Lys^{27}, Asp^{30}), and *cyclo*(Glu^{22}, Lys^{26}), designed in order to lock these putative salt bridges irreversibly. The peptides were assembled on MBHA resin using Boc/Bzl chemistry for the *C*-terminal portion and Fmoc/Bzl chemistry subsequently. Side-chain cyclization was achieved at an intermediate stage, after incorporation of the residues to be linked, bearing Boc and *t*Bu side-chain protecting groups. After their deprotection (TFA–CH$_2$Cl$_2$, 7:13) and neutralization (DIEA–CH$_2$Cl$_2$, 1:9), cyclization of the peptides not containing Arg within the lactam occurred with TBTU/HOB*t*/DIEA (8 equiv. each) for 2 h. In the case of the Arg(Mtr)-containing peptide, it was necessary to adopt a modified protocol because Arg was partially deblocked during the TFA cleavage. The same authors reported the synthesis of similar cyclic lactams, based on the sequence hPTH (1–31), using a different protection strategy (see later).

Similarly, Schiller's protocol was employed by Aldrich and coworkers [48] for the preparation of lactam analogues of DynA-(1–13), namely *cy-*

$clo(\text{D}\text{Asp}^2$, $\text{Xaa}^5)\text{DynA-}(1-13)$, $\text{Xaa} = \alpha,\beta$-diaminopropionic acid (Dap), α,γ-diaminobutyric acid (Dab), and Orn. The linear precursors were constructed on MBHA resin (0.23 mmol/g) using Fmoc/Bzl chemistry, with the exception of the two residues to be cyclized, which were Boc or *t*Bu protected on the side chains. After the usual TFA deprotection, cyclization was attempted with TBTU/HOBt/DIEA (4:4:6 equiv.) followed by HF cleavage to obatin in all cases two major peptide products. The by-products were characterized as the tetramethylguanidinium (Tmg) Schiff base derivatives to the ω-amino group of Dap, Dab, and Orn, respectively. In a comparative study [49], BOP and DIPCDI/HOBt were found to give the desired cyclic peptide in much better yield, although 3 days of reaction were required. However, the relative difficulty of cyclization reactions was found to be highly dependent on amino acid sequence and ring size. Comparison of the results for the cyclizations of different peptides indicated that the more difficult the cyclization, the larger the proportion of guanidinium by-products formed. The same group described the synthesis of several other cyclic lactam analogues of DynA-(1–13) of general structure $cyclo(\text{D}\text{Asp}^i$, $\text{Dap}^{i+3})\text{DynA-}(1-13)\text{NH}_2$, $i = 3$, 5, and 6 [50]. Syntheses were achieved by procedures similar to those described previously, featuring BOP- or PyBOP-mediated cyclization, which requried 2–5 days, yielding the desired cyclic peptides as the major component (50–60% by HPLC).

A second general approach for the synthesis of side chain–to–side chain cyclic peptides, described by Felix et al. [51], was based on Boc/Bzl chemistry for the construction of the peptide and Fmoc/OFm protections for the side chains to be selectively deprotected before cyclization. Accordingly, linear peptides were assembled on BHA resin (0.7 mmol/g) using Boc/Bzl chemistry and Fmoc and OFm groups on amino and carboxy side chains that must be cyclized, respectively. The authors showed that these base-labile functions are sufficiently stable to the tertiary amine (DIEA–CH_2Cl_2, 1:19) used to achieve neutralization after each Boc deprotection step. Selective removal of Fmoc and OFm was achieved with piperidine–DMF (1:4), followed by cyclization and final HF cleavage (Scheme 8). Several activation agents were systematically tested for the on-resin ring closure reaction of the model peptide $cyclo^{3,12}[\text{Ala}^{15}]$-GRF-(1–29)-$\text{NH}_2$, (GRF, gonadotropin releasing factor) including BOP, which was first proposed as cyclization mediator by these authors. The best cyclization procedures involved BOP (3 equiv.) in DMF containing 1.5% DIEA ($\geq$6 equiv.) for 2 h at 20°C. The use of DCC/HOBt (6 equiv. each) proceeded sluggishly and was 25–55% complete in 24 h, and the use of bis-(2-oxo-3-oxazolidinyl)phosphinic chloride (BOP-Cl) or diphenylphosphoryl azide (DPPA) resulted in even less cyclization. Essentially the same protocol was exploited for the preparation of the more constrained analogue $cyclo^{8,12}[\text{Asp}^8, \text{Ala}^{15}]$-GRF-(1–29)-$\text{NH}_2$ [52].

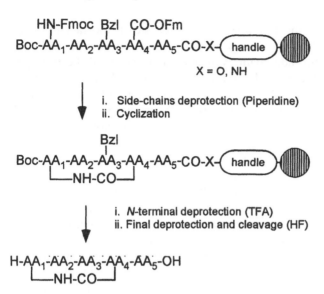

Scheme 8 Synthesis of side chain–to–side chain cyclic peptides according to Felix et al. [51].

Several applications of both Schiller's and Felix's methods are due to Hruby and coworkers [53], who reported the synthesis of cyclic lactam analogues of α-melanocyte-stimulating hormone (α-MSH) [54–56], glucagon [57], and dynorphin A [58]. For the preparation of these peptides MBHA resin was loaded in the range 0.35–0.45 mmol/g to minimize intermolecular cyclizations. The side-chain protection strategy followed substantially Felix's protocol; cyclization was achieved using BOP/DIEA (6:8 equiv.) in DMF or NMP for 2–15 h at room temperature. The synthesis of $cyclo^{5,10}$[Nle^4, Asp^5, DPhe^7, Lys^{10}]-α-MSH-(4–10)-NH$_2$ was performed both in solution and on the solid phase. The latter method was found to provide easier workup and higher yield of the cyclic peptide (55–60% vs. 30–40% using the solution method) [55]. The superiority of BOP to DIPCDI/HOBt and DPPA as a cyclization reagent was also confirmed [53].

A different strategy for the preparation of side chain–to–side chain cyclic peptides, based on the previously described cyclization–cleavage reaction enabled by the oxime resin, was described by Taylor and coworkers [59–62]. The synthetic protocol, which enables the preparation of fully protected cyclic peptides to be used in fragment condensations, is characterized by the following steps: (1) side-chain attachment of Boc-Asp(OH)-OPac (Pac, phenacyl) to the oxime resin; (2) peptide chain elongation using Boc/Bzl chemistry; (3) use of an amine protecting group of a third orthogonality for the N^ε function of the Lys residue to be closed in the lactam bridge (Fmoc or Trt); (4) selective cleavage of the N^ε protecting group with concomitant side chain–to–side chain cyclization and release of the fully protected peptide from the resin by intramolecular aminolysis; (5) selective cleavage of the C-terminal Pac protecting group to provide the C-terminal free, side chain fully protected, cyclic peptide (Scheme 6B). In one example Boc-Glu-OPac was attached to an oxime resin (0.066 mmol/g) through the side chain and the linear peptide Boc-Lys(2ClZ)-Leu-Lys(Trt)-Glu(OBzl)-Leu-Lys(2ClZ)-Glu(oxime resin)-OPac was assembed by fragment condensation. After selective removal of the Trt function from Lys^3 side chain with TFA–TFE–CH$_2$Cl$_2$ (2:19:19) and neutralization with DIEA–CH$_2$Cl$_2$ (1:19), cyclization–cleavage was achieved in CH$_2$Cl$_2$ (3 days) in the presence of CH$_3$COOH (10 equiv.). The resultant protected cyclic peptide, obtained in 60% yield, was subsequently used for the construction of a 21-residue tricyclic amphiphilic model peptide by segment condensation in solution [59].

In later studies [63,64] a different strategy for the preparation of side chain–to–side chain cyclic peptides on oxime resin, based on backbone cyclization, was devised. Accordingly, the preparation of $cyclo^{17,21}$-[Lys^{17}, Asp^{21}]-hCT(17–21) (hCT, human calcitonin) was achieved starting with the attachment of Boc-Lys^{18}(2ClZ)-OH to the oxime resin (0.47 mmol/g), followed by Boc cleavage and coupling of Fmoc-Lys^{17}(Boc)-OH. The lactam bridge was subsequently formed after Boc cleavage and side chain–to–side chain coupling with Boc-Asp^{21}(OH)-OAl. The backbone was then completed via successive coupling of Boc-His^{20}(Bom)-OH and Boc-Phe^{19}-OH to Asp^{21}. Cleavage of the Boc group enabled N^α-to-C^α cyclization, with concomitant cleavage of the fully protected cyclic peptide allyl ester. In a comparative study, this cyclization strategy, in combination with C^α-allyl ester protection was found to improve significantly the yields and purity of $i,i + 4$ lactambridged peptide as compared with the direct side chain–to–side chain cyclization on oxime resin reported previously.

Use of the allyl chemistry for the selective protection of the side chains to be linked in the lactam was first proposed by Lyttle and Hudson [65]. The hGRF (human gonadotropin releasing factor) analogue $cyclo^{3,12}$-GRF-(1–29)-NH$_2$, previously synthesized by Felix (see earlier), was assembled on PAL-PEG-PS resin using N^α-Fmoc amino acids with tBu side-chain protections, except that Asp^3 and Lys^{12} were incorporated in the linear chain as Fmoc-Asp(OAl)-OH and Fmoc-Lys(Aloc)-OH, respectively. Allyl functions were removed by treatment of the fully protected peptide-resin with Pd(PPh$_3$)$_4$ and cyclization was achieved by treatment with BOP (yield >50%) (Scheme 9). This approach was subsequently used to generate two series of eight cyclic structures immobilized on a cellulose support [66]. Analogues of two 15-mer model peptides were constructed with a lactam bridge linking

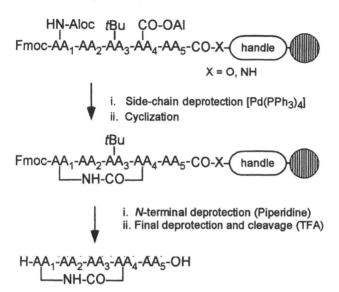

Scheme 9 Synthesis of side chain–to–side chain cyclic peptides according to Lyttle and Hudson [65].

carboxy and amino side chains of glutamic acid and ornithine, respectively, inserted five amino acids apart and "walked" through the structure of the two peptides.

Andreu and coworkers [67,68] compared solid-phase versus solution cyclization methods for the preparation of a 22-mer cyclic peptide with a lactam bridge involving 18 residues (60-atom cycle), designed to mimic the antigenic site A of foot-and-mouth disease virus. The solid-phase synthesis, performed essentially following Felix's protocol, suffered from poorer yield but required fewer purification steps than the solution approach, and therefore global yields were comparable. However, when the same synthesis was repeated on a PEG-PS resin (substitution level 0.2 mmol/g) using the Fmoc/ tBu/allyl protection strategy described in the previous example, the main product was a cyclodimer. This unexpected result was tentatively attributed to the higher flexibility of the PEG-PS resin.

An example of an application of allyl-based methods for the preparation of side chain–to–side chain lactam peptides using a solid-phase strategy was published by Willick and coworkers [69]. These authors reported the synthesis of hPTH full-length analogues with an $i, i + 4$ lactam bridge designed to mimic a salt bridge, as previously reported by the same authors using shorter C-terminal fragments of the same peptide (see earlier). Syntheses were performed on PEG-PS resin (0.2 mmol/g) using Fmoc/tBu chemistry and a continuous-flow strategy. Aloc and Al protections were used on the amino and carboxy side chains to be linked, respectively. After completion of the C-terminal segment to be cyclized, the peptide-resin was removed from the column and the allyl groups were removed with Pd(PPh$_3$)$_4$ in CH$_2$Cl$_2$–CH$_3$COOH–NMM (92.5:5:2.5) for 6 h at 20°C under argon. The deprotection mixture was washed using DEDT and NMM (0.5% each) in DMF, and side chain–to–side chain cyclization was achieved with HOAt/ NMM (1:2 equiv.) in DMF for 14 h at 20°C. Next, the peptide resin was repacked into the column and the synthesis completed. Cleavage was achieved using reagent K.

A special case of side chain–to–side chain cyclization on the solid phase, involving the use of a preformed side-chain linked protected dipeptide, has been reported by Mendre et al. [70]. These authors prepared a cyclic lactam peptide designed to mimic the inner loop of endothelin, with replacement of the two disulfide bridges by amide bonds: cyclo^{1,8}(Dpr-Ser-βAla-βAla-Val-Tyr-Phe-Asp-His-Leu-Asp-Ile-Ile-Trp). The lactam bridge was obtained by using a preformed building block prepared by linking through the side chains Fmoc-Asp-OH and Z-Dpr-CAM (CAM, carboxyamidomethyl ester). This approach avoids any possibility of aspartimide formation resulting either from Asp side-chain activation or from the repetitive piperidine-mediated Fmoc deprotection. The solid-phase assembly of the linear peptide was carried out by a conventional continuous-flow procedure (Fmoc/ tBu chemistry); the dipeptide building block was introduced by double coupling (TBTU activation; 2 × 1 h). After removal of the N-terminal Fmoc group, the CAM ester was hydrolyzed with NaOH (3 equiv.) in iPrOH–H$_2$O (7:3) for 1 h at room temperature and the peptide-resin was neutralized with CH$_3$COOH. Cyclization was obtained with TBTU/HOBt/DIEA (3:3:6 equiv.) in DMF (3 × 2 h) and the final peptide was obtained by deprotection–cleavage with reagent K, followed by removal of the Z group by TFA–TFMSA–thioanisole–1,2-ethanedithiol. Interestingly, this protocol would allow the synthesis of branched peptides (see later) if the Z protecting group is replaced by a truly orthogonal one.

An example of side chain–to–side chain cyclization between two Lys residues has been reported by Pawlak et al. [71]. For the preparation of a cyclic analogue of enkephalinamide, the linear precursor Boc-Tyr(2BrZ)-DLys(Fmoc)-Gly-Phe-Lys(Fmoc)-resin was prepared on MBHA resin. After removal of the two Fmoc protecting groups, cyclization was performed on the solid support by reaction of bis(4-nitrophenyl)carbonate (1 equiv. for 20 h + 0.03 equiv. for 2 days) with the free side-chain amino groups of the two Lys residues, yielding a ureido group incorporating the two side-chain amino functions. Next, Boc removal (TFA) followed by HF treatment yielded the desired peptide.

IV. SIDE CHAIN TO END

The preparation of peptides bearing a lactam bridge between the *N*-terminal amino function and the carboxy function of a side chain can be achieved on the solid support using strategies conceptually identical to those previously reported for side chain–to–side chain cyclization (Scheme 10). The first applications of this principle are due independently to Ho and Plaué. Ho et al [72] reported the synthesis of a deaminodicarba analogue of eel calcitonin (eCT) in which the *N*-terminal 1,7-disulfide bridge is replaced by a single spanning α-aminosuberic acid (Asu) residue with the carboxy function of the side chain linked to the *N*-terminal amino function of the peptide. The synthesis was performed using Fmoc/Bzl chemistry on MBHA resin; the Asu residue was introduced as the corresponding *t*Bu ester and the *N*-terminal Ser as an N^{α}-Boc derivative. TFA treatment exposed the functional groups for cyclization, which was achieved with BOP/HOBt/DIEA (6 equiv. each) in DMF for 24 h in greater than 90% yield.

A. C-terminal anchoring
P
CO-OB"
P'
B'-NH
CO-X
handle
X = O, NH
i. Removal of B' and B"
ii. Cyclization
iii. Final deprotection and cleavage
B. Side-chain anchoring (cyclization/cleavage strategy)
handle
O
C=O
P
P'
A-NH
CO-O-P"
i. Removal of A
ii. Cyclization/cleavage (intramolecular aminolysis)
iii. Final deprotection (in solution)

Scheme 10 General strategies for the synthesis of side chain–to–*N*-terminal cyclic peptides.

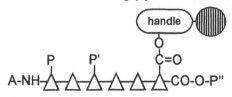

An interesting paper by Plaué [73] described the synthesis of a series of cyclic peptides related to region 139–147 of the hemagglutinin of influenza virus. Sequences were X-Lys-Arg-Gly-Pro-Gly-Ser-Asp-Phe-Y-Tyr-NH$_2$, where X = Cys, Ser and Y = Asp, Lys(ε-succinyl). In the so-called large-loop peptides, the succinyl unit closed the bridge with the N-terminal amino group. They were assembled by Boc/Bzl chemistry on MBHA resin (0.1–0.6 mmol/g) using Boc-Lys(Fmoc)-OH as a precursor for the site of cyclization. Following removal of the N-terminal Boc group, succinic anhydride and DIEA (3 equiv. each) were added for 15 min and the side-chain Fmoc group was subsequently removed with piperidine–DMF (1:1) for 3 + 7 min. Cyclizations were then achieved with either BOP/HOBt/DIEA (3:3:7.5 equiv.) in DMF for 2 h or DIPCDI/HOBt (3 equiv. each) in DMF for 3 × 24 h. Using the latter method, the yield of cyclic monomer peptide versus cyclodimer was investigated over the substitution level range 0.1–0.6 mmol/g, and the amount of cyclodimer increased from 8 to 19%. The more rapid activation obtained with BOP caused more oligomerization, but it was tested only at the higher resin substitution level (i.e., 0.6 mmol/g). Plaué also described the synthesis of "small-loop" peptides, not including the succinyl unit, again using Fmoc/Bzl chemistry on MBHA resin and Fmoc-Asp(tBu)-OH and N^α-Boc terminal amino acid as cycle precursors. Removal of the N-terminal Boc group and the tBu group blocking the side chain of Asp with TFA–CH$_2$Cl$_2$ (13:20) for 45 min and neutralization by DIEA–DMF (1:9, 2 × 2 min) were followed by cyclization with DIPCDI/HOBt (3 equiv. each) in DMF for 48 h, indicating more facile folding of the small loop.

The previously reported cyclization–cleavage stragegy on oxime resin was also applied to the preparation of side chain–to–head cyclic peptides (Scheme 10B) by Nishino et al. [74], who described the synthesis of the cyclic portion of [Asu^{1,7}]-eCT (see earlier). Boc-Asu-NHNH-Z was linked to the oxime resin through the side chain (0.4 mmol/g) and the peptide chain elongated using Boc/Bzl chemistry. After removal of the Boc group from the N-terminus, cyclization–cleavage was achieved with the previously discussed method in 24 h, yielding the fully protected cyclic peptide. The same strategy also enabled the preparation of the cyclic portion of [Asu^{1,6}]-oxytocin.

A different orthogonal protection strategy was reported by Marlowe [75], who prepared a model peptide cyclized between the N-terminus and an Asp side chain: *cyclo*(Glu-Ser-Thr-Arg-Pro-Met-Asp-NH$_2$). The third level of orthogonal protection was obtained using a trimethylsilylethyl (TMSE) ester, which is selectively deprotected by fluoride ions. Accordingly, Fmoc-Asp(OTMSE)-OH was loaded onto Rink amine resin and the linear precursor peptide was constructed using Fmoc/tBu chemistry. Treatment with 1 M tetrabutylammonium fluoride in DMF for 20 min removed both the TMSE and the N-terminal Fmoc groups and cyclization was achieved using BOP/DIEA in DMF for 4 h. Side chain deprotection and cleavage with reagent K yielded the desired cyclic peptide.

The use of allyl esters as the third level of orthogonal protection in the solid-phase synthesis of side chain–to–head cyclic peptides was reported by Tumelty et al. [76]. During the preparation of cyclic heptapeptides according to Kates' protocol, these authors observed that the linear peptides were partially modified by some of the coupling reagents used rather than forming the cyclic product, as also reported by others (see earlier). To overcome this problem it was proposed that the immobilized peptide be activated while the N^α-protecting group was still present; subsequent removal of this group from the carboxyl-activated intermediate would lead to cyclization. Accordingly, Fmoc-Glu(OAl)-OH was attached to a PEG-PS resin through either a Wang or 4-(4-hydroxymethyl-3-methoxyphenoxy)butyric acid (HMPB) linker. The sequence Phe-Gly-Gly-Phe-Ala-Gly was next assembled using Fmoc chemistry, except that the final residue was incorporated with either a photolabile (PLG) or a hyper–acid-labile (Ddz) group as its N^α-protection. The allyl group was removed with Pd(PPh$_3$)$_4$ in CHCl$_3$–morpholine (9:1) containing PPh$_3$ for 2 h. The unmasked side chain was activated by Pfp ester formation, achieved using an excess of Pfp in TFA–pyridine–DMF (2:1:1, 60 equiv.) for 2 h. After deprotection of the N-terminal amino function [either photolysis at 365 nm for 45 min or treatment with TFA–CH$_2$Cl$_2$ (1:49) for 30 min], upon cleavage from their respective resins the crude peptides contained 50–65% of the monomeric cyclized product when the photolabile groups were used and 95% when the hyper–acid-labile group was used. For comparison, the same sequence prepared according to Kates' protocol and cyclized with PyBOP/HOBt gave 87% crude product.

A side chain–to–N-terminal cyclic pentapeptide prepared on the solid phase was used by Eichler et al. [77] for the construction of a cyclic peptide template combinatorial library. Accordingly, the linear precursor Fmoc-Lys(Dde)-Lys(Boc)-Lys(Boc)-Glu(OAl)-Gly-resin was constructed on PAM polystyrene resin (substitution level 0.53 mmol/g). After removal of the allyl and Fmoc protecting groups, cyclization was achieved using PyBOP/HOBt/DIEA activation. Subsequently, chemical diversity was introduced by acylating the ε-amino groups of the Lys residues using 10 carboxylic acids in addition to the 20 proteinogenic amino acids.

An example of a solid-phase synthesis of a side chain–to–C-terminal cyclic peptide has also been published (Scheme 11). Lee et al. [78] exploited the "side-chain resin attachment" principle and Kates' protocol to prepare bacitracin A, a cyclic dodecapeptide antibiotic, $cyclo$(H-Δ^2Thi-Leu-DGlu-Ile-

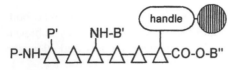

i. Removal of B' and B"
ii. Cyclization
iii. Final deprotection and cleavage

Scheme 11   General strategy for the synthesis of side chain–to–C-terminal cyclic peptides.

Lys-DOrn-Ile-DPhe-His-DAsp-Asn), Δ^2Thi, 2[1'(S)-2'(R)-methylbutyl]-4(R)-carboxy-Δ^2-thiazoline. The synthesis was performed on an automatic synthesizer using the PAL-PS resin, to which Fmoc-Asp(OH)-OAl was attached through the side chain in order to obtain the required Asn residue upon acidic cleavage. The linear precursor was assembled by sequential couplings of Fmoc/tBu-protected amino acids, with the exception of the Lys residue, which was side-chain protected with the Aloc group. After the coupling of the Leu residue, the allyl protecting groups were removed with Pd(PPh₃)₄ in CHCl₃–CH₃COOH–NMM (75:4:8) for 4 + 12 h at room temperature under argon and the cyclization was achieved with PyBOP/HOBt/DIEA (2:2:6 equiv.) in NMP for 1 day. Subsequently, N-terminal Fmoc removal was followed by the double coupling of the Boc-protected thiazoline dipeptide. This step was carried out after the cyclization procedure in order to avoid the possibility that the sulfur-containing thiazoline moiety would inhibit the palladium-catalyzed removal of the allyl groups. Finally, TFA cleavage yielded the desired cyclic peptide in 24% overall yield after HPLC purification.

A similar protocol was used by Flouzat et al. [79] for the preparation of a series of cyclic tripeptides of general structure *cyclo*(H-Lys-Trp-Asp), where the side chain of Lys was connected to the C-terminal carboxy function. The syntheses were performed using an automated continuous-flow strategy, starting from commercial Fmoc-Asp(OPAC-PEG-PS)-OAl resin (0.16 mmol/g). After sequential coupling of Fmoc-Trp-OH and Boc-Lys(Fmoc)-OH and Al and Fmoc removal in this order, following standard procedure, cyclization was achieved using PyBOP/HOBt/NMM (3:3:9 equiv.) for 1 h (recycle rate 10 mL/min). Cleavage with TFA–anisole–1,2-ethanedithiol–CH₂Cl₂ (60:2.5:2.5:35) gave the desired cycle tripeptide (50% HPLC purity; yield 22% after preparative HPLC). Analogues containing DLys and/or DTrp were also obtained with comparable yields (8–22%) and purity (50–88%), and the series of analogues based on DAsp were prepared by cyclization in solution, with yields and purity slightly higher (13–36% and 52–83%, respectively).

V. BRANCHED

The synthesis of branched cyclic peptides on a solid support requires the availability of four levels of orthogonal protections. An example of such a synthesis was reported by Bloomberg et al. [80], who used a combination of Fmoc/tBu chemistry with allyl and Dde side-chain protections for Glu and Lys, respectively. The first branch was assembled on PAL-PEG-PS resin by sequential coupling of Fmoc-Lys(Dde)-OH, Fmoc-Ala-OH, Fmoc-Glu(OAl)-OH, and Fmoc-Cys(Trt)-OH and N-terminal acetylation. The Dde protection was then removed with 1.5% hydrazine in DMF and the second branch (Fmoc-Gly-Gly-Glu-Lys-Thr-Arg-Asn-Gln-Met-Gly) was added by Fmoc/tBu chemistry, with a C-terminal–to–side chain connection. Subsequent removal of the allyl group blocking the Glu side chain with Pd(PPh$_3$)$_4$ and of the Fmoc group blocking the N-terminus of the second branch was followed by HBTU/HOBt/NMM-mediated cyclization (N-terminal–to–side chain connection). Finally, the crude branched peptide was deprotected and cleaved with TFA–1,2-ethanedithiol–phenol–H$_2$O–tri(iso-butyl)silane (92:3:2:2:1) for 2 h at 25°C (Scheme 12).

A distinct example of a branched peptide was synthesized by Zhao and Felix [81] using three levels of orthogonal protection (Boc/Bzl/Fmoc). Accordingly, the designated "extended lactam ring" was obtained using Boc/Bzl chemistry to construct the main branch on a BHA resin. An Fmoc-protected basic residue (i.e., Fmoc-Orn-OH) was incorporated and an Fmoc-Gly-OH residue was coupled to the side chain. Next, the main chain was continued using Boc/Bzl chemistry and the acidic residue necessary to close the lactam bridge was incorporated as a side-chain Fm ester. After deprotection of Fmoc and Fm groups, cyclization was carried out with BOP/DIEA (3:6 equiv.) for 48 h. The main chain was subsequently further elongated and the final product was obtained after HF cleavage. Similarly, a "reverse-extended lactam ring" was obtained by initially incorporating the acidic component (i.e., Boc-Asp(OFm)-OH) in the main branch. The second branch was elongated with H-Gly-OFm (BOP-activated double coupling), followed by insertion of Boc-Lys(Fmoc)-OH in the main branch. After Fmoc and Fm removal, the lactam ring was closed to yield the dicyclic GRF analogue, [Ala^{15}]-dicyclo-(Asp^8-[Gly]-Orn^{12})-(Orn^{21}-[Gly]-Asp^{25})-GRF(1–29)-NH$_2$.

A similar approach published by Zhang and Taylor [82] was based on a combination of Boc/Bzl and Fmoc/tBu methods as well as the use of a preformed pseudodipeptide building block, which was previously proposed

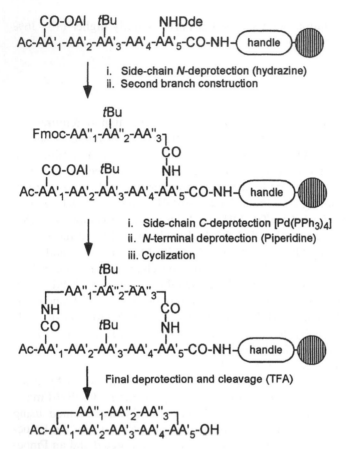

Scheme 12 Synthesis of branched cyclic peptides according to Bloomberg et al. [80].

by Mendre (see earlier). Accordingly, the first branch of the peptide was constructed on Merrifield resin using Boc/Bzl chemistry, except that an Asp residue was side-chain protected as its Fm ester, to provide the following sequence: Boc-Glu(OBzl)-Leu-Gln-Gln-Lys(ClZ)-Leu-Asp(OFm)-Glu(OBzl)-Leu-Lys(ClZ)-Gln-resin. Subsequently, the building block Boc-Lys(Fmoc-Ala)-OH (*C*-terminal–to–side chain connection) was introduced, followed by further elongation of the first branch by Boc chemistry. After Fmoc and OFm selective removal, the cycle was closed (*N*-terminal–to–side chain connection) using BOP/DIEA activation (1.5 equiv. each, 2–4 h at room temperature). The final product was obtained by treatment with TFA (*N*-terminal deprotection) and HF (side-chain deprotection and cleavage).

The yield of monomeric cyclic peptide was strongly influenced by the solvent mixture used in the cyclization reaction; the best results were obtained with DMSO–NMP (1:4) (65% monomer when LAla was used as the second branch; 94% monomer when DAla was used as the second branch). Further complexity was created by introduction of a "tripodal side-chain bridge," where the Ala residue of the second branch was replaced by 1,3-diaminopropionic acid (Dap). Dap was side chain–to–side chain linked to an Asp residue of the first branch to form a bicyclic structure. The synthesis was performed by introducing allyl protection as a further dimension of orthogonality according to the following scheme: (1) construction of the resin-linked first branch Boc-Glu(OBzl)-Leu-Asp(OFm)-Gln-Lys(ClZ)-Leu-Asp(OAl)-Glu(OBzl)-Leu-Lys(ClZ)-Gln-resin; (2) coupling of the building block Boc-Lys[Aloc-Dap(Fmoc)]-OH; (3) further elongation of the first branch by coupling of Boc-Leu-OH and Boc-Lys(ClZ)-OH; (4) Fmoc and OFm removal; (5) first cyclization [BOP/DIEA 1.5 equiv. each, in DMSO–NMP (1:4), 2 h at room temperature]; (6) Aloc and Al removal [Pd(Ph$_3$)$_4$]; (7) second cyclization (conditions as before); (8) N-terminal Boc removal; (9) HF deprotection and cleavage.

VI. BACKBONE TO BACKBONE

A special class of homodetic cyclic peptides was introduced by Gilon et al. [83]; the cycle is formed between two amide nitrogens in the peptide backbone, interconnected via a bridge consisting of alkyl groups and an amide bond. This approach was initially used to prepare a series of cyclic analogues of the N-terminal hexapeptide of substance P (H-Arg-Phe-Phe-Gly-Leu-Met-NH$_2$) with alkyl bridges of variable length of general structure —CO—(CH$_2$)$_m$—CO—NH—(CH$_2$)$_n$— (m = 2, 3, 4; n = 2, 3, 6) connecting the N-terminal amino function with the nitrogen of the amide bond between Phe and Gly [84]. The peptides were prepared on MBHA resin using Boc chemistry for the first two residues (Leu and Met); next, the coupling of Fmoc-N^α-[ω-(Boc-amino)-alkyl]Gly-OH did not require any special procedure. Fmoc chemistry was used for subsequent construction of the peptide; the coupling of Fmoc-Phe-OH to the N^α-(ω-alkyl)Gly building block was repeated three times, using BOP activation. After removal of the N-terminal Fmoc group, the peptide was reacted with the appropriate dicarboxylic acid spacer (as its anhydride) with a catalytic amount of DMAP. Then the Boc protecting group of the N^α-(ω-alkyl)Gly was removed and the peptide was cyclized on the resin (BOP/DIEA activation; a few hours to 3 days, depending on the ring size); final deprotection and cleavage (HF) yielded the expected cyclopeptide. A similar protocol was used by the same authors [85] to prepare a library of 60 backbone-bicyclic substance P analogues containing both a lactam and a disulfide bridge. Similarly, a library of backbone cyclic peptides designed to mimic functionally the nuclear localization signal region of the human immunodeficiency virus type 1 matrix protein was also reported [86]. In this case, an amino acid residue Axx was inserted in the alkyl bridge [—CO—$(CH_2)_m$—CO—Axx—NH—$(CH_2)_n$—] connecting the N-terminus and the Lys-Gly amide bond of the sequence H-Lys-Lys-Lys-Gly-Lys-Cys-NH$_2$. The preparation of a backbone cyclic somatostatin analogue was described by the same authros [87].

VII. HOMODETIC CYCLIC PEPTIDES— EXPERIMENTAL SECTION

General methods for the removal of orthogonal protecting groups

1. *Fmoc/Fm*: The resin is washed successively with CH_2Cl_2, methanol, and DMF (2 × 1 min each), then treated with piperidine– DMF 1:4 (3 + 7 min) and washed again as before [17].
2. *Boc/tBu*: These protecting groups are removed by treatment with TFA–CH_2Cl_2 (1:1), 2 + 30 min, followed by neutralization using DIEA–CH_2Cl_2 (1:9), 2 + 10 min, and washing with CH_2Cl_2 and DMF, 3 × 1 min each [45].
3. *Aloc/Al*: In-batch procedure: The resin is washed with DMF, 5 × 0.5 min; treated with Pd(PPh$_3$)$_4$ (5 equiv.) in THF–DMSO–0.5 M HCl–morpholine (2:2:1:0.1) under Ar for 180 min; and washed successively with THF, DMF, CH_2Cl_2, DIEA–CH_2Cl_2 (1:19), CH_2Cl_1, 3 × 2 min each, then sodium N,N-dimethyldithiocarbamate (0.03 M in DMF), 3 × 15 min, DMF, 5 × 2 min, CH_2Cl_2, 3 × 2 min, and DMF, 3 × 1 min [31].

   Automated side-chain deblocking of the allyl group under continuous-flow conditions is performed by placing a vial containing Pd(PPh$_3$)$_4$ (125 mg) under Ar in the amino acid module following the last amino acid in the sequence. The catalyst is dissolved to a final concentration of 0.14 M in $CHCl_3$–CH_3COOH– NMM (37:2:1), delivered to a 0.3 mmol/g substituted peptidyl-resin (500 mg), and recycled through the column for 120 min. The column is washed with a solution of 0.5% DIEA and 0.5% sodium N,N-diethyldithiocarbamate in DMF (10 min, 6 mL/min) and DMF (10 min, 6 mL/min) [27].
4. *Dmb*: Resins containing this protecting group were washed with CH_2Cl_2, 5 × 30 s, and then treated with TFA–CH_2Cl_2 (1:99), 6 × 5 min, and washed with CH_2Cl_2, 5 × 30 s [22].

5. *Dde*: In continuous-flow conditions in the Dde group is removed by flowing a 1.5% solution of hydrazine in DMF through the resin for 9 min [80].

REFERENCES

1. SA Kates, NA Solé, F Albericio, G Barany. In: C Basava, GM Anantharamaiah, eds. Peptides: Design, Synthesis, and Biological Activity. Boston: Birkhauser, 1994, pp 39–58.
2. C Blackburn, SA Kates. Methods Enzymol 289:175–198, 1997.
3. M Fridkin, A Patchornik, E Katchalski. J Am Chem Soc 87:4646–4648, 1965.
4. E Flanigan, GB Marshall. Tetrahedron Lett 27:2403–2406, 1970.
5. E Giralt, R Eritja, C Navalpotro, E Pedroso. An Quim 80:118–122, 1984.
6. WF DeGrado, ET Kaiser. J Org Chem 45:1295–1300, 1980.
7. G Ösapay, A Profit, JW Taylor. Tetrahedron Lett 31:6121–6124, 1990.
8. N Nishino, M Xu, H Mihara, T Fujimoto. Tetrahedron Lett 33:1479–1482, 1992.
9. N Nishino, J Hayashida, T Arai, H Mihara, Y Ueno, H Kumagai. J Chem Soc Perkin Trans I 939–946, 1996.
10. M Xu, N Nishino, H Mihara, T Fujimoto, N Izumiya. Chem Lett 191–194, 1992.
11. LS Richter, JYK Tom, JP Burnier. Tetrahedron Lett 35:5547–5550, 1994.
12. JA Camarero, GJ Cotton, A Adeva, TW Muir. J Peptide Res 51:303–316, 1998.
13. PE Dawson, TW Muir, I Clark-Lewis, SBH Kent. Science 266:776–779, 1994.
14. S Jackson, W DeGrado, A Dwivedi, A Parthasarathy, A Higley, J Krywko, A Rockwell, J Markwalder, G Wells, R Wexler, S Mousa, R Harlow. J Am Chem Soc 116:3220–3230, 1994.
15. BH Lee. Tetrahedron Lett 38:757–760, 1997.
16. SS Isied, CG Kuehn, JM Lyon, RB Merrifield. J Am Chem Soc 104:2632–2634, 1982.
17. P Rovero, L Quartara, G Fabbri. Tetrahedron Lett 32:2639–2642, 1991.
18. C Taylor, RJ Quinn, P Alewood. Bioorg Med Chem Lett 6:2107–2112, 1996.
19. A Tromelin, MH Fulachier, G Mourier, A Menez. Tetrahedron Lett 33:5197–5200, 1992.
20. ML Valero, E Giralt, D Andreu. Tetrahedron Lett 37:4229–4232, 1996.
21. K Burgess, D Lim, SA Mousa. J Med Chem 39:4520–4526, 1996.
22. JS McMurray. Tetrahedron Lett 32:7679–7682, 1991.
23. JS McMurray, CA Lewis, NU Obeyesekere. Peptide Res 7:195, 1994.
24. HF Brugghe, HAN Timmermans, LMA VanUnen, GJ TenHove, G VanDe-Werken, JT Poolman, P Hoogerhout. Int J Peptide Protein Res 43:166–172, 1994.
25. A Trzeciak, W Bannwarth. Tetrahedron Lett 33:4557–4560, 1992.
26. SA Kates, NA Solé, CR Johnson, D Hudson, G Barany, F Albericio. Tetrahedron Lett 34:1549–1552, 1993.

27. SA Kates, SB Daniels, F Albericio. Anal Biochem 212:303–310, 1993.

28. D Delfoge, M Art, B Gillon, M Dieu, E Delaive, M Raes, J Remacle. Anal Biochem 242:180–186, 1996.

29. J Alsina, F Rabanal, E Giralt, F Albericio. Tetrahedron Lett 35:9633–9636, 1994.

30. J Alsina, C Chiva, M Ortiz, F Rabanal, E Giralt, F Albericio. Tetrahedron Lett 38:883–886, 1997.

31. KJ Jensen, J Alsina, MF Songster, J Vagner, J Alsina, F Albericio, G Barany. J Am Chem Soc 120:5441–5452, 1998.

32. G Sabatino, M Chelli, S Mazzucco, M Ginanneschi, AM Papini. Tetrahedron Lett 40:809–812, 1999.

33. D Andreu, S Ruiz, C Carreno, J Alsina, F Albericio, MA Jimenez, N DeLa-Figuera, R Herranz, MT Garcia-Lopez, R Gonzales-Muniz. J Am Chem Soc 119:10597–10586, 1997.

34. KH Wiesmuller, S Feiertag, B Fleckenstein, S Kienle, D Stoll, M Herrmann, G Jung. In: G Jung, ed. Combinatorial Peptide and Nonpeptide Libraries. Weinheim, Germany: VCH, 1996, pp 221–223.

35. K Darlak, P Romanovskis, AF Spatola. In: RS Hodges, JA Smith, eds. Peptides: Chemistry, Structure and Biology. Proceedings of the 13th American Peptide Symposium. Leiden, The Netherlands: ESCOM, 1994, pp 981–983.

36. AF Spatola, K Darlak, P Romanovskis. Tetrahedron Lett 37:591–594, 1996.

37. JJ Wen, AF Spatola. J Peptide Res 49:3–14, 1997.

38. H Mihara, S Yamabe, T Niidome, H Aoyagi, H Kumagai. Tetrahedron Lett 36:4837–4840, 1995.

39. I Sanders, MC Allen, IM Matthews, MC Campbell. Bioorg Med Chem Lett 7:2603–2606, 1997.

40. H Franzyk, MK Christensen, RM Jorgensen, M Meldal, H Cordes, S Mouritsen, K Bock. Bioorg Med Chem 5:21–40, 1997.

41. PW Schiller, TMD Nguyen, J Miller. Int J Peptide Protein Res 25:171–177, 1985.

42. PW Schiller, TMD Nguyen, C Lemiueux, LA Maziak. J Med Chem 28:1766–1771, 1985.

43. PW Schiller, TMD Nguyen, LA Maziak, BC Wilkes, C Lemieux. J Med Chem 30:2094–2099, 1987.

44. PW Schiller, G Weltrowska, TMD Nguyen, C Lemieux, NN Chung, BJ Marsden, BC Wilkes. J Med Chem 34:3125–3132, 1991.

45. PW Schiller, TMD Nguyen, C Lemieux. Tetrahedron 44:733–743, 1988.

46. W Neugebauer, G Willick. In: CH Schneider, AN Eberle, eds. Peptides 1992. Proceedings of the 22nd European Peptide Symposium. Leiden, The Netherlands: ESCOM, 1993, pp 395–396.

47. W Neugebauer, L Gagnon, J Whitfield, GE Willick. Int J Peptide Protein Res 43:555–562, 1994.

48. S Arttamangkul, TF Murray, GE DeLander, JV Aldrich. J Med Chem 38:2410–2417, 1995.

49. SC Story, JV Aldrich. Int J Peptide Protein Res 43:292–296, 1994.

50. S Arttamangkul, JE Ishmael, TF Murray, DK Grandy, GE DeLander, BL Kieffer, JV Aldrich. J Med Chem 40:1211–1218, 1997.

51. AM Felix, CT Wang, EP Heimer, A Fournier. Int J Peptide Protein Res 31: 231–238, 1988.

52. AM Felix, EP Heimer, CT Wang, TJ Lambros, A Fournier, TF Mowles, S Maines, RM Campbel, BB Wegrzynski, V Toome, D Fry, VS Madison. Int J Peptide Protein Res 32:441–454, 1988.

53. VJ Hruby, F AlObeidi, DG Sanderson, DD Smith. In: R Epton, ed. Innovation and Perspectives in Solid Phase Synthesis: Peptides, Polypeptides and Oligonucleotides. Birmingham, UK: SPCC, 1990, pp 197–203.

54. EE Sugg, AML Castrucci, ME Hadley, G VanBinst, VJ Hruby. Biochemistry 27:8181–8188, 1988.

55. F AlObeidi, AML Castrucci, ME Hadley, VJ Hruby. J Med Chem 32:2555–2561, 1989.

56. VJ Hruby, D Lu, SD Sharma, AL Castrucci, RA Kesterson, FA AlObeidi, ME Hadley, RD Cone. J Med Chem 38:3454–3561, 1995.

57. R Dharanipragada, D Trivedi, A Bannister, M Siegel, D Tourwe, N Mollova, K Schram, VJ Hruby. Int J Peptide Protein Res 42:68–77, 1993.

58. FDT Lung, N Collins, D Stropova, P Davis, HI Yamamura, F Porreca, VJ Hruby. J Med Chem 39:1136–1141, 1996.

59. G Ösapay, JW Taylor. J Am Chem Soc 112:6046–6051, 1990.

60. G Ösapay, JW Taylor. J Am Chem Soc 114:6966–6973, 1992.

61. M Bouvier, JW Taylor. J Med Chem 35:1145–1155, 1992.

62. C Bracken, J Gulyas, JW Taylor, J Baum. J Am Chem Soc 116:6431–6432, 1994.

63. A Kapurniotu, JW Taylor. Tetrahedron Lett 34:7031–7034, 1993.

64. A Kapurniotu, JW Taylor. J Med Chem 38:836–847, 1995.

65. MH Lyttle, D Hudson. In: JA Smith, JE Rivier, eds. Peptides: Chemistry, Structure and Biology. Proceedings of the 12th American Peptide Symposium. Leiden, The Netherlands: ESCOM, 1992, pp 583–584.

66. MH Lyttle, COA Berry, HO Villar, MH Hocker, LM Kauvar. In: RS Hodges, JA Smith, eds. Peptides: Chemistry, Structure and Biology. Proceedings of the 13th American Peptide Symposium. Leiden, The Netherlands: ESCOM, 1994, pp 1009–1011.

67. ML Valero, JA Camarero, A Adeva, N Verdaguer, I Fita, MG Mateu, E Domingo, E Giralt, D Andreu. Biomed Peptides Proteins Nucleic Acids 1:133–140, 1995.

68. JA Camarero, JJ Cairo, E Giralt, D Andreu. J Peptide Sci 1:241–250, 1995.

69. JR Barbier, W Neugebauer, P Morley, V Ross, M Soska, JF Whitfield, G Willick. J Med Chem 40:1373–1380, 1997.

70. C Mendre, R Pascal, B Calas. Tetrahedron Lett 35:5429–5432, 1994.

71. D Pawlak, NN Chung, PW Schiller, J Izdebski. J Peptide Sci 3:277–281, 1997.

72. P Ho, D Slavazza, D Chang, K Bassi, K Chang. In: JE Rivier, GR Marshall, eds. Peptides: Chemistry, Structure and Biology. Proceedings of the 11th American Peptide Symposium. Leiden, The Netherlands: ESCOM, 1990, pp 993–995.

73. S Plaué. Int J Peptide Protein Res 35:510–517, 1990.

74. N Nishino, M Xu, H Mihara, T Fujimoto, M Ohba, Y Ueno, H Kumagai. J Chem Soc Chem Commun 180–181, 1992.

75. CK Marlowe. Bioorg Med Chem Lett 3:437–440, 1993.

76. D Tumelty, D Vetter, VV Antonenko. J Chem Soc Chem Commun 1067–1068, 1994.

77. J Eichler, AW Lucka, RA Houghten. Peptide Res 7:300–307, 1994.

78. J Lee, JH Griffin, TI Nicas. J Org Chem 61:3893–3986, 1996.

79. C Flouzat, F Marguerite, F Croizet, M Percebois, A Monteil, M Combourieu. Tetrahedron Lett 38:1191–1194, 1997.

80. GB Bloomberg, D Askin, AR Gargaro, MJA Tanner. Tetrahedron Lett 34:4709–4712, 1993.

81. Z Zhao, AM Felix. Peptide Res 7:218–222, 1994.

82. W Zhang, JW Taylor. Tetrahedron Lett 37:2173–2176, 1996.

83. C Gilon, D Halle, M Chorev, Z Selinger, G Byk. Biopolymers 31:745–750, 1991.

84. G Byk, D Halle, I Zeltser, G Bitan, Z Selinger, C Gilon. J Med Chem 39: 3174–3178, 1996.

85. G Bitan, I Sukhotinsky, Y Mashriki, M Hanani, Z Selinger, C Gilon. J Peptide Res 49:421–426, 1997.

86. A Friedler, N Zakai, O Karni, YC Broder, L Baraz, M Kotler, A Loyter, C Gilon. Biochemistry 37:5616–5622, 1998.

87. C Gilon, M Huenges, B Matha, G Gellerman, V Hornik, M Afargan, O Amitay, O Ziv, E Feller, A Gamliel, D Shohat, M Wanger, O Arad, H Kessler. J Med Chem 41:919–929, 1998.

8

Disulfide Formation in Synthetic Peptides and Proteins: The State of the Art

David Andreu and Ernesto Nicolás
University of Barcelona, Barcelona, Spain

Disulfide bridges are key structural features of many peptides and proteins, with a fundamental role in the folding and stabilization of their bioactive conformations [1–3]. The unique reactivity of cysteine and the selectivity requirements inherent in correct cystine pairing combine to make disulfide-containing peptides demanding synthetic targets. Interest in disulfide formation has not been limited to the reproduction of the patterns of natural structures; rather, it has often extended to the engineering of artificial disulfide bonds into natural or de novo–designed peptide molecules for different purposes: conformational restriction, mimetization of active sites, immunoconjugation, etc. The substantial literature in this field has been reviewed in depth in several publications [4–9]. The variety of methods described to form intramolecular (cyclic) and intermolecular (homo- and heteromeric) disulfides can be conveniently grouped into three main approaches [5]: (1) disulfides formed by oxidation of free thiol precursors, (2) disulfides formed from symmetrically *S*-protected precursors, and (3) directed disulfide formation from unsymmetrically protected precursors. The first two approaches are applicable to monomeric peptides with single or multiple disulfides, as well as parallel bis-cystine homodimers. Approach (3) is particularly suited for antiparallel bis-cystine homodimers and for heteromeric disulfides in general. We will use this classification in our review of relevant contributions to the chemistry of disulfide formation during the years 1995–1999.

I. DISULFIDES BY OXIDATION OF THIOL PRECURSORS

Treatment of a polythiol precursor at high dilution under a variety of oxidizing conditions is the simplest and remains the most favored approach to disulfide formation whenever the synthetic target is the native structure of a natural protein or peptide. The process is generally facilitated by the use of thiol–disulfide exchange conditions that ultimately favor correct pairing or folding into thermodynamically stable conformations. The number of successful examples of application of this methodology has increased (Table 1) to include structures with four or even five intramolecular disulfide bonds.

Two routes to the precursor polythiols are generally employed. The most straightforward is direct acidolysis of peptide-resin precursors with all Cys residues protected by either Meb [14,19] or Trt [12,13,16,20] groups, depending on whether Boc- of Fmoc-based chemistries are used for the synthesis, respectively. Alternatively, polythiol precursors can be obtained by metal-assisted removal of S-Acm protections, usually by means of silver triflate [10,11] or mercury acetate [15,17,18]. Transformation of these precursors into cystine residues has been achieved in either of two ways: (1) air oxidation, most often in the presence of disulfide exchange–promoting agents (see later) or (2) use of dimethyl sulfoxide (DMSO) [21,22]. The latter approach is particularly adequate for direct oxidation of Ag or Hg thiolates into disulfides, avoiding conversion to the intermediate free SH form by treatment with conventional reducing agents such as dithiothreitol (DTT), BME, or phosphines. Other previously reported thiol oxidation conditions, such as use of potassium ferricyanide or diethyl azodicarboxylate [5], do not seem to have found application in recent work.

On the other hand, two newly described methods of Cys oxidation deserve mention, although further refining may be required before they are applied to complex multiple disulfide targets. One is charcoal surface–promoted catalytic oxidation [23], which reportedly produces superior results to either (1) or (2) in the preceding paragraph, particularly in the case of small, strained disulfides. The method has been demonstrated only for single cystine residues and its possibilities for multiple disulfide systems remain to be ascertained. Another oxidation condition, related to the DMSO-based procedures, has been found as a result of Met sulfoxide reduction with NH_4I–TFA. When the reduction is conducted in the presence of unprotected Cys residues, concomitant Cys pairing is observed [24,25]. Some intermediate species involved in sulfoxide reduction are presumed to cause thiol activation, which is followed by attack of an unactivated thiol group to give the disulfide. This approach, which is compatible with the presence of protecting groups (e.g., Acm) at other Cys residues, has been found particularly useful for preparing single disulfide homodimers without the risk of Met(O) formation inherent in other oxidation methods. Its potential application to multiple disulfide arrays is under investigation.

The addition of glutathione (GSH–GSSG) or cysteine–cystine to air oxidation mixtures promotes thiol–disulfide exchanges among transient folding intermediates and ultimately favors the formation of the thermodynamically most stable product (often although not always coincident with the correctly folded one). This practice is becoming practically routine in the synthesis of complex multiple disulfide structures, as shown by several examples in Table 1. Most recent accounts use a 1:10 ratio of reduced to oxidized forms of the reagent. Because practically no differences in efficiency are observed between glutathione and cysteine (see, e.g., Ref. 11), the latter additive, which is simpler and cheaper, will presumably gain progressive acceptance.

The outcome of thiol oxidation reactions can also be significantly influenced by environmental factors. Interesting examples are the effect of trifluoroethanol on the formation of the antiparallel dimeric protein uteroglobin [24] or on the relative amounts of cyclic monomer, parallel or antiparallel dimer, or trimer forms of a designed palindromic α-helical peptide [26]. The fluorocarbon solvent, known for its gas-solubilizing ability, is thought to increase the amount of available oxygen in the oxidation reaction.

In conclusion, the oxidation of free thiol precursors remains the method of choice for preparing multiple disulfide peptides, in particular the synthetic versions of natural structures. It requires only one type of protecting group for Cys and allows Cys pairing under mild conditions that often favor nativelike folding.

II. DISULFIDES BY OXIDATION OF SYMMETRICALLY PROTECTED PRECURSORS

Despite the relative success of the methods for disulfide formation from free thiols just described, there are cases in which simultaneous oxidation of all Cys residues in the protein molecule either does not provide satisfactory Cys pairing or encounters other difficulties. These situations require a sequential approach to disulfide formation, in which the Cys residues to be paired are selectively protected, deprotected, and oxidized, ideally under conditions leaving other Cys residues or disulfides unaffected. The full potential of these gradual deprotection–oxidation schemes is achieved when as many of the operations as possible are performed on the solid phase, thus diminishing the number of intermediate purification steps. Elegant descriptions of strat-

Table 1 Recent Examples of Multiple Disulfide Peptides Prepared from Free Thiol Precursors

Peptide	Length	Cys connectivity	Synthetic chemistry	Disulfide formation method	Reference
Stromal cell–derived factor 1	67	9–34, 11–50	Fmoc/tBu	(1) Acid deprotection of $Cys^{9,34}$(Trt) and air oxidation (2) $Cys^{11,50}$(Acm) AgOTf deprotection and DMSO oxidation	10
Amaranth α-amylase inhibitor	32	1–18, 8–23, 17–31	Fmoc/tBu	AgOTf deprotection of Acm_6 precursor and air oxidation (+cysteine/cystine or GSH/GSSG)	11
PMP-D2	35	4–19, 17–27, 14–32	Fmoc/tBu	Acid deprotection of Trt_6 precursor and air oxidation	12
Decorsin	39	7–15, 17–27, 22–38	Fmoc/tBu	Acid deprotection of Trt_6 precursor and air oxidation (+GSH–GSSG and protein disulfide isomerase)	13
C1r protein (EGF-like module)	53	7–26, 22–35, 38–50	Boc/benzyl	Acid deprotection	14
Dendrototoxin	60	7–57, 16–40, 32–53	Boc/benzyl[a]	$Hg(OAc)_2$ deprotection of Acm_6 precursor and air oxidation (+GSH/GSSG)	15

Name	Size	Disulfides	Chemistry	Method	Ref.
Maurotoxin	34	3–24, 9–29, 13–19, 31–34	Fmoc/tBu	Acid deprotection of Trt_8 precursor and air oxidation	16
ω-Agatoxin IV A	48	4–20, 12–25, 19–36, 27–34	Boc/benzyl	Hg(OAc)$_2$ deprotection of Acm_8 precursor and air oxidation (+GSH/ GSSG)	17
Human midkine	121	14–39, 23–48, 30–52, 62–94, 72–104	Boc/benzyl[a]	Hg(OAc)$_2$ deprotection of Acm_{10} precursor and air oxidation (+GSH/ GSSG)	18
Human relaxin	32 (A) + 24 (B)	A10–A15, A11–B11, A24–B23	Boc/benzyl	(1) Acid deprotection of A(Meb$_4$) and B(Meb$_2$) chains (2) Joint air oxidation of A and B chains (2 mM urea, 1.5 mM DTT)	19
Mabinlin II	33 (A) + 72 (B)	A5–B21, A18–B10, B11–B59, B23–B67	Fmoc/tBu[a]	(1) Acid deprotection of A(Trt$_4$) and B(Trt$_4$) chains (2) Joint air oxidation of A and B chains	20

[a]Peptide chain built by fragment condensation in solution.

egies of this type have appeared [9,27]. In contrast, the synthetic chemistry involved in the preparation of orthogonally Cys-protected peptides anchored to solid-phase supports through compatible linkers is often very demanding. This is probably the reason for the relatively low number of recent synthetic accounts using this type of approach (Table 2).

The choice of the gradual approach to disulfide formation is often dictated by difficulties encountered in the handling of the free polythiol. For instance, the tetrathiol precursors of the relatively short, double-disulfide peptides guanylin and uroguanylin [28] were practically insoluble in aqueous basic medium and thus attempts at oxidative folding resulted in inactive materials. In contrast, products with correct Cys pairing and activity similar to that of their natural counterparts were obtained by a two-step procedure: the first disulfide was formed by air oxidation of two cysteines (following acidolytic Trt removal) and the second one by iodine treatment of the other two Acm-protected positions.

Another elegant example of the successful application of gradual disulfide formation is provided by the complex cyclic peptides circulin B and cyclopsychotride (CPT) [29], which, in addition to the end-to-end cycle, display three internal disulfides folded into a cystine knot pattern. In this particular case, DMSO-promoted simultaneous oxidation of the hexathiol precursor gave very poor yields of the target compounds. Presumably, the mobility restriction posed by the starting end-to-end cyclic arrangement causes a substantial reduction in the conformational space available for adequate folding. Selective Acm protection of two Cys residues, however, eased most difficulties by reducing from 15 to 3 the number of possible cystine pairings. The correctly folded, partially protected regioisomer could then be readily identified, purified, and converted into the desired structure by iodine-promoted deprotection–oxidation.

In contrast to these two successful reports, a somewhat sobering account of poor rewards in multistep disulfide formation is provided by the systematic exploration of the various regioselective approaches to the three disulfides of antimicrobial peptide PMP-D2 [12]. Two combinations of Cys protecting groups (Trt, Acm, Mob and Trt, Mmt, Acm) and four different routes to the target compound were evaluated. The results indicated that neither strategy was as efficient as the synthetically less demanding simultaneous oxidation of the hexathiol derived from an all-Trt protected peptide resin precursor.

In these and other accounts [30] of multistep disulfide formation, the merits of the Acm-iodine method [32] (or CNI in lieu of iodine [31]) are clearly demonstrated and, by extension, the versatility of the Acm group for different strategies of disulfide formation is firmly established.

Table 2 Recent Examples of Peptides Prepared by Deprotection–Oxidation of *S*-Protected Precursors

Peptide	Length	Cys connectivity	Synthetic chemistry	Disulfide formation method	Reference
Uroguanylin	16	4–12, 7–15	Fmoc/tBu	(1) $Cys^{7,15}$(Trt) acid deprotection and air oxidation (2) $Cys^{4,12}$(Acm) iodine-promoted oxidation	28
Circulin B	31 (cyclic)	1–16, 3–20, 8–25	Boc/benzyl	(1) $Cys^{1,8,16,25}$(Meb) acid deprotection and DMSO oxidation (2) $Cys^{3,20}$(Acm) iodine-promoted oxidation	29
PMP-D2	35	4–19, 17–27, 14–32	Fmoc/tBu	(1) $Cys^{4,19}$(Mmt) very mild (2% TFA) acid deprotection and on-resin oxidation (2) $Cys^{17,27}$(Trt) acid deprotection and DMSO oxidation (3) $Cys^{14,32}$(Acm) iodine-promoted oxidation	12
Human protein S EGF 1–like domain	41	5–18, 10–27, 29–38	Boc/benzyl	(1) $Cys^{5,18,29,38}$(Meb) acid deprotection and air oxidation (2) $Cys^{10,27}$(Acm) iodine-promoted oxidation	30
Human factor VII EGF 2–like domain	47	10–21, 17–31, 33–46	Fmoc/tBu	(1) $Cys^{10,21,33,46}$(Trt) acid deprotection and air oxidation (+GSH/GSSG) (2) $Cys^{17,31}$(Acm) CNI-promoted oxidation	31

III. DIRECTED UNSYMMETRIC DISULFIDE FORMATION AND RELATED APPROACHES

A third general method of disulfide formation that is particularly valuable in preparing heteromeric disulfide-linked structures [5,9] is based on the attack of a Cys S-nucleophile (usually the thiol) on another Cys residue protected as the heterodisulfide of an electron-withdrawing group. Among several Cys-protecting groups that fit this description, the Npys group has received most attention. Its dual role as protector (Boc chemistry compatible, HF stable) and S-activator and its ability to replace directly (by means of Npys chloride) other protecting groups such as Acm clearly make it one of the most valuable disulfide formation tools available. A synthesis of hetero-trimeric collagen peptides [33,34] provides an impressive example of the potential of this type of chemistry in the preparation of complex multimeric structures.

Another closely related application of this type of chemistry is an elegant method of intramolecular disulfide formation on the solid phase using a polymer-bound version of Ellman's reagent [35]. Although similar conceptual schemes had been previously applied to solid-phase heterodimer formation [36,37] from single Cys-containing fragments, in this account the solid-phase chemistry has been especially adapted to favor intramolecular reaction and minimize irreversibly resin-bound intermediates.

IV. SELECTED EXPERIMENTAL PROTOCOLS

The following descriptions cover procedures not described in similar detail in preceding review chapters or significantly improved in recent publications. A more comprehensive group of general protocols can be found in Refs. 5 and 9.

A. Removal of S-Acm with Ag(I), Followed by DMSO Oxidation

Adapted from Ref. 10.

1. Dissolve 1.78 μmol Cys(Acm)-protected peptide and 0.02 mL an-isole in 2 mL neat TFA.
2. Chill the solution to 4°C in an ice bath, and add solid AgOTf [molecular weight (MW) 256.9, 50 eq./Acm].
3. After 2 h reaction, precipitate peptide with 30 mL ice-cold dry Et$_2$O; wash precipitate with same solvent (3 $\times$ 20 mL).

4. Redissolve the residue in 17 mL 50% DMSO–1 M aqueous HCl (v/v) at room temperature.
5. After 7 h, filter the AgCl precipitate and dilute filtrate with 150 mL water.
6. Purify by preparative HPLC as required.

B. Disulfide Formation Mediated by Glutathione–Protein Disulfide Isomerase

Adapted from Ref. 13.

1. Prepare a solution of protein disulfide isomerase (0.4 μg/mL) in 0.1 M Tris-HCl, pH 8.5 buffer containing 1 mM EDTA, 0.2 M NaCl, 2 mM GSH, and 1.5 mM GSSG.
2. Dissolve the prepurified, reduced peptide (0.2 mg/mL) in the buffer.
3. Monitor folding and eventually purify by reverse-phase HPLC.

C. Charcoal Surface–Assisted Intramolecular Disulfide Cyclization

Adapted from Ref. 23.

1. Dissolve prepurified, reduced peptide to a concentration of 1 mg/ mL in water; adjust pH to 7.5–8 with 5% aqueous ammonia as required.
2. Add an equal weight of granulated charcoal (powder not recommended!) to the peptide solution.
3. Shake gently at room temperature for 2–6 h, monitoring the reaction by HPLC or Ellman analysis.
4. Upon completion, filter off charcoal, acidify filtrate, lyophilize, and purify by HPLC as required.

D. Simultaneous Methionine Reduction–Disulfide Formation

Adapted from Refs. 24 and 25.

1. Dissolve 3.6 μmol peptide (one Cys residue) in 1.8 mL neat TFA (2 mM) and chill the solution to 4°C in an ice bath.
2. Add solid NH$_4$I (MW 145, 10 eq./sulfoxide) and Me$_2$S (MW 62.1, 1 eq./NH$_4$I) with vigorous magnetic stirring of the suspension.
3. After 3 h (based on HPLC monitoring) quench the reaction by dilution with 10 mL water.

4. Extract with CCl_4 (3 × 25 mL) to remove iodine.
5. Remove volatiles under vacuum, dissolve the residue in 20 mL 10% AcOH, and lyophilize.
6. Redissolve the peptide in minimal amount of CH_3CN–water (0.05% of TFA), apply to preparative reverse-phase HPLC column, and purify as required.

ABBREVIATIONS

Acm, acetamidomethyl; Boc, *tert*-butyloxycarbonyl; EDTA, ethylenediaminotetraacetic acid; Fmoc, 9-fluorenylmethyloxycarbonyl; GSH, glutathione (H-γ-Glu-Cys-Gly-OH); GSSG, glutathione disulfide; Meb, 4-methylbenzyl; Mmt, 4-methoxytrityl; Mob, 4-methoxybenzyl; Npys, 3-nitro-2-pyridinesulfenyl; Tf, trifluoromethanesulfonyl; TFA, trifluoroacetic acid; Tris, tris(hydroxymethyl)aminomethane; Trt, trityl (triphenylmethyl).

ACKNOWLEDGMENTS

Work carried out in the authors' laboratories has been supported by Generalitat de Catalunya (CERBA), DGICYT (PB95-1131; PB97-9873), and the European Union (FAIR5-CT97-3577).

REFERENCES

1. JM Thornton. J Mol Biol 151:261–287, 1981.
2. SF Betz. Protein Sci 2:1551–1558, 1993.
3. RT Raines. Nat Struct Biol 4:424–427, 1997.
4. EE Büllesbach. Kontakte (Darmstadt) 1:21–29, 1992.
5. D Andreu, F Albericio, NA Solé, MC Munson, M Ferrer, G Barany. Methods Mol Biol 35:91–169, 1994.
6. Y Kiso, H Yajima. In: B Gutte, ed. Peptides: Synthesis, Structures and Applications. San Diego: Academic Press, 1995, pp 61–81.
7. L Moroder, D Besse, HJ Musiol, S Rudolph-Böhner, F Siedler. Biopolymers 40:207–234, 1996.
8. P Lloyd-Williams, F Albericio, E Giralt. Chemical Approaches to the Synthesis of Peptides and Proteins. Boca Raton, FL: CRC Press, 1997, pp 209–236.
9. I Annis, B Hargittai, G Barany. Methods Enzymol 289:198–221, 1997.
10. H Tamamura, F Matsumoto, K Sakano, A Otaka, T Ibuka, N Fujii. Chem Commun 151–152, 1998.
11. V Lozanov, C Guarnaccia, A Patthy, S Foti, S Pongor. J Peptide Res 50:65–72, 1997.

12. C Kellenberger, H Hietter, B Luu. Peptide Res 8:321–327, 1995.
13. P Polverino de Laureto, E Scaramella, V de Filippis, O Marin, M-G Doni, A Fontana. Protein Sci 7:433–444, 1998.
14. J-F Hernandez, B Bersch, Y Pétillot, J Gagnon, GJ Arlaud. J Peptide Res 49: 221–231, 1997.
15. H Nishio, T Inui, Y Nishiuchi, CLC de Medeiros, EG Rowan, AL Harvey, E Katoh, T Yamazaki, T Kimura, S Sakakibara. J Peptide Res 51:355–364, 1998.
16. R Kharrat, K Mabrouk, M Crest, H Darbon, R Oughideni, M-F Martin-Eauclaire, G Jacquet, M El-Ayeb, J van Rietschoten, H Rochat, J-M Sabatier. Eur J Biochem 242:491–498, 1996.
17. J Najib, T Letailleur, J-C Gesquière, A Tartar. J Peptide Sci 2:309–317, 1996.
18. T Inui, J Bódi, S Kubo, H Nishio, T Kimura, S Kojima, H Maruta, T Muramatsu, S Sakakibara. J Peptide Sci 2:28–39, 1996.
19. JD Wade, F Lin, D Salvatore, L Otvos Jr, GW Tregear. Biomed Peptides Proteins Nucleic Acids 2:27–32, 1996.
20. M Kohmura, Y Ariyoshi. Biopolymers 46:215–223, 1998.
21. JP Tam, CR Wu, W Liu, JW Zhang. J Am Chem Soc 113:6657–6662, 1991.
22. A Otaka, T Koide, A Shide, N Fujii. Tetrahedron Lett 32:1223–1226, 1991.
23. R Volkmer-Engert, C Landgraf, J Schneider-Mergener. J Peptide Res 51:365–369, 1998.
24. T Ferrer, E Nicolás, E Giralt. Lett Peptide Sci 6:165–172, 1999.
25. M Vilaseca, E Nicolás, F Capdevila, E Giralt. Tetrahedron 54:15273–15286, 1998.
26. M Royo, MA Contreras, E Giralt, F Albericio, M Pons. J Am Chem Soc 120: 6639–6650, 1998.
27. MC Munson, G Barany. J Am Chem Soc 115:10203–10210, 1993.
28. J Klodt, M Kuhn, UC Marx, S Martin, P Rösch, W-G Forssmann, K Adermann. J Peptide Res 50:222–230, 1997.
29. JP Tam, Y-A LU. Protein Sci 7:1583–1592, 1998.
30. TM Kackeng, PE Dawson, SBH Kent, JH Griffin. Biopolymers 46:53–63, 1998.
31. M Husbyn, L Örning, KS Sakariassen, PM Fischer. J Peptide Res 50:475–482, 1998.
32. B Kamber, A Harmann, K Eisler, B Riniker, H Rinker, P Sieber, W Riettel. Helv Chim Acta 63:899–915, 1980.
33. J Ottl, L Moroder. J Am Chem Soc 121:653–661, 1999.
34. J Ottl, L Moroder. Tetrahedron Lett 40:1487–1490, 1999.
35. I Annis, L Chen, G Barany. J Am Chem Soc 120:7226–7238, 1998.
36. B Ponsati, M Ruiz-Gayo, E Giralt, F Albericio, D Andreu. J Am Chem Soc 112:5345–5347, 1990.
37. PT Lansbury Jr, I Sucholeiki. J Org Chem 58:1318–1324, 1993.

9

Solid-Phase Convergent Approaches to the Synthesis of Native Peptides and Proteins

Paul Lloyd-Williams and Ernest Giralt
University of Barcelona, Barcelona, Spain

Although several impressive linear solid-phase syntheses of proteins have been achieved [1], methodological improvements are continually allowing larger peptides to be routinely considered. However, homogeneous large target molecules are probably best produced chemically by convergent strategies that rely on the union of peptide segments. A key advantage of this approach is that such segments can be purified and characterized, providing a series of analytical data on the nature of the material being synthesized and facilitating the purification of the final product. Most modern convergent peptide synthesis strategies are based to some extent on the use of solid-phase methods because of the advantages in economics and rapidity that these present. Several reviews [2–4] of such strategies have already appeared; here we present an updated report on practical, useful methods.

I. CONVERGENT SOLID-PHASE PEPTIDE SYNTHESIS

In *convergent solid-phase peptide synthesis* (CSPPS), protected peptide segments are prepared by modifying the standard chemistry used for solid-phase peptide synthesis in such a way that the protected peptide segment rather than the free peptide can be cleaved from the solid support after synthesis. These protected peptide segments are then usually purified by chromatography before coupling on a solid support thus allowing the amino acid se-

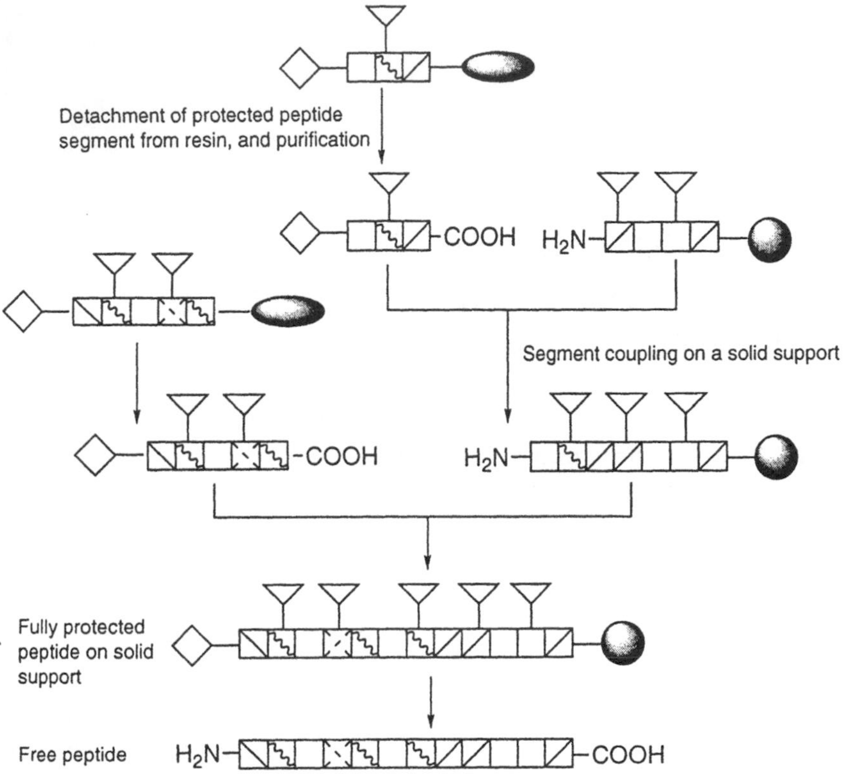

Scheme 1 Convergent solid-phase peptide synthesis (CSPPS). Amino acids are represented by squares, the protected α-amino group by a diamond, and the protected side-chain functional groups by triangles.

quence of the desired peptide or protein to be constructed. The procedure is represented in Scheme 1.

A. Solid-Phase Synthesis of Protected Peptide Segments

Protected peptide segments are required because to couple them in an unambiguous manner, all reactive functional groups except those required for amide bond formation usually must be protected. (See Section II, however, for alternatives that allow unprotected peptide segments to be coupled.)

Modern approaches are based on modification of the peptide-resin anchorage so that the bond that links the peptide to the solid support can be cleaved selectively. This requires a fine adjustment of the chemistry involved, but this aspect of peptide research has been extensively investigated and many different types of anchorages have been described. Stability to all of the conditions necessary to effect synthesis of the peptide is a prerequisite, as is the ability to detach the protected segment from the solid support in high yield under mild conditions and without racemization at the C-terminal amino acid.

A variety of different chemical processes can be used but there must be a high degree of compatibility between the protecting groups of the segment and the peptide-resin anchorage. The most important methods are discussed here.

1. Acidolytic Cleavage of Protected Peptides from the Solid Support

If acidolysis is to be used for cleavage, then all of the protecting groups of the peptide must be stable to these conditions. Thus, strong acids such as liquid hydrogen fluoride are impractical and cleavage must occur under milder acidolysis conditions. The development of CSPPS has stimulated the search for more acid-labile peptide-resin anchorages. Modification of the chemical structure of the solid support can increase its lability to acid, and a series of highly acid-labile resins and linkers are now available. These are compatible only with 9-fluorenylmethoxycarbonyl/t-butyl (Fmoc/tBu) SPPS because the use of acidic conditions is avoided throughout; highly acid-labile solid supports are not compatible with t-butyloxycarbonyl/benzyl (Boc/Bzl) methods. The most commonly used handles and resins that are cleaved under mild acidolytic conditions are discussed here. Generally, these present the considerable advantage in practical terms of providing a very clean and simple cleavage reaction with easy isolation of the peptide.

a. Highly Acid-Labile Handles and Resins. Handles and resins that can be cleaved by extremely mild acidolysis currently appear to be the most promising option for the solid-phase preparation of protected peptide segments. The trityl-based resin **1** was developed specifically for use in CSPPS and is among the most acid labile of all solid supports [5–11]. Cleavage can be accomplished in excellent yield by treatment with less than 1% trifluoroacetic acid (TFA) or a mixture of acetic acid (HOAc) and trifluoroethanol (TFE) in dichloromethane (DCM). An alternative cleavage reagent is hexafluoroisopropanol (HFIP) in DCM [12], which has the advantage of avoiding contamination of the protected peptide segment with a carboxylic acid. Such contamination can lead to capping of the free N^α-amino group of protected peptide segments in coupling reactions [13].

Several trityl-based handles, which can be attached to a range of functionalized solid supports, have also been described [14–20]. A representative
example is **2**, which has been for the synthesis of a 94-residue protein [21].
(See Section I.D.2.) Trityl-based handles and resins have now been established as the most generally applicable and satisfactory solution to the construction of protected peptide segments with Fmoc/*t*Bu peptide synthesis.

A number of other useful highly acid-labile handles are also available.
Most of these linkers are formed by introducing electron-donating substituents into the benzyl groups that serve as the peptide-resin anchor. The first
handle of this type, 4-hydroxymethyl-3-methoxyphenoxyacetic acid (HMPA),
3, was introduced by Sheppard and Williams [22,23]. A closely related and
more widely used alternative is 4-(4-hydroxymethyl-3-methoxyphenoxy)
butyric acid [24–26] (HMPB), **4**.

Such handles, when attached to suitable resins, provide highly acid-
sensitive solid supports for peptide synthesis [27,28]. Fmoc/*t*Bu-protected
peptides can be cleaved, in the case of **4**, by repetitive treatments with 1%
TFA in DCM for 6–8 min at room temperature. These conditions do not
affect *t*Bu-based protecting groups.

Another highly acid-labile resin, **5**, known as SASRIN (super-acid-
sensitive resin) [29–31], has also found widespread application, and other,
even more acid labile, handles and resins are available [32,33]. Cleavage of
peptides from **5** can be accomplished in high yield by treatment of peptide-

resin with 1% TFA in DCM, whereas treatment with 0.1% TFA suffices for cleavage from the hyper-acid-labile (HAL) linker **6**.

Protected peptide segments are often required with Gly or Pro at the C-terminus in order to reduce the risk of epimerization on segment coupling. (See Section I.C.5.) The presence of Pro in this position tends to favor the formation of diketopiperazines on incorporation of the third amino acid of the synthesized peptide with concomitant reduction of the level of resin functionalization. This unwanted cyclization can normally be controlled by using special Fmoc deprotection protocols. The use of shorter exposures to piperidine in N,N-dimethylformamide (DMF) [34] or tetrabutylammonium fluoride (TBAF) quenched with methanol (MeOH) [35] has been reported to alleviate the problem. The steric hindrance to cyclization normally ensures that diketopiperazine formation is minimal in trityl-based resins and handles even when Pro is at the C-terminus, constituting a further advantage of this solid support.

b. Representative Experimental Procedure for the Cleavage of Protected Peptides from High Acid-Labile Resins.   This text was adapted from Ref. 4. The optimal protocol for the cleavage of protected peptides from trityl-, SASRIN-, and HMPB-based solid supports is to use a continuous flow procedure, which can be effectively applied only when the resin is a polyethylene glycol–grafted polystyrene (PEG-PS) support. The uneven shrinkage of polystyrene (PS) resins in TFA–DCM under moderate pressure makes these supports incompatible with the continuous flow approach. A general alternative for all classes of solid supports is a batchwise procedure of several short cleavage steps with neutralization of TFA in the effluent.

The peptide-resin (0.5 g) was preswollen in a 10-mL polypropylene syringe fitted with a polyethylene disk and a stopcock with washes of DCM (5 mL, 3 × 10 min). Cleavage was carried out at 25°C, by alternating washes of 1% TFA in DCM and DCM (2.5 mL of each solution, 10 × 20 s) and filtering into MeOH–pyridine (9:1) (20 mL). This operation was repeated three times. For peptides containing Trp and Cys, 5% 1,2-ethanedithiol (EDT) was added to the TFA solution. The combined cleavage portions were concentrated to one third of the original volume, and the peptide was precipitated with H_2O (usually about one half of the organic volume). The precipitate was filtered and dried in vacuum over P_2O_5. Using these conditions, all common protecting groups including the Trt of His are stable.

Detachment of protected peptides from chlorotrityl-resins [9] using HFIP–DCM (1:4) can also be performed in a polypropylene syringe fitted with a polyethylene disk and a stopcock. Quantitative cleavage is obtained after 3 min at 25°C. Neutralization of the effluents with MeOH–pyridine (9:1) is advisable when His(Trt) is present. The combined filtrates were evaporated to dryness.

2. Nucleophile- and Base-Mediated Cleavage of Protected Peptides from the Solid Support

Similarly to the compatibility of highly acid-labile resins and handles for Fmoc/tBu synthesis only base-labile solid supports can normally be used for the Boc/Bzl approach. The conditions necessary for Fmoc group removal would lead to cleavage of a base-labile handle or resin.

a.   Cleavage of the Peptide-Resin Anchorage by a β-Elimination Reaction.   Base-mediated cleavage of peptides from solid supports incorporating fluorene-based handles [36,37] such as **7, 8,** and **9** occurs by a β-elimination mechanism. Solid supports incorporating such handles are completely stable to acidic condition. The handle **7**, *N*-[(9-hydroxymethyl)-2-fluorenyl]succinamic acid (HMFS), can be used to form solid supports that are more stable than either **8** or **9** with respect to premature cleavage of the peptide during synthesis. This is probably because the link between the fluorene nucleus and the side chain is an electron-donating *N*-amide group. Resins incorporating **7** are much more stable to solutions of *N,N*-diisopropylethylamine (DIEA) in DMF, so loss of peptide chains from the support during Boc/Bzl synthesis is avoided. However, lability to secondary amines such as piperidine is maintained, allowing protected segments to be released from the resin in high yield using 20% morpholine in DMF for 2 h [36,37]. These conditions do, however, provoke partial or complete loss of the *N*-formyl group of Trp and the BrZ group of Tyr. Such loss can be avoided by use of the cyclohexyloxycarbonyl and 3-pentyl groups, respectively. Handle **7** currently provides perhaps the best option for the synthesis of protected peptide segments by Boc/Bzl chemistry, and its use has been demonstrated in complex syntheses [38].

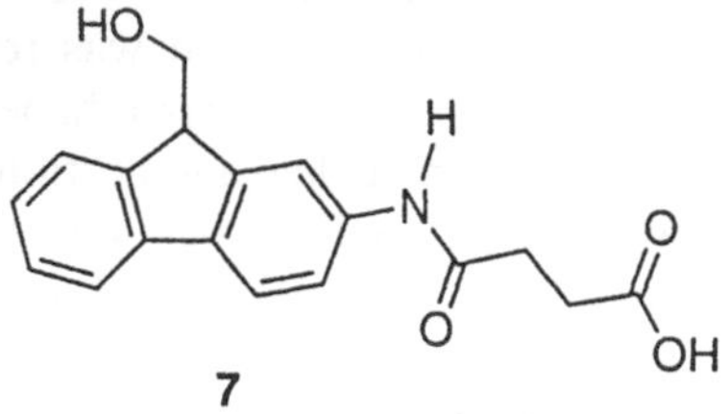

7

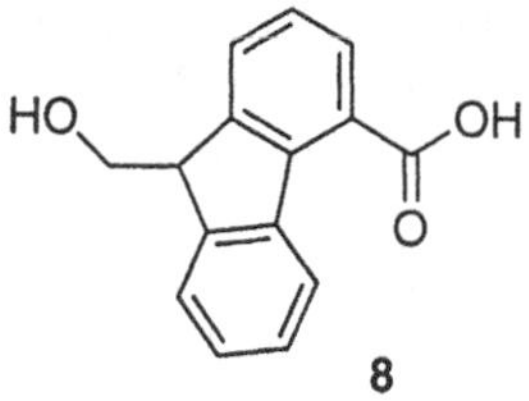

8

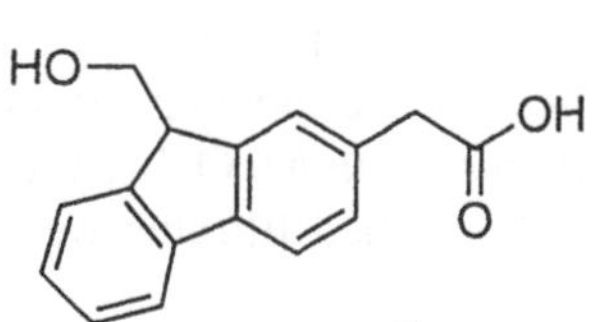

9

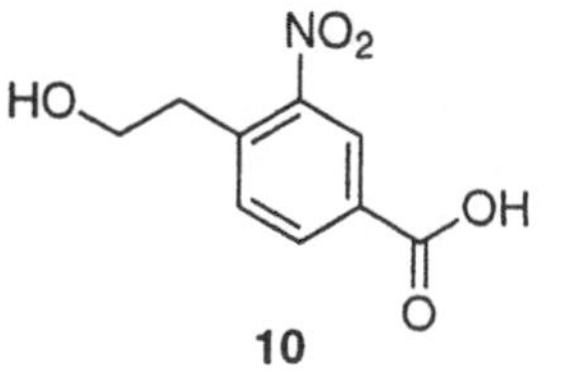

10

Several other handles and resins allow cleavage by base-catalyzed β-elimination reactions [39–42]. The 2-(2-nitrophenyl)ethyl handle **10** has been used for the synthesis of protected peptides and nucleopeptides [43,44]. Cleavage can be accomplished in high yield on treatment with 0.1 M 1,8-diazabicyclo[5.4.0] undec-7-ene (DBU) in dioxane, or 20% piperidine in DMF, for 2 h at room temperature.

b. *Representative Experimental Procedure for the Cleavage of Protected Peptides from Fluorenylmethyl-Resins Based on the HMFS Handle* **7**.   This text was adapted from Ref 37. Peptide-resin (1 g) was preswollen in a 20-mL polypropylene syringe fitted with a polyethylene disk and a stopcock with washes of DMF (10 mL, 3 × 10 min). Cleavage was carried out by treatment of peptide-resins with a freshly prepared solution of morpholine–DMF (1:4) (10 mL) for 2 h at 25°C with occasional agitation. After filtration, the resin was washed with DMF (3 × 5 mL) and the combined cleavage portions were evaporated to dryness under high vacuum.

c. *The Kaiser Oxime Resin.*   The base-labile resin that has been most used in peptide synthesis in the *p*-nitrobenzophenone oxime resin **11** developed by Kaiser and coworkers [45–47]. This resin has been used extensively for the synthesis of Boc/Bzl-protected peptide segments.

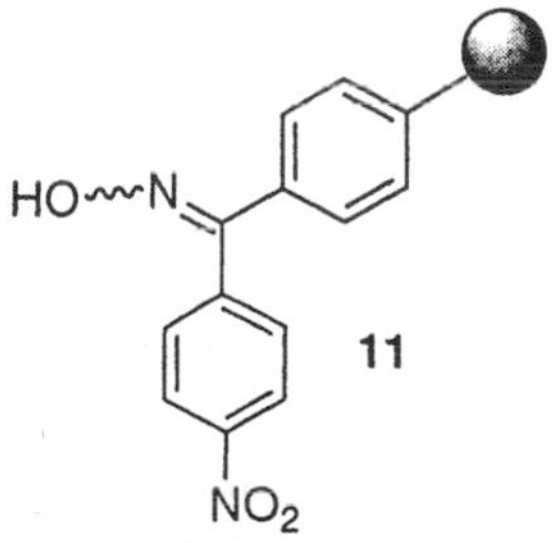

Various methods have been used to cleave protected peptides from the resin, including hydrazinolysis, ammonolysis, or aminolysis using a suitable amino acid ester [48–52]. Probably the most useful procedure, however, is transesterification of the peptide-resin with hydroxypiperidine. This initially forms the hydroxypiperidine ester of the protected peptide, which, after treatment with zinc in HOAc, furnishes the corresponding free carboxylic acid. This method does not lead to epimerization at the *C*-terminal amino acid or to loss of acid-sensitive protecting groups. Cleavage yields are usually high.

The main problem associated with use of Kaiser oxime resin is its lability to nucleophiles. Loss of peptide from the resin can be provoked by the free N^α-amino group of the growing peptide chain on neutralization with base, after acidolytic removal of the N^α-Boc group. Chain elongation on this resin is usually carried out with in situ neutralization in order to reduce such loss. This is particularly important in the coupling of the third amino acid, where diketopiperazine formation may occur.

d. Representative Experimental Procedure for the Cleavage of Protected Peptides from Oxime-Based Resins. This text was adapted from Ref. 49. In a silylated screw-cap tube reaction vessel with a Teflon-lined cap, a sintered glass frit, and a stopcok, protected peptide-oxime resin (1 g) was pre-swollen with washes of DCM (10 mL, 3 × 10 min). The cleavage was carried out by treatment of the peptide-resin with hydroxypiperidine (3 equiv., freshly recrystallized from hexane) for 3–16 h at 25°C with mechanical shaking. The resin was filtered and washed with DCM, DMF, and MeOH (3 × 5 mL in each case). The combined filtrates were evaporated under vacuum. The residue was triturated with ether or hexane to give the crude peptide 1-piperidyl ester. This was then dissolved in HOAc–H_2O (9:1) (15 mL) and Zn dust (30 equiv.) was added. After vigorous stirring for 15–30 min at 25°C, Zn was removed by filtration and washed with HOAc–H_2O (9:1) (3 × 5 mL). The combined filtrates were evaporated under vacuum. Gel-filtration chromatography using Sephadex LH-60 in DMF is advisable for removal of residual Zn and of by-products originating from hydroxypiperidine.

3. Photolytic Cleavage of Protected Peptides from the Solid Support

Photolysis [53] of the peptide-resin bond is a mild, noninvasive technique that is, in principle, compatible with both the Boc/Bzl and Fmoc/*t*Bu approaches, creating an attractive cleavage technique for use in CSPPS. Although a range of possibilities exist for introducing photolability in a solid support, the most widely used in peptide synthesis are nitrobenzyl or phenacyl anchoring linkages.

a. Nitrobenzyl Resins. Several nitrobenzyl-based solid supports have been employed for the synthesis of protected peptide segments [54–60]. The most useful are resins incorporating the 4-bromomethyl-3-nitrobenzoic acid handle **12**. A typical example is 3-nitro-4-bromomethylbenzhydrolyamido-polystyrene (known as Nbb-resin), which is formed by attaching **12** to a benzhydrylamine resin. This support is fully compatible with Boc/Bzl SPPS and partially compatible with the Fmoc/*t*Bu approach. The peptide–handle bond is not completely stable to treatment with piperidine, so the synthesis of long sequences by Fmoc/*t*Bu synthesis is not advisable [61]. A nitroveratryl handle **13** has been quite extensively used for the elaboration of small-molecule libraries [62].

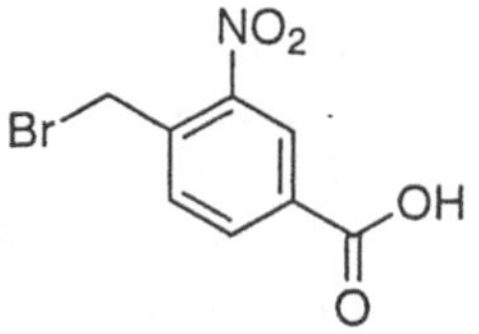

In order to use such supports, the first amino acid must be incorporated as its cesium carboxylate [63], displacing the benzyl bromide by nucleophilic substitution. This can be accomplished either by esterifying the amino acid onto the handle in solution to give a performed handle [64] or by attaching the residue to a handle previously incorporated on the resin. The coupling of the third amino acid of the sequence (regardless of whether a Boc/Bzl or Fmoc/tBu approach is used) must be carried out under conditions that minimize the production of diketopiperazines [65]. Formation of these cyclic dipeptides is more of a problem in Fmoc/tBu synthesis and can be avoided on **13** only by coupling the second and third amino acids as a protected dipeptide [34]. In the Boc/Bzl approach, the problem can be minimized by using specially designed coupling protocols for the third amino acid involving in situ neutralization [66,67]. The coupling of all other amino acids is carried out by standard protocols.

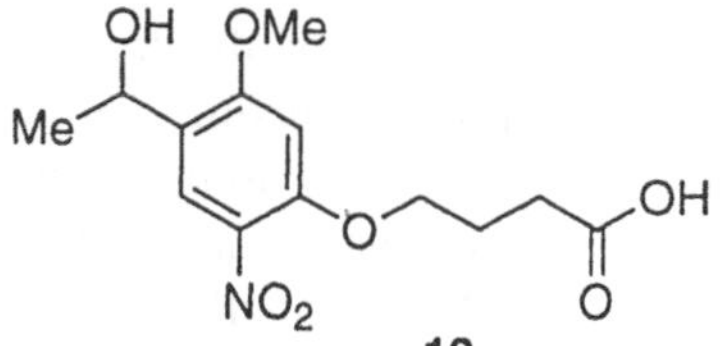

Photolysis of a suspension of the peptide-resin in a mixture of DCM and TFE by irradiation at 360 nm then detaches the peptide from the solid support in moderate to good yields. All commonly used protecting groups are stable to the cleavage conditions.

b. Phenacyl Resins. An alternative to nitrobenzyl-based resins is provided by resins such as **14** or handles such as **15** or **16**. These allow the formation of a photolabile α-methylphenacyl ester [68] anchoring linkage on treatment with the cesium carboxylate of the first amino acid of the sequence to be constructed. Phenacyl-based resins [69–74] are, however, compatible only with the Boc/Bzl approach because the peptide-resin anchor is not stable to the basic conditions used in Fmoc/tBu synthesis.

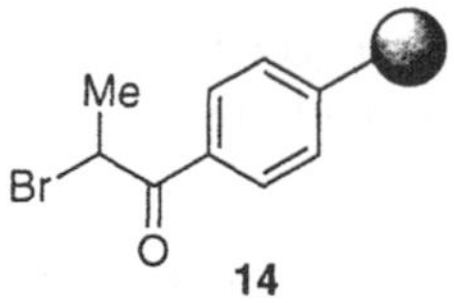

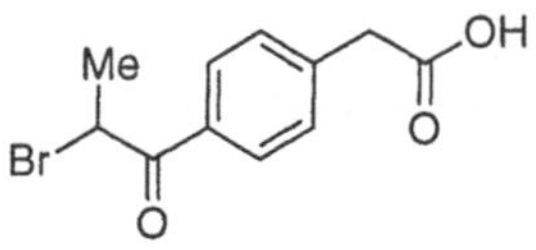

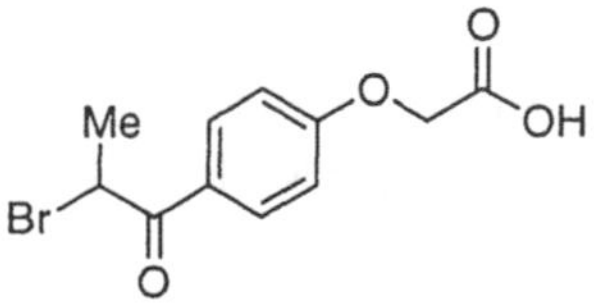

Phenacyl-based solid supports are subject to the same side reactions as occur when the phenacyl ester is used as a *C*-terminus protecting group in solution synthesis. Racemization of the first amino acid can occur, especially in the case of Pro. Incorporation of the second amino acid can be hampered by cyclization of the free amino group of the first residue onto the carbonyl group of the resin, forming Schiff bases. Diketopiperazine formation may also compete with the coupling of the third amino acid, although this can normally be overcome by using methods similar to those used for nitrobenzyl resins. Occasionally, in difficult cases, these side reactions can be avoided only by incorporating previously prepared di- or even tripeptides on the resin [70].

Photolysis is carried out in a manner similar to that for nitrobenzyl resins, and cleavage yields of protected peptides are comparable for both types of support.

c. Representative Experimental Procedure for the Cleavage of Protected Peptides from Photolabile Resins. This text was adapted from Ref. 75. To prevent resin adhering to its walls, a two-neck cylindrical reaction vessel was silylated before photolysis by rinsing three or four times with trimethylsilyl chloride (Me$_3$SiCl)–toluene (1:9), followed by washing with absolute ethanol (EtOH) and drying. Peptide-resin (0.5–0.7 g) was suspended in TFE–DCM or toluene (1:4) (100 mL) in the reaction vessel. Before photolysis, the peptide-resin suspension was degassed by evacuation at water-pump pressure and purging with Ar three times in succession. The resin was photolyzed at 360 nm in a Rayonnet RPR-100 apparatus for 1–15 h, maintaining vigorous magnetic stirring and keeping the temperature below 40°C, using the incorporated fan. The reaction crude was filtered and the resin washed with TFE–DCM (1:4), DCM, DMF, and MeOH (3 × 5 mL in each case). The combined filtrates were then evaporated to dryness.

4. Cleavage of Protected Peptides from Allyl-Functionalized Resins

Allyl-based solid supports are potentially very useful because the mild cleavage conditions are, in principle, compatible with both Boc/Bzl and Fmoc/*t*Bu peptide synthesis. Furthermore, when they are used with the latter approach, an orthogonal [64,76,77] scheme for the preparation of protected peptide segments is achieved. Useful allyl-functionalized polymers are provided by incorporation of the handle **17, 18,** or **19** into suitable resins [78–80].

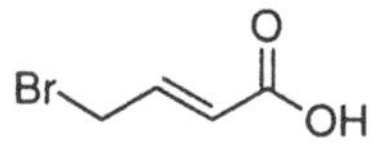

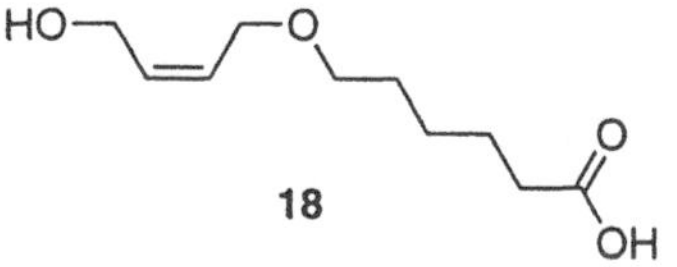

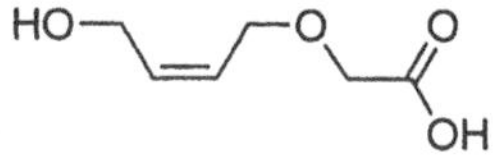

The general requirements for the cleavage of peptides include a palladium catalyst and a severalfold excess, relative to the degree of ally substitution of the resin, of a suitable nucleophile or "allyl acceptor." Two general procedures have been developed. The first incorporates tetrakis(triphenylphosphine) palladium, $(Ph_3P)_4Pd$, as a catalyst and 1-hydroxybenzotriazole, morpholine, N,N-dimethylbarbituric acid as a nucleophile [81,82]. For Fmoc-protected peptides the nucleophile used as an allyl acceptor must be carefully chosen to be compatible with the base-labile protecting groups; N-methylaniline (NMA) can give good results [83,84]. In an alternative hydrostannolytic cleavage procedure [85,86], the peptide-resin is treated with dichlorobis(triphenylphosphine)palladium (II), $Pd(PPh_3)_2Cl_2$, and tributyltin hydride in the presence of a proton donor. This method is particularly useful in the case of Fmoc-containing peptides because the conditions involved do not provoke loss of this protecting group. For both cleavage procedures, the conditions are mild and yields are usually high, although careful control of the reaction conditions is often required.

a. Representative Experimental Procedure for the Cleavage of Protected Peptides from Allyl-Based Resins. This was adapted from Ref. 84.

> Method A. Peptide-resin (100 mg, ~0.075 mmol) and crystalline, yellow $Pd(Ph_3)_4$ (30 mg, 0.026 mmol) were suspended in a previously degassed mixture of DMSO–THF–0.5 M HCl (2:2:1) (5 mL) in a silylated two-neck round-bottom flash and stirred vigorously under Ar. The dissolved catalyst gave a yellow-colored suspension. NMA (385 μL, 3.5 mmol) was added and the mixture stirred under Ar for 12 h at 25°C. Filtration, followed by washing the resin with DMF and TCM (3 × 1 mL in each case), and solvent removal under high vacuum gave the crude peptide. For Met-containing peptides, TCM–HOAc (9:1) can be used as a solvent.

> Method B. Peptide-resin (100 mg, ~0.075 mmol) and $PdCl_2(Ph_3P)_2$ (2 mg, 2.5 μmol) were suspended in a previously degassed mixture of DMF–DCM (1:1) (5 mL) in a silylated two-neck round-bottom flask and stirred vigorously under Ar. The dissolved catalyst gave a yellow-colored suspension. Tributyltin hydride (75 μL, 0.25 μmol) in DCM (1 mL) was added over 30 min and the mixture stirred for a further 10 min at 25°C. After filtration, the resin was washed with DMF–DCM (1:1) (3 × 1 mL), and the filtrate extracted repeatedly (4 × 5 mL) with pentane in order to eliminate tin by-products. HCl (1 M, 5 mL) was added to convert the tin carboxylate to the peptide free acid, followed by H_2O (5 mL) to precipitate the crude peptide, which was isolated by centrifugation and filtration.

In this cleavage procedure gel-filtration chromatography through Sephadex LH-60 with DMF as an eluent is advisable in order to remove completely any tin by-products.

B. Purification of Protected Peptide Segments

The purification of the protected peptide intermediates is an important aspect of the CSPPS strategy to ensure homogeneous molecular species, free from single-residue deletion peptides and other impurities. However, a prerequisite for any chromatographic purification is adequate solubility of the material to be purified in a solvent compatible with the procedure. Protected peptides exhibit unpredictable, but generally poor, solubility in water and in most of the commonly used organic solvents, which makes them difficult to purify and causes some of the most serious problems in CSPPS.

1. Enhancement of the Solubility of Protected Peptide Segments

The low solubility of protected peptide segments has prompted investigation to enhance methods for facilitating purification and segment coupling reactions. Poor solubility is thought to be primarily due to intermolecular association by hydrogen bonding, leading to the formation of β-sheet–like secondary structures [87], similar to structures thought to cause "difficult sequences" in linear SPPS [88]. Disruption of the hydrogen bonding that stabilizes these secondary structures is necessary in order to solubilize protected peptides. Improved solubility can be achieved when certain solvents or solvent mixtures are used. Alternatively, the structure of the protected peptide segments may be modified to reduce the formation of β-sheet–like structures, rendering them more soluble.

a. Structural Modification.   Model studies of peptides containing Pro, **20**, demonstrated that the presence of tertiary amide bonds at central positions in the segment led to better solubility [89]. For target peptides lacking a regular distribution of Pro, pseudoprolines, **21**, may be incorporated in the segments by the formation of oxazolidines or thiazolidines from Ser, Thr, or Cys [90–94]. Regeneration of the original residue is straightforward in the case of Ser- and Thr-derived oxazolidines, although it can be more difficult for Cys-derived thiazolidines. Another alternative is to introduce tertiary amides by the protection of peptide bonds [95–102]. Hmb amide bond protection, **22**, has been used to great effect in CSPPS. It confers much improved solubility upon protected peptide segments, significantly facilitating their purification, as demonstrated by the synthesis of the Tau protein of Alzheimer's disease [21,103–105]. (See Section I.D.2.)

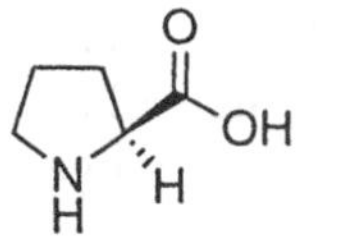

This type of structural modification, however, is not always practical because it tends to depend to some extent on the presence of specific residues in the desired sequence. As a consequence, efforts have been devoted to developing more generally applicable strategies based on the protection of the C- or N-terminals or the side chains with "solubilizing" groups. The Sulfmoc group **23**, described by Merrifield and Bach [106], was the first of a series of solubilizing N^α-protecting groups that were evaluated for use in the purification of peptides. Other removable chromatographic probes, based on derivatives of the Fmoc group, have been reported [107], as have non–Fmoc-based groups [108].

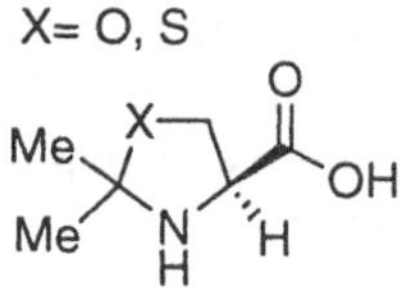

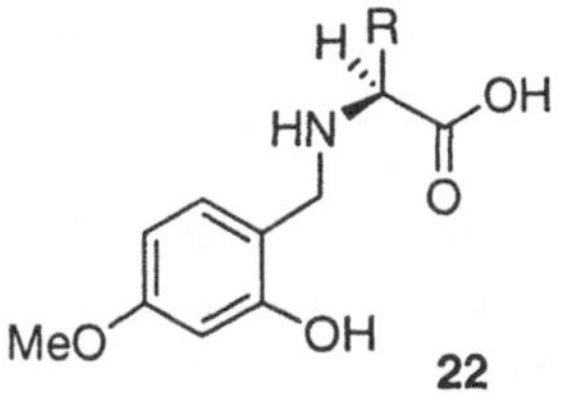

Various possibilities are available for the enhancement of solubility by modification of side-chain protection. An example is the use of picolyl groups as side-chain protection instead of the more common benzyl groups. This produces protected peptides **24** that are generally more polar, have lower retention times on high-performance liquid chromatography (HPLC), and are more soluble in acetic acid [109]. Such peptides may be amenable to purification by cation exchange chromatography, as may those containing Lys residues, which can be reversibly protected [110,111].

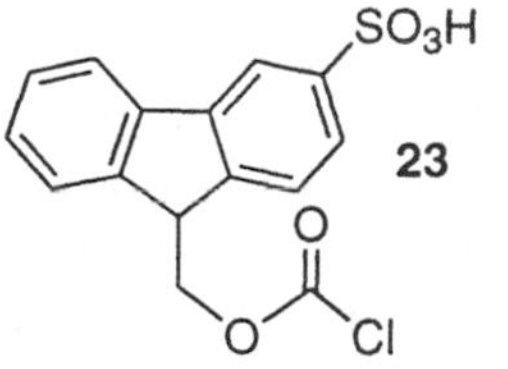

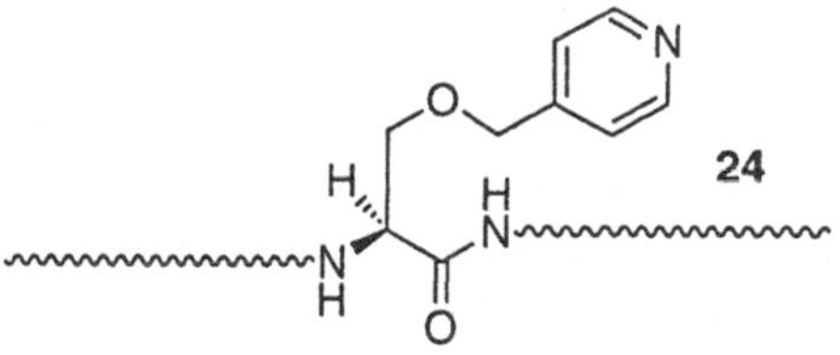

b. *Use of Special Solvents or Additives to Enhance Solubility.* The structural modification of peptides and the use of special protection schemes in order to enhance solubility are promising approaches and are topics of considerable current interest. However, the use of special solvent systems and/or additives has the advantage that it can allow the initially synthesized, unmodified protected peptide segment to be dissolved. Protected peptide segments are usually only poorly soluble in aqueous media and in most of the commonly used organic solvents but many are at least moderately soluble in dipolar aprotic solvents such as DMF, dimethyl sulfoxide (DMSO), or *N*-methylpyrrolidinone (NMP). In addition to these, attention has focused on the use of TFE and HFIP [112] as well as mixed solvent systems such as hexamethylphosphorotriamide (HMPA)–DMSO, HFIP–EtOH–DCM, and HFIP–DMF [113,117]. The solubility of peptides in nonpolar solvents such as tetrahydrofuran (THF) is improved by the addition of inorganic salts [118]. This has been used in the purification of protected peptide segments [119] and is an interesting and potentially useful development in this area of research.

2. Purification Methods

Even in favourable cases, the purification of a synthetic peptide can take considerable time and effort. The purification of protected peptide intermediates is complicated by problems of poor solubility and is often laborious. Assuming that the protected peptide can be dissolved in some suitable solvent, purification can be carried out using one or more of the chromatographic techniques that are applicable to peptide molecules. Frequently, however, quite drastic modifications of the normal operating conditions are necessary if a protected peptide is to be purified to homogeneity [10,120–122]. Two techniques in particular, deserve special mention because of their usefulness in the purification of protected peptide molecules.

a. Gel Filtration.   Gel filtration [123,124] separates molecules on the basis of molecular size—larger molecules elute more rapidly than smaller ones, which are retarded by the stationary phase to a greater extent [125–127]. The technique is used routinely in the purification of peptides but its resolving power is somewhat limited. Purification of protected peptides often requires the use of dipolar aprotic solvents such as DMF, DMSO, or NMP.

b. Reversed-Phase High-Performance Liquid Chromatography.   The importance of HPLC in the peptide field cannot be overemphasized and the development and refinement of this technique have exerted a profound influence on peptide analysis, characterization, and synthesis [128]. In reversed-phase HPLC, retention of the sample by the column occurs through hydrophobic interaction with the column support. Elution is carried out in such a way as to decrease the ionic nature or to increase the hydrophobicity of the eluant so that it competes for the hydrophobic groups on the column.

The analysis of protected peptides by reversed-phase HPLC often requires injection into the system as solutions in a polar aprotic solvent such as DMF. Perfectly acceptable analytical chromatograms are usually obtained with the loading solvent appearing as a broad peak at the beginning of the chromatogram. For the purification of protected peptide segments at the preparative level, addition of another solvent such as 2-propanol or DMF to the eluants may be necessary to ensure adequate solubility [13,120,121,129,130]. Some increase in viscosity and column pressure is observed and ultraviolet (UV) detection is also complicated owing to the strong absorbance of DMF between 220 and 270 nm. If a suitable chromophore such as the Fmoc group or the Tyr residue is present, detection at 300 or 280 nm may be possible; if not, the monitoring can be carried out by analytical HPLC. Some hydrophobic segments that are soluble in TCM or TCM–TFE mixtures can be purified by reversed-phase HPLC, loading a TCM solution onto the column and eluting successively with MeOH–H$_2$O, MeOH, and MeOH–TCM. For protected peptides having only a free C-terminal carboxylic acid, an ion-pairing reagent is often not required and no deleterious effect on recovery of chromatographic efficiency is observed. In cases in which separation efficiency is improved by using a reagent such as propionic acid, or HOAc, care must be taken to remove them completely from the peptide prior to coupling; otherwise capping of amino groups may be observed. Purification of His(Trt) peptides should be done in the presence of 0.1% pyridine in order to avoid loss of the Trt group. Such losses can occur even when ACN–H$_2$O mixtures without a carboxylic acid modifier are used [131].

Normal-phase LC is a less potent alternative [132,133] and has been recommended only when impurities in the crude were derived from the loss of side-chain protecting groups. If the crude product exhibits a significant amount of deletion peptides, then purification by reversed-phase HPLC techniques is required [131]. In this case, C$_8$ or even C$_4$ columns are the most suitable for peptides of ~10 residues. For longer peptides the less polar diphenyl-based phase is recommended.

Other techniques that can be useful for the purification of protected peptide segments include normal-phase column chromatography on silica gel and medium-pressure liquid chromatography (MPLC) in both normal and reversed-phase modes [2,3].

3. Determination of Covalent Structure

The characterization of a peptide after purification is necessary to establish whether the desired peptide and not some structural modification has been isolated. Characterization of peptide molecules is not always straightforward owing to the particular type of structural complexity that these molecules present. Characterizations are best realized by mass spectrometry, especially in the fast atom bombardment (FAB) mode [132–136], although electrospray and matrix-assisted laser desorption–time of flight (MALDI-TOF) modes can also be useful, as well as by nuclear magnetic resonance (NMR) spectroscopy. In the latter technique, the interpretation of the NMR spectra of large protected peptides can be complicated. One-dimensional spectra are not normally sufficient and more sophisticated two- and even three-dimensional experiments are usually required for rigorous assignments [137]. Extensive use should be made both of mass spectrometric and NMR techniques for a comprehensive characterization of protected peptide segments.

C.   Solid-Phase Coupling of Protected Peptide Segments

Although the coupling of a protected peptide segment on a solid support is analogous to that of a single amino acid, the former is more demanding for several reasons. There is a large risk of epimerization at the C-terminus and consequently protected peptides are often chosen to have either Gly or Pro at this position. In linear SPPS, excesses of urethane-protected amino acids can be used to drive the coupling reaction to completion. Alternatively, in CSPPS the use of excesses of a protected peptide segment whose synthesis may have required considerable investment in time and effort must be considered carefully. There is also the question of solubility. It is often difficult to achieve acceptably high concentrations of a peptide segment in the coupling medium. This tends to lower the yield of the coupling reaction and it is necessary to repeat until coupling is complete and the segment is consumed.

Even with these difficulties, protected peptide segments can be coupled on solid supports with good results but to maximize the efficiency of the process the following factors must be considered.

1.   The Solid Support

The nature of the solid support can have a significant effect on the outcome of segment coupling reactions. Not surprisingly, the solid supports that give the best results are those that also perform best in standard SPPS. Studies have shown that polystyrene and polyacrylamide resins give higher segment coupling yields and shorter reaction times than other types of support, such as controlled pore glass [138]. PEG-PS also gives similarly good results in peptide segment couplings [60].

In addition to the resin type, the degree of functionalization is important. The optimal level for segment coupling is lower than that normally used in linear SPPS, with values in the range 0.04–0.2 mEq/g giving the best results [139].

2.   Synthesis Strategy

In addition to normal chain elongation in the $C \rightarrow N$ direction, solid-phase segment coupling has been carried out by $N \rightarrow C$ elongation [140–144]. For stepwise SPPS $N \rightarrow C$ chain elongation suffers from the disadvantage that the risks of epimerization of the C-terminus of the resin-bound peptide in the amino acid coupling steps are much greater. However, in CSPPS, although activation of the resin-bound carboxyl groups may be correspondingly more difficult, epimerization of the activated C-terminal amino acid is no more a risk than it is in standard C $\rightarrow$ N elongation. One of the possible advantages of N $\rightarrow$ C chain elongation in segment coupling is that, because the activated C-terminus of the peptide remains attached to the resin, the excess of protected peptide in solution may be recycled. Nonetheless, N $\rightarrow$ C chain elongation has not been established and is rarely used; almost all SPPS is carried out in the C $\rightarrow$ N direction.

When designing a convergent solid-phase synthesis strategy for a particular peptide, one of the factors that must be considered is the size of the segments to be coupled. If the criterion of choosing only protected peptide segments having Pro or Gly at the C-terminus is strictly observed, inevitably, some segments will be rather long. The danger of working with long protected peptide segments is that they may be too insoluble to be purified or to provide sufficiently concentrated solutions for the coupling reaction [145]. A compromise must be established between a longer, poorly soluble segment having Gly or Pro at the C-terminus or a shorter, more soluble segment that lacks these residues at the C-terminus and will be much more prone to epimerize on coupling. A related consideration is that peptide segments should be chosen to have Pro at the C-terminus or in the middle of a sequence. As previously discussed (See Section I.B.1.a), Pro in the middle of a protected peptide segment can confer better solubility, whereas at the C-terminus it is resistant to epimerization. One possibility is to choose segments having Pro at the C-terminus and to introduce tertiary amide bonds into the peptide via Hmb peptide bond protection. (See Section I.D.2.)

3. Incorporation of the First Segment

The first segment of the target molecule is perhaps best synthesized directly on the resin by linear SPPS. Although a previously synthesized, purified, and characterized peptide segment can, in principle, be attached to the resin, yields are not always high, especially when this must be done by esterification. An example of this method may be found in Section I.D.4. The advantage of stepwise synthesis is that it is rapid, but a drawback is that the peptide cannot be purified. If the segment is quite long, the possibility of having deletion or truncated peptides in the C-terminal region is increased. If the segment is relatively short, however, modern SPPS usually allows it to be constructed in an essentially pure state.

Irrespective of the incorporation of the first segment, it is often accomplished in a strategy to reduce the substitution level of the resin [52,146]. For stepwise synthesis the first amino acid can be loaded using less than the amount required to react with all of the active sites on the resin. The resin is then acetylated to block the remaining amino groups and synthesis is continued under normal conditions. If a protected peptide is to be loaded, the substitution level of the resin can be reduced when an internal reference amino acid is incorporated.

4. Coupling Methods

Even assuming that a reasonable excess of the segment to be coupled can be used and that there are no solubility problems, a segment coupling reaction may require many hours or several days for completion. It is important, therefore, to treat each coupling reaction independently. If good results are to be obtained, careful attention is required to the different variables such as reagents, reaction time, concentration of the soluble components, temperature, and efficiency of agitation of the resin.

Coupling methods for protected peptide segments have evolved in an analogous manner to those used for the coupling of single amino acids. Early procedures included the azide and oxidation–reduction protocols. These were superseded by the use of carbodiimides, usually in the presence of additives such as hydroxysuccinimide (HOSu) or 1-hydroxybenzotriazole (HOBt). Contemporary methods for effecting the solid-phase coupling of peptide segments are based on the use of phosphonium or uronium salts in the presence of HOBt. New reagents based upon 1-hydroxy-7-azabenzotriazole (HOAt) [147,148] such as N-[(dimethylamino)-1H-1,2,3-triazolo-[4,5b]pyridin-1-ylmethylene]-N-methylmethanaminium hexafluorophosphate N-oxide (HATU) show much promise [131,149] and have been used successfully in solid-phase segment couplings. (See Section I.D.3.)

5. Side Reactions

Perhaps the most important side reaction is epimerization at the activated C-terminal amino acid of the segment. For Gly, epimerization is not possible and is thought to be minimal for Pro under normal coupling conditions. For all other amino acids, however, epimerization may occur to a greater or lesser extent [150,151]. Peptide segments with C-terminal amino acids that are neither Gly nor Pro must, therefore, be coupled in such a way that epimerization is minimized. Furthermore, analytical methods that are sufficiently sensitive to detect the presence of diastereomeric peptides in quite small quantities are necessary in order to be able to quantify the amount of epimerization that has occurred in a given segment coupling [152–154].

Other side reactions may have their origins in the instability of side-chain protecting groups during coupling reactions that may take considerable time to go to completion. One example is the formation of pGlu when Glu is the *N*-terminal amino component. The stability of side-chain protecting groups to the coupling conditions should be carefully checked.

6. Monitoring of the Coupling Reaction

Monitoring of solid-phase segment coupling reactions is not necessarily straightforward. The qualitative ninhydrin test is useful for determining whether or not unreacted amino groups remain on the resin. As the length of the peptide chain increases, however, the test becomes less sensitive. Amino acid analysis can also be used to determine the extent of segment coupling reactions. However, as the peptide attached to the solid support becomes longer, the information becomes more limited. It can be difficult to judge the extent of incorporation of a new segment if it contains residues that are already present in the peptide. If several of these residues are already present, the problem is correspondingly more difficult. When both of these techniques give ambiguous results, solid-phase Edman sequencing can be a useful and accurate alternative for determining the yields of segment couplings [60,155,156]. Alternatively, during the course of a synthesis the weight of the peptide-resin increases and, in principle, can be used to monitor segment couplings. This method has the advantage of simplicity and its merits are documented [157,158]. However, it is often not a reliable or accurate guide to coupling yields. It should be used only as supporting evidence when the coupling yield is determined using other techniques. The best method for monitoring segment coupling reactions is to remove an aliquot of the resin and to release the peptide from the support. The product can then be analyzed by one or more physical techniques such as HPLC, capillary electrophoresis, mass spectrometry, or nuclear magnetic resonance.

D. Examples of Convergent Solid-Phase Peptide Synthesis

1. Prothymosin α

Prothymosin α, a protein consisting of 109 residues, is one of the largest peptides that has been synthesized using a CSPPS strategy [159]. Its synthesis demonstrates the potential of CSPPS for the provision of large peptides and proteins. The segments spanning the 1–75 sequence of the protein were synthesized on trityl resin **1** and, after cleavage, were purified by chromatography on silica gel, eluting with TCM–DCM mixtures. The *C*-terminal segment, consisting of 34 amino acid residues and corresponding to the 76–109 sequence of the protein, was synthesized by stepwise Fmoc/*t*Bu-SPPS. The initial functionalization level of resin **1** was between 0.4 and 0.6 mEq/

g. The protected peptide segments were then coupled to the 76–109 peptide-resin, as shown in Scheme 2.

A fivefold excess of each peptide segment–HOBt–N,N'-dicyclohex-ylcarbodiimide (DCC) (1:1.5:1) in DMSO was used and reactions were complete after 6–18 h. The segment corresponding to the 1–10 sequence of the protein could be coupled in only 55% yield. All other segment couplings, however, proceeded in very high yields as judged by HPLC analysis of material obtained by cleavage of aliquots of peptide-resin taken throughout the synthesis. Prothymosin α was obtained in 11% overall yield and the synthetic material had biological activity identical to that of the natural protein.

2.   The 3-Repeat Region of Human Tau-2

The Tau protein is associated with Alzheimer's disease, although its precise role in the onset and development of the condition is unknown. The 3-repeat region of this protein, from Asp[158] to Leu[251], a 94-residue peptide, has been synthesized by CSPPS using the versatile Hmb amide bond protecting group. As with the protection of selected amide bonds dramatically reducing the level of on-resin aggregation, protected peptide segments with Hmb-derived protection at strategic points are much more soluble. This increase in solubility has two important consequences. First, the purification of the protected peptide segments is facilitated and conventional chromatographic methods can be used. Second, but equally important, the increased solubility allows more concentrated solutions to be incorporated in segment coupling reactions. This promotes more efficient couplings with shorter reaction times and higher yields.

The problem of the incorporation of the first segment of the protein onto the solid support in a pure state was solved in a very elegant manner as outlined in Scheme 3. Fmoc-β-Ala was esterified onto the pentafluoro-phenol active ester of handle 2, and this performed handle was attached to a polyamide resin giving solid support 25. After removal of the Fmoc group, the HMPAA handle [160] incorporating Fmoc leucine was attached, giving double-handle-containing solid support 26. The first segment was then elaborated on this support, incorporating Hmb amide bond protection at Lys[245]. When chain elongation was complete the peptide-resin was acetylated, giving 27. Treatment with 0.75% TFA in DCM selectively cleaved the trityl-based handle, giving the protected peptide HMPAA ester 28. Good solubility of this species was ensured by its Hmb amide bond protection. After purification by chromatography on silica gel, it was attached to a polyamide resin, giving 29, and the synthesis was carried out to completion. The substitution level of the peptide resin at this point was 0.048 mmol peptide/g peptide-resin.

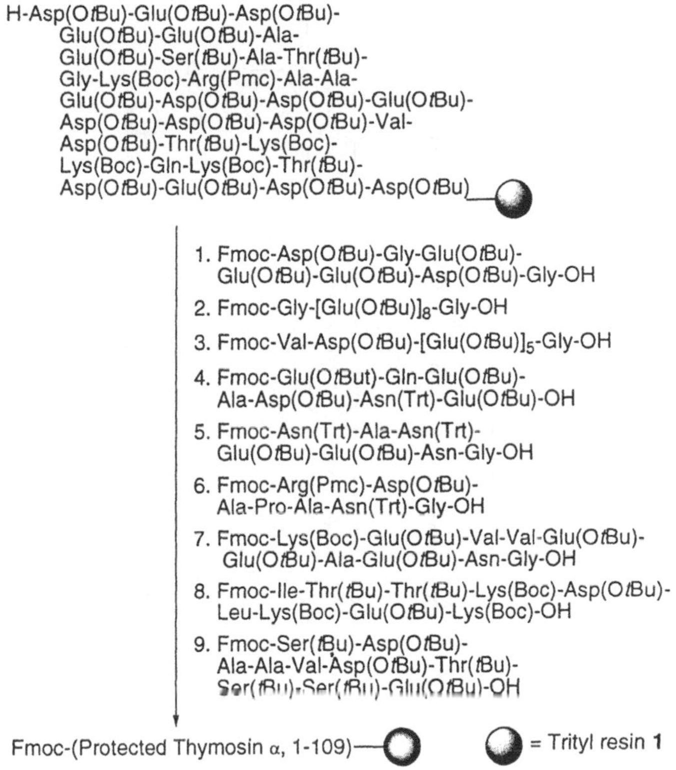

H-Asp(O*t*Bu)-Glu(O*t*Bu)-Asp(O*t*Bu)-
    Glu(O*t*Bu)-Glu(O*t*Bu)-Ala-
    Glu(O*t*Bu)-Ser(*t*Bu)-Ala-Thr(*t*Bu)-
    Gly-Lys(Boc)-Arg(Pmc)-Ala-Ala-
    Glu(O*t*Bu)-Asp(O*t*Bu)-Asp(O*t*Bu)-Glu(O*t*Bu)-
    Asp(O*t*Bu)-Asp(O*t*Bu)-Asp(O*t*Bu)-Val-
    Asp(O*t*Bu)-Thr(*t*Bu)-Lys(Boc)-
    Lys(Boc)-Gln-Lys(Boc)-Thr(*t*Bu)-
    Asp(O*t*Bu)-Glu(O*t*Bu)-Asp(O*t*Bu)-Asp(O*t*Bu)⟷◯

           1. Fmoc-Asp(O*t*Bu)-Gly-Glu(O*t*Bu)-
              Glu(O*t*Bu)-Glu(O*t*Bu)-Asp(O*t*Bu)-Gly-OH

           2. Fmoc-Gly-[Glu(O*t*Bu)]$_8$-Gly-OH

           3. Fmoc-Val-Asp(O*t*Bu)-[Glu(O*t*Bu)]$_5$-Gly-OH

           4. Fmoc-Glu(O*t*But)-Gln-Glu(O*t*Bu)-
              Ala-Asp(O*t*Bu)-Asn(Trt)-Glu(O*t*Bu)-OH

           5. Fmoc-Asn(Trt)-Ala-Asn(Trt)-
              Glu(O*t*Bu)-Glu(O*t*Bu)-Asn-Gly-OH

           6. Fmoc-Arg(Pmc)-Asp(O*t*Bu)-
              Ala-Pro-Ala-Asn(Trt)-Gly-OH

           7. Fmoc-Lys(Boc)-Glu(O*t*Bu)-Val-Val-Glu(O*t*Bu)-
              Glu(O*t*Bu)-Ala-Glu(O*t*Bu)-Asn-Gly-OH

           8. Fmoc-Ile-Thr(*t*Bu)-Thr(*t*Bu)-Lys(Boc)-Asp(O*t*Bu)-
              Leu-Lys(Boc)-Glu(O*t*Bu)-Lys(Boc)-OH

           9. Fmoc-Ser(*t*Bu)-Asp(O*t*Bu)-
              Ala-Ala-Val-Asp(O*t*Bu)-Thr(*t*Bu)-
              Ser(*t*Bu)-Ser(*t*Bu)-Glu(O*t*Bu)-OH

Fmoc-(Protected Thymosin α, 1-109)⟷◯   ● = Trityl resin **1**

Scheme 2   Convergent solid-phase synthesis of prothymosin α.

    All of the other protected peptide segments required for the synthesis were also prepared on trityl-based solid supports of the type **25** (except that the *C*-terminal amino acid of the segment in question was incorporated rather than β-Ala). In each case, following chain elongation, the peptide-resins were acetylated so that backbone amide protection was in the form of AcHmb groups. Cleavage of the protected peptides occurred in high yield on treatment of the peptide-resins with 0.75% TFA in DCM.

    Removal of the Fmoc group of peptide-resin **29** with piperidine also deacetylated the resin-bound backbone protecting groups, giving peptide resin **30**. If Fmoc removal is performed to leave the AcHmb groups unaffected, *O*- to *N*-acyl migration can occur on segment coupling, effectively capping the resin bound N^α-amino group and causing low yields. Segment couplings were carried out with 2 equivalents of peptide segment using benzotriazol-1-yl-*N*-oxy-tris(dimethylamino)phosphonium hexafluorophos-

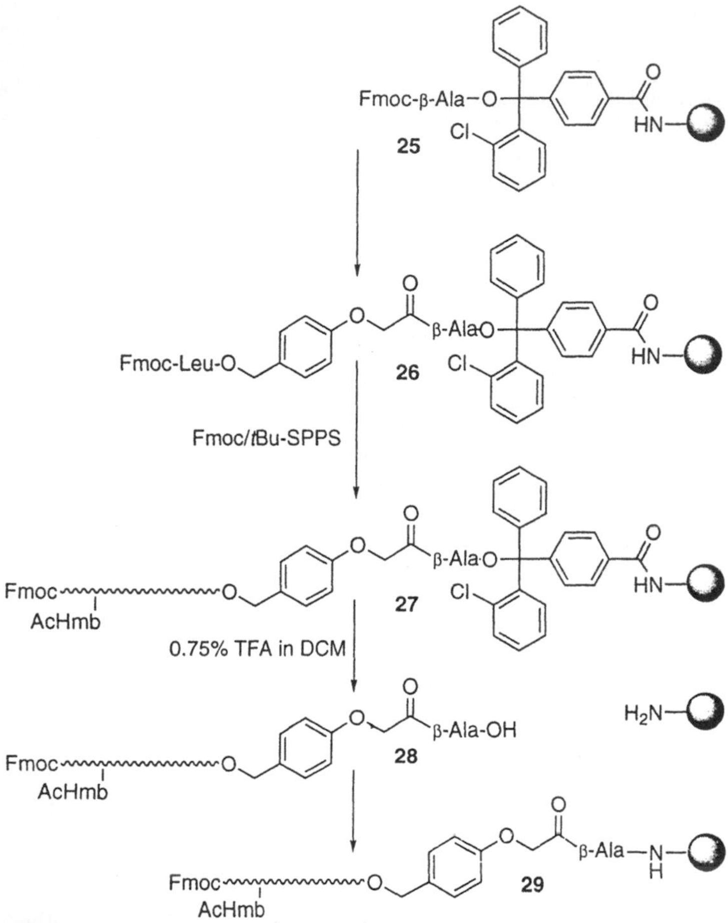

Scheme 3 Attachment of the first protected peptide to the reins in the synthesis of the 3-repeat region of human tau-2.

phate (BOP) reagent in a minimum volume of DMF and were usually complete within 6 h. Even quite large (up to 21 residues) protected peptide segments could be coupled in high yield. After each coupling step the resin was acetylated, capping any residual resin-bound amino-groups and re-acetylating the resin-bound Hmb groups. Once chain assembly was complete, treatment of peptide-resin **31** with piperidine removed the N^α-Fmoc group and deacetylated all backbone protection. Acidolytic cleavage with TFA containing scavengers then gave the crude protein fragment **32**, which

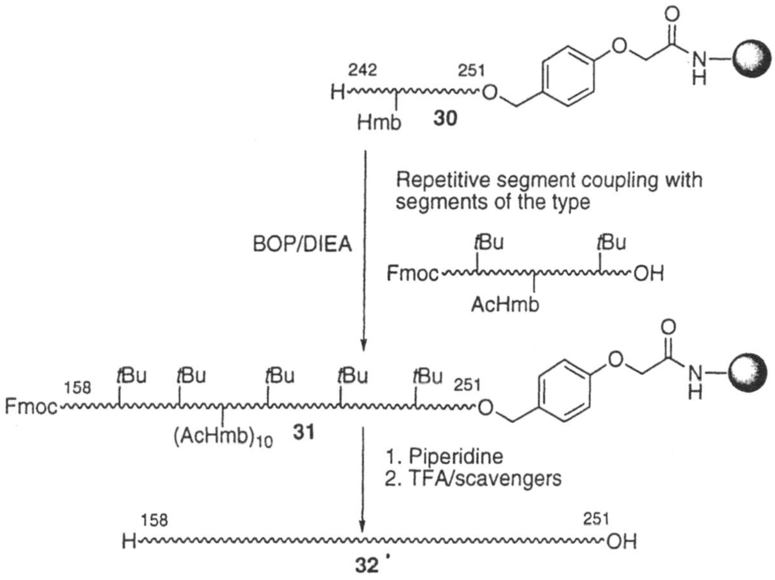

Scheme 4 Convergent solid-phase synthesis of the 3-repeat region of human tau-2.

was sufficiently soluble to allow its purification by reversed-phase HPLC in an H_2O and ACN mixture. The synthesis is outlined in Scheme 4.

The use of backbone-amide bond protection in CSPPS provides a possible solution to the problem of poor segment solubility, allowing the facile purification of protected peptide segments and improving the yields in coupling reactions. These factors clearly surpass the slight disadvantage of the modified coupling protocols required for the incorporation of Hmb-protected amino acids into peptides.

3. The *N*-Terminal Repeat Region of γ-Zein

An example from the authors' own laboratory [131] illustrates the particular suitability of CSPPS for the preparation of peptides with repetitive sequences. The *N*-terminal region of the maize γ-zein protein consists of eight conserved repeats of the sequence Val-His-Leu-Pro-Pro-Pro. If the *N*-terminal region of this protein is synthesized by CSPPS, then only the one protected peptide segment, corresponding to the monomer, need be prepared. Sequential coupling of the monomer then gives the desired protein fragment. Its synthesis is outlined in Scheme 5.

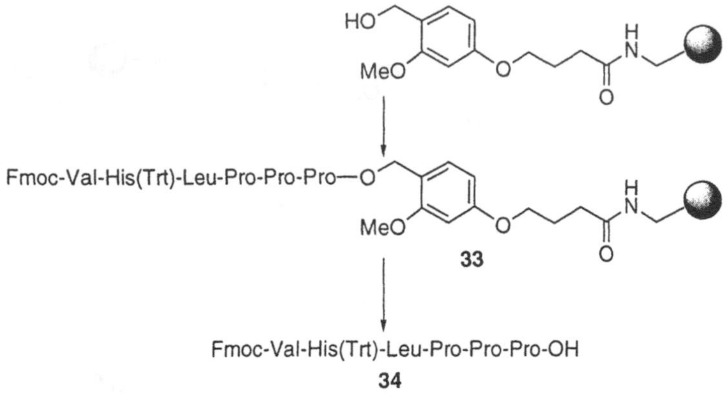

Scheme 5   Solid-phase synthesis of the Fmoc-Val-His(Trt)-Leu-Pro-Pro-Pro-OH monomer.

The peptide was elaborated on a 4-methylbenzhydrylamine resin incorporating the highly acid-labile HMPB handle **4**. Three residues of Phe were incorporated between the solid support and the handle as an internal standard. The first Pro residue was anchored by esterification using *N,N'*-diisopropylcarbondiimide (DIPCDI) in the presence of 4-dimethylamino-pyridine (DMAP) in a yield greater than 98%. The second two Pro residues were incorporated as the Fmoc-Pro-Pro-OH dipeptide because sequential incorporation led to unacceptably high amounts of diketopiperazine formation [161,162]. (See Chapter 2, Section IV.B.1.a.) The remaining amino acids were added using standard coupling protocols. Treatment of peptide-resin **33** with 1% TFA in DCM led to cleavage of peptide **34** in 91% yield. This peptide was purified by reversed-phase MPLC using H_2O and ACN eluants containing 0.1% pyridine in order to avoid deprotection of the His(Trt) group. Sequential couplings of this protected segment then allowed the repetitive sequence of the *N*-terminal region of maize γ-zein to be synthesized, as shown in Scheme 6.

The Val-His(Trt)-Leu-Pro-Pro-Pro sequence was first synthesized on an aminomethyl resin incorporating the 3-(4-hydroxymethyl)-phenoxy)-propionic acid handle [27] and three Phe residues as internal standard, **35**. Peptide-resin **36** was synthesized using the same methods as those for the protected segment **34**. Removal of the *N*-terminal Fmoc group from **36**, followed by HATU-mediated segment coupling and a capping step, then gave the dimer **37** (*n* = 2). Higher oligomers were synthesized by repetition of these steps until the desired octamer had been produced. Removal of the Fmoc group and cleavage with TFA in DCM (1:1) containing 3% H_2O then furnished the desired repetitive protein fragment, **38**.

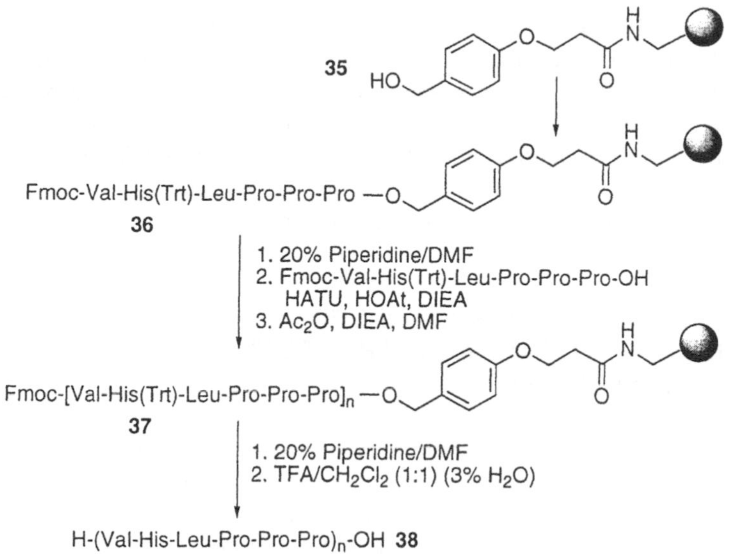

Scheme 6 Convergent solid-phase synthesis of the N-terminal repeat region of γ-zein.

4. Typical Experimental Procedure of the Solid-Phase Coupling of Protected Peptides

This procedure is adapted from Refs. 131 and 155. For the couplings that occur for up to 4 h, a polypropylene syringe fitted with a polythene disk and stopcock can be used. Occasional manual stirring may be sufficient. However, for longer couplings, a silylated screw-cap tube reaction vessel with a Teflon-lined cap, a sintered glass frit, and a stopcock and continuous mechanical shaking is preferable. Coupling should first be tested on a small amount of peptide-resin (2–3 μmol) in a 1-mL syringe before carrying out the large-scale experiment. The process of dissolving the protected peptide may need to be started several hours before initiating the coupling reaction and may require continuous agitation or ultrasound and even gentle heating. The most concentrated protected peptide solution possible should be prepared. The excess of peptide segment required depends on the coupling, but between 2 and 5 equivalents are most widely used. Protected peptide resins are preswollen in DMF (1 mL/0.1 g of resin, three times for 10 min plus once for 1 h) After removal of the N^{α}-protecting group with the appropriate reagent [TFA–DCM (4:6, v/v) for Boc chemistry or piperidine–DMF (1:4, v/v) for Fmoc chemistry, with an extended reaction time of 1 h] and washings, a mixture of the protected peptide, coupling reagent (carbodiimide or aminium or phosphonium salt, 1 equivalent with respect to the peptide), HOBt or HOAt (1 equivalent), and base (1 equivalent), if aminium or phosphonium salts are used, is added to the resin at 0°C. If carbodiimide-based couplings are carried out, some researchers prefer to perform a step of preactivation of the protected peptide with the carbodiimide and HOBt or HOAt for 10–60 min at 0°C. Acetylation, when required, is done with acetic anhydride–DIEA (1;1 v/v, 30 equivalents) in the minimum amount of DMF.

II. SYNTHESIS OF NATIVE PROTEINS BY CONDENSATION OF MINIMALLY PROTECTED OR UNPROTECTED PEPTIDES

The problems caused by the poor solubility of protected peptide segments, inherent both in the CSPPS strategy and in classical synthesis in solution, stimulated the investigation of other convergent strategies based on the coupling of minimally protected or even completely unprotected peptide segments. These are much more soluble in aqueous solvent systems, allowing them to be purified to homogeneity by the standard chromatographic techniques used for the purification of peptides. The coupling of such peptide segments is a delicate and demanding project because the multitude of free functional groups can lead to various side reactions that can compete with the desired amidation. In order to achieve this process chemically, the *C*-terminal carboxylic acid on one of the peptides must be activated without affecting any of the other carboxyl groups present in either peptide. In addition, all other amino groups except the one involved in coupling must either be blocked or deactivated; otherwise they will participate in acylation reactions, leading to the formation of branched peptides. However, methods are available and the approach is one of the most promising general strategies currently available for the assembly of proteins.

A. Coupling of Minimally Protected Peptide Segments

The most favorable method of the ones currently receiving attention appears to be that of Aimoto and coworkers [163–169], which uses the specific activation of *C*-terminal thioesters of minimally protected peptide segments for differentiating the carboxyl group of the *C*-terminus from those of the side chains. The chemistry involved is based on previous work by Blake, Li, and Yamashiro [170–177]. The fundamental strategy is to divide the target molecule into two (sometimes more) large segments, one of which has a thiocarboxyl group at the *C*-terminus. Both of these segments can be constructed using Boc/Bzl SPPS; thioester peptide-resin linkages are stable to the conditions used. They are then detached from the resin with strong acid so that most side-chain protecting groups are also removed, ensuring that even quite large segments are soluble in aqueous solvent systems. The approach is outlined in Scheme 7.

The *S*-alkyl thioester of glycine **40** is synthesized by reaction of the thioacid **39** with Boc-Gly-OH and is then loaded onto a *p*-methylbenzhydrylamine resin using conditions similar to those used for loading a normal amino acid. Boc/Bzl peptide synthesis is then carried out on **41** until the desired sequence has been assembled. The last amino acid is incorporated with N^α-2,2,2-trichloroethoxycarbonyl (Troc) or isonicotinoyloxycarbonyl (*i*Noc) protection (both stable to hydrogen fluoride treatment), giving the peptide-resin **42**. After hydrogen fluoride–mediated cleavage, the peptide *S*-

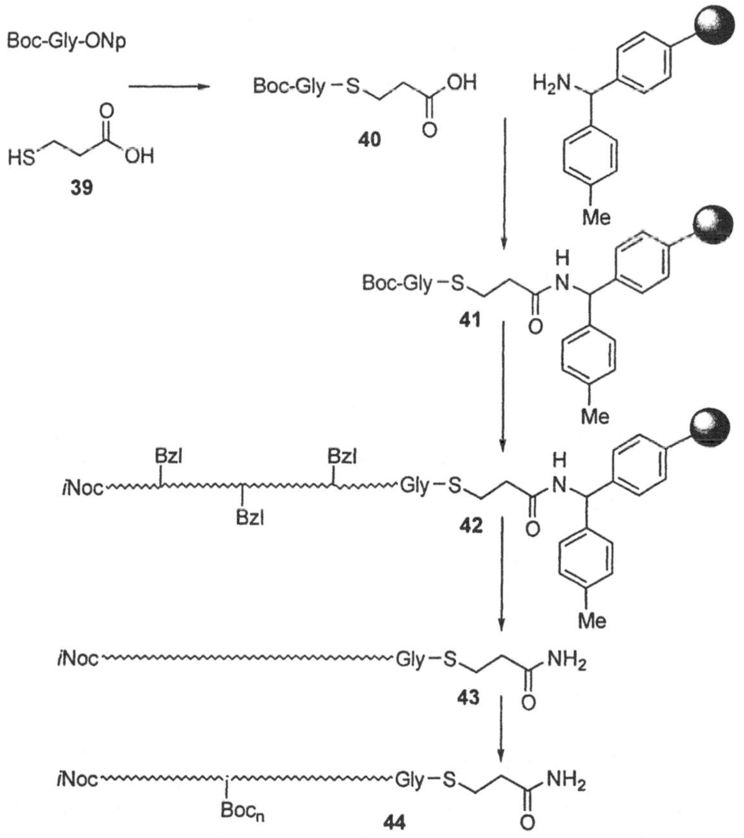

Scheme 7 Synthesis of partially protected peptide *C*-terminal thioesters.

alkylthioester **43** is treated with *tert*-butylsuccinimidyl carbonate (BocOSu) to protect the side chains of any Lys residues it might contain, giving minimally protected peptide **44**. This can then be selectively activated at the *C*-terminus using silver salts and converted to the corresponding *p*-nitrophenyl active ester for coupling with the N^α-amino group of another minimally protected peptide segment.

Such an approach has been applied successfully in the synthesis of the 90-amino-acid DNA-binding protein in *Bacillus stearothermophilus* [165].

B. Coupling of Unprotected Peptides

If the bond used to join two peptides is not an amide, then, at least in principle, protecting groups can be eliminated. Each of the segments can be designed so that each has a unique and complementary reactive functional group that remains completely unprotected. The peptides can then be joined by a specific chemical reaction, although very strict selectivity is required. The linking of peptide segments via bonds that are not amides is often called "chemical ligation" [178], although "ligation" or "chemical ligation" is also used to describe the union of free peptides by amide bond formation, especially performed enzymatically. Here we will interpret chemical ligation as the covalent joining of peptide segments by any bond that is not an amide.

The chemical ligation of peptide segments can be accomplished to form stable nonnatural structures in which the peptide backbone still consists predominantly of amide bonds but now incorporates some other covalent linkage at intervals coinciding with the ligation sites. Such structures are referred to as "backbone engineered" and can exhibit the full biological activity of the natural protein. The most important of these methods for the synthesis of backbone-engineered proteins were developed by Schnölzer and Kent [179,180] and by Offord and Rose and colleagues [181–184].

The chemical ligation strategies demonstrate that large unprotected peptides can be joined chemoselectively in high yield giving chemically well-defined molecules [185–189]. This has prompted the investigation of new methods for the formation of amide bonds between peptides so that native structures can be produced. Contemporary work has focused on the use of a prior chemical ligation step to bring the peptides together before amide bond formation occurs.

1. Template-Assisted Coupling

The original concept was outlined several decades ago independently by Brenner et al. [190–195] and Wieland et al. [196] but it has been applied only recently in complex peptide synthesis. The conversion of "template-assisted'' amide bond formation into a practical general method for peptide synthesis has proved to be very challenging and at present has not been achieved in an entirely satisfactory manner despite the application of various approaches [197–199]. Among the problems that must be overcome for the method to be useful for the coupling of large peptide segments is attaching components on the template. This could conceivably be just as difficult to achieve as the coupling of the two segments under more conventional conditions.

The most promising current variation of the strategy is that proposed by Kemp [200] known as "thiol capture ligation." The method involves amide bond formation between two segments, brought together on 4-hydroxy-6-mercaptodibenzofuran template **45**, by chemical ligation through disulfide formation [201,202]. Both of the peptide segments involved are synthesized by solid-phase methods; the second segment must have an N-terminal Cys residue in order to permit its chemical ligation by disulfide formation to the peptide-template. With the two peptides attached on the template, as in **45**, base-catalyzed intramolecular acyl transfer can occur, forming an amide bond that leads to desired target sequence **46**. Any remaining protecting groups including the template are then removed before target molecule **47** is purified. The intramolecular acyl transfer is shown in Scheme 8.

Thiol capture ligation is an elegant approach to complex peptide synthesis. It has been proved in model systems, including those with unprotected His, Lys, and Arg residues [203,204], and in the construction of a complex 39-residue peptide [205]. There are many advantages of this type of ligation as opposed to conventional segment coupling methods; for example, the solubility of the intermediates in aqueous solvent systems is improved owing to the lack of side-chain protection. Second, the intramolecular acyl transfer reaction is fast, usually being complete within a few hours and sometimes within minutes. Third, the use of excesses of one of the segments is avoided. However, the strategy does not constitute a general approach to peptide synthesis because it is applicable only in cases in which there is a convenient distribution of Cys residues. In addition, the chemistry is rather intricate and some, albeit minimal, protection of certain amino acid side chains is necessary. Because the technique has not been extensively incorporated in synthetic design, it is unknown whether this strategy is a general method for coupling peptide segments.

2. Native Chemical Ligation

Kemp's thiol capture ligation demonstrates that amide bond formation, even between large peptide segments with only moderate activation, can be rapid

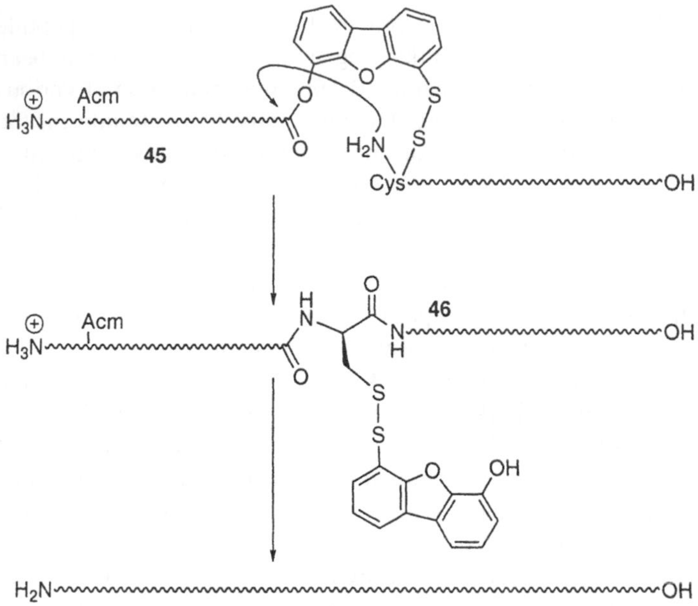

Scheme 8 Template-assisted intramolecular acyl transfer to give an amide bond between the two peptide segments.

when the reacting amino and carboxyl termini are held in close proximity. An alternative approach, which does not rely on the use of a template, has been devised [206–209]. This method involves the chemical ligation of two unprotected segments by thioester formation, followed by the creation of an amide bond at this position. For this strategy, one of the peptides must have cysteine as its *N*-terminus and the other a thioester *C*-terminus. In model experiments, using H-Cys-OH and pentapeptide thiobenzyl ester **48** (R = Bzl), initial chemical ligation gives cysteine thioester **49**. Amide bond formation to yield hexapeptide **50** then occurs. The proposed mechanism of such coupling is shown in Scheme 9.

The initially formed chemical ligation product **49** spontaneously rearranges, giving an amide bond at the ligation site, hence the name "native chemical ligation." The rapid rearrangement of **49** is presumably due to the favorable geometric arrangement of the α-amino group with respect to the thioester. Alternative chemical methods for the formation of the initial ligation product have also been investigated by Tam et al. [210–212].

Such amide bond formation can also be applied to the coupling of larger peptides. The *N*-terminal segment is synthesized on a suitable solid

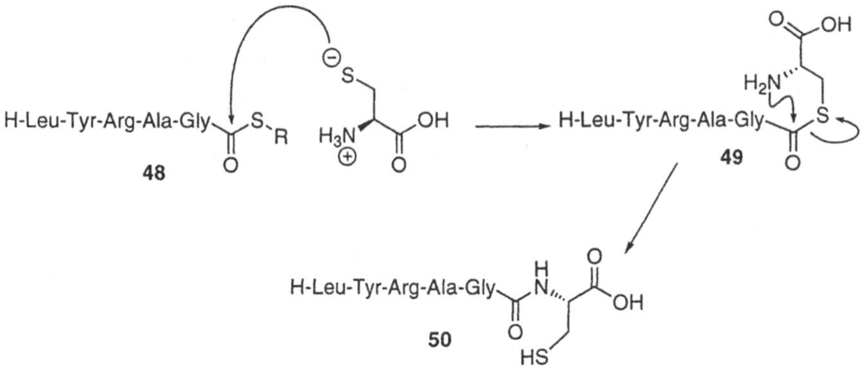

Scheme 9 Native chemical ligation.

support and has a thioglycine residue at its *C*-terminus. After cleavage, the thiocarboxyl peptide is allowed to react with an alkyl halide to give the required thioester [213]. Model studies demonstrated that the use of better thioester leaving groups led to faster ligation reactions [214]. The second segment, which must have Cys as its *N*-terminus, as in Kemp's thiol capture ligations, can be synthesized by standard solid-phase procedures. Once purified, the segments are ready for chemical ligation. Amide bond formation occurs spontaneously as in the model studies outlined earlier. Protection of functional groups other than the *C*-terminus thioester is unnecessary, and the method tolerates the presence of unprotected Lys or Cys residues in either of the segments to be coupled. One of the molecules that native chemical ligation has been used to synthesize is interleukin-8, a 72-amino-acid protein [206]. The synthesis is outlined in Scheme 10.

The *N*-terminal peptide segment, which has two internal unprotected Cys residues in addition to unprotected Lys residues, was synthesized on a suitable solid support. After cleavage from the resin, treatment with benzyl

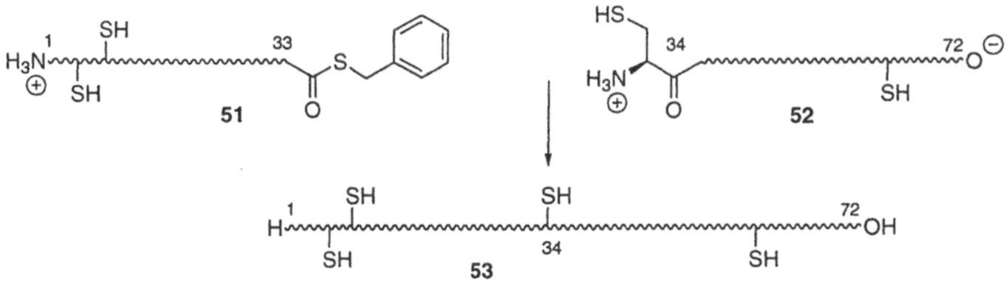

Scheme 10 Synthesis of human interleukin-8 by native chemical ligation.

bromide furnished thioester **51**. The second segment **52** was assembled by standard SPPS and both peptides were purified by reversed-phase chromatography. Native chemical ligation of **51** and **52** was accomplished at pH 7.6 in phosphate buffer in the presence of 6 M guanidine hydrochloride, affording desired sequence **53**. Oxidation followed by purification gave material identical to natural interleukin-8.

Muir and coworkers [215] demonstrated that native ligation can be carried out on solid supports, eliminating the need to manipulate peptide thioacid and peptide thioester intermediates in solution. This in turn reduces handling losses and improves overall yield. The key to solid-phase native chemical ligation is the global detachment of all peptide protecting groups while maintaining a thioester anchorage to the resin. This step requires fine control of the conditions of the hydrogen fluoride deprotection reaction. The approach is summarized in Scheme 11.

Optimized Boc/Bzl peptide synthesis is carried out on a polyethylene glycol–polyamide resin to which the first amino acid is anchored as a thioes-

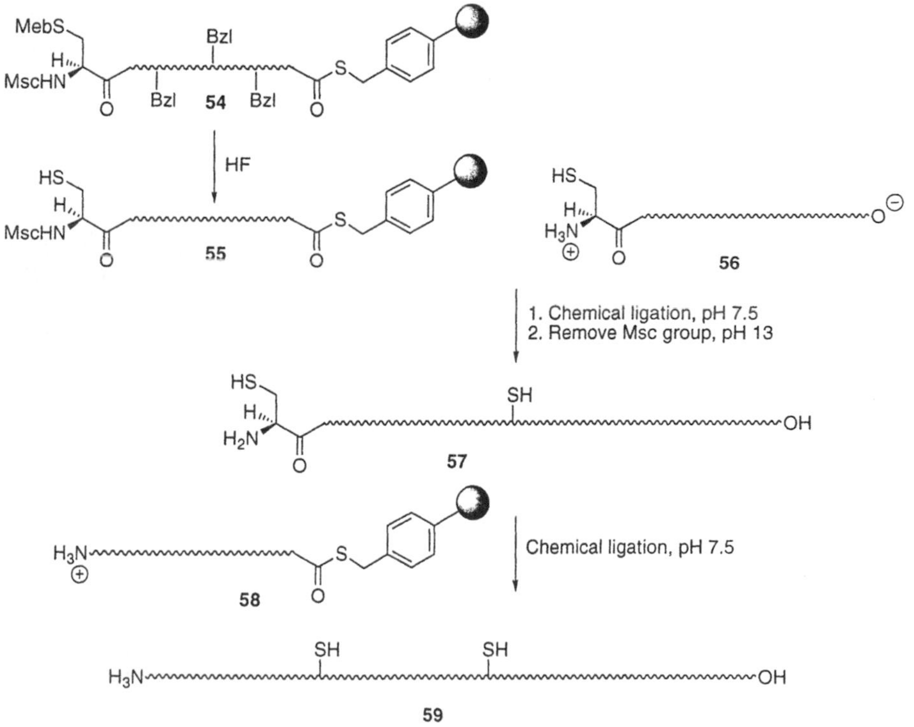

Scheme 11 Solid-phase chemical ligation.

ter giving peptide resin **54**. The last amino acid of **54** is manipulated to introduce the acid-stable 2-(methylsulfonyl)ethoxycarbonyl Msc group at the *N*-terminus. Hydrogen fluoride treatment of **54** then yields peptide resin **55**, in which all side-chain protecting groups have been removed. Solid-phase native ligation of peptide-resin **55** with the *N*-terminal cysteine peptide **56** prepared by standard solid-phase methods leads to peptide coupling with concomitant detachment of the peptide from the solid support. Posterior base treatment to remove the Msc group gives **57**. This peptide can be applied in a second solid-phase native ligation step with the peptide-resin **58**, prepared in an analogous manner to **55**, affording desired protein molecule **59**, which is then purified.

The rate of solid-phase native ligation is accelerated considerably by the introduction of aromatic thiols into the reaction medium, with peptide coupling complete in 3 to 4 h. Such techniques have been used to assemble several medium-sized (up to 79 residues in length) polypeptides in acceptable yield and excellent purity.

Muir et al. [217,218] also reported a convenient in vitro chemical ligation strategy that allows folded recombinant proteins to be joined together. This strategy permits segmental, selective isotopic labeling of the product [219].

3. Typical Experimental Procedures for the Solid-Phase Native Ligation of Peptides

a. Preparation of Unprotected Peptides Anchored to Solid Supports by Thioester Linkages. These procedures are adapted from Ref. 215. The desired peptides are synthesized by optimized Boc solid-phase peptide synthesis [216] on a 3-mercaptopropionamide–polyethylene glycol–poly-(*N,N*-dimethylacrylamide) copolymer support (HS-PEGA). Once the desired sequence has been constructed, the peptide resin is treated with liquid hydrogen fluoride for 1 h at 0°C in the presence of 4% v/v *p*-cresol as the sole carbocation scavenger. [His(Bom)-containing peptides also require the addition of 4% v/v resorcinol to the reaction mixture.] Less than 2% cleavage from the resin is observed with this procedure. Soluble crude peptide products are precipitated and washed with anhydrous diethyl ether before being dissolved in degassed aqueous ACN (10–50%) and lyophilized. The unprotected peptide-resin was washed with anhydrous diethyl ether, 50% ACN in 0.1% aqueous TFA and dried. If required, small aliquots (1–2 mg) of this peptide-resin can be cleaved using 50% ACN in 0.5 M sodium hydroxide solution at 0°C for 1 min and/or benzyl thiol (2% v/v) in ACN–sodium phosphate buffer (1:1 v/v) at pH 7.5 for 2–3 h and then analyzed by HPLC and mass spectrometry.

b. Solid-Phase Ligation. The unprotected peptide-resin (1 equivalent) and the *N*-terminal cysteine peptide (1 equivalent) were swollen in freshly degassed 6 M guanidinium hydrochloride, 0.1 M sodium phosphate, pH 7.5 buffer containing thiophenol (2% v/v) and benzyl thiol (1% v/v) and the vessel was shaken for 3 h at room temperature. The supernatant was removed after centrifugation and the resin washed with the ligation buffer. The washes and supernatant were combined and the ligation product purified by semipreparative HPLC.

ACKNOWLEDGMENTS

Work on convergent solid-phase peptide synthesis in our laboratory is supported by grants from CICYT (PB95-1131 and 2FD97-0267), Generalitat de Catalunya [Grup Consolidat (1995SGR494)], and Centre de Referència en Biotecnologia.

REFERENCES

1. P Lloyd-Williams, F Albericio, E Giralt. Chemical Approaches to the Synthesis of Peptides and Proteins. Boca Raton, FL: CRC press, 1997.
2. P Lloyd-Williams, F Albericio, E Giralt. Tetrahedron 49:11065–11133, 1993.
3. H Benz. Synthesis 337–358, 1994.
4. F Albericio, P Lloyd-Williams, E Giralt. Methods Enzymol 289:313–336, 1997.
5. K Barlos, D Gatos, S Kapolos, G Papaphotiu, W Schäfer, Y Wenqing. Tetrahedron Lett 30:3947–3950, 1989.
6. K Barlos, D Gatos, J Kallitsis, G Papaphotiu, P Sotiriu, Y Wenqing, W Schäfer. Tetrahedron Lett 30:3943–3946, 1989.
7. K Barlos, O Chatzi, D Gatos, G Stavropoulos. Int J Peptide Protein Res 37: 513–520, 1991.
8. K Barlos, D Gatos, G Papaphotiou, W Schäfer. Liebigs Ann Chem 215–220, 1997.
9. J Bodi, H SuliVargha, K Ludanyi. Tetrahedron Lett 38:3293–3296, 1997.
10. T Kaise, W Worlter. J Prakt Chem Chem Ztg 339:371–380, 1997.
11. D Gatos, S Patrianaku, O Hatzi, K Barlos. Lett Peptide Sci 4:177–184, 1997.
12. R Bollhagen, M Schmiedberger, K Barlos, E Grell. J Chem Soc Chem Commun 2559–2560, 1994.
13. M Gairí, P Lloyd-Williams, F Albericio, E Giralt. Int J Peptide Protein Res 46:119–133, 1995.
14. A van Vliet, RHPH Smulders, BH Rietman, GI Tesser. In: R Epton. ed. Innovation and perspectives in Solid Phase Synthesis. Peptides, Polypeptides and Oligonucleotides. Canterbury, England: Intercept, 1992, pp 475–477.

15. A van Vliet, RHPH Smulders, BH Rietman, IF Eggen, G van der Werben, GI Tesser. In: CH Schneider, AN Eberle, eds. In: Peptides 1992. Proceedings of the 22nd European Peptides Symposium, Interlaken, Switzerland. Leiden: ESCOM, 1993, pp 279–280.
16. L Zhang, W Rapp, C Goldhammer, E Bayer. In: R Epton, ed. Innovation and Perspectives in Solid Phase Synthesis. Peptides, Proteins and Nucleic Acids. Biological and Biomedical Applications, Oxford. Birmingham: Mayflower Worldwide, 1994, pp 717–722.
17. P White. In: R Epton, ed. Innovation and Perspectives in Solid Phase Synthesis. Peptides, Proteins and Nucleic Acids. Biological and Biomedical Applications, Oxford. Birmingham: Mayflower Worldwide, 1994, pp 701–704.
18. BH Rietman, RHPH Smulders, IF Eggen, A van Vliet, G van de Werben, GI Tesser. Int J Peptide Protein Res 44:199–206, 1994.
19. CC Zikos, NG Ferderigos. Tetrahedron Lett 35:1767–1768, 1994.
20. A van Vliet, DTS Rijkers, GI Tesser. In: HLS Maia, ed. Peptides 1994. Proceedings of the 23rd European Peptides Symposium, Braga, Portugal. Leiden: ESCOM, 1995, pp 267–268.
21. M Quibell, LC Packman, T Johnson, J Am Chem Soc 117:11656–11668, 1995.
22. RC Sheppard, BJ Williams. J Chem Soc Chem Commun 587–589, 1982.
23. RC Sheppard, BJ Williams. Int J Peptide Protein Res 20:451–454, 1982.
24. A Flörsheimer, B Riniker. In: E Giralt, D Andreu, eds. Peptides 1990. Proceedings of the 21st European Peptides Symposium, Platja D'Aro, Spain. Leiden: ESCOM, 1991, pp 131–133.
25. B Riniker, A Flörsheimer, H Fretz, P Sieber, B Kamber. Tetrahedron 49:9307–9320, 1993.
26. B Riniker, A Flörscheimer, H Fretz, B Kamber. In: CH Schneider, AN Eberle eds. Peptides 1992. Proceedings of the 22nd European Peptides Symposium, Interlaken, Switzerland. Leiden: ESCOM, 1993, pp 34–35.
27. F Albericio, G Barany. Int J Peptide Protein Res 26:92–97, 1985.
28. J Alsina, E Giralt, F Albericio. Tetrahedron Lett 37:4195–4198, 1996.
29. M Mergler, R Tanner, J Gosteli, P Grogg. Tetrahedron Lett 29:4005–4008, 1988.
30. M Mergler, R Nyfeler, R Tanner, J Gosteli, P Grogg. Tetrahedron Lett 29:4009–4012, 1988.
31. AM Felix, Z Zhao, T Lambros, M Ahmad, W Liu, D Daniewski, J Michaleswky, EP Heimer. J Peptide Res 52:155–164, 1998
32. H Rink. Tetrahedron Lett 28:3787–3790, 1987.
33. F Albericio, G Barany. Tetrahedron Lett 32:1015–1018, 1991.
34. E Pedroso, A Grandas, X de las Heras, R Eritja, E Giralt. Tetrahedron Lett 27:743–746, 1986.
35. M Ueki, M Amemiya. Tetrahedron Lett 28:6617–6620, 1987.
36. F Rabanal, E Giralt, F Albericio. Tetrahedron Lett 33:1775–1778, 1992.
37. F Rabanal, E Giralt, F Albericio. Tetrahedron Lett 15:1449–1458, 1995.
38. Y Nishiuchi, T Inui, H Nishio, J Bódi, T Kimura, FI Tsuji, S Sakakibara. Proc Nat Acad Sci U S A 95:13549–13554, 1998.

39. GI Tesser, JTWARM Buis, ETM Wolters, EGAM Bothé-Helmes. Tetrahedron 32:1069–1072, 1976.
40. JTWARM Buis, GT Tesser, RJF Nivard. Tetrahedron 32:2321–2325, 1976.
41. R Schwyzer, E Felder, P Failli. Helv Chim Acta 67:1316–1327, 1984.
42. SB Katti, PK Misra, W Haq, KB Mathur. J Chem Soc Chem Commun 843–844, 1992.
43. J Robles, E Pedroso, A Grandas. J Org Chem 59:2482–2486, 1994.
44. J Robles, E Pedroso, A Grandas. Nucleic Acids Res 23:4151–4161, 1995.
45. WF DeGrado, ET Kaiser. J Org Chem 47:3258–3261, 1982.
46. ET Kaiser. Acc Chem Res 22:47–54, 1989.
47. ET Kaiser, H Mihara, GA Laforet, JW Kelly, L Walters, MA Findeis, T Sasaki. Science 243:187–192, 1989.
48. RB Scarr, MA Findeis. Peptide Res 3:328–241, 1990.
49. SH Nakagawa, ET Kaiser. J Org Chem 48:678–685, 1983.
50. S Nakagawa, HSH Lau, FJ Kézdy, ET Kaiser. J Am Chem Soc 107:7087–7092, 1985.
51. T Sasaki, ET Kaiser. J Am Chem Soc 111:380–381, 1989.
52. T Sasaki, MA Findeis, ET Kaiser. J Org Chem 56:3159–3168, 1991.
53. VNR Pillai. Synthesis 1–26, 1980.
54. DH Rich, SK Gurwara. J Chem Soc Chem Commun 610–611, 1973.
55. DH Rich, SK Gurwara. J Am Chem Soc 97:1575–1579, 1975.
56. E Giralt, F Albericio, D Andreu, R Eritja, P Martin, E Pedroso. An Quim 77C:120–125, 1981.
57. E Bayer, M Dengler, B Hemmasi. Int J Peptide Protein Res 25:178–186, 1985.
58. A Ajayaghosh, VNR Pillai. J Org Chem 52:5714–5717, 1987.
59. A Ajayaghosh, VNR Pillai. Tetrahedron 44:6661–6666, 1988.
60. N Kneib-Cordonier, F Albericio, G Barany. Int J Peptide Protein Res 35:527–538, 1990.
61. F Albericio, E Nicolás, J Josa, A Grandas, E Pedroso, E Giralt, C Granier, J van Rietschoten. Tetrahedron 43:5961–5971, 1987.
62. CP Holmes, DG Jones. J Org Chem 60:2318–2319, 1995.
63. BF Gisin. Helv Chim Acta 56:1476–1482, 1973.
64. G Barany, F Albericio. J Am Chem Soc 107:4936–4942, 1985.
65. E Giralt, R Eritja, E Pedroso. Tetrahedron Lett 22:3779–3782, 1981.
66. K Suzuki, K Nitta, N Endo. Chem Pharm Bull 23:222–224, 1975.
67. M Gairí, P Lloyd-Williams, F Albericio, E Giralt. Tetrahedron Lett 31:7363–7366, 1990.
68. JC Sheehan, K Umezawa. J Org Chem 38:3771–3774, 1973.
69. SS Wang. J Org Chem 41:3258–3261, 1976.
70. FS Tjoeng, JP Tam, RB Merrifield. Int J Peptide Protein Res 14:262–274, 1979.
71. FS Tjoeng, GA Heavner. J Org Chem 48:355–359, 1983.
72. D Bellof, M Mutter. Chimia 39:317–320, 1985.
73. NA Abraham, G Fazal, JM Ferland, S Rakhit, J Gauthier. Tetrahedron Lett 32:577–580, 1991.

74. RLE Furlan, EG Mata, OA Mascaretti. J Chem Soc Perkins Trans I:355–358, 1998.
75. P Lloyd-Williams, M Gairi, F Albericio, E Giralt. Tetrahedron 49:10069–10078, 1993.
76. G Barany, RB Merrifield. J Am Chem Soc 99:7363–7365, 1977.
77. F Albericio, G Barany. Int J Peptide Protein Res 30:177–205, 1987.
78. H Kunz, B Dombo. Angew Chem Int Ed Engl 27:711–713, 1988.
79. B Blankemeyer-Menge, R Frank. Tetrahedron Lett 29:5871–5874, 1988.
80. F Guibé, O Dangles, G Balavoine, A Loffet. Tetrahedron Lett 30:2641–2644, 1989.
81. H Kunz, In: R Epton, ed. Innovation and Perspectives in Solid Phase Synthesis and Related Technologies. Peptides, Polypeptides and Oligonucleotides. Macro-organic Reagents and Catalysts, Oxford. Birmingham: SPCC (UK), 1990, pp 371–378.
82. P Lloyd-Williams, G Jou, F Albericio, E Giralt. Tetrahedron Lett 32:4207–4210, 1991.
83. SA Kates, SB Daniels, F Albericio. Anal Biochem 212:303–310, 1993.
84. P Lloyd-Williams, A Merzouk, F Guibé, F Albericio, E Giralt. Tetrahedron Lett 35:4437–4440, 1994.
85. O Dangles, F Guibé, G Balavoine, S Lavielle, A Marquet. J Org Chem 52:4984–4993, 1987.
86. A Loffet, N Galeotti, P Jouin, B Castro, F Guibé, O Dangles, G Balavoine. In: JE Rivier, GR Marshall, eds. Peptides. Chemistry, Structure and Biology. Proceedings of the 11th American Peptides Symposium, La Jolla, CA. Leiden: ESCOM, 1990, pp 1015–1016.
87. C Toniolo, GM Bonora, EP Heimer, AM Felix. Int J Peptide Protein Res 30:232–239, 1987.
88. RCdL Milton, SCF Milton, PA Adams. J Am Chem Soc 112:6039–6046, 1990.
89. C Toniolo, GM Bonora, M Mutter. VNR Pillai. Makromol Chem 182:2007–2014, 1981.
90. T Haack, M Mutter. Tetrahedron Lett 33:1589–1592, 1992.
91. T Haack, A Zier, A Nefzi, M Mutter. In: CH Schneider, AN Eberle, eds. Peptides 1992. Proceedings of the 22nd European Peptides Symposium, Interlaken, Switzerland. Leiden: ESCOM, 1993, pp 595–596.
92. T Haack, A Nefzi, D Dhanapal, M Mutter. In: R Epton, ed. Innovation and Perspectives in Solid Phase Synthesis. Peptides, Proteins and Nucleic Acids. Biological and Biomedical Applications, Oxford. Birmingham: Mayflower Worldwide, 1994, pp 521–524.
93. T Wöhr, M Mutter. Tetrahedron Lett 36:3847–3848, 1995.
94. T Wöhr, F Wahl, A Nefzi, B Rohwedder, T Sato, XC Sun, M Mutter. J Am Chem Soc 118:9218–9227, 1996.
95. F Weygand, W Steglich, J Bjarnason, R Akhtar, NM Khan. Tetrahedron Lett 3483–3487, 1966.
96. F Weygand, W Steglich, J Bjarnason. Chem Ber 101:3642–3648, 1968.

97. S Isokawa, I Tominaga, T Asakura, M Narita. Macromolecules 18:878–881, 1985.
98. M Narita, K Ishikawa, H Nakano, S Isokawa. Int J Peptide Protein Res 24: 14–24, 1984.
99. J Blaakmeer, T Tijsse-Keasen, GI Tesser. Int J Peptide Protein Res 37:556–564, 1991.
100. R Bartl, K-D Klöppel, R Frank. In: JA Smith, JE Rivier, eds. Peptides. Chemistry and Biology. Proceedings of the 12th American Peptides Symposium, Boston. Leiden: ESCOM, 1992, pp 505–506.
101. R Bartl, K-D Klöppel, R Frank. In: CH Schneider, AN Eberle, eds. Peptides 1992. Proceedings of the 22nd European Peptides Symposium, Interlaken, Switzerland. Leiden: ESCOM, 1993, pp 277–278.
102. H Eckert, C Seidel. Angew Chem Int Ed Engl 25:159–160, 1986.
103. T Johnson, M Quibell, D Owen, RC Sheppard. J Chem Soc Chem Commun 369–372, 1993.
104. M Quibell, WG Turnell, T Johnson. In: R Epton, ed. Innovation and Perspectives in Solid Phase Synthesis. Peptides, Proteins and Nucleic Acids. Biological and Biomedical Applications, Oxford. Birmingham: Mayflower Worldwide, 1994, pp 653–656.
105. M Quibell, T Johnson. In: HLS Maia, ed. Peptides 1994. Proceedings of the 23rd European Peptides Symposium, Braga, Portugal. Leiden: ESCOM, 1995, pp 173–174.
106. RB Merrifield, AE Bach. J Org Chem 43:4808–4816, 1978.
107. H Anzinger, M Mutter, E Bayer. Angew Chem Int Ed Engl 18:686–687, 1979.
108. S Funakoshi, H Fukuda, N Fujii. Proc Natl Acad Sci U S A 88:6981–6985, 1991.
109. J Rizo, F Albericio, G Romero, C García-Echeverría, J Claret, C Muller, E Giralt, E Pedroso. J Org Chem 53:5386–5389, 1988.
110. K Suzuki, Y Sasaki, N Endo. Chem Pharm Bull 24:1–9, 1976.
111. J Rizo, F Albericio, E Giralt, E Pedroso. Tetrahedron Lett 33:397–400, 1992.
112. M Narita, S Honda, H Umeyama, S Obana. Bull Chem Soc Jpn 61:281–284, 1988.
113. M Narita, S Honda, S Obana. Bull Chem Soc Jpn 2:342–344, 1989.
114. M Narita, H Umeyama, S Isokawa, S Honda, C Sasaki, H Kakei. Bull Chem Soc Jpn 62:780–785, 1989.
115. M Narita, H Umeyama, T Yoshida. Bull Chem Soc Jpn 62:3582–3586, 1989.
116. H Kuroda, Y-N Chen, T Kimura, S Sakakibara. Int J Peptide Protein Res 40: 294–299, 1992.
117. L Zhang, C Goldhammer, B Henkel, F Zühl, G Panhaus, G Jung, E Bayer. In: R Epton, ed. Innovation and Perspectives in Solid Phase Synthesis. Peptides, Proteins and Nucleic Acids. Biological and Biomedical Applications, Oxford. Birmingham: Mayflower Worldwide, 1994, pp 711–716.
118. D Seebach, A Thaler, AK Beck. Helv Chim Acta 72:857–867, 1989.
119. K Halverson, PE Fraser, DA Kirschner, PT Lansbury. Biochemistry 29:2639–2644, 1990.

120. P Lloyd-Williams, F Albericio, E Giralt. Int J Peptide Protein Res 37:58–60, 1991.
121. P Lloyd-Williams, F Albericio, M Gairí, E Giralt. In: R Epton, ed. Innovation and Perspectives in Solid Phase Synthesis and Combinatorial Libraries. Peptides, Proteins and Nucleic Acids. Small Molecule Organic Chemical Diversity, Edinburgh. Birmingham: Mayflower Scientific, 1996, pp 195–200.
122. D Gatos, C Tzavara, P Athanassopoulos, K Barlos. Lett Peptide Sci 51:194–200, 1998.
123. B Lindquist, T Storgards. Nature 175:511–513, 1955.
124. J Porath, P Flodin. Nature 183:1657–1659, 1959.
125. J Porath. Biochim Biophys Acta 39:193–207, 1960.
126. P Flodin. J Chromatogr 5:103–115, 1961.
127. P Andrews. Biochem J 91:222–233, 1964.
128. MTW Hearn. Methods Enzymol 104:190–212, 1984.
129. GE Tarr, JW Crabbe. Anal Biochem 131:99–107, 1983.
130. BS Welinder, HH Sorenson. J Chromatogr 537:181–199, 1991.
131. I Dalcol, F Rabanal, M-D Ludevid, F Albericio, E Giralt. J Org Chem 60:7575–7581, 1995.
132. TA Lyle, SF Brady, TM Ciccarone, CD Colton, WF Paleveda, DF Veber, RF Nutt. J Org Chem 52:3752–3759, 1987.
133. M Barber, RS Bordoli, RD Sedgwick, AN Tyler. J Chem Soc Chem Commun 325–327, 1981.
134. DJ Surman, JC Vickerman. J Chem Soc Chem Commun 324–325, 1981.
135. A Grandas, E Pedroso, A Figueras, J Rivera, E Giralt. Biomed Environ Mass Spectrom 15:681–684, 1988.
136. C Celma, E Giralt. Biomed Environ Mass Spectrom 19:235–239, 1990.
137. TL James, NJ Oppenheimer, eds. Nuclear Magnetic Resonance, Part C, Methods in Enzymology. Vol 239. San Diego: Academic Press, 1994.
138. F Albericio, M Pons, E Pedroso, E Giralt. J Org Chem 54:360–366, 1989.
139. E Krambovitis, G Hatzidakis, K Barlos. J Biol Chem 273:10874–10879, 1998.
140. AM Felix, RB Merrifield. J Am Chem Soc 92:1385–1391, 1970.
141. R Matsueda, H Maruyama, E Kitazawa, H Takahagi, T Mukaiyama. Bull Chem Soc Jpn 46:3240–3247, 1973.
142. T Mukaiyama, K Goto, R Matsueda, M Ueki. Tetrahedron Lett 5293–5296, 1970.
143. R Matsueda, H Maruyama, E Kitazawa, H Takahagi, T Mukaiyama. J Am Chem Soc 97:2573–2575, 1975.
144. H Maruyama, R Matsueda, E Kitasaura, H Takahagi, T Mukaiyama. Bull Chem Soc Jpn 49:2259–2267, 1976.
145. H Rink, W Born, JA Fischer. In: JE Rivier, GR Marshall, eds. Peptides. Chemistry, Structure and Biology. Proceedings of the 11th American Peptides Symposium, San Diego. Leiden: ESCOM, 1990, pp. 1041–1042.
146. H Yajima, Y Kiso. Chem Pharm Bull 22:1087–1094, 1974.
147. L Carpino, A El-Faham, F Albericio. Tetrahedron Lett 35:2279–2282, 1994.

148. LA Carpino, A El-Fahan, CA Minor, F Albericio. J Chem Soc Chem Commun 201–203, 1994.
149. I Dalcol, M Pons, MD Ludevid, E Giralt. J Org Chem 61:6775–6782 (1996).
150. NL Benoiton, K Kuroda. Int J Peptide Protein Res 17:197–204, 1981.
151. NL Benoiton, FMF Chen. Can J Chem 59:384–389, 1981.
152. E Bayer, E Gil-Av, WA König, S Nakapartksin, J Oró, W Parr. J Am Chem Soc 92:1738–1740, 1970.
153. M Goodman, P Keogh, H Anderson. Bioorgan Chem 6:239–247, 1977.
154. R Steinauer, FMF Chen, NL Benoiton. J Chromatogr 325:111–126, 1985.
155. A Grandas, F Albericio, J Josa, E Giralt, E Pedroso, JM Sabatier, J van Rietschoten. Tetrahedron 45:4637–4648, 1989.
156. JC Hendrix, KJ Halverson, PT Lansbury. J Am Chem Soc 114:7930–7931, 1992.
157. D Lelievre, Y Trudelle, F Heitz, G Spach. Int J Peptide Protein Res 33:379–385, 1989.
158. K Noda, S Terada, N Mitsuyasu, M Waki, T Kato, N Izumiya. Mem Fac Sci Kyushu Univ Ser B 7:189–201, 1970.
159. K Barlos, D Gatos, W Schäfer. Angew Chem Int Ed Engl 30:590–593, 1991.
160. E Atherton, CJ Logan, RC Sheppard. J Chem Soc Perkin Trans I 538–546, 1981.
161. M Rothe, J Mazánek. Liebigs Ann Chem 439–459, 1974.
162. M Rothe, M Rott, J Mazánek. In: A Loffet, ed. Peptides 1976. Proceedings of the 14th European Peptides Symposium, Wépion, Belgium. Brussels: Editions de l' Université de Bruxelles, 1976, pp 309–318.
163. S Aimoto, N Mizoguchi, H Hojo, S Yoshimura. Bull Chem Soc Jpn 62:524–531, 1989.
164. H Hojo, S Aimoto. Bull Chem Soc Jpn 64:111–117, 1991.
165. H Hojo, S Aimoto. Bull Chem Soc Jpn 65:3055–3063, 1992.
166. H Hojo, YD Kwon, S Aimoto. In: A Suzuki, ed. Peptide Chemistry 1991. Proceedings of the 29th symposium on Peptide Chemistry, Tokyo. Osaka: Protein Research Foundation, 1992, pp 115–120.
167. Y Kwon, R Zhang, H Hojo, S Aimoto. In: N Yanaihara, ed. Peptides 1992. Proceedings of the 2nd Japan Symposium on Peptide Chemistry, Shizuoka, Japan. Leiden: ESCOM, 1993, pp 58–60.
168. S Aimoto, Y Kwon, H Hojo. In: N Yanaihara, ed. Peptide Chemistry 1992. Proceedings of the 2nd Japan Symposium on Peptide Chemistry, Shizuoka, Japan. Leiden: ESCOM, 1993, pp 54–57.
169. S Aimoto, H Hojo. In: Y-C Du, JP Tam, Y-S Zhang, eds. Peptides. Biology and Chemistry. Proceedings of the 1992 Chinese Peptide Symposium, Hangzhou, China. Leiden: ESCOM, 1993, pp 273–277.
170. J Blake. Int J Peptide Protein Res 17:273–274, 1981.
171. J Blake, CH Li. Proc Natl Acad Sci U S A 78:4055–4058, 1981.
172. J Blake, CH Li. Proc Natl Acad Sci U S A 80:1556–1559, 1983.
173. J Blake, M Westphal, CH Li. Int J Peptide Protein Res 24:498–504, 1984.
174. J Blake. Int J Peptide Protein Res 27:191–200, 1986.

175. J Blake, D Yamashiro, K Ramasharma, CH Li. Int J Peptide Protein Res 28: 468–476, 1986.
176. D Yamashiro, CH Li. Int J Peptide Protein Res 31:322–334, 1988.
177. H-C Cheng, D Yamashiro. Int J Peptide Protein Res 38:70–78, 1991.
178. TW Muir, SBH Kent. Curr Opin Biotechnol 4:420–427, 1993.
179. M Schnölzer, SBH Kent. Science 256:221–225, 1992.
180. M Schnölzer, SBH Kent. In: CH Schneider, AN Eberle, eds. Peptides 1992, Proceedings of the 22nd European Peptides Symposium, Interlaken, Switzerland. Leiden: ESCOM, 1993, pp 237–238.
181. K Rose, LA Vilaseca, R Werlen, A Meunier, I Fisch, RML Jones, RE Offord. Bioconj Chem 2:154–159, 1991.
182. I Fisch, G Künzi, K Rose, RE Offord. Bioconj Chem 3:147–153, 1992.
183. HF Gaertner, K Rose, R Cotton, D Timms, R Camble, RE Offord. Bioconj Chem 3:262–268, 1992.
184. HF Gaertner, RE Offord, R Cotton, D Timms, R Camble, K Rose. J Biol Chem 269:7224–7230, 1994.
185. S Futaki, T Ishikawa, M Niwa, K Kitagawa, T Yagami. Bioorg Med Chem 5:1883–1891, 1997.
186. M Mutter, G Tuchscherer. Cell Mol Life Sci 53:851–863, 1997.
187. LS Zhang, TR Torgerson, XY Liu, S Timmons, AD Colosia, J Hawiger, JP Tam. Proc Natl Acad Sci U S A 95:9184–9189, 1998.
188. O Melnyk, M Bossus, D David, C Rommens, H Gras-Masse. J Peptide Res 52:180–184, 1998.
189. JP Tam, QT Yu. Biopolymers 46:319–327, 1998.
190. M Brenner, JP Zimmermann, J Wehrmüller, P Quitt, A Hartmann, W Schneider, U Beglinger. Helv Chim Acta 40:1497–1517, 1957.
191. M Brenner, JP Zimmermann. Helv Chim Acta 40:1933–1939, 1957.
192. M Brenner, J Wehrmüller. Helv Chim Acta 40:2374–2383, 1957.
193. M Brenner, W Hofer. Helv Chim Acta 44:1794–1798, 1961.
194. M Brenner, W Hofer. Helv Chim Acta 44:1798–1801, 1961.
195. M Brenner. In: HC Beyerman, A van de Linde, W Maassen van den Brink, eds. Peptides. Proceedings of the 8th European Peptides Symposium, Noordwijk, The Netherlands. Amsterdam: North-Holland, 1967, pp 1–7.
196. T Wieland, E Bokelmann, L Bauer, HU Lang, H Lau. Liebigs Ann Chem 583:129–149, 1953.
197. S Sasaki, M Shionoya, K Koga. J Am Chem Soc 107:3371–3372, 1985.
198. C Gennari, F Molinari, U Piarulli. Tetrahedron Lett 31:2929–2932, 1990.
199. JCHM Wijkmans, JH van Boom, W Bloemhoff. Tetrahedron Lett 34:7123–7126, 1993.
200. DS Kemp. Biochemistry 20:1793–1804, 1981.
201. DS Kemp, NG Galakatos. J Org Chem 51:1821–1829, 1986.
202. DS Kemp, NG Galakatos, B Bowen, K Tam. J Org Chem 51:1829–1838, 1986.
203. DS Kemp, RI Carey. Tetrahedron Lett 32:2845–2848, 1991.
204. N Fotouhi, NG Galakatos, DS Kemp. J Org Chem 54:2803–2817, 1989.
205. DS Kemp, RI Carey. J Org Chem 58:2216–2222, 1993.

206. PE Dawson, TW Muir, I Clark-Lewis, SBH Kent. Science 266:776–779, 1994.
207. WY Lu, MA Strarovasnik, SBH Kent. FEBS Lett 429:31–35, 1998.
208. J Wilken, SBH Kent. Curr Opin Biotech 9:412–426, 1998.
209. LP Miranda, PF Alewood. Proc Natl Acad Sci U S A 96:1181–1186, 1999.
210. JP Tam, WF Cunningham-Rundles, BW Erickson, RB Merrifield. Tetrahedron Lett 4001–4004, 1977.
211. JP Tam, Y-A Lu, C-F Liu, J Shao. Proc Natl Acad Sci U S A 92:12485–12489, 1995.
212. C-F Liu, C Rao, JP Tam. Tetrahedron Lett 37:933–936, 1996.
213. X Li, T Kawakami, S Aimoto. Tetrahedron Lett 39:8669–8672, 1998.
214. PE Dawson, MJ Churchill, MR Ghadiri, SBH Kent. J Am Chem Soc 119: 4325–4329, 1997.
215. JA Camarero, GJ Cotton, A Adeva. TW Muir. J Peptide Res 51:303–316, 1998.
216. M Schnölzer, P Alewood, A Jones, D Alewood, SBH Kent. Int J Peptide Protein Res 40:180–193, 1992.
217. TW Muir, S Sondhi, PA Cole. Proc Natl Acad Sci U S A 95:6705–6710, 1998.
218. K Severinov, TW Muir. J Biol Chem 273:16205–16209, 1998.
219. R Xu, B Ayers, D Cowburn, TW Muir. Proc Natl Acad Sci U S A 96:388–393, 1999.

10
Preparation of Glyco-, Phospho-, and Sulfopeptides

Carlos García-Echeverría
Novartis Pharma Inc., Basel, Switzerland

I. GLYCOPEPTIDES

A. Introduction

Many peptides and proteins in nature carry carbohydrate moieties [1,2] that are covalently bound to amino acid side chains through *N*- or *O*-glycosidic linkage.* The oligosaccharide chain of these biopolymers affects the properties of the parent compound in many and diverse ways (e.g., uptake, excretion, proteolytic stability, solubility, conformation) and has a significant impact on their biological function. As a logical consequence of these important effects, synthetic and biological studies of glycopeptides and glycoproteins have been very attractive areas of research. A comprehensive book on glycopeptides and related compounds has appeared [4] and several reviews have also described the progress in the development of methods for the solid-phase synthesis of glycopeptides [5–11].

The glycopeptide assembly process relies mainly on the established methods of solid-phase peptide chemistry. However, the synthetic strategy has to be adapted to accommodate the chemical sensitivities of the carbohydrate moiety, whether the saccharide is present during chain elongation or the carbohydrate moiety is incorporated after chain assembly is complete. Of the different reported strategies for the preparation of glycopeptides [12],

---

*In addition to these anchor points, the glycan can be C-linked to tryptophan. An aldohexopyranosyl residue was found C-glycosidically linked to the C-2 atom of the indole side chain of a tryptophan residue in human ribonuclease (RNase) US [3].

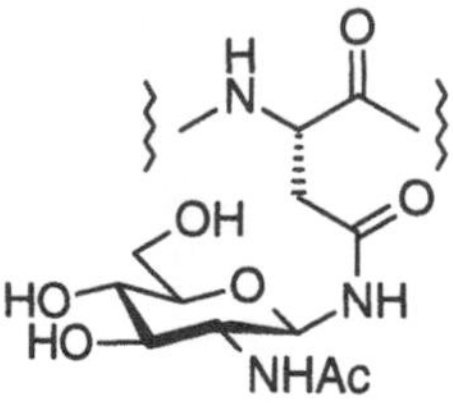

R= H, Ser
R= Me, Thr

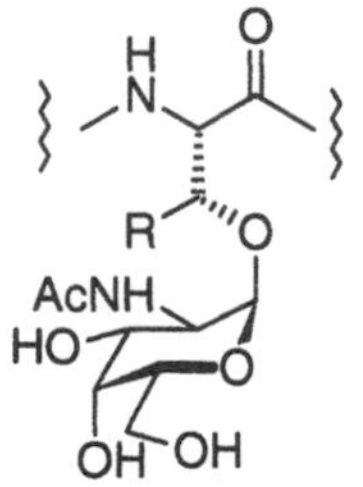

1   **2**

Figure 1   Typical linkages between glycans and peptides.

this section will focus on contributions in the area of solid-phase synthesis of glycopeptides by the building block approach, which is the most common and versatile strategy for the synthesis of N- and O-glycopeptides. In the building block approach, the suitably protected derivative of the appropriate amino acid is first linked to a saccharide derivative. In N-linked glycosylation, the oligosaccharide chains are attached via a core N-acetyl-D-glucosamine (GlcNAc) residue to asparagine in a β-N-glycosidic linkage (Fig. 1, **1**). In O-linked glycosylation, the oligosaccharide chains are linked to serine or threonine* through an α-anomeric linkage of a core N-acetyl-D-galactosamine (GalNAc) residue (Fig. 1, **2**). The resulting glycosylated amino acid is then incorporated in the growing resin-bound peptide. The synthesis of glycopeptides on the solid phase by the building block approach is an interactive combination of both peptide and carbohydrate chemistries. This strategy requires a proper choice of the protecting groups of the building blocks as well as the solid support, linker, coupling protocol, and deprotection conditions. These topics are reviewed here.

B. Solid Supports and Linkers

The synthesis of glycopeptides have been performed manually as well as by using automated or semiautomated peptide synthesizers on poly(ethyleneglycol)–poly(N,N-dimethylacrylamide) copolymer (PEGA) [13,16–22], polystyrene–1% divinylbenzene copolymer [23–37], kieselguhr-supported

---

*The saccharide moiety can also be O-linked to 5-hydroxylysine, 4-hydroxyproline, or tyrosine [1,13,14]. 2-Acetamido-2-deoxy-β-D-glycopyranose O-glycosydically linked to serine or threonine has been reported as a novel type of posttranslational glycosylation [15].

poly(N,N-dimethylacrylamide) [17,22,38–48], and polystyrene–poly(ethyleneglycol) graft copolymer [22,39,49,50] solid supports. In the case of coupling of a large building block [e.g., N^{α}-Fmoc-Asn(Gal$_3$GlcNAc$_3$Man$_3$Glc NAc$_2$)-OH], the PEGA resin has been selected [17]. This solid support provides good swelling capacity in N,N-dimethylformamide and adequate pore size to allow large glycosylated amino acids to have good access to the growing resin-bound peptide [17].

Depending on the method used, the preceding solid supports have been derivatized with a particular linker to anchor the growing peptide chain and obtain the required functionality at the C-terminus. For 9-fluorenylmethoxylcarbonyl (Fmoc) chemistry, the 4-alkoxybenzyl alcohol resin (Fig. 3, **3**), or Wang resin [24–27,32,51–53], the 4-hydroxymethylbenzoic acid (HMBA) linker (Fig. 2, **4**) [16], the 4-hydroxymethylphenoxyacetic acid (HMPAA) linker (Fig. 2, **5**) [17,42], the 2-chlorotrityl chloride resin (Fig. 2, **6**) [54] and the "super-acid-sensitive" resin (SASRIN) (Fig. 2, **7**) [55] have been used in the synthesis of glycopeptides having a C-terminal carboxylic acid. To obtain glycopeptides with C-terminal carboxamides, the 4-(2', 4'-dimethoxyphenylaminomethyl)-phenoxymethyl resin (Fig. 2, **8**), or Rink amide resin [13,18,21,29,31,38] (Fig. 2, **8**), and the 5-(4-aminomethyl-3,5-dimethoxyphenoxy)-valeric acid (PAL) linker (Fig. 2, **9**) [20,22,30,33,34,39,43–46,49] have been employed. The cleavage from the resin support to provide fully or partially protected glycopeptides is accomplished under mild acid conditions compatible with most saccharide-containing peptides. The selection of the cleavage–deprotection conditions and scavenger cocktail is dictated by the linker and the protecting groups present in the sequence. Representative examples of cleavage conditions for the preceding linkers are listed in Table 1, but these are only guidelines and an optimized protocol for each specific glycopeptide is usually required.

The Merrifield resin (Fig. 3, **10**) in combination with standard t-butyloxycarbonyl (Boc)-chemistry was used in the early 1980s by Lavielle and collaborators [56,57] for the solid-phase synthesis of glycosylated analogues of somatostatin. Although the desired glycopeptides were obtained, the use of liquid hydrogen fluoride for the cleavage–deprotection from the Merrifield resin is too harsh for most of the glycosidic linkages. Another variant using Boc chemistry is the Kaiser–DeGrado oxime resin (Fig. 3, **11**), which was employed by Bauman and collaborators [58] to obtain glycopeptides in fully protected form for further chemical manipulation. In this case, the acetate salt of glycosylated serine derivative **12** was used to displace a 3-mer protected peptide from the resin (Fig. 4). From synthetic studies with single amino acids attached to the polymer, it appears that the efficiency of this approach is related to steric effects and low yields are expected for hindered C-terminal β-branched amino acids.

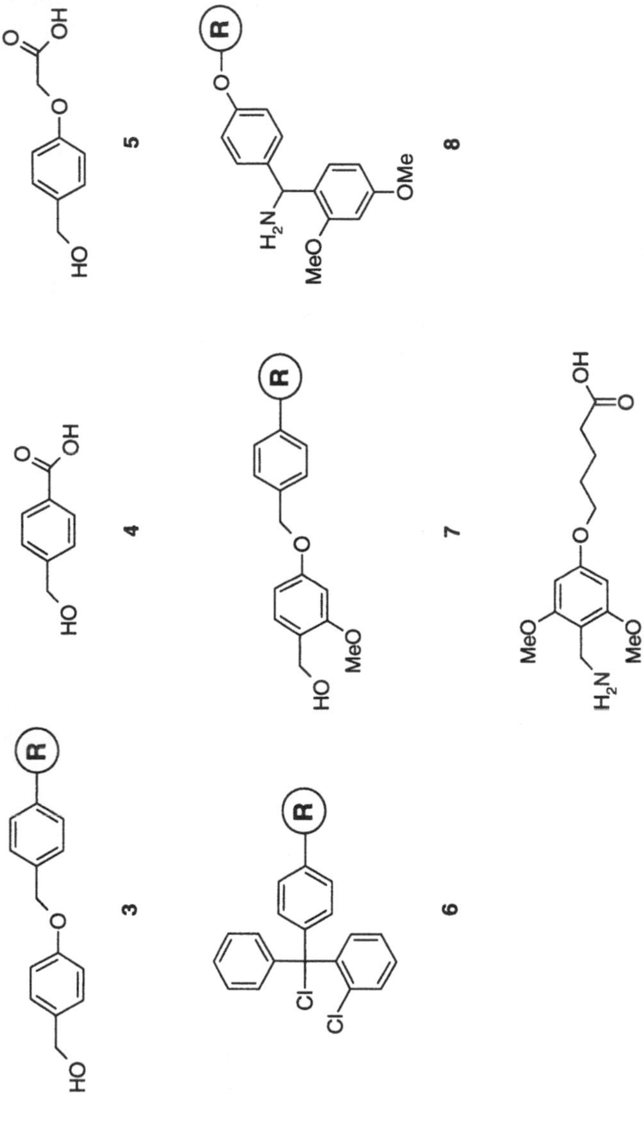

Figure 2 Linkers used in the solid-phase synthesis of glycopeptides by Fmoc methodology.

Table 1 Linkers and Cleavage Conditions Used in the Solid-Phase Synthesis of Glycopeptides by Fmoc Methodology

Linker	Cleavage conditions	References
Wang Resin	TFA–phenol–1,2-ethanedithiol–thioansiole–water (42.5:2.9:1:2:2)	[24]
	TFA–water–1,2-ethanedithiol (92.6:4.9:2.5)	[25,26]
	TFA–water (19:1)	[27,53]
	TFA–phenol–thioanisole–water (40:2:1:1)	[32]
	TFA–water–phenol–anisole–1,2-ethanedithiol (40:2:2.8:2:1)	[51]
	TFA–water–thioanisole–1,2-ethanedithiol (87.5:5:5:2.5)	[52]
HMBA	TFA–water (19:1) followed by 0.1 M NaOH	[16]
HMPAA	TFA–water (19:1)	[17,42]
2-Cholorotrityl	DCM–TFE–AcOH (7:2:1)	[54]
SASRIN	TFA–DCM (1:99)	[55]
Rink amide	TFA–water (19:1)	[13,18,21,38]
	TFA–water–thioanisole–1,2-ethanedithiol (87.5:5:5:2.5)	[29]
	TFA–DCM–water (8:16:1)	[31]
PAL	TFA–water (19:1)	[20,22,30,39, 43–46]
	TFA–DCM (7:3)	[33,34]
	TFA–water–thioanisole–1,2-ethanedithiol (87.5:5:5:2.5)	[49]

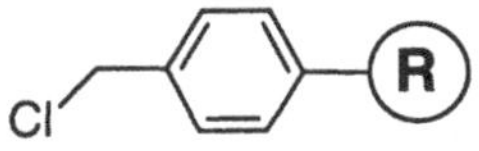

10

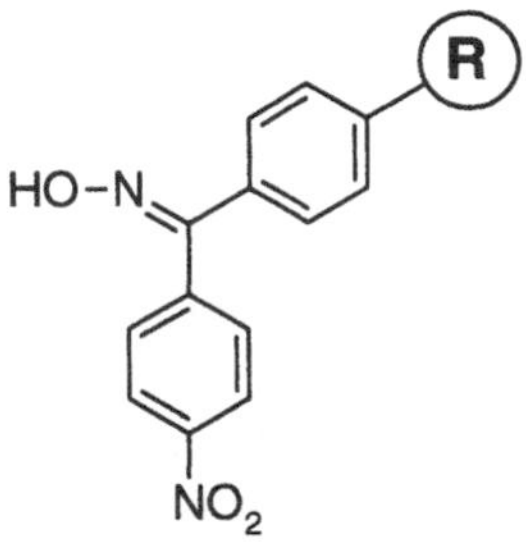

11

Figure 3 Linkers used in the solid-phase synthesis of glycopeptides by Boc methodology.

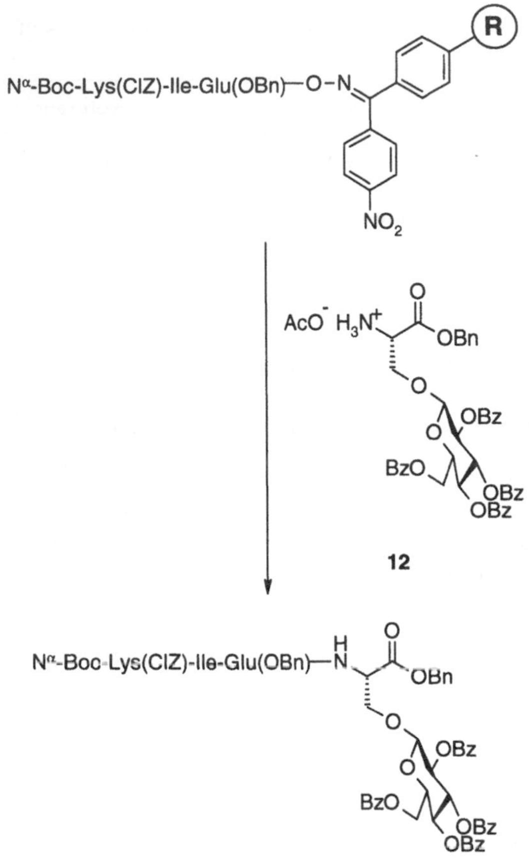

Figure 4 Synthesis of fully protected *O*-linked glycopeptide using the Kaiser–DeGrado oxime resin.

Mild and selective release of glycopeptides from the solid support in the absence of acidic or basic conditions has been accomplished by Kunz and collaborators [19,28,37,59–65] using the allylic anchoring principle. The (*E*)-17-hydroxy-4,7,10,13-tetraoxy-15-heptadecenoyl (HYCRON)-linker (Fig. 5, **13**) [19,28,60,61] is a revised allylic anchor and was developed to overcome undesired side reactions observed from the original HYCRAM (Fig. 5, **14**) and β-HYCRAM (Fig. 5, **15**) [37,59,62–65] resins in the Fmoc methodology [66]. The allyl ester linkage between the assembled glycopeptide and the resin anchor is cleaved by a Pd(0)-catalyzed allyl transfer reaction to a weakly basic nucleophile such as *N*-methylaniline. This approach, which has been used successfully for *N*- and *O*-linked glycopeptides, is

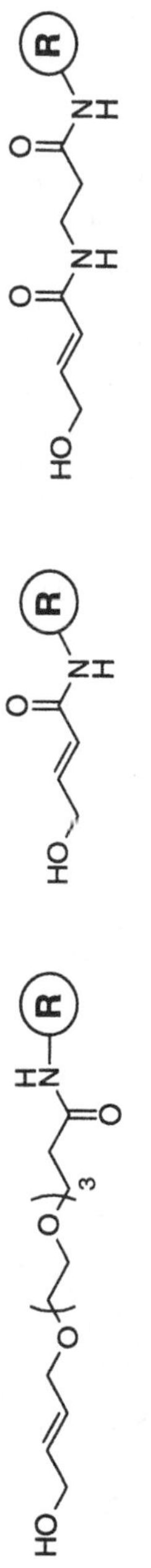

13 14 15

Figure 5 Allylic linkers suitable for solid-phase glycopeptide synthesis.

orthogonal to a variety of protecting group systems and adds another degree of flexibility in the protection scheme.

C. Coupling Protocols

Different coupling conditions have been employed to incorporate protected glycosylated amino acids on the solid phase. Carbodiimides (e.g., DCC; Fig. 6, **16**) with 1-hydroxybenzotriazole (HOBt) (Fig. 6, **17**) used as an auxiliary agent have been extensively employed for in situ activation of protected glycosylated amino acids in solid-phase synthesis [24–27,32,34,49,63,67]. Other coupling agents such as BOP (Fig. 6, **18**) [51,] BOP–HOBt (1:1) [31], TBTU (Fig. 6, **19**) [17,28,54] (Fig. 6, **20**) [23], PyBOP (Fig. 6, **21**) [53], and PfPyU (Fig. 6, **22**) [19] in the presence of N-N-diisopropylethylamine or N-methylmorpholine have also been used to activate the carboxyl function of protected glycoamino acids.

A large number of protected glycosylated amino acids have been prepared as their pentafluorophenyl esters. Originally introduced into glycopeptide chemistry by Meldal and Jensen [48], this group has a double function. It masks the carboxyl function of the amino acid during the glycosylation derivatization and later acts as an effective carboxyl-activating group for the

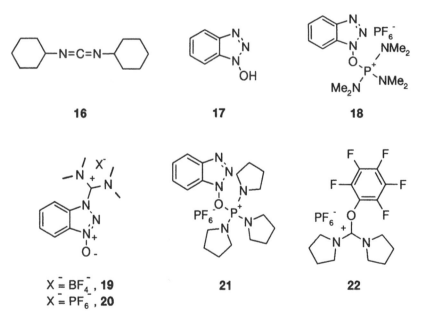

Figure 6   Coupling agents used for the incorporation of glycosylated amino acids.

incorporation of the protected building block in the growing resin-bound peptide [13,18,20–22,29,30,38,39,41,43–46,48–50,68–72]. The reactivity of the pentafluorophenyl ester is further enhanced by the addition of 3-hydroxy-4-oxo-3,4-dihydro-1,2,3-benzotriazine (HODhbt; Fig. 7, **23**) [13,18,21,22,30,38,39,43–46,48,69–71]. This reagent increases the rate of coupling, presumably by in situ formation of the more activated HODhbt ester from the pentafluorophenly ester. In addition, this auxiliary nucleophile allows the progress of the peptide bond formation to be followed visually. A bright yellow color of the ion pair, formed between resin-bound unreacted amino groups and ionized HODhbt (Fig. 7), fades away during the acylation. Under continuous-flow conditions, the progress of the acylation reaction can be monitored with a solid-phase spectrophotometer at 440 nm [13,39,45]. However, real-time monitoring with HODhbt was found to be problematic in one case because of some persistent coloration of the PEGA solid support [13].

Because of the efforts invested in the synthetic preparation of the protected glycosylated amino acids, it is recommended that the smallest excess be used in each coupling. Use of a 1- to 1.5-fold excess of the protected glycosylated amino acid, with respect to the loading of the solid support, is in many cases sufficient to obtain complete or high couplings even when the glycopeptide contains two vicinal glycosylated residues [27]. When a higher excess of a valuable building block is required because of sterically hindered amino acids, steric interaction, or chain aggregation, the excess of the glycosylated building block used in the coupling step could be recovered. For example, 70% of the excess of N^{α}-Fmoc-3-O-[2,3,6-tri-O-acetyl-4-O-(2,3,4,6-tetra-O-acetyl-α-D-galactopyranosyl)-β-D-galactopyranosyl]-L-serine was recovered after aqueous workup of the reaction solution and flash column chromatography [29].

The use of a low excess of the building block requires careful control of the progress of the acylation reaction. Besides the 3-hydroxy-4-oxo-3,4-dihydro-1,2,3-benzotriazine reagent **23**, the incorporation of the building block can be monitored by addition of Bromophenol Blue (0.05% of the resin loading) [52,67,73] or Violet Acid 17 [20] to the reaction vessel, as well as the ninhydrin test [74]. If the coupling is carried out by in situ formation of the 1-hydroxybenzotriazole ester, an excess of 1-hydroxybenzotriazole is required for reliable monitoring with Bromophenol Blue.

D. Fmoc Deprotection

Fmoc chemistry has been the preferred methodology for solid-phase glycopeptide synthesis because mild acidic conditions can be used for the final deprotection and cleavage. After some initial controversy about the use of

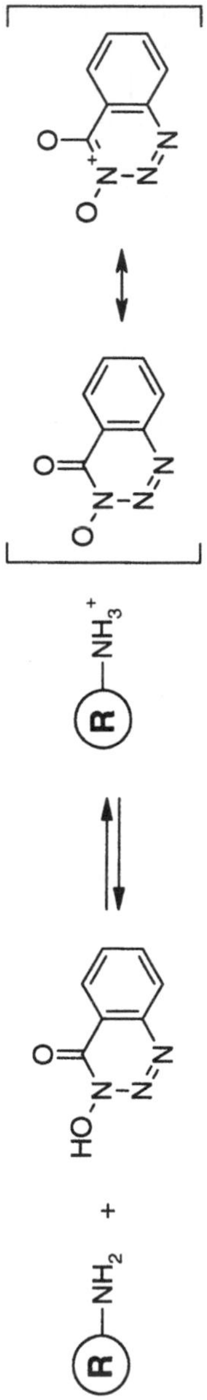

Figure 7 Monitoring the progress of peptide bond formation with HODhbt.

piperidine for Fmoc-deprotection of glycopeptides, studies have shown that neither piperidine nor 1,8-diazabicyclo[5.4.0]undec-7-ene DBU causes β-elimination or epimerization of carbohydrate moieties linked to serine and threonine [39,49,67]. If such side reactions occur, the weaker base morpholine p$K_a^{\text{morpholine}}$ 8.3 versus p$K_a^{\text{piperidine}}$ 11.1) can be used. However, Fmoc cleavage with 50% morpholine is less efficient and can result in poor crude products, particularly for long glycopeptides that are more prone to internal aggregation during the synthesis [41,67]. Thus, the long T-cell antigenic glycopeptides from hen egg lysozyme [75] or glycopeptides from human immunodeficiency virus (HIV) glycoprotein 120 (gp120) and mucins [76] were incompletely deprotected even with prolonged treatment with morpholine after the onset of aggregation.

Treatment with DBU (1 to 2.5% solution in DMF or DMA) can cause succinimide formation from cyclization of aspartyl residues [77] (Fig. 8) and decomposition when silyl-based groups are used for the protection of carbohydrate hydroxy groups [78].

E. Azido Group Reduction on the Solid Support

The azide strategy in the solid-phase synthesis of O-glycopeptides was initially applied by the Paulsen group for the synthesis of mucin glycopeptides [30,44,45]. In this approach, the glycoside synthesis with the 2-azido-2-deoxy-D-galactose donor (Fig. 9, **24**) is carried out with the pentafluorophenyl ester of the amino acid (serine or threonine), which, after derivatization, is incorporated directly on the solid support, After completion of peptide assembly, the azido group is reduced and converted into an acetamido group by treatment of the resin-bound peptide with freshly distilled thioacetic acid [gas–liquid chromatography (GLC), >99.5%] (Fig. 9). The progress of the reduction of the azido group is followed by monitoring the complete disappearance of the azide absorption band at ν 2117 cm^{-1} in the

Pr: protecting group

Figure 8 DBU-mediated succinimide formation from cyclization of aspartyl residues.

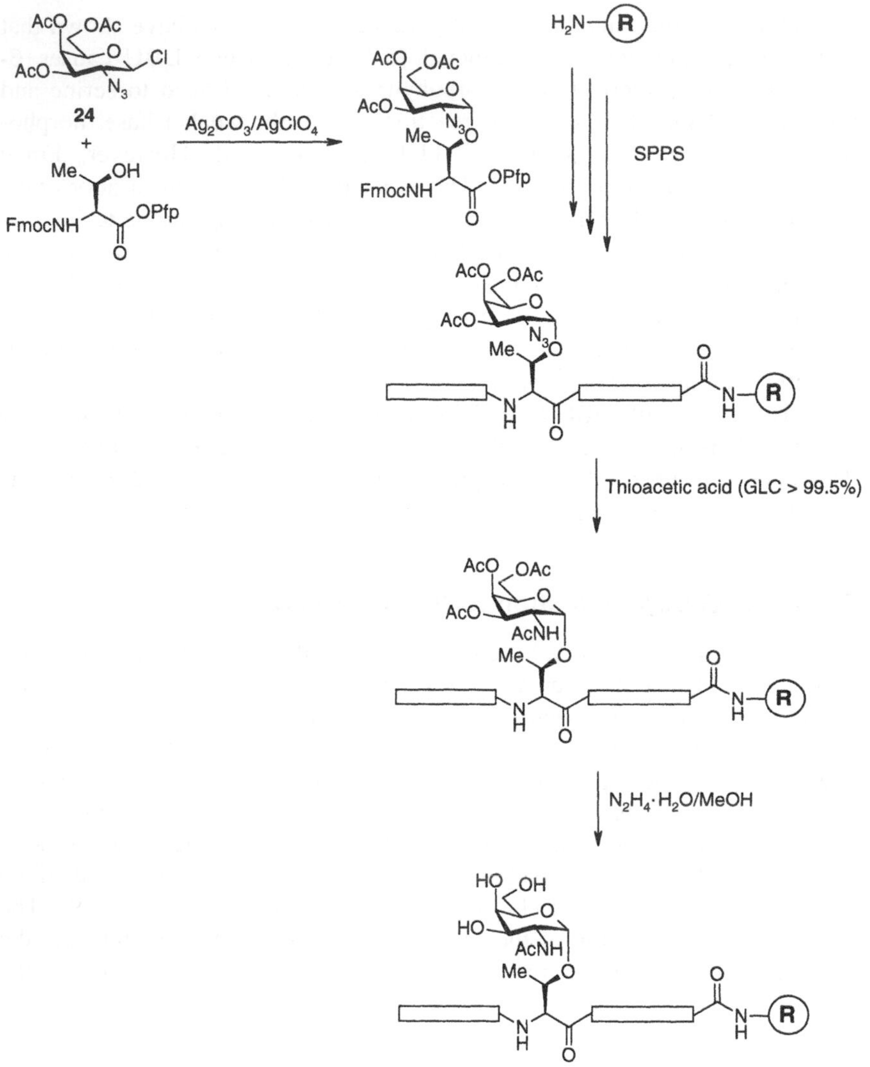

Figure 9 The azide strategy for solid-phase synthesis of *O*-glycopeptides.

infrared spectrum of a KBr pellet containing resin (~150 mg of KBr for 1 mg of resin). After the azide transformation, the resin-bound glycopeptides have been often deacetylated with hydrazine hydrate in methanol (1:7 v/v) [30,39,43,44]. The azide procedure has also been used in a multiple-column solid-phase glycopeptide synthesis [38,43] and by other laboratories [20,53].

However, several side reactions have been reported. Arsequell et al. [53] found that it was necessary to maintain the N-terminus of the resin-bound peptide protected with Fmoc during the azide transformation to avoid unwanted N-acetylation. Another side reaction during the reduction of the azido group was reported by Rio-Anneheim et al. [38]: 10–15% of the corresponding 2-deoxy-2-thioacetamido-α-D-galactopyranose derivative of mucin O-glycopeptides was isolated after solid-phase reduction of the azido group and preparative reversed-phase high performance liquid chromatographic (HPLC) purification of the crude mixtures. In this case, the reduction of the azido group lasted up to 8 days.

F.   Glycan Protection Strategy

The glycosidic linkages of common mono- and disaccharides with unprotected hydroxyl groups are sufficiently stable to allow the cleavage–deprotection of the peptides with trifluoroacetic acid provided this treatment is performed for a limited period of time (≤ 2 h) and in the absence of water (50,79]. In addition, no acylation of the free hydroxyl groups of the sugars is expected to occur during the coupling step using standard activating agents (e.g., pentafluorophenyl esters, DCC **16**/HOBt **17**, symmetrical anhydrides). This approach has been used by different groups for the synthesis of N- and O-glycopeptides [32,42,50,72], but, in spite of these examples, protection of the carbohydrate moieties of glycopeptide should be considered as a suitable precaution for most glycopeptides (e.g., glycosides of sialic acid and 6-deoxy-L-fucose) in order to prevent decomposition during acid-promoted cleavage and deprotection.

Acetates are a common choice for the protection of the carbohydrate hydroxy groups. O-Acetyl groups stabilize the glycosidic bonds under the acidic conditions used for side-chain deprotection and cleavage from the resin and are removed at the last synthetic step [19–21,28,38,43,45,51,60,80]. It was anticipated that β-elimination of carbohydrates linked to serine and threonine and epimerization of peptide stereocenters might constitute serious problems on removal of O-acetyl protecting groups from glycopeptides under alkaline conditions [11,18,82]. An investigation of conditions used for the deacetylation of glycopeptides has revealed that neither β-elimination nor epimerization was encountered in the treatment of a model 3-mer peptide containing a β-D-galactosyl residue linked to threonine (Fig. 10, **25**) with the following deprotection conditions: (1) hydrazine hydrate in methanol (1:5) for 3 h at 20°C, (2) saturated methanolic ammonia for 6 h at 20°C, or (3) 6 mM methanolic sodium methoxide for 30 min at 20°C [83]. Under more severe conditions (120 mM methanolic sodium methoxide for 2.5 h at 20°C), which are required for removal of O-benzoyl groups [42], β-elimination occurred slowly and was accompanied by slight (<5%) epimerization.

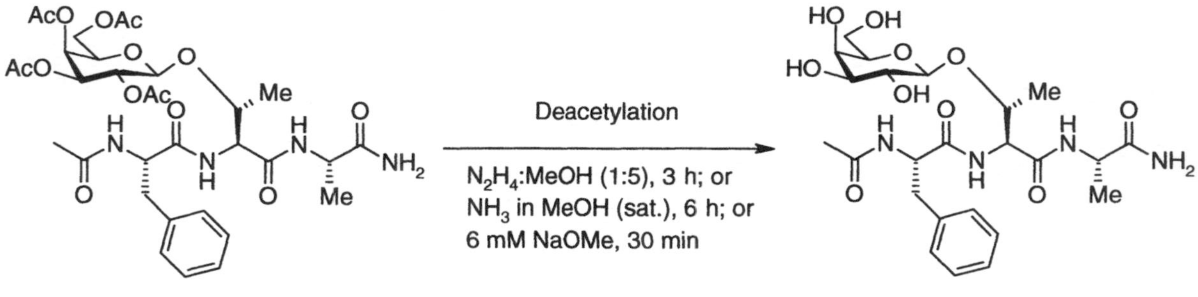

Figure 10 Deacetylation of the 3-mer model *O*-glycopeptide did not cause *β*-elimination or epimerization of threonine.

Other conditions have also been used for the removal of the O-acetyl groups from the sugar moieties. The deacetylation of a 26-mer peptide containing the uncommon O-fucosyl-threonine linkage (Fig. 11, **26**) was successfully carried out by using N,N-diethylmethylamine in water at a constant pH of 9.0 [51]. Similar conditions were used for the O-acetyl deprotection of a glycopeptide derivative of peptide T [63]. The N,N-diethylmethylamine base can be efficiently removed after deacetylation by evaporation under reduced pressure, which avoids side reactions [21] and allows convenient purification of the crude product. Alternatively, the O-acetyl protecting groups can be removed with hydrazine hydrate in methanol while the peptide remains anchored to the resin (Fig. 9) [30,33,34,43,44]. Even though O-acyl protection is commonly employed in glycopeptide synthesis, deacetylation of glycopeptides has occasionally been reported to be accompanied by various side reactions, including β-elimination [84], carbohydrazide formation [85], partial formylation [13] and cysteine-catalyzed degradation of the peptide backbone [76].

As an alternative to the O-acyl protection of glycopeptides, the trimethylsilyl (TMS) group was used for protection of the hydroxy groups of 1-amino-alditols [78,86] and glycosylamines that were N-glycosydically linked to asparagine (Fig. 12, **27–29**) [86]. The more stable *tert*-butyldimethylsilyl (TBDMS) and *tert*-butyldiphenylsilyl (TBDPS) protective groups have been used for the glycosylated 5-hydroxynorvaline building block (Fig. 12, **30**) employed in the preparation of a glycopeptide analogue of type II collagen [52]. The cleavage cocktail (TFA–water–thioanisole–1,2-ethanedithiol, 87.5:5:5:2.5) removed the silyl and isopropylidene protective groups without affecting the glycosidic bond. The *tert*-butyldimethylsilyl protective group has also been employed in the synthesis of a glycopeptide from HIV gp120 containing the Tn [α-D-GalNAc-(1→O)-Thr] epitope (Fig. 12, **31**) [84]. Regardless of these examples, these protective groups cannot be used for labile deoxygenated saccharides [10].

O-Benzoyl protection of the glycosyl donor was proved to give increased yields in glycosylations [26,87], but the removal of this group demands prolonged treatment with base [42,88] that may result in side reactions (e.g., β-elimination and epimerization of peptide stereocenters) [42,83].

Benzyl ethers are removed by hydrogenolysis [24–26,54], and their use is therefore restricted to glycopeptides that lack cysteine and methionine. In addition, precautions have to be taken in the hydrogenation of tyrosine-containing glycopeptides. Thus, the tyrosine side chain of a glycopeptide fragment of the interleukin-8 (IL-8) receptor was reduced to cyclohexanone (51%) during the debenzylation step [27]. No effort was made to minimize this side reaction, which was attributed to the long reduction time (6 days). Partial debenzylation of the O-linked disaccharide side chain O-(2,3,4,6-

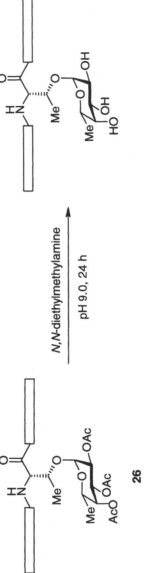

Figure 11   Example of deacetylation using *N,N*-diethylmethylamine in water.

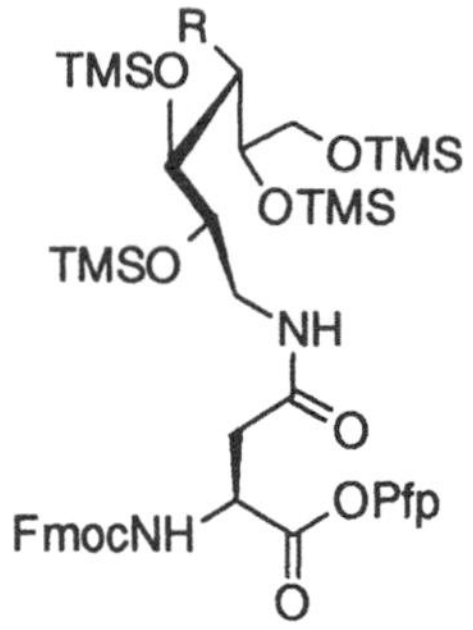

R= (OTMS)₄-β-D-Gal-O **27**
R= (OTMS)₄-β-D-Glc-O **28**
R= (OTMS)₄-α-D-Glc-O **29**

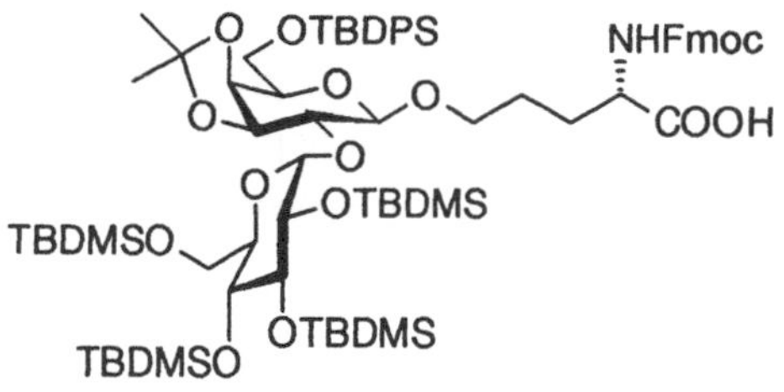

30

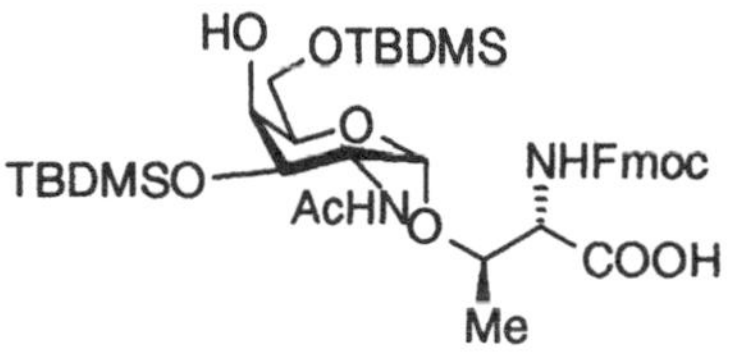

31

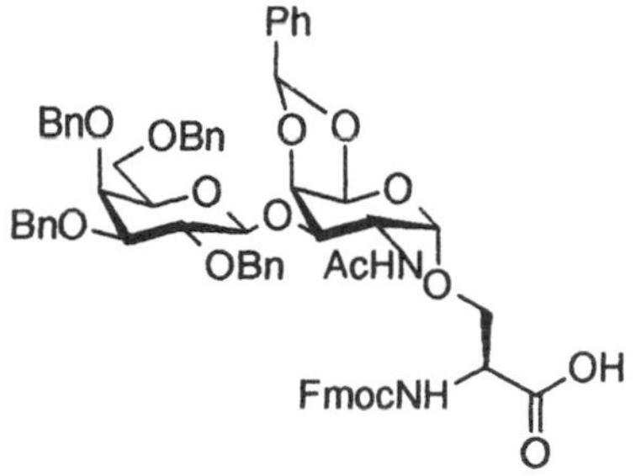

32

Figure 12 Examples of protected-glycosylated amino acids.

tetra-*O*-benzyl-β-D-galactopyranosyl-(1→3)-2-acetamido-4,6-*O*-benzyli-dene-2-deoxy-α-D-galactopyranosyl) (Fig. 12, **32**) was reported during the cleavage–deprotection of a 27-mer glycopeptide (asialo-[Ala¹⁸]-B-chain of human α2HS glycoprotein) with TFA and a scavenger mixture [24]. Debenzylation after treatment of the resin-bound peptide with TFA–water–1,2-ethanediitiol (92.6:4.9:2.5) was also observed by Guo et al. [26] in the synthesis of a CD52 glycopeptide carrying an *N*-linked pentasaccharide.

II. PHOSPHOPEPTIDES

A. Introduction

Because of the important role of phosphorylation and dephosphorylation in the regulation of cellular processes [89–92], phosphopeptides have received much attention as useful biological and biochemical tools for studying and elucidating the role of these posttranslational cellular events. The increased interest in protein phosphorylation and the need to have access to phospho-peptides for different applications have resulted in substantial efforts to de-velop methods for the preparation of these modified peptides. Three methods are currently used for the chemical synthesis of phosphopeptides on the solid phase: (1) the building block or preassembly approach, where fully or par-tially protected phosphoamino acid derivatives are incorporated stepwise onto the growing peptide resin; (2) the postassembly or global phosphory-lation approach, where the free hydroxyl groups are phosphorylated after chain assembly is complete; and (3) the use of side-chain-unprotected phos-phoamino acids. The following discussion provides general information about the procedures employed in each approach and highlights their merits and drawbacks. The selection of the synthetic strategy will depend on the sequence and final application of the phosphopeptide.

B. The Building Block or Preassembly Approach

This method involves the incorporation of fully or partially protected phos-phoamino acid derivatives into the desired site of the growing resin-bound peptide. A number of studies have been reported in which this strategy has been utilized to synthesize phosphopeptides of significant length and func-tional complexity, especially phosphotyrosine-containing peptides. For sim-plicity, the syntheses of phosphotyrosyl peptides and phosphoserine- and phosphothreonine-containing peptides are presented separately. The syn-thetic schemes employed to obtain the protected synthons are not discussed, but adequate references are provided.

1. Phosphotyrosine-Containing Peptides

The synthesis of phosphotyrosine-containing peptides by Boc methodology has been accomplished mainly by using methyl [93–96] and benzyl [97,98] protected phosphotriester derivatives. Unfortunately, these protecting groups do have some limitations. Incomplete incorporation of the N^α-Boc-Tyr(PO$_3$Me$_2$)-OH building block and partial demethylation and dephosphor-ylation during peptide assembly were encountered during the solid-phase synthesis of a 14-mer fragment of the regulatory domain of pp60^{c-src} [93].

Monodebenzylation of the resin-attached Tyr(PO$_3$Bzl$_2$) residue oc-curred under Boc deprotection conditions, but this side reaction did not preclude the synthesis of a 13-mer phosphotyrosyl peptide in good yield [98]. A major S-benzylsulfonium salt by-product (66 to 83% of the crude material as determined from the HPLC chromatogram peaks) was identified when N^{α}-Boc-Tyr(PO$_3$Bzl$_2$)-OH was used in the synthesis of methionine-containing peptides derived from the native PDGF-β sequence [97]. This by-product was converted quantitatively into the desired phosphopeptide by catalytic hydrogenation using 20% palladium/carbon in methanol.

In addition to the preceding synthetic problems, anhydrous hydrogen fluoride, often used for the final cleavage–deprotection step in Boc methodology, was found to be incompatible with phosphotyrosine-containing peptides because the aryl oxygen-to-phosphorus bond is not completely stable to this acid [99,100]. To overcome this important drawback, several deprotection procedures currently used in peptide Boc chemistry were investigated by Otaka and collaborators [101]. As a result of this study, an optimized two-step deprotection protocol that minimizes dephosphorylation and other side reactions (e.g., migration of phosphate groups or alkylation of methionine and tryptophan residues) was proposed. The new procedure involves incorporation of the O,O-dimethyl–protected phosphorylated amino acid derivative and, after peptide assembly is complete, removal of the protecting groups using a two-step deprotection method consisting of a high-acid treatment [for 4 μmol of protected peptide resin, 1 M TMSOTf–thioanisole in TFA (2.0 mL), m-cresol (100 μL), EDT (100 μL) for 1.5 h at 4°C] and a low-acid treatment [first-step reagent + DMS − TMSOTf (1 mL, 30:20 to 40:10, v/v) for 2 to 4 h at 4°C]. This synthetic method was verified by preparing a 19-mer MAP-kinase derived peptide containing two phosphoamino acid residues as well as methionine and tryptophan. The purified peptide was obtained in 24% yield based on the protected peptide resin loading.

The milder conditions for N^{α}-amino temporary deprotection and final cleavage–deprotection from the solid support in Fmoc methodology have promoted more diversified work in the synthesis of protected phosphotyrosine derivatives. Thus, methyl [94,102–104], benzyl [104–109], $tert$-butyl [104,105,109–111], dimethylamine [112], propylamine [113], isopropylamine [113], (methydiphenylsilyl)ethyl [114], and ally [104] O,O-protected phosphotyrosine derivatives have been described. Using a test pp60^{s-src}-derived peptide sequence in the continuous-flow method, Kitas et al. [104] reported no significant differences in the reactivities of the methyl-, benzyl-, $tert$-butyl- and allyl-protected synthons or in the incorporation of subsequent residues. However, several drawbacks associated with some of the derivatives have been reported.

Rapid monodealkylation was observed during N^{α}-Fmoc deprotection of methyl- and benzyl-protected phosphates by the customary piperidine–DMF solution treatment [103,108]. Although this side reaction does not seem to compromise peptide purity, it could be decreased but not completely suppressed by employing the nonnucleophilic base 1,8-diazabicyclo[5.4.0]undec-7-ene (DBU) (1 to 2.5% solution in DMF or DMA) [104,108,109], albeit at the risk of causing succinimide formation (Fig. 8) [109] from cyclization of aspartyl residues [77].

Complete removal of the methyl groups of Tyr(PO_3Me_2)-containing peptides requires harsh acidolytic or silylytic treatments [see the preceding discussion on N^α-Boc-Tyr(PO_3Me_2)-OH]. For peptides synthesized using the Fmoc mode, the deprotection and release of the resin-bound peptide have been performed mainly by adding trimethylsilyl bromide (TMSBr) to standard cleavage cocktails [e.g., TMSBr–thioanisole–TFA–m-cresol (16.1: 14.8:93.8:1) for 16 h at 4°C] [104]. Alternatively, demethylation of O,O-dimethyl-phosphotyrosyl peptides prior to the TFA-mediated side-chain deprotection and cleavage from the solid support can be carried out by means of trimethylsilyl iodide (TMSI) in acetonitrile (Fig. 13) [115]. This method has been demonstrated to be suitable for sequences containing arginine, histidine, methionine, or tryptophan residues.

a.   Dealkylation of O,O-Dimethyl-Phosphotyrosyl Peptides on Solid Support

1. Prior to the incorporation of the O,O-dimethyl-phosphotyrosyl synthon, the synthesis of the desired peptide is performed using standard Fmoc protocols. Coupling of N^α-Fmol-Tyr(PO_3Me_2)-OH is achieved by first dissolving the protected amino acid (2 equiv.), DIEA (2.2 equiv.), and TPTU (2 equiv.) in NMP, waiting 3 min for preactivation, and adding the mixture to the resin preswollen in NMP.

2. The reaction vessel is mixed for at least 90 min. The incorporation of the amino acid is monitored using the ninhydrin test [74]. After coupling, the vessel is drained and the peptide resin washed three times with NMP.

3. After the incorporation of the preceding building block and for the rest of the synthesis, the removal of the Fmoc group is performed with a 2.5% solution of DBU in DMA for 10 min. **Caution:** DBU can cause succinimide formation from cyclization of aspartyl residues.

4. Synthesis then resumes until the entire peptide chain is assembled. After the incorporation of the last amino acid, the peptide resin is preswollen in MeCN. Then a 20-fold molar excess of TMSI in MeCN is added and the reaction vessel is mixed for 4 h at room temperature. After this reaction time, the vessel is drained and the peptide resin is washed with MeCN, MeOH, DMA, and DCM (three times each). To ensure clean and effective demethylation, the use of freshly prepared trimethylsilyl iodide or commercially available reagent packaged in ampules is recommended.

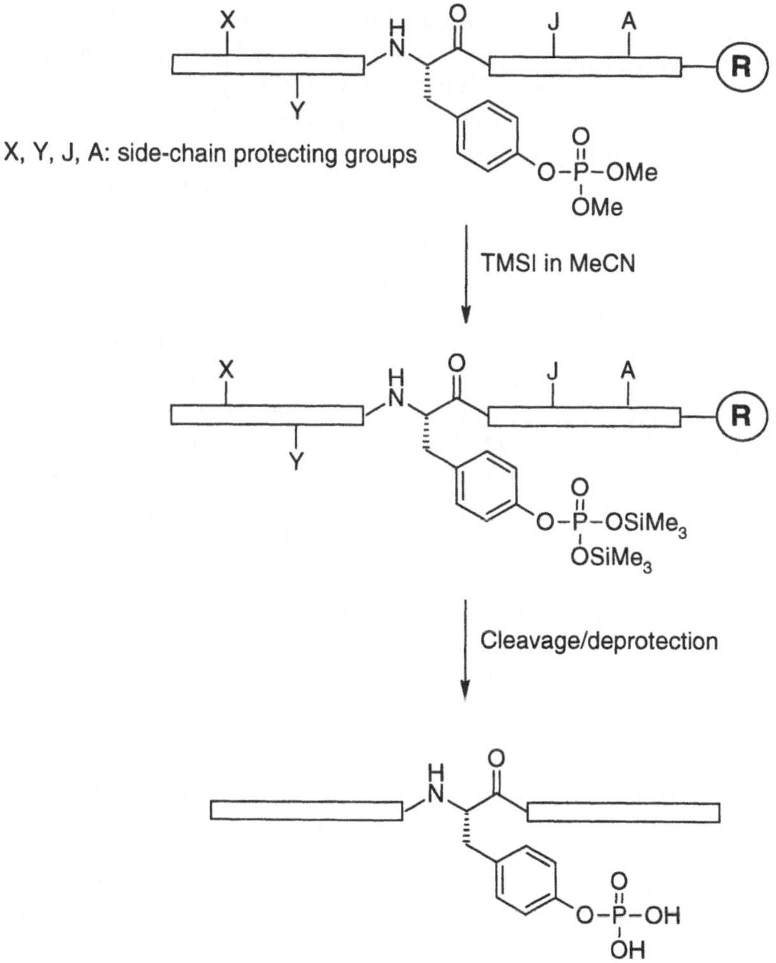

Figure 13 Dealkylation of *O,O*-dimethyl-phosphotyrosyl peptides on solid support.

5. After the completion of the dealkylation reaction, the peptide resin is ready for final cleavage and deprotection. As usual, the scavenger cocktail is dictated by the linker used and the protecting groups present in the sequence.

In contrast to the methyl ester, the benzyl phosphonate group is labile to mild acidolytic conditions. Cocktails of TFA–anisole, water, or phenol (19:1) or TFA–water–anisole–thioanisole–1,2-ethanedithiol (82.5:5:5:5:

2.5; reagent K) have been used [104–109]. In an evaluation of protected Fmoc-phosphotyrosine derivatives, Valerio et al. [105] reported minor dephosphorylation (2 to 3%) of Tyr(PO$_3$Bzl$_2$) during peptide assembly. A higher value (~7% by weight) was previously observed by Kitas et al. [108] in the synthesis of a tridecapeptide.

The monobenzyl protected N^α-Fmoc-Tyr[PO(OH)(OBzl]-OH derivative has been introduced [116]. This building block is incorporated using standard coupling agents (e.g., TBTU **19** or PyBOP **21**) and is suitable for automated synthesis. No complications in the coupling of subsequent residues seem to arise from the presence of the monoprotected phosphate in the growing peptide resin. The benzyl group is removed during the course of the standard TFA-based simultaneous cleavage–deprotection reaction.

The *tert*-butyl phosphate protection was considered ideally suited for use in Fmoc chemistry [104,109,110,117]. This protecting group is not susceptible to nucleophilic attack during Fmoc deprotection and its removal is readily effected during the acidolytic cleavage of the peptide from the solid support. Unfortunately, N^α-Fmoc-Tyr(PO$_3$tBu$_2$)-OH is not stable on storage because of autocatalyzed decomposition. In spite of this problem, it was the preferred derivative for use in Fmoc solid-phase peptide synthesis in an evaluation of several N^α-Fmoc-Tyr(PO$_3$R$_2$)-OH derivatives (R = *t*Bu, Bzl, and H) [105]. The same study revealed that the tyrosyl di-*tert*-butyl phosphorodiester is subject to minor dephosphorylation through the action of nucleophilic scavengers used in the acidolytic cleavage–deprotection treatment.

The hydrolytic lability of the phosphorus–nitrogen bond under acidic conditions has inspired the development of several amide-based protecting groups (e.g., Tyr[PO(NHR)$_2$, R = *n*-Pr and *i*-Pr] [113] or Tyr[PO(NMe$_2$)$_2$] [112]). Decomposition (~79%) was observed in stability studies when N^α-Z-Tyr[PO(NH*n*-Pr)$_2$]-OBzl was treated with 2 equiv. of tetrabutylammonium fluoride (TBAF) hydrate. Therefore, the use of TBAF for Fmoc deprotection should be avoided for this protected synthon.* Differences in the acid-mediated cleavage of the amide-based phosphate protecting groups have been reported. The deprotection of Tyr[PO(NH*n*-Pr)$_2$] with TFA–water (19:1) is faster than for Tyr[PO(NMe$_2$)$_2$] with TFA–water (9:1), which requires more than 10 h for completion. No significant differences between propyl and isopropyl were observed by Ueki et al. [113], although the propyl protecting group is more stable on storage and more convenient for coupling.

---

*Tetrabutylammonium fluoride was initially used by Arendt et al. [118] for the deprotection of diphenyl phosphotriesters of serine and threonine while the peptide was still attached to the resin. Later, it was determined that TBAF attacks the phosphoester bond, giving a mixture of phosphorylated and unphosphorylated peptides. As a general precaution, the use of TBAF for Fmoc deprotection should be avoided.

Allyl- [104] and (methyldiphenylsilyl)ethyl [114] O,O-protected Fmoc phosphotyrosine derivatives have also been described and their use illustrated with the synthesis of several peptides. The allyl protecting group adds another degree of flexibility to the protection scheme because it is orthogonal to Fmoc/tBu. Mild and effective removal of the allyl group is accomplished in solution by overnight treatment under an argon atmosphere with Et_3N- HCOOH (1:1, v/v), PPh_3, and $Pd(PPh_3)_4$ in 1,4-dioxane–THF–water (1:1:1) [104]. Alternatively, the deprotection can be carried out on the solid support before the TFA-based cleavage–deprotection treatment.

The silyl-based (methyldiphenylsilyl)ethyl protecting group is removed by treatment with TFA–phenol–Et_3SiH (95:3:2) for 2 h at room temperature, and its application was illustrated by the synthesis of three peptides, one of them containing cysteine and methionine [114].

2. Phosphoserine- and Phosphothreonine-Containing Peptides

Originally, the synthesis of phosphoserine- and phosphothreonine-containing peptides was generally carried out by Boc [119,120] or Aloc [121,122] methods to avoid the ready β-elimination of the bis-protected phosphate group under basic conditions in Fmoc methodology [123].

The allyl [124], benzyl [119,125,126], 3-chlorobenzyl [127], cyclohexyl [126,127], cyclopentyl [127], 2,6-dimethylphenyl [128], methyl [101,120], 2-methylphenyl [128], phenyl [118,119,122,129], and 2-$tert$-butylphenyl [128] groups have been used for O,O-phosphate protection of Boc-phosphoamino acids in solid-phase peptide synthesis.

The allyl and phenyl groups require a two-step reaction for final deprotection. In addition, the use of the phenyl-based groups, which can be removed by catalytic hydrogenation over PtO_2, is restricted to peptides devoid of aromatic or sulfur-containing amino acids. For the phenyl group, incomplete deprotection in the hydrogenation step has been reported for large peptides [121].

As for the phosphotyrosine-containing peptides, the final cleavage from the resin with hydrogen fluoride has been shown to cause complications in certain peptide sequences [128,129]. Extensive dephosphorylation of the Ser/Thr(PO_3Ph_2)-residue in a model peptide occurred using three commonly employed hydrogen fluoride cleavage conditions: (1) anisole–thiocresol–dimethyl sulfide–HF (1:1:13:13), 0°C, 2 h, then anisole–HF (1:19), 0°C, 45 min ("low–high HF"); (2) anisole–HF (1:19), 0°C, 45 min; and (3) neat HF, 0°C, 45 min [129]. The extent of dephosphorylation was higher for the low–high procedure and was dependent of the HF contact time. In addition, the Ser(PO_3Ph_2) residue underwent dephosphorylation at a slightly higher rate than the Thr(PO_3Ph_2) residue. These results contradict the data reported by Arendt et al. [118] on the synthesis of a 7-mer phosphopeptide using Boc-diphenylphosphono esters of serine and threonine. No dephosphorylated peptides were detected in the crude products after cleavage of the peptide-resins with liquid HF–anisole.

In search of protective groups more stable during the hydrogen fluoride treatment, Tsukamoto et al. [128] examined several substituted phenyl groups in the synthesis of a 13-mer phosphoserine-containing peptide. The 2-methyl- and 2,6-dimethylphenyl groups decreased but did not completely suppress dephosphorylation during HF treatment. Effective suppression of dephosphorylation was obtained with the 2-*tert*-butylphenyl group, but the stability of substituted phenyl groups with HF treatment appears to be related to the difficulty of their final cleavage by hydrogenolysis. Thus, the reductive cleavage of the 2-*tert*-butylphenyl group was slow, required the use of large amounts of PtO_2, and gave considerable amounts of side products.

Monodebenzylation of the Thr(PO$_3$Bzl$_2$) residue (10–30% yield) was observed in solution under Boc deprotection conditions [125,130]. This side reaction, as with the Tyr(PO$_3$Bzl$_2$) analogue (see Section II.B.1), did not preclude the solid-phase synthesis of a 12-mer Cys-(EGFR[649–659]) peptide [125]. For this particular example, the simultaneous cleavage and deprotection of the peptide from the phenylacetamidomethyl (PAM) linker **33** was performed using 1 M TFMSA in TFA in the presence of thioanisole [130,131]. As an alternative to the TFA-labile benzyl group, Wakamiya et al. [132,133] proposed the cyclohexyl group, but removal of this group at the end of the synthesis requires a hard acid treatment for an extended period of time. To overcome this problem, the cyclopentyl and 3-chlorobenzyl groups were proposed by the same authors [134]. Both protective groups are sufficiently stable to trifluoroacetic acid and are more easily removed than the cyclohexyl group by treatment with TFMSA and scavengers in TFA. A comparative study of the cyclopentyl and cyclohexyl protecting groups using a phosphoserine-containing peptide related to heat shock protein HSP27A, has been published [127].

33

The Ser/Thr(PO$_3$Me$_2$)-protected amino acids have been rarely used because the harsh acid treatment required for complete removal of protecting groups can result in severe side reactions. As discussed previously (Section

II.B.1), Otaka et al. [101,120] have proposed a two-step deprotection protocol consisting of high-acid (S_N1/S_N2) and low-acid (S_N2) reagent systems for the deprotection of O,O-dimethyloxyphosphinyl amino acid derivatives. Using these optimized deprotection conditions, several phosphopeptides in 69–81% yields, based on the loading of the protected peptide resins, were obtained [101].

As already described, the use of Fmoc-protected serine or threonine phosphotriester building blocks was precluded because of the high susceptibility of the phosphotriester functionality to undergo β-elimination under the basic conditions required for Fmoc group removal on the solid support.* Two independent groups have reported a novel and practical procedure based solely on Fmoc solid-phase synthesis. The monobenzyl protected N^α-Fmoc-Ser[PO(OH)(OBzl)]-OH and N^α-Fmoc-Thr[PO(OH)(OBzl)]-OH [135] are used as synthons in standard Fmoc SPPS [116,123,127,136,137]. The phosphodiester group is no longer sensitive to β-elimination and no complication in the coupling of subsequent residues seems to result from the use of the monobenzyl phosphate. The monobenzyl-protected phosphoamino acids are incorporated using BOP **18**–HOBt **17** (1:1), TBTU **19**, HBTU **20**, or TPTU **34** and are suitable for automated Fmoc synthesis.

34

A combined Fmoc/Aloc building block strategy has been applied with success for the synthesis of phosphoserine-containing peptides using N^α-Aloc-Ser(PO$_3$Al$_2$)-OH as the key component [138,139]. After incorporation of the protected synthon at the N-terminus of a sequence prepared by standard Fmoc solid-phase synthesis, the allyl protecting groups are removed and a protected fragment is coupled to complete the target sequence (Fig. 14). The cleavage of the allyl groups is performed by treatment of the resin suspended in dichloromethane with a premixed solution of trimethylsilylazide and tetrabutylammonium trihydrate, followed by addition of Pd(PPh$_3$)$_4$

---

*The stability of protected phosphoamino acid derivatives in 20% piperidine in DMF was examined in detail by Wakamiya et al. [123]. This study revealed that β-elimination occurs in the diphenyl protected phosphate, but for the Bzl(4-NO$_2$), Bzl, and Me groups the cleavage of one protecting group to produce the O-monoprotected phosphono derivatives occurs faster than the β-elimination.

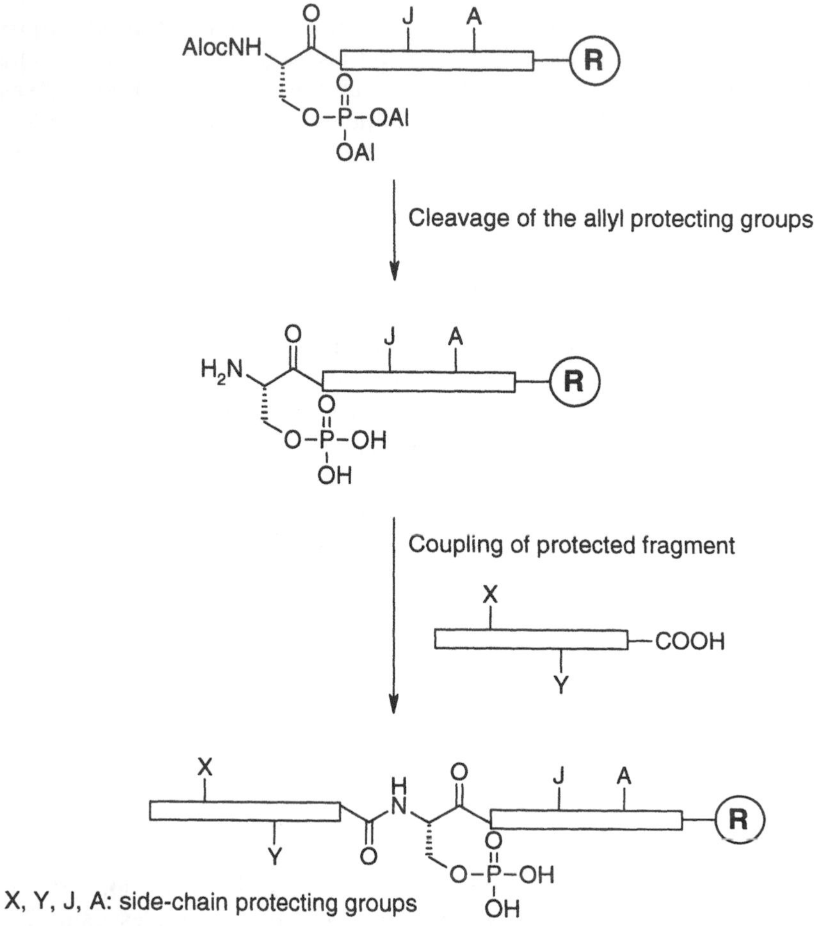

Figure 14 Fmoc/Aloc strategy for the synthesis of a resin-bound peptide containing an unprotected phosphate group.

under argon (Fig. 14). This strategy is particularly suitable in cases in which a resin-bound peptide containing a free phosphate group is required.

C. The Postassembly or Global Phosphorylation Approach

The postassembly approach involves the phosphorylation of side-chain unprotected hydroxyl groups on the solid support (Fig. 15). The global phosphorylation on the resin offers the advantage that the unphosphorylated and

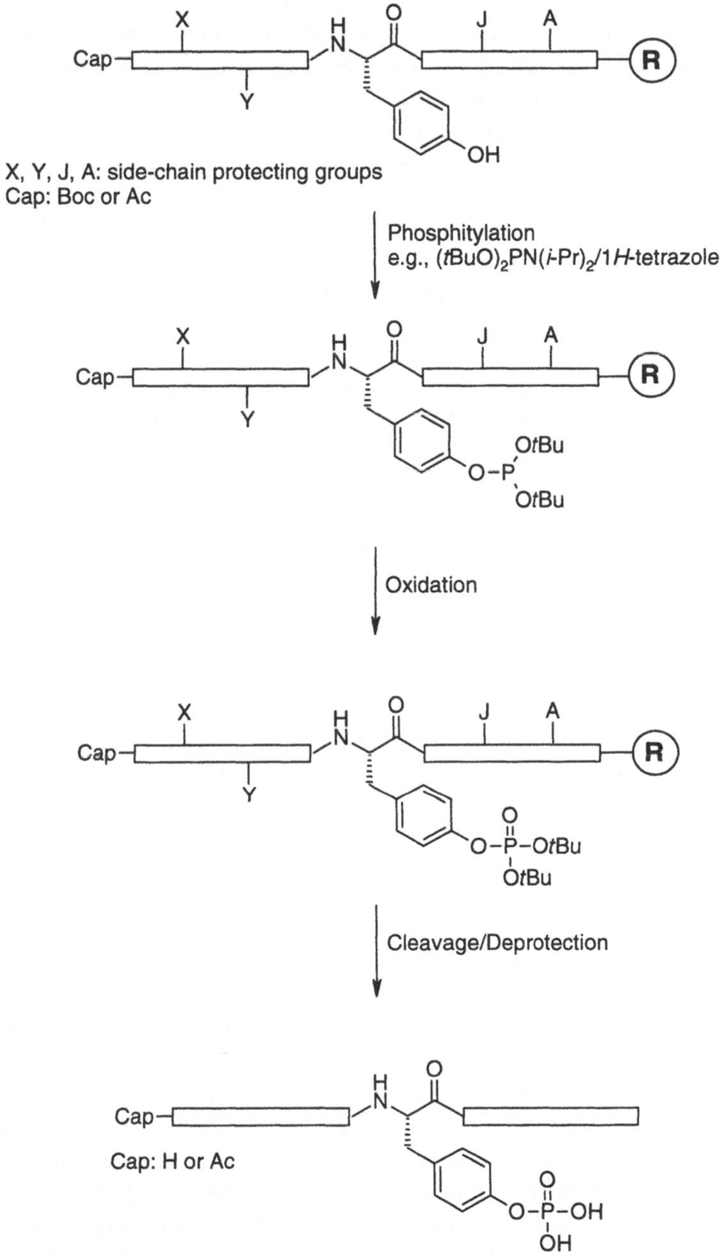

Figure 15 Solid-phase synthesis of a phosphotyrosyl peptide by the postassembly approach.

phosphorylated sequence can be obtained from one batch. Because the functionalization of the free hydroxyl group is performed at the end of solid-phase assembly, this approach allows the use of an array of different phosphoryl protecting groups, even those that are not fully compatible with the conditions of the elongation cycles. The amino acid to be phosphorylated on the solid support is generally incorporated without side-chain protection using pentafluorophenyl or 3-hydroxy-4-oxo-3,4-dehydro-1,2,3-benzotriazine esters **23**, DCC **16**–HOBt **17**, BOP **18**, TBTU **19**, HBTU **20**, PyBOP **21**, or TPTU **34** as coupling agents. However, there is always an increasing risk of side reactions with every coupling step after incorporation of unprotected side-chain hydroxyl groups, particularly for longer sequences having the residue to be phosphorylated close to the *C*-terminus. To mask this functional group during peptide assembly, selectively protected derivatives such as N^α-Fmoc-Ser(TBDMS)-OH [140] or N^α-Fmoc-Ser(Trt)-OH [141,142] have been used. The *tert*-butyldimethylsilyl (TBDMS) group is selectively cleaved with fluoride (2.0 equiv. of TBAF hydrate in DMA at room temperature for 1 h), unmasking the serine hydroxyl to be phosphorylated. The trityl group is removed from serine by exposure to solutions of 1% TFA–5% triethylsilane (3 × 3 min) in DCM or 4% TCA in DCM (10 × 2 min) [**Caution:** these conditions can also cause partial detritylation of Cys(Trt)].

If the target peptide has a free *N*-terminal group, this functionality is protected with Boc, either by direct incorporation of the final residue as the appropriate Boc/*t*Bu derivative or by acylation of the free *N*-terminus with Boc$_2$O. With this synthetic approach, the *N*-terminal protecting group is removed when the entire peptide resin is deprotected and cleaved (Fig. 15). Protecting the *N*-terminus with the Boc group prevents undesired reactions during the phosphorylation step and eliminates exposure to a final treatment of piperidine to remove the *N*-terminal Fmoc group.

The free resin-bound hydroxyl group is functionalized by means of phosphochloridates, *H*-phosphonates, or phosphoramidites (Table 2).

Phosphorochloridic acid dibenzyl ester (Fig. 16, **35**) is a sluggish reagent [143], and it can yield in addition to the target peptide a great number of unidentified by-products [144]. Better results have been obtained with phosphorochloridic acid dipentafluorophenyl ester (Fig. 16, **36**) [145].

The use of benzyl *H*-phosphonate (Fig. 16, **37**) was illustrated with the synthesis of a 5-mer phosphoserine-containing peptide [146]. The reagent was activated by pivaloyl chloride in pyridine, followed by in situ oxidation with 1% iodine (w/v) in pyridine–water (98:2 v/v). Phosphoramidites are considered to be superior to the other phosphitylating reagents and have been extensively used for the synthesis of peptides of significant functional complexity. Kitas et al. [104] did not observe differences in reactivity for phosphoramidites **d**, **i**–**k** (Table 2), but the use of this type of

Table 2 Reagents for the Synthesis of Phosphopeptides by Phosphorylation on the Solid Support

Entry	Reagent	Method	References
a	Phosphorochloridic acid dibenzyl ester	Fmoc	[143,144]
b	Phosphorochloridic acid dipentafluorophenyl ester	Fmoc	[145]
c	Benzyl H-phosphonate	Fmoc	[146]
d	Dibenzyl N,N-diisopropylphosphoramidite	Fmoc	[104,140,147, 148–150]
e	Di-t-butyl N,N-diethylphosphoramidite	Fmoc	[117,142,144, 148,151 152, 153–155, 156–158]
		Boc/Fmoc	[159]
f	Dibenzyl N,N-diethylphosphoramidite	Boc/Fmoc	[159]
g	Di-4-chlorobenzyl N,N-diisopropylphosphoramidite	Fmoc	[160,161]
h	Di[2-(trimethylsilyl)ethyl] N,N-diisopropylphosphoramidite	Fmoc	[162]
i	Di-t-butyl N,N-diisopropylphosphoramidite	Fmoc	[104]
j	Dimethyl N,N-diethylphosphoramidite	Fmoc	[104]
k	Diallyl N,N-diisopropylphosphoramidite	Fmoc	[104]

reagent is not exempt from side reactions. A derivative with two peptide fragments linked through a phosphodiester bridge and an H-phosphonate by-product was reported by Hoffmann et al. [144] upon use of dibenzyl N,N-diisopropylphosphoramidite (Fig. 16, **38**) in a chemical phosphorylation study. The H phosphonate derivative (Fig. 17, **41**) was initially identified and described by Ottinger et al. [158], who unsuccessfully tried to eliminate this by-product by changes in oxidation conditions (e.g., temperature, reaction time, and/or reagents). The resolution of H-phosphonate formation has been reported by Perich [163]. The H-phosphonate by-product, which was originally attributed by Ottinger et al. to incomplete oxidation of the phosphite triester [158], results from 1H-tetrazole acid-mediated decomposition of the intermediate di-tert-butyl or dibenzyl phosphite triester and rearrangement of the resultant hydroxy phosphite diester. The side reaction is rectified by using iodine in THF–pyridine–water (1:1:1 v/v/v) for the oxidation step. This system converts the H-phosphonate to the corresponding

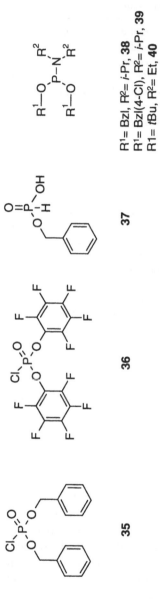

Figure 16 Examples of reagents for the phosphorylation of hydroxy groups on solid support.

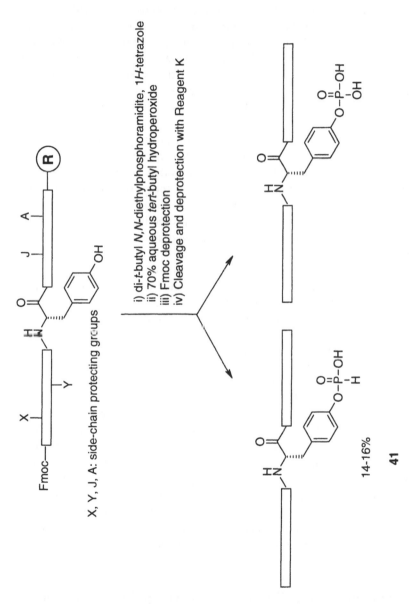

Figure 17 *H*-phosphonate by-product identified in the solid-phase synthesis of a phosphotyrosyl peptide by the postassembly approach.

phosphodiester, which upon acidolytic treatment during the cleavage–deprotection step gives the target dihydrogen phosphate functionality.

After phosphitylation, the initially formed phosphoric acid triester is oxidized to the corresponding phosphotriester. The oxidation is effected using *m*-chloroperbenzoic acid, 1 M iodine in THF–2,6-dimethylpyridine–water (40:10:1 v/v/v) or *tert*-butyl hydroperoxide. This oxidation step might cause side reactions in peptides containing oxidation-sensitive amino acids such as methionine, cysteine, and tryptophan. A study by Bannwarth et al. [148] has demonstrated that peptides containing the preceding residues can be phosphorylated by the postassembly approach if iodine–THF–2,6-dimethylpyridine–water is used for the oxidation of the phosphoric acid triester to the phosphotriester. Alternatively, allyloxycarbonyl protection on the indole moiety of tryptophan [164] is recommended when oxidation is performed under acidic conditions. However, N^α-Fmoc-Trp(Aloc)-OH requires the use of DBU for Fmoc removal because significant cleavage of the Aloc group was detected under standard Fmoc deprotection conditions (20% piperidine in DMF). The Aloc group is removed from the indole following synthesis with Pd(PPh$_3$)$_4$ [148,164] or piperidine–water [148].

The cleavage and simultaneous deprotection of the resin-bound peptide after phosphorylation are performed in accordance with the linker, sequence, and protecting groups of the phosphate moiety (see Section II.B).

The phosphorylation efficiencies of this approach vary widely depending on the reactivity of the hydroxyl group and steric constraints. Both effects are highly sequence dependent and cannot be predicted. To reduce the potential for steric hindrance, the phosphorylation of an unprotected tyrosine residue was carried out by conventional methods immediately after introduction of the amino acid (Fig. 18) [156]. This intraassembly or "on-line" phosphorylation strategy allowed the synthesis of a phosphotyrosine-containing peptide (Stat 91[695–708]), which was inaccessible by the building block or "classical" postassembly phosphorylation procedures. The on-line phosphorylation of tyrosine was also used in the syntheses of a 9-mer model peptide and a 17-mer Fcγ receptor–derived peptide [157]. The isolated crude phosphopeptides showed less by-product formation than with the standard global phosphorylation method. Steric constraints have also been overcome by adopting a segment condensation strategy [142,155] or by using the *N*-(2-hydroxy-4-methoxybenzyl (Hmb) backbone amide-protecting group [165].

1. General Protocol for Phosphorylation Using Phosphoramidites

1. The synthesis of the desired peptide is performed using standard Fmoc protocols. The residue to be phosphorylated is incorporated

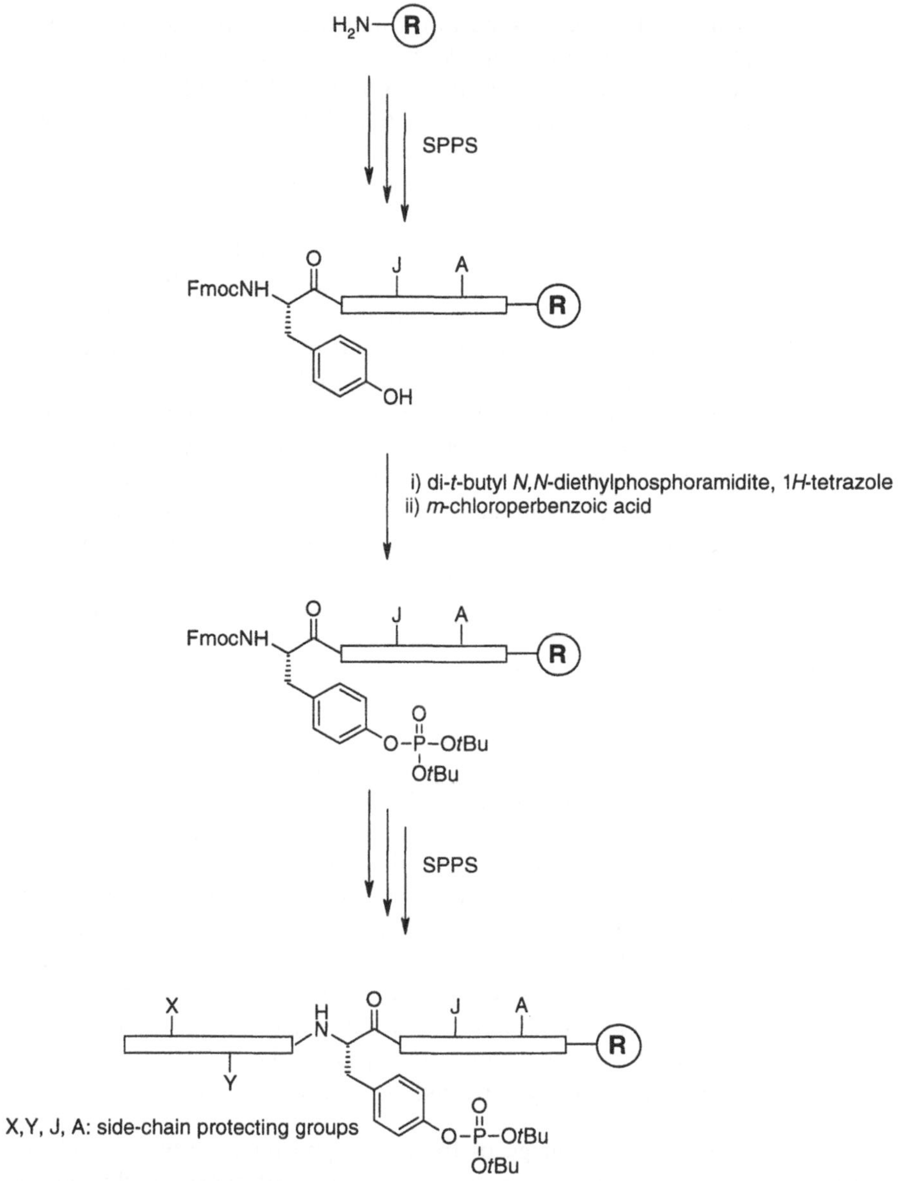

Figure 18 Interassembly or "on-line" phosphorylation strategy.

without side-chain protection using the coupling agents reported above. If the target peptide has a free N-terminal group, the last amino acid is coupled as the appropriate N^{α}-Boc/tBu derivative or, after Fmoc deprotection, the free N-terminus is acylated with Boc_2O (10 equiv.) in DMF for at least 1 h.

2. When the entire peptide is assembled, the resin is dried under high vacuum overnight.

3. The resin-bound peptide is washed several times with anhydrous DMF and the reaction vessel is sealed with a rubber septum and flushed with a stream of dry argon delivered via a needle.

4. Anhydrous DMF is added to a vial of DNA grade 1H-tetrazole using a dry syringe. An aliquot (50-fold molar excess of 1H-tetrazole) is transferred with a syringe to a vial containing dibenzyl N,N-diisopropylphosphoramide (25-fold molar excess), and the resulting solution is mixed and added, also via syringe, to the reaction vessel. Additional anhydrous DMF may have to be added to ensure complete coverage of the resin. **Caution:** Phosphoramides are moisture and air sensitive, and the manipulation of these reagents must be carried out under an atmosphere of dry argon and using dry syringes.

5. After 2 h, the reaction vessel is drained and the peptide resin washed four times with DMF.

6. Iodine (1 M) in THF–2,6-dimethylpyridine–water (40:10:1 v/v/v) (25-fold molar excess) is added and the reaction vessel is mixed for 30 min. After this time, the vessel is drained and the peptide resin is washed with DMF and diethyl ether (three times each) and dried in vacuum. The peptide resin is ready for final cleavage and deprotection. As usual, the scavenger cocktail is dictated by the linker used and the protecting groups present in the sequence.

7. The phosphopeptide is purified by ion-exchange chromatography or preparative HPLC on a reversed-phase column. The separation of phosphopeptide from nonphosphorylated impurities can be carried out by affinity chromatography on Fe^{2+} Chelex gel (iminodiacetic acid epoxy activated Sepharose 6B) [118].

One important advantage of the global phosphorylation method over the other approaches is that it allows the synthesis of phosphorothioate peptides from the same batch. The solid-phase synthesis of these modified peptides can be accomplished using di-4-chlorobenzyl N,N-diisopropylphosphoramidite **39** and 1H-tetrazole followed by sulfurization using phenylacetyl disulfide in dry DMA. This method was applied to the synthesis of a 15-mer O-phosphorothioylserine–containing peptide using the Wang linker

3 in a kieselguhr-supported poly(N,N-dimethylacrylamide) solid support [161]. The optimal cleavage–deprotection conditions after phosphorothioylation involved the use of a mixture of TMSBr–thiophenol–m-cresol–TFA (1.1:1:1:6.4) for 1 h at 0°C under an argon atmosphere.

A similar strategy was used by Tegge [154] for the synthesis of thiophosphorylated peptides related to the EGFR[1168–1180] sequence. Global phosphorylation of the unprotected hydroxyl amino acids (serine, threonine, and tyrosine) on the solid support was carried out with di-t-butyl N,N-diethylphosphoramidite **40** and 1H-tetrazole followed by in situ sulfurization with dibenzoyl tetrasulfide in THF. If possible, this phosphoramidite should be avoided for the introduction of the phosphite moiety in the synthesis of phosphorothioate peptides because S-$tert$-butylation was detected during the cleavage step. Whereas the serine- and threonine-thiophosphate group was stable to the deprotection conditions (TFA containing 3% triisobutylsilane and 2% water), tyrosine-thiophosphate was hydrolytically labile under acidic conditions.

A similar result was observed by Kitas et al. [150] in the synthesis of an insulin receptor derived O-thiophosphotyrosyl peptide (IR[1142–1153]). In this example, the thiophosphorylation reaction was performed using dibenzyl N,N-diisopropylphosphoramidite **38** for the phosphitylation step followed by oxidation with a 5 M solution of elemental sulfur in CS_2–pyridine (1:1) or a 15% solution of tetraethylthiuram disulfide in acetonitrile. Stability measurements showed that the thiophosphotyrosyl functionality may be more stable at extreme pH than in solvents of low acid strength (e.g., 0.1% TFA in water). Purifications performed using a gradient of acetonitrile in 0.1 M triethylammonium acetate (pH 7.5) partially overcome the initial stability problems observed during the purification step with a 0.1% TFA$_{aq}$–EtOH gradient.

D. Side-Chain Unprotected Phosphoamino Acids

The side-chain unprotected N^α-Fmoc-Tyr(PO_3H_2)-OH derivative has been applied for the syntheses of single and multiple phosphotyrosine-containing peptides [105,158,166–170] and in most cases has given satisfactory results. However, pyrophosphate formation has been observed for some peptide sequences. Thus, the incorporation of the preceding synthon into peptides with two consecutive phosphotyrosine residues can result in the formation of an intramolecular pyrophosphate bond (Fig. 19, **42**) [171]. This side reaction occurs during subsequent coupling steps and is promoted by extended and/ or repeated exposure to coupling agents. Intermolecular pyrophosphate bond formation between two side-chain unprotected phosphotyrosine residues has also been reported (Fig. 19, **43**) [172]. Under identical coupling conditions,

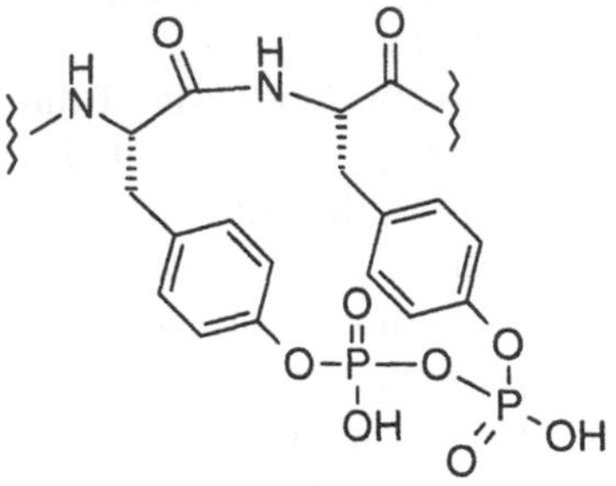

Figure 19   Intra- and intermolecular pyrophosphate by-products.

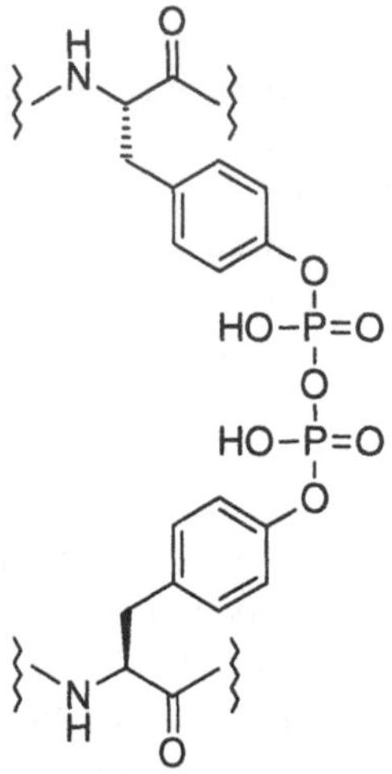

the extension of this side reaction appears to be sequence dependent and its occurrence is unpredictable. In order to minimize the formation of the intermolecular pyrophosphate bridge, the side-chain unprotected phosphotyrosine can be incorporated using BOP **18**/HOBT **17**-NMM in the presence of a chaotropic salt (0.4 M LiCl in NMP) [173].

1.   Incorporation of N^α-Fmoc-Tyr(PO$_3$H$_2$)-OH

1.   The synthesis of the desired peptide sequence is performed using a standard Fmoc strategy until the incorporation of N^α-Fmoc-Tyr(PO$_3$H$_2$)-OH.

2.   Coupling is achieved by first dissolving N^α-Fmoc-Tyr(PO$_3$H$_2$)-OH (3 equiv.), NMM (7 equiv.), and BOP/HOBt (1:1; 3 equiv.) in 0.4 M LiCl in NMP, then waiting 3 min for preactivation, and adding the mixture to the resin preswollen in 0.4 M LiCl in NMP. **Caution:** BOP must be handled with caution because of the respiratory toxicity of the hexamethylphosphotriamide by-product formed upon amino acid activation.

3.   The reaction vessel is mixed for at least 90 min. After this reaction time, the vessel is drained and a double coupling is carried out using the same conditions.

4. All subsequent amino acids are incorporated using the protocol described. Double couplings have frequently been found to be required to reach ninhydrin-negative end points for the remainder of the synthesis.

N^α-Boc-Tyr(PO$_3$H$_2$)-OH has been applied with success in the synthesis of tyrosine phosphorylated peptides 4 to 10 residues in length [174]. The most effective procedure for the phosphopeptide syntheses includes the formation of the sodium salt of the derivative after the neutralization step, addition of 1-hydroxybenzotriazole to catalyze coupling, and removal of the Boc group with 50% trifluoroacetic acid in dichloromethane for 15 min at room temperature. For this building block, no pyrophosphate by-products have been reported.

Side-chain unprotected phosphoserine has been used in the synthesis by the Fmoc strategy of a phosphopeptide related to heat shock protein, i.e., HSP27-(87–92) [H-Ser-Gly-Val-Ser(PO$_3$H$_2$)-Glu-Ile-OH] [123,137]. Problems in yield and purity were noted, as well as the formation of an intermolecular pyrophosphate by-product (16% in the crude mixture) [123]. Contrary to the phosphotyrosine case described before, no attempts to minimize or, at best, abolish this side reaction have been reported.

E. Synthesis of Peptides Containing Phosphoamino Acid Mimetics

The cellular and in vivo use of natural phosphoamino acid–containing peptides has severe limitations because, in addition to the poor chemical stability of the O-phosphate group, they are subjected to dephosphorylation by phosphatases and are poorly absorbed into cells. These circumstances and the potential therapeutic use of compounds containing enzymatically and chemically stable phosphoamino acid isosteres have motivated the synthesis of phosphoamino mimetics. Figure 20 shows some selected examples of phosphotyrosine (**44** [175–177], **45** [176], **46** [178], **47** [179], **48–49** [177], **50** [180], **51** [177], **52** [181], **53** [182], **54** [183], **55–56** [184,185], **57** [186]), phosphoserine (**58** [187], **59** [188], **60** [189,190], and phosphothreonine (**61** [191]) mimetics. These building blocks have been incorporated into peptides and peptidomimetics by solid-phase synthesis using standard procedures [175,179,182,183,189,192–198].

For the phosphonate synthons in the Fmoc chemistry, the *tert*-butyl group clearly appears to be the most suitable phosphonate protective group because it is readily removed during the acidolytic cleavage of the peptide from the solid support. The methyl or ethyl ester hydrolysis requires severe acidolytic conditions, often over extended periods of time, leading in some

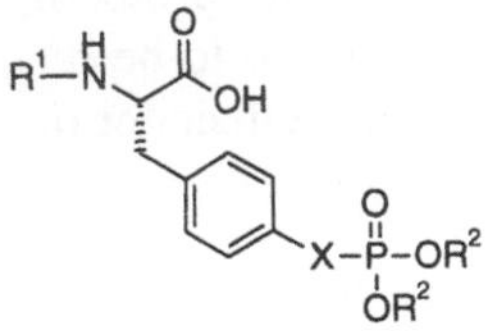

R¹	R²	X	
Fmoc	tBu	CH₂	44
Fmoc	Me	CH₂	45
Boc	Et	CH₂	46
Boc	Me	CH₂	47
Fmoc	tBu	CHF	48
Fmoc	tBu	CF₂	49*
Boc	Et	CF₂	50
Fmoc	tBu	CHOH	51

*This compound proved to be extremely unstable.

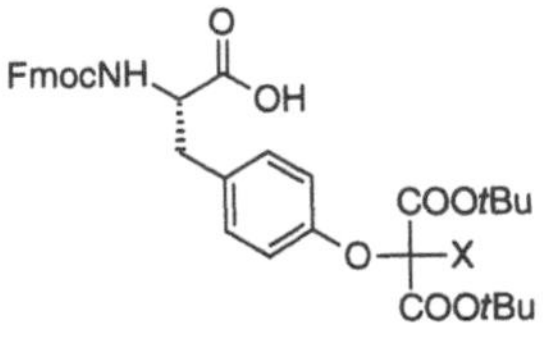

X= H **52**
X= F **53**

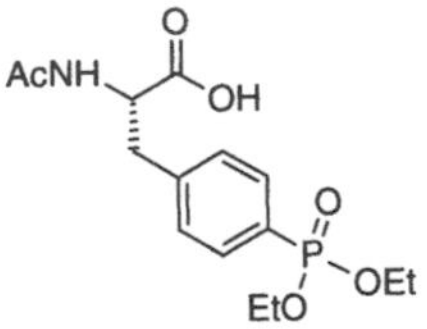

R= tBu **55**
R= Et **56**

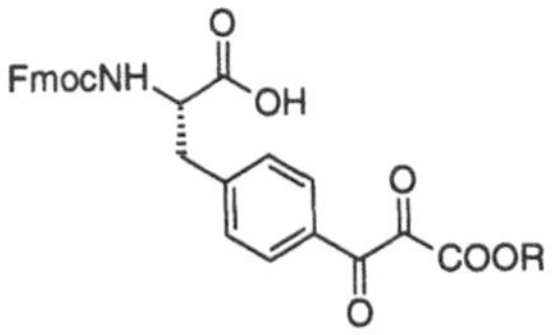

54

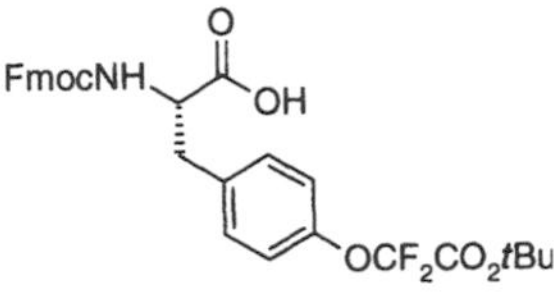

57

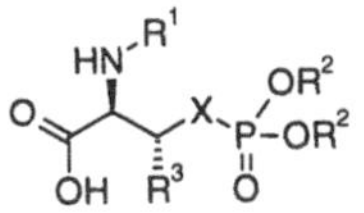

R¹	R²	R³	X	
Fmoc	Al	H	CH₂	58
Boc	Me	H	CH₂	59
Boc	Et	H	CF₂	60
Fmoc	Al	CH₃	CH₂	61

Figure 20 Selected examples of protected phosphoamino acid mimetics.

cases to the formation of by-products or degradation of the target peptide. As an alternative to these harsh acidolytic conditions, trimethylsilyl iodide (TMSI) in acetonitrile was proposed by Green [193]. This reagent provides clean and rapid dealkylation of *O,O*-diethylphosphonomethylphenyllananine–containing peptides after the TFA-mediated side-chain deprotection and cleavage from the solid support.

1. **Dealkylation of Phosphonate Diethyl Ester for the Preparation of 4-Phosphonomethylphenylalanine–Containing Peptides**

    1. After completion of the synthesis, the peptide resin is simultaneously cleaved and deprotected. The cleavage conditions are dictated by the linker used and the types of protecting groups present

in the sequence. An optimized protocol for each specific phospho-peptide is usually required.

2. The filtrate from the cleavage reaction is precipitated in diisopro-pyl ether–petroleum ether (1:1 v/v; 20 mL per 1 mL of cleavage mixture) at 0°C, and the precipitate is collected by filtration.

3. The dried, partially protected crude peptide is dissolved or sus-pended in acetonitrile at room temperature. TMSI (100- to 150-fold molar excess) is added dropwise to the stirred solution and the reaction is monitored by analytical HPLC. To ensure clean and effective deprotection, the use of freshly prepared trimethylsilyl iodide or commercially available reagent packaged in ampules is recommended.

4. The reaction is terminated by evaporation to dryness followed by the addition of 2 N AcOH. The aqueous solution is extracted with diethyl ether, concentrated, and directly applied to the MPLC–HPLC purification system. Depending on the scale, the solution has to be first lyophilized and then purified by MPLC on a re-versed-phase or ion-exchange column.

III. SULFOPEPTIDES

A. Introduction

Sulfation of amino acids* is a ubiquitous posttranslational modification of peptides and proteins [200]. Interest in the functional importance of this posttranslational event is relatively new and therefore synthetic methods for obtaining sulfated peptides have been developed only recently and have mainly focused on the preparation of sulfated tyrosine–containing peptides.

B. Synthesis of *O*-Sulfotyrosine-Containing Peptides

Two main strategies can be distinguished for the synthesis of $Tyr(SO_3H)$-containing peptides. The first involves incorporation of tyrosine-O-sulfate derivatives into the protected peptide resins followed by a global deprotec-tion–cleavage step conducted under optimized conditions to avoid or min-imize desulfuration. In the second approach, the tyrosine-containing peptide, obtained after peptide-resin cleavage, is regioisomerically sulfated in solu-tion with the aid of orthogonally removable protecting groups.

The direct solid-phase synthesis of sulfated tyrosine–containing pep-tides can be achieved by using N^{α}-Fmoc-Tyr(SO_3X)-OH (X = Na^+ or

---

*To the best of our knowledge, the *O*-sulfonate modification of serine and threonine is so far not found in nature, but the existence of alkyl-sulfatases has been considered to support its occurrence [199].

$Ba_{1/2}^{2+}$) as a building block for peptide chain assembly [201–205]. The tyrosine-O-sulfate derivative is incorporated using standard agents and coupling conditions (e.g., BOP **18**/HOBt **17**, PyBOP **21**, or as a pentafluorophenyl ester). The use of a handle labile to weak acid conditions (e.g., 2-chlorotrityl **6**)* is highly advisable [201,202]. This will allow, upon completion of the chain assembly, a two-step cleavage–deprotection protocol to minimize the deterioration of the sulfate moiety of Tyr(SO₃H) (Fig. 21). The weak acid treatment should allow nearly quantitative detachment from the solid support with retention of all or a significant portion of the side-chain protecting groups and with negligible loss of the sulfate group. The conditions for the latter acid treatment can then be optimized to minimize deterioration of the sulfate group of tyrosine. The selection of the cleavage–deprotection conditions and scavenger cocktail is dictated by the linker used and the protecting groups present in the sequence, but general recommendations can be formulated.† Sulfur-containing scavengers, such as thioanisole, dimethyl sulfide, or 1,2-ethanedithiol, accelerate decomposition of O-sulfate-tyrosine to tyrosine in trifluoroacetic acid media, but water, m-cresol, and 2-methylindole have little effect on the decomposition [204]. The reaction temperature is also a decisive determinant of the stability of Tyr(SO₃H) during the final acid treatment [204]. In these circumstances, a 90% aqueous TFA-based reagent (90% aqueous TFA–m-cresol or 90% aqueous TFA–m-cresol–2-methylindole) [204]‡ for tryptophan-containing peptides at 4°C has been judged suitable as a final deprotection reagent [201,203,204]. However, generally a number of conditions have to be examined to find the optimal compromise between acid-promoted loss of sulfate and incomplete deprotection.

In the sulfation of tyrosine in solution after peptide chain assembly, the side chain of the tyrosine residues not to be sulfated in multiple tyrosine-containing peptides and the hydroxyl groups of serine and/or threonine, which are known to be sulfated in preference to the phenolic hydroxyl group of tyrosine [208], have to be protected. In addition to the hydroxyl groups,

---

* Alternatively, the [[9-][(9-fluorenylmethyloxycarbonyl)amino]xanthen-3-yl]oxy]acetic acid linker (Fig. 22, **62**) has been applied successfully to the synthesis of the O-sulfate-tyrosine cholecystokinin octapeptide (residues 26–33) [206]. In this case, the cleavage–deprotection was carried out in a single step using TFA–DCM–H₂O (1:18:1) for 15 min at 45°C. The overall yield was 71% and the target peptide represented >95% of the total crude material. The acid-labile PAL-linked resin **9** [203,204] and the 4-succinylamino-2,2′,4′-trimethoxybenzhydrylamine (SAMBHA) linker (Fig. 22, **63**) [205] have also been used in this approach.
† Sodium salt formation increased the acid stability of the O- and S-sulfate groups. Therefore, the peptide resin is preferentially washed with 1 M aqueous NaBr/DMF (1:2 v/v) [207] or a methanolic NaOH solution before the cleavage [205].
‡ 50 equiv. of m-cresol and 25 equiv. of 2-methylindole with respect to the peptide [204].

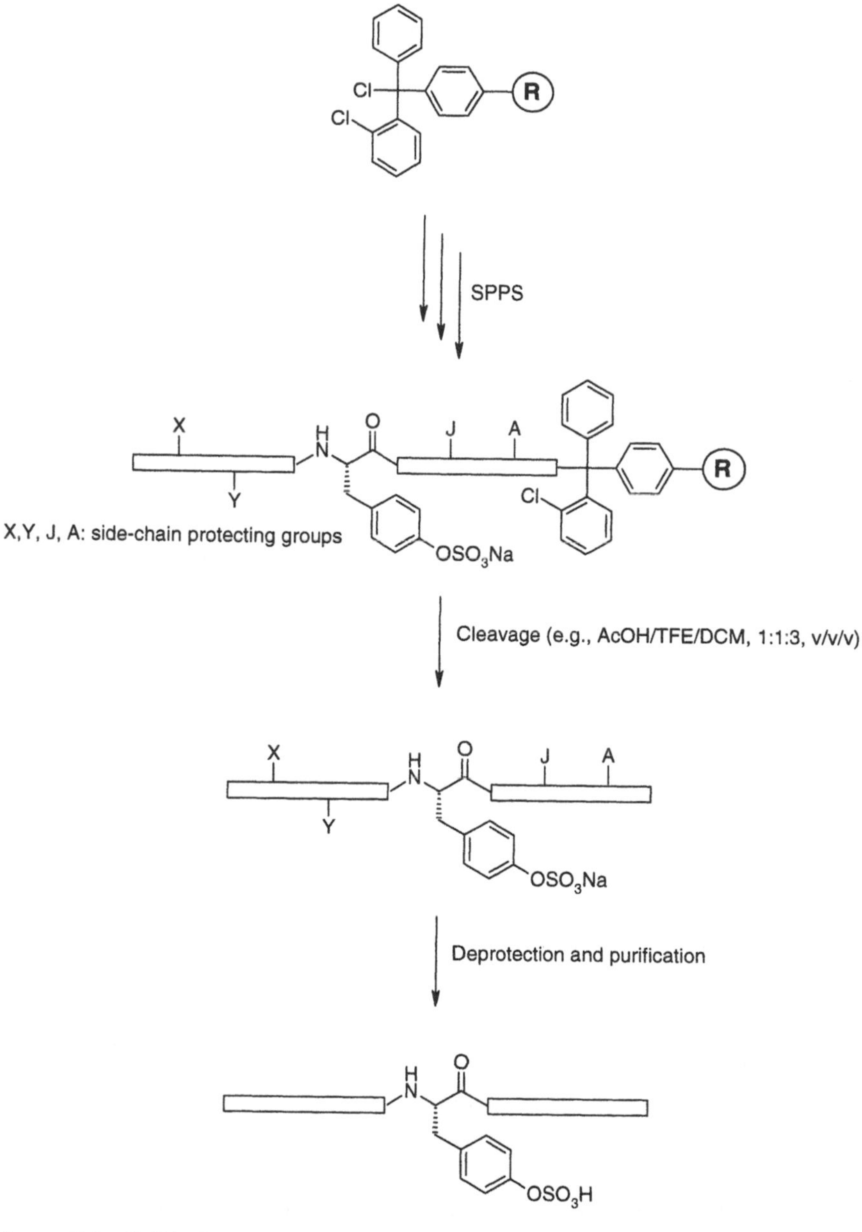

Figure 21 Solid-phase synthesis of a sulfotyrosine-containing peptide using the 2-chlorotrityl resin.

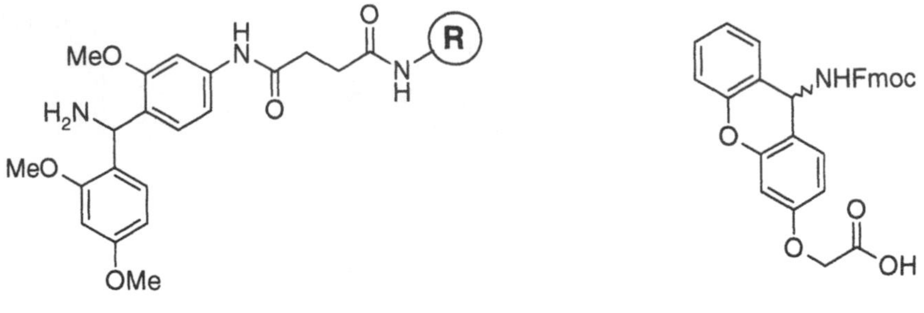

62   **63**

Figure 22 Linkers suitable for solid-phase sulfopeptide synthesis.

the *N*-terminal amino group and the side chains of lysine and/or ornithine have to be protected during sulfation in order to avoid possible sulfamic acid formation. To overcome these problems, the trifluoroacetic stable *p*-(methylsulfinyl)benzyl (Msib) and *p*-methylsulfinyl)benzyloxycarbonyl (Msz) groups (Fig. 23) are used for the protection of hydroxyl and amino functions, respectively [203,204,209,210]. The tyrosine residues not to be sulfated are incorporated using N^α-Fmoc-Tyr(Msib)-OH and the tyrosine residues to be sulfated are coupled using N^α-Fmoc-Tyr(*t*Bu)-OH. Upon incorporation of the last amino acid, the *N*-terminal Fmoc group is removed and replaced by the Msz group. After peptide-resin cleavage, partially protected peptides containing the Msib or Msz protecting groups are treated with the DMF–SO_3* complex in the presence of 1,2-ethanedithiol to achieve the sulfation of free tyrosine residues and the reduction of the Msib or Msz groups to TFA-labile *p*-(methylthio)benzyl (Mtb) or *p*-(methylthio)benzyloxycarbonyl (Mtz) groups. Final deprotection of the Mtb or Mtz groups is achieved with a 90% aqueous TFA-based reagent (e.g., 90% aqueous TFA–*m*-cresol or 90% aqueous TFA–*m*-cresol–2-methylindole for tryptophan-containing peptides [203,204].[†]

1. Synthesis of *O*-Sulfated Tyrosine-Containing Peptides Using Msib and Msz Protecting Groups

1. Introduction of the respective amino acid derivatives and manipulations of the peptide chain assembly are performed using standard Fmoc protocols. The tyrosine residue to be sulfated is incor-

---

*The DMF–SO_3 complex has been shown to have greater ability for sulfation than the commonly used pyridine–SO_3 complex [211].
[†]50 equiv. of *m*-cresol and 25 equiv. of 2-methylindole with respect to the peptide [204].

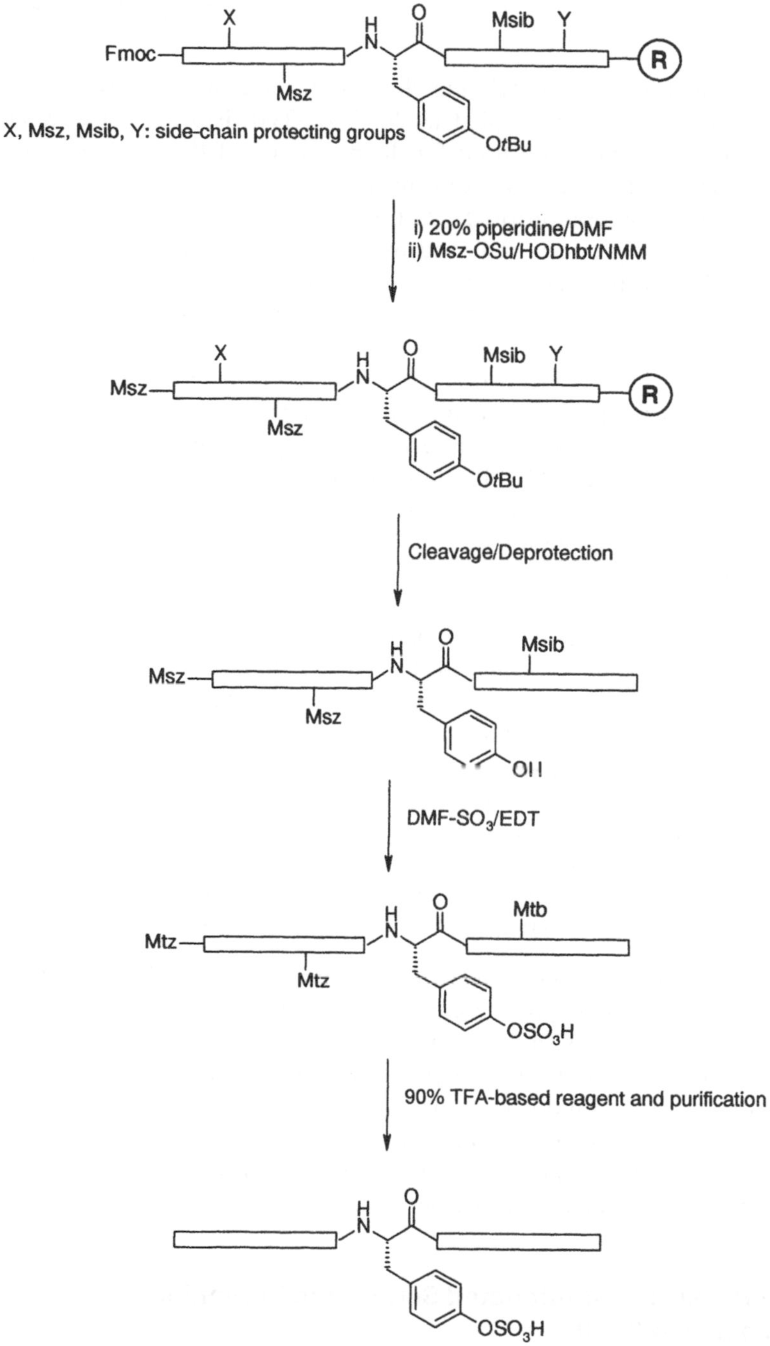

Figure 23 Solid-phase synthesis of a sulfotyrosine-containing peptide using Msib or Msz protecting groups.

porated with *tert*-butyl side-chain protection and the tyrosine not to be sulfated is coupled as N^α-Fmoc-Tyr(Msib)-OH. The Msib and Msz groups are used for the protection of the side-chain hydroxyl and amino functions, respectively.

2. After incorporation of the last amino acid, the *N*-terminal Fmoc protecting group is removed by treatment with 20% piperidine–DMF and the peptide-resin is acylated with Msz-OSu (6 equiv.) in the presence of HODhbt (6 equiv.) and NMM (6 equiv.). The reaction is monitored using standard methods.

3. Following the completion of the acylation reaction, the peptide resin is ready for cleavage. As usual, the scavenger cocktail is dictated by the linker employed and the protecting groups present in the sequence. The filtrate from the cleavage reaction is precipitated in diisopropyl ether–petroleum ether (1:1 v/v; 20 mL per 1 mL of cleavage mixture) at 0°C, and the precipitate is collected by filtration.

4. The dried, partially protected crude peptide is dissolved in DMF–pyridine (4:1 v/v) and treated with DMF–SO$_3$ (125-fold molar excess) in the presence of 1,2-ethanedithiol (100-fold molar excess). After about 15 h at room temperature, the solution is applied to a size exclusion chromatography column and eluted with DMF. The target compound is collected and the solvent evaporated to dryness. After lyophilization, the crude compound is treated with a 90% TFA-based reagent at 4°C. The reaction time of this step is optimized by monitoring the acidolytic treatment of a small aliquot of the *O*-sulfated peptide and analysis of the synthetic products by analytical HPLC and mass spectrometry.

For the two above approaches, it is recommended that the purification of the crude *O*-sulfated tyrosine-containing peptides be performed by reversed-phase HPLC or MPLC using 0.1 M ammonium acetate buffer (pH 6.5) in the gradient system. Lower pH values (e.g., with application of the most common 0.1% trifluoroacetic-based buffers) can induce slow desulfuration during chromatography, concentration of the samples, and lyophilization. The ammonium acetate solution, although volatile, had to be lyophilized several times to gain salt-free pure peptides.

C. Synthesis of *O*-Sulfonated Serine- and Threonine-Containing Peptides

A literature survey reveals a paucity of papers concerning the synthesis of peptides containing *O*-sulfonated serine and threonine residues. The preparation of a model peptide [H-Ala-Arg-Gly-Ala-Xxx-Gly-OH; Xxx = Ser (SO_3H) or Thr (SO_3H)] was performed using N^α-Fmoc-Ser/Thr(SO_3Na)-OH as building blocks and DCC **16**/HOBt **17** as coupling agents [199]. A relatively small amount of the nonsulfated peptide was identified by analytical HPLC after cleavage–deprotection with 95% aqueous TFA. Identical conditions were used for the continuous-flow synthesis of [Ser$(SO_3H)^3$, Ahx^{17}]–vasoactive intestinal peptide (VIP) and related analogues on a low loaded kieselguhr-supported poly-N,N-dimethylacrylamide) resin with the Rink amide linker **8** [212].

The sulfation of serine and threonine in solution after solid-phase peptide chain assembly can be accomplished with DMF–SO_3 in DMF–pyridine (4:1 v/v) at 4°C [213] (see also Section III.B) or by exploiting a side reaction observed during the removal of the 2,2,5,7,8-pentamethyl-chroman-6-sulfonyl (Pmc) group from arginine with TFA under nonaqueous conditions [199,214,215]. The arylsulfonyl transfer reaction [214] has obvious practical limitations because of the sequence requirement and potential side reactions related to the lack of scavengers in the cleavage–deprotection mixture.

D. Synthesis of *S*-Sulfocysteine–Containing Peptides

In addition to O-sulfated tyrosine, serine, or threonine, *S*-sulfocysteine–containing peptides have been known for a long time. The *S*-sulfo derivatives, which are also called Bunte salts, have been used for different purposes, including cysteine protection [207,216], direct cysteine bridging [217], and peptide purification by ion-exchange chromatography [216]. The *S*-sulfocysteine derivatives can be prepared from the corresponding peptides via oxidative sulfitolysis [217] or direct incorporation in the growing peptide chain in solution or the solid phase [207,216]. The preparation of N^α-Fmoc-Cys(SSO_3Na)-ONa has been described and its use as a protecting group for cysteine was illustrated with the synthesis on the solid phase of Arg^8-vasopressin [207,216] and a phosphorylase kinase β-subunit (PKS) fragment (420–436) [207]. As with the O-sulfated analogues, partial *S*-sulfonate removal occurs during the TFA treatment. For the Arg^8-vasopressin example, higher homogeneity in the crude peptide was obtained when p-cresol was used as the only scavenger in the cleavage–deprotection solution [216]. Alternatively, a washing step with 1 M aqueous–NaBr–DMF (1:2 v/v) before cleavage and the use of an Et$_3$SiH–TFA (0.7:1 or 13:87) cleavage–deprotection mixture have been proposed to increase the quality of the crude compound [207]. After purification of the intermediate by ion-exchange chromatography or reversed-phase chromatography, the *S*-sulfonate group, which has a strong infrared band at 1030–1050 cm^{-1}, is removed in quantitative yield with tributylphosphine [216] or a 4% solution of DTT in water [207].

ABBREVIATIONS

Ac, acetyl; Ahx, aminohexyl; Al, allyl; Aloc, allyloxycarbonyl; Bn, benzyl; Boc, *tert*-butyloxycarbonyl; Boc$_2$O, di-*tert*-butyl-dicarbonate; BOP, 1*H*-benzotriazol-1-yloxy-tris(dimethylamino) phosphonium hexafluorophosphate; Bz, benzoyl; Bzl, benzyl; Bzl(4-Cl), 4-chlorobenzyl; Bzl(4-NO$_2$) 4-nitrobenzyl; ClZ, 2-chlorobenzyloxycarbonyl; DBU, 1,8-diazabicyclo[5.4.0]undec-7-ene; DCC, *N,N'*-dicyclohexylcarbodiimide; DCM, dichloromethane; DIEA, *N,N*-diisopropylethylamine; DMA, *N,N*-dimethylacetamide; DMF, *N,N*-dimethylformamide; DMS, dimethyl sulfide; DNA, deoxyribonucleic acid; DTT, dithiothreitol; EDT, 1,2-ethanedithiol; EGFR, epidermal growth factor receptor; Fmoc, fluorenyl-9-methoxycarbonyl; Gal, galactose; GalNAc, *N*-acetyl-D-galactosamine; GLC, gas–liquid chromatography; Glc, glucosamine; GlcNAc, *N*-acetyl-D-glucosamine; HBTU, 2-(1*H*-benzotriazole-1-yl)-1,1,3,3-tetramethyluronium hexafluorophosphate; HIV, human immunodeficiency virus; Hmb, *N*-(2-hydroxy-4-methoxybenzyl); HMBA, 4-hydroxymethylbenzoic acid; HMPAA, 4-hydroxymethylphenoxyacetic acid; HOBt, 1-hydroxybenzotriazole; HODhbt, 3-hydroxy-4-oxo-3,4-dihydro-1,2,3-benzotriazine; HPLC, high-performance liquid chromatography; HSP, heat shock protein; HYCRAM, hydroxycrotonylaminomethylpolystyrene; HYCRON, (*E*)-17-hydroxy-4,7,10,13-tetraoxy-15-heptadecenoyl; IL, interleukin; *i*-Pr, isopropyl; IR, insulin receptor; MAP, mitogen-activated protein; MPLC, medium-pressure liquid chromatography; Msib, *p*-(methylsulfinyl) benzyl; Msz, *p*-(methylsulfinyl)benzyloxycarbonyl; Mtb, *p*-(methylthio)benzyl; Mtz, *p*-(methyl)benzyloxycarbonyl; NMM, *N*-methylmorpholine; NMP, *N*-methyl-2-pyrrolidone; *n*-Pr, propyl; OSu, succinimide ester; PAL, 5-(4-aminomethyl-3,5-dimethoxyphenoxy)-valeric acid; PAM, phenylacetamidomethyl; PDGF, platelet-derived growth factor; PEGA, poly(ethyleneglycol)-poly(*N,N*-dimethylacrylamide) copolymer; Pfp, pentafluorophenyl; PfPyU, *N,N,N',N'*-bis(tetramethylene)-*O*-pentafluorophenyluronium hexafluorophosphate; Ph, phenyl; PKS, phosphorylase kinase β-subunit; Pmc, 2,2,5,7,8-pentamethyl-chroman-6-sulfonyl; PyBOP, benzotriazole-1-yl-oxy-tris-pyrrolidino-phosphonium hexafluorophosphate; SAMBHA, 4-succinylamino-2, 2',4'-trimethoxybenzhydrylamine; SASRIN, super-acid-sensitive resin; SPPS, solid-phase peptide synthesis; TBAF, tetrabutylammonium fluoride; TBDMS, *tert*-butyldimethylsilyl; TBDPS, *tert*-butyldiphenylsilyl; TBTU, 2-(1*H*-benzotriazole-1-yl)-1,1,3,3-tetramethyluronium tetrafluoroborate; *t*Bu, *tert*-butyl; TCA, trichloroacetic acid; TFA, trifluoroacetic acid; TFE, 2,2,2-trifluoroethanol; TFMSA, trifluoromethanesulfonic acid; THF, tetrahydrofuran; TMS, trimethylsilyl; TMSBr, trimethylsilyl bromide; TMSI, trimethylsilyl iodide; TMSOTf, trimethylsilyl trifluoromethanesulfonate; TPTU, O-(1,2-dihydro-2-oxo-1-pyridyl)-1,1,3,3-tetramethyluronium tetrafluoroborate; Trt, trityl; Stat, signal transducer and activator of transcription; VIP, vasoactive intestinal peptide; Z, benzyloxycarbonyl.

REFERENCES

1. H Lis, N Sharon. Eur J Biochem 218:1–27, 1993.
2. J Montreuil. Adv Carbohydr Chem Biochem 37:157–223, 1980.
3. J Hofsteenge, DR Müller, T de Beer, A Loeffler, WJ Richter, JFG Vliegenthart. Biochemistry 33:13524–13530, 1994.
4. DG Large, CD Warren, eds. Glycopeptides and Related Compounds. New York: Marcel Dekker, 1997.
5. M Meldal, PMS Hilaire. Curr Opin Chem Biol 1:552–563, 1997.
6. T Norberg, B Luening, J Tejbrant. Methods Enzymol 247:87–106, 1994.
7. M Meldal. Curr Opin Struct Biol 4:710–718, 1994.
8. M Meldal. In: YC Lee, RT Lee, ed. Neoglycoconjugates: Preparation and Applications. San Diego: Academic Press, 1994, pp 145–198.
9. H Kunz. In: C Basava, GN Anantharamaiah, eds. Recent Developments in the Synthesis of Glycopeptides. Boston: Birkhaeuser, 1994, pp 69–79.
10. J Kihlberg, M Elofsson, LA Salvador. Methods Enzymol 289:221–245, 1997.
11. H Kunz, WKD Brill. Trends Glycosci Glycotechnol 4:71–82, 1992.
12. DG Large, IJ Bradshaw. In: DG Large, CD Warren, eds. Chemical Synthesis of the Peptide Moiety of Glycopeptides. New York: Marcel Dekker, 1997, pp 295–325.
13. AM Jansson, KJ Jensen, M Meldal, J Lomako, WM Lomako, CE Olsen, K Bock. J Chem Soc Perkin Trans 1 1001–1006, 1996.
14. K Bock, J Schuster-Kolbe, E Altman, G Allmaier, B Stahl, R Christian, UB Sleytr, P Messner. J Biol Chem 269:7137–7144, 1994.
15. CR Torres, GW Hart. J Biol Chem 259:3308–3317, 1984.
16. U Tedebark, M Meldal, L Panza, K Bock. Tetrahedron Lett 39:1815–1818, 1998.
17. E Meinjohanns, M Meldal, H Paulsen, RA Dwek, K Bock. J Chem Soc Perkin Trans 1 549–560, 1998.
18. E Meinjohanns, A Vargas-Berenguel, M Meldal, H Paulsen, K Bock. J Chem Soc Perkin Trans 1 2165–2175, 1995.
19. J Habermann, H Kunz. Tetrahedron Lett 39:265–268, 1998.
20. J Rademann, RR Schmidt. Carbohydr Res 269:217–225, 1995.
21. MK Christensen, M Meldal, K Bock, H Cordes, S Mouritsen, H Elsner. J Chem Soc Perkin Trans 1 1299–1310, 1994.
22. S Peters, T Bielfeldt, M Meldal, K Bock, H Paulsen. J Chem Soc Perkin Trans 1 1163–1171, 1992.

23. K Witte, O Seitz, C-H Wong. J Am Chem Soc 120:1979–1989, 1998.
24. Y Nakahara, T Ogawa. Carbohydr Res 292:71–81, 1996.
25. Z-W Guo, Y Nakahara, T Ogawa. Angew Chem Int Ed Engl 36:1464–1466, 1997.
26. Z-W Guo, Y Nakahara, T Ogawa. Bioorg Med Chem 5:1917–1924, 1997.
27. Z-W Guo, Y Nakahara, T Ogawa. Carbohydr Res 303:373–377, 1997.
28. O Seitz, H Kunz. Angew Chem Int Ed Engl 34:803–805, 1995.
29. M Elofsson, S Roy, B Walse, J Kihlberg. Carbohydr Res 246:89–103, 1993.
30. H Paulsen, T Bielfeldt, S Peters, M Meldal, K Bock. Liebigs Ann Chem 369–379, 1994.
31. R Polt, L Szabo, J Treiberg, Y Li, VJ Hruby. J Am Chem Soc 114:10249–10258, 1992.
32. M Gobbo, L Biondi, F Filira, B Scolaro, R Rocchi, T Piek. Int J Peptide Protein Res 40:54–61, 1992.
33. E Bardaji, JL Torres, P Clapes, F Albericio, G Barany, RC Rodriguez, MP Sacristan, G Valencia. J Chem Soc Perkin Trans 1 1755–1759, 1991.
34. E Bardaji, JL Torres, P Clapes, F Albericio, G Barany, G Valencia. Angew Chem 102:311–313, 1990.
35. H Paulsen, G Merz, S Peters, U Weichert. Liebigs Ann Chem 1165–1173, 1990.
36. H Paulsen, G Merz, U Weichert. Angew Chem 100:1425–1427, 1988.
37. H Kunz, B Dombo. Angew Chem 100:732–734, 1988.
38. S Rio-Anneheim, H Paulsen, M Meldal, K Bock. J Chem Soc Perkin Trans 1 1071–1080, 1995.
39. M Meldal, T Bielfeldt, S Peters, KJ Jensen, H Paulsen, K Bock. Int J Peptide Protein Res 43:529–536, 1994.
40. H Paulsen, S Peters, T Bielfeldt, M Meldal, K Bock. Carbohydr Res 268:17–34, 1995.
41. DM Andrews, PW Seale. Int J Peptide Protein Res 42:165–170, 1993.
42. KB Reimer, M Meldal, S Kusumoto, K Fukase, K Bock. J Chem Soc Perkin Trans 1 925–932, 1993.
43. H Paulsen, T Bielfeldt, S Peters, M Meldal, K Bock. Liebigs Ann Chem 381–387, 1994.
44. T Bielfeldt, S Peters, M Meldal, K Bock, H Paulsen. Angew Chem 104:881–883, 1992.
45. S Peters, T Bielfeldt, M Meldal, K Bock, H Paulsen. Tetrahedron Lett 33:6445–6448, 1992.
46. S Peters, T Bielfeldt, M Meldal, K Bock, H Paulsen. Tetrahedron Lett 32:5067–5070, 1991.
47. L Biondi, F Filira, M Gobbo, B Scolaro, R Rocchi. Int J Peptide Protein Res 37:112–121, 1991.
48. M Meldal, KJ Jensen. J Chem Soc Chem Commun 483–485, 1990.
49. M Elofsson, B Walse, J Kihlberg. Int J Peptide Protein Res 47:340–347, 1996.
50. L Urge, DC Jackson, L Gorbics, K Wroblewski, G Graczyk, L Otvos Jr. Tetrahedron 50:2373–2390, 1994.
51. H Hietter, M Schultz, H Kunz. Synlett 1219–1220, 1995.

52. J Broddefalk, K-E Bergquist, J Kihlberg. Tetrahedron Lett 37:3011–3014, 1996.

53. G Arsequell, JS Haurum, T Elliott, RA Dwek, AC Lellouch. J Chem Soc Perkin Trans 1 1739–1745, 1995.

54. H Zhang, Y Wang, W Voelter. Tetrahedron Lett 36:8767–8770, 1995.

55. B Luening, T Norberg, C Rivera-Baeza, J Tejbrant. Glycoconj J 8:450–455, 1991.

56. S Lavielle, NC Ling, R Saltman, RC Guillemin. Carbohydr Res 89:229–236, 1981.

57. S Lavielle, NC Ling, RC Guillemin. Carbohydr Res 89:221–228, 1981.

58. AC Bauman, JS Broderick, RM Dacus, DA Grover, LS Trzupek. Tetrahedron Lett 34:7019–7022, 1993.

59. B Liebe, H Kunz. Angew Chem Int Ed Engl 36:618–621, 1997.

60. B Liebe, H Kunz. Helv Chim Acta 80:1473–1482, 1997.

61. M Gewehr, H Kunz, A Lauterbach, B Loehr, O Seitz. Use of the allylic HYCRON anchor in solid phase peptide synthesis. Proceedings of Innovation and Perspectives in Solid Phase Synthesis. Fourth International Symposium, Edinburgh, 1995, pp 117–120.

62. W Kosch, J Maerz, KVD Bruch, H Kunz. Solid phase synthesis of glyco-peptide antigens on HYCRAM resin utilizing the allylic anchor principle. Proceedings of Innovation and Perspectives in Solid Phase Synthesis. Third International Symposium, Oxford, 1993, pp 267–272.

63. W Kosch, J Maerz, H Kunz. React Polym 22:181–194, 1994.

64. H Kunz, B Dombo, W Kosch. Solid-phase synthesis of peptides and glyco-peptides on resins with allylic anchoring groups. Proceedings of the Twenty European Peptide Symposium, Tübingen, 1988, pp 154–156.

65. H Kunz, W Kosch, J Maerz, M Ciommer, W Guenther, C Unverzagt. Synthesis of glycopeptides in solution and on solid phase. Proceedings of Innovation and Perspectives in Solid Phase Synthesis. Second International Symposium, Canterbury, 1991, pp 171–178.

66. H Kunz, M Schultz. In: DG Large, CD Warren. Recent Advances in the Synthesis of Glycopeptides. New York: Marcel Dekker, 1997, pp 23–78.

67. J Kihlberg, T Vuljanic. Tetrahedron Lett 34:6135–6138, 1993.

68. MK Christensen, M Meldal, K Bock. J Chem Soc Perkin Trans 1 1453–1460, 1993.

69. I Christiansen-Brams, M Meldal, K Bock. J Chem Soc Perkin Trans 1 1461–1471, 1993.

70. M Meldal, K Bock. Tetrahedron Lett 31:6987–6990, 1990.

71. Am Jansson, M Meldal, K Bock. Tetrahedron Lett 31:6991–6994, 1990.

72. L Otvos Jr, L Urge, M Hollosi, K Wroblewski, G Graczyk, GD Fasman, J Thurin. Tetrahedron Lett 31:5889–5892, 1990.

73. M Flegel, RC Sheppard. J Chem Soc Chem Commun 536–538, 1990.

74. E Kaiser, RL Colescott, CD Bossinger, PI Cook. Anal Biochem 34:595–598, 1970.

75. M Meldal, S Mouritsen, K Bock. ACS Symp Ser 519:19–33, 1993.

76. T Vuljanic, K-E Bergquist, H Clausen, S Roy, J Kihlberg. Tetrahedron 52: 7983–8000, 1996.

77. M Bodanszky, J Martinez. Synthesis 333–356, 1981.

78. I Christiansen-Brams, M Meldal, K Bock. Tetrahedron Lett 34:3315–3318, 1993.

79. L Urge, L Otvos Jr, E Lang, K Wroblewski, I Laczko, M Hollosi. Carbohydr Res 235:83–93, 1992.

80. H Paulsen, S Peters, T Bielfeldt, M Meldal, K Bock. Carbohydr Res 268:17–34, 1995.

81. K Wakabayashi, W Pigman. Carbohydr Res 35:3–14, 1974.

82. H Kunz, Angew Chem 99:297–311, 1987.

83. P Sjölin, M Elofsson, J Kihlberg. J Org Chem 61:560–565, 1996.

84. M Elofsson, LA Salvador, J Kihlberg. Tetrahedron 53:369–390, 1997.

85. S Peters, TI Lowary, O Hindsgaul, M Meldal, K Bock. J Chem Soc Perkin Trans 1 3017–3022, 1995.

86. I Christiansen-Brams, AM Jansson, M Meldal, K Breddam, K Bock. Bioorg Med Chem 2:1153–1167, 1994.

87. PJ Garegg, T Norberg. Acta Chem Scand Ser B B33:116–118, 1979.

88. M Elofsson, J Broddefalk, T Ekberg, J Kihlberg. Carbohydr Res 258:123–133, 1994.

89. R Seger, EG Krebs. FASEB J 9:726–735, 1995.

90. EG Krebs. Trends Biochem Sci 19:439, 1994.

91. HV Rickenberg, BH Leichtling. In: PD Boyer, EG Krebs, eds. Enzymes. 3rd ed. Orlando, FL: Academic Press, 1998, pp 419–455.

92. L Pike, EG Krebs. In: PM Conn, ed. Receptors. Orlando, FL: Academic Press, 1986, pp 93–134.

93. E-S Lee, M Cushman. J Org Chem 59:2086–2091, 1994.

94. JW Perich. Methods Enzymol 201:234–245, 1991.

95. RM Valerio, PF Alewood, RB Johns, BE Kemp. Int J Peptide Protein Res 33: 428–436, 1989.

96. RM Valerio, PF Alewood, RB Johns, BE Kemp. Tetrahedron Lett 25:2609–2612, 1984.

97. K Ramalingam, SR Eaton, WL Cody, JA Loo, AM Doherty. Lett Peptide Sci 1:73–79, 1994.

98. Z Tian, C Gu, RW Roeske, M Zhou, RL Van Etten. Int J Peptide Protein Res 42:155–158, 1993.

99. EA Kitas, JW Perich, RB Johns, GW Tregear. Tetrahedron Lett 29:3591–3592, 1988.

100. JW Perich, RB Johns. J Org Chem 54:1750–1752, 1989.

101. A Otaka, K Miyoshi, M Kaneko, H Tamamura, N Fujii. J Org Chem 60: 3967–3974, 1995.

102. EA Kitas, RB Johns, CN May, GW Tregear, JD Wade. Peptide Res 6:205–210, 1993.

103. EA Kitas, JW Perich, JD Wade, RB Johns, GW Tregear. Tetrahedron Lett 30: 6229–6232, 1989.

104. EA Kitas, R Knorr, A Trzeciak, W Bannwarth. Helv Chim Acta 74:1314–1328, 1991.

105. RM Valerio, AM Bray, NJ Maeji, PO Morgan, JW Perich. Lett Peptide Sci 2:33–40, 1995.

106. JW Perich, F Meggio, RM Valerio, RB Johns, La Pinna, EC Reynolds. Bioorg Med Chem 1:381–388, 1993.

107. A Chavanieu, H Naharisoa, F Heitz, B Calas, F Grigorescu. Bioorg Med Chem Lett 1:299–302, 1991.

108. EA Kitas, JD Wade, RB Johns, JW Perich, GW Tregear. J Chem Soc Chem Commun 338–339, 1991.

109. W Bannwarth, EA Kitas. Helv Chim Acta 75:707–714, 1992.

110. JW Perich, M Ruzzene, LA Pinna, EC Reynolds. Int J Peptide Protein Res 43:39–46, 1994.

111. JW Perich, EC Reynolds. Int J Peptide Protein Res 37:572–575, 1991.

112. H-G Chao, B Leiting, PD Reiss, AL Burkhardt, CE Klimas, JB Bolen, GR Matsueda. J Org Chem 60:7710–7711, 1995.

113. M Ueki, J Tachibana, Y Ishii, J Okumura, M Goto. Tetrahedron Lett 37:4953–4956, 1996.

114. H-G Chao, MS Bernatowicz, PD Reiss, GR Matsueda. J Org Chem 59:6687–6691, 1994.

115. H Fretz. Lett Peptide Sci 3:343–348, 1997.

116. P White, J Beythien. Preparation of phosphoserine, threonine and tyrosine containing peptides by the Fmoc methodology using pre-formed phospho-amino acid building blocks. Proceedings of Innovation and Perspectives in Solid Phase Synthesis. Fourth International Symposium, Edinburgh, 1995, pp 557–560.

117. JW Perich, ND Le, EC Reynolds. Tetrahedron Lett 32:4033–4034, 1991.

118. A Arendt, K Palczewski, WT Moore, RM Caprioli, JH McDowell, PA Hargrave. Int J Peptide Protein Res 33:468–476, 1989.

119. JW Perich, Methods Enzymol 201:225–233, 1991.

120. A Otaka, K Miyoshi, PP Roller, TR Burke Jr, H Tamamura, N Fujii. J Chem Soc Chem Commun 387–389, 1995.

121. JM Lacombe, F Andriamanampisoa, AA Pavia. Int J Peptide Protein Res 36:275–280, 1990.

122. N Mora, JM Lacombe, AA Pavia. Tetrahedron Lett 34:2461–2464, 1993.

123. T Wakamiya, K Saruta, J-I Yasuoka, S Kusumoto. Synthetic study of phosphopeptides based on the Fmoc-mode pre-phosphorylation strategy. Proceedings of the Thirty-Second Symposium on Peptide Chemistry, Fukuoka, 1994, pp 5–8.

124. Y Ueno, F Suda, Y Taya, R Noyori, Y Hayakawa, T Hata. Tetrahedron Lett 5:823–826, 1995.

125. T Wakamiya, K Saruta, S Kusumoto, K Nakajima, K Yoshizawa-Kumagaye, S Imajoh-Ohmi, S Kanegasaki. Chem Lett 1401–1404, 1993.

126. H Mihara, H Kuwahara, T Niidome, H Kanegae, H Aoyagi. Chem Lett 399–400, 1995.

127. T Wakamiya, R Togashi, T Nishida, K Saruta, J-I Yasuoka, S Kusumoto, S Aimoto, YK Kumagaye, K Nakajima, K Nagata. Bioorg Med Chem 5:135–145, 1997.

128. M Tsukamoto, R Kato, K Ishiguro, T Uchida, K Sato. Tetrahedron Lett 32:7083–7086, 1991.

129. JW Perich, E Terzi, E Carnazzi, R Seyer, E Trifilieff. Int J Peptide Protein Res 44:305–312, 1994.

130. N Fujii, A Otaka, O Ikemura, M Hatano, A Okamachi, S Funakoshi, M Sakurai, T Shioiri, H Yahima. Chem Pharm Bull 35:3447–3452, 1987.

131. N Fujii, S Funakoshi, T Sasaki, H Yahima. Chem Pharm Bull 25:3096–3098, 1977.

132. T Wakamiya, K Saruta, J Yasuoka, S Kusumoto. Bull Chem Soc Jpn 68:2699–2703, 1995.

133. T Wakamiya, K Saruta, S Kusumoto. Acid-stable phosphate-protection for phosphoamino acid in the synthesis of phosphopeptides. Proceedings of the Second Japan Symposium on Peptide Chemistry, Shizuoka, 1992, pp 39–41.

134. T Wakamiya, R Togashi, K Saruta, S Kusumoto, S Aimoto, K Yoshizawa-Kumagaye, K Nakajima. New phosphoryl protecting groups for synthesis of phosphopeptides by Boc method. Proceedings of the Thirty-Third Symposium on Peptide Chemistry, Japan, 1995, pp 17–20.

135. T Wakamiya, T Nishida, R Togashi, K Saruta, J-I Yasuoka, S Kusumoto. Bull Chem Soc Jpn 69:465–468, 1996.

136. T Vorherr, W Bannwarth. Bioorg Med Chem Lett 5:2661–2664, 1995.

137. T Wakamiya, K Saruta, J Yasuoka, S Kusumoto. Chem Lett 1099–1102, 1994.

138. G Shapiro, D Buchler, C Dalvit, P Frey, M Del Carmen Fernandez, B Gomez-Lor, E Pombo-Villar, U Stauss, R Swoboda, C Waridel. Bioorg Med Chem 5:147–156, 1997.

139. G Shapiro, D Buechler, C Dalvit, M Fernandez, B Gomez-Lor, E Pombo-Villar, U Stauss, R Swoboda. Bioorg Med Chem Lett 6:409–414, 1996.

140. G Shapiro, R Swoboda, U Stauss. Tetrahedron Lett 35:869–872, 1994.

141. K Barlos, D Gatos, S Koutsogianni, W Schäfer, G Stavropoulos, Y Wenging. Tetrahedron Lett 32:471–474, 1991.

142. H Sakamoto, H Kodama, Y Higashimoto, M Kondo, MS Lewis, CW Anderson, E Appella, K Sakaguchi. Int J Peptide Protein Res 48:429–442, 1996.

143. L Otvos Jr, I Elekes, VMY Lee. Int J Peptide Protein Res 34:129–133, 1989.

144. R Hoffmann, WO Wachs, RG Berger, H-R Kalbitzer, D Waidelich, E Bayer, W Wagner-Redeker, M Zeppezauer. Int J Peptide Protein Res 45:26–34, 1995.

145. P Hormozdiari, D Gani. Tetrahedron Lett 37:8227–8230, 1996.

146. E Larsson, B Lüning. Tetrahedron Lett 35:2737–2738, 1994.

147. DM Andrews, J Kitchin, PW Seale. Int J Peptide Protein Res 38:469–475, 1991.

148. W Bannwarth, E Küng, T Vorherr. Bioorg Med Chem Lett 6:2141–2146, 1996.

149. M Rodriguez, R Crosby, K Alligood, T Gilmer, J Berman. Lett Peptide Sci 2:1–6, 1995.

150. EA Kitas, E Küng, W Bannwarth. Int J Peptide Protein Res 43:146–153, 1994.
151. JW Perich. Int J Peptide Protein Res 40:134–140, 1992.
152. G Staerkaer, MH Jakobsen, CE Olsen, A Holm. Tetrahedron Lett 32:5389–5392, 1991.
153. MW Pennington. Methods Mol Biol 35:195–200, 1994.
154. W Tegge. Int J Peptide Protein Res 43:448–453, 1994.
155. H Mostafavi, S Austermann, W-G Forssmann, K Adermann. Int J Peptide Protein Res 48:200–207, 1996.
156. WDF Meutermans, PF Alewood. Tetrahedron Lett 37:4765–4766, 1996.
157. JW Perich. Lett Peptide Sci 3:127–132, 1996.
158. EA Ottinger, LL Shekels, DA Bernlohr, G Barany. Biochemistry 32:4354–4361, 1993.
159. JW Perich, RB Johns. Tetrahedron Lett 29:2369–2372, 1988.
160. DBA De Bont, JH van Boom, RMJ Liskamp. Tetrahedron Lett 31:2497–2500, 1990.
161. DBA De Bont, WJ Moree, JH van Boom, RMJ Liskamp. J Org Chem 58: 1309–1317, 1993.
162. HG Chao, MS Bernatowicz, CE Klimas, GR Matsueda. Tetrahedron Lett 34: 3377–3380, 1993.
163. JW Perich. Lett Peptide Sci 5:49–55, 1998.
164. T Vorherr, A Trzeciak, W Bannwarth. Int J Peptide Protein Res 48:553–558, 1996.
165. T Johnson, LC Packman, CB Hyde, D Owen, M Quibell. J Chem Soc Perkin Trans 1 719–728, 1996.
166. N Sotirellia, TM Johnson, ML Hibbs, IJ Stanley, E Stanely, AR Dunn, H-C Cheng. J Biol Chem 270:29773–29780, 1995.
167. Q Xu, J Zheng, D Cowburn, G Barany. Lett Peptide Sci 3:31–36, 1996.
168. C García-Echeverría, P Furet, B Gay, H Fretz, J Rahuel, J Schoepfer, G Caravatti. J Med Chem 41:1741–1744, 1998.
169. P Furet, B Gay, C García-Echeverría, J Rahuel, H Fretz, J Schoepfer, G Caravatti. J Med Chem 40:3551–3556, 1997.
170. C García-Echeverría, C Stamm, R Wille, D Arz, B Gay. Lett Peptide Sci 4: 49–53, 1997.
171. EA Ottinger, Q Xu, G Barany. Peptide Res 9:223–228, 1996.
172. C García-Echeverría. Lett Peptide Sci 2:93–98, 1995.
173. C García-Echeverría. Lett Peptide Sci 2:369–373, 1996.
174. G Zardeneta, D Chen, ST Weintraub, RJ Klebe. Anal Biochem 190:340–347, 1990.
175. SE Shoelson, S Chatterjee, M Chaudhuri, TR Burke Jr. Tetrahedron Lett 32: 6061–6064, 1991.
176. K Baczko, W-Q Liu, BP Roques, C Garbay-Jaureguiberry. Tetrahedron 52: 2021–2030, 1996.
177. TR Burke Jr, MS Smyth, M Nomizu, A Otaka, PR Roller. J Org Chem 58: 1336–1340, 1993.

178. C Garbay-Jaureguiberry, D Ficheux, BP Roques. Int J Peptide Protein Res 39:523–527, 1992.

179. M Cushman, ES Lee. Tetrahedron Lett 33:1193–1196, 1992.

180. MN Qabar, J Urban, M Kahn. Tetrahedron 53:11171–11178, 1997.

181. B Ye, TR Burke Jr. Tetrahedron Lett 36:4733–4736, 1995.

182. TR Burke Jr, B Ye, M Akamatsu, H Ford, X Yan, HK Kole, G Wolf, SE Shoelson, PP Roller. J Med Chem 39:1021–1027, 1996.

183. CJ Stankovic, N Surendran, EA Lunney, MS Plummer, KS Para, A Shahripour, JH Fergus, JS Marks, R Herrera, SE Hubbell, C Humblet, AR Saltiel, BH Stewart, TK Sawyer. Bioorg Med Chem Lett 7:1909–1914, 1997.

184. H Fretz. Tetrahedron Lett 37:8475–8478, 1996.

185. H Fretz. Tetrahedron Lett 37:8479–8482, 1996.

186. H Fretz. Tetrahedron 54:4849–4858, 1998.

187. G Shapiro, D Buechler, V Ojea, E Pombo-Villar, M Ruiz, H-P Weber. Tetrahedron Lett 34:6255–6258, 1993.

188. G Tong, JW Perich, RB Johns. Aust J Chem 45:1225–1240, 1992.

189. A Otaka, K Miyoshi, TR Burke Jr, PP Roller, H Kubota, H Tamamura, N Fujii. Tetrahedron Lett 36:927–930, 1995.

190. DB Berkowitz, Q Shen, J Maeng. Tetrahedron Lett 35:6445–6448, 1994.

191. M Ruiz, V Ojea, G Shapiro, H-P Weber, E Pombo-Villar. Tetrahedron Lett 35:4551–4554, 1994.

192. B Ye, M Akamatsu, SE Shoelson, G Wolf, S Giorgetti-Peraldi, X Yan, PP Roller, TR Burke Jr. J Med Chem 38:4270–4275, 1995.

193. OM Green. Tetrahedron Lett 35:8081–8084, 1994.

194. M Nomizu, A Otaka, TR Burke Jr, PP Roller. Tetrahedron 50:2691–2702, 1994.

195. SM Domchek, KR Auger, S Chatterjee, TR Burke Jr, SE Shoelson. Biochemistry 31:9865–9870, 1992.

196. A Otaka, TR Burke Jr, MS Smyth, M Nomizu, PP Roller. Tetrahedron Lett 34:7039–7042, 1993.

197. M Nomizu, A Otaka, MS Smyth, SE Shoelson, RD Case, TR Burke Jr, PP Roller. Synthesis and structure of SH2 binding peptides containing phosphonomethyl-phenylalanine and analogs. Proceedings of the Thirty-First Symposium on Peptide Chemistry, Japan, 1993, pp 281–284.

198. G Shapiro, D Buechler, A Enz, E Pombo-Villar. Tetrahedron Lett 35:1173–1176, 1994.

199. E Jaeger, P Rücknagel, HA Remmer, G Jung. Synthesis and properties of peptides containing O-sulfonated serine and threonine residues. Proceedings of the Eighth FRG–USSR Symposium on Chemistry of Peptides and Proteins, Aachen, 1991, pp 115–124.

200. WB Huttner. Annu Rev Physiol 50:363–376, 1988.

201. K Kitagawa, C Aida, H Fujiwara, T Yagami, S Futaki, Tetrahedron Lett 38: 599–602, 1997.

202. K Kitagawa, C Aida, H Fujiwara, T Yagami, S Futaki. Efficient solid-phase synthesis of sulfated tyrosine containing peptides using 2-chlorotrityl resin

and its application to gastrin/cholecystokinin peptides. Proceedings of the Thirty-Four Symposium on Peptide Chemistry, Japan, 1996, pp 21–24.

203. K Kitagawa, S Futaki, T Yagami, S Sumi, K Inoue. Int J Peptide Protein Res 43:190–200, 1994.

204. T Yagami, S Shiwa, S Futaki, K Kitagawa. Chem Pharm Bull 41:376–380, 1993.

205. B Penke, L Nyerges. Peptide Res 4:289–295, 1991.

206. Y Han, SL Bontems, P Hegyes, MC Munson, CA Minor, SA Kates, F Albericio, G Barany. J Org Chem 61:6326–6339, 1996.

207. I Maugras, J Gosteli, E Rapp, R Nyfeler. Novel cysteine derivatives and their use in peptide synthesis. Proceedings of the Twenty-Third European Peptide Symposium, Braga, 1994, pp 161–162.

208. N Fujii, S Futaki, S Funakoshi, K Akaji, H Morimoto, R Doi, K Inoue, M Kogire, S Sumi, M Yun, T Tobe, M Aono, M Matsuda, H Narusawa, M Moriga, H Yajima. Chem Pharm Bull 36:3281–3291, 1988.

209. S Futaki, T Taike, T Akita, K Kitagawa. Tetrahedron 48:8899–8914, 1992.

210. S Futaki, T Taike, T Akita, K Kitagawa. J Chem Soc Chem Commun 523–524, 1990.

211. S Futaki, T Taike, T Yagami, T Ogawa, T Akita, K Kitagawa. J Chem Soc Perkin Trans 1 1739–1744, 1990.

212. HA Remmer, E Jaeger, P Rücknagel, T Abdel-Razek, SI Said, Structure activity studies on VIP III: Synthesis and properties of [Ser(SO$_3$H)^3, Ahx^{17}]VIP, [Ahx^{17}]PACAP-27 and the VIP/PACAP-hybrid [Ahx^{17}, Ala^{24}, Ala^{25}, Val^{26}]VIP-(1–27)-amide. Proceedings of the Twenty-Third European Peptide Symposium, Braga, 1994, pp 367–368.

213. I Imajo, Y Yamagata, H Katoh, S Satomura, K Nakamura. Preparation of acidic peptides for use in isolation and analysis of living body components. European Patent Application. EP 755941 A2 970129, 1997.

214. E Jaeger, HA Remmer, G Jung, J Metzger, W Oberthür, KP Rücknagel, W Schäfer, J Sonnenbichler, I Zetl. Biol Chem Hoppe-Seyler 374:349–362, 1993.

215. E Jaeger, G Jung, HA Remmer, P Rücknagel. Formation of hydroxy amino acid-O-sulfonates during removal of the Pmc-group from arginine residues in SPPS. Proceedings of the Twelfth American Peptide Symposium, Cambridge, 1991, pp 629–630.

216. I Maugras, J Gosteli, E Rapp, R Nyfeler. Int J Peptide Protein Res 45:152–156, 1995.

217. JM Swan. Nature 180:643–645, 1957.

11
Oligonucleotide Synthesis

Laurent Bellon and Francine Wincott
Ribozyme Pharmaceuticals, Inc., Boulder, Colorado

I. INTRODUCTION

The demand of the scientific community for synthetic oligonucleotides has grown exponentially over the past decade. Fortunately, the abundant source of DNA oligonucleotide primers has satisfied the tremendous needs of the genome sequencing efforts, functional genomics, or other polymerase chain reaction (PCR)–based detection methods. Oligonucleotides also have widespread use in the development of therapeutics and diagnostic applications, including chip-based DNA microarrays. Significant advances in structural biology and biochemistry have been achieved through concomitant advances in DNA and RNA chemistry. For instance, the current state of the art in ribozyme research, including crystal structures, would not have been possible without the accompanying improvements in RNA synthesis. Research on the many roles of nucleic acids has, in the past, been hindered by limited means of producing such biologically relevant molecules [1–4]. Although enzymatic methods existed, protocols that allowed one to probe structure–function relationships were limited. Only uniform postsynthetic chemical modification [5] or site-directed mutagenesis [6] was available. Fortunately, oligonucleotide synthesis by the phosphoramidite method has greatly increased our understanding of DNA and RNA. Site-specific introduction of modified nucleotides at any position in a given oligonucleotide has now become routine, allowing easy chemical probing of define functionalities.

The presence of a single hydroxyl at the 2′-position of the ribofuranose ring has been the major reason that research in the RNA field has lagged so far behind comparable DNA studies. Progress has been made in improving methods for DNA synthesis that have enabled the production of large amounts of antisense oligodeoxynucleotides for therapeutic applications. Only recently have similar gains been achieved for oligoribonucleotides [7–9].

The chasm between DNA and RNA syntheses is due to the added difficulty of identifying orthogonal protecting groups for the 5'- and 2'-hydroxyls. However, protecting group compatibility is not solely relevant to the 2'- and 5'-OH groups but applies to the heterocyclic base and phosphate protecting groups as well.

Solid-phase synthesis of oligonucleotides using the phosphoramidite method follows the iterative process outlined in Fig. 1. A solid support with an attached nucleoside is subjected to removal of the protecting group on the 5'-hydroxyl. The incoming amidite is coupled to the growing chain in the presence of an activator. Any unreacted 5'-hydroxyl is capped and the newly created phosphite triester linkage is then oxidized to provide the desired phosphotriester bond. The process is then repeated until an oligomer of the desired length results. The actual reagents used may vary according to the 5'- and 2'-protecting groups.

From the generic synthesis wheel (Fig. 1), it is clear that oligonucleotide synthesis is a multistep elongation process that entails reactions between specific nucleophilic (5'-hydroxyl) and electrophilic (3'-activated phosphoramidite azolide intermediate) moieties. These two reactive species need to be transiently protected so that they can be unveiled only during the critical coupling step. A typical nucleotide scaffold exhibits additional reactive centers such as the exocyclic amino groups, present at positions N_2, N_4, and N_6 of the heterocyclic bases guanine, cytidine, and adenine, respectively; the anionic phosphodiester backbone; or the additional 2'-hydroxyl, present in the RNA series. All of these reactive groups need to be permanently protected during synthesis to minimize undesired reactions and to allow efficient elongation of the oligomer.

Following synthesis, the crude oligonucleotide, bound to the solid support, needs to be further processed to remove all protecting groups. This is typically a single-step (DNA synthesis) or a two-step (RNA synthesis) process that entails cleavage of the oligomer from the support and deprotection of the base and phosphate blocking groups, followed by removal of the 2'-hydroxyl protecting groups when required. In all cases it is imperative that indiscriminate removal of the protecting groups does not occur. This is particularly an issue for RNA deprotection in the classic situation wherein the first step is base mediated. In this case, if the 2'-hydroxyl present in oligoribonucleotides is revealed under these conditions, strand scission will result because of attack of the vicinal hydroxyl group on the neighboring phosphate backbone. Two other concerns that are prevalent in RNA synthesis but do not occur in DNA assembly are the propensity for 2',3'-phosphate mi-

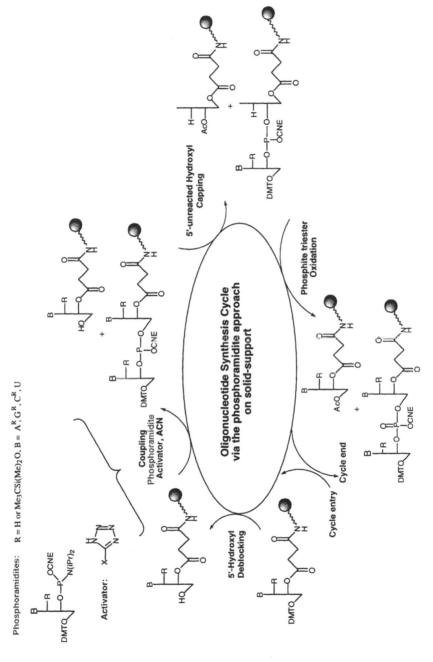

Figure 1   Oligonucleotide synthesis on solid support via the phosphoramidite approach.

gration to provide undesired 2'–5' linkages and the susceptibility of oligo-ribonucleotides to degradation by ribonucleases. The latter fact has led many researchers to develop 2'-OH protecting groups that can remain in place until the oligomer is required for the desired experiment.

II. PROTECTING GROUPS

Choosing the appropriate protecting groups is the key to successful oligo-nucleotide synthesis. Although the 2'- and 5'-protecting groups are obvi-ously the most critical and will be discussed in depth, the importance of the base and phosphate blocking groups cannot be ignored. Therefore, the var-ious protecting groups to be used must be orthogonal to account for synthesis and deprotection scheme requirements. Some base and phosphate groups are not stable to conditions required for the repetitive removal of the 5'-group during each nucleotide addition cycle. In other cases, the protecting group may not be stable to the conditions required to synthesize the monomers.

A. Nucleobase Protecting Groups

1. Acyls

For oligonucleotide synthesis, the standard base protecting groups, used ever since Khorana and coworkers [10] introduced them in 1963, are benzoyl for A and C and isobutyryl for G. Despite the fact that these groups can easily be removed with 30% aqueous ammonium hydroxide at 55°C for 5 h, these deprotection conditions are not easily amenable to proper deprotection of alkali-sensitive modified 2'-deoxyribonucleotides. Moreover, many of the 2'-OH protecting groups used in RNA synthesis are not stable to this harsh reagent. Milder conditions have been developed that are more amenable to deprotection of base-sensitive oligonucleotides, such as the use of 3:1 NH$_4$OH–EtOH [11] or EtOH–NH$_3$ [12]. However, long incubation times are still required: ~16 h at elevated temperatures (55–65°C), to effect com-plete deprotection.

Consequently, more base-labile acyl groups were developed such as phenoxyacetyl, **1** (PAC) [13], *t*-butylphenoxyacetyl, **2** (TAC) [14,15], and isopropylphenoxyacetyl, **3** (*i*PrPAC) (Fig. 2). All can be removed in 1–4 h at 55°C with 3:1 NH$_4$OH–EtOH. Because the oligomer is not exposed to severe deprotection conditions for prolonged periods, better yields of higher quality product result. More recently, a faster deprotection protocol, entailing the use of aqueous methylamine (MA) has been reported for DNA [16] and RNA [17]. Incubation times have been reduced to 10 min at 65°C. When compared with other RNA deprotection methods, treatment with this reagent

Figure 2 Base protecting groups typically used for the synthesis of oligoribonucleotides.

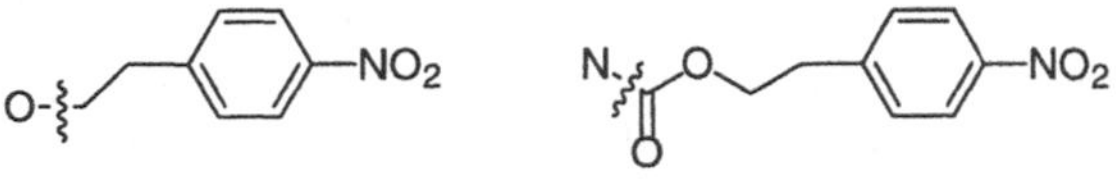

gave greater full length product than the standard protocol using 3:1 NH_4OH–EtOH [7]. The only requirement is that acetyl is used as the protecting group for cytidine because of a well-documented transamination reaction [18].

2. Amidines

Another strategy used to protect the exocyclic amino groups is amidine [19] chemistry, which is used mainly for ribopurine phosphoramidites because the N_4-acetyl protection of cytidine appears quite optimal. This chemistry which includes acetamidine [19] and dialkylformamidine [20] moieties, **4**, (Fig. 2), allows cleavage in 2–3 h at 55°C with ammonium hydroxide–ethanol (3:1). The main advantage of the dimethylformamidine protecting group when used in conjunction with the 2'-deoxyribophosphoramidite series [21], is that it confers increased resistance toward depurination occurring during the detritylation step. However, since oligoribonucleotides are inherently less sensitive to depurination, the phenoxyacetyl chemistry has been preferred on the scale of commercial manufacturing.

3. β-Eliminating

Another family of base protecting groups that has found some favor for DNA–RNA synthesis comprises the 2-(4-nitrophenyl), **5** (Npe), 2-(4-nitrophenyl)ethoxy carbonyl, **6** (Npeoc), groups [22,23] (Fig. 2). They are stable to both weak acids and bases yet can be readily removed with a nonnucleophilic base such as 1,8-diazabicyclo[5.4.0.]undec-7-ene (DBU). Use of these

β-eliminating blocking groups is usually indicated when a base-labile 5'-protecting group is present. Similarly, the 9-fluorenylmethoxycarbonyl (Fmoc) [24–26] group has been introduced on the exocyclic amines of dA and dG and deprotected under mild basic conditions [25,27].

4. Miscellaneous

Aside from the popular base-labile acyl protecting groups, a number of alternative orthogonal nucleobase protecting groups have been studied, mostly for very specific applications. Such applications include synthesis of oligonucleotide phosphotriesters and phosphoramidates, carboxylic ester–containing oligonucleotide for postsynthesis conjugation, and, more generally, use when the base lability of a particular functionality is a concern. Interestingly, methylphosphotriester oligonucleotides have been prepared by the phosphoramidite method using the pent-4-enoyl group for the protection of the N_6, N_2, and N_4 amino groups of adenine, guanine, and cytidine phosphoramidites, respectively [28]. Deprotection can easily be achieved using a mixture of 2% iodine in pyridine–water [28] or pyridine–methanol [29] (98:2) at room temperature for 30 min. The use of an N-pent-4-enoyl protecting group implies that the phosphitetriester oxidation into the corresponding phosphotriester linkage has to be conducted with an iodine-free formulation such as 1 M *tert*-butylhydroperoxide in toluene.

B. Phosphate Protecting Groups

Ever since the development of the modified phosphoramidite method [30] that made use of methyl-*N,N*-dimethylaminophosphine, the 2-cyanoethyl group, 7 [31] (Fig. 3), has been the main choice for permanent protection of the phosphitetriester–phosphotriester linkages during oligonucleotide synthesis because of ease of removal via a β-elimination under mildly basic conditions. A number of alternative allyl, alkyl, haloalkyl, and diversely substituted aryl or alkylaryl protecting groups have been studied and reviewed extensively by Beaucage and Iyer [32]. The methyl group introduced by Beaucage and Caruthers [30] in the DNA series was further studied by Usman et al. [11] in the RNA series. Despite the higher reactivity of methyl phosphoramidite as compared with the 2-cyanoethyl analogue, this group requires the use of soft nucleophiles as provided by noxious thiophenolate ions [33] to prevent nucleophilic attack on the phosphorus center leading to strand scission. Further development of less toxic and malodorous mercaptans allowed more user-friendly methyl phosphate deprotection protocols [34,35].

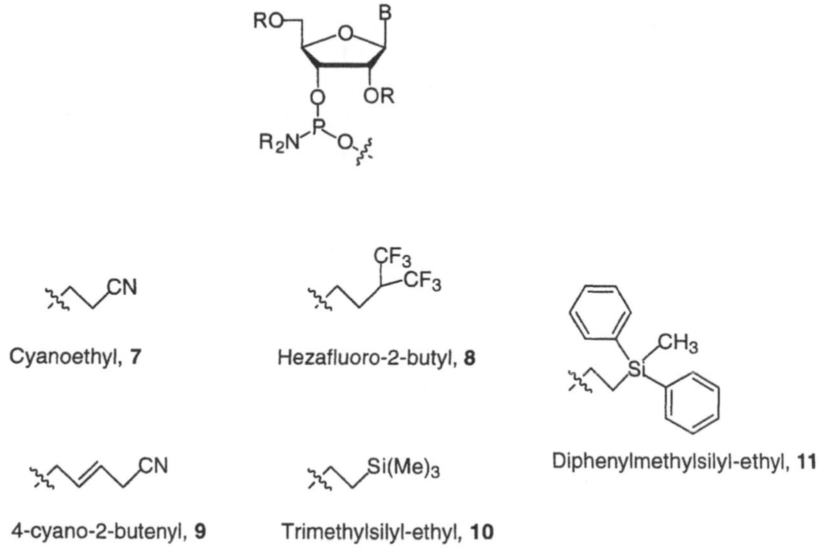

Cyanoethyl, 7

Hezafluoro-2-butyl, 8

Diphenylmethylsilyl-ethyl, 11

4-cyano-2-butenyl, 9

Trimethylsilyl-ethyl, 10

Figure 3 Phosphoramidite protecting groups used in oligonucleotide synthesis.

Other useful phosphate protecting groups that operate through a mildly basic β-elimination route include the p-nitrophenylethyl (NPE) **5** group [22] (Fig. 2) and the hexafluoro-2-butyl (HFB) group, **8** [36] (Fig. 3).

An alternative to the β-elimination cleavage uses a δ-elimination mechanism by substituting the 2-cyanoethyl moiety by a 4-cyano-2-butenyl functionality, **9** [37]. This group is released conveniently with concentrated aqueous ammonium hydroxide for 2 h at room temperature.

Alternatively, a β-fragmentation mechanism has been applied in the design of 2-(trimethylsilyl)ethyl, **10** (TSE) [38], and 2-(diphenylmethylsilyl)ethyl, **11** (DPSE) [39] (Fig. 3). The DPSE group can be removed either under mild ammonium hydroxide conditions or by the use of tetrafluorosilane, whereas TSE appears to require fluoride ions to initiate the β-fragmentation cleavage.

C. 5′-Hydroxyl Protecting Group

1. Acid-Labile 5′-Hydroxyl Protecting Group

a. Trityl. Alkoxy-substituted trityl groups are the most favored groups for protection of 5′-hydroxyl because of ease of introduction on the nucleoside scaffold and ease of deprotection under acidic conditions. The acid sensitivity of the trityl ether linkage is driven by the electron-donating properties of the substituents of the aromatic rings of the trityl moiety [40]. This led to the investigation of 4-methoxytrityl (MmTr) and especially 4,4'-di-methoxytrityl, **12** (DmTr) [40], which is still routinely used in oligonucleotide synthesis (Fig. 4). These groups are cleaved efficiently with 3% trichloroacetic acid or 2% dichloroacetic acid in apolar solvents such as dichloromethane or dichloroethane. Alternatively, Lewis acids such as zinc bromide in nitromethane–water (99:1) [41] or dichloromethane–isopropanol (85:15) [42] can be used instead of the conventional protic haloacetic acids responsible for the depurination of N_6-benzoyl deoxyadenosine residues. The trityl moiety constitutes an elegant template for designing application-specific trityl-based 5'-hydroxyl protecting groups used, for instance, in photo-cross-linking experiments [43] or to provide a more hydrophobic handle for reverse-phase purification [44,45].

b. Pixyl. Another alternative to the popular 4,4'-dimethoxytrityl group is the 9-phenylxanthen-9-yl group, **13** (Pixyl, Px) [46] (Fig. 4), which has comparable lability toward acidic conditions such as tosic acid in chloroform–methanol (95:5) [47].

2. Non–Acid-Labile 5'-Hydroxyl Protecting Groups

The rationale for the development of non–acid-labile 5'-hydroxyl protecting groups stems from the notorious potential depurination of N_6-benzoyl deoxyadenosine residues under trichloroacetic acid conditions and the need for orthogonal acid-labile protecting groups for the 2'-hydroxyl in the RNA series.

a. 2-Dansylethoxycarbonyl (Dnseoc). In the RNA series, one approach to resolving the incompatibility of 2'-acetal protecting groups with the standard acid-labile 5'-protecting groups, DMT and Px, was the development of a new 5'-blocking group, 2-dansylethoxycarbonyl, **14** (Dnseoc) [48]. This base-labile group can be readily removed using dilute DBU in 140 s [49]. As a result, the more stable Npe phosphate protecting group is required in place of the traditional cyanoethyl group and Npe and Npeoc are used for base protection. Oligomers of 20 nucleotides were synthesized, and in special cases 40-mers have been constructed. It was determined that the optimal results were obtained with the N,N-diethylphosphoramidite rather than the N,N-diisopropyl analogue. Deprotection of the oligoribonucleotide first required treatment with 0.5–1M DBU for 10 h to remove the Npe and Npeoc groups, followed by cleavage of the support with NH_3 for 200 min. The oligomer could be stored at this point or exposed to acid to remove the Mthp group.

b. Levulinyl. Oligoribonucleotides as long as 21-mers were synthesized with a 5'-O-levulinyl group, **15** [50], in combination with 2'-O-tetrahydro-

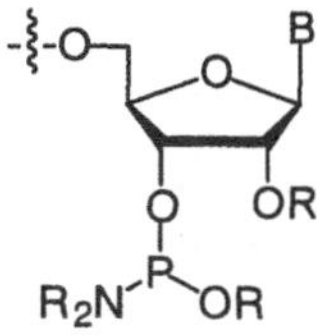

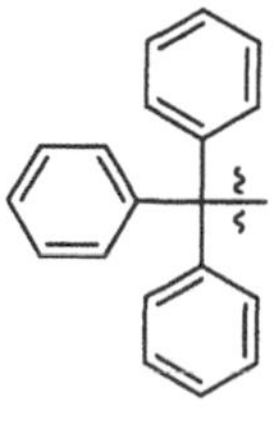

DMT, **12**

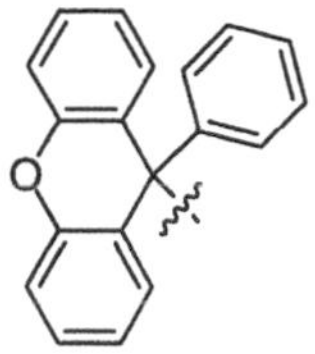

Px, **13**

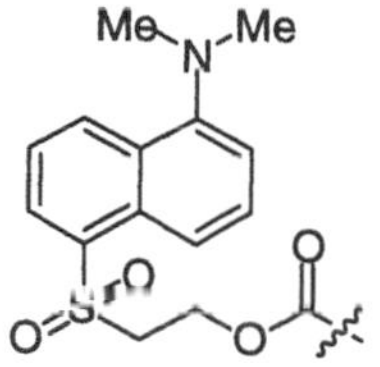

Dnseoc, **14**

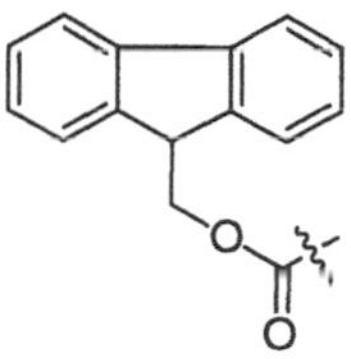

Fmoc, **16**

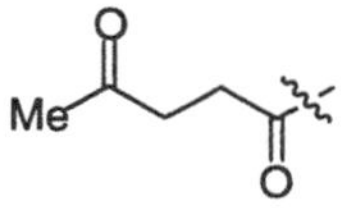

Levulinyl, **15**

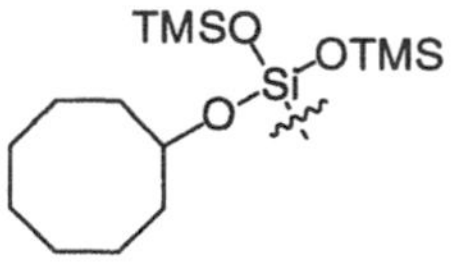

SIL, **17**

Figure 4 5′-Hydroxyl protecting groups used in oligonucleotide synthesis.

furanyl (Thf) protection [51]. The levulinyl group is removed during solid-phase synthesis with 0.5 M hydrazine in 10 min. Following ammonia treatment, the base and phosphate deprotected oligomer is then treated with 0.01 N HCl (pH 2) for 24 h to effect removal of the 2′-acetal. Although no base modification is observed, there are some drawbacks to this scheme. As is the case for a number of alternative 5′-protecting groups, removal of the levulinyl groups cannot be easily monitored. Furthermore, because of the prolonged time required for full removal of the levulinyl group, cycle times are very long. Finally, introduction of the levulinyl group to the 5'-position of the required monomers is not selective, which reduces yields.

c.   9-Fluorenylmethoxycarbonyl (Fmoc).   The lability of acetal groups to iterative acidic treatment led to the development of the 9-fluorenylmethoxycarbonyl group, **16** (Fmoc), as an alternative to the DMT group for 5'-hydroxyl protection [52]. The Fmoc group is readily introduced in the 5'-position of the 2'-Mthp protected nucleosides. Furthermore, release of the Fmoc group can be detected by ultraviolet (UV) spectroscopy, allowing the researcher to quantitate the coupling efficiency of the synthesis. During solid-phase synthesis, the 5'-Fmoc is removed with 0.1 M DBU in ACN in 2 min. The oligomer is cleaved from the support and base and phosphate groups are removed with ammonia treatment. At this point the 2'-protected oligoribonucleotide can be purified if desired. After careful analysis it was shown that all internucleotidic linkages were 3'–5' and that no base modification occurred. Oligomers containing 20 residues have been successfully synthesized using this combination of protecting groups.

Ogawa et al. [53] substituted the acid-labile 1-(isopropoxy) ethyl group (IPE) for the Mthp group at the 2'-OH position in addition to using Fmoc as the 5'-hydroxyl protecting group. The desired nucleosides were prepared from the corresponding Markiewicz protected intermediates in a four-step procedure in good yields. Removal of the Fmoc group during solid-phase synthesis was accomplished with 0.1 M piperidine in ACN in 2 min, and coupling was effected with tetrazole for 20–25 min. Following deprotection with ammonia for 6–12 h at 55°C, the IPE group was removed at pH 2.0 in 3 h at room temperature. Again, no 2'-5' isomerization or base modification was observed under these conditions. Oligomers consisting of 21 residues were reported with this combination of protecting groups.

d.   5'-Bis(Trimethylsiloxy)-Cyclooctyloxy Silyl Ether (SIL)/2'-Bis(2-Acetoxyethoxy) Methyl Orthoester (ACE).   A completely different approach to 5'-protection was reported by Scaringe et al. [54] wherein a silyl ether is utilized for 5'-protection in tandem with a 2'-orthoester. The bis(trimethylsiloxy)-cyclooctyloxy silyl ether, **17** (SIL), can be removed with fluoride ion under conditions that will not affect an acid-labile 2'-protecting group. The bis(2-acetoxyethoxy) methyl orthoester, **31** (ACE) (Fig. 5), is a convertible protecting group that is stable to all synthesis conditions but is modified during base deprotection of the oligoribonucleotide. The resulting bis(2-hydroxyethoxy)methyl orthoester is 10 times more acid labile than the original protecting group. The required monomers can be produced in four steps from the Markiewicz protected nucleosides in overall yields of 45–55%. Because cyanoethyl groups are not compatible with repeated exposure to fluoride ion, the methyl N,N-diisopropylphosphoramidite is used. The 5'-silyl group is removed in 35 s with 1.1 M HF in TEA−DMF. Coupling is complete after 90 s with S-ethyltetrazole as the activator with coupling yields reported as >99%. Once the oligomer has been synthesized, deprotection of the methyl phosphate group is effected with disodium-2-carbamoyl-2-cyanoethylene-1,1-dithiolate [35] for 10 min, followed by treatment with 40% MA in water at 55°C for 10 min. The 2'-protected oligomer can then be analyzed, purified if necessary, and stored. To remove the modified 2'-orthoester, the oligoribonucleotide is heated to 55°C for 10 min in pH 3 buffer, followed by incubation for 10 min at 55°C at pH 7.7−8.0. This final step cleaves any remaining 2'-formyl groups that result from the orthoester deprotection. The synthesis of oligomers up to 36 residues in length has been reported. Careful analysis of the resulting oligomers showed that there was no base modification and no indication of 2'-5' migration. Furthermore, appropriate molecular weights and enzymatic activities were observed for the oligomers that were synthesized. In comparisons of identical oligoribonucleotides synthesized using 5'-DMT/2'-TBDMS chemistry versus this new protocol, better yields were obtained with the 5'-SIL/2'-ACE process [54].

D. 2'-Hydroxyl Protecting Group

The most common paradigm has been to adapt DNA synthesis to RNA. As a result, a 2'-hydroxyl protecting group must be identified that is compatible with DNA protecting groups but can easily be removed once the oligomer is synthesized. Because of constraints placed by the existing amide protecting groups on the bases and the 5'-trityl (DMT) group [or in some cases the 9-(p-anisyl)xanthen-9-yl (Px) group], the 2'-blocking group must be stable to both acid and base. In addition, the group must be inert to the oxidizing and capping reagents. Although the most widely used 2'-hydroxyl protecting group is the TBDMS ether, many others have been explored because of the longer coupling times required when the bulky 2'-TBDMS substituent is used. Additional alternatives not discussed in this chapter can be found in reviews [32,55].

In the early 1990s the 5'-DMT/2'-Fpmp combination [56,57] showed great promise. However, since that time there have been very few reports of successful RNA syntheses using this protocol, although these monomers are commercially available. The results obtained with the o-nitro- [58] and p-nitrobenzyloxymethyl [59] groups also appeared quite encouraging. Since the initial reports, these 2'-protecting groups do not appear to have been readily accepted by researchers. Other 2'-protecting groups that are similar to TBDMS and Fpmp in their compatibility with current DNA synthesis protocols are the convertible protecting groups HIFA [60] and NEFE [23].

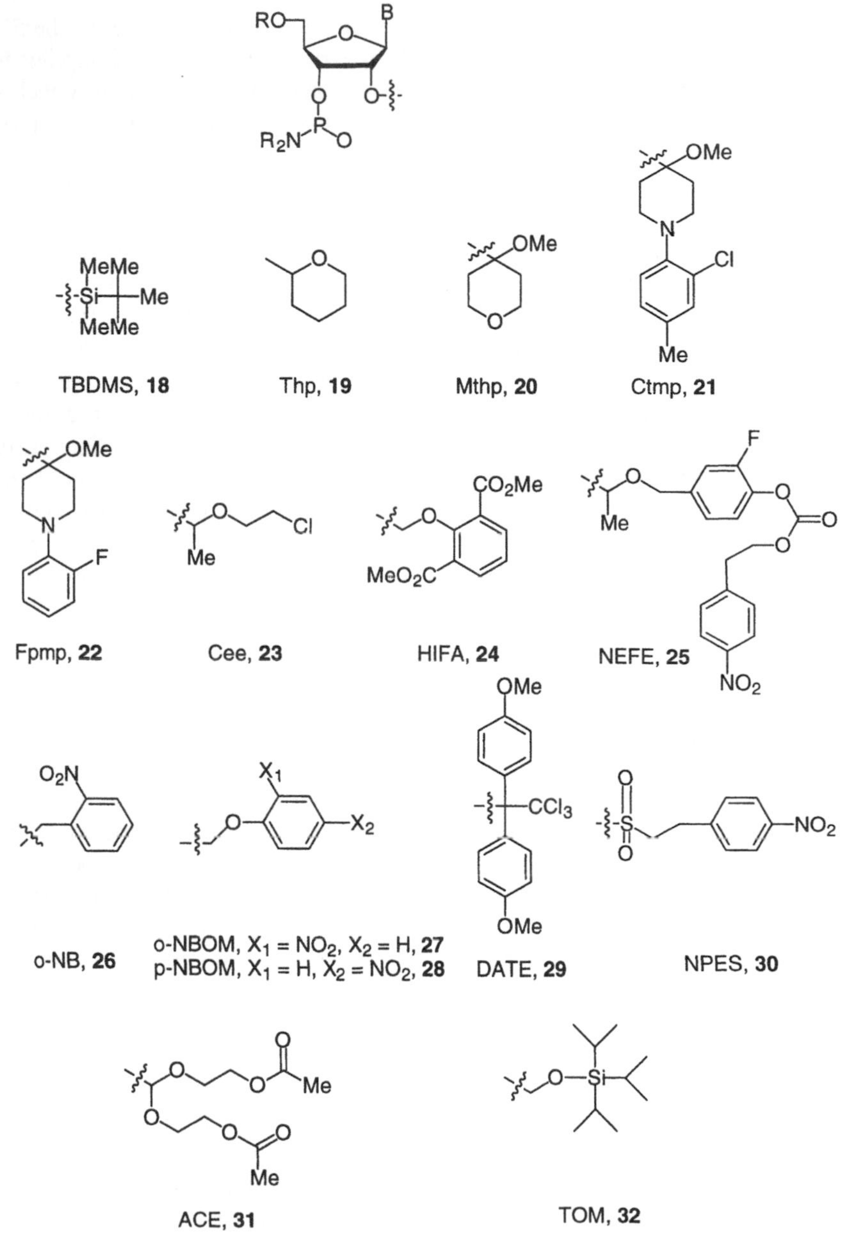

Figure 5   RNA synthesis compatible 2′-hydroxyl protecting groups.

These acetal-derived groups appear interesting, but there have been few reports since the initial publications.

1. *t*-Butyldimethylsilyl (TBDMS)

By far the most popular 2′-protecting group is the *t*-butyldimethylsilyl group, **18** (TBDMS), developed principally by Ogilvie and coworkers [11]. Synthesis of the nucleotides can be quite readily accomplished in good yields [12]. The chemistry utilized in the construction of the oligoribonucleotides is completely compatible with the DNA synthesis cycle, thereby allowing the simple construction of DNA–RNA chimeras. In addition, many of the earlier disadvantages of this protecting group no longer exist. Although coupling times are not as short as with DNA, through the use of *S*-ethyl tetrazole coupling times have been reduced from 30 to 5 min [7]. Furthermore, the development of new deprotection protocols has not only reduced the incubation times but also greatly increased the quality of the product to the extent that contaminating 2′–5′ linkages can be eliminated completely.

Currently, the 5′-DMT/2′-TBDMS combination constitutes the benchmark for the synthesis of oligoribonucleotides. Although many other methods for the synthesis of RNA have been developed, none have gained the popularity of the TBDMS chemistry. Advances in the use of this silyl chemistry in both synthesis [7,8] and deprotection [7,8,61] strategies have made it an even more viable approach to the production of oligoribonucleotides.

2. Acetals

Because of concerns about conversion of the desired 3′–5′ internucleotidic linkages to 2′–5′ linkages, acid-labile acetals were considered to be the ideal 2′-protecting groups. They are stable to alkaline conditions and can be hydrolyzed with dilute acids, so there are no residual reagents to complicate purification. Furthermore, the oligonucleotide can be isolated with the 2′-protecting group intact, allowing one to store the oligonucleotide in a nuclease-resistant form.

A number of different acetals have been investigated. The tetrahydropyranyl, **19** (Thp), and methoxytetrahydropyranyl, **20** (Mthp), groups have proved to be unstable to the conditions required for iterative removal of the dimethoxytrityl (DMT) group [62,63]. Although some successful syntheses have resulted from the use of these acetals, they have been limited to very short oligomers. As a result, aryl-substituted piperidines were developed. The 1-(2-chloro-4-methylphenyl)-4-methoxy-piperidin-4-yl (Ctmp) group, **21**, was first investigated by Reese et al. [64]. However, because the reagent required for introduction of the Ctmp group into the monomer was difficult to prepare, the 1-(2-fluorophenyl)-4-methoxypiperidin-4-yl group **22** (Fpmp), was developed as an alternative. This aryl-substituted piperidine is easily introduced to provide the required monomer [65] and is more stable to acidic hydrolysis than the previously described acetals. Consequently, the Fpmp group can be paired with either a DMT or Px group at the 5′-position. Early reports indicated that room temperature deprotection at pH 2.0 for 20 h was optimal [56]. However, it has since been determined that the rate of acid-catalyzed hydrolysis of the internucleotidic linkages is sequence dependent. To avoid hydrolytic cleavage and migration, the best conditions for removal of the Fpmp group are 24 h at pH 3 at room temperature [57].

Another approach to acetal protection of the 2′-hydroxyl function led to the development of the 1-(2-chloroethoxy)ethyl group, **23** (Cee) [66,67]. Oligoribonucleotides up to 20 residues in length have been prepared using the Cee group. This protecting group is stable under the acidic conditions required for removing the trityl group during synthesis, yet is reportedly removed postsynthetically with 0.01 N HCl (pH 2.0) at room temperature for 8–30 h.

More recently, a new 2′-protecting group was reported: the 2-hydroxyisophthalate formaldehyde acetal, **24** (HIFA) [60]. This is a convertible protecting group that, as the bis ester, is stable to acidic treatment during synthesis but is altered upon treatment with ammonia to a bis acid that is then more labile in acid. The half-life for the deprotection of the resulting bis acid is ~390 min at pH 3, as compared with 166 min for Fpmp cleavage under the same conditions. At this time only UpU and UpG dimers and the corresponding uridine phosphoramidite have been synthesized.

Finally, Pfleiderer et al. [23] have designed a new acetal for the solid-phase synthesis of oligoribonucleotides that is used successfully in conjunction with a 5′-*O*-DMT group. As in the case of the HIFA protecting group, this acetal, 1-{4-[2-(4-nitrophenyl)ethoxycarbonyloxy]-3-fluorobenzyloxy}-ethyl, **25** (NEFE), is also a convertible, or "protected protecting" group. For the nucleoside, upon cleavage of the ethoxy carbonyl group, the resulting acetal can be removed in 4 h by acid hydrolysis as compared with 24 h for the "protected" acetal. The best results were obtained when this acetal was used with the Npe/Npeoc base protecting groups using a 20-min coupling time. Treatment with DBU removed the base and phosphate protecting groups as well as the 2-[4-(nitrophenyl)ethoxycarbonyl group from the acetal. After cleavage from the support, the acetal was reportedly removed with 0.5% AcOH at room temperature in 18 h.

3. Photolabile

Another approach to the protection of the ribonucleoside 2′-hydroxyl is the use of photolabile protecting groups. The advantages of this strategy are many. The protecting groups are completely orthogonal as they are resistant to both acid and base and, as a result, remain intact throughout synthesis and base deprotection. Furthermore, introduction onto the monomer is accomplished quite readily without any migration. Originally, the o-nitrobenzyl group, **26** (o-NB) [68,69], was the photolabile protecting group of choice. Coupling of the amidite was accomplished in 15 min with tetrazole as the activator or 2.5 min with 5-p-nitrophenyltetrazole in conjunction with the methyl phosphate protecting group [70]. Following base hydrolysis of the oligomer, irradiation with the long-wave UV light occurs at pH 3.5 for 1 h in solutions that have been purged with N_2. Use of higher pH resulted in the formation of side products.

Because the o-nitrobenzyl group (in conjunction with the cyanoethyl phosphate group) requires extended coupling times [71], an alternative photolabile group was devised: the o-nitrobenzyloxymethyl group, **27** (o-NBOM) [58]. It was postulated that the extended arm present in this group might ease steric crowding, thereby reducing coupling times. Synthesis of the monomers proceeds similarly to that of the 2'-O-TBDMS amidites. However, unlike silyl-protected ribonucleotides, these amidites required only a 2-min coupling time. Following a standard basic deprotection protocol [pyridine–NH$_4$OH (1:4) 50°C, 24 h], the 2'-protected oligomers were exposed to long-wave UV light for 4.5 h at room temperature (pH 3.7) to remove the o-nitrobenzyloxymethyl groups. More recently, the p-nitrobenzyloxymethyl group, **28** (p-NBOM), has been described [59]. This protecting group behaves almost identically to the o-nitro version with regard to monomer synthesis and coupling times; however, it can be removed with (tetrabutylammonium fluoride (TBAF) in 24 h at room temperature.

4. 1,1-Dianisyl-2,2,2-Trichloroethyl (DATE)

Klosel et al. [72] have described a completely new protecting group; the 1,1-dianisyl-2,2,2-trichloroethyl group, **29** (DATE). This β-haloalkyl group is stable to acid and base, yet cleavable under mild, neutral conditions via reductive fragmentation. Furthermore, there is no tendency toward migration between 2'- and 3'-hydroxyls. Only the synthesis of the uridine phosphoramidite and the corresponding UpT dimer has been described. The synthesis of the monomer is fairly straightforward; the product is provided in five steps from uridine. To form the dimer, a 15-min coupling time is required. Following treatment with ammonia, the dimer was exposed to lithium cobalt(I) phthalocyanine and phenol in MeOH (O_2 free) for 14 h at room temperature. The reaction mixture was then quenched with buffer, and analysis showed that the backbone was intact and that no 2'–5' linkages were present.

5. *p*-Nitrophenylethyl Sulfonyl (NPES)

The *p*-nitrophenylethyl sulfonyl group, **30** (NPES), has also been proposed as a 2′-protecting group [48]. The advantages of this sulfonate-derived group are acid stability and lack of 2′–3′ migration. This protecting group works best when coupled with the Npeoc and β-cyanoethyl groups for base and phosphate protection, respectively. Treatment with DBU results in removal of all protecting groups. Unfortunately, this scheme is not compatible with uridine that is not protected at O-4 because of a propensity for anhydro nucleoside formation. As a result, protection of O-4 with a *p*-cyanoethyl group was explored. This group can also be removed upon exposure to DBU, but only at an elevated temperature (50°C).

6. 2′-Bis(2-Acetoxyethoxy) Methyl Orthoester, **31** (ACE)

See Section II.C.2.d for a discussion.

7. Triisopropylsilyloxymethyl (TOM)

Among the recent developments in the 2′-hydroxyl specific protecting groups, the 2′-*O*-triisopropylsilyloxymethyl, group, **32** (TOM), developed by Weiss and Pitsch [73,74] from Xeragon A.G. (Zurich, Switzerland), presents a number of features that render it extremely appealing for efficient oligoribonucleotide synthesis. Because of the additional oxymethyl group separating the 2′-hydroxyl from the triisopropylsilyl group, there is much reduced steric hindrance between the activated phosphoramidite at the 3′-*O*-position and the 2′-OH protecting group. This makes it possible to shorten the coupling time to ~120 s while maintaining high coupling efficiency with 5-benzylthio-1-*H*-tetrazole. The synthesis of 2′-TOM protected nucleosides is quite straightforward because the well-known 2′–3′ isomerization [75] pathway under weakly basic conditions with TBDMS chemistry cannot occur. Deprotection of 2′-*O*-TOM crude oligoribonucleotides can be accomplished using standard methylamine- and fluoride-based conditions. More specifically, the 2′-*O*-TOM deprotection makes use of a fluoride-mediated silylether cleavage reaction followed by formaldehyde elimination with weakly basic aqueous workup.

III. SOLID SUPPORTS

The solid-phase synthesis of oligonucleotides by the phosphoramidite approach requires the use of a solid support that is functionalized with an appropriate nucleoside. Solid supports are typically macroporous structures with functional groups present at the inner surface of the matrix. A solid support allows proper filtration of all chemicals and solvents (dichloromethane, acetonitrile) involved in oligonucleotide synthesis. Nonswellability of solid supports is essential because swelling in organic solvent may inhibit rapid diffusion of reagents through the support. A solid support is composed of a polymeric matrix, an appropriate linker, and a defined chemistry of attachment between the linker and the 3'-nucleoside of the oligonucleotide to be synthesized. Clearly, all of these constituents are important for successful oligonucleotide synthesis.

A. Supports

Over the years, a variety of stationary matrices have been investigated. Most notably, these solid supports are composed of silica, polystyrene, or miscellaneous polymeric lattices.

1. Silica-Based Supports

Silica-based supports were recognized early as suitable supports for efficient oligonucleotide synthesis, mainly because of their rigid nonswellable characteristics. Silica gels, such as Kieselgel 60 [76], Vydak TP [76,77], Fractosil [78,79] or Porasil [80], were all porous supports ranging from 200 to 1000 Å in pore size that allowed only modest yields by today's standards of relatively short oligonucleotides. From this precursor work were inferred important conclusions regarding the role of the length and chemistry of the spacer–linker. Glass solid supports of controlled pore size were also introduced in the early 1980s with the pioneering work of Koster [81] and Gough [82] and their coworkers. Advantages of controlled pore glass or CPG are that the accessibility of functional groups is not dependent on the swellability of the support or choice of the organic solvent, the incompressible rigid structure allows high flow rates of reagents and solvents, and the narrow pore size distribution can be varied over a wide range of porosity. Because of the macroporous structure of CPG, oligonucleotide growth is strictly limited by the pore size, because the functional groups (i.e., nucleosides loaded on the CPG) are available at the inner surface of cavity [83,84]. Hence, for long oligonucleotides 50 nucleotides in length or more, high-porosity CPG of 1000 Å may be preferred, whereas CPG of 500 Å is typically used for shorter sequences [85]. The CPG porosity also governs the loading of the functional nucleoside in such a way that higher loadings (>50 μmol/g) can be achieved only with smaller pore sizes ($\leq$700 Å), which are optimal for limited oligonucleotide lengths [86]. However, excellent syntheses of RNA on CPG on scales of $\sim$100 μmol have been described [9].

2. Polystyrene-Based Supports

Polystyrene-based resin developed by Merrifield and colleagues [87] for peptide and protein chemistry has long constituted a state-of-the-art solid support [88]. Application to the oligonucleotide area has been more challenging
because the classical low cross-linking of the polystyrene moieties by 1; or
2% divinylbenzene during the copolymerization allowed swelling characteristics that were suboptimal for the organic solvents used in oligonucleotide
chemistry. Aminomethyl polystyrene supports were used to derivatize succinyl nucleoside [89] and to produce oligonucleotides of various lengths
[90,91]. However, the isolated yield of 82% for an octamer [92] or 43.5%
for an 18-mer DNA oligonucleotide [93] showed the limitations of these
types of low cross-linked copolymers of polystyrene.

   To circumvent some of the drawbacks of 1% divinylbenzene polystyrene [84], optimized polystyrene solid supports based on a high degree
(50%) of cross-linking with divinylbenzene were developed to confer nonswelling and rigidity characteristics to the polystyrene beads [84]. Such polystyrene resins have been shown to perform optimally for the synthesis of
RNA on small scales [7,8] using nucleoside loadings of ~ 30 μmol/g.

   Grafts of 1% divinylbenzene cross-linked polystyrene with polyethylene glycol ($n = 70$) have also been shown to confer good solvatation in
acetonitrile because of the polyethyleneoxy linker [86]. This support, known
as Tentagel, swells well in polar or apolar nonprotic solvents such as dichloromethane or acetonitrile [86] and has been shown to be efficient in the
synthesis of large amounts (up to 200 μmol) of RNA [94]. It is of interest
to note that retention of water by hydrogen bonding interaction with the
polyethyleneglycol component of Tentagel does not appear to constitute a
problem, whereas McCollum and Andrus [84] described the activated phosphoramidite quenching effect of water retained on CPG.

3. Miscellaneous Supports

A variety of other resins have been investigated with the goal of perfecting
solid-supported oligonucleotide synthesis. A polyacrylamide support was
used for the production of hexadecanucleotides [95] and kieselguhr–poly-
dimethylacrylamine was used in the synthesis of a 27-mer oligonucleotide
[96]. Synthesis of a 101-mer DNA was reported using a methacrylate-based
support copolymerized with vinyl alcohol [97]. A new type of support makes
use of sintered amino derivatized polyethylene membranes (Porex X-4920)
coupled to polymer colloid composed of styrene, divinylbenzene. Crude purity in the 80% range was observed on a 15-mer DNA [98].

   Mixed supports made of a polytetrafluoroethylene (PTFE) core and a
graft coating of polystyrene have been shown to compare well with com-

mercially available CPG or polystyrene supports [99]. An interesting feature of this PTFE-PS mixed support is that the final nucleoside loading can be easily controlled by varying the degree of polystyrene grafting, allowing loading up to 160 μmol/g with 12–17% polystyrene. Averaged stepwise yields in excess of 99% were obtained for oligonucleotide lengths ranging from 34- to 146-mers using long-chain alkylamine PTFE-PS loaded with ~80 μmol/g nucleoside [100].

Finally, oligonucleotide synthesis of a soluble polymeric support that can be crystallized from the reaction mixture for all the necessary washing steps has been reported [101,102]. This support, based on polyethylene glycol 5000, was loaded at high levels of 3'-O-succinyl nucleosides (~160 μmol/g) but allowed only a modest ASWY of 95–97% of octamers. A newer generation of soluble solid supports, based on polymeric assembling of N-acryloylmorpholine, was described as an alternative to the PEG 5000 matrix [103].

B. Spacer–Linker

The spacer between the polymeric matrix and the leading conveniently functionalized 3'-nucleoside of the oligonucleotide to be synthesized has been extensively studied since the inception of solid phase–supported oligonucleotide synthesis. Most of the work described in the literature focused primarily on optimizing the spacer length and chemical structure on silica-based solid supports. Early spacer structures, used in conjunction with silica gel solid supports, were relatively simple. Treatment of silica gel with (3-aminopropyl)triethoxysilane yields aminopropyl-silica, which is further reacted with succinic anhydride to afford **33** (Fig. 6). Support **33** is subsequently loaded with the 3'-hydroxyl of the 3'-leading nucleoside through classical N,N-dicyclohexylcarbodiimide activation [77]. A long-chain alkyl amine (LCAA)–CPG solid support, **34**, was successfully used by Adams [104] and Gough [82] and their coworkers in conjunction with phosphoramidite and phosphotriester chemistries, respectively. The LCAA–CPG solid support is still the most widely used commercially available silica-based support. Indeed, the structural features of the LCAA spacer concur with the length and chemical structure requirements that emerged from the thorough investigations conducted independently by Katzhendler [79,80], van Aerschot [78], and Arnold [105] and their colleagues.

Katzhendler et al. [79,80] reported the effect of various spacers on the synthesis of oligonucleotides. Most notably, long aliphatic polyureid spacers and polyethylene glycol spacers, **35**, gave ASWY in excess of 99.5%. It was inferred not only that the spacer length is important for successful oligonucleotide synthesis but also that a high number of carbonyl groups in the

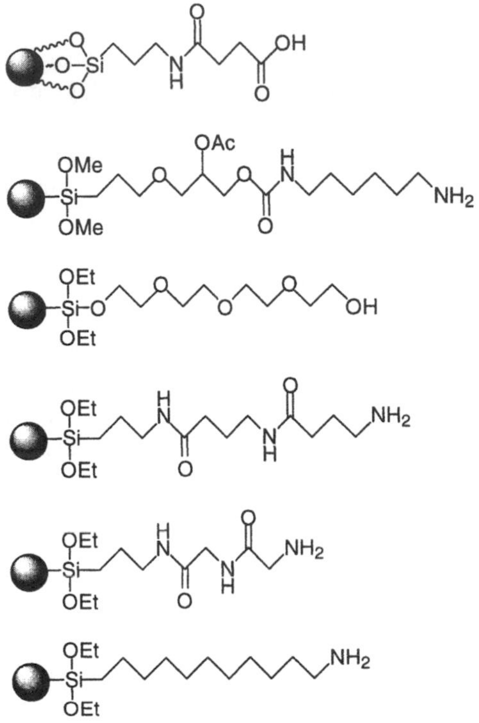

Aminopropylsuccinyl silica, **33**

LCAA-CPG, **34**

TEG-CPG, **35**

bis-(γ-aminobutyroyl) Fractosyl, **36**

bis-glycine Fractosyl, **37**

aminoundecyl Fractosyl, **38**

Figure 6 Linker structures used in solid supports for DNA–RNA synthesis.

spacer structure may promote hydrogen bondings. This phenomenon could lead to folded conformations of the spacer structure that could decrease the overall yield. Conversely, the tetraethyleneglycol spacer, **35**, despite a shorter length, gave high yields in oligonucleotide synthesis, presumably because of the gauche–gauche effects of the oxygen atoms of the glycol moiety that maintain the conformation of the spacer in a fully extended form.

Similar observations were made by van Aerschot et al. [78] using bis-(γ-aminobutyroyl, **36**, or bis-glycine, **37**, units on an aminopropylated Fractosyl support or aminoundecyl derivatized Fractosyl, **38**. The bis-(γ-aminobutyroyl)- aminoundecyl-linked supports were not as effective as the bis-glycine-linked support, presumably because of the formation of folded structures due to intramolecular bonds or hydrophobic interactions, respectively. Interestingly, Arnold et al. [105] described the synthesis of DNA in good yields of 30-mers using a solid support composed of a long aliphatic spacer made by successive addition of succinic anhydride and 1,10-decanediol on aminopropylated CPG. These results imply that the hydrophobic interaction [78] responsible for suboptimal folding of aliphatic linkers proposed by van Aerschot et al. may not be so prevalent.

C.   Chemistry of Attachment of Leader 3′-Nucleoside

The chemistry of attachment of the leader 3′-nucleoside to an amino-linked support is of importance in oligonucleotide chemistry because it allows efficient release of the oligonucleotide from the support when the synthesis is complete. This section will focus on linker structures that generate unmodified 3′-hydroxyl–containing oligonucleotides as opposed to the wealth of "specialty linkers" that have been developed to introduce a variety of functional groups (phosphates, thiols, etc.) at the 3′-terminus of the oligonucleotide.

The succinyl linkage, **39** (Fig. 7), that was introduced in the early 1980s [76,77,95] remains the most widely used linkage to date. Its facile introduction on the 3′-hydroxyl of the leader nucleoside [106] and the easy coupling of the resulting 3′-succinyl derivative to an amino-linked solid support [83,107,108] are favorable features that justified the widespread use of this linker. In addition, this linker provides a convenient deprotection protocol involving aqueous ammonia or a wide range of alkylamines.

A urethane linker, **40**, was introduced by Sproat and Brown [109] as an alternative to the succinyl linkage and is prepared by reacting LCAA-CPG with tolylene-2,6-diisocyanate. A 19-mer oligonucleotide constructed on this urethane-linked CPG support was successfully deprotected by extended aqueous ammonia treatment at 56°C for 48 h.

In order to deprotect base-sensitive oligonucleotide derivatives efficiently, an oxalyl linkage, **41**, was developed to allow expedited cleavage of newly synthesized oligonucleotides under mild conditions [110].

Oxalyl-CPG is obtained by reacting oxalyl chloride and 1,2,4-triazole with a 3′-hydroxyl nucleoside and LCAA-CPG. Cleavage of the oxalyl linker can be effected by a variety of mild reagents including triethylamine, propylamine–DCM, 5% ammonium hydroxide in methanol, or 0.5 M cesium fluoride in methanol for 15 min. Alternatively, 0.5 M DBU in either pyridine, dichloromethane, or dioxane was shown to effect full cleavage of the oxalyl linkage in less than 30 min at room temperature [111].

Silyl-based linkers were investigated to effect a nonbasic fluoride-mediated cleavage of the oligonucleotide from the solid support. The siloxyl, **42** [112,113], or disiloxyl, **43** [114], backbones were grafted on CPG or polystyrene supports. Release of the oligonucleotide was achieved by use of 0.5 M tetrabutylammonium fluoride in THF or DMF. Such conditions render the use of this linker structure less compatible with conventional 2′-*O*-TBDMS RNA chemistry.

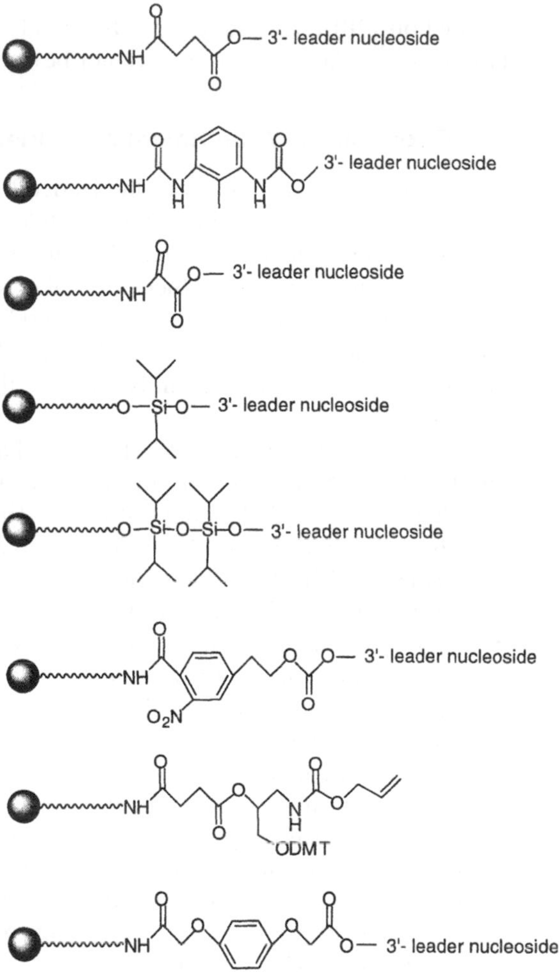

Figure 7 3'-Leader nucleoside chemistry of attachment onto amino-linked solid support.

Strongly basic nonnucleophilic conditions such as 0.5 M DBU in dioxane or pyridine were used by Eritja et al. [115] to effect the β-elimination cleavage of a 4-ethyl-3-nitrobenzoyl carbonate linker, **44**, allowing complete deprotection of oligonucleotides synthesized with p-nitrophenylethyl–based protecting groups under totally ammonia-free conditions [116].

An elegant universal linker was designed by Lyttle et al. [117]. A 1-O-(4,4'-dimethoxytrityl)-2-O-succinoyl-3N-allyloxycarbo propane linker is immobilized on amino-propyl CPG, **45**. After palladium-mediated deprotection of the 3-amino group, a mild triethylammonium treatment catalyzes the nucleophilic attack of the amine functionality on the adjacent 3′-phophotriester followed by elimination of the oligonucleotide.

Photolabile CPG supports, based on the *ortho*-nitrobenzyl moiety, have been designed by Greenberg et al. [118,119] to provide orthogonal oligonucleotide cleavage conditions suitable for alkali-sensitive oligonucleotides.

A "Q- or HDQA-linker," **46**, based on the hydroquinone-*O,O*′-diacetic acid linker was developed by Pon and Yu [120]. This HDQA linker is more labile than the succinyl linker but more stable than the oxalyl linker as evidenced by the mere 5-min deprotection time required to effect close to quantitative cleavage of the leader nucleoside from either Q-linked CPG or polystyrene-based supports under concentrated ammonium hydroxide conditions. Milder deprotection reagents, such as 5% ammonium hydroxide, potassium carbonate, or *t*-butylamine, were also shown to be very effective in cleaving this linker [121]. This HDQA linker in conjunction with a modified capping procedure allowed CPG regeneration after oligonucleotide synthesis, hence significantly reducing solid support–associated cost in large-scale oligonucleotide manufacturing [122].

IV. ACTIVATION CHEMISTRY

The acidic activation step that allows rapid conversion of the stable N,N-diisopropylamino phophoramidite to an azolide intermediate is critical to commercial phosphoramidite chemistry. This highly reactive transient species is then substituted by the free 5′-hydroxyl of the nucleotide bound to the support to give rise to the phosphite triester internucleotidic bond, which is further oxidized to the acid-stable pentavalent phosphotriester linkage. 1-H-Tetrazole, **47** (Fig. 8), constitutes the activator of choice DNA synthesis, allowing extremely short coupling times, typically on the order of 30 s. With this activator, average stepwise yields (ASYs) of 99% are regularly achieved with DNA. In contrast, reaction times as long as 1 h have been reported in the RNA series, although contact times in the range of 10–30 min have usually been achieved with 1-H-tetrazole as the activator [11,12]. The typically more sluggish reaction rate obtained in the RNA series is the result of the sterically hindered 2′-protecting group. Changes in 1-H-tetrazole concentration and/or coupling time result in additional side products and not increased coupling yields.

The use of substituted tetrazoles [123] as activators for DNA and RNA synthesis has been reported [124,125]. Dramatic improvements in oligoribonucleotide synthesis are achieved when more acidic activators are used in

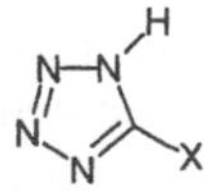

X = H: 1-H-tetrazole, **47**
X = S-Et: 5-ethylthio-tetrazole, **49**

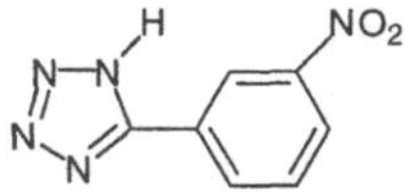

5-(3-nitrophenyl)-tetrazole, **48**

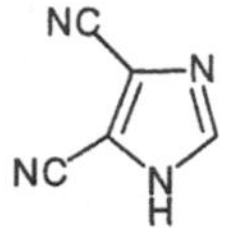

4,5-dicyanoimidazole, **50** 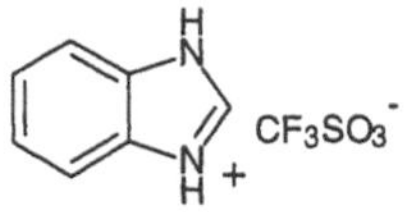   benzyimidazolium triflate, **51** 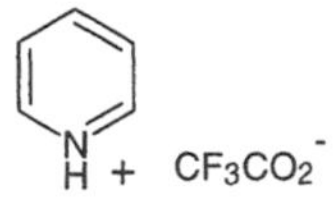   pyridinium trifluoroacetate, **52**

Figure 8 Activators commonly used for the synthesis of oligoribonucleotides according to phosphoramidite chemistry.

place of the standard 1-H tetrazole (pK_a = 4.8). Such activators include 5-(p-nitrophenyl)-1H-tetrazole, **48** (pK_a = 3.7) [126] which allows coupling times as short as 6 min, and 5-ethylthio-1H-tetrazole, **49** (pK_a = 4.28) [7,127], with reported coupling times of 5–8 min, using less equivalents of activator than tetrazole. These tetrazole derivatives are supposedly more efficient at catalyzing the protonation step of the trivalent phosphorus. Once protonated, this electrophilic phosphorus center reacts with a tetrazole molecule that displaces the N,N-disoproylamine moiety. Consequently, 5-ethyl-thio-1H-tetrazole (SET) has become the preferred activator for oligoribonucleotides, used in a number of reports describing successful RNA synthesis. However, the acidic advantage of the ethylthio-tetrazole is counterbalanced on larger oligonucleotide synthesis scales because of concomitant acidic activator–mediated detritylation of the nucleoside phosphoramidite during the extended coupling time required to obtain satisfactory coupling efficiency [128]. This problem initiated the development of less acidic but more nucleophilic activators as exemplified by 4,5-dicyanoimidazole, **50** (pK_a = 5.2) [9], or benzimidazolium triflate, **51** (pK_a = 4.5) [129]. Recently, the use of pyridinium trifluoroacetate, **52** (pK_a = 5.2), was reported as an inexpensive and efficient alternative to regular 1H-tetrazole for DNA synthesis [130]. Although no mechanistic studies have been presented, these activators presumably accelerate the coupling reaction because of the increased nucleophilicity of their conjugated base while maintaining sufficient acidity to protonate the phosphorus center.

Another approach to optimizing the coupling step in oligonucleotide synthesis has been to modify the dialkylamine or transient protecting group component of the phosphoramidite group. In most cases the N,N-diisopropylamino group [131] is utilized for the synthesis of oligonucleotides. This group constitutes a good compromise between the reactivity of the corresponding phosphoramidite, under tetrazole conditions, and its stability in acetonitrile because of the increased steric hindrance conferred by the diisopropylamine group at the phosphorus center [104]. A number of different protecting groups for the P–N linkage present in the phosphoramidite moiety have been investigated, including N,N-dimethylamino [30], N-morpholino [31,132], and N,N-diethylamino [104] groups.

The use of lower alkyl substituted phosphoramidites, such as diethylamino instead of diisopropylamino, has been reported to improve coupling yields in RNA synthesis when measured by dimethoxytrityl cation quantitation [133]. Despite the good coupling conditions obtained, these compounds have not been used extensively because of their instability, and the O-2-cyanoethyl, N,N-diisopropyl combination is most frequently used for DNA and RNA synthesis.

V. ROUTINE DNA SYNTHESIS

A. Synthesis of Oligodeoxynucleotides

The elongation cycle for automated oligonucleotide synthesis is composed of the classical detritylation, coupling, capping, and oxidation steps as delineated in Fig. 1. Clearly, each of these steps is critical for successful DNA synthesis. Detritylation, which allows cleavage of the 5'-hydroxyl protecting group, is typically performed by a 3% formulation of trichloroacetic acid ($pK_a = 0.7$) in dichloromethane. Worth noting is the sensitivity of N_6-benzoyl-2'-deoxyadenosine to string acidic conditions leading to depurination. This may be particularly true during large-scale synthesis (>200 μmol) that require extended detritylation treatment. Hence, the preferred detritylation solution at these scales is classically composed of 2% dichloroacetic acid ($pK_a = 1.5$) in dichloromethane to limit undesired depurination reactions. Alternatively, the use of *tert*-butylphenoxyacetyl as the protecting group for the exocyclic amino functionality of 2'-deoxyadenosine greatly enhanced the depurination half-life of 2'-deoxyadenosine under acidic conditions [15].

The coupling reaction involves the activation of the N,N-diisopropyl phosphoramidite by an acidic activator. In the DNA series, the most commonly used activator is 1-H-tetrazole ($pK_a = 4.8$). The coupling reaction is typically performed in a dissociating polar nonprotic solvent such as acetonitrile to facilitate the nucleophilic displacement of the tetrazolide moiety. Following coupling, oxidation and capping steps proceed to oxidize the resulting phosphite triester linkage and cap the unreacted 5′-hydroxyls of the abortive oligonucleotides, respectively.

The order in which these steps are conducted has been studied. Oxidation–capping, capping–oxidation, and capping–oxidation–capping cycles have been investigated in conjunction with a possible branching reaction occurring during oligonucleotide synthesis at the unprotected O-6 position of guanidylate residues [134,135]. Capping followed by oxidation appears to provide a synthesis advantage when guanosine-rich oligonucleotides are produced [11]. However, it is still ambiguous whether this type of cycle provides a clear advantage, as exemplified by the report that a cap–ox–cap cycle was shown to give slightly better yields in straight DNA synthesis [15].

B. Deprotection of Oligodeoxynucleotides

The deprotection of crude oligodeoxynucleotides traditionally requires a basic treatment that allows the concomitant removal of the exocyclic acyl protecting groups, β-elimination of the 2-cyanoethyl phosphate protecting group, and cleavage of the succinic ester bond that links the oligonucleotide to the solid support (Fig. 9).

1. Nucleobase Deprotection

In the early days of oligonucleotide chemistry, the heterocyclic amino functions of the nucleobases were almost exclusively protected via the classical

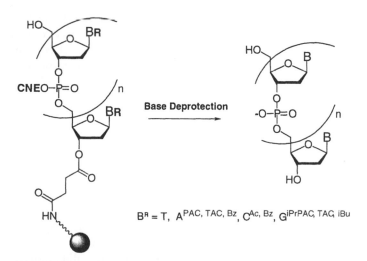

Figure 9 Base deprotection of oligodeoxynucleotide.

benzoyl (dA and dC) and isobutyryl (dG) groups. These protecting groups are efficiently removed by using concentrated ammonium hydroxide in the DNA series. However, deprotection of more complexed oligonucleotides, including base-sensitive phosphoramidites such as 5-bromo-2′-deoxyuridine (BrdU), necessitates the use of a 3:1 mixture of ammonium hydroxide and ethanol for 12 to 16 h at room temperature to minimize the undesired nucleophilic displacement of the halogen group. To limit the exposure of specialty oligodeoxynucleotides to extended basic treatment, more base-labile amino protecting groups were developed. Amidine chemistries, described in Section II.A.2, and phenoxyacetyl (PAC), described in Section II.A.1, constitute the main alternatives to standard benzoyl and isobutyryl groups.

C. Oligodeoxynucleotide Synthesis Protocol

This protocol describes automated chemical synthesis of oligodeoxynucleotide by means of the phosphoramidite method. The following procedure was developed for the ABI 394 DNA–RNA synthesizer at 0.2 μmol.

Materials list

Empty synthesis columns for 0.2-μmol-scale syntheses (Applied Biosystems Inc., ABI, Foster City, CA).
Controlled pore glass derivatized with 5′-O-DMT-3′-O-succinyl deoxyribonucleoside (Glen Research, Sterling, VA).
DNA phosphoramidites:
5′-O-DMT-N^6-(phenoxyacetyl)-adenosine-3′-O-(β-cyanoethyl-N,N-diisopropylamino) phosphoramidite (Amersham-Pharmacia)
5′-O-DMT-N^2-(isopropylphenoxyacetyl)-guanosine-3′-O-(β-cyanoethyl-N,N-diisopropylamino) phosphoramidite (Amersham-Pharmacia)
5′-O-DMT-N^4(acetyl)-cytidine-3′-O-(β-cyanoethyl-N,N-diisopropylamino) phosphoramidite (Glen Research)
5′-O-DMT-thymidine-3′-O-(β-cyanoethyl-N,N-diisopropylamino) phosphoramidite (Amersham-Pharmacia)
All phosphoramidites are diluted on the synthesizer to 0.1 M in acetonitrile, using automated protocols.
Ancillary reagents: 3% TCA in methylene chloride (ABI); cap B: 16% N-methyl imidazole in THF (ABI); cap A: 10% acetic anhydride/ 10% 2,6-lutidine in THF (ABI); 16.9 mM I_2, 49 mM pyridine, 9% water in THF (Perspective); Burdick & Jackson (Muskegon, MI) synthesis grade acetonitrile. A choice of 5-ethylthio-1-H-tetrazole solution (0.25 M in acetonitrile), prepared from the solid obtained from American International Chemical (Matick, MA), or 1-H-tetrazole

solution (0.45 M in acetonitrile) obtained from Glen Research (Sterling, VA), may be used for DNA synthesis.

Steps

1. Load an empty synthesis column with ~8 mg of the DNA CPG solid support (~25 μmol/g) corresponding to the first nucleotide at the 3'-end of the oligonucleotide.
2. Perform synthesis on ABI 394 synthesizer according to the cycle outlined in Table 1.
3. At the end of the synthesis, perform a manual detritylation cycle, if desired, to remove the trityl group at the 5'-end of the oligonucleotide.
4. Remove synthesis column from synthesizer and dry under a stream of argon or in a vacuum desiccator for 10–15 min.

Table 1 Summary of a 0.2-μmol-Scale Cycle for Oligodeoxynucleotide Synthesis on ABI 394 DNA–RNA Synthesizer[a]

Function	Time (s)	Function	Time (s)
1. ACN to waste	3	18. Iodine to colume	4
2. ACN to column	10	19. Repeat steps 13 and 4	
3. Argon reverse flush	8	20. Wait	15
4. Argon block flush	4	21. ACN to column	10
5. Activator to waste	1.7	22. Argon flush to waste	4
6. Amidite+Act to column	1.2	23. ACN to column	10
7. Push to column	NA	24. Repeat steps 3 and 4	
8. Wait	10	25. Repeat steps 21, 23, and 24	
9. Argon flush to waste	0.1	26. TCA/DCM to column	6
10. Wait	5	27. Wait	5
11. Argon flush to waste	0.1	28. Argon trityl flush	5
12. Repeat steps 10 and 11 six times		29. Repeat steps 26–28	
13. ACN to waste	4	30. ACN to column	10
14. Repeat steps 3 and 4		31. Argon trityl flush	5
15. Cap A&B to column	4	32. Repeat steps 2, 3, and 4	
16. Wait	5	33. End	
17. Repeat steps 13 and 14			

[a]Delivery flow rates are ~3.1 mL/min for phosphoramidites and activators and ~3.6 mL/min for all other reagents.

D. Oligodeoxynucleotide Deprotection Protocol

1. Oligodeoxynucleotide Deprotection with Concentrated NH₄OH

This protocol describes a deprotection scheme using 30% aqueous ammonium hydroxide to cleave and deprotect the oligonucleotide from the solid support.

Materials list

29% ammonium hydroxide (J.T. Baker, Phillipsburg, NJ)

Steps

1. Transfer the dried solid support from the synthesis column to a 4-mL glass-screw-top vial with Teflon-lined lid (Wheaton).
2. Add 1 mL of ammonium hydroxide to the vial, screw cap on tightly, and place in a heat block at 65°C for 4 h.
3. Remove vial from heat block, place in a room temperature block, and place in −20°C freezer until cooled (~30 min).
4. Decant the solution into a 14-mL Falcon tube. Dry aqueous ammonium hydroxide solution in a Speed-Vac (~2.5 h on medium heat).
5. The deprotected oligonucleotide can then be precipitated by re suspending the pellet in 0.5 mL of MilliQ water and adding 25 μL of 3 M NaOAc, followed by 1 mL of *n*-BuOH.
6. Cool mixture to −20°C for 2 h to overnight and centrifuge at 4°C, 10,000 rpm for 30 min.
7. Decant solution and wash the pellet with 70% EtOH. Centrifuge at 4°C, 10,000 rpm for 10 min and decant. Dry oligonucleotide pellet.

2. Oligodeoxynucleotide Deprotection with Aqueous Methylamine

This protocol describes a deprotection scheme using aqueous methylamine as a substitute for ammonium hydroxide.

Additional materials list

40% aqueous methylamine (Aldrich)

Steps

1. Transfer the oligoribonucleotide attached to the solid support from the synthesis column to a 4-mL glass screw-top vial (Wheaton).
2. Add 1 mL of 40% aqueous methylamine to the vial, screw cap on tightly, and place in a heat block at 65°C for 10 min.

3. Remove vial from heat block, place in a room temperature block, and place in −20°C freezer until cooled (~20 min).
4. Decant the solution into a 14-mL Falcon tube. Dry aqueous ammonium hydroxide solution in a Speed-Vac (~2.5 h on medium heat).
5. Return to step 5 of Section IV.D.1.

E. Analysis

Anion-exchange high-performance liquid chromatography (HPLC) and capillary gel electrophoresis (CGE) are the best methods for accurately assessing the percentage of full-length oligomer present in a fully deprotected crude sample.

For HPLC analyses, we recommend the use of a Dionex NucleoPac PA-100 column, 4 × 250 mm, at 50°C, with $NaClO_4$ buffers. This system gives optimal resolution of full-length product material versus abortive truncations.

Method

Analytical anion-exchange HPLC analysis: Hewlett Packard 1090 HPLC.
Dionex NucleoPac PA-100 column, 4 × 250 mm, at 50°C.
Flow rate 1.5 mL/min
Buffer A, 20 mM $NaClO_4$; buffer B, 300 mM $NaClO_4$ (Fluka).
Universal gradient as follows:

Time (min)	% B
0.00	0.0
40.00	100.0
42.00	100.0
46.00	0.0

For CGE analyses, good resolution is obtained using a coated capillary and refillable gel eCAP single-stranded DNA (ssDNA) 100-R kit commercially available from Beckman Instruments (Fullerton, CA) with Tris-borate and 7 M urea as the running buffer and an effective capillary length of 208 mm. The crude or purified oligonucleotide is typically desalted by membrane dialysis against water for 30 min. A 0.1 OD_{260}/mL solution is electrokinetically injected for 10 s at −10 kV. General running conditions make use of a voltage of −9.3 kV and a run time of 30 min.

1. Purification

Anion-exchange chromatography is an efficient purification tool for a variety of oligonucleotide lengths. The following example applies to the purification of a 30- to 35-mer oligonucleotide synthesized at a scale of 2.5 μmol.

Method

Milli Q water.

Dionex NucleoPac PA-100 22 × 250 mm column.

Buffer A, 10 mM $NaClO_4$; buffer B, 300 mM $NaClO_4$.

SepPak cartridge (C_{18}).

CH_3CN, CH_3CN–MeOH–H_2O (1:1:1), and Milli Q water.

The crude material was diluted to 5 mL with ribonuclease (RNase)-free water and injected onto a Dionex NucleoPac PA-100 22 × 250 mm column with 100% buffer A (10 mM $NaClO_4$).

To elute the oligonucleotide, a gradient from 100 to 150 mM $NaClO_4$ at a rate of 15 mL/min for a Dionex NucleoPac anion-exchange column was used.

Fractions were analyzed by HPLC and those containing full-length product ~80% by peak area were pooled for desalting.

The pooled fractions were applied to a C_{18} SepPak cartridge that was prewashed successively with CH_3CN (10 mL), CH_3CN–MeOH–H_2O (1:1:1) (10 mL), and MilliQ H_2O (20 mL).

Following sample application, the cartridge was washed with MilliQ H_2O (10 mL) to remove the salt.

Product was then eluted from the column with CH_3CN–MeOH–H_2O (1:1:1) (10 mL) and dried.

VI. ROUTINE RNA SYNTHESIS

A. Synthesis of Oligoribonucleotides

As with oligodeoxyribonucleotides, the chemical synthesis of oligoribonucleotides on a solid support is routinely performed via the phosphoramidite method. However the additional 2'-hydroxyl moiety, present on the ribofuranosyl sugar, requires suitable protection during oligoribonucleotide synthesis. Among the various protecting groups available to the chemist, trialkylsilyl ethers [136–138] and particularly the *tert*-butyldimethylsilyl ether [139] (TBDMS) have been the most extensively studied because the fluoride-mediated silyl ether deprotection is quite orthogonal to the other acid- and base-labile protecting groups commonly used in oligonucleotide chemistry. Because of the steric bulk induced by the TBDMS group, coupling of

2'-*O*-TBDMS–protected ribophosphoramidites to a growing oligonucleotide chain is notoriously more sluggish than that of their 2'-deoxy analogues.

The elongation cycle for automated oligoribonucleotide synthesis is also composed of the classical detritylation, coupling, capping, and oxidation steps. One important difference regarding the detritylation step is the possibility of using higher concentration of haloacids because of the reduced sensitivity of the ribofuranosylnucleotide to acid-mediated depurination. The oxidation and capping steps do not differ significantly from those of classical oligodeoxyribonucleotide synthesis. Conversely, the coupling step has been the focus of much attention, and a number of phosphate protecting groups [11,38,39,140–142] and different dialkylamines [30,31,133] have been developed with the aim of faster coupling. Although some of these modifications have been commercialized, the 2-cyanoethyl and *N,N*-diisopropylamino combination [30,31,132] is used predominantly in the field. Using this chemistry, the 2'-*O*-TBDMS ribophosphoramidite coupling time is typically much longer than that of the 2'-deoxy series for synthesis scales ranging from 0.2 to 2.5 μmol.

A number of activators have been investigated to increase the quality of RNA synthesis (see Section V). Figure 10 shows anion-exchange chromatograms of a 36-mer oligoribonucleotide synthesized via the 2'-*O*-TBDMS–protected ribophosphoramidites according to the synthesis protocol delineated in Section V.C. using either 5-ethylthio-tetrazole, 1-*H*-tetrazole or 4,5-dicyanoimidazole as the activator. From this comparative analysis, 5-ethylthio-tetrazole still appears to provide the most full-length product.

Currently, 5'-*O*-DMT/2'-*O*-TBDMS [7,8,11,12] is the benchmark for the synthesis of oligoribonucleotides. The suitably protected amidites are commercially available and quality products can be produced on a reasonable scale. However, RNA synthesis chemistry using the 2'-TBDMS group has not yet reached the level achieved by DNA synthesis despite advances in the use of this silyl chemistry in both the synthesis [7–9] and deprotection [7,8,61] steps. As a result, the search for improved protocols or new approaches continues.

Among the developments in 2'-hydroxyl specific protecting groups, the 2'-*O*-triisopropylsilyloxymethyl (TOM) group, developed by Weiss and Pitsch [73,74], constitutes in our opinion the most suitable protecting group for efficient RNA synthesis. These 2'-*O*-TOM–protected phosphoramidites are commercially available from Glen Research (Sterling, VA) in the United States and from Xeragon A.G. in Europe. Weiss and Pitsch [73,74] recommend the use of 5-benzylthio-tetrazole for the activation step. However, we have found that high-quality material can be obtained with 5-ethylthio-tetrazole as the activator at a fraction of the coupling time required for conventional 2'-*O*-TBDMS–protected RNA synthesis (120 s vs. 450 s). Figure

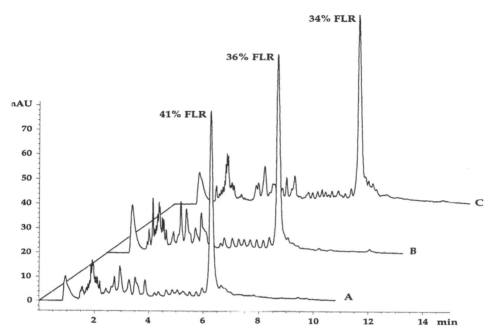

Figure 10   HPLC chromatograms of a 36-mer oligoribonucleotide (5′-GUU UUC CCU GAU GAG GCC GAA AGG CCG AAA UUC UCC-3′) synthesized at the 0.2-μmol scale according to the basic synthesis protocol and deprotected using the second alternative "one-pot" deprotection protocol. A, 0.25 M SET; B, 0.5 M DCI; C, 0.45 M TET. Dionex NucleoPac PA-100 22 × 250 mm column at 50°C. Buffer A (1 mM Tris, 20 mM NaClO₄), buffer B (1 mM Tris, 300 mM NaClO₄). Gradient 40% B to 70% B in 12 min, flow rate, 1.5 mL/min.

11 shows anion-exchange chromatograms of a 33-mer oligoribonucleotide synthesized via either 2′-O-TBDMS–protected ribophosphoramidites or 2′-O-TOM–protected ribophosphoramidites according to the synthesis protocol delineated in Section V.C. using 5-ethylthio-tetrazole as the activator. From this comparative analysis, the triisopropylsilyloxymethyl protecting group allows better RNA synthesis yields while lowering the coupling time by more than 70%.

Currently, none of the synthetic methods that have been designed specifically for RNA synthesis are commercially available. The SIL–ACE protocol [54], however, appears very attractive. The quality of the product is excellent and oligomers of up to 36 residues have been synthesized. Currently, none of these phosphoramidites are readily available, although efforts are under way to commercialize the SIL–ACE process.

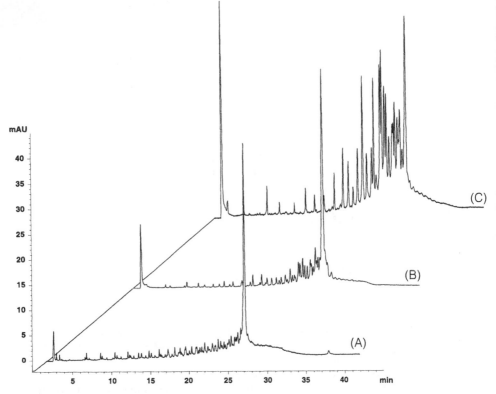

Figure 11 HPLC chromatograms of a 33-mer oligoribonucleotide (5′- AGA UUA UUG AUA CGC UGA AUU AGG ACA UAG AUC-3′) synthesized at the 0.2-μmol scale according to the basic synthesis protocol and deprotected using the one-pot deprotection protocol (see Section V.D.3). A, 120 s 2′-O-TOM; B, 450 s 2′-O-TBDMS; C, 120 s 2′-O-TBDMS. Dionex NucleoPac PA-100 22 × 250 mm column at 50°C. Buffer A (1 mM Tris, 20 mM NaClO$_4$), buffer B (1 mM Tris, 300 mM NaClO$_4$). Gradient 0% B to 100% B in 40 min, flow rate 1.5 mL/min.

B. Deprotection of 2′-O-Silyl–Protected Oligoribonucleotides

The deprotection of crude oligoribonucleotides traditionally requires a basic treatment that concomitantly allows transamination of the exocyclic acyl protecting groups, β-elimination of the 2-cyanoethyl phosphate protecting group, and cleavage of the succinic ester bond that links the oligoribonucleotide to the solid support. The base deprotection is by far the most rate-limiting step for these three reactions. Once these reactions are accomplished, a fluoride treatment removes the substituted silyl ether protecting group at the 2'-hydroxyl of the ribofuranose ring (Fig. 12).

1. Nucleobase Deprotection

Base deprotection of oligoribonucleotides typically requires milder conditions than those used in the DNA series. The main reason for this is the possible premature deprotection of the 2'-hydroxyl during the basic treatment, resulting in extensive oligoribonucleotide chain cleavage through the 2'-alcoholate attack of the phophodiester bond. The presence of the 2'-O-TBDMS group in the RNA series necessitates the use of ethanolic ammonia [143] or a 3:1 mixture of ammonium hydroxide and ethanol for 12 to 16 h at 55°C to minimize the undesired concomitant cleavage of the 2'-O-TBDMS group [144,145]. To circumvent this unwanted side reaction and shorten the basic deprotection time, two main strategies were investigated over the past decade.

On the one hand, alternative basic deprotection cocktails were developed using various combinations of hydrazine, ethanolamine, and alcohol [146] or incorporating the use of more nucleophilic alkylamines [7,16]. The latter method favors 40% aqueous methylamine in place of or in addition to 30% ammonium hydroxide, which cleaves the acyl protecting groups present on the nucleobases in a few minutes at 65°C or in over 1 h at room temperature.

On the other hand, more base-labile amino protecting groups were developed, thereby limiting the exposure of the 2'-O-TBDMS group to extended basic treatment. Aside from the amidine chemicals described in Section II.A.2, the main chemical used in RNA synthesis is phenoxyacetyl (PAC) [13,14,147], mainly used for the ribopurine phosphoramidites because the N_4-acetyl protection of cytidine is quite optimal. Phenoxyacetyl protecting groups and *tert*-butyl or isopropyl-phenoxyacetyl derivatives [15] are considerably more labile under basic conditions. Typically, they can be quantitatively cleaved from the exocyclic amino functions after 15 min to 1 h incubation at 65°C using ammonium hydroxide–ethanol (3:1). Alternately, 2 to 4 h at room temperature also allows complete deprotection.

Clearly, the use of the "fast deprotecting groups" such as the PAC family does not preclude the concomitant use of more nucleophilic alkylamines. Indeed, combining these two methods allows expedited base-labile protecting group removal as exemplified in the first alternative deprotection protocol, where 10 min incubation at 65°C has been shown to be sufficient without premature deprotection of the 2'-O-TBDMS or base modification.

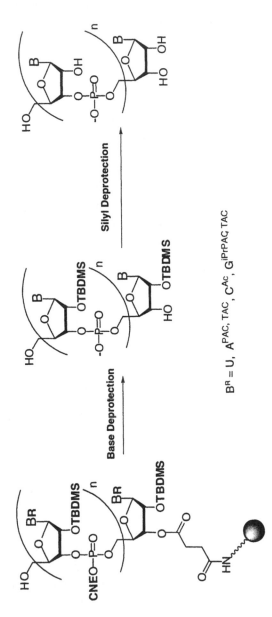

Figure 12 Two-step deprotection of oligoribonucleotide.

2. 2'-O-Silyl Deprotection

The fluoride-sensitive *tert*-butyldimethylsilyl group allows efficient orthogonal deprotection of the 2'-hydroxyl functionalities present in RNA. By a large extent, the use of fluoride-based reagents is the preferred methodology to remove the 2'-O-TBDMS group, although Sekine and coworkers [148] reported the application of an acid-catalyzed desilylation scheme to oligoribonucleotides. After completion of the base and phosphotriester deprotection followed by evaporation of the basic solution, addition of tetrabutylammonium fluoride (TBAF) in THF to the partially deprotected RNA cleaves the 2'-O-TBDMS group at room temperature within 24 h. However, solubility of the RNA in this apolar mixture hampers the efficient deprotection of longer oligoribonucleotides. To circumvent this problem, RNA solubilization in either dimethyl sulfoxide [149] or ethanol−water (1:1) [12] has been investigated. However, the notorious water sensitivity of the TBAF reagent [150] and the subsequent necessary desalting step after the quenching reaction have led to the development of other fluoride-based reagents.

Triethylamine trihydrofluoride (TEA 3HF) has been shown to be a superior reagent either neat [151,152] or in combination with polar nonprotic solvents such as dimethylformamide [153] or *N*-methylpyrrolidinone [7]. This reagent promotes 2'-hydroxyl deprotection within 30−90 min at 65°C or 4−8 h at room temperature. Using this desilylating reagent, the time-consuming desalting step may be replaced by a sodium acetate−1-butanol precipitation [7] procedure that is not compatible with the high organic content of the TBAF in THF. Further application of neat TEA 3HF in a one-pot deprotection procedure [61] using a mixture of anhydrous ethanolic methylamine as described in Section V.D.3 allows expedited RNA deprotection without detectable 2'-5'-phosphodiester migration [61]. This procedure requires a quenching step with ammonium bicarbonate, if one desires to retain the trityl group on the oligonucleotide and subsequent desalting−purification on a reverse-phase cartridge because the ethanol and dimethyl sulfoxide prevent RNA precipitation from butanol.

C. Oligoribonucleotide Synthesis Protocols

This protocol describes automated chemical synthesis of oligoribonucleotides by the phosphoramidite method. The following procedure was developed for the ABI 394 DNA−RNA synthesizer at the 0.2-μmol scale, although it can be modified to utilize any standard synthesizer.

Materials list

Empty synthesis columns for 0.2-μmol-scale syntheses (Applied Biosystems Inc., ABI, Foster City, CA)

Aminomethyl polystyrene (RNA Primers) derivatized with 5'-O-DMT-2'-O-TBDMS-3'-O-succinyl ribonucleoside (Pharmacia, Milwaukee, WI).

RNA phosphoramidites (Pharmacia):

5'-O-DMT-N^6-(phenoxyacetyl)-2'-O-TBDMS-adenosine-3'-O-(β-cyanoethyl-N,N-diisopropylamino) phosphoramidite

5'-O-DMT-N^2-(isopropylphenoxyacetyl)-2'-O-TBDMS-guanosine-3'-O-(β-cyanoethyl-N,N-diisopropylamino) phosphoramidite

5'-O-DMT-N^4-(acetyl)-2'-O-TBDMS-cytidine-3'-O-(β-cyanoethyl-N,N-diisopropylamino) phosphoramidite

5'-O-DMT-2'-O-TBDMS-uridine-3'-O-(β-cyanoethyl-N,N-diisopropylamino) phosphoramidite

All phosphoramidites are diluted on the synthesizer to 0.1 M in acetonitrile, using automated protocols.

Ancillary reagents: 3% TCA in methylene chloride (ABI); cap B: 16% N-methyl imidazole in THF (ABI); cap A: 10% acetic anhydride, 10% 2,6-lutidine in THF (ABI); 16.9 mM I_2, 49 mM pyridine, 9% water in THF (Perseptive); Burdick & Jackson (Muskegon, MI) synthesis grade acetonitrile. A choice of 5-ethylthio-1-H-tetrazole solution (0.25 M in acetonitrile), prepared from the solid obtained from American International Chemical (Natick, MA), or 4,5-dicyanoimidazole solution (0.5 M in acetonitrile), prepared from the solid obtained from Nexstar Technology Products (Boulder, CO) or 1-H-tetrazole solution (0.45 M in acetonitrile) obtained from Glen Research (Sterling, VA), may be used for RNA synthesis.

Steps

1. Load an empty synthesis column with ~8 mg of the RNA primer solid support (~25 μmol/g) corresponding to the first nucleotide at the 3'-end of the oligoribonucleotide.
2. Perform synthesis on ABI 394 synthesizer according to the cycle outlined in Tables 2 and 3.
3. At the end of the synthesis, perform a manual detritylation cycle, if desired, to remove the trityl group at the 5'-end of the oligonucleotide.
4. Remove synthesis column from synthesizer and dry under a stream of argon or in vacuum desiccator for 10–15 min.

D. Oligoribonucleotide Deprotection Protocols

Deprotection of oligoribonucleotides according to the three protocols presented in Sections V.D.1–3 is quite straightforward. However, the one-pot

Table 2 Equivalents to the Synthesis Scale, Volumetric Amounts, and Wait Time for the Main Components of the Synthesis Process

Reagents	Equivalents	Amounts (μL)	Wait time (s)
Phosphoramidites	15	31	465
Activator { SET	38.7	31	465
Activator { DCI	80	31	465
Activator { TET	70	31	465
Acetic anhydride	655	124	5
N-Methylimidazole	1245	124	5
TCA	700	732	10
Iodine	20.6	244	15

Table 3 Summary of a 0.2-μmol-Scale Cycle for Oligoribonucleotide Synthesis on ABI 394 DNA–RNA Synthesizer[a]

Function	Time (s)	Function	Time (s)
1. ACN to waste	3	18. Iodine to column	4
2. ACN to column	10	19. Repeat steps 13 and 4	
3. Argon reverse flush	8	20. Wait	15
4. Argon block flush	4	21. ACN to column	10
5. Activator to waste	1.7	22. Argon flush to waste	4
6. Amidite+Act to column	1.2	23. ACN to column	10
7. Push to column	NA	24. Repeat steps 3 and 4	
8. Wait	450	25. Repeat steps 21, 23, and 24	
9. Argon flush to waste	0.1	26. TCA/DCM to column	6
10. Wait	45	27. Wait	5
11. Argon flush to waste	0.1	28. Argon trityl flush	5
12. Repeat steps 10 and 11 six times		29. Repeat steps 26–28	
13. ACN to waste	4	30. ACN to column	10
14. Repeat steps 3 and 4		31. Argon trityl flush	5
15. Cap A&B to column	4	32. Repeat steps 2, 3, and 4	
16. Wait	5	33. End	
17. Repeat steps 13 and 14			

[a]Delivery flow rates are ~3.1 mL/min for phosphoramidites and activators and ~3.6 mL/min for all other reagents.

deprotection protocol should not be used in conjunction with RNA synthesized on controlled pore glass because of the inherent incompatibility between TEA 3HF and the silyl ethers bonds that are part of the CPG. All basic solutions used are composed of gaseous amines dissolved in water or ethanol, and freshly opened bottles will ensure that the effective concentration of the amine (29% NH_4OH or 40% $MeNH_2$ in water) is close to its nominal value. Typically, reagent bottles should be replaced every 2 weeks if opened on a regular basis. This may be especially true for the TBAF solution in THF because a low water content is critical for efficient desilylation with this reagent. The 65°C deprotection times are calculated for 2 mL of basic solution and need to be extended if larger amounts of reagents are to be used.

At a small scale (<10 μmol), the quality of oligoribonucleotides generated from the basic and alternative deprotection protocols does not vary much. Therefore, one should consider the timing requirement as an important parameter in selecting a particular protocol. However, because oligoribonucleotide synthesis often incorporates one or more modified synthons that may be more or less sensitive to the basic or fluoride treatments, the chemical compatibility should be examined carefully before selecting the optimal deprotection method.

Finally, deprotected oligoribonucleotides are highly sensitive to nuclease degradation. Therefore, gloves should be worn at all times, and sterile disposable containers and MilliQ water should be used to limit potential exposure to nucleases.

1. Oligoribonucleotide Deprotection with $NH_4OH/ETOH$ and TBAF

This protocol describes a deprotection scheme using a 3:1 cocktail of concentrated ammonium hydroxide–ethanol to cleave the oligoribonucleotide from the solid support, perform β-elimination of the cyanoethyl phosphodiester protecting group, and cleave the acyl-based exocyclic amine protecting groups. A subsequent treatment with tetrabutylammonium fluoride effects the cleavage of the *tert*-butyldimethylsilyl protecting group on the 2'-hydroxyl functionality.

Materials list

29% ammonium hydroxide (J.T. Baker, Phillipsburg, NJ)
Absolute ethanol
Tetrabutyl ammonium fluoride 1.0 M in THF (Aldrich, Milwaukee, WI)
EtOH-MeCN-H_2O (3:1:1)

Qiagen desalt: Qiagen500 column (Chatsworth, CA), 50 mM triethylammonium bicarbonate (TEAB) pH 7.8, 2M TEAB pH 7.8

Steps

Caution: It is important to cool the sample vial in step 4 before opening the screw cap to avoid loss of content.

1. Transfer the dried solid support from the synthesis column to a
   4-mL glass screw-top vial with a Teflon-lined lid (Wheaton).
2. Prepare a solution of ammonium hydroxide–ethanol (3:1).
3. Add 4 mL of ammonium hydroxide–ethanol (3:1) to the vial,
   screw cap on tightly, and place in a heat block at 65°C for 4 h.
4. Remove vial from heat block, place in a room temperature block,
   and place in −20°C freezer until cooled (~30 min).
5. Decant the solution into a 14-mL Falcon tube. Add 1 mL of
   EtOH–MeCN–H₂O (3:1:1), vortex well, and allow support to
   settle. Decant wash and add to deprotection solution. Repeat
   wash two more times.

*Because of the presence of the hydrophobic 2'-O-TBDMS groups on
the RNA, this organic wash helps increase the recovery yield.*

6. Dry combined supernatants in the Falcon tube in a Speed-Vac
   (~2.5 h on medium heat).
7. Add 1 mL of TBAF 1.0 M in THF to the dried RNA in the
   Falcon tube and place at room temperature for 24 h.
8. Quench desilylation reaction with 9 mL of 50 mM TEAB and
   refrigerate at 4°C until desalting step (see following).
9. Load the quenched reaction in TEAB onto a Qiagen 500 anion-
   exchange cartridge that was prewashed with 10 mL of 50 mM
   TEAB.
10. After the loaded cartridge is washed with 10 mL of 50 mM
    TEAB, the RNA is eluted with 10 mL of 2 M TEAB and dried
    to a white powder in a Speed-Vac.

2. Oligoribonucleotide Deprotection with Aqueous Methylamine and TEA.3HF

This protocol describes a deprotection scheme using aqueous methylamine
and triethylamine trihydrofluoride as alternative reagents to effect nucleobase, 2'-hydroxyl, and phosphodiester deprotection.

Additional materials list

40% aqueous methylamine (Aldrich)

Triethylamine trihydrofluoride (TEA 3HF), *N*-methyl-pyrrolidinone (NMP)
Triethylamine (TEA) (Aldrich)
EtOH–MeCN–H$_2$O (3:1:1)
Sodium acetate (Fluka, Ronkonkoma, NY), *n*-butanol, 70% ethanol
Ammonium bicarbonate (Fluka)

Steps

1. Transfer the oligoribonucleotide attached to the solid support from the synthesis column to a 4-mL glass screw-top vial (Wheaton).
2. Add 1 mL of 40% aqueous methylamine to the vial, screw cap on tightly, and place in a heat block at 65°C for 10 min.
3. Remove vial from heat block, place in a room temperature block, and place in −20°C freezer until cooled (~20 min).
4. Decant the solution into a 14-mL Falcon tube. Add 1 mL of EtOH–MeCN–H$_2$O (3:1:1), vortex well, and allow support to settle. Decant wash and add to deprotection solution. Repeat wash two more times.
5. Dry combined supernatants in the Falcon tube in a Speed-Vac (~2.5 h on medium heat).
6. Add 0.3 mL of TEA 3HF–NMP–TEA (1.5:0.75:1) to the dried RNA in the Falcon tube, cap tube, and place in a heat block at 65°C for 90 min; then bring to room temperature.
7. If oligonucleotide has been synthesized trityl-off, proceed to quench and desalting steps according to steps 8–10 in Section V.D.1.
8. Alternatively, the solution can be precipitated directly from the desilylation reaction by adding 25 μL of 3 M NaOAc, followed by 1 mL of *n*-BuOH.
9. Cool mixture to −20°C for 2 h to overnight and centrifuge at 4°C, 10,000 rpm for 30 min.
10. Decant solution and wash the pellet with 70% EtOH. Centrifuge at 4°C, 10,000 rpm for 10 min and decant. Dry oligoribonucleotide pellet.

This precipitation procedure cannot be applied to the TBAF procedure outlined in Section V.D.1 because of the high organic content of the desilylation reaction.

11. Alternatively, if a trityl-on deprotected oligoribonucleotide is required, quench the desilylation reaction with 5 mL of 1.5 M

ammonium bicarbonate, pH 8.0. Proceed to HPLC analysis or purification.

3. "One-Pot" Oligoribonucleotide Deprotection with Anhydrous Methylamine and Neat TEA 3HF

This protocol describes an expedited deprotection using anhydrous ethanolic methylamine and triethylamine trihydrofluoride as alternative reagents to effect nucleobase, 2'-hydroxyl, and phosphodiester deprotection.

Additional materials list

33% ethanolic methylamine (Fluka)
Anhydrous dimethyl sufloxide (DMSO) (Aldrich)
C_{18} SepPak cartridge (Waters, Millford, MA)

Steps

1. Transfer the dried solid support from the synthesis column to a 4-mL glass screw-top vial (Wheaton).
2. Add 0.8 mL of 33% ethanolic methylamine–DMSO (1:1) to the vial, screw cap on tightly, and place in a heat block at 65°C for 15 min. DMSO is useful for solubilizing the partially deprotected oligoribonucleotide and helps prevent alkaline hydrolysis of the fully deprotected RNA.
3. Remove vial from heat block and place in a room temperature block.
4. Add 0.1 mL of neat TEA 3HF, vortex well, and transfer to a 65°C heating block for 15 min. The solution usually gels out after the addition of TEA 3HF. Cool sample vial at room temperature and the place at −20°C for 10 min.
5. Quench the reaction with 1 mL of 1.5 M ammonium bicarbonate, pH 7.5. **Caution:** *If sample is not cool enough, the addition of ammonium bicarbonate solution may lead to significant effervescence.* Allow support to settle and decant the supernatant. Proceed to HPLC analysis–purification or to a SepPak desalting.
6. Apply the quenched solution to a C_{18} SepPak cartridge that was prewashed successively with CH_3CN (10 mL), CH_3,CN–MeOH–H_2O (1:1:1) (10 mL), and RNase-free H_2O (20 mL).
7. Following sample application, wash the cartridge with RNase-free H_2O (10 mL) to remove all salts.
8. Elute product from the column with CH_3CN–MeOH–H_2O (1:1:1) (10 mL) and dry in a Speed-Vac. *This desalting step will detritylate a trityl-on deprotected oligoribonucleotide. The precipita-*

tion procedure (steps 8–10 in Section V.D.2) cannot be applied to the one-pot deprotection protocol because of the high organic content.

E. Oligoribonucleotide Analysis High-Performance Liquid Chromatography

Refer to Section IV.E.1–2 for methods of analysis and purification of oligoribonucleotides because they are very similar to those for oligodeoxyribonucleotides. RNase-free MilliQ water and sterile containers should be used at all times when processing oligoribonucleotides.

VII. SUMMARY

This chapter contains updated methodologies for chemically synthesizing oligonucleotides according to the prevalent phosphoramidite method. Various critical parameters including protecting group strategy, solid support and linker chemistries, activation methods, and deprotection protocols are delineated. Oligoribonucleotide synthesis is discussed using the well-established 2′-*O*-*t*-butyldimethylsilyl/cyanoethyl-*N,N*-disopropyl phosphoramidite chemistry and other promising approaches. This chapter also describes updated deprotection protocols allowing expedited deprotection times.

ACKNOWLEDGMENTS

The authors gratefully acknowledge Lara Maloney, Kuyler Jones, and Victor Mokler for their efforts in oligonucleotide synthesis, deprotection, and analysis as well as Dr. Nassim Usman for continuous support.

REFERENCES

1. Cech T. Ribozyme engineering. Curr Opin Struct Biol 2:605–609, 1992.
2. Francklyn C, Schimmel PR. Aminoacylation of RNA minihelixes with alanine. Nature 337:478–481, 1989.
3. Cook KS, Fisk GJ, Hauber J, Usman N, Daly TJ, Rusche JR. Characterization of HIV-1 REV protein: Binding stoichiometry and minimal RNA substrate. Nucleic Acids Res 19:1577–1583, 1991.
4. Gold L. Posttranscriptional regulatory mechanisms in *Escherichia coli*. Annu Rev Biochem 57:199–233, 1988.
5. Karaoglu D, Thurlow DL. A chemical interference study on the interaction of ribosomal protein L11 from *Escherichia coli* with RNA molecules con-

taining its binding site from 23S rRNA. Nucleic Acids Res 19:5293–5300, 1991.

6. Johnson KA, Benkovic SJ. In: Sigman DS, Boyer PD, eds. The Enzymes. Vol 19. San Diego: Academic Press, 1990, pp 159–211.

7. Wincott F, DiRenzo A, Shaffer C, Grimm S, Tracz D, Workman C, Sweedler D, Gonzalez C, Scaringe S, Usman N. Synthesis, deprotection, analysis and purification of RNA and ribozymes. Nucleic Acids Res 23:2677–2684, 1995.

8. Sproat B, Colonna F, Mullah B, Tsou D, Andrus A, Hampel A, Vinayak R. An efficient method for the isolation and purification of oligoribonucleotides. Nucleosides Nucleotides 14:255–273, 1995.

9. Vargeese C, Carter J, Yegge J, Krivjansky S, Settle A, Kropp E, Peterson K, Pieken W. Efficient activation of nucleoside phosphoramidites with 4,5-dicyanomidazole during oligonucleotide synthesis. Nucleic Acids Res 26:1046–1050, 1998.

10. Schaller H, Weiman G, Lerch B, Khorana HG. The stepwise synthesis of specific deoxypolynucleotides. J Am Chem Soc 85, 3821–3827, 1963.

11. Usman N, Ogilvie KK, Jiang M-Y, Cedergren RJ. Automated chemical synthesis of long oligoribonucleotides using 2'-O-silylated ribonucleoside 3'-O-phosphoramidites on a controlled-porc glass support: Synthesis of a 43-nucleotide sequence similar to the 3'-half molecule of *Escherichia coli* formylmethionine tRNA. J Am Chem Soc 109:7845–7854, 1987.

12. Scaringe SA, Francklyn C, Usman N. Chemical synthesis of biologically active oligoribonucleotides using β-cyanoethyl protected ribonucleoside phosphoramidites. Nucleic Acids Res 18:5433–5341, 1990.

13. Wu T, Ogilvie KK, Pon RT. N-Phenoxyacetylated guanosine and adenosine phosphoramidities in the solid-phase synthesis of oligoribonucleotides: Synthesis of a ribozyme sequence. Tetrahedron Lett 34:4249–4252, 1988.

14. Chaix C, Duplaa AM, Molko D, Téoule R. Solid-phase synthesis of the 5'-half of the initiator t-RNA from *B. subtilis*. Nucleic Acids Res 17:7381–7393, 1989.

15. Sinha ND, Davis P, Usman N, Pérez J, Hodge R, Kremsky J, Casale R. Labile exocyclic amine protection in DNA, RNA and oligonucleotide analog synthesis facilitating N-deacylation, minimizing depurination and chain degradation. Biochimie 75:13–23, 1993.

16. Reddy MP, Hanna NB, Farooqui F. Fast cleavage and deprotection of oligonucleotides. Tetrahedron Lett 35:4311–4314, 1994.

17. Reddy MP, Hanna NB, Farooqui F. Methylamine deprotection provides increased yield of oligoribonucleotides. Tetrahedron Lett 36:8929–8932, 1995.

18. Reddy MP, Farooqui F, Hanna NB. Elimination of transamination side product by the use of dCAc methylphosphonamidite in the synthesis of oligonucleoside phosphoramidites. Tetrahedron Lett 37:8691–8694, 1996.

19. McBride LJ, Kierzek R, Beaucage SL, Caruthers, MH. Amidine protecting groups in oligonucleotide synthesis. J Am Chem Soc 108:2040–2048, 1986.

20. Vinayak R, Anderson P, McColum C, Hampel A. Chemical synthesis of RNA using fast oligonucleotide deprotection chemistry. Nucleic Acids Res 20:1265–1269, 1992.

21. Vu H, McColum C, Jacobson K, Theisen P, Vinayak R, Spiess E, Andrus A. Fast oligonucleotide deprotection phosphoramidite chemistry for DNA synthesis. Tetrahedron Lett 31:7269–7272, 1990.

22. Himmelsbach F, Schulz BS, Trichtinger T, Charubala R, Pfleiderer W. The *p*-nitrophenylethyl (Npe) group, a versatile new blocking group for phosphate and aglycone protection in nucleosides and nucleotides. Tetrahedron 40:59–72, 1984.

23. Pfleiderer W, Matysiak S, Bergman F, Schnell R. Recent progress in oligonucleotide synthesis. Acta Biochim Pol 43:37–44, 1996.

24. Webb T, Matteucci M. Hybridization triggered cross-linking of deoxyoligonucleotides. Nucleic Acids Res 14:7661–7674, 1986.

25. Kuijpers W, Huskens J, Koole L, van Boeckel C. Synthesis of a well defined phosphate methylated DNA fragment: The application of potassium carbonate in methanol as deprotecting agent. Nucleic Acids Res 18:5197–5201, 1990.

26. Zhou Y, Chladek S, Romano L. Synthesis of oligonucleotides containing site specific carcinogen adducts. J Org Chem 59:556–563, 1994.

27. Scalfi-Happ C, Happ E, Chladek S. New approach to the synthesis of 2′,3′-*O*-aminoacyl oligoribonucleotides related to the 3′-terminus of aminoacyl transfer ribonucleic acid. Nucleosides Nucleotides 6:345–348, 1987.

28. Iyer RP, Yu D, Ho N-H, Devlin T, Agrawal S. Methylphosphotriester oligonucleotides: Facile synthesis using *N*-pent-4-enoyl nucleoside phosphoramidites. J Org Chem 60:8132–8133, 1995.

29. Iyer RP, Yu D, Ho N-H, Devlin T, Habus Y, Agrawal S. *N*-Pent-4-enoyl nucleosides: Application in the synthesis of support-bound and free oligonucleotide analogs by the *H*-phosphonate approach. Tetrahedron Lett 37:1539–1542, 1996.

30. Beaucage S, Caruthers, M. Deoxynucleoside phosphoramidites. A new class of key intermediates for deoxypolynucleotide synthesis. Tetrahedron Lett 37:1859–1862, 1981.

31. Sinha N, Biernat J, McManus J, Koster H. Use of cyanoethyl-*N,N*-dialkylamine-*N*-morpholino phosphoramidite of deoxynucleosides for the synthesis of DNA fragments simplifying deprotection and isolation of the final product. Nucleic Acids Res 12:4539–4557, 1984.

32. Beaucage SL, Iyer RP. Advances in the synthesis of oligonucleotides by the phosphoramidite approach. Tetrahedron 48:2223–2311, 1992.

33. Daub GW, van Tamalen EE. Synthesis of oligoribonucleotides based on the facile cleavage of methyl phosphotriesters intermediates. J Am Chem Soc 99:3526–3527, 1977.

34. Andrus A, Beaucage SL. 2-Mercaptobenzothiazole: An improved reagent for the removal of methylphosphate protecting groups from oligodeoxynucleotide phosphotriesters. Tetrahedron Lett 29:5479–5482, 1988.

35. Dahl BH, Bjergarde K, Henriksen L, Dahl O. A highly reactive odourless substitute for thiophenol/triethylamine as a deprotection reagent in the synthesis of oligonucleotides and their analogues. Acta Chem Scand 44:639–641, 1990.

36. Kim S-G, Eida K, Takaku H. Use of the hexafluoro-2-butyl protecting group in the synthesis of DNA fragments via the phosphoramidite approach on solid-support. Bioorg Med Chem Lett 5:1663–1666, 1995.

37. Ravikumar VT, Cheruvallath ZS, Cole DL. 4-Cyano-2-butenyl group: A new type of protecting group in oligonucleotide synthesis via phosphoramidite approach. Tetrahedron Lett 37:6643–6646, 1996.

38. Wada T, Sekine M. TSE as a phosphate protecting group in oligonucleotide synthesis. Tetrahedron Lett 35:757–760, 1994.

39. Ravikumar VT, Cole DL. DPSE: A versatile protecting group for oligonucleotide synthesis. Gene 149:157–161, 1994.

40. Smith M, Rammler DH, Goldberg IH, Khorana HG. Studies on polynucleotides. Specific synthesis of the C3'-C5' interribonucleotide linkage. J Am Chem Soc 84:430–440, 1962.

41. Josephson S, Lagerholm E, Palm G. Automatic synthesis of oligodeoxynucleotides and mixed oligodeoxynucleotides using the phosphoramidite method. Acta Chem Scand B38:539–545, 1984.

42. Ikatura K, Rossi JJ, Wallace RB. Synthesis and use of synthetic oligonucleotides. Annu Rev Biochem 53:1755–1769, 1984.

43. Bidaine A, Berens C, Sonveaux E. The phototrityl group. Photocrosslinking of oligonucleotides to BSA. Bioorg Med Chem Lett 6:1167–1170, 1996.

44. Gortz IIH, Seliger H. New hydrophobic protecting groups for the chemical synthesis of oligonucleotides. Angew Chem Int Ed Engl 20:680–681, 1981.

45. Seliger H, Schmidt G. Derivatization with the 4-decyloxytrityl group as aid in the affinity chromatography of oligo- and poly-nucleotides. J Chromatogr 397:141–151, 1987.

46. Chattopadhyaya JB, Reese CB. The 9-phenylxanthen-9-yl protecting group. J Chem Soc Chem Commun 639–640, 1978.

47. Chattopadhyaya JB, Reese CB. Chemical synthesis of a tridecanucleoside dodecaphosphate sequence of SV40 DNA. Nucleic Acids Res 8:2039–2043, 1980.

48. Pfister M, Farkas S, Charubala R, Pfleiderer W. Recent progress in oligoribonucleotide synthesis. Nucleosides Nucleotides 7:595–600, 1988.

49. Bergmann F, Pfleiderer W. Solid-phase synthesis of oligoribonucleotides using the 2-dansylethoxycarbonyl group for 5'-hydroxy protection. Helv Chim Acta 77:481–500, 1994.

50. Iwai S, Ohtsuka E. 5'-Levulinyl and 2'-tetrahydrofuranyl protection for the synthesis of oligoribonucleotides by the phosphoramidite approach. Nucleic Acids Res 16:9443–9456, 1988.

51. Iwai S, Yamada E, Asaka M, Hayase Y, Inoue H, Ohtsuka E. A new solid-phase synthesis of oligoribonucleotides by the phosphoro-p-anisidate method using tetrahydrofuranyl protection of 2'-hydroxyl groups. Nucleic Acids Res 15:3761–3772, 1987.

52. Lehmann C, Xu YZ, Christodoulou C, Tan ZK, Gait MJ. Solid-phase synthesis of oligoribonucleotides using 9-fluorenylmethoxycarbonyl (Fmoc) for 5'-hydroxyl protection. Nucleic Acids Res 17:2379–2390, 1989.

53. Ogawa T, Hosaka H, Makita T, Takaku H. Solid-phase synthesis of oligori-bonucleotides using 5'-9-fluorenylmethoxycarbonyl and 2'-1-(isopropoxyl)ethyl protection. Chem Lett 1169–1172, 1991.

54. Scaringe SA, Wincott FE, Caruthers MH. Novel RNA synthesis method using 5'-O-silyl-2'-O-orthoester protecting groups. J Am Chem Soc 120:11820–11821, 1998.

55. Gait MJ, Pritchard C, Slim G. Oligoribonucleotide synthesis. In: Eckstein F, ed. Oligonucleotides and Analogues, a Practical Approach. Oxford: Oxford University Press, 1991, pp 25–48.

56. Rao MV, Reese CB, Schehlmann V, Yu PS. Use of the 1-(2-fluorophenyl)-4-methoxypiperidin-4-yl (Fpmp) protecting group in the solid-phase synthesis of oligo- and poly-ribonucleotides. J Chem Soc Perkin Trans I 43–55, 1993.

57. Capaldi DC, Reese CB. Use of the 1-(2-fluorophenyl)-4-methoxypiperidin-4-yl (Fpmp) and related protecting groups in oligoribonucleotide synthesis: Stability of internucleotide linkages to aqueous acid. Nucleic Acids Res 22:2209–2216, 1994.

58. Schwartz ME, Breaker RR, Asteriadis GT, deBear JS, Gough GR. Rapid synthesis of oligoribonucleotides using 2'-O-(o-nitrobenzyloxymethyl)–protected monomers. Bioorg Med Chem Lett 2:1019–1024, 1992.

59. Gough GR, Miller TJ, Mantick NA. p-Nitrobenzyloxymethyl: A new fluoride-removable protecting group for ribonucleoside 2'-hydroxyls. Tetrahedron Lett 37:981–982, 1996.

60. Rastogi J, Usher D. A new 2'-hydroxyl protecting group for the automated synthesis of oligoribonucleotides. Nucleic Acids Res 23:4872–4877, 1995.

61. Bellon L, Workman C. Deprotection of RNA. Application Serial No 09-164,964, Docket 236–249.

62. Reese CB, Skone PA. Action of acid on oligoribonucleotide phosphotriester intermediates. Effect of released vicinal hydroxy functions. Nucleic Acids Res 13:3501–3504, 1985.

63. Christodoulou C, Agrawal S, Gait MJ. Incompatibility of acid-labile 2' and 5' protecting groups for solid-phase synthesis of oligoribonucleotides. Tetrahedron Lett 27:1521–1522, 1986.

64. Reese CB, Serfinowska HT, Zappia G. An acetal group suitable for the protection of 2'-hydroxy functions in rapid oligoribonucleotide synthesis. Tetrahedron Lett 27:2291–2294, 1986.

65. Beijer B, Sulston I, Sproat BS, Rider P, Lamond AI, Neuner P. Synthesis and applications of oligoribonucleotides with selected 2'-O-methylation using the 2'-O-[1-(2-fluorophenyl)-4-methoxypiperidin-4-yl] protecting group. Nucleic Acids Res 18:5143–5151, 1990.

66. Sakatsume O, Yamaguchi T, Ishikawa M, Hirao I, Miura K, Takaku H. Solid-phase synthesis of oligoribonucleotides by the phosphoramidite approach using 2'-O-1-(2-chloroethoxy)ethyl protection. Tetrahedron 47:8717–8728, 1991.

67. Sakatsume O, Ogawa T, Hosaka H, Kawashima M, Takaki M, Takaku H. Synthesis and properties of non-hammerhead RNA using 1-(2-chloroethoxy)ethyl group for the protection of 2'-hydroxyl function. Nucleosides Nucleotides 10:141–153, 1991.

68. Ohtsuka E, Fujiyama K, Ikehara M. Studies on transfer ribonucleic acids and related compounds. XL. Synthesis of an eicosaribonucleotide corresponding to residues 35–54 of tRNAfMet from *E. coli*. Nucleic Acids Res 9:3503–3522, 1981.

69. Hayes JA, Brunden MJ, Gilham PT, Gough GR. High-yield synthesis of oligoribonucleotides using *o*-nitrobenzyl protection of 2'-hydroxyls. Tetrahedron Lett 26:2407–2410, 1985.

70. Tanaka T, Tamatsukuri S, Ikehara M. Solid-phase synthesis of oligoribonucleotides using *o*-nitrobenzyl protection of 2'-hydroxyl via a phosphite triester approach. Nucleic Acids Res 14:6265–6279, 1986.

71. deBear JS, Hayes JA, Koleck MP, Gough GR. A universal glass support for oligonucleotide synthesis. Nucleosides Nucleotides 6:821–830, 1987.

72. Klosel R, Konig S, Lehnhoff S, Karl RM. The 1,1-dianisyl-2,2,2-trichloroethyl group as a 2'-hydroxyl protection of ribonucleotides. Tetrahedron 52:1493–1502, 1996.

73. Pitsch S, Weiss P, Luzzi J. Ribonucleoside derivative and methods for preparing the same. Swiss Patent Application 01 937/97, August 18, 1997.

74. Wu X, Pitsch S. Synthesis and base-pairing properties of oligoribonucleotide analogues containing a metal-binding site attached to β-D-allofuranosyl cytidine. Nucleic Acids Res 26:4315–4323, 1998.

75. Jones JJ, Reese CB. Migration of *t*-butyldimethylsilyl protecting groups. J Chem Soc Perkin Trans I 2762–2764, 1979.

76. Ogilvie KK, Nemer MN. Silica gel as solid-support in the synthesis of oligoribonucleotide. Tetrahedron Lett 21:4159–4162, 1980.

77. Matteucci MD, Caruthers MH. Synthesis of deoxyoligonucleotides on a polymeric support. J Am Chem Soc 103:3185–3191, 1981.

78. van Aerschot A, Herdewijn P, Vanderhaeghe H. Silica gel functionalised with different spacers as solid-support for oligonucleotide synthesis. Nucleosides Nucleotides 7:75–90, 1988.

79. Katzhendler J, Cohen S, Rahanim E, Weisz M, Ringel I, Deutsch A. The effect of spacer linkage and solid-support on the synthesis of oligonucleotides. Tetrahedron 45:2777–2792, 1989.

80. Katzhendler J, Cohen S, Weisz M, Ringel I, Camerini-Oetrio RD, Deutsch A. Spacer effect on the synthesis of oligodeoxynucleotides by the phosphite method. React Polym 6:175–187, 1987.

81. Koster H, Stumpe A, Wolter A. Polymer support oligonucleotide synthesis: Rapid and efficient synthesis of oligodeoxynucleotides on porous glass support using the triester approach. Tetrahedron Lett 24:747–750, 1983.

82. Gough GR, Brunden MJ, Gilham PT. Recovery and recycling of synthetic units in the construction of oligodeoxyribonucleotide on solid-support. Tetrahedron Lett 24:747–750, 1983.

83. Pon RT, Usman N, Ogilvie KK. Derivatization of controlled pore glass beads for solid-phase oligonucleotide synthesis. Biotechniques 6:768–775.

84. McCollum C, Andrus A. An optimized polystyrene support for rapid efficient oligonucleotide synthesis. Tetrahedron Lett 32:4069–4072, 1991.

85. Escavitch JW, McBride LJ, Eadie JS. Effects of pore diameter on the support bound synthesis of long oligonucleotide. In: Bruzik KS, Stee WJ. Biophosphates and Their Analogues. Synthesis, Structure, Metabolism and Activity. Amsterdam: Elsevier, 1987, pp. 65–70.

86. Wright P, Lloyd D, Rapp W, Andrus A. Large scale synthesis of oligonucleotides via phosphoramidite nucleosides and a high loaded polystyrene support. Tetrahedron Lett 34:3373–3376, 1993.

87. Mitchell AR, Kent SB, Erickson B, Merrifield RB. Preparation of aminomethyl-polystyrene. Tetrahedron Lett 42:3795–3798, 1976.

88. Bayer E. Towards the chemical synthesis of proteins. Angew Chem Int Ed Engl 30:113–129, 1991.

89. Montserrat FX, Grandas A, Pedroso E. Predictable and reproducible yields in the anchoring of DMT nucleosides succinates to highly loaded aminoalkyl polystyrene resins. Nucleosides Nucleotides 12:967–971, 1993.

90. van der Marel GA, Marugg JE, de Vroom E, Wille G, Tromp M, van Boeckel CA, van Boom JH. Phosphotriester synthesis of DNA fragments on cellulose and polystyrene solid supports. Rec Trav Chim Pays Bas 101:234–246, 1982.

91. Kume A, Iwase R, Sekine M, Hata T. Cyclic diacyl groups for protection of deoxyadenosine in oligonucleotide synthesis. Nucleic Acids Res 1:8525–8538, 1984.

92. Bardella F, Giralt E, Pedroso E. Polystyrene supported synthesis by the phosphite triester approach: An alternative for the large scale synthesis of small oligonucleotides. Tetrahedron Lett 31:6231–6234, 1990.

93. Ito H, Ike Y, Ikuta S, Itakura K. Solid-phase synthesis of polynucleotides: Further studies on polystyrene co-polymers for the solid-support. Nucleic Acids Res 10:1755–1769, 1982.

94. Tsou D, Hampel A, Andrus A, Vinayak R. Large scale synthesis of oligoribonucleotides on high-loaded polystyrene HLP support. Nucleosides Nucleotides 14:1481–1492, 1995.

95. Miyoshi KI, Huang TH, Itakura K. Solid-phase synthesis of polynucleotides: Synthesis of polynucleotides with defined sequences by the block coupling phosphotriester approach. Nucleic Acids Res 8:5491–5505, 1980.

96. Gait MJ, Matthes HW, Singh M, Sproat BS, Titmas RC. Rapid synthesis of oligodeoxyribonucleotides. Nucleic Acids Res 10:6243–6248, 1982.

97. Reddy MP, Michael MA, Farooqui F, Girgis NS. New and efficient solid support for synthesis of nucleic acids. Tetrahedron Lett 35:55771–5774, 1994.

98. Devivar RV, Koontz SL, Peltier WJ, Pearson JE, Guillory TA, Fabricant J. A new solid-phase support for oligonucleotide synthesis. Bioorg Med Chem Lett 9:1239–1242, 1999.

99. Birch-Hirschfeld E, Foldes-Papp Z, Guhrs KH, Seliger H. A versatile support for the synthesis of oligonucleotides of extended length and scale. Nucleic Acids Res 22:1760–1761, 1994.

100. Birch-Hirschfield E, Foldes-Papp Z, Guhrs KH, Seliger H. Oligonucleotide synthesis on polystyrene grafted poly(tetrafluoroethylene) support. Helv Chim Acta 79:137–150, 1996.

101. Bonora GM, Scremin CL, Colonna FP, Garbesi A. HELP (high efficiency liquid phase) new oligonucleotide synthesis on soluble polymeric support. Nucleic Acids Res 18:3155–3159, 1990.

102. Scremin CL, Bonora GM. Liquid phase synthesis of phosphorothioate oligonucleotides on polyethylene glycol support. Tetrahedron Lett 34:4663–4666, 1993.

103. Bonora GM, Baldan A, Schiavon O, Ferruti P, Veronese FM. Poly(N-acryolylmorpholine) as a new soluble support for the liquid phase synthesis of oligonucleotides. Tetrahedron Lett 37:4761–4764, 1996.

104. Adams SP, Kavka KS, Wykes EJ, Holder SB, Gallupi GT. Hindered dialkylamino nucleoside phosphate reagents in the synthesis of two DNA 51-mers. J Am Chem Soc 105:661–663, 1983.

105. Arnold L, Tocik Z, Bradkova E, Hostomsky Z, Paces V, Smrt J. Automated chloridite and amidite synthesis of oligodeoxyribonucleotides on a long chain support using amidine protected purine nucleosides. Collect Czech Chem Commun 54:523–532, 1989.

106. Pon RT. Protocols for oligonucleotides and analogs. Methods Mol Biol 20:465–496, 1993.

107. Damha MJ, Giannaris PA, Zabarylo S. An improved procedure for derivatization of controlled pore glass beads or solid-phase oligonucleotide synthesis. Nucleic Acids Res 18:3813–3821, 1990.

108. Kumar P, Sharma AK, Sharma P, Garg BS, Gupta KC. Express protocol for functionalization of polymer supports for oligonucleotide synthesis. Nucleosides Nucleotides 15:879 888, 1996.

109. Sproat BS, Brown DM. A new linkage for solid-phase synthesis of oligodeoxyribonucleotides. Nucleic Acids Res 13:2979–2987, 1985.

110. Alul RH, Singman CN, Zhang G, Letsinger RL. Oxalyl-CPG: A new labile support for synthesis of sensitive oligonucleotide derivatives. Nucleic Acids Res 19:1527–1532, 1991.

111. Avino A, Garcia RG, Diaz A, Albericio F, Eritja R. A comparative study of supports for the synthesis of oligonucleotides without using ammonia. Nucleosides Nucleotides 15:871–1889, 1996.

112. Holmberg L. U.S. patent 5,589,586, 1996.

113. Routledge A, Wallis MP, Ross KC, Fraser W. A new deprotection strategy for automated oligonucleotide synthesis using a novel silyl-linked solid-support. Bioorg Med Chem Lett 5:2059–2064, 1995.

114. Kwiatkowski M, Nilson M, Landegren U. Synthesis of full-length oligonucleotides: Cleavage of apurinic molecules on a novel support. Nucl Acids Res 24:4632–4638, 1996.

115. Eritja R, Robles J, Avino A, Albericio F, Pedroso E. A synthetic procedure for the preparation of oligonucleotides without ammonia and its application to the synthesis of oligonucleotides containing O-4-alkyl thymidines. Tetrahedron Lett 48:4171–4182, 1992.

116. Avino AM, Eritja R. Use of NPE protecting groups for the preparation of oligonucleotides without using nucleophiles during the final deprotection. Nucleosides Nucleotides 13:2059–2069, 1994.

117. Lyttle MH, Hudson D, Cook RM. A new universal linker for solid-phase DNA synthesis. Nucleic Acids Res 24:2793–2798, 1996.

118. Greenberg MM, Gilmore JL. Cleavage of oligonucleotides from solid-phase supports using o-nitrobenzyl photochemistry. J Org Chem 59:746–753, 1994.

119. Venkatesan H, Greenberg MM. Improved utility of photolabile solid-phase synthesis supports for the synthesis of oligonucleotides containing 3'-hydroxyl termini. J Org Chem 61:525–529, 1996.

120. Pon RT, Yu S. Hydroquinone-O,O'-diacetic acid as a more labile replacement for succinic acid linkers in solid-phase oligonucleotide synthesis. Tetrahedron Lett 38:3327–3330, 1997.

121. Pon RT, Yu S. Hydroquinone-O,O'-diacetic acid (Q-linker) as a replacement for succinyl and oxalyl linker arms in solid-phase oligonucleotide synthesis. Nucleic Acids Res 25:3629–3635, 1997.

122. Pon RT, Yu S, Guo Z, Sanghvi Y. Multiple oligodeoxynucleotide syntheses on a reusable solid-phase CPG support via the hydroquinone-O,O'-diacetic acid (Q-linker) linker arm. Nucleic Acids Res 27:1531–1538, 1999.

123. Leiber E, Enkoju T. Synthesis and properties of 5-(substituted) mercaptotetrazoles. J Org Chem 26:4472–4479, 1961.

124. Andrus A, Beaucage S, Ohms J, Wert K. American Chemical Society Meeting, New York, Organic Division, Abstract 333, 1986.

125. Vinayak R, Ratmeyer L, Wright P, Andrus A, Wilson D. Chemical synthesis of biologically active RNA using labile protecting groups In: Epton R, ed. Innovations and Perspectives in Solid-Phase Synthesis. Birmingham: Mayflower Worldwide, 1994, pp 45–50.

126. Sproat B, Lamond A, Beijer B, Neuner P, Ryder U. Highly efficient chemical synthesis of 2'-O-methyloligoribonucleotides and terabiotinylated derivatives: Novel probes that are resistant to degradation by RNA or DNA specific nucleases. Nucleic Acids Res 17:3373–3386, 1989.

127. Vinayak R, Andrus A, Mullah B, Tsou D. Advances in the chemical synthesis and purification of RNA. Nucleic Acids Symp Ser 33:123–125, 1995.

128. Krotz A, Klopchin P, Walker K, Srivatsa S, Cole D, Ravikumar V. On the formation of longmers in phosphorothioate oligodeoxyribonucleotide synthesis. Tetrahedron Lett 38:3875–3878, 1997.

129. Hayakawa Y, Kataoka M, Noyori R. Benzimidazolium triflate as an efficient promoter for nucleotide synthesis via the phosphoramidite method. J Org Chem 61:7996–7997, 1996.

130. Sanghvi Y. Importance of backward integration in the manufacture of oligonucleotides. IBC's Third International Conference Oligonucleotide Technology, San Diego, May 5–6, 1999.

131. McBride LJ, Caruthers MH. An investigation of several deoxynucleoside phosphoramidites useful for synthesizing deoxynucleotides. Tetrahedron Lett 24:245–248, 1983.

132. Sinha N, Biernat J, Koster H. Cyanoethyl N,N-dialkylamino N-morpholinomonochloro phosphoamidites. New phosphitylating agents facilitating ease of deprotection and work-up of synthesized oligonucleotides. Tetrahedron Lett 24:5843–5846, 1983.

133. Lyttle MH, Wright PB, Sinha ND, Bain JD, Chamberlin AR. New nucleoside phosphoramidites and coupling protocols for solid-phase RNA synthesis. J Org Chem 56:4608–4615, 1991.

134. Pon RT, Usman N, Damha MJ, Ogilvie KK. Prevention of guanine modification and chain cleavage during the solid-phase synthesis of oligonucleotides using phosphoramidite derivatives. Nucleic Acids Res 14:6453–6470, 1986.

135. Pon RT, Damha MJ, Ogilvie KK. Modification of guanine bases by nucleoside phosphoramidite reagents during the solid-phase synthesis of oligonucleotides. Nucleic Acids Res 13:6447–6465, 1985.

136. Ogilvie KK, Theriault N, Sandana K. Synthesis of oligoribonucleotides. J Am Chem Soc 99:7741–7743, 1977.

137. Ogilvie KK, Beaucage SL, Entwistle DW, Thompson EA, Quilliam MA, Westmore JB. Alkylsilyl groups in nucleoside and nucleotide chemistry. J Carbohydr Nucleosides Nucleotides 3:197–227, 1976.

138. Wincott F, Usman N. 2'-(Trimethylsilyl)ethoxymethyl protection of the 2'-hydroxyl group in oligoribonucleotide synthesis. Tetrahedron Lett 35:6827–6830, 1994.

139. Usman N, Pon RT, Ogilvie KK. Preparation of ribonucleoside 3'-O-phosphoramidites and their application to the automated solid-phase synthesis of oligonucleotides. Tetrahedron Lett 26:4567–4570, 1985.

140. Schwarz M, Pfleiderer W. Solution synthesis of fully protected thymidine dimers using various phosphoramidites. Tetrahedron Lett 24:5513–5516, 1984.

141. Hamamoto S, Takaku H. New approach to the synthesis of deoxyribonucleoside phosphoramidite derivatives. Chem Lett 1401–1404, 1986.

142. Kayakawa Y, Wakabayashi S, Kato H, Noyori R. The allylic protection in solid-phase oligonucleotide synthesis. An efficient preparation of solid-anchored DNA oligomers. J Am Chem Soc 112:1691–1696, 1990.

143. Lyttle M. Chain cleavage during deprotection of RNA synthesized by the 2'-O-trialkylsilyl protection strategy. Nucleosides Nucleotides 12:95–106, 1993.

144. Stawinski J, Stromberg R, Thelin M, Westman E. Studies on the t-butyldimethylsilyl group as 2'-O-protection in oligoribonucleotide synthesis via the H-phosphonate approach. Nucleic Acids Res 16:9285–9298, 1988.

145. Wu T, Ogilvie KK, Pon RT. Prevention of chain cleavage in the chemical synthesis of 2'-O-silylated oligoribonucleotides. Nucleic Acids Res 17:3501–3517, 1989.

146. Polushin N, Pashkova I, Efimov V. Rapid deprotection procedures for synthetic oligonucleotides. Nucleic Acids Res Symp Ser 24:49–50, 1991.

147. Chaix C, Molko D, Teoule R. The use of labile base protecting groups in oligoribonucleotide synthesis. Tetrahedron Lett 30:71–74, 1989.

148. Kawahara S, Wada T, Sekine M. Unprecedented mild acid-catalyzed desilylation of 2'-O-tert-butyldimethylsilyl group from chemically synthesized oligoribonucleotides intermediates via neighboring group participation of the internucleotidic phosphate residue. J Am Chem Soc 118:9461–9468, 1996.

149. Gasparutto D, Livache T, Bazin H, Duplaa AM, Guy A, Khorlin A, Molko

D, Roget D, Teoule R. Chemical synthesis of a biologically active natural tRNA with its minor bases. Nucleic Acids Res 20:5159–5166, 1992.

150. Hogrefe R, McCaffrey A, Borodzine L, McCampbell E, Vaghefi M. Effect of excess water on the desilylation of oligoribonucleotides using tetrabutylammonium fluoride. Nucleic Acids Res 21:4739–4741, 1993.

151. Pirrung MC, Shuey SW, Lever DC, Fallon L. A convenient procedure for the deprotection of silylated nucleosides and nucleotides using triethylamine trihydrofluoride. Tetrahedron Lett 4:1345–1346, 1994.

152. Westman E, Stromberg R. Removal of *t*-butyldimethylsilyl protection in RNA synthesis. Triethylamine trihydrofluoride is a more reliable alternative to tetrabutylammonium fluoride. Nucleic Acids Res 22:2430–2431, 1994.

153. Vinayak R, Andrus A, Mullah B, Tsou D. Advances in the chemical synthesis and purification of RNA. Nucleic Acids Symp Ser 33:123–125, 1995.

12
Synthesis of Oligonucleotide-Peptide Conjugates and Nucleopeptides

Ramon Eritja
Consejo Superior de Investigaciones Cientificas, Barcelona, Spain

I. INTRODUCTION

Covalent bonds between nucleic acids and amino acids, peptides, or proteins occur naturally in the genomes of certain RNA and DNA viruses. In these molecules, the protein–nucleic acid linkage occurs through a 5′-phospho-diester bond to serine, tyrosine, or threonine [1,2]. Other naturally occurring examples of RNA–amino acid conjugates are the aminoacyl–transfer RNAs (tRNAs), in which the amino acids are esterified to the 2′- or 3′-hydroxyl group of tRNA [3,4]. These naturally occurring conjugates are called nucleoproteins and *nucleopeptides*. Moreover, interest in *oligonucleotide-peptide conjugates* (also called oligonucleotide-peptide chimeras or oligonucleotide-peptide hybrids) has grown. They are chimeras constituted by oligonucleotides that are connected to peptide sequences and are produced to add some of the biological or biophysical properties of peptides to synthetic oligonucleotides. They are different from peptide nucleic acids (PNAs) because PNAs are not formed by natural amino acids but are oligonucleotide analogues with modified backbones. They are also different from nucleopeptides because they are not found in nature. From the synthetic point of view, nucleopeptides and oligonucleotide-peptide conjugates are similar. In fact, nucelopeptides may be considered a particular case of oligonucleotide-peptide conjugates in which the linkage between peptide and oligonucleotide is performed through the side chain or the carboxylic function of particular amino acids.

In most cases, nucleopeptides have been prepared to be used as model compounds for the study of the biological properties of nucleoproteins [5–12], whereas oligonucleotide-peptide conjugates have found use in several fields. The introduction of multiple nonradioactive labels was one of the first applications described for oligonucleotide-peptide conjugates [13–17], together with their use as sequence-specific artificial nucleases [18–28]. Furthermore, oligonucleotide-peptide conjugates may become important in the field of antisense oligonucleotides. Conjugation of oligonucleotides to poly-lysine [29–32], basic [33] or hydrophobic [34] peptides, fusogenic peptides [35,36], signal peptides [37–47], and penetratin [48,49] has been performed in order to increase cellular uptake or intracellular delivery of antisense oligonucleotides. Moreover, binding oligonucleotides to certain peptide sequences improves binding to complementary DNA [50–55], hybridization speed [56,57], binding to RNA [58] and proteins [59], and nuclease resistance [60]. Finally, peptide sequences have been used for the preparation of encoded combinatorial libraries [61–63].

The preparation of oligonucleotide-peptide conjugates and nucleopeptides presents an interesting challenge because the standard protection schemes used in peptide and oligonucleotide synthesis are not compatible. For example, amide-type protecting groups are used for the protection of nucleobases. These protecting groups are removed by ammonia under conditions that could hydrolyze nucleopeptide bonds [5–12] or provoke unwanted side reactions such as racemization or aspartimide formation. In contrast, all standard protection schemes in solid-phase peptide synthesis utilize acid treatments, which could provoke partial depurination of DNA [64,65]. In order to overcome these problems two different approaches have been described: the postsynthetic conjugation approach and the stepwise solid-phase synthesis approach (Table 1). In the postsynthetic conjugation approach the oligonucleotide and the peptide are built in a separate support using standard protocols but are conveniently functionalized to be linked after synthesis. In the stepwise solid-phase approach the oligonucleotide-peptide conjugate is prepared in a single support using special protecting groups and modified protocols that minimize unwanted side reactions. Examples of both approaches will be described in this chapter.

II. THE POSTSYNTHETIC CONJUGATION APPROACH

One of the simplest solutions to the problem of chemical incompatibility between oligonucleotides and peptides is to avoid the issue by preparing both components separately using peptide and oligonucleotide standard methodologies followed by linking both compounds together. Prior to the

Table 1 Different Methods Described for the Preparation of Oligonucleotide-Peptide Conjugates

No. and name	Remarks
1. Postsynthetic conjugation	Peptides and oligonucleotides are prepared using standard protocols. Special groups are introduced for conjugation.
1a. Thiol-oligonucleotide + maleimido-peptide	Maleimido-peptide and thiol-oligonucleotides are deblocked just before use.
1b. Maleimido-oligonucleotide + thiol-peptide	Most commonly used. Maleimido-oligonucleotide is prepared from 5'-amino-oligonucleotide. Introduction of a thiol group in peptides is performed by adding a Cys residue.
1c. Oligonucleotide with ribonucleoside at 3' + poly-Lys	Special protocol to attach poly-Lys and proteins.
2. Stepwise solid-phase synthesis	Peptides and oligonucleotides are prepared in the same support. Peptides are usually linked at the 3' end.

conjugation, attachment of certain functional groups to oligonucleotides and peptides to ensure a unique and specific linkage between both molecules is required. Several protocols have been described, most of which are based on the specific reactivity of the thiol groups (Fig. 1). Some protocols introduce a thiol group on the oligonucleotide and a thiol reactive group in the peptide (Fig. 1, strategy a), whereas others introduce the thiol reactive group in the oligonucleotide and the thiol group in the peptide (Fig. 1, strategy b).

In strategy a, the thiol reactive groups added to the peptide are maleimido [37–39,45], bromoacetyl [42], chloromethyl ketone [59], or thiol [18–22,33,56,57] groups. The introduction of thiol groups in oligonucleotides is usually accomplished at the 5' end using a special phosphoramidite carrying a protected thiol group [66–68] or at the 3' end using a special support containing a disulfide bond that is cleaved during oligonucleotide deprotection [65,68]. Some protocols activate the thiol-oligonucleotide with 2,2'-dithiopyridine prior to the conjugation reaction [18–22,33,35,56,57,69], especially when the linkage between oligonucleotide and peptide is a disulfide bond. However, this activation is not needed when thiol-oligonucleotides are reacted with maleimido- [37–39,45,68], bromoacetyl- [42], or chloromethyl ketone-peptides [59].

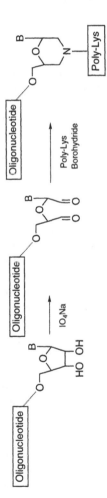

Figure 1 The postsynthetic conjugation approach.

In strategy b, a thiol reactive group such as maleimido [44,50,53] or iodoacetamido [41,54] is introduced at the 5' end of the oligonucleotide. Typically these groups are added by reaction of an oligonucleotide containing an amino group at the 5' end [66,70] with commercially available bifunctional cross-linking agents such as succinimidyl-4-(N-maleimidomethyl) cyclohexane-1-carboxylate (SMCC) or succinimidyl iodoacetate. Maleimido and iodoacetamido groups are not stable to ammonia, and for this reason the direct introduction of these groups in oligonucleotides is not possible. The thiol group for the peptide is provided by the side chain of a cysteine residue.

Finally, oligonucleotides have been conjugated to poly-Lys by adding a ribonucleotide at the 3' end of the oligonucleotide [29,32]. Periodate oxidation of the *cis*-diol function of the ribonucleotide followed by cyanoborohydride reductive amination with the amino groups of poly-Lys yields the desired conjugates (strategy c).

In this chapter examples of the different strategies used for the preparation of oligonucleotide-peptide conjugates are described.

A. Synthesis of Oligonucleotide-Peptide Conjugates Using Strategy A: Preparation and Conjugation of Maleimido-Peptides with Thiol-Oligonucleotides

1. Solid-Phase Synthesis of Peptides Carrying Maleimido Groups

The solid-phase synthesis of peptides carrying maleimido groups is similar to conventional procedures for preparing peptides. We used the solid-phase methodology on polystyrene supports with the 9-fluorenylmethoxycarbonyl (Fmoc)/t-butyl (tBu) (see Chapter 2). In order to avoid having an ionizable carboxyl group at the C-terminal position of the peptide, an amide function is introduced in the peptide using peptide amide linker (PAL) or Rink linkers (see Chapter 5). The sequence is assembled on an approximately 0.2-mmol scale (0.4–0.5 g of support). Fmoc groups were removed by 20% piperidine in dimethyl formamide (DMF). Four equivalents of Fmoc-protected amino acid over the resin capacity were preactivated (5 min) with equimolar amounts of 1-hydroxybenzotriazole (HOBt) and 2-(1H-benzotriazol-1-yl)-1,1,3,3-tetramethyluronium hexafluorophosphate (HBTU) together with N,N-diisopropylethylamine (DIEA) (8 equivalents) in DMF and added to the resin. After 25 min the completion of the coupling reaction was checked with the ninhydrin test. If the ninhydrin test was positive after 1 h, the coupling reaction was repeated. At the last step 6 equivalents of 3-maleimidobenzoic acid N-hydroxysuccinimide ester (MBS) and 6 equivalents of

HOBt in a small amount of DMF were added. The maleimido-peptide was deprotected and released from the resin by treatment with trifluoroacetic acid (TFA)–H_2O (19:1) for 4 h at room temperature. The resin was filtered and washed several times with H_2O. The combined solutions were evaporated under reduced pressure. The residue was treated three times in diethyl ether and concentrated to dryness to eliminate the excess of TFA. The peptide was analyzed by high-performance liquid chromatography (HPLC) and characterized by mass spectrometry.

We have observed that the maleimido-containing peptides are not completely stable in aqueous solutions. After 3–4 days in neutral aqueous solutions, the purified peptides were transformed to a complex mixture of products formed by reaction of the maleimido groups with the side-chain amino groups. To avoid this problem, the peptidyl-resin is stored in the freezer and small aliquots of the resin (5–10 mg per 20–30 OD units at 260 nm of crude oligonucleotide) are deprotected just before the coupling with the thiol-containing oligonucleotide.

2. Synthesis of Thiol-Containing Oligonucleotides and Conjugation with Maleimido-Peptides

Oligonucleotide sequences were synthesized using an automatic DNA synthesizer (see Chapter 11). The standard β-cyanoethyl protected phosphoramidite method on controlled pore glass (CPG) supports was used (0.2 or 1 μmol scale). For the introduction of the thiol group at the 5′ or at the 3′ end of the oligonucleotide, commercially available (Glen Research) phosphoramidite or CPG supports were used (Fig. 2). Oligonucleotides were removed from the solid support by treatment with a solution (1 mL) of 0.1 M dithiothreitol (DTT) in concentrated ammonia at 55°C overnight. The solution was concentrated to dryness. The residue was dissolved in sterile

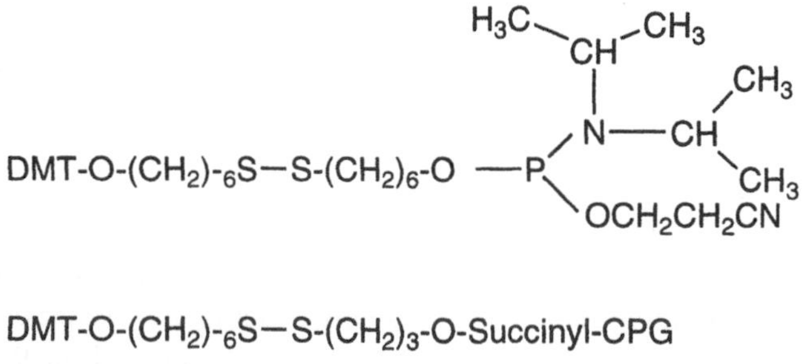

DMT-O-(CH$_2$)-$_6$S—S-(CH$_2$)$_3$-O-Succinyl-CPG

Figure 2 Chemical structures of the commercially available phosphoramidite and solid support used for the preparation of thiol-oligonucleotides.

water and the excess DTT was removed by using an NAP-10 column (Pharmacia). The oligonucleotide was eluted with 1.5 mL of sterile water. This solution was collected directly in buffer A (0.5 mL), which contained a 10-fold excess of the maleimido-peptide recently prepared. To adjust the solution to pH 6.5, some drops of 1 M aqueous triethylammonium acetate (TEAA) solution (pH 6.5) were added. The reaction can be followed by HPLC and is usually complete in less than 1 h. When the reaction was complete the reaction mixture was concentrated to dryness, dissolved in sterile water, and purified by HPLC using one of the following columns: Nucleosil 120 C_{18} (250 × 4 mm) or PRP-1 (Hamilton, 305 × 7 mm). Buffer A was 5% acetonitrile in 0.1 M aqueous TEAA, pH 6.5. Buffer B was 70% acetonitrile in 0.1 M aqueous TEAA, pH 6.5. A 30-min gradient from 0% B to 80% B was used. Elution was followed by reading the absorbance at 260 nm (Fig. 3).

Oligonucleotide-peptide conjugates are characterized by UV spectrophotometry, amino acid analysis [71], nucleoside analysis after enzymatic degradation and mass spectrometry [72,73].

B. Synthesis of Oligonucleotide-Peptide Conjugates Using Strategy B: Preparation and Conjugation of Maleimido-Oligonucleotides with Thiol Peptides

Peptide sequences were assembled using standard Fmoc protocols as described before. A cysteine residue was added at either the *C*- or the *N*-terminal position. For this purpose, Fmoc-L-Cys(Trt)-OH or Fmoc-L-Cys(MMT)-OH was used. After synthesis, the peptide-resin was dried under vacuum and treated with trifluoroacetic acid–triisopropylsilane–H_2O (95:2.5:2.5) for 3 h. Solvents were removed and the residue was purified by HPLC.

Oligonucleotide sequences were synthesized on a DNA synthesizer using standard protocols on 0.2-μmol scale. The primary amino group was introduced at the 5′ end using *N*-MMT-6-aminohexanol phosphoramidite (Glen Research). After ammonia deprotection, MMT-NH-oligonucleotides were purified with cartridge oligonucleotide purification (COP) (Cruachem) cartridges essentially as described by the manufacturer with minor modifications. Removal of the MMT group on the cartridges was performed with a 5-min treatment with 2% aqueous solution of trifluoroacetic acid and the elution of oligonucleotides from the cartridges was performed with CH_3CH–H_2O (7:13). Oligonucleotides were converted to the sodium salt form by passage of an aqueous solution of oligonucleotide through a small column (30 × 6 cm) containing a cation-exchange resin (Dowex 50 × 2, sodium form). The 5′-amino oligonucleotides were dissolved in 0.5 mL of sodium carbonate buffer (pH 9.0) and 1 mg of SMCC dissolved in 0.5 mL of CH_3CN

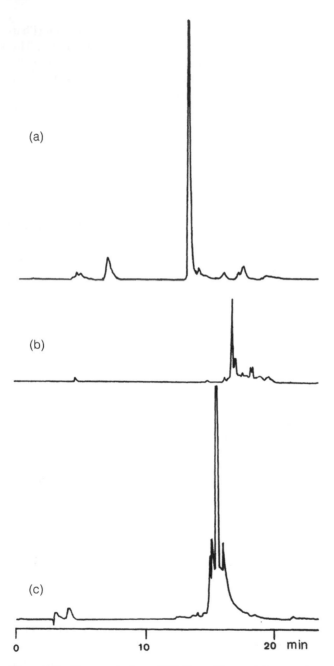

Figure 3 Reversed-phase HPLC profiles at 260 nm of (a) thiol-hexanucleotide, (b) maleimido-Ala-Glu-Gln-Lys-Leu-Ile-Ser-Glu-Glu-Asp-Leu-Asn-CONH$_2$, and (c) oligonucleotide-peptide conjugate (see conditions in the text).

was added. The mixtures were incubated at 37°C overnight and concentrated to dryness. The residues were dissolved in H_2O and desalted on a Sephadex G-25 column (NAP-10, Pharmacia). Fractions containing the oligonucleotides were pooled, concentrated, and purified by HPLC as described in Section II.A. 5'-Maleimido-oligonucleotides were isolated in 50% yield and stored dry at -20°C.

An aliquot of the thiol-peptide (~ 1 μmol, 10 equiv.) was dissolved in 0.2 mL of 0.1 M triethylammonium acetate buffer, pH 6.5, mixed, and the solution was added to a solution of 5'-maleimido-oligonucleotide (0.1 μmol) in the same solution (0.2 mL). The reaction was allowed to occur for 2–3 h at room temperature and the product was purified as described in Section II.A.

C. Synthesis of Oligonucleotide-Peptide Conjugates Using Strategy C: Preparation of Oligonucleotides Carrying a Ribonucleoside at the 3'-End and Conjugation with Poly-Lys [32]

Oligonucleotides carrying a ribonucleotide at the 3' end were prepared using special RNA supports in which 2'-O-acetylribonucleotides are attached to CPG via a succinyl linkage. These supports are commercially available (Glen Research). Oligonucleotide sequences were assembled on a DNA synthesizer using a 0.2-μmol-scale cycle. Standard ammonia deprotection yielded the desired oligonucleotides carrying the ribonucleoside at the 3' end, which were purified by HPLC as described previously.

Purified oligonucleotides (80–100 nmol) were dissolved in 0.1 mL of 20 mM sodium acetate (pH 4.4) and 4.6 μmol of sodium metaperiodate was added. The reaction was stirred for 30 min at 0°C in the dark. An equal volume of poly-Lys (Sigma, mean 15 kd) 80–100 nmol in 2 M NaCl, 0.2 M sodium borate buffer (pH 8.4), and 100 μmol sodium cyanoborohydride were added. The mixture was incubated overnight at 20°C and then loaded on a Sephadex G-50 column equilibrated with 0.5 M NaCl, 20 mM sodium acetate buffer (pH 6.0). Each fraction was assayed for its oligonucleotide-poly-Lys content by absorbance at 260 nm and by protein assay. The conjugates were stored at -80°C.

III. THE STEPWISE SOLID-PHASE SYNTHESIS APPROACH

One of the drawbacks of the postsynthetic conjugation approach is the amount of labor involved during the preparation of the oligonucleotide-peptide conjugates. The stepwise synthesis of the oligonucleotide-peptide conjugates on the same support increases the efficiency of the synthesis by reducing the number of manipulations. On the other hand, the conventional methodologies for the solid-phase synthesis of oligonucleotides and peptides are not compatible, and for this reason special protecting groups and modified deprotection protocols should be used. Most of the stepwise syntheses described in the references produce oligonucleotides containing the peptide moiety at the 3' end [9,10,13–17,26–28,34,36,46,60]. In this strategy the peptide is first assembled using Boc-amino acids containing base labile protecting groups for the protection of trifunctional amino acid side chains [9,10,27,28,46,60]. The repetitive acid treatments used for the removal of the t-butyloxycarbonyl (Boc) group were performed without the presence of the oligonucleotide. In addition, the base-labile groups (Fmoc, Fm, TFA, Tos, Dnpe) were removed concomitantly with the nucleotide protecting groups. Some research laboratories prefer use of Fmoc-amino acids instead of Boc-amino acids for the peptide moiety [13–17,26,34,36]. When the peptide sequence contains trifunctional amino acids, acid-labile (Boc) groups are used for the protection of side chains removed via TFA–1,2-ethanedithiol (9:1) treatment [13,14,34]. In order to avoid the strong acid treatment, Truffert et al. [26] used the Dde group for amino side-chain protection of lysine. The Dde group is stable to piperidine but removed with hydrazine or ethanolamine [26].

In order to attach via chemical bonds the oligonucleotide and the peptide fragment, different compounds have been described (Fig. 4). These molecules contain a carboxylic acid to react with the amino-terminal group of the peptide and a hydroxyl function as a site to assemble the oligonucleotide. The simplest case of the connecting linkers is the side chain of the amino acids containing hydroxyl functions (Ser, Thr, Tyr, and Hse) [9,10,28,60].

Finally, some oligonucleotides carrying small peptides at the 5' end have also been prepared [63,74]. In this example, 5'-amino-2',5'-dideoxythymidine (Fig. 4) was incorporated at the 5' end of the oligonucleotide, generating an amino group from which peptide synthesis begins [63,74]. Peptide assembly was performed with Fmoc-amino acids but protection of the trifunctional amino acids was not addressed.

Another important feature of the synthesis of oligonucleotide-peptide conjugates is the choice of a solid support. Polystyrene supports are used in peptide synthesis and CPG supports are incorporated for oligonucleotide synthesis. Most reports describe the use of CPG for the synthesis of oligonucleotide-peptide conjugates [13,14,26,27,36,63,74]. However, in some cases low coupling yields have been reported with CPG and alternative supports have been described, such as Teflon [34], polystyrene (PS) [9,10,28,60], and polyethylene glycol–polystyrene (PEG-PS) grafts [46]. Some of the low coupling yields reported on CPG are related to the method

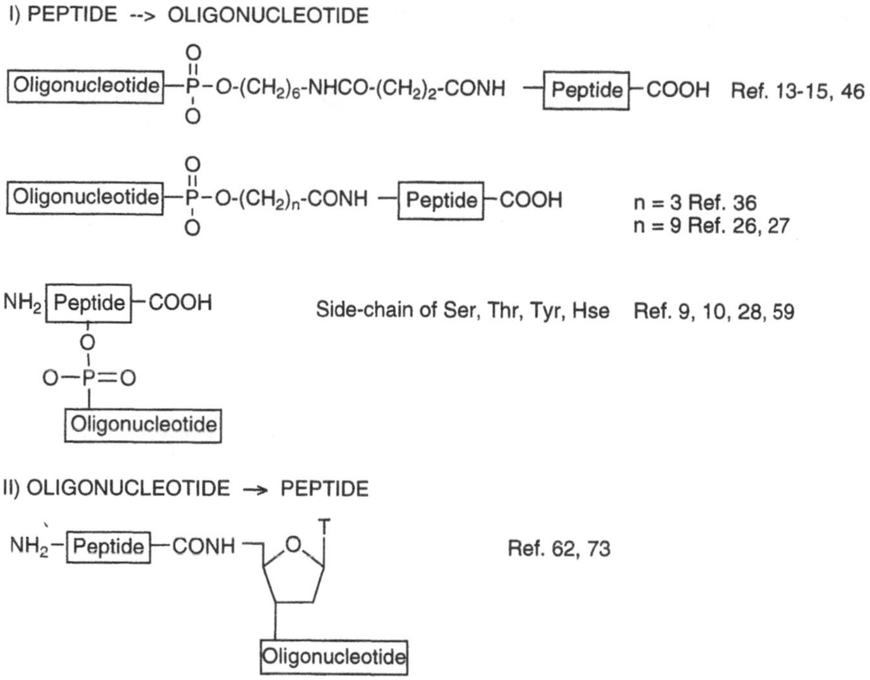

I) PEPTIDE --> OLIGONUCLEOTIDE

II) OLIGONUCLEOTIDE → PEPTIDE

Figure 4 Linker molecules described for connecting the oligonucleotide and the peptide part in the stepwise solid-phase approach.

of activation for the amino acid or to an intrinsic characteristic of a given CPG batch [63]. Coupling reactions with symmetrical anhydrides and some active esters are less efficient than coupling reactions mediated by HBTU/ HOBt [63]. On the other hand, coupling reactions on Teflon, PS, and PEG-PS are less dependent on the activation method during peptide synthesis but are less efficient in oligonucleotide synthesis. The low efficiency of these supports during oligonucleotide synthesis is compensated by increasing the coupling time and changing the solvent to dissolve the phosphoramidites [9,28,46,60].

The linkage between the oligonucleotide-peptide conjugate and the solid support is usually a base-labile linker inspired by the oligonucleotide synthesis field (Fig. 5). Linkers from the peptide field are acid labile and not compatible with the oligonucleotide moiety. Most the linkers used to anchor the peptide moiety have an ester function that is cleaved with concentrated ammonia [13–17,46], ethylenediamine [34], ethanolamine [26], sodium hydroxide [27], or tetrabutylammonium fluoride (TBAF)

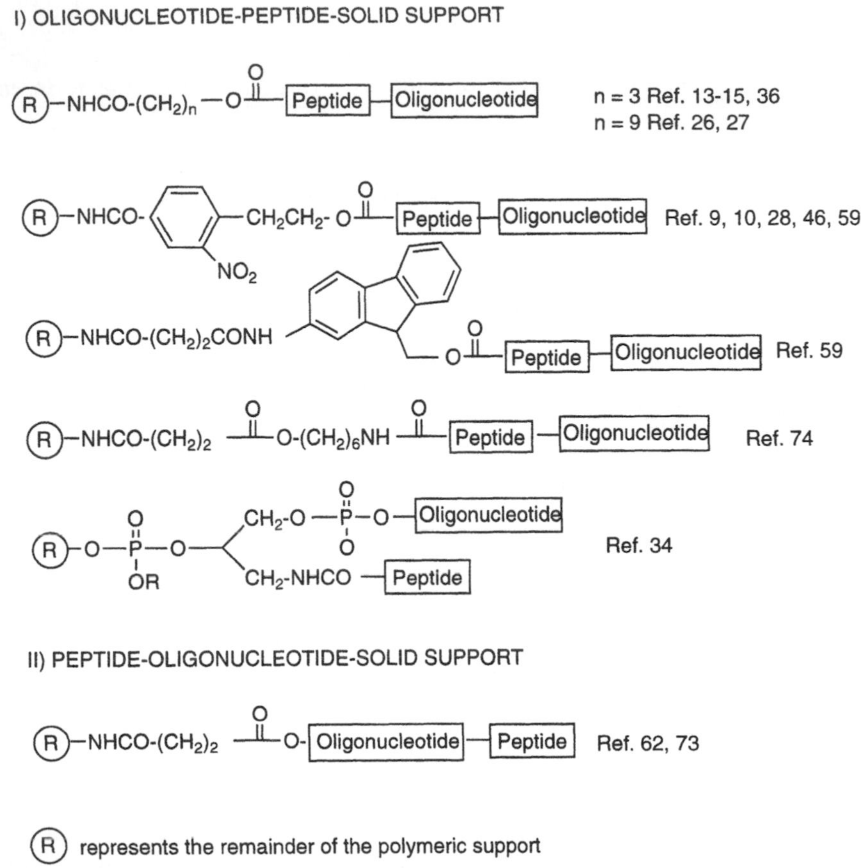

I) OLIGONUCLEOTIDE-PEPTIDE-SOLID SUPPORT

$n = 3$ Ref. 13-15, 36
$n = 9$ Ref. 26, 27

Ref. 9, 10, 28, 46, 59

Ref. 59

Ref. 74

Ref. 34

II) PEPTIDE-OLIGONUCLEOTIDE-SOLID SUPPORT

Ref. 62, 73

(R) represents the remainder of the polymeric support

Figure 5 Linkers used to anchor the oligonucleotide-peptide conjugate during the stepwise solid-phase synthesis.

[9,10,28,59]. In most cases, treatment of traditional peptide handles with ammonia and primary amines gives a mixture of the C-terminal amide and carboxylate. The use of NaOH or TBAF produces the pure carboxylate peptides, but an additional ammonia treatment is required for the complete deprotection of the nucleobases. The handle described for PNA synthesis, 6-[(4-methoxyphenyl)-diphenylmethylamino]hex-1-yl succinylamido linker, is an interesting alternative to produce peptide-oligonucleotide conjugates. This linker is easily cleaved by concentrated ammonia, yielding oligonucleotide-peptide conjugates having the N-(6-hydroxyhexyl)amide group at the C-terminus [75].

Finally, some side reactions produced during the stepwise synthesis of oligonucleotide-peptide conjugates have been described. Accounting for the large number of functional groups that can interact and the complexity of the molecules, oligonucleotide-peptide conjugates constructed by the stepwise approach should be carefully analyzed. One of the possible side reactions is racemization, which may occur during deprotection under basic conditions. Although racemization has not been rigorously excluded during synthesis and cleavage, analytical data such as HPLC traces, capillary electrophoresis, and nuclear magnetic resonance (NMR) spectra of some conjugates do not suggest undesired diastereoisomers [36,63]. The use of phenoxyacetyl or *t*-butylphenoxyacetyl (Expedite) groups for the protection of nucleobases has been shown to be deleterious for oligonucleotide-peptide conjugates [60,63]. Migration of protecting groups from nucleobases to the terminal amino group has been observed [63]. Conventional protecting groups such as isobutyryl, benzoyl, and dimethylformamidino groups solved this problem. Alkylation of lysine residues by the acrylonitrile formed during phosphate deprotection has been described [63]. The addition of 3% cresol as a scavenger to the deprotection mixture is recommended to avoid this side reaction [63]. Fragmentation of the phosphodiester backbone has also been observed, particularly when basic amino acids were present in the peptide component. This undesired reaction has been suggested to occur via nucleophilic attack on phosphotriester intermediates during deprotection [63]. This side reaction can be minimized by using allyloxycarbonyl as a phosphate protecting group [63].

A. Synthesis of Oligonucleotide-Peptide Conjugates Using the Stepwise Solid-Phase Synthesis Approach

The route to the synthesis of oligonucleotide-peptide conjugates is illustrated in Fig. 6 [46]. Using the base-labile linker described by Will et al. [75] attached to a PEG-PS support, the peptide is first assembled. Boc-amino acids protected with the base-labile Fmoc and Fm groups are used as coupling units. A 6-hydroxyhexanoate linker was added to connect the peptide fragment to the oligonucleotide [26,27,36] by converting the last amino group of the peptide to a dimethoxytrityl (DMT)-protected hydroxyl function. The oligonucleotide component is assembled using the standard (isobutyryl, benzoyl) 2'-deoxyribonucleoside 3'-O-(2-cyanoethyl) phosphoramidites. Following linear assembly, some of the protecting groups (Fmoc, Fm, cyanoethyl) are removed with 0.5 M 1,8-diazabicyclo[5.4.0]undec-7-ene (DBU). The conjugate is liberated from the solid support by concentrated ammonia with concomitant removal of the nucleobase protecting groups.

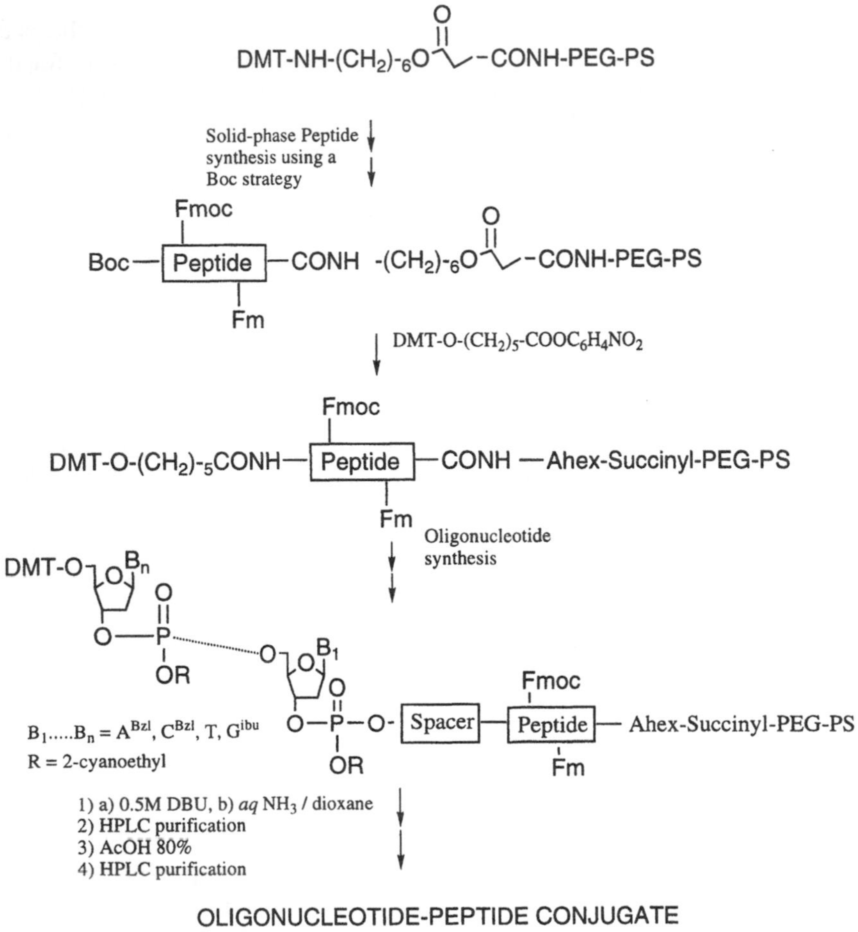

Figure 6   Stepwise synthesis of oligonucleotide-peptide conjugates.

1.   Peptide Syntheses

Peptide sequences were synthesized via the solid phase on a 20 to 50-mmol scale using a homemade manual synthesizer and amino-PEG-PS (0.19 mmol/g, PerSeptive Biosystems) as the solid support. The handle, 6-DMT-aminohex-1-yl hemisuccinate [75], was anchored to the amino-PEG-PS-resin as described previously [76]. The loading of the support was measured by the absorbance of the DMT cation and was 0.1 mmol/g. The elongation of the peptide chain was carried out in DMF using a 5-fold excess of Boc-amino acid, 10-fold excess of DIEA, and 5-fold excess of benzotriazole-1-

yl-oxy-tris-pyrrolidino-phosphonium hexafluorophosphate (PyBOP) for 1 h. The protecting groups of the side chains were Fmoc for lysine and Fm for glutamic and aspartic acids (Novabiochem and Bachem). For the protection of histidine, tosyl and dinitrophenyl groups are recommended [27,28]. In order to remove the Boc group, the supports were treated with TFA–CH$_2$Cl$_2$ (3:7) for 30 min and neutralized with DIEA–CH$_2$Cl$_2$ (1:19). After the addition of the last amino acid, p-nitrophenyl 6-dimethoxytrityloxyhexanoate [36] (5-fold excess, in DMF, 1 h) was coupled to the peptide support. The resin was washed with DMF, dichloromethane, and methanol. Finally, it was dried and stored for further use in oligonucleotide synthesis.

2. Synthesis of Oligonucleotide Part, Deprotection, and Purification of Conjugates

Oligonucleotides were synthesized on a Perkin Elmer–Applied Biosystems 392 DNA Synthesizer using standard β-cyanoethyl–protected phosphoramidites on a 1-μmol scale. The phosphoramidites were dissolved in dry dichloromethane, giving 0.1 M solutions. A slightly modified program was used. Coupling time was increased to 5 min, capping and oxidation times were increased to 1 min, and detritylation times were increased to 2 min (4 $\times$ 30 s). Following chain elongation, the last DMT protecting group was not removed. The average coupling yield was >98%.

The solid supports containing the conjugates were initially treated with a solution of 0.5 M DBU in CH$_3$CN for 5 min. The supports were washed with CH$_3$CN and dried. The resulting supports were treated with concentrated aqueous ammonia–dioxane (10:1) at 50°C overnight. After filtration of the solid supports, the solutions were evaporated to dryness, the residue dissolved in water, and the conjugates purified using standard two-step HPLC purification (Fig. 7). HPLC conditions were as follows: column, Nucleosil 120-10 C$_{18}$ (250 $\times$ 4 mm) or PRP-1 (305 $\times$ 7 mm); a 20-min linear gradient from 15% B to 80% B and 5 min 80% B; flow rate 1 mL/min. Solvent A was 5% acetronitrile in 100 mM triethylammonium acetate (TEAA), pH 6.5, and solvent B 70% acetonitrile in 100 mM TEAA, pH 6.5. In the first step, truncated sequences were separated from the DMT-containing product, and in the second step the conjugates were isolated after removal of the DMT group with 80% aqueous acetic acid (30 min).

Characterization was performed via amino acid analysis after 6 N HCl hydrolysis, nucleoside analysis after enzyme digestion, and mass spectrometry. The homogeneity of the conjugates was analyzed by ion-exchange chromatography. HPLC conditions were as follows: column, Mono Q HR 5/5 (Pharmacia); flow rate 0.5 mL/min; solvent A 10 mM NaOH, 0.3 M NaCl, pH 12.5; solvent B, 10 mM NaOH, 0.9 M NaCl, pH 12.5; a 50-min linear gradient from 10% B to 60% B.

(a)

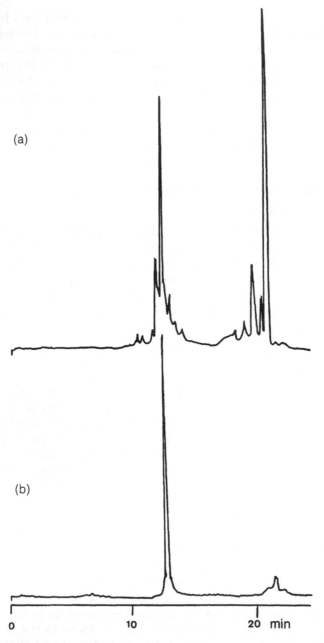

(b)

0   10   20 min

Figure 7 HPLC purification of a hexanucleotide-peptide conjugate prepared using the stepwise approach. (a) DMT-on purification. The desired DMT-oligonucleotide-peptide elutes at 22 min. (b) DMT-off purification (see conditions in the text). The HPLC profiles are measured at 260 nm.

IV. SUMMARY

Nucleopeptides and oligonucleotide-peptide conjugates will play an important role in the future of drug discovery. It has been shown that peptide sequences linked to oligonucleotides enhance some of the desired properties of oligonucleotides such as hybridization to complementary strands [50–55], speed of hybridization [56,57] and nuclease resistance [60]. Moreover, peptide sequences have been used for the introduction of multiple nonradioactive labels in oligonucleotides [13–17], for the preparation of sequence-specific artificial nucleases [18–28], and as molecular bar codes in oligonucleotide combinatorial libraries [61–63].

Two main routes can be used for the preparation of oligonucleotide-peptide conjugates. At present, the postsynthetic conjugation approach is the best choice for the preparation of small amounts of oligonucleotide-peptide conjugates in which the peptide fragment is a protein or a large peptide. The stepwise solid-phase approach will be useful for the construction of large amounts of conjugates carrying short peptide sequences. Several protocols have been described for the stepwise synthesis of oligonucleotide-peptide conjugates, but there is still room for improvements as well as a need for new protocols to assemble complex structures.

During the preparation of this chapter some recent improvements were described. For example, in the postsynthetic conjugation approach, L-cysteine(S-thiobutyl) [77] and N^{α}-tert-butyldithioethylcytosine [78] have been introduced for the preparation of thiol-oligonucleotides, as well as the conjugation of thiol-peptides to oligonucleotide phosphorothioates [79]. In the stepwise solid-phase synthesis approach, the use of the Fmoc group for the side chain of arginine and the formyl group for Trp [80] has been proposed. Finally, the use of protected peptide fragments is emerging as a promising approach for the preparation of oligonucleotide-peptide conjugates [81]. However, protocols for the preparation of conjugates carrying disulfide bridges or circular peptides remain a challenge for synthetic chemists.

ACKNOWLEDGMENTS

I would like to thank my collaborators Beatriz García de la Torre, Anna Aviñó, Mónica Escarceller, Anna Pons, Fernando Albericio, Ramon Güimil Garcia, Viviane Adam, Marten Wiersma, Dirk Gottschling, Hartmut Seliger, Gema Tarrasón, Jaume Piulats, and Ester-Tula Saison-Behmoaras for their support and the mass spectrometry group at EMBL leaded by M. Mann and M. Wilm for help in the characterization of the conjugates.

REFERENCES

1. AB Vartapetian, AA Bogdanov. Prog Nucleic Acids Res 34:209–251, 1987.
2. BA Juodka. Nucleosides Nucleotides 3:445–483, 1984.
3. VW Cornish, D Mendel, PG Schultz. Angew Chem Int Ed Engl 34:621–633, 1995.
4. SA Robertson, JA Ellman, PG Schultz. J Am Chem Soc 113:2722–2729, 1991.
5. E Kuyl-Yeheskiely, CM Tromp, AWM Lefeber, GA van der Marel, JH van Boom. Tetrahedron 44:6515–6523, 1988.
6. CM Dreef-Tromp, EMA van Dam, H van den Elst, GA van der Marel, JH van Boom. Nucleic Acids Res 18:6491–6495, 1990.
7. CM Dreef-Tromp, H van den Elst, JE van den Boogaart, GA van der Marel, JH van Boom. Nucleic Acids Res 20:2435–2439, 1992.
8. CM Dreef-Tromp, JCM van der Maarel, H van den Elst, GA van der Marel, JH van Boom. Nucleic Acids Res 20:4015–4020, 1992.
9. J Robles, E Pedroso, A Grandas. J Org Chem 59:2482–2486, 1994.
10. J Robles, E Pedroso, A Grandas. Nucleic Acids Res 23:4151–4161, 1995.
11. V Jungmann, H Walmann. Tetrahedron Lett 39:1139–1142, 1998.
12. BP Zhao, GB Panigrahi, PD Sadowski, JJ Krepinsky. Tetrahedron Lett 37:3093–3096, 1996.
13. J Haralambidis, L Duncan, GW Tregear. Tetrahedron Lett 28:5199–5202, 1987.
14. J Haralambidis, L Duncan, K Angus, GW Tregear. Nucleic Acids Res 18:493–499, 1990.
15. J Haralambidis, K Angus, S Pownall, L Duncan, M Chai, GW Tregear. Nucleic Acids Res 18:501–505, 1990.
16. G Tong, JM Lawlor, GW Tregear, J Haralambidis. J Org Chem 58:2223–2231, 1993.
17. G Tong, JM Lawlor, GW Tregear, J Haralambidis. J Am Chem Soc 117:12151–12158, 1995.
18. DR Corey, PG Schultz. Science 238:1401–1403, 1987.
19. RN Zuckermann, PG Schultz. J Am Chem Soc 110:6592–6594, 1988.
20. RN Zuckermann, DR Corey, PG Schultz. J Am Chem Soc 110:1614–1615, 1988.
21. DR Corey, D Pei, PG Schultz. Biochemistry 28:8277–8286, 1989.
22. D Pei, DR Corey, PG Schultz. Proc Natl Acad Sci U S A 87:9958–9962, 1990.
23. NN Polushin, B Chen, LW Anderson, JS Cohen. J Org Chem 58:4606–4613, 1993.
24. JK Bashkin, JK Gard, AS Modak. J Org Chem 55:5125–5132, 1990.
25. JK Bashkin, RJ McBeath, AS Modak, KR Sample, WB Wise. J Org Chem 56:3168–3176, 1991.
26. JC Truffert, O Lorthioir, U Asseline, NT Thuong, A Brack. Tetrahedron Lett 35:2353–2356, 1994.
27. JC Truffert, U Asseline, A Brack, NT Thuong. Tetrahedron 52:3005–3016, 1996.
28. M Beltran, E Pedroso, A Grandas. Tetrahedron Lett 39:4115–4118, 1998.

29. M Lemaitre, B Bayard, B LeBleu. Proc Natl Acad Sci U S A 84:648–652, 1987.
30. G Degols, JP Leonetti, C Gagnor, M Lamaitre, B Lebleu. Nucleic Acids Res 17:9341–9350, 1989.
31. JP Leonetti, G Degols, B LeBleu. Bioconjug Chem 1:149–153, 1990.
32. JP Leonetti, B Rayner, M Lemaitre, C Gagnor, PG Milhaud, JL Imbach, B Lebleu. Gene 72:323–332, 1988.
33. E Vives, B LeBleu. Tetrahedron Lett 38:1183–1186, 1997.
34. CD Juby, CD Richardson, R Brousseau. Tetrahedron Lett 32:879–882, 1991.
35. JP Bongartz, AM Aubertin, PG Milhaud, B LeBleu. Nucleic Acids Res 22: 4681–4688, 1994.
36. S Soukchareun, GW Tregear, J Haralambidis. Bioconjug Chem 6:43–53, 1995.
37. NJ Ede, GW Tregear, J Haralambidis. Bioconjug Chem 5:373–378, 1994.
38. JJ Hangeland, JT Levis, YC Lee, POP Ts'o. Bioconjug Chem 6:695–701, 1995.
39. K Arar, M Monsigny, R Mayer. Tetrahedron Lett 34:8087–8090, 1993.
40. SB Rajur, CM Roth, JR Morgan, ML Yarmush. Bioconjug Chem 8:935–940, 1997.
41. MW Reed, D Fraga, DE Schwartz, J Scholler, RD Hinrichsen. Bioconjug Chem 6:101–108, 1995.
42. K Arar, AM Aubertin, AC Roche, M Monsigny, R Mayer. Bioconjug Chem 6: 573–577, 1995.
43. E Bonfils, C Depierreux, P Midoux, NT Thuong, M Monsigny, AC Roche. Nucleic Acids Res 20:4621–4629, 1992.
44. A Chollet. Nucleosides Nucleotides 9:957–966, 1990.
45. R Eritja, A Pons, M Escarceller, E Giralt, F Albericio. Tetrahedron 47:4113–4120, 1991.
46. BG de la Torre, A Aviñó, G Tarrason, J Piulats, F Albericio, R Eritja. Tetrahedron Lett 35:2733–2736, 1994.
47. P Seibel, J Trappe, G Villani, T Klopstock, S Papa, H Reichmann. Nucleic Acids Res 23:10–17, 1995.
48. CM Troy, D Derossi, A Prochiantz, LA Green, ML Shelanski. J Neurosci 16: 253–261, 1996.
49. B Allinquant, P Hantraye, P Mailleux, K Moya, C Bouillot, A Prochiantz. J Cell Biol 128:919–927, 1995.
50. JG Harrison, S Balasubramanian. Nucleic Acids Res 26:3136–3145, 1998.
51. CH Tung, KJ Breslauer, S Stein. Bioconjug Chem. 7:529–531, 1996.
52. Z Wei, CH Tung, T Zhu, WA Dickerhof, KJ Breslauer, DE Georgopoulos, MJ Leibowitz, S Stein. Nucleic Acids Res 24:655–661, 1996.
53. T Zhu, Z Wei, CH Tung, WA Dickerhof, KJ Breslauer, DE Georgopoulos, MJ Leibowitz, S Stein. Antisense Res Dev 3:265–275, 1993.
54. Z Wei, CH Tsung, T Zhu, S Stein. Bioconjug Chem 5:468–474, 1994.
55. CH Tung, MJ Rudolph, S Stein. Bioconjug Chem 2:464–465, 1991.
56. DR Corey. J Am Chem Soc 117:9373–9374, 1995.
57. SV Smulevitch, CG Simmons, JC Norton, TW Wise, DR Corey. Nat Biotechnol 14:1700–1704, 1996.
58. CH Tung, J Wang, MJ Leibowitz, S Stein. Bioconjug Chem 6:292–295, 1995.

59. Y Lin, A Padmapriya, KM Morden, SD Jayasena. Proc Natl Acad Sci U S A 92:11044–11048, 1995.
60. J Robles, M Maseda, M Beltrán, M Concernau, E Pedroso, A Grandas. Bioconjug Chem 8:785–788, 1997.
61. J Nielsen, S Brenner, KD Janda. J Am Chem Soc 115:9812–9813, 1993.
62. S Brenner, RA Lerner. Proc Natl Acad Sci U S A 89:5381–5383, 1992.
63. CN Tetzlaff, I Schwope, CF Bleczinski, JA Steinberg, C Richert. Tetrahedron Lett 39:4215–4218, 1998.
64. SL Beaucage, RP Iyer. Tetrahedron 48:2223–2311, 1992.
65. SL Beaucage, RP Iyer. Tetrahedron 49:10441–10488, 1993.
66. SL Beaucage, RP Iyer. Tetrahedron 49:1925–1963, 1993.
67. BA Connolly, P Rider. Nucleic Acids Res 13:4485–4502, 1985.
68. BG de la Torre, AM Aviñó, M Escarceller, M Royo, F Albericio, R Eritja. Nucleosides Nucleotides 12:993–1005, 1993.
69. R Zuckermann, D Corey, P Schultz. Nucleic Acids Res 15:5305–5321, 1987.
70. M Manoharan. In: ST Crooke, B Lebleu, eds. Antisense Research and Applications. Boca Raton, FL: CRC Press, 1993, pp 303–349.
71. T Zhu, Y Peng, H Lackland, S Stein. Anal Biochem 214:585–587, 1993.
72. ON Jensen, S Kulkarni, JV Aldrich, DF Barofsky. Nucleic Acids Res 24:3866–3872, 1996.
73. K Berlin, RK Jain, C Tetzlaff, C Steinberg, C Richert. Chem Biol 4:63–77, 1997.
74. F Bergmann, W Bannwarth. Tetrahedron Lett 36:1839–1842, 1995.
75. DW Will, G Breipohl, D Langmer, J Knolle, E Uhlmann. Tetrahedron 51:12069–12082, 1995.
76. KC Gupta, P Kumar, D Bhatia, AK Sharma. Nucleosides Nucleotides 14:829–832, 1995.
77. S Soukchareun, J Haralambidis, G Tregear. Bioconj Chem 9:466–475, 1998.
78. D Gottschling, H Seliger, G Tarrasón, J Piulats, R Eritja. Bioconj Chem 9:831–837, 1998.
79. M Antopolsky, E Azhayeva, U Tengvall, S Auriola, I Jaaskelainen, S Ronkko, P Honkakoshi, A Urtti, H Lonnberg, A Azhayev. Bioconjug Chem 10:598–606, 1999.
80. V Marchan, L Debethune, M Beltran, J Robles, I Travesset, G Fabregas, E Pedroso, A Grandas. Nucleosides Nucleotides 18:1493–1494, 1999.
81. S Peyrottes, B Mestre, F Burlina, MJ Gait. Tetrahedron 54:12513–12522, 1998.

13

The Chemistry of Peptide Nucleic Acids

Michael Egholm and Ralph A. Casale
PE Biosystems, Framingham, Massachusetts

I. INTRODUCTION

A peptide nucleic acid (PNA) is an oligonucleotide mimic in which the (deoxy)ribose–phosphate backbone has been replaced by an *N*-(?-amino-ethyl)glycine unit and the nucleobases have been attached through methyl-enecarbonyl linkages (Fig. 1) [1,2]. PNA mirrors the hybridization properties of nucleic acids originally detailed by Watson and Crick in 1953 [3] and demonstrates significant advantages under specific conditions [4]. This function is being employed in a number of applications in molecular biology, diagnostics, and therapeutics [5–12]. As a consequence, a demand has grown for PNA oligomers, and hence their preparation and chemical properties are the subjects of this chapter.

Oligonucleotide analogues have commonly been considered as "nucleic acids" even in the absence of an acidic group, e.g., methyl phosphonates. In a formal sense, "peptide nucleic acids" are neither peptides nor composed of nucleic acids, nor should they be confused with peptide/protein oligonucleotide conjugates as described in the previous section. The PNA monomeric unit contains features of both amino acids and nucleosides. The four common base portions of nucleosides—adenyl, cytosyl, guanidyl, and thymidyl—are tethered to the PNA backbone, which carries the functionality of common amino acids. Amide bonds then consecutively link these monomer units. The term polyamide is more chemically appropriate; thus an alternative name is polyamide nucleoside analogue, which is still abbreviated PNA.

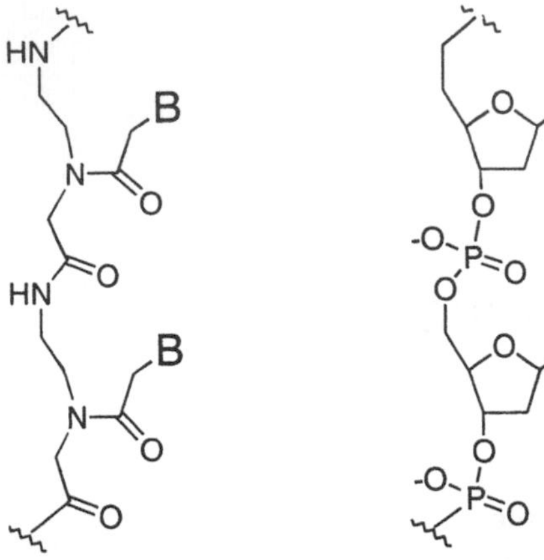

Figure 1   Structure of PNA and DNA molecules.

The sequence of PNA oligomers is written from amine terminus to carboxy terminus. The conventional peptide symbols H- to designate amine terminus and -NH$_2$ for carboxy-terminal amide may be included but are not common. The amine terminus of PNA is analogous to the 5'-hydroxyl of DNA. Although PNA has been found to hybridize to DNA and RNA in both the parallel and antiparallel modes, optimal binding is antiparallel to the oligonucleotide target [4].

Many analogues of PNA, including structures with a more "peptide-like" character, have been prepared, but to date only compounds that have the original monomeric unit in PNA retain superior hybridization properties. Minor side-chain substitutions such as charge and hydrophilicity are accepted and can be used to modulate the properties of PNA [13]. Most interestingly, "true" *peptide* nucleic acids lack the ability to hybridize specifically with oligonucleotides; i.e., the peptide backbone conformation cannot adapt to the structure required for Watson–Crick base pairing but may form homoduplexes [14,15].

II. CHEMICAL PROPERTIES OF PEPTIDE NUCLEIC ACIDS

The behavior and properties of PNA oligomers should be considered prior to a discussion of the oligomer synthesis. PNA is constructed with a poly-amide backbone and oligomers are prepared by traditional solid-phase peptide synthesis (SPPS) techniques. The terminal amino group of the PNA

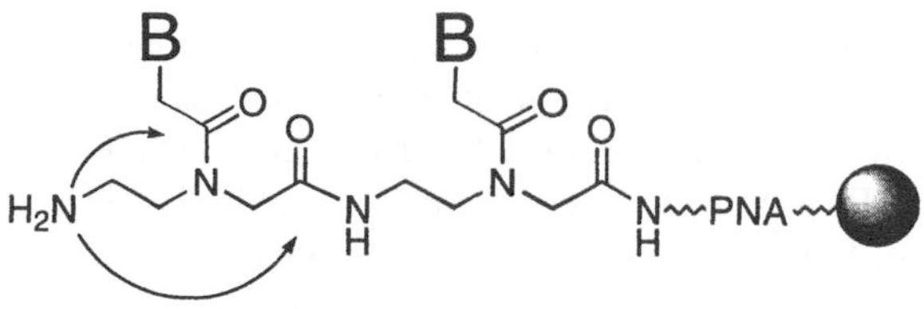

Figure 2 Potential transacylation pathways.

monomeric unit is positioned at the end of an ethylene unit as opposed to alpha to a carboxyl group. As a result, the amine is more basic and less hindered compared with α-amino acids, and coupling is significantly more efficient. In addition, PNA monomers are achiral and racemization of the α-carbon is not a concern during coupling. Factors deleterious to coupling efficiency are that protected PNA monomers are much larger than similarly protected amino acids (molecular weight 384–742) and the terminal amino group can form a strong internal hydrogen bond with either of the two adjacent amides in the growing PNA oligomer. Although there is no equivalent to complex α-helix formation in peptide chemistry, the synthesis is more difficult through homopurine stretches (adenine and guanine) and through repeats of the same base (three or more). This is presumably due to stacking of the protected nucleobases.

Probably the most precarious aspect of PNA synthesis is a facile rearrangement of the N-terminal unit in the PNA oligomer. The N-terminal amino group under basic conditions will attack either the carboxymethyl unit that links the nucleobase to the backbone or the interunit amide linkage and release a ketopiperazine derivative (Fig. 2). The cleavage is so efficient that this side reaction can be used advantageously to sequence PNA oligomers [16]. The rearrangement also has important implications for oligomer synthesis as well as handling and storage of PNA oligomers (see later).

Finally, nucleobases are excellent acceptors for the electrophilic agents generated in deprotection and cleavage steps following the oligomer assembly. Simple cation scavengers such as m-cresol and/or thioanisole should be added to quench any undesired side reaction during this final step.

III. OLIGOMER SYNTHESIS

A. t-Boc Chemistry

SPPS based on the tert-butoxycarbonyl (t-Boc) protection strategy applied to PNA synthesis provides a method for optimal yield and quality of oligomer. The first PNA syntheses of homothymine oligomers were carried out with the *t*-Boc/benzyloxycarbonyl (Z) protection scheme and utilized "active" pentafluorophenyl (Pfp) esters [2]. The synthetic method relies on the varied acid lability of the two protecting groups. The *t*-Boc group is removed with trifluoroacetic acid (TFA), and the Z group requires hydrofluoric acid (HF) or trifluoromethanesulfonic acid (TMSA) to effect cleavage. No loss of the Z groups on the nucleobases is observed during the *t*-Boc removal with TFA. In situ activation of the corresponding carboxylic acids of the monomers was employed when the synthesis was expanded to include all four nucleobases (Fig. 3) [17–19].

PNA assembly with *t*-Boc synthesis chemistry may be employed by either manual or automated methods (Fig. 4). Manual synthesis based on the *t*-Boc chemistry yields PNA of very high quality but is labor intensive and requires a chemist trained in *t*-Boc SPPS. Manual synthesis has the advantage of being easily monitored at each monomer inclusion and also allows great flexibility of scale.

Automation of *t*-Boc SPPS is available on a few commercial peptide synthesizers but is not widely applied because this method uses harsher chemicals than other available technologies. PNA is prepared at a scale more common to oligonucleotide than peptide synthesis; however, the TFA used in the synthesis cycles has prevented effective conversion of commercial DNA synthesizers to PNA assembly by the *t*-Boc method. The implemen-

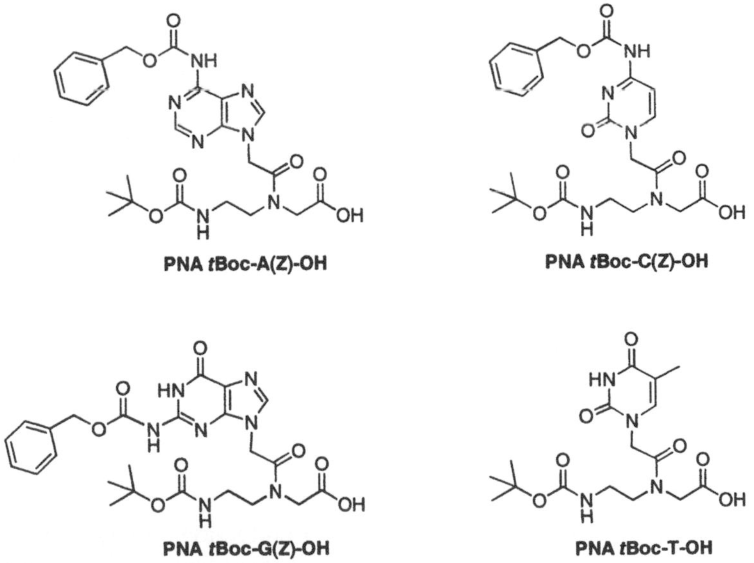

PNA *t*Boc-A(Z)-OH

PNA *t*Boc-C(Z)-OH

PNA *t*Boc-G(Z)-OH

PNA *t*Boc-T-OH

Figure 3   *t*-Boc/Z-protected PNA monomers.

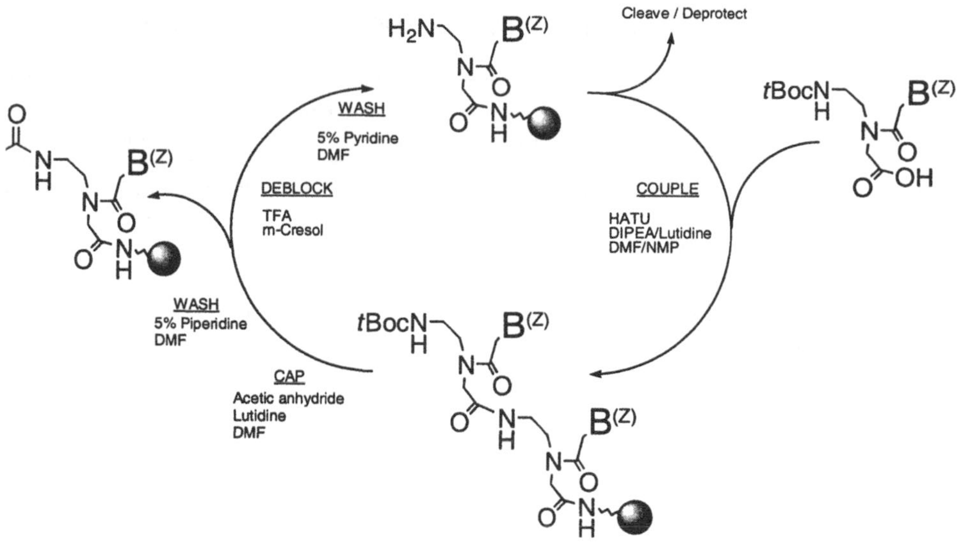

Figure 4 *t*-Boc/Z synthesis cycle.

tation of this chemistry on the ABI 433A allows for the automated synthesis of PNA [20]. The instrument employs a batch mode with a 3-mL vortexing reaction vessel and this system has available protocols for preparing PNA on 5-, 10-, and 20-μmol scales (Table 1). The automated synthesis cycle is essentially identical to that used in manual synthesis with some minor variances detailed in the following text. Protocols listed at the end of this section detail the synthesis procedures used for manual synthesis as well as the cleavage–deprotection method used for both styles of preparation.

1. Solid Support

A cross-linked polystyrene bead functionalized with 4-methylbenzhydrylamine (MBHA) is commonly used for PNA prepared by *t*-Boc/Z chemistry. This resin provides *C*-terminal amides upon final cleavage. A loading of 0.15 mmol/g or less is optimal for PNA synthesis; at higher loading significant failures are observed beyond eight and nine units that have been attributed to interstrand aggregation during synthesis. The *C*-terminal monomer unit may be precoupled to a commercially available MBHA resin to obtain the desired loading. This process is referred to as downloading (Protocol 1).

Table 1 Automated Preparation of PNA via *t*–BOC/Z Synthesis Chemistry on a 433A, 5—μmol Scale

Step	Reagent (bottle position)	Module	Time (min)	Comment
Deprotection	TFA–*m*–cresol 19:1 (2)	B	8.6	Removal of *t*-Boc amine protection, washes.
Wash	DCM (9)	G	1	Wash resin with DCM.
Wash	NMP (10)	D	1	Wash resin with NMP.
Read cartridge, Preactivation	PNA monomer 0.2 M DIEA 0.4 M (7) HATU 0.19 M (8)	A	3.2	Activate monomer for coupling. **Note**: Preactivation is performed in the monomer cartridge.
Coupling	Preactivated monomer	F	35	Transfer and coupling of preactivated monomer to extending PNA oligomer.
Capping	Acetic anhydride– lutidine–DMF 5:6:89	C	1.5	Acetylation of uncoupled PNA oligomer
Piperidine wash	NMP–piperidine 9:1	d	1.5	Removal of undesired acylation.
Wash	NMP	d	1.5	Rinse resin with NMP.
Total cycle time			54 min	

Protocol 1. Downloading of MBHA Resin for PNA Preparation via *t*-Boc/Z Chemistry.   The following procedure is for loading 3 g of MBHA resin with a PNA monomer to a final range of 0.10–0.15 mmol/g and may be scaled appropriately.

Prepare the following solutions:

0.45 mmol PNA monomer as a 0.2 M *N*-methyl-2-pyrrolidone (NMP) solution

0.45 mmol HATU as a 0.2 M NMP solution

0.90 mmol DIEA as a 0.4 M solution in NMP

Dimethyl formamide (DMF)–acetic anhydride–lutidine 89:5:6 (cap solution)

Dichloromethane (DCM)–*N*,*N*-diisopropylethylamine (DIEA) (19:1)

MBHA resin, 3.0 g (>0.2 mmol/g), is added to a filtered assembly suitable for manual peptide synthesis. DCM is added and the resin is allowed to swell for 1 h. DCM should completely cover the resin. DCM is removed by vacuum filtration and the (DCM/DIEA) solution is added to neutralize the resin. After 1 min the base solution is removed by filtration and this treatment is repeated. The resin is washed with DCM and air dried by vacuum filtration for 5 min.

The NMP solutions of monomer, HATU, and DIEA are mixed. The monomer is activated for 2 min and then added to the resin. The monomer is coupled to the resin for 1 h and the NMP solution is removed by filtration. The resin is washed with NMP. The cap solution is added to the loaded resin, allowed to react for 10 min, and removed by filtration. The cap treatment is repeated twice, and then the resin is washed with NMP, DCM–DIEA (19:1), and finally DCM. The capping treatment is repeated if the resin demonstrates a positive Kaiser test. Resin loading may be determined by quantitative Kaiser test after removal of the t-Boc protection on the monomer.

2. Deblocking

Removal of the t-Boc blocking group from the amine terminus of the extending PNA is accomplished with TFA–m-cresol (19:1). The m-cresol effectively scavenges the $tert$-butyl cations formed in the deprotection, preventing electrophilic attack of the nucleobases. A 5-min treatment was determined adequate to deblock each of the monomers fully even for the first resin-bound monomer, which demonstrates slower deprotection rates [19]. TFA treatment provides a protonated amine that is inactive to the acyl migrations mentioned previously. The terminal amine must be neutralized prior to coupling the next monomer. This is accomplished with in situ neutralization during automated synthesis or with a brief wash with DMF–pyridine (19:1) in manual preparation.

3. Activation and Coupling

PNA was originally prepared with a pentafluorophenyl (Pfp) ester of the t-Boc-T monomer [2]. Consequent studies employed in situ activation of the corresponding carboxylic acids (Fig. 3). A variety of coupling agents were investigated and ultimately concluded that N-[(dimethylamino)-1H-1,2,3-triazol[4,5-b]pyridin-1-ylmethylene]-N-methylmethanaminium hexafluorophosphate. N-oxide (HATU) provided the highest overall efficiency [20]. Activation of the monomers with HATU requires a base component to deprotonate the carboxylic acid moiety of the monomer. A mixture of strong and weak tertiary amines has proved effective for uronium salt–based activation of the monomer and subsequent coupling to the solid phase [21]. The monomer and base may be prepared as a mixed solution, which allows a higher concentration following the addition of coupling agent (HATU). For manual synthesis, a solution of 0.2 M monomer, 0.3 M collidine, and 0.2 M N-methyl N-dicyclohexylamine (MDCHA) in N-methylpyrolidinone (NMP) is mixed with 0.19 M HATU to affect activation. Automated protocols on the 433A add both a base and an activator solution sequentially to a predissolved monomer, resulting in a lower overall concentration of activated intermediate. The monomer is used in slight molar excess to the activator and a short activation time (1–2 min) is recommended to avoid tetramethylguanidinium capping of the terminal amine by the activator [20,22]. Manual and automated syntheses both use ~5 equiv. of activated monomer in the coupling protocol. A Kaiser test may be used to evaluate the coupling efficiency in manual synthesis. Repeating the coupling followed by capping is recommended in the case of a positive ninhydrin test.

4. Capping

The acetylation of uncoupled oligomer, also referred to as capping, is performed during each cycle of the stepwise synthesis. Capping simplifies the purification of the full-length PNA and is highly recommended although not absolutely necessary. A number of effective acetic anhydride–containing solutions have been applied in the literature. The DMF–acetic anhydride–lutidine (89:5:6) mixture is currently preferred and has been found to have a longer shelf life (months to years) than acetic anhydride and bases such as DIEA or pyridine.

Acetic anhydride may also react with the protected exocyclic amino groups of the nucleobases. The product of this reaction is a potent acetylating agent itself and is stable to the TFA deprotection procedure. The result is an observed "self-capping" of the PNA in the neutralization step after TFA treatment (unpublished results). The inclusion of a DMF–piperidine (19:1) wash following the capping protocol nullifies the acetyl-nucleobase intermediate.

Protocol 2. Preparing an Automated t-Boc PNA Synthesis on the 433A. The 433A software contains protocols and instructions for 5-, 10-, and 20-μmol PNA synthesis with t-Boc chemistry. These syntheses are performed in a 3-mL reaction vessel.

The bottle positions for PNA synthesis on the 433A are as follows:

1. NMP–piperidine (9:1)
2. TFA–*m*-cresol (19:1)
4. DMF–acetic anhydride–lutidine (89:5:6) (cap solution)
7. 0.4 M DIEA in DMF
8. 0.19 M HATU in DMF
9. DCM
10. NMP

A flow test for each bottle position should be performed before each run. Download the appropriate synthesis protocol to the instrument. The sequence and run information are prepared and downloaded to the instrument. Solutions of PNA monomers are prepared in NMP at 0.2 M. For a 5-μmol synthesis, 125 μL of the monomer solutions is added to the corresponding cartridge and set on the instrument according to the sequence. The synthesis is run as detailed in Table 1.

Protocol 3. Manual Synthesis of PNA.   Prepare the following solutions:

TFA–*m*-cresol (19:1)
DMF–pyridine (19:1)
0.2 M monomer, 0.3 M collidine, and 0.2 M MDCHA in NMP
0.19 M HATU in NMP
DMF–acetic anhydride–lutidine (89:5:6)
DMF–piperidine (19:1)

PNA synthesis may be carried out using apparatus suitable for the manual preparation of peptides and minimally consisting of a reaction vessel fitted with a filter. Materials other than glass, polyethylene, or polypropylene may not stand up to the harsh chemical treatments and should be avoided. In each step of the synthesis where reagents are applied, the reaction vessel should be either agitated or mixed with a glass rod to remove air bubbles and ensure full exposure to the chemistry. PNA should be prepared on an MBHA resin downloaded by coupling of the first monomer as described in Protocol 1. DCM should be added to the support and the resin allowed to swell for 1 h prior to initiating synthesis.

Deblocking: The synthesis support is washed with DCM. TFA–*m*-cresol (19:1) is added and should fully cover the resin. After 5 min the TFA solution is removed by filtration and the deblock treatment is repeated. After removal of the TFA, the resin is washed consecutively with DCM, DMF, and DCM, followed by DMF–pyridine (19:1) to neutralize the resin before coupling.

Preactivation and coupling: Monomer solution (0.5 equiv., 0.2 M monomer, 0.3 M collidine, and 0.2 M MDCHA in NMP) is combined with an equal volume of 0.19 M HATU in NMP to activate the monomer. After a 2-min preactivation, the combined solution is added to the synthesis support and allowed to couple for 15 min. After removal of the monomer-containing solution by filtration, the resin is washed consecutively with DMF, DCM, and DMF. Coupling should be repeated if the resin fails to provide a negative Kaiser test.

Capping: The capping solution [DMF–acetic anhydride–lutidine (89: 5:6)] is added to the resin. After a 1-min treatment the capping solution is removed and the resin is washed with DMF. A 1-min treatment with DMF–piperidine (19:1) is performed to remove undesired acyl moieties. After washing with DMF and DCM, the cycle is repeated.

At the conclusion of oligomerization, the PNA is cleaved from the support and deprotected as detailed in Protocol 4.

Protocol 4. Cleavage of PNA from the Solid Support and Deprotection of the Nucleobases (*t*-Boc chemistry). *5–μmol-scale synthesis*: Transfer the PNA-containing resin from the column housing to an Eppendorf tube containing a filtration insert (Millipore Ultrafree-MC PTFE 0.2 μm or equivalent). TFA–*m*-cresol–thioanisole–TFMSA (400 μL, 6:1:1:2) is added to the resin and the assembly is briefly vortexed. At the conclusion of cleavage (2 h) the Eppendorf assembly is centrifuged (maximum 2000 rpm) for 5 min. The cleavage solution is transferred to a 15-mL centrifuge tube. TFA (200 μL) is added and the assembly is again spun down to rinse the support; repeat for a total of two resin washes. The filtration insert is discarded and the TFA washes are added to the cleave mix in the centrifuge tube. Diethyl ether (5 mL) is added to the cleave mix and the tube is vortexed. The precipitated PNA is concentrated by centrifugation (maximum 2000 rpm) for 5 min and the ether solution is carefully decanted. The crude PNA is washed with diethyl ether (2 × 1 mL), vortexed to suspend PNA, and concentrated by centrifugation. The ether wash is removed by decanting. The PNA pellet is placed on a heating block at 37°C for 15 min to dry. Crude PNA is reconstituted for analysis by the addition of 0.1% TFA in H$_2$O (500 μL).

B. Fmoc/Bhoc Chemistry

Because PNAs contain four principal monomer units and are typically prepared on the micromole scale, preparation on oligonucleotide synthesis plat-
forms was preferred. These instruments are advantageous over peptide synthesizers with respect to more effective handling of small volumes and fixed
reservoirs for a small number of building blocks. Although the t-Boc chemistry provides PNAs of unmatched quality, it has not been amenable for
adoption to such systems. A synthesis cycle incorporating milder chemical
treatments has therefore been developed for synthesis on this platform.

Based on developments in solid-phase peptide synthesis, an alternative
protection method would incorporate the 9-fluorenylmethylcarbonyl (Fmoc)
group. Removal of Fmoc protection may be accomplished with mild chemistry, which is compatible with oligonucleotide synthesizers. Initial reports
of PNA synthesis based on this method used the Fmoc protection group for
the primary amines while retaining the Z group for protection of the exocyclic amino groups of the nucleobases [23]. Harsh conditions, however, are
still required for the removal of the Z group. Furthermore, the resulting
monomers are not sufficiently soluble for preparation of stock solutions that
would remain on the instrument for extended periods. Application of milder
chemistry in the final cleavage–deprotection stage of the synthesis has the
additional advantage of facile preparation of a wider range of labeled compounds and conjugates. This issue was addressed by seeking alternative protection for the nucleobases and compatible synthesis handle on the solid
support.

A protection group for the exocyclic amino groups has to provide
sufficient protection during synthesis, be removed readily under cleavage
conditions, and render sufficient solubility to the resulting monomers. The
benzhydryloxycarbonyl (Bhoc) group [24], fulfills all of these criteria and
is readily introduced by employing essentially the same chemical procedures
as used for the preparation of the Z-protected nucleobase intermediates (unpublished results). Fmoc- and Bhoc-protected PNA monomers (Fig. 5) can
be maintained at 0.2 M in NMP for weeks. The Bhoc group may be removed
at the conclusion of the oligomerization with neat TFA in less than a minute.
The Fmoc/Bhoc synthesis chemistry has allowed the development of rapid
synthesis cycles with short cleavage and deprotection times. The overall
preparation time is therefore significantly shortened, thus giving researchers
better access to PNA.

We based our synthesis protocol development on a PE Biosystems
(formerly PerSeptive Biosystems) dual-column Expedite 8909 nucleic acid
synthesis system that has the benefit of employing calibrated reagent delivery and a microfluidic plate. These features allow very accurate volumes

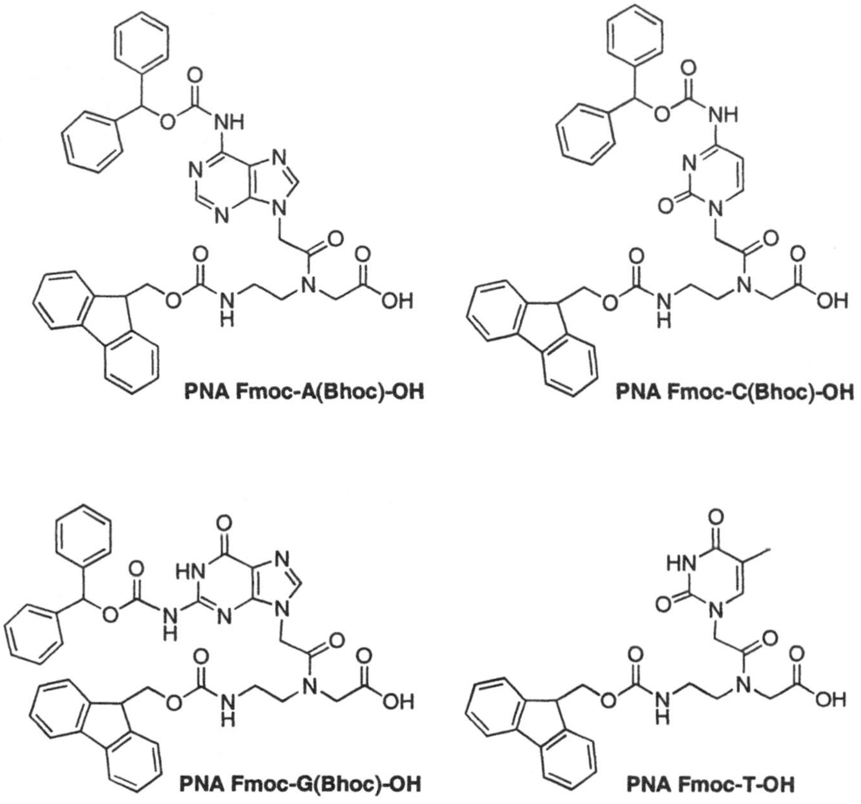

Figure 5 Fmoc/Bhoc-protected PNA monomers.

and mixing of small amounts of reagents (16-μL pulses). Preparation of PNA on the Expedite 8909 is generally performed at a 2-μmol scale (Table 2), although a 25-μmol protocol is also available. PNA has also been prepared on an Expedite 8909 outfitted with a multiple oligomer synthesis system (MOSS), which allows the preparation of 16 2-μmol oligomers with a single instrument setup. As in t-Boc/Z PNA synthesis, Fmoc/Bhoc-based PNA assembly (Fig. 6) involves repetitive cycles of deprotection, activation–coupling, and capping. The individual steps of this sequence have been optimized and are discussed in the following.

1. Solid Support

A number of supports have been investigated with regard to yield and purity of PNAs prepared by Fmoc/Bhoc synthesis chemistry [25]. Optimal results

Table 2 Fmoc PNA Synthesis Protocol

Train	Step	Reagent	Time (min)	Comment
A	Deprotection	DMF–piperidine (4:1)	2.0	Fmoc removal
A	Wash	DMF–gas–DMF–gas	1.0	
B	Preactivation	Monomer, HATU, base mix	2.5	Activation of monomer
B	Coupling	Monomer, HATU, base mix	5.0	Extension of oligomer
B	Wash	DMF–gas–DMF–gas	0.75	
A	Capping	Acetic anhydride–lutidine– DMF (5:6:89)	3.0	Acetylation of uncoupled oligomer
A	Wash	DMF–gas–DMF–gas	1.5	
	Total cycle time in dual-column mode:		16.5 min	2× coupling stage

were obtained with polyethylene glycol–polystyrene (PEG-PS) resins utilizing either the XAL or PAL synthesis handles (Fig. 7). These supports provide PNA containing a carboxy-terminal amide, similar to the MBHA resin used in the *t*-Boc synthesis. The PEG groups provide a "solution-like" environment for the applied chemistry. Furthermore, PEG-PS resins maintain a more consistent volume during synthesis and are compatible with flow-through instruments, such as the Expedite 8909, in which the capacity of the column assembly is fixed (see Chapter 1).

2. Deblocking

The removal of the Fmoc group from the amine terminus during chain elongation was the largest concern in the development of this synthesis method. The potential for side reactions involving the amine terminus and the acyl moieties of the nucleobase tether or the interunit linkage has been described in the introduction (Fig. 2). The removal of the Fmoc protection is performed in the highly basic solution of DMF–piperidine (4:1), which may induce transacylation. In *t*-Boc chemistry the terminal amine is inactive to rearrangement during deprotection due to protonation.

The rate of Fmoc removal was initially studied by measuring the absorbance at 301 nm in the effluent from a 2-μmol synthesis. Each monomer was found to be fully deprotected within 30 s as determined by return to baseline absorbance. The extent of the acyl transfer rearrangement was ascertained by preparing a series of 5-mers, XACGT. The sequence design was based on the assumption that the rate of rearrangement would be different for the four monomers X=A, C, G, and T. Each oligomer was exposed to DMF–piperidine (4:1) for times ranging from 1 to 24 h and the fraction

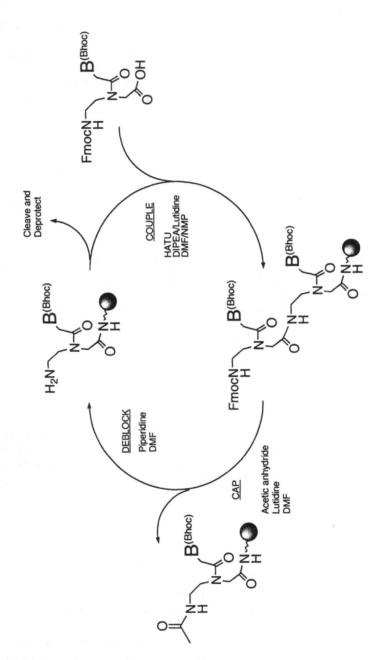

Figure 6   Fmoc/Bhoc synthesis cycle.

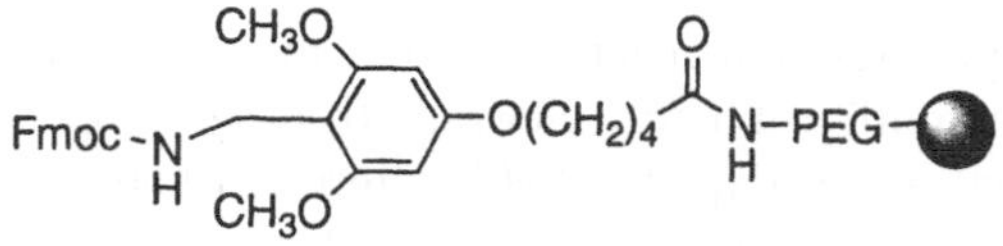

Fmoc-PAL-PEG-PS

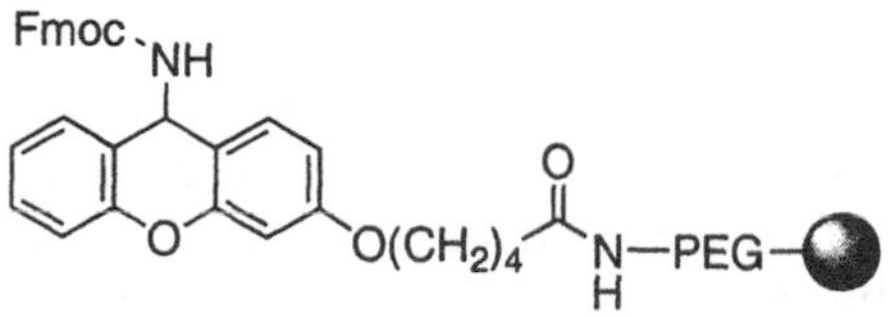

Fmoc-XAL-PEG-PS

Figure 7   Fmoc-PAL and Fmoc/XAL-PEG-PS.

of rearranged product, as determined by reversed-phase high-performance liquid chromatography (RP-HPLC), was plotted versus time. The results for the transacylation at the amine terminus indicated the order A > G > T, with no observed rearrangement for the C monomer. Interpolation of these data implied that in the original deblocking program of 10 min, transfer of the nucleobase to the primary amine of the A monomer could be as high as 3.4% but was reduced to <0.65% for a 1-min treatment. The corresponding values for G and T were 1.2% and 1.1%, respectively, for 10 min and <0.2% rearrangement over 1 min for both residues. Interpolation of the data does not consider the time required for the actual removal of the Fmoc group, which would be expected to reduce these values further. Therefore, the side reaction is not a significant issue with the reduced deprotection times used in automated synthesis.

Fmoc removal from PNA monomers is more rapid than from amino acids, presumably because of reduced hindrance in the vicinity of the amine. The PNA synthesis protocol on the Expedite 8909 (Table 2) employs a deprotection time of 2 min to ensure complete N-Fmoc removal of PNA monomers and allow efficient incorporation of amino acids into the sequence. Deblocking the synthesis handles Fmoc-PAL and Fmoc-XAL is also slow compared with PNA monomers. Longer initial deprotection of universal supports is incorporated into the synthesis protocols of the Expedite 8909 to avoid C-terminal deletion products.

A consequence of the acyl migration reaction is that the N-terminal protection should not be removed until just before cleavage from the support. Selection of the Fmoc-on option during Expedite 8909 setup will leave the final protection intact. Removal of this Fmoc group should be performed just before final cleavage–deprotection and is accomplished with a designated prime cycle (per instrument manual). Manual preparation is *not* recommended with Fmoc PNA chemistry as the reaction (deprotection) times are less easily controlled.

As a practical concern, because of the facile Fmoc removal from PNA monomers, caution must be exercised when choosing the solvating liquid. High-grade NMP is recommended for monomer solvation because DMF generally contains more trace amines and has demonstrated shorter solution lifetimes in our hands. Typically, PNA monomers are prepared as 0.2 M solutions in NMP, which are stable for at least 2 weeks.

3. Coupling

Optimization of the coupling step for Fmoc-based methods was based on the experience gained in the development of the *t*-Boc protocols. HATU is maintained as the activator of choice and this selection was not revisited. Activation requires a base component to proceed (see Chapter 6). The base solution is composed of 0.2 M DIEA and 0.3 M lutidine. This combination of amines proved more effective than a 0.5 M solution of either base. A 2.5-min preactivation step is included in the instrument protocol and a slight excess of monomers over activator (0.2 M to 0.19 M) is used in the coupling step. This prevents undesired tetramethylguanidine (TMG) capping, derived from the reaction of HATU directly with the primary amine of the growing PNA chain. The TMG-capped product is easily identified by MALDI-TOF mass spectroscopy; traces are readily observed. DNA synthesizers do not typically utilize activation chambers, common on peptide synthesis instruments. To accomplish preactivation, three components must be combined and adequately mixed in the instrument lines before delivery to the reaction chamber containing the solid support. The solenoid-driven reagent system of the 8909 allows this multiple simultaneous controlled delivery.

The Fmoc synthesis protocol delivers ~ 8 equiv. of activated monomer for a 2-μmol preparation. An activation–coupling time of 7.5 min (8.25 min including washing steps) is used to ensure the robustness of the synthesis. The Expedite 8909 is designed to perform two simultaneous syntheses. The instrument carries out activation–coupling on one reaction vessel (column A) and the remainder of the synthesis cycle (capping, deblocking) is performed on the second (column B). The activation–coupling stage determines the overall cycle time in the dual-column mode, as the activation–coupling procedure is more time consuming than the concurrent capping and deblocking steps.

4. Capping

As previously described in the section on *t*-Boc chemistry, capping is recommended in each synthesis cycle to ease the purification of full-length PNA product. A solution consisting of acetic anhydride–lutidine–DMF (5:6:89) effectively acetylates any amines that failed to extend in the coupling cycle. The piperidine wash step, which was useful in the *t*-Boc protocols for elimination of products derived from acylation of the exocyclic amino groups of the nucleobases, is accomplished during every consequent Fmoc deprotection.

5. Washing

A wash–gas–wash–gas procedure was placed at the end of each stage of the synthesis cycle. This operation more effectively cleaned the PEG-PS support of residual reagents and resulted in improved synthesis performance. DMF is used as the wash solvent for each stage. The PNA protocol also incorporates an optional wash at the end of a synthesis. This wash uses either dichloromethane (DCM) or methanol to rinse the resin so that it may be rapidly dried and more easily removed from the column housing. A DCM resin wash is recommended for long-term storage of PNA on a solid support. Other than this optional wash, no chlorinated solvents are used in Fmoc PNA synthesis.

Protocol 5. Performing a Synthesis. *Expedite 8909 equipped with PNA upgrade software:* Monomer solutions are prepared by adding PNA diluent (NMP, 3.25 mL) to the prepackaged dry monomers (0.7 mmol) and gently shaking for 10 min to provide 0.2 M solutions. The Fmoc-AEEA-OH spacer ("O" monomer) is similarly prepared at 0.2 M in NMP with the addition of 2.4 mL of PNA diluent to the prepackaged spacer. Activator solution is prepared by adding DMF (13.5 mL) to the prepackaged HATU to a final concentration of 0.19 M. The prepared solutions and additional reagents are loaded onto the Expedite 8909 as described in the user's guide. After entering the desired sequence and priming the instrument chemistry, a prepackaged 2-μmol column housing containing Fmoc-PAL-PEG-PS or Fmoc-XAL-PEG-PS is fitted onto the instrument and the synthesis is begun.

The synthesis should be performed Fmoc-on. The amine terminal Fmoc group is removed with a special prime cycle just before final cleavage–deprotection.

---

6. Cleavage and Final Deprotection

Following automated assembly, an "off instrument" workup consisting of final deprotection and cleavage from the synthesis support is required. PNA is both deprotected and freed from the solid support by treatment with TFA–*m*-cresol (19:1) (Protocol 6). The use of a scavenger is required because the electron-rich aromatic rings of the nucleobases may become alkylated by cations generated by the TFA treatment. The removal of the Bhoc protection is complete within 1 min, releasing the exocyclic amines of the nucleobases and producing benzhydryl cations that are scavenged by the *m*-cresol.

The TFA treatment also provides release of the PNA from the solid support. The PAL synthesis handle requires treatment with TFA–*m*-cresol (19:1) for 90 min to complete cleavage. The XAL linker was designed for cleavage of protected peptide fragments and may be cleaved with prolonged treatment with low TFA concentrations or with neat TFA in minutes [26]. Release of the PNA from XAL-PEG-PS with TFA–*m*-cresol (19:1) is complete within 5 min. Although both the PAL- and XAL-linked resins provide PNA of similar high quality, the XAL handle is preferred in our laboratory as the total workup time is considerably shortened. Unlike the solid supports common for DNA synthesis, the linkers are universal and do not require attachment of the first monomer unit to be performed off line. The PNA product is isolated from the cleavage mixture simply by precipitation with diethyl ether.

Protocol 6. Cleavage of PNA from the Solid Support and Deprotection of the Nucleobases (Fmoc Chemistry). *2-μmol-scale synthesis*: The PNA-containing resin is transferred from the column housing to an Eppendorf tube containing a filtration insert (Millipore Ultrafree-MC PTFE 0.2 μm or equivalent). TFA–*m*-cresol (200 μL, 19:1) is added to the resin and the assembly is briefly vortexed. At the conclusion of cleavage (5 and 90 min for XAL- and PAL-linked resins, respectively), the Eppendorf assembly is centrifuged (maximum 2000 rpm) for 5 min. An additional 200 μL of the cleavage cocktail is added and the assembly is again spun down to rinse the support. The filtration insert is discarded and diethyl ether (1 mL) is added to the cleave mix and vortexed. The precipitated PNA is concentrated by centrifugation (maximum 2000 rpm) for 5 min and the ether solution is carefully decanted. The crude PNA is washed with diethyl ether (2 × 1 mL), vortexed to suspend PNA, concentrated by centrifugation, and the ether wash decanted. Place on a heating block at 37°C for 10 min to dry the PNA pellet. Crude PNA is reconstituted for analysis by the addition of 0.1% TFA in H_2O (200 μL).

C. MMT Chemistry

Many of today's molecular biology techniques are based on the manipulation of enzymatic processes. Although PNA has enhanced hybridization properties, it is inert to enzyme activity. Therefore, it has been reasoned that PNA–DNA chimeras might be useful constructs to introduce biological functionality [27]. Preparation of such complexes is complicated by the incompatibility of oligonucleotides with the acid treatments of the previously described chemistries. Synthesis efforts focused on creating a "DNA-like" chemistry. Such a method would incorporate protection of the primary backbone amine with a group that is readily removed with dilute acid and ammonia-labile protection for the exocyclic amines of the nucleobases. Successful preparation has utilized the acid-labile MMT group for protection of the backbone amine and the standard DNA nucleobase protection, i.e., benzoyl on A and C and iso-butyryl on G [28]. Substituted benzoyl nucleobase protection has also been reported [29]. Employing these monomers, PNA–DNA and DNA–PNA chimeras may be constructed provided that appropriate linkers are used at the transition between the PNA and DNA oligomer. For an overview of this chemistry and the properties of the chimera see the review by Uhlmann et al. [12].

IV. ANALYSIS AND PURIFICATION OF PNA

A. Optical Density

PNA products released from the solid support are analyzed for purity by reverse-phase HPLC and MALDI-TOF mass spectroscopy (Fig. 8). Reconstitution of the PNA precipitate for analysis is best accomplished in a solution of 0.1% TFA in H_2O as described earlier. The quantity of PNA produced is measured in absorbance units as a total optical density (OD) at 260 nm. The OD values provide a rapid early evaluation of the synthesis performance and should range between 80 and 180 for a 2-μmol preparation. This measurement varies with the length and purine content of the sequence. The OD is directly related to the concentration of the solution by the Beer–Lambert law $A = \varepsilon c L$ where A is the absorbance at 260 nm (OD), c is the concentration, L is the path length, and ε is the molar extinction coefficient.

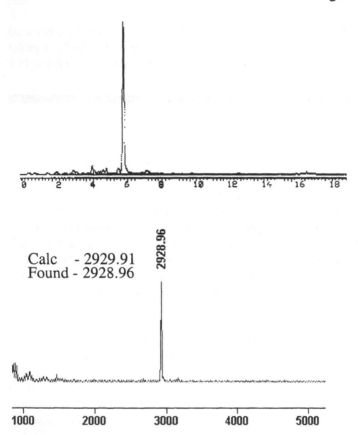

Figure 8 HPLC and MALDI-TOF MS of a biotinylated PNA 10-mer prepared by Fmoc/Bhoc synthesis chemistry. Biotin labeling was performed prior to cleavage from the solid phase.

This relation constant may be calculated from the PNA sequence by employing the following extinction coefficients.

Protocol 7. Measuring the Optical Density (OD) of PNA.   Crude PNA is constituted as described earlier in 0.1% TFA in H_2O. A 1:2500 dilution of the crude sample is prepared in two steps.

1.   Dilute the crude sample 1:10: 4 μL of crude PNA in 36 μL of 18 M H_2O.
2.   Dilute the 1:10 solution by 1:250: 4 μL of 1:10 solution in 996 μL of 18 M H_2O.

The OD_{Total} is measured in a UV spectrometer at 260 nm in a prezeroed 1 mL cuvette with a 1-cm path length.

$$OD_{Total} = Abs_{260} * D * V, \quad \text{where } D = \text{dilution (2500) and}$$

$$V = \text{sample volume (mL)}$$

Example: A 2-μmol Fmoc synthesis is dissolved in 0.1% TFA in H_2O (200 μL). After performing the dilutions an absorbance reading of 0.224 is recorded at 260 nm.

$$OD_{Total} = Abs_{260} * D * V = 0.224 \text{ AU/mL} * 2500 * 0.2 \text{ mL} = 112 \text{ AU}$$

B. Analytical HPLC

Reverse-phase HPLC is performed with UV detection at 260 nm using a method common in peptide analysis. Both a Waters Delta-Pak 5 (C_{18}, 300 Å, 3.9 × 190 mm) and a YMC AQ 3 (C_{18}, 120 Å, 4 × 50 mm) are routinely used in our laboratories. The column is warmed to 55°C to avoid aggregation of the PNA. Solution A is 0.05% TFA in H_2O and solution B is 0.04% TFA in 1:1 acetonitrile–methanol. An elution gradient of 19:1 to 13:7 over 35 min at a flow rate of 1.2 mL/min is used. For best results, 0.05 to 0.2 ODs of PNA are injected for analysis.

C. HPLC Purification

Purification of PNAs is best accomplished by preparative reverse-phase HPLC. This is accomplished analogously to the analytical method. Conditions are 55°C on a YMC AQ 3 (C_{18}, 120 Å, 10 × 30 nm). An elution gradient of 19:1 to 13:7 over 35 min at a flow rate of 7.5 mL/min is used. Analytical HPLC and MS are used to analyze collected fractions. Purified samples are quantified by OD measurement and freeze dried for long-term storage.

D. MALDI-TOF Mass Spectroscopy

Mass spectroscopy is performed on a PerSeptive Biosystems Voyager DE MALDI-TOF system using a sinapinic acid matrix and insulin as an internal standard [30].

Protocol 8. MALDI-TOF (Matrix Assisted Laser Desorption Ionization-Time-of-Flight) Mass Spectroscopy of PNA. Prepare the following solutions:

1. Matrix solution: 10 mg/mL sinapinic acid in 0.1% TFA in H_2O–ACN (2:1)
2. Insulin solution: 1.0 mg/mL in 0.1% TFA in H_2O.

Each week prepare a fresh solution of 1:19 insulin in matrix solution: 25 μL insulin solution is added to 475 μL matrix solution.

Crude PNA: Dilutions of 1:10, 1:100, and 1:1000 of crude PNA in matrix–insulin solution are prepared by performing a series dilution of 1 μL crude PNA into three consecutive portions of 9 μL matrix–insulin solution. The MS sample plate is spotted with 1 μL of each the 1:100 and 1:1000 dilutions, with the latter typically providing the best results.

Purified PNA: 1 μL of the purified PNA fraction collected from the HPLC and 1 μL of the matrix–insulin solution are spotted on the MS sample plate.

The MS sample plate is dried under high vacuum for 5 min prior to analysis on the MALDI-TOF instrument. Internal calibration may be performed using insulin 5734.59 daltons and sinapinic acid 225.22 daltons.

V. LABELING OF PNA

Fluorescent labels and other reporter groups are easily attached to PNA. Fluorescence, specific binding, or antibody interaction of reporter groups is routinely used in all PNA and DNA hybridization-based assays, and thus a majority of the prepared PNAs carry a label or affinity tag. A large number of labeling agents are specifically designed to react with free amines and hence are easily attached to the *N*-terminus of PNA. Thiol-specific labels may also be attached to PNAs by incorporating a cysteine in the sequence. Labeling of PNA may be performed either on resin or in solution after cleavage–deprotection. The decision depends on the stability of the label compound to the PNA cleavage conditions. The three most common labels —fluorescein (FAM), rhodamine (TAMRA), and biotin—are stable to both *t*-Boc and Fmoc cleavage conditions and may be coupled to the PNA while still on resin. Additional acid-sensitive labels, which may be attached to PNA after cleavage, include digoxigenin, cyanine dyes, BODIPY, Oregon green, and alkaline phosphatase. Some of these labels are resistant to the TFA cleavage of the Fmoc methodology but not stable to TFMSA or HF used for *t*-Boc chemistry.

A. AEEA-OH Spacer

Similarly to the attachment of reporter groups to DNA, the labeling of PNA is best accomplished with the addition of a linker to the amine terminus. In the case of DNA, a linker is often necessary to introduce the functionality, such as an amine, required for the labeling reaction. PNA may be directly labeled on the amine terminus, but a spacer moiety is often preferred to separate the reporter group from the sequence. this is particularly necessary with biotin labels, as the large biotin–streptavidin complex will block a portion of the sequence information if not adequately spaced. Quenching of fluorescence by the nucleobases can also be minimized by the introduction of a spacer moiety. It is recommended that two spacers be incorporated between the PNA sequence and label or reporter group.

An 8-amino-3,6-dioxaoctanoic acid moiety, designated as the O monomer or AEEA-OH, is the commonly used spacer and is available suitably protected for use with both Fmoc and *t*-Boc chemistries (Fig. 9). The O spacer often improves the product yields particularly for solution labeling. This is partially due to the hydrophilic character of the O spacer, which increase the aqueous solubility of PNA oligomers. The placement of an O spacer at the amine terminus effectively prevents acyl transfer rearrangement, a side reaction observed with the standard monomer in basic solutions. This addition of an *N*-terminal O spacer to the sequence also allows Fmoc-on purification of the PNA oligomer because no transacylation reaction occurs during the Fmoc removal (see earlier).

B. Labeling of PNA on Resin

Labeling with biotin, fluorescein, rhodamine, and succinimidyl-4-(*N*-mal-eimidomethyl)-cyclohexane-1-carboxylate (SMCC) as well as other less common reporter groups is accomplished while the PNA is still immobilized on the resin. These labels are stable to the cleavage and final deprotection chemistry employed by both *t*-Boc and Fmoc synthesis methods (Fig. 8). Removal of the *N*-terminal protection must be performed before labeling if the synthesis was done with the final protecting group remaining on the solid support (often the default for Fmoc chemistry). The automated synthesizers described earlier contain subroutines for this *N*-terminal deprotection. Labeling solutions may be prepared according to Table 3.

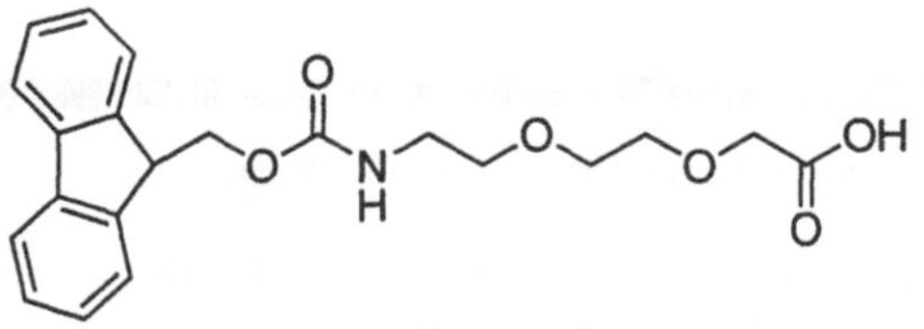

Figure 9 Structure of the O spacer.

Table 3 On-Resin Labeling (2-μmol Synthesis)

Label	Labeling reagent	Weight (mg)	DMF (μL)	DIEA (μL)	Time (min)
Biotin[a]	Biotin–Pfp	10	300	6.6	30
Fluorescein	5(6)–FAM–NHS	10.7	300	11	60
Rhodamine	5(6)–TAMARA–NHS	12.5	300	11	60
SMCC	SMCC–NHS	10.0	300	11	60

[a]The use of low-quality or old TFA in the PNA cleavage cocktails has been observed to result in oxidation of biotin.

Protocol 9. On-Resin Labeling (2-μmol Synthesis).   With a 1-mL disposable syringe draw up the labeling solution, attach the synthesis column to the syringe tip, and inject enough reagent to saturate the resin. Attach a second 1-mL syringe to the other side of the column and carefully push the reagent back and forth (approximately 10 times). Allow the labeling to proceed for the time indicated in Table 3. Intermittently push the labeling solution gently back and fourth to enhance labeling efficiency. When the reaction time has passed, discard the labeling solution. Rinse the resin by pushing DMF (3 × 1 mL) followed by DCM (or methanol, 2 × 1 mL) through the column housing with a 1-mL syringe. Dry the resin under vacuum and remove from the column housing before proceeding to cleavage–deprotection.

C. Labeling of PNA in Solution

For reporter groups sensitive to the reagents used in cleavage–deprotection of the PNA, labeling of the PNA in solution postsynthetically is performed (Fig. 10). The PNA should be purified (typically by reverse-phase HPLC) before attempting solution labeling. This is customarily accomplished as follows:

Protocol 10. Solution Labeling (General Procedure).   Prepare:

Solution A: 1 mg of labeling agent dissolved in 110 μL DMF.
Solution B: Dilute 10–15 OD PNA to 100 μL with ultrapure water.
Solution C: 1 mL water–1-methylmorpholine (NMM) (9:1).

(a)

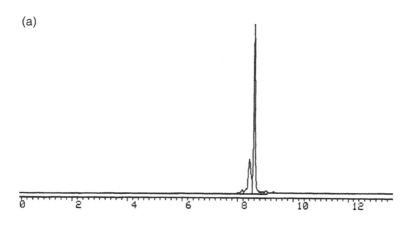

(b)

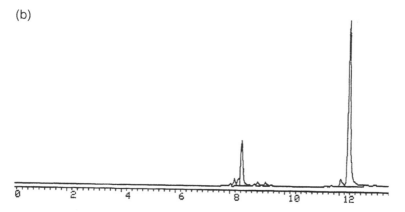

Figure 10 HPLC analysis of solution labeling PNA with digoxigenin. (a) PNA alone; (b) PNA combined with label as in Protocol 10.

Control 1: Unlabeled PNA: 2 μL solution B in 100 μL 0.1% TFA in H$_2$O for HPLC analysis.

Control 2: Labeling reagent: 2 μL solution A in 100 μL 0.1% TFA in H$_2$O for HPLC analysis (partial hydrolysis of NHS esters is often observed).

Labeling reaction: Combine 100 μL of solution A and the remaining 98 μL of solution B. Add small aliquots of solution C (up to 25 μL) and ensure that the PNA is maintained in solution and the pH is 7.0 or above.

Vortex the solution thoroughly. Allow the reaction to proceed for 1 h (in the dark if the reagent is light sensitive).

The labeling reaction should be followed by reverse-phase HPLC and compared with the control solutions. Dual-wavelength monitoring is often helpful with fluorescent labels. Upon confirming that the labeling reaction was successful, purify the labeled PNA immediately. Otherwise, freeze the solution, lyophilize (in the dark if the label is light sensitive), and store the dried material at $-20°C$ until purification can be performed.

VI. MODIFIED PNAs

A. Nucleobase Alterations

PNA may be systematically modified to incorporate varied functionality. The two distinct portions of PNA, nucleobase and backbone, can both accommodate alterations. Nucleobase modifications affect a PNA's ability to bind DNA in a sequence-specific manner. A pseudoisocytosine nucleobase has been successfully incorporated in PNAs designed to bind DNA as a triple helix [31]. This nucleobase mimics a protonated cytosine and effectively binds DNA in the Hoogsteen mode.

B. Alternative Backbones

In the PNA backbone three glycine units are readily recognized. To explore the limits of the PNA structure, amino acids other than glycine have been incorporated in the monomer unit at each of these positions [32–37]. Such modifications allow fine-tuning of the properties of PNA including charge incorporation and hydrophobicity. These modified oligomers have been analyzed for DNA binding affinity and solubility. Generally, only substitutions with D-amino acids at the carboxy position of aminoethyl glycine leads to PNAs that have retained or slightly increased affinity for complementary oligonucleotides. It has been speculated that such modifications of the backbone could increase cellular uptake, but this remains to be proved.

There has also been an extensive search for backbone structures other than aminoethyl glycine that may support Watson–Crick base pairing to complementary DNA and RNA. Although many structures have been pursued, little success has been observed. Structures that are based on peptides (i.e., true *peptide* nucleic acids) have been reported. These derivatives are virtually unable to hybridize nucleic acids but some do display the ability to form homoduplexes [14,15,38–42].

It should be added that the emergence of PNA has resulted in a resurgence of research on the introduction of amide (or carbamate) linkages between the ribose units of nucleic acids first described by Halford and Jones in 1968 [43]. One such structure called HNA, hexose nucleic acid, demonstrates the ability to hybridize to nucleic acids [44]. These types of oligonucleotide mimics have been reviewed [45].

C. PNA-Peptide Conjugates

Amino acids and even peptides can be incorporated directly during PNA synthesis using either t-Boc or Fmoc synthesis methods. The most common amino acid additions are lysine and cysteine. Lysine can provide additional aqueous solubility to PNA and carboxy-terminal lysine with orthogonal protection may also be used for doubly labeled PNA preparations. Cysteine allows the attachment of thiol-specific labels and further derivatized thiol- or maleimide-linked reagents.

Peptide conjugates introduce functionality beyond sequence hybridization to PNA. These conjugate biomolecules, for example, may provide a metal binding site [46] as well as the seven-amino-acid Kemptide sequence that can be radiolabeled with ^{32}P using protein kinase A [47]. Alternatively, ^{32}P can be introduced by conjugation of a deoxynucleoside 3'-phosphate to the terminal amino followed by phosphorylation of the free 5'-hydroxyl group with T4 polynucleotide kinase and ATP [48]. Other examples include His_6 tags that facilitate strong binding to nickel-NTA columns [49]. Preparation of these constructs is typically by direct synthesis. In general, the peptide is prepared initially followed by PNA elongation on the solid support.

Carrier peptides represent another interesting class of molecules to which PNA has been conjugated. These peptides have been found to traverse the cell membrane and are capable of actively transporting attached constructs [50]. PNA conjugated to these biomolecules by a disulfide linkage has been shown to be internalized in the cytoplasm and dissociated from the carrier peptides, presumably by reduction of the linkage. These internalized PNAs have demonstrated antisense inhibition of messenger RNA in the model systems studied.

VII. SUMMARY

The preparation of PNA oligomers represent challenges derived from the chemistry of both peptides and nucleic acids. In this chapter we described in detail the two solid-phase synthesis methods incorporating the t-Boc/Z and Fmoc/Bhoc protection schemes. The monomers for these chemistries are available commercially and are in widespread use.

The *t*-Boc/Z chemistry provides oligomers of the highest quality and allows the preparation of PNA by manual synthesis. The shortcomings of this method, however, include the use of extremely harsh chemicals and long cycle times. The Fmoc/Bhoc protection method provides PNA with a shortened cycle time and has been adapted to a DNA synthesis platform. Custom PNA suppliers have adopted this method for more rapid oligomer production.

PNA is readily labeled for use in a variety of molecular biology applications. The versatile chemistry of PNA has also allowed the preparation of PNA-peptide conjugates and PNA–DNA chimeras. The PNA base unit itself has incorporated backbone and nucleobase modifications to affect solubility and binding characteristics. The application of solid-phase synthesis methods to the preparation of PNA provides researchers with an impressive new molecular biology tool.

REFERENCES

1. P Nielsen, M Egholm, R Berg, O Buchardt. Science 254:1497–1500, 1991.
2. M Egholm, O Buchardt, P Nielsen, R Berg. J Am Chem Soc 114:1895–1897, 1992.
3. J Watson, F Crick. Nature 171:737–738, 1953.
4. M Egholm, O Buchardt, L Christensen, C Behrens, S Freier, D Driver, R Berg, S Kim, B Nordén, P Nielsen. Nature 365:566–568, 1993.
5. O Buchardt, M Egholm, R Berg, P Nielsen. TIBTECH 11:384–386, 1993.
6. B Hyrup, P Nielsen. Bioorg Med Chem 4:5–23, 1996.
7. D Corey. TIBTECH 15:224–229, 1997.
8. T Koch, M Borre, M Naesby, H Batz, H Ørum. Nucleosides. Nucleotides 16: 1171–1174, 1997.
9. H Knudsen, P Nielsen. Anticancer Drug 8:113–118, 1997.
10. P Nielsen, H Ørum. In: Molecular Biology: Current Innovations and Future Trends. Part 2. Horizon Scientific Press, 1995, pp 73–89.
11. P Nielsen. Antisense Therapeutics. Vol 4. Leiden: SECOM Science Publishers, 1996, pp 76–84.
12. E Uhlmann, A Peyman, G Breipohl, D Will. Angew Chem Int Ed Engl 37: 2797–2823, 1998.
13. A Puschl, S Sforza, G Haaima, O Dahl, P Nielsen. Tetrahedron Lett 39:4707–4710, 1998.
14. U Diederichsen, H Schmitt. Tetrahedron Lett 37:475–478, 1996.
15. U Diederichsen. Bioorg Med Chem Lett 8:165–168, 1998.

16. L Christensen, R Fitzpatrick, B Gildea, B Warren, J Coull. Solid Phase Synthesis. Peptides, Proteins, and Nucleic Acid. Birmingham: Mayflower Worldwide, 1994, pp 149–156.

17. M Egholm, P Nielsen, O Buchardt, R Berg. J Am Chem Soc 114:9677–9678, 1992.

18. K Dueholm, M Egholm, C Behrens, L Christensen, H Hansen, T Vulpius, K Petersen, R Berg, P Nielsen, O Buchardt. J Org Chem 59:5767–5773, 1994.

19. L Christensen, R Fitzpatrick, B Gildea, K Petersen, H Hansen, T Koch, M Egholm, O Buchardt, P Nielsen, J Coull, R Berg. J Peptide Sci 3:175–183, 1995.

20. T Koch, H Hansen, P Andersen, T Larsen, H Batz, K Otteson, H Oerum. J Peptide Res 49:80–88, 1997.

21. L Carpino, A El-Faham. J Org Chem 59:695–698, 1994.

22. M Schnolzer, P Alewood, A Jones, D Alewood, S Kent. Int J Peptide Protein Res 40:180–193, 1992.

23. S Thomson, J Josey, R Cadilla, M Gaul, C Hassman, M Luzzio, A Pipe, K Reed, KD Ricca, R Wiethe, S Noble. Tetrahedron 51:6179–6194, 1995.

24. R Hiskey, J Adams. J Am Chem Soc 87:3969–3973, 1965.

25. R Casale, C Paul, I Jensen, M Moyer, S Kates, M Egholm. Proceedings of the 5th Solid Phase Synthesis and Combinatorial Libraries Symposia, in press.

26. Y Han, S Bontems, P Hegyes, M Munson, C Minor, S Kates, F Albericio, G Barany. J Org Chem 61:6326–6339,1996.

27. E Uhlmann. Biol Chem 379:1045–1052, 1998.

28. F Bergmann, W Bannwarth, S Tam. Tetrahedron Lett 36:6823–6826, 1995.

29. D Will, G Breipohl, D Langner, J Knolle, E Uhlmann. Tetrahedron 51:12069–12082, 1995.

30. J Butler, P Jiang-Baucom, M Huang, P Belgrader, J Girard. Anal Chem 68:3283–3287, 1996.

31. M Egholm, L Christensen, K Dueholm, O Buchardt, J Coull, P Nielsen. Nucleic Acids Res 23:217–222, 1995.

32. B Hyrup, M Egholm, M Rolland, R Berg, P Nielsen, O Buchardt. J Chem Soc Chem Commun 518–519, 1993.

33. B Hyrup, M Egholm, P Nielsen, P Wittung, B Nordén, O Buchardt. J Am Chem Soc 116:7964–7970, 1994.

34. B Hyrup, M Egholm, O Buchardt, P Nielsen. Bioorg Med Chem Lett 6:1083–1088, 1996.

35. K Dueholm, P Nielsen. New J Chem 21:19–31, 1997.

36. G Haaima, A Lohse, O Buchardt, P Nielsen. Angew Chem Int Ed Engl 35:1939–1942, 1996.

37. E Lesnik, F Hassman, J Barbeau, K Teng, K Weiler, R Griffey, S Freier. Nucleosides Nucleotides 16:1775–1779, 1997.

38. U Diederichsen. Bioorg Med Chem Lett 7:1743–1746, 1997.

39. U Diederichsen. Angew Chem Int Ed Engl 36:1886–1889, 1997.

40. U Diederichsen, H Schmitt. Angew Chem Int Ed Engl 37:302–305, 1998.

41. U Diederichsen, H Schmitt. Eur J Org Chem 827–835, 1998.

42. U Diederichsen. Angew Chem Int Ed Engl 37:2273–2276, 1998.

43. M Halford, A Jones. Nature 217:638–640, 1968.
44. R Goodnow, S Tam, D Pruess, W McComas. Tetrahedron Lett 38:3199–3202, 1997.
45. K Weisz. Angew Chem Int Ed Engl 36:2592–2594, 1997.
46. M Footer, M Egholm, S Kron, J Coull, P Matsudaira. Biochemistry 35:10673–10679, 1996.
47. T Koch, M Naesby, P Wittung, M Jørgensen, C Larsson, O Buchardt, C Stanley, B Nordén, P Nielsen, H Ørum. Tetrahedron Lett 36:6933–6936, 1995.
48. I Kozlov, P Nielsen, L Orgel. Bioconj Chem 9:415–417, 1998.
49. H Ørum, P Nielsen, M Jørgensen, C Larsson, C Stanley, T Koch. Biotechniques 19:472–480, 1995.
50. M Pooga, U Soomets, M Hallbrink, A Valkna, K Saar, K Rezaei, U Kahl, J Hao, X Xu, Z Wiesenfeld-Hallin, T Hokfelt, T Bartfai, U Langel. Nat Biotechnol 857–861, 1998.

14
Solid-Phase Synthesis of Oligosaccharides

Jacques Y. Roberge
Bristol-Myers Squibb, Princeton, New Jersey

I. INTRODUCTION

The tremendous biological importance of oligosaccharides and glycoconjugates [1,2] has been recognized for many years. Besides acting as a source of stored energy and a structural element of life,* the oligosaccharides and glycoconjugates play a very significant role in cell–cell and cell–pathogen interactions [5–19]. In order to understand their exact roles fully, it is necessary to obtain molecular tools in the form of well-defined fragments of those biopolymers [20,21]. It is possible to isolate them from natural sources, but it is a tedious and difficult process that often yields very small amounts of material [22]. The current understanding of the biochemistry of the other two major classes of biopolymers, proteins and nucleotides, was largely achieved by the ability to synthesize oligosaccharides efficiently and rapidly.

One of the most significant accomplishments toward an understanding of molecular biology was the development of automated peptide synthesis on solid supports [23] initially by Merrifield [24]; this was later applied to oligonucleotide assembly, and both are now performed on commercially available synthesizers. Although the initial attempts to synthesize oligosaccharides on solid polymeric supports were made in the late 1960s, the synthetic challenge of developing general regio- and stereospecific methods for the assembly of complex glycosidic linkages has prevented the availability

---

*For reviews of the enzymes involved in oligosaccharide biosynthesis, see Refs. 3 and 4.

of an automated oligosaccharide synthesizer. The discovery of new, efficient solution-phase oligosaccharide synthetic strategies [25,26] has provoked a flurry of activity in the field of solid-phase preparation of oligosaccharides. The solid-phase synthesis of oligosaccharides was first reviewed in 1980 by Fréchet [27,28], one of the pioneers in the field. Sixteen years later, Danishefsky and Roberge [29] reviewed the field, and many new reviews have appeared covering numerous aspects of the solid-phase synthesis of oligosaccharides [30–34]. With some of the exciting new methods available, the goal of assembling oligosaccharides in an automated fashion is approaching [35,36].

The potential of solid-phase synthesis in combinatorial chemistry has been the focus of many reviews and will not be discussed here [37]. Polysaccharide-based drugs are still not favored by most pharmaceutical companies. They are metabolically very labile and also suffer from poor oral absorption in intact form. In addition, binding of polysaccharides to their targets often requires multivalent interactions to achieve significant effects [38]. Perhaps the greatest value of these molecules is as tools for better understanding the biology of glycoconjugates. Glycomimetics [39] or compound libraries using sugars as scaffolds [31–34] may prove to be more successful for generating new therapeutic agents. These areas will not be discussed in detail because they have been the subjects of recent reviews [39]. The objective is to provide the reader with a broad view of the field but in enough detail to permit the synthesis of novel oligosaccharide constructs. The experimental procedures for key transformations are presented when available.

The difficulties associated with assembling oligosaccharides on solid supports arise from a monomeric unit that has more than two connecting sites, resulting in much higher levels of ramification than with amino acids or nucleotides. Adding to this complexity, the stereochemical outcome of bond formation at the anomeric center must be defined to avoid obtaining intractable mixtures. Protecting groups must be used to synthesize oligosaccharides, but they must be very carefully selected as they often have a profound influence on the reactivity of the donor–acceptor pair and modulate the stereochemistry of the products. Finally, monitoring reactions on solid supports presents a difficult challenge [40–42].

Two general strategies are used to assemble oligosaccharides on polymeric supports: the glycoside acceptor can be attached to the support via the reducing end (the anomeric carbon) or the glycoside donor may be attached to the support via the nonreducing end (any of the other carbons of the carbohydrate) (Figs. 1 and 2). The former approach is the most com-

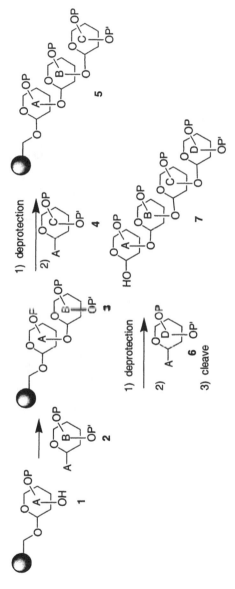

Figure 1 Acceptor attached to the resin (reducing end).

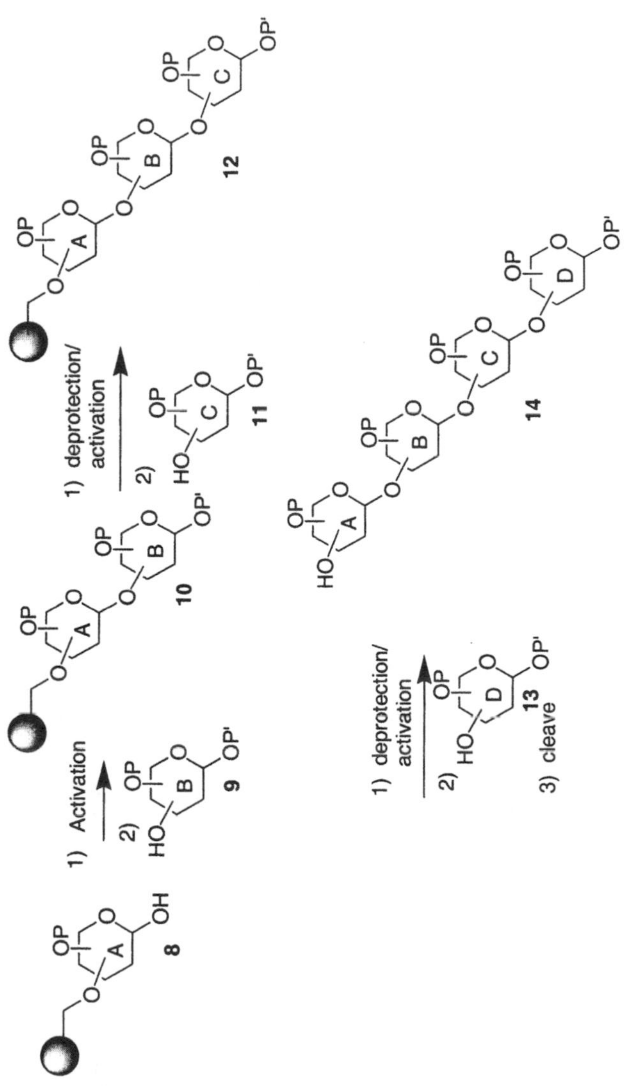

Figure 2 Donor attached to the resin (nonreducing end).

monly used presently (Fig. 1). Polymer-bound acceptor **1** is treated with the donor **2** to give disaccharide **3**. The protecting group of compound **3** is removed and reiteration of the sequence leads to trisaccharide **5**, tetrasaccharide **7**, and so on.

Because glycoside donors are usually reactive and unstable, this approach is advantageous, allowing the use of an excess of the donor to form glycosidic bonds with the polymer-supported acceptor. To avoid formation of side products when using this strategy, it is often necessary to repeat the coupling step to drive the reaction toward completion or to derivatize the unreacted alcohol groups to avoid the formation of products having sequence deletions (e.g., A-C-D instead of A-B-C-D). The deletion products are generally difficult to separate because of their similarity to the desired products. A disadvantage of using an excess of the unstable donor is that it may be difficult to recover, adding to the expense of the synthesis. Another major challenge of this approach is the synthesis of a variety of donors with orthogonal protecting groups that can be removed selectively and efficiently.

Glycosidic bond formation using a supported donor is not often used, although it has the advantage that an excess of the acceptor can be used and the unreacted acceptor can usually be recovered without any decomposition (Fig. 2). Oligosaccharides are assembled by activating the polymer-bound sugar **8** and coupling it to acceptor **9** to give the product **10**, from which the anomeric protecting group must be selectively removed in order to undergo another coupling step with **11** to **12**. Reiteration of the sequence with acceptor **13** will give **14**. Most glycopeptides and glycolipids are attached via the anomeric position of the oligosaccharide. The method shown in Fig. 2 is preferred for the preparation of glycopeptide and glycolipid libraries, coupling a given glycoside to an array of peptide and lipid building blocks. A drawback of this method is that an orthogonal protecting group strategy is necessary, which requires labor-intensive synthesis of the various building blocks. The anomeric hydroxyl protecting groups must be removable in high yield and the activation procedures very efficient, otherwise a capping must be added to prevent the formation of deletion sequences. Alternatively, orthogonal anomeric activation groups can be installed, obviating the need for anomeric protecting groups [43]. The most significant problem with this approach is that it is not always possible to repeat a coupling step to drive the reaction to completion as the reactive donor may decompose on the solid support, leading to chain termination. The glycal assembly method, which is described in the next section, circumvents many of these difficulties and provides an elegant solution to the problem of using polymer-bound donors.

II. SOLID-PHASE SYNTHESIS OF OLIGOSACCHARIDES

A. Nonreducing End Attached to the Resin

1. Glycal Assembly Method

The glycal assembly method for the solid-phase synthesis of oligosaccharides was developed by Danishefsky and coworkers working initially at Yale University and later at the Memorial Sloan-Kettering Cancer Center and Columbia University. The work has been reviewed [50–54]. For early glycosylation work using glycals see [44–49]. The glycal assembly method is one of the very few successful solid-phase procedures for the synthesis of oligosaccharides in which the glycoside donor is attached to the resin (Fig. 3). Activation of glycal **15** with an appropriate reagent results in the formation of the glycoside donor **16**, which is then reacted with the free hydroxyl of glycal **17** with or without a catalyst. The resulting (1→6)-disaccharide **18** is activated and coupled to another acceptor **19**, producing a (1→4)-linked trisaccharide, **20**. Reiteration of this process yields oligosaccharides that can be linked to lipids, peptides, proteins, or other molecules to generate a variety of new products.

The use of glycals greatly simplifies the formation of oligosaccharides on the solid support in many ways. Over the years, a great number of efficient methods have been developed to activate glycals and couple them to other glycosides. Glycal building blocks with orthogonal protecting groups are usually less difficult to synthesize than other glycosides because the former have two fewer hydroxyls [55–58]. There is enough difference in the reactivity of the glycal hydroxyls that selective protection of each group

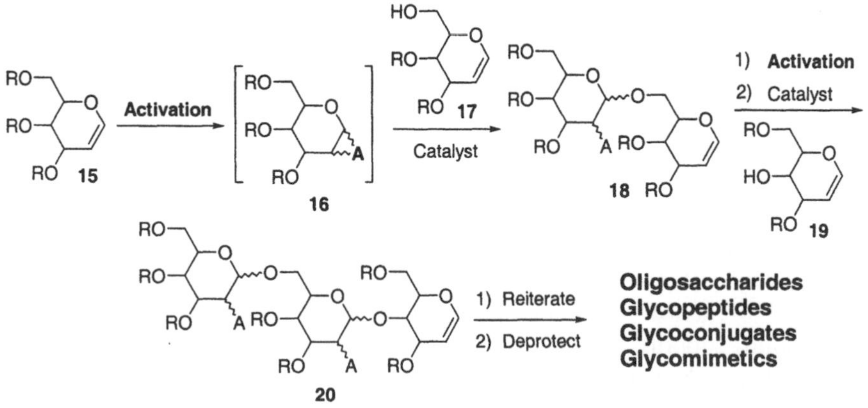

Figure 3 Glycal assembly method.

is easily achieved. An increasing number of selectively protected glycals are now commercially available from Sussex Research Laboratories or Aldrich [59]. A drawback of this method is that α-linked glycosides are more difficult to access from glycals than β-linked sugars using the glycal epoxide opening reaction. β-Mannosides are also difficult to obtain directly from glycals, although they may be formed by oxidizing the 2-hydroxyl resulting from the epoxide opening and then reducing the ketone to give the mannoside hydroxyl [60–62]. The free hydroxyl resulting from the epoxide opening is available, in a differentiated form, for coupling with glycoside donors including glycals. Epoxides from glucals are not as good donors as the galactal carbonate epoxides because they are more unstable and the rate of their decomposition competes with the glycosidic linkage formation. A practical solution to this problem was published [63] and will be discussed later. 2,6-Dideoxy sugar libraries were prepared using glycals [64].

a. The Glycal Assembly Methods: Early Results. The strength of the glycal assembly method for synthesis of oligosaccharides is its simplicity and the ease with which oligosaccharides are built. A detailed discussion of the evolution of the glycal assembly method was published in 1996 [53]. The initial application of glycal assembly of oligosaccharides to solid-phase synthesis [65] incorporated a chlorodiphenylsilyl-derived polystyrene resin, **22** (R = Ph) (Fig. 4), which was first described by Chan and Huang [66]. This resin later exhibited some lability with nucleophiles and was replaced by the chlorodiisopropylpolystyryl resin **22** (R = iPr) [67,68]. Galactal carbonate **23**, an excellent donor for the glycal assembly method, was attached in good yield to the silyl resin to give **24** and was then epoxidized with 3,3-dimethyldioxirane [69]. Opening of the epoxide with excess galactal carbonate **23** and zinc chloride yielded disaccharide **25** in good overall yield after cleavage with buffered tetrabutylammonium fluoride (TBAF). The epoxide opening occurred with good control and only one isomer was observed. The oxidation–epoxide opening procedure was easily iterated to produce, after buffered TBAF cleavage, tri-, tetra-, penta-, and hexasaccharides in very good overall yields (**26–29** respectively, Figs. 4 and 5). Glycoside acceptors with secondary hydroxyls have also been used for the epoxide opening (**24** → **31**). In this example, two secondary hydroxyls were free to react but only the one at C-3 coupled with the glycal to give **31**. Disaccharide **31** has two free hydroxyl groups and was used to prepare the precursor to the Lewis[b] blood group determinant **33** by bis-fucosylation [70].

The final products obtained using an iterative glycal epoxidation–epoxide opening sequence are typically very easily separable from the side products. Simple chromatography using silica gel usually gives pure compounds. It has been proposed that glycal epoxides that failed to react with the glycoside acceptor are hydrolyzed and further oxidized in the subsequent

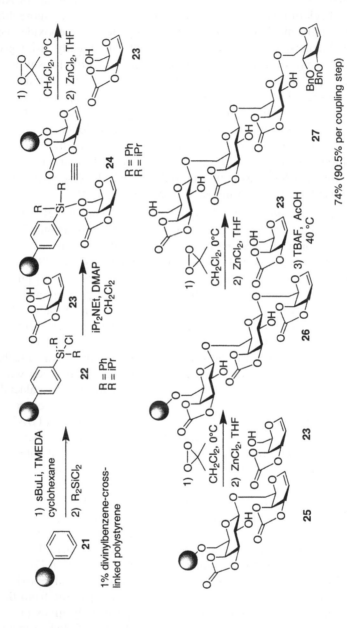

Figure 4 Synthesis of a tetrasaccharide using glycals.

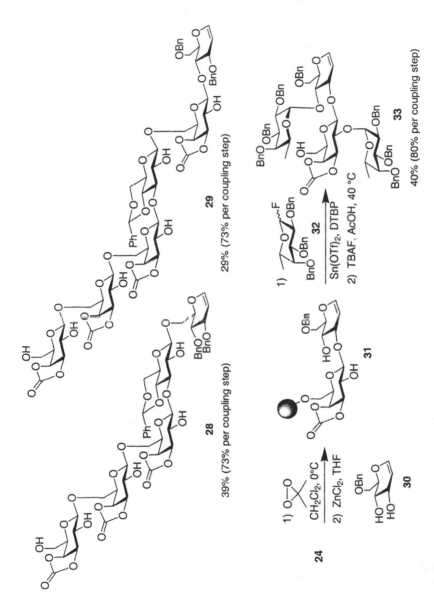

Figure 5 Examples of oligosaccharides built using glycals.

glycal activation step, rendering those products much more polar than the desired product. This desirable effect has been called "the self-policing feature of the glycal assembly method" [53]. The reaction can be monitored by applying magic angle spinning nuclear magnetic resonance (MAS-NMR) techniques to a solvent-swelled sample of resin [42].

Experimental. McClure flask: The McClure flask (Fig. 6) is commonly used for solid-phase syntheses in Danishefsky's group [67]. It has a medium porosity glass frit attached to a side arm. The advantage of using this flask is that the chemistry can be performed under an inert atmosphere with heating or cooling if necessary without any additional special equipment. The stirring is typically done with a slow-spinning magnetic stirrer. This may sometimes cause grinding of the resin, resulting in clogging or loss of product through the filter. Orbital shaking may be used or the design of the stirrer bar may be changed to minimize grinding; i.e., suspend the stirrer from a glass rod hanging from the cap or to have the stirrer bar balanced on a glass tip at the bottom of the flask. We have performed up to 10 chemical steps with the same resin in a McClure flask without substantial clogging of the filter.

Preparation of the resin: Polystyrene-divinylbenzene (1%) copolymer **21** (Fluka) was washed [71] to removed shorter polystyrene components and remaining monomers and reagents with each of the following solutions at 60–80°C for 30–60 min: NaOH (1 N), HCl (1 N), NaOH (2 N)–dioxane (1:2), HCl (2 N)–dioxane (1:20, water, dimethyl formamide (DMF). The resin was then washed at room temperature with HCl (2 N) in methanol, water, methanol, methanol–dichloromethane (1:3), and methanol–dichloromethane (1:10) and the resin was dried at 50–70°C under reduced pressure. The washed polystyrene resin (13 g) was suspended under nitrogen in anhydrous cyclohexane (80 mL) in a 250-mL polymer synthesis flask. Tetramethylethylenediamine (TMEDA, 20 mL, 132.5 mmol) and *n*-butyllithium (2.0 M in cyclohexane, 80 mL, 160 mmol) were successively added and the mixture was stirred at 65°C for 4 h. The dark burgundy–colored resin was filtered under nitrogen and washed with anhydrous cyclohexane (2 × 100

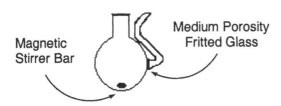

Figure 6 McClure's polymer synthesis flask.

mL), stirring 20 min after each addition of solvent. Anydrous benzene (100 mL) was added to the lithiated polymer followed by diisopropylsilyl chloride (20 mL). The mixture was stirred at room temperature for 3 h and the resin was filtered and washed with dry benzene (3 × 100 mL), stirring 20 min after each addition of solvent. The resin **22** was dried under reduced pressure to give a light brown powder.

Loading of the resin: A solution of glycal **23** (1 g, 5.8 mmol) in di-chloromethane (10 mL) and N,N-diisopropylethylamine (1 mL, 5.7 mmol) was added, under nitrogen, to the resin **22** (R = iPr, 1 g) and 4-dimethylam-inopyridine (DMAP) (10 mg, cat.) and the mixture was stirred at room temperature for 72 h. The mixture was filtered under nitrogen and the resin was rinsed with dichloromethane (2 × 50 mL), methanol–N,N-diisopropyl-ethylamine–dichloromethane (1:1:18, 10 mL, stirred 1 h), dichloromethane (2 × 50 mL) (with 20 min of stirring before draining each solvent), tetrah-ydrofuran (THF) (2 × 50 mL), acetone (2 × 50 mL), dimethyl sulfoxide (DMSO) (2 × 50 mL), acetone (2 × 50 mL), and THF. The resin was dried under reduced pressure to give **24** as a light cream powder.

Cleavage from the resin: Tetrabutylammonium fluoride (1.0 M in THF, 0.2 mL, 0.2 mmol) and acetic acid (1.0 M in THF, 0.1 mL, 0.1 mmol) were added to a suspension of resin **24** (50 mg) in THF (1 mL). The mixture was stirred at 40°C for 18 h, the resin was drained and washed with THF (3 × 5 mL), and the combined solvents were concentrated. The residual oil was purified by chromatography on silica gel (methanol–ether, 1:19) to give **23** as a colorless gum (8.0 mg, 46 μmol) giving a loading of 0.92 mmol/g for the resin **24**.

Preparation of dimethyldioxirane: Dimethyldioxirane was prepared us-ing a modification of the procedure of Murray and Jeyaraman [69]. In a three-neck 5-L flask equipped with a mechanical stirrer, an air condenser attached to a vacuum adapter, and a receiving flask immersed in a dry ice–acetone bath was mixed sodium bicarbonate (120 g, 1.42 mol), water (330 mL, 18.3 mol), and acetone (240 mL, 3.27 mol). Oxone (375 g, 0.61 mol) was added in two portions, 10 min apart, with vigorous stirring. Vacuum (~10 mm Hg) was applied for about 2 h, during which time the light yellow solution of dimethyldioxirane was collected. This solution was dried with stirring over powdered calcium sulfate for 20 min at 0°C. We have observed that the use of calcium sulfate as a drying agent yielded more concentrated solutions of dimethyldioxirane than the magnesium sulfate used by Murray and Jeyaraman. The final light yellow solution (~0.1 M, ~200 mL) was stored over molecular sieves (3 Å, ~20 g) at −20°C for up to a week. Higher concentrations of dimethyldioxirane can be maintained if the solution is stored over molecular sieves that have been reactivated after being used for storage of dimethyldioxirane.

Epoxidation of galactal carbonate: A solution of dimethyldioxirane (~0.1 M in acetone, 50 mL) was added to a cold (0°C) suspension of resin **24** (900 mg) in dichloromethane (20 mL). The mixture was stirred 1.5 h at 0°C, the solvents were drained, and the procedure was repeated until the dimethyldioxirane solution remained faintly yellow. The solvents were drained and the resin was dried under reduced pressure to give the polymer-bound epoxide. The epoxide was suspended in a solution of anhydrous THF (10 mL) and glycal **23** (1.0 g, 5.8 mmol) and the mixture was cooled to 0°C. A solution of anhydrous zinc chloride (1.0 M in THF, 1 mL) was added and the mixture was allowed to warm to room temperature in 8 h. The resin was drained, washed with THF (4 × 20 mL), and dried under reduced pressure to give **25** as a colorless powder.

b. Solid-Phase Synthesis Using Thioethyl Glycosyl Donors. As described previously, galactal carbonate epoxides react readily with nucleophiles to give good yields of products that are essentially stereochemically pure. Epoxide opening of glucals with less reactive or hindered nucleophiles is not as efficient in terms of selectivity and yields. A solution to this problem was achieved using a high-yielding three-step procedure (Fig. 7) [63,72]. The method uses thioethyl glycosides that are potent glycosyl donors when treated with thiophilic reagents [73]. 3,4-Dibenzylglucal **34** was loaded on diisopropylsilyl resin **22** and **35** and oxidized with 3,3-dimethyloxirane as described previously to give the epoxide **36**. The epoxide was opened by treatment with ethanethiol in the presence of a catalytic amount of trifluoroacetic anhydride. The resulting hydroxyl group of the polymer-bound thioglycoside **37** was acylated with pivaloyl chloride to give **38**. The bulky pivaloyl group acted as a neighboring participating group, ensuring good stereochemical control, as well as being unlikely to promote the formation of orthoester side products [72]. Glycosylation was achieved by treatment of the thioethylglycoside **38** with 2,6-dibenzylglucal **39** in the presence of methyl triflate and the nonnucleophilic base di-*tert*-butylpyridine (DTBP) to give disaccharide **40**. The pivaloyl group was removed by reduction with diisobutylaluminum hydride (DIBAL) and the free hydroxyl was coupled to glucosyl donor **41** to give the branched trisaccharide **42**. The final protected sugar **43** was obtained in 59% overall yield from **35** after cleavage with buffered TFA. The development of thioethyl glycosylation on the solid support greatly enhanced the scope of the solid-phase glycal assembly method.

Experimental. *Opening of glycal epoxide with ethylthiol* [63]: A suspension of polymer-bound glycal **35** (1.0 g) in dichloromethane (20 mL) was treated with dimethyldioxirane as described previously to give epoxide **36**. This resin was suspended in a mixture of dichloromethane (10 mL) and ethanethiol (10 mL), cooled to −78°C, and treated with trifluoroacetic an-

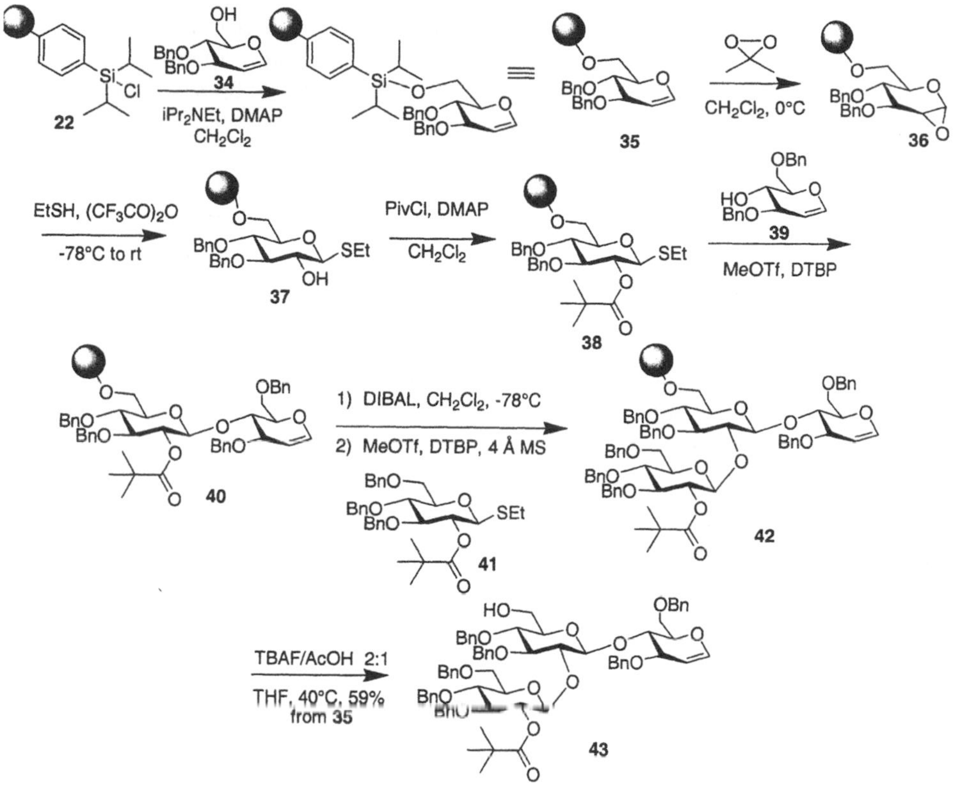

Figure 7 Thioethylglycoside donors from glycals.

hydride. The mixture was warmed to room temperature over 8 h and the polymer was washed with THF (4 × 20 mL) and dried under reduced pressure to give **37** as a colorless powder.

Pivaloylation: Pivaloyl chloride (0.35 mL, 2.8 mmol) was added to a suspension of resin **37** (1.00 g, 0.56 mmol) and DMAP (0.68 g, 5.6 mmol) in dichloromethane (10 mL). After 4 h at room temperature, the mixture was drained, washed with acetone (4 × 20 mL) and THF (3 × 20 mL), and dried under reduced pressure to give **38** as a colorless powder.

Coupling with glycal acceptor: A suspension of resin **38** (200 mg, 0.12 mmol), glucal **39** (196 mg, 0.6 mmol), DTBP (539 μl, 2.4 mmol), and molecular sieves 4 Å (200 mg) in dichloromethane (10 mL) was stirred at room temperature for 10 min. The suspension was cooled to 0°C, followed by dropwise addition of methyl trifluoromethanesulfonate (272 μL, 2.4

mmol). The mixture was stirred for 2 h at 0°C, stirred at room temperature for 8 h, and then triethylamine (1 mL) was added before filtering the resin. The resin was suspended in acetone (3 × 20 mL), stirred for several minutes, and then allowed to settle to remove the molecular sieves. The resin was washed with DMSO (2 × 20 mL), dichloromethane (2 × 20 mL), and THF (2 × 20 mL) and dried under reduced pressure to give **40**.

Pivaloyl reduction: The pivaloyl group was removed by addition of DIBAL (1 M in toluene, 0.7 mL) to a cold (−78°C) suspension of resin **40** (217 mg, 0.14 mmol) in dichloromethane (15 mL). The mixture was stirred at this temperature for 5 h and a saturated solution of sodium potassium tartrate was added. The mixture was warmed to room temperature and the solvents were drained. The resin was washed with water (2 × 20 mL), acetone (2 × 20 mL), and dichloromethane (2 × 20 mL) and dried under reduced pressure.

c.   Convergent Solid-Phase Synthesis of Glycopeptides [74].   One of the major challenges of glycopeptide synthesis [75] is to develop a set of chemical reactions and protecting groups that are compatible for the assembly of both the peptide and oligosaccharide moieties [76]. Two major approaches have been developed to prepare glycopeptides on a solid support.

The most commonly used strategy for the synthesis of glycopeptides on a solid support is derived from normal solid-phase synthesis of peptides (see Chapter 3). The approach is shown schematically in Fig. 8. A modified amino acid having the desired glycoside appendage is prepared in solution from the activated glycoside **44** and the amino acid **45**. The glycosides are usually attached to the hydroxyl group side chain of serine or threonine for *O*-linked glycopeptides or to the amide side chain of asparagine for *N*-linked glycopeptides. The glycosylated amino acid **46** is coupled to a growing peptide attached to the resin **47**. The product **48** can be further elongated and cleaved from the resin to give **49**. This method works well for the synthesis of glycopeptides with multiple glycoslyated amino acids, usually *O*-linked, such as the mucins [8]. One drawback of this approach is that the yields for the coupling of glycosylated amino acids to the growing peptide tend to drop as the number of attached glycosides increases.

The second approach for glycopeptide synthesis initially builds the oligosaccharide component on the solid support (**50**, Fig. 9). Then the peptide **52**, prepared in solution or on the solid support (**51**), is convergently coupled to the polymer-bound oligosaccharide to give **53**. The peptide can be further extended and the whole assembly can be cleaved from the solid support (**53**→**54**→**55**). One advantage of using the glycal assembly method to synthesize oligosaccharides on a solid support is that the reducing end of the sugar is readily available for conjugation to peptides, lipids, or other molecules. This feature allows the rapid addition of nonsugar components

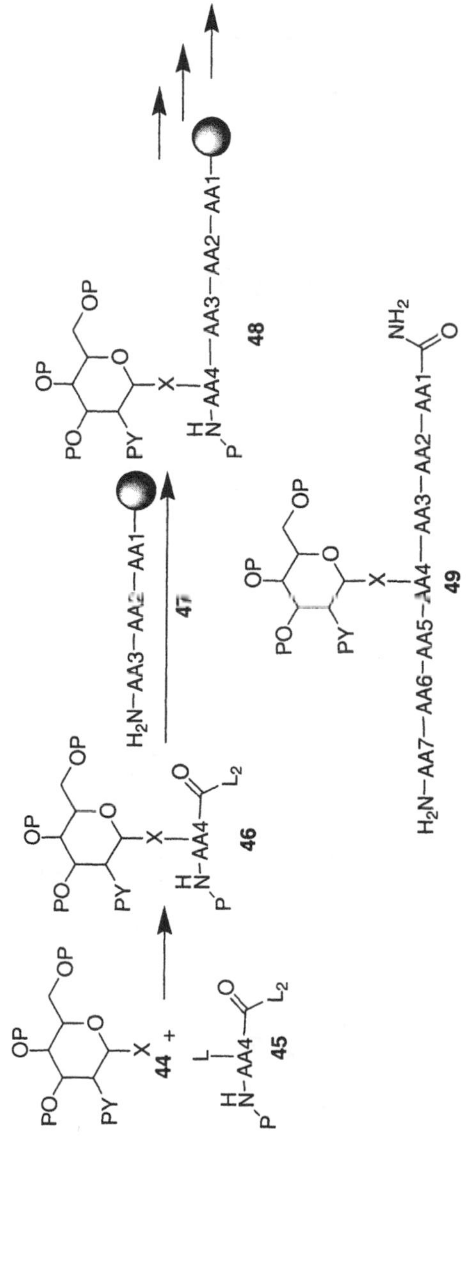

Figure 8 Glycopeptides using glycosylated amino acid building blocks.

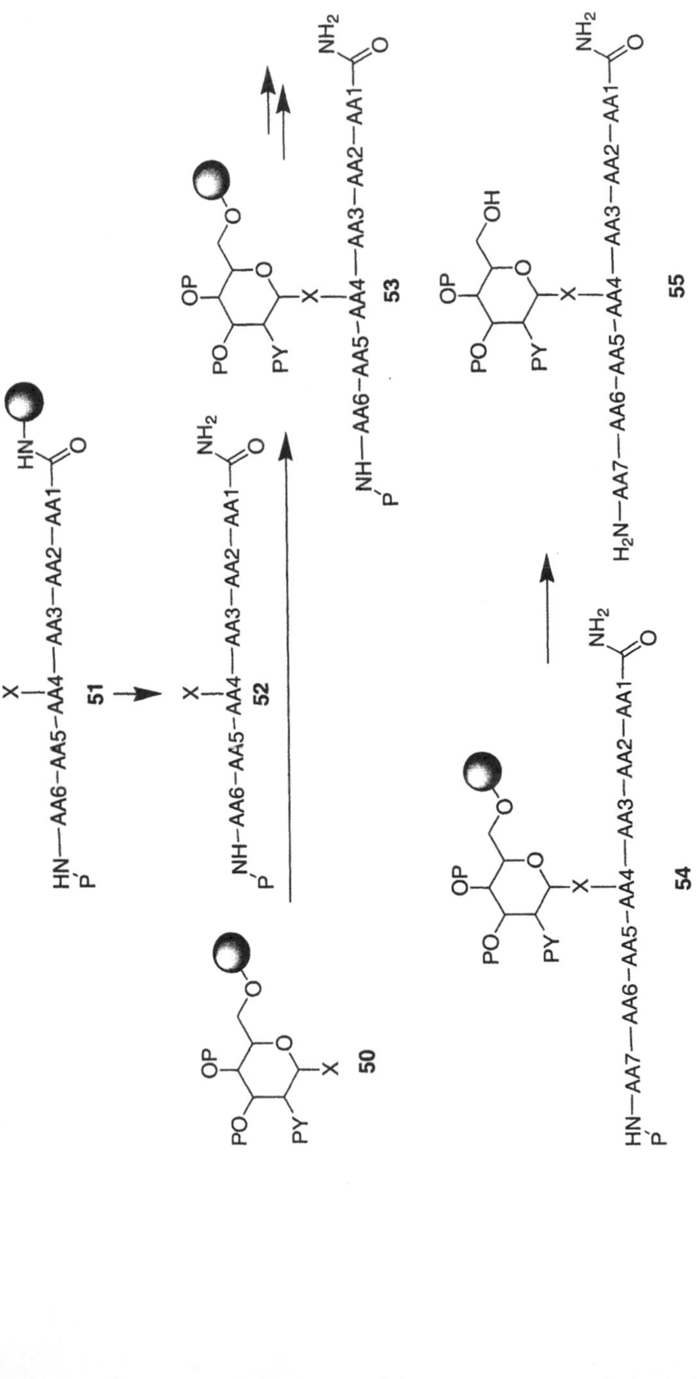

Figure 9 Glycopeptides from polymer-bound glycosides.

to a give oligosaccharide construct. Because *N*-linked glycopeptides do not usually have multiple repeats of glycosylated amino acids in the peptide sequence, these glycopeptides were a good initial target for the application of the glycal assembly method [50,51,77,78]. When the oligosaccharide is attached to the solid support via the reducing end, one can build the desired fragments of complex *N*-linked glycans but the sugars have to be cleaved from the resin before coupling to the peptide can take place [79].

A branched trisaccharide-pentapeptide was prepared on a solid support using the glycal assembly method (Fig. 10) [80]. Coupling of 3,6-dibenzyl-glucal **39** to the epoxide derived from polymer-bound galactal carbonate **22** gave disaccharide **56**. Treatment of the disaccharide with iodonium bis(sym-collidine) perchlorate [81] and 9-anthracenesulfonamide (AnthSO$_2$NH$_2$) [82] gave the desired diaxial product **57**. The use of 9-anthracenesulfonamide was key to the successful preparation of glycopeptides on a solid support. The anthracene-SO$_2$ fragment is easily removable under mild conditions after it has served its function. It is also very soluble in THF and gives clean, high-yielding iodosulfonamidation reactions on the solid phase. Product **57** was converted to the azide **58** with tetrabutylammonium azide, probably via an aziridine intermediate. Fucosylation of the free hydroxyl of **58** with the anomeric fluoride **32** in the presence of tin triflate [83] gave the precursor to H–type II blood group determinant trisaccharide **59**. This compound was acylated and converted in a one-pot reaction to the anomeric 1-amino-2-*N*-acetylglucosamine **60** by treatment with propanedithiol and *N,N*-diisopropylethylamine. The anomeric amine was coupled to the free carboxyl side chain of the aspartic acid residue of peptide **61** in the presence of 2-iso-butoxy-1-isobutoxycarbonyl-1,2-dihydroquinoline (IIDQ) [84] to give glycopeptide **62**. The protecting group scheme for this peptide was based on the pioneering work of Kunz' group [26,85]. Cleavage from the resin **62** with HF-pyridine gave the desired glycopeptides **63** and **64**.

Experimentals. *Iodosulfonamidation and azide displacement*: Freshly prepared iodonium(bis)collidine perchlorate [46] (397 mg, 0.847 mmol) was added to a cold (0°C) suspension of dried 9-anthracenesulfonamide (217.9 mg, 0.847 mmol) and glycal **56** in anhydrous THF (3.0 mL) in a polymer synthesis flask. The mixture was protected from light, stirred at 0°C for 6 h, and allowed to warm to 5°C. The reaction was quenched with a solution of ascorbic acid (5 g, 22 mL THF, 3 mL H$_2$O). The resin was drained; washed with THF–H$_2$O (1:1, 3 × 20 mL), THF (3 × 20 mL), and CH$_2$Cl$_2$ (3 × 20 mL); and dried under vacuum overnight to give 270 mg of resin **57**. This resin was swollen in THF (2.5 mL) and tetrabutylammonium azide (180 mg, 0.762 mg) was added. The suspension was stirred at room temperature for 21 h and the resin was filtered, washed with THF (3 × 20 mL)

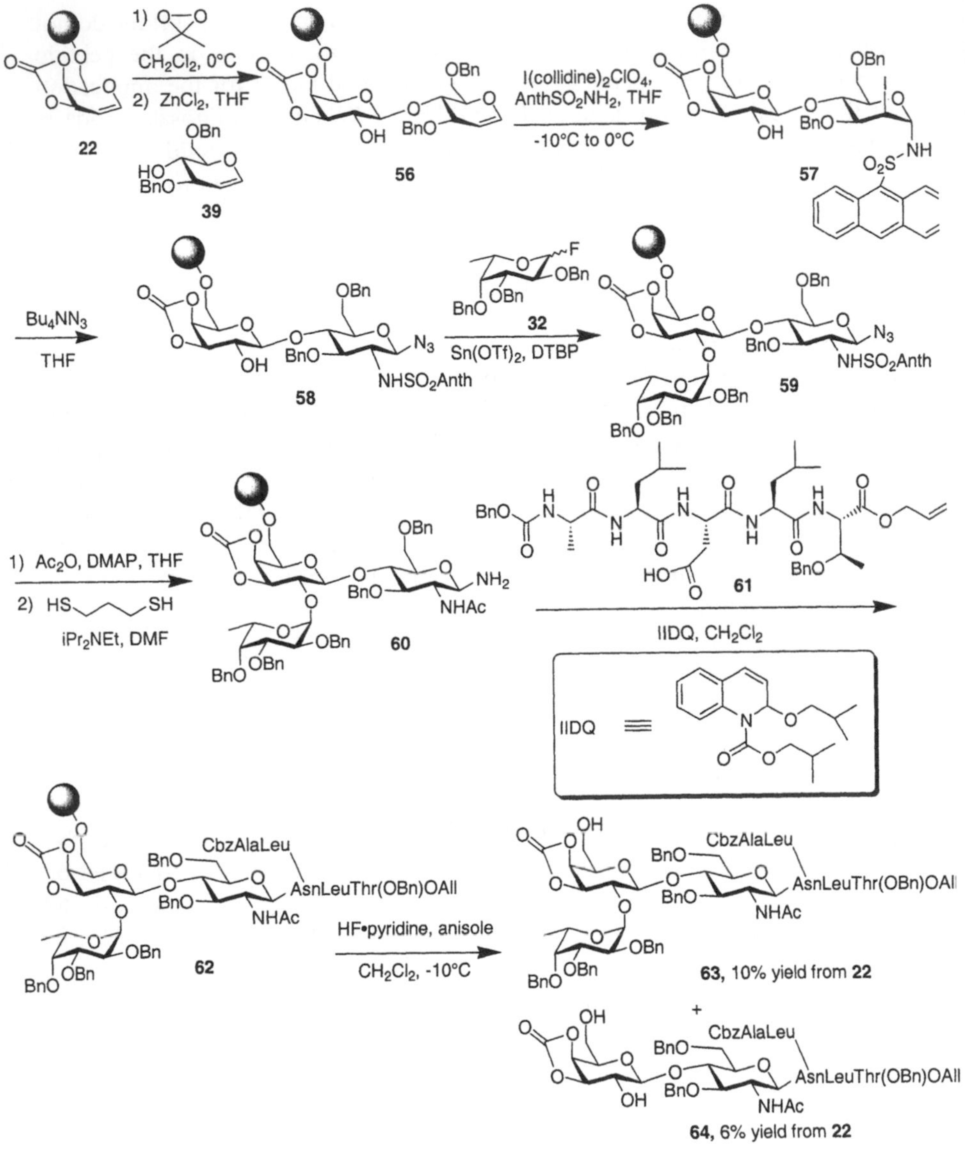

Figure 10 Glycopeptides from glycals.

and CH_2Cl_2 (3 × 20 mL), and dried under vacuum to give 280 mg of resin **58**.

Acylation and reduction: Acetic anhydride (200 µL, 2.12 mmol) was added to a suspension of resin **59** (130 mg, ~0.05 mmol) in THF (2.0 mL) followed by a solution of DMAP (166 mg, 1.36 mmol) in THF (2.0 mL). The reaction was shielded from light and stirred for 22 h. The solvents were drained and the resin was washed with THF (3 × 20 mL) and dichloro-methane (3 × 20 mL) and dried under reduced pressure. This resin was suspended in DMF (3.0 mL), 1,3-propanedithiol (1 mL, 9.96 mmol) and *N,N*-diisopropylethylamine (1 mL, 5.74 mmol) were added, and the resin was stirred for 15 h, during which time the color of the resin faded. The solvents were drained and the resin was washed with DMF (3 × 20 mL), THF (4 × 20 mL), and dichloromethane (2 × 5 mL) and dried under vac-uum to give the resin **60** (104 mg).

Glycopeptide formation with IIDQ: A solution of pentapeptide **61** (100 mg, 0.11 mmol) in dichloromethane (5 mL) and DMF (250 µL) was added to a suspension of resin **60** (104 mg) in dichloromethane (5 mL), IIDQ (29 µL, 0.1 mmol) was added, and the mixture was stirred for 25 h. The solvents were drained and the resin was washed with dichloromethane (5 × 5 mL), THF (5 mL), and dichloromethane (3 × 5 mL) and dried under vacuum to give resin **62** (100 mg).

Cleavage with HF·pyridine: In a Teflon tube, resin **62** (93 mg) was suspended in a mixture of dichloromethane (2 mL) and anisole (2 µL, 0.018 mmol) and was cooled to −10°C (ice–acetone bath) and HF·pyridine (200 µL, ~7 mmol HF) was added. After 2 h at −10°C, a cold (−10°C) mixture of phosphate buffer (32 mM, 5 mL) and ethyl acetate (10 mL) was added and the mixture was shaken vigorously. The mixture was extracted with ethyl acetate (4 × 20 mL) and the combined organic layers were washed with pH 8 brine (10 mL, brine–NaHCO$_3$ saturated 9:1) and dried over so-dium sulfate. The resulting white powder was purified using RP-18 reverse-phase silica with methanol–water (7:3 → 9:1) as eluant to give the desired trisaccharide-pentapeptide **63** (4.5 mg, 10% yield) and the disaccharide-pen-tapeptide **64** (2.4 mg, 6% yield).

2. Other Solid-Phase Syntheses with the Glycoside Attached to the Resin at the Nonreducing End

a. Orthogonal Glycosylation Strategy. Ito et al. [43] have assembled tri- and tetrasaccharides using an orthogonal glycosylation strategy and a soluble polyethylene glycol (PEG) support [86,87]. Anomeric thiomethyl glycosides were used and were activated in an orthogonal fashion from glycosyl fluo-rides (Fig. 11). The PEG-supported glycoside **65** was prepared in six steps

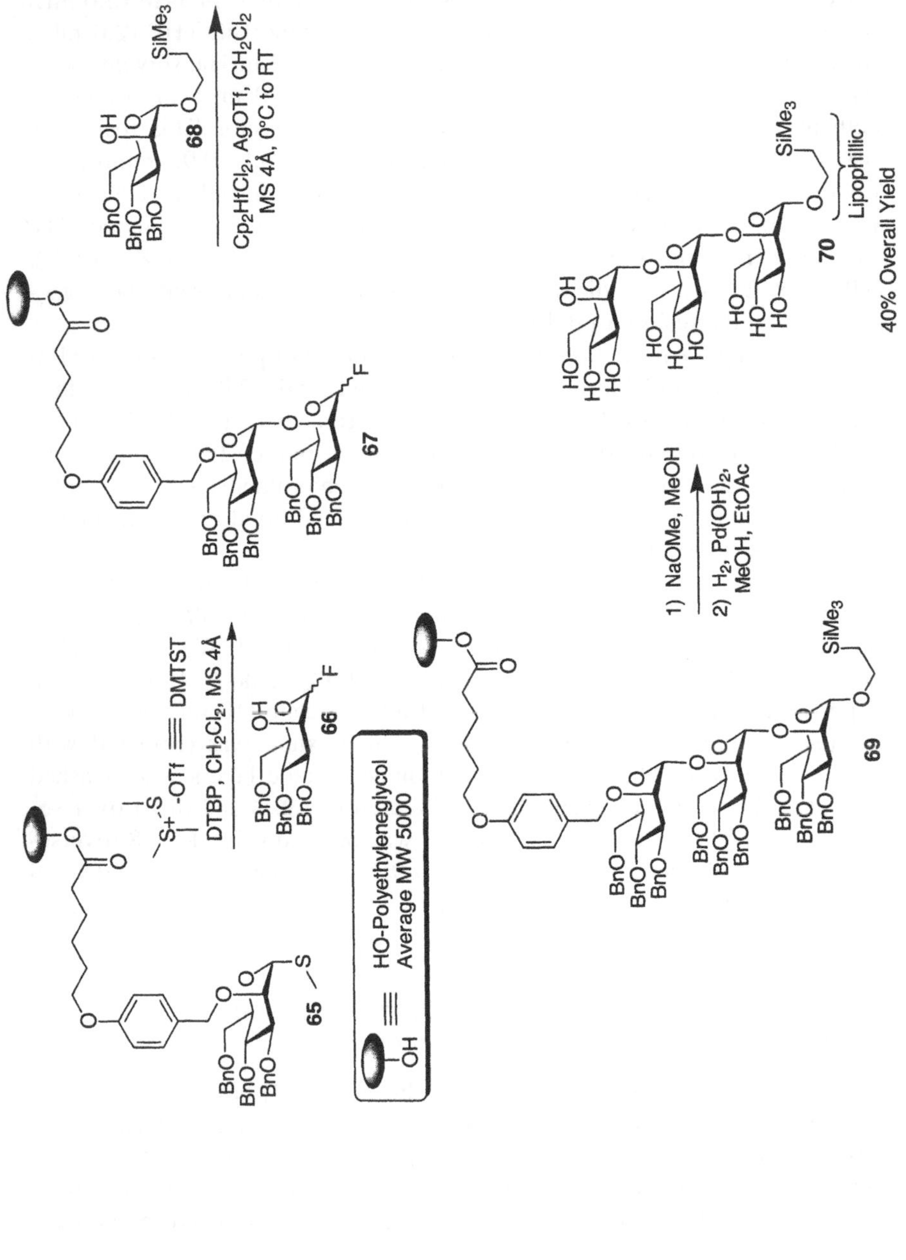

Figure 11 Orthogonal glycosylation.

from readily available starting materials [86]. The selectively protected mannosyl fluoride **66** was glycosylated in the presence of buffered dimethyl(methylthio)sulfonium triflate (DMTST) to give **67**. The disaccharide **67** was then treated under Suzuki activation [88] with the mannosyl derivative **68**, which had the lipophilic group 2-(trimethylsilyl)ethyl at the anomeric position [89,90], to give **69**. The PEG resin was removed by methanolysis and the benzyl protecting groups and linker were cleaved using hydrogenolysis to give final product **70**. Compound **70** had a hydrophobic tail and was easily purified from more polar by-products by reverse-phase chromatography. Using this approach, a set of anomeric thiomethyl and fluoride building blocks could be prepared and used to assemble a variety of oligosaccharides on PEG resin.

The use of soluble polymeric supports combines some of the advantages of solution-phase and solid-phase chemistry. The reaction kinetics are very similar to those of solution-phase reactions but the products are usually easily separated from the reagents by precipitation of the polymer and filtration. Alternatively, the soluble resin can be recovered using size exclusion chromatography. NMR and other spectroscopic methods can readily be used to monitor the reactions. Combinatorial libraries may be prepared by split and pool techniques using soluble supports, but lengthy deconvolution may be required to determine the structures of active compounds because current tagging techniques are not applicable to soluble PEG resins. One other liability of soluble supports is that the maximum loading is often low compared with that of cross-linked polystyrene resins. For example, the PEG resin used by Ito and coworkers has an average molecular weight of 5000, which would correspond to a loading of approximately 0.2 mmol/g, significantly lower than the ~1 mmol/g usually obtained using cross-linked polystyrene supports. PEG resin is also partially labile to repeated exposure to some glycosylation conditions [33].

b. Diethylalkylsilyl Linker. A series of disaccharides was prepared using glycoside donors attached to the resin at the nonreducing end via a diethylalkylsilyl linker [91,92]. The silyl chloride [93] **71** (Fig. 12) was reacted with the anomeric thiophenyl glycoside **72** to give the polymer-bound glycoside donor **73**. The donor was activated with either DMTST or N-iodosuccinimide-triflic acid in the presence of 4-Å molecular sieves and DTBP to give disaccharide **75**. The final product **76** was obtained in quantitative mass balance and high purity after cleavage from the resin with hot aqueous acetic acid in THF using either activating procedure. The glycosylation worked well when both glycosides were armed (having either type of protecting group, as in the previous example) or when only one of the two was disarmed (with all acyl protecting groups) (**77, 78**) [94]. The reaction failed when both donor and acceptor were disarmed (**79**). Other donor–activator

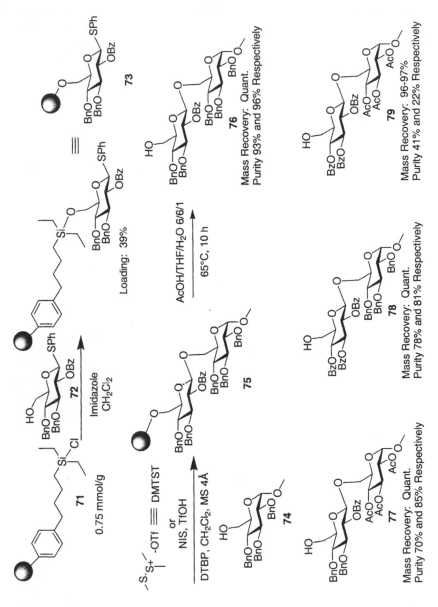

Figure 12 Disaccharides from diethylsilyl resin.

combinations were also tested. Anomeric fluoroglycoside and bis-cyclopentadienyl hafnium bis-triflate as an activator gave good glycosylation yields, although the use of a combination of anomeric trichloroacetimidates with triflic acid or boron fluoride etherate was less satisfactory and anomeric sulfoxide with a variety of activators did not react at all.

c. Others. Many other examples of polymer-bound saccharide glycosylations attached at the nonreducing end have used solution-based trichloroacetimidate donors [95]. 6-Deoxyoligosaccharides were prepared using a polystyrene-based sulfonate linker attached at the 6-position to selectively protected glycosides and glycals [96]. Di- and trisaccharides were assembled using standard coupling conditions and the products were cleaved with iodide ions (**80**) (Fig. 13).

Heparan sulfate–like oligomers **81** were assembled on a soluble PEG resin using repeating disaccharide building blocks (Fig. 14) [97]. The coupling efficiency was about 95% and, after deprotection and cleavage from the resin, the products were sulfated. The structure of the products was confirmed by ^{1}H NMR and MALDI–mass spectrometry.

With the goal of developing a solid-phase, semiautomated synthesis of nucleotide–glycoside hybrids as antisense reagents, Iadonisi and coworkers [98,99] compared the yields of oligosaccharides obtained using three different solid supports with trichloroacetimidate donors (Fig. 15). Using similar reaction conditions with trimethylsilyl triflate activation, controlled pore glass (CPG) gave the best yields of disaccharide **82**, followed closely by polystyrene; Tentagel gave the lowest yields. CPG, which is commonly used for automated synthesis of oligonucleotides, has the advantage of being rigid and easily washed after each coupling. Its major drawback is its very low loading capacity, which makes it an impractical support. By comparison, the Tentagel resin has more than seven times the loading capacity of CPG and polystyrene can hold 33 times more material than CPG.

B. Reducing End Attached to the Resin

1. Enzymatic Synthesis [29]

a. Solid-Phase Synthesis of a Trisaccharide-Octapeptide. Wong and coworkers have been developing methods for solid-phase syntheses of oligosaccharides and glycopeptides for a number of years [100–103]. This group published a procedure for the solid-phase enzymatic synthesis of a base- and acid-labile *O*-linked trisaccharide-octapeptide and a solution-phase synthesis of an *O*-linked sialyl-Lewis-X-octapeptide [104,107]. The peptide substrate was built on a CPG support via a palladium-labile HYCRON linker (Fig. 16). This linker is both acid and base stable and can be cleaved using nearly

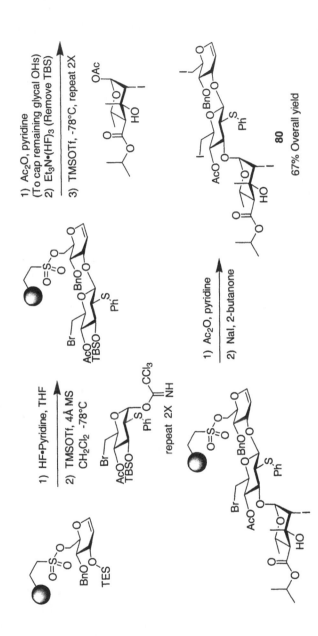

Figure 13 2-Deoxyglycosides using a sulfonate linker.

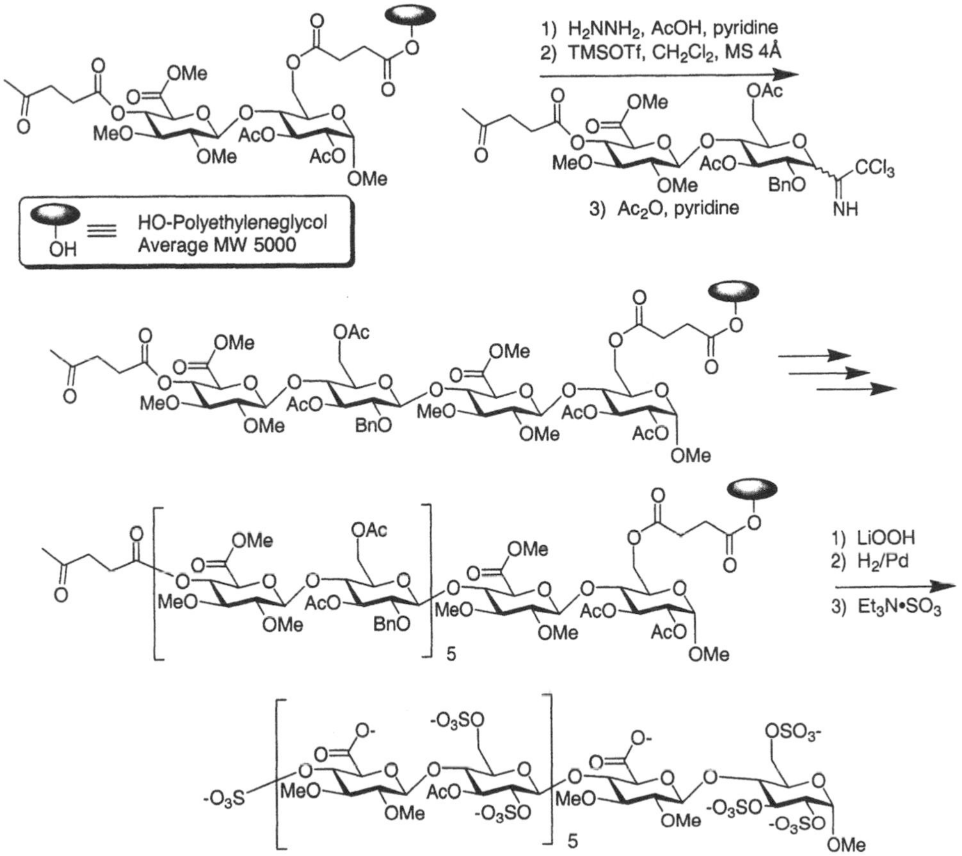

Figure 14 Synthesis of heparan sulfate–like oligomers.

neutral conditions with palladium catalysis. This linker was used for the synthesis of the N-acetylglucosylated peptide substrates using an Fmoc/tBu protecting group strategy with all of the protecting groups for both the peptidic and glycosidic fragments removable while the glycopeptide was anchored to the solid support. Although the CPG support gives poorer yields for peptide synthesis [105], it is a better substrate for enzymatic reactions than the polyacrylamide–polyethylene glycol resin commonly used for peptide synthesis [106].

A preformed alanine-HYCRON construct **83** was coupled with the CPG beads using O-benzotriazol-l-yl-N,N,N',N'-tetramethyluronium hexafluorophosphate (HBTU) and N-methylmorpholine (NMM) to give **84**. The

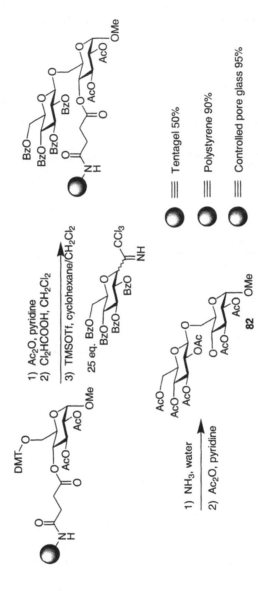

Figure 15   Glycosylation using control pore glass, polystyrene, and Tentagel supports.

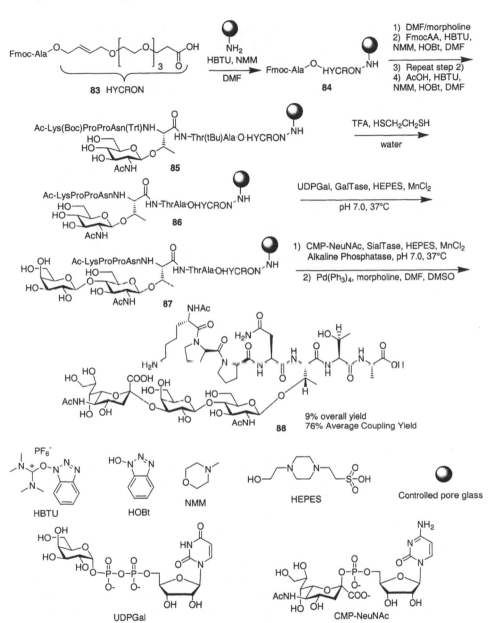

Figure 16 Enzymatic synthesis of glycopeptides.

Fmoc group was removed with morpholine and the next amino acid was coupled. The second threonine fragment in the octapeptide sequence, which had an unprotected glucosamine side chain, was coupled using the same conditions. The couplings were repeated until all eight amino acids were on the support. Finally, the last amine was acylated to give the *N*-acetylglucos-amine-octapeptide **85**. The acid-labile protecting groups were removed by treatment with TFA and glycopeptide **86** was coupled to uridine diphosphate galactoside (UDPGal) in the presence of galactosyl transferase (GalTase) to give **87**. The sialyl group was then installed with cytidine 5'-monophospho-sialic acid (CMP-NeuNAc) in the presence of sialyl transferase (SialTase) and alkaline phosphatase to prevent product inhibition by hydrolyzing the CMP to cytidine. The final product **88** was obtained in 9% overall yield. Because SialTase has a lower specific activity than GalTase, side reactions were responsible for lower yields, mainly the hydrolysis of the glycopeptide from the HYCRON linker. Surprisingly, the product found in the supernatant solution from this reaction was purer than the product obtained from the final palladium-catalyzed cleavage. The authors speculated that this hydro-lysis might be due to a special conformation of the sialylated glycopeptide. Although the yields were low, the synthesis of the glycopeptide was completed in only 9 days.

b. Enzymatic Synthesis of Sialyl-Lewis-X-Tetrasaccharide (**92**) *on Sephar-ose Matrix.* Blixt and Norberg [108] have used enzymatic oligosaccharide synthesis to prepare the Lewis-X-tetrasaccharide conjugates **91** and **92** (Fig. 17) on a Sepharose matrix [109]. The tetrasaccharide was assembled using available enzymes and sugar nucleotides and the final products were re-moved from the support by cleaving the disulfide linkage to Sepharose with dithiothreitol (DTT). The yield of the synthesis was excellent, producing the desired products in 57% overall yield after purification by size exclusion chromatography. The key to this success was the use of a long spacer be-tween the support and the initial saccharide substrate. The spacer was easily and efficiently assembled from commercially available materials. For the galactosylation of polymer-bound 2-*N*-acetyl-2-deoxyglucosamine, a spacer length of 59 to 71 atoms was optimal for all the polymer-bound-glycosides to be readily accessible to the enzymes.

Methods for the enzymatic synthesis of oligosaccharide have been de-veloped and are now becoming an attractive alternative to nonenzymatic preparation. The advantages of this protocol are numerous; in particular, great specificity of the enzymes precludes the use of protecting groups. The stereochemical and regiochemical outcomes of the reactions are also very well controlled. The drawbacks are that the number of available enzymes is still small, although that number is rapidly increasing, and the sugar nucle-otides are expensive. One major problem is that the method lacks the flex-

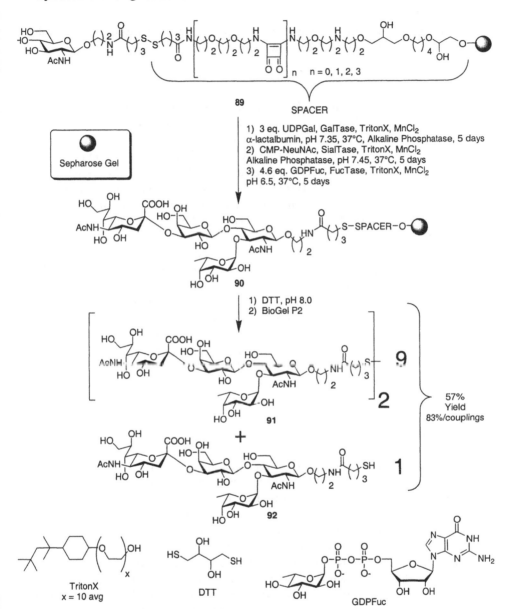

Figure 17 Enzymatic synthesis of tetrasaccharides on Sepharose matrix.

ibility of nonenzymatic syntheses that allow the preparation of "natural-like" and "non-natural-like" saccharides.

2. Chemical Synthesis

a. Polymer-Supported Synthesis of Oligosaccharides Using Trichloroacetimidates. Schmidt and coworkers [79,110] have demonstrated the usefulness of trichloroacetimidates for the synthesis of oligosaccharides on a solid support by preparing a branched pentasaccharide fragment of complex-type glycans [111]. Their efficient approach took advantage of the pseudosymmetry of the final product **100** (Fig. 18) to perform two double glycosylations, assembling a pentasaccharide in only six steps!* The initial differentially protected trichloroacetimidate-mannose **93** was loaded on the thiol resin **94** to give **95**. This resin linker was stable to the activation conditions required for glycosidic bond formation with trichloroacetimidates. The resulting thioglycoside could later be cleaved with thiophilic reagents. The benzoates at C-6 and C-3 were removed with sodium methoxide and the resulting diol was mannosylated with **96** to give the branched trisaccharide **97**. The reactions were monitored by cleaving an aliquot of resin and recording their matrix-assisted laser desorption ionization time-of-flight mass spectra (MALDI-TOF MS) or by chromatographic techniques. Another cycle of deprotection–mannosylation with **98** yielded the polymer-bound pentasaccharide **99**, which was cleaved from the resin by treatment with di-*tert*-butylperoxide, N-iodosuccinimide, and methanol to give **100** in 38% overall yield.

b. Polymer-Supported Synthesis of Oligosaccharides Using Thioglycosides. Nicolaou et al. [112] have prepared one of the largest oligosaccharide via solid-phase assembly. Their approach was an improvement over their previous work [113] and allowed the synthesis of building blocks on a solid support that could be reused to prepare larger oligosaccharides. The use of a photocleavable linker and a 4-hydroxybenzoate spacer inserted between the photolinker and the glycoside had many advantages. The photolinker is stable to most glycosylation reaction conditions and is cleaved selectively using mild conditions. Isolation of the cleaved final products is simplified because the stereochemistry of the 4-hydroxybenzoate spacer group is already defined (see **107** → **108**). Also, the polymer-bound glycoside can be cleaved from the resin to give thioglycosides that are readily activated for glycosylation (see **105** → **106**). Using their block-type solid-phase strategy, the dodecasaccharide **108** was assembled in only 12 chemical steps using

---

*Double glycosylation has previously been accomplished on the solid support; see Ref. 70.

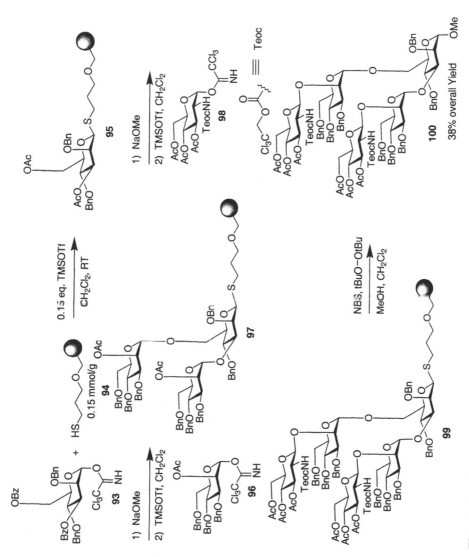

Figure 18 Solid-phase synthesis of pentasaccharides using trichloroacetimidates.

the same trisaccharide building block three times (Fig. 19). The trisaccharide building block was assembled in five steps from the polymer-bound saccharide **101**. Glycosylation of **101** with the thioglycoside **102** in the presence of dimethylthiomethylsulfonium triflate (DMTST) gave the disaccharide **103** after removal of the Fmoc protecting group. Addition of thiophenylglucoside **104** to **103** required the presence of tetramethylurea to give the trisaccharide **105**. This sugar was cleaved with light but, more important, treatment of this material with S-trimethylsilylthiophenol, zinc iodide, and tetrabutyl ammonium iodide gave the thioglycoside building block **106** in 76% yield. This demonstration was limited to the formation of (β1,6) and (β1,3) glucosidic bonds, but the method will probably be expanded to include other glycosides and types of linkages. The overall yield for the synthesis was acceptable at 10%, representing an average of about 82% yield per step.

c.   Polymer-Supported Synthesis of Oligosaccharides Using n-Pentenyl Glycosides and Anomeric Sulfoxide.   The anomeric sulfoxides have been used extensively for the synthesis of oligosaccharides on a solid support. These anomeric sulfoxides are very reactive and provide one of the most general glycosylation methods currently available. This reactivity becomes a handicap when constructing compounds on a solid support because low temperatures are required. More recently, the *n*-pentenyl glycosides have been used on the solid support and appeared to work effectively. These two techniques have been well reviewed [30] and will not be discussed.

III. CONCLUSION

Solid-phase synthesis of oligosaccharides is becoming a mature science, tackling more complex and relevant synthetic problems. An increasing number of promising techniques and strategies have been tested on the solid support and it is now possible to compare the results and decide the optimal approach for a given problem. With the development of analytical techniques compatible with the solid support (infrared, MAS-NMR, MS) it is not possible to follow reactions on the solid support with relative ease. Many of the glycosylation methods are also compatible and/or orthogonal, and in the future we will see a larger number of syntheses done using more than one type of glycosylation reaction. The research will probably move from developing the methods on the solid support to preparing oligosaccharides uniquely for their biological relevance. We are getting closer to the time when the syntheses of complex oligosaccharides will be established well enough to be conducted in most laboratories.

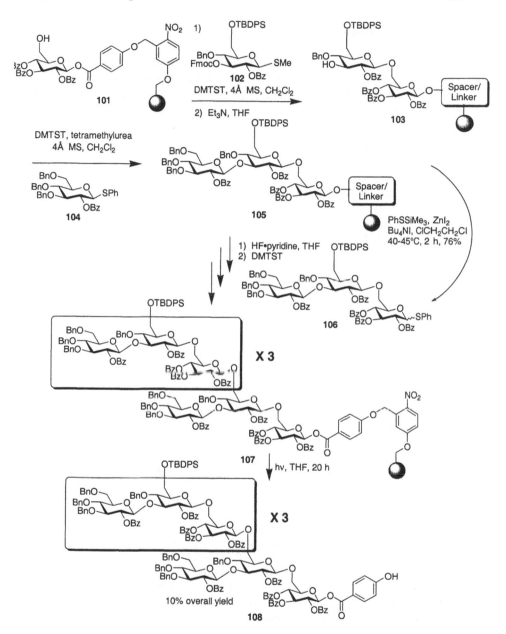

Figure 19 Solid-phase synthesis of dodecasaccharide.

ACKNOWLEDGMENTS

I would like to thank Dr. Jana Pika for her help assembling the bibliography, editing the text, and taking care of our son Marc while I was typing this chapter. I would also like to thank Drs. Prashant Deshpande and Ashok Purandare for their valuable comments and suggestions.

REFERENCES

1. Many of the references in this chapter were listed in M Lebl, Z Leblova. Dynamic database of references in molecular diversity. Internet *http:// www.5z.com.*
2. For a description of the systematic nomenclature of carbohydrates, see HJ Allen, EC Kisailus, eds. Glycoconjugates: Composition, Structure and Function. New York: Marcel Dekker, 1992.
3. DH van den Eijden, DH Joziasse. Curr Opin Struct Biol 3:711–721, 1993.
4. S Natsuka, JB Lowe. Curr Opin Struct Biol 4:683–691, 1994.
5. RA Dwek. Chem Rev 96:683–720, 1996.
6. YC Lee, RT Lee. Neoglycoconjugates: Preparation and Applications. San Diego: Academic Press, 1994.
7. YC Lee, RT Lee. Acc Chem Res 28:321–327, 1995.
8. PV den Steen, PM Rudd, RA Dwek, G Opdenakker. Crit Rev Biochem Mol Biol 33:151–208, 1998.
9. G Ragupathi, TK Park, S Zhang, IJ Kim, L Graber, S Adluri, KO Lloyd, SJ Danishefsky, PO Livingston. Angew Chem Int Ed Engl 36:125–128, 1997.
10. T Feizi. Nature 314:53–57, 1985.
11. S-I Hakomori. Cancer Res 45:2405–2414, 1985.
12. T Feizi. Curr Opin Struct Biol 3:701–710, 1993.
13. BK Hayes, GW Hart. Curr Opin Struct Biol 4:692–696, 1994.
14. CA Ryan. Proc Natl Acad Sci U S A 91:1–2, 1994.
15. A Kobata. Acc Chem Res 26:319–324, 1993.
16. G Opdenakker, PM Rudd, CP Ponting, RA Dwek. FASEB J 7:1330–1337, 1993.
17. T Feizi, D Bundle, Curr Opin Struct Biol 4:673–676, 1994.
18. A Varki. Proc Natl Acad Sci U S A 91:7390–7397, 1994.
19. JC Castro-Palomino, G Ritter, SR Fortunato, S Reinhardt, LJ Old, RR Schmidt. Angew Chem Int Ed Engl 36:1998–2001, 1997.
20. H Paulsen. GBF Monogr Ser Vol 8. Chem Synth Mol Biol 115–135, 1987.
21. C-H Wong, Acc Chem Res 32:376–385, 1999.
22. Y Zhao, SBH Kent, BT Chait. Proc Natl Acad Sci U S A 94:1629–1633, 1997.
23. For a discussion of the phenomenon occuring in polymer beads used in solid-phase synthesis, see P Hodge. Chem Soc Rev 26:417–424, 1997.
24. RB Merrifield. J Am Chem Soc 85:2149, 1963.
25. H Paulsen. Angew Chem Int Ed Engl 29:823–938, 1990.

26. H Kunz. Angew Chem Int Ed Engl 26:294–308, 1987.
27. JM Fréchet. In: P Hodge, DC Sherrington, eds. Polymer-Supported Reactions in Organic Synthesis. New York: Wiley, 1980, pp 407–434.
28. JMJ Fréchet. Tetrahedron 37:663–683, 1981.
29. SJ Danishefsky, JY Roberge. In: DG Large, CD Warren, eds. Glycopeptides and Related Compounds: Chemical Synthesis, Analysis and Applications. New York: Marcel Dekker, 1997, pp 245–294.
30. For a general review of the field, see HMI Osborn, T Khan. Tetrahedron 55: 1807–1850, 1999.
31. MJ Sofia. Drug Discov Today 1:27–34, 1996.
32. MJ Sofia. In: EM Gordon, JF Kerwin Jr, eds. Combinatorial Chemistry and Molecular Diversity in Drug Discovery. 1998, pp 243–269.
33. Z-W Wang, O Hindsgaul. Glycoimmunology 2:219–236, 1998.
34. MJ Sofia, DJ Silva. Curr Opin Drug Discov Dev 2:365–376, 1999.
35. Z Zhang, IR Ollmann, X-S Ye, R Wischnat, T Baasov, C-H Wong. J Am Chem Soc 121:734–753, 1999.
36. S Borman. Chem Eng News, March 30–31, 1999.
37. NK Terrett. Combinatorial Chemistry. Oxford Chemistry Masters. Oxford: Oxford University Press, 1998.
38. M Mammen, S-K Choi, GM Whitesides. Angew Chem Int Ed Engl 37:2754–2794, 1998.
39. Y Chapleur. Carbohydrate Mimics. Concepts and Methods. Weinheim: Wiley-VCH, 1998.
40. TY Chan, R Chen, M Sofia, BC Smith, D Glennon. Tetrahedron Lett 38: 2821–2824, 1997.
41. T Kanemitsu, O Kanie, C-H Wong. Angew Chem Int Ed Engl 37:3415–3418, 1998.
42. PH Seeberger, X Beebe, GD Sukenick, S Pochapsky, SJ Danishefsky. Angew Chem Int Ed Engl 36:491–493, 1997.
43. Y Ito, O Kanie, T Ogawa. Angew Chem Int Ed Engl 35:2510–2512, 1996.
44. RU Lemieux, B Fraser-Reid. Can J Chem 42:532–538, 1964.
45. RU Lemieux, B Fraser-Reid. Can J Chem 43:1460–1475, 1965.
46. RU Lemieux, AR Morgan. Can J Chem 43:2190–2198, 1965.
47. J Thiem, P Ossowski. J Carbohydr Chem 3:287–313, 1984.
48. J Thiem, A Prahst, T Wendt. Liebigs Ann Chem 1044–1056, 1986.
49. J Thiem. In: Trends in Synthetic Carbohydrate Chemistry. Washington DC: American Chemical Society, 1989, pp 131–149.
50. SJ Danishefsky, JT Randolph, JY Roberge, KF McLure, RB Ruggeri. ACS Polym Prepr 35:977–978, 1994.
51. SJ Danishefsky, JT Randolph, JY Roberge, KF McLure, RB Ruggeri. Schering Lect Ser 25:7–22, 1995.
52. SJ Danishefsky, JY Roberge. Pure Appl Chem 67:1647–1662, 1995.
53. SJ Danishefsky, MT Bilodeau. Angew Chem Int Ed Engl 35:1380–1419, 1996.
54. PH Seeberger, SJ Danishefsky. Acc Chem Res 31:685–695, 1998.
55. SJ Danishefsky. Chemtracts Org Chem 2:273–297, 1989.

56. DB Berkowitz, SJ Danishefsky, GK Schulte. J Am Chem Soc 114:4518–4529, 1992.

57. RP Spencer, CL Cavallaro, J Schwartz. J Org Chem 64:3987–3995, 1999 and references therein.

58. W Holla. Angew Chem Int Ed Engl 28:220–221, 1989.

59. Sussex Research Laboratories Inc. 100 Sussex Drive, Suite 129, Ottawa, Canada, K1A 0R6, (613) 993-4402.

60. KK-C Liu, SJ Danishefsky. J Org Chem 59:1895–1897, 1994.

61. KK-C Liu, SJ Danishefsky. J Org Chem 59:1893–1894, 1994.

62. A Fürstner, I Konetzki. Tetrahedron Lett 39:5721–5724, 1998.

63. C Zheng, PH Seeberger, SJ Danishefsky. J Org Chem 63:1126–1130, 1998.

64. M Izumi, Y Ichikawa. Tetrahedron Lett 39:2079–2082, 1998.

65. SJ Danishefsky, KF McLure, JT Randolph, RB Ruggeri. Science 260:1307–1309, 1993.

66. TH Chan, WQ Huang. J Chem Soc Chem Commun 909–911, 1985.

67. JT Randolph, KF McLure, SJ Danishefsky. J Am Chem Soc 117:5712–5719, 1995.

68. For the use of siylene linker for the protection of hindered glycals hydroxyls, see KA Savin, JCG Woo, SJ Danishefsky. J Org Chem 64:4183–4186, 1999.

69. RW Murray, R Jeyaraman. J Org Chem 50:2847–2853, 1985.

70. JT Randolph, SJ Danishefsky. Angew Chem Int Ed Engl 33:1470–1473, 1994.

71. MF Farrall, JMJ Fréchet. J Org Chem 41:3877–3882, 1976.

72. PH Seeberger, M Eckhardt, CE Gutterridge, SJ Danishefsky. J Am Chem Soc 119:10064–10072, 1997.

73. PJ Garegg. Adv Carbohydr Chem Biochem 52:179–266, 1997.

74. For an introduction to the biological relevance of glycoproteins, see OP Bahl. In: HJ Allen, EC Kisailus, eds. Glycoconjugates: Composition, Structure and Function, New York: Marcel Dekker, 1992, pp 1–12.

75. M Meldal, K Bock. Glycoconj J 11:59–63,1994.

76. M Meldal. Curr Opin Struct Biol 4:710–718, 1994.

77. FE McDonald, SJ Danishefsky. J Org Chem 57:7001–7002, 1992.

78. JY Roberge, X Beebe, SJ Danishefsky, Science 269:202–204, 1995.

79. J Rademann, A Geyer, RR Schmidt. Angew Chem Int Ed Engl 37:1241–1245, 1998.

80. JY Roberge, X Beebe, SJ Danishefsky. J Am Chem Soc 120:3915–3927, 1998.

81. RU Lemieux, S Levine. Can J Chem 42:1473–1480, 1964.

82. AJ Robinson, PB Wyatt. Tetrahedron 49:11329–11340, 1993.

83. KC Nicoloau, CW Hummel, Y Iwabuchi. J Am Chem Soc 114:3126–3128, 1992.

84. Y Kiso, H Yajima. J Chem Soc Chem Commun 942–943, 1972.

85. H Kunz. Pure Appl Chem 65:1223–1232, 1993.

86. Y Ito, O Kanie, T Ogawa. J Am Chem Soc 116:12073–12074, 1994.

87. For a report on the recent development of the use of soluble polymers for synthesis, see CW Harwig, DJ Gravert, KD Janda. Chemtracts Org Chem 12:1–26, 1999.

88. K Suzuki, KH Maeta, T Matsumoto. Tetrahedron Lett 30:4853–4856, 1989.
89. P Stangier, MM Palcic, DR Bundle. Carbohydr Res 267:153–159, 1995.
90. Y Ito, JC Paulson. J Am Chem Soc 115:1603–1605, 1993.
91. T Doi, M Sugiki, H Yamada, T Takahashi, JA Porco Jr. Tetrahedron Lett 40: 2141–2144, 1999.
92. H Yamada, H Harada, T Takahashi. J Am Chem Soc 116:7919–7920, 1994.
93. Y Hu, JA Porco Jr, JW Labadie, OW Gooding, BM Trost. J Org Chem 63: 4518–4521, 1998.
94. DR Mootoo, P Konradsoon, U Udodong, B Fraser-Reid. J Am Chem Soc 110:5583–5584, 1988.
95. Z-G Wang, SP Douglas, JJ Kripinsky, Tetrahedron Lett 37:6985–6988, 1996.
96. JA Hunt, WR Roush, J Am Chem Soc 118:9998–9999, 1996.
97. CM Dreef-Tromp, HAM Willems, P Westerduin, P van Veelen, CAA van Boeckel. Bioorg Med Chem Lett 7:1175–1180, 1997.
98. M Adinolfi, G Barone, L De Napoli, A Iadonisi, G Piccialli. Tetrahedron Lett 39:1953–1956, 1998.
99. M Adinolfi, G Barone, L De Napoli, A Iadonisi, G Piccialli. Tetrahedron Lett 37:5007–5010, 1996.
100. M Schuster, P Wang, JC Paulson, C-H Wong. J Am Chem Soc 116:1135–1136, 1994.
101. RL Halcomb, H Huang, C-H Wong. J Am Chem Soc 116:11315–11322, 1994.
102. C-H Wong, RL Halcomb, Y Ichikawa, T Kajimoto. Angew Chem Int Ed Engl 34:521–546, 1995.
103. K Witte, O Seitz, C-H Wong. J Am Chem Soc 120:1979–1989, 1998.
104. O Seitz, C-H Wong. J Am Chem Soc 119:8766–8776, 1997.
105. F Albericio, M Pons, E Pedroso, E Giralt. J Org Chem 54:360–366, 1989.
106. U Slomczynska, F Albericio, F Cardenas, E Giralt. Biomed Biochim Acta 50: S67–S73, 1991.
107. A Giannis. Angew Chem Int Ed Engl 33:178–180, 1994.
108. O Blixt, T Norberg. J Org Chem 63:2705–2710, 1998.
109. O Blixt, T Norberg. J Carbohydr Chem 16:143–154, 1997.
110. J Rademann, RR Schmidt, Tetrahedron Lett 37:3989–3990, 1996.
111. See also H Shimizu, Y Ito, O Kanie, T Ogawa. Bioorg Med Chem Lett 6: 2841–2846, 1996.
112. KC Nicolaou, N Watanabe, J Li, J Pastor, N Winssinger. Angew Chem Int Ed Engl 37:1559–1561, 1998.
113. KC Nicolaou, N Winssinger, J Pastor, F DeRoose. J Am Chem Soc 119:449–450, 1997.

15

Solid-Phase Synthesis of Heterocyclic Compounds from Amino Acids and Linear Peptides

Adel Nefzi, John M. Ostresh, and Richard A. Houghten
Torrey Pines Institute for Molecular Studies, San Diego, California

I. INTRODUCTION

The explosively growing field of combinatorial chemistry clearly has its roots in Merrifield's solid-phase synthetic approaches [1]. All aspects of this approach for the assembly of peptides were extensively developed in the 20 years following Merrifield's initial publication [2–5]. The concept of rapidly constructing large numbers of compounds in parallel was first presented by Geysen et al. (pins) [6] and Houghten (free resin in mesh packets—"tea-bags") [7] in 1984 and 1985, respectively. As is true for all combinatorial strategies, these two seminal parallel synthesis methods were first applied to the synthesis of peptides. They are now the most widely used means of preparing combinatorial libraries of all types, including peptidomimetics [8–10], oligonucleotides [11,12], and oligosaccharides [13–15]. The success of all solid-phase synthetic approaches is indicated by the capability to drive reactions on polymer supports to completion (often >99.8%), the ability to remove readily excess reagents and starting materials, and ease of automation. The design and development of strategies for the synthesis of individual and combinatorial libraries of heterocyclic compounds have been areas of intense research over the past 5 years. It is of note that Leznoff [16,17] and Crowley and Rapoport [18] carried out the solid-phase synthesis of heterocycles from 1972 to 1977, but this powerful approach remained virtually untapped for 15 years until the solid-phase synthesis of libraries of diazepines was reported by Bunin and Ellman in 1992 [19].

Several general syntheses of heterocyclic compounds from α-amino acids or dipeptidomimetics have been reported [20]. Sardina and Rapoport [21] reviewed the enantiospecific synthesis in solution of heterocycles from α-amino acids. For the solid-phase construction of heterocyclic combinatorial libraries, the use of four key techniques must be considered: (1) the synthesis of parallel arrays of individual compounds, generally involving the preparation of compounds from all combinations of the building blocks used; (2) the systematic preparation of mixtures of compounds, also involving all combinations of the building blocks used; (3) the synthesis of soluble or immobilized versions of the individual compounds and mixtures in (1) and (2); and (4) the deconvolution methods used to identify the most active compounds from a library. Here we review the use of amino acids and linear peptides for the solid-phase assembly of heterocyclic compounds, in particular, the use in our laboratory of dipeptides and peptidomimetics. The preparation of mixtures of compounds and their deconvolution using the positional scanning approach and the preparation of combinatorial libraries of cyclic ureas, cyclic thioureas, hydantoins, and thiohydantoins are also presented.

II. TECHNIQUES

A. Mixtures Versus Individual Compound Arrays

There exists an intense ongoing debate on the relative merits of the preparation and screening of individual compound arrays versus mixtures of the same compounds. The central issue is the balance between time and cost. An example is the time and cost to prepare and screen 10,000 to 30,000 individual variations of a particular pharmacophore (this is currently the typical size of a library) versus the same 10,000 to 30,000 compounds as mixtures, with each mixture containing 10, 100, 1000, etc. compounds. For illustrative purposes, if a library of 27,000 compounds is prepared (30 different building blocks at each of the three positions: $30 \times 30 \times 30 = 27,000$), then an individual compound array format would require the preparation, screening, storing, etc. of 27,000 individual compounds. Alternatively, the same 27,000 compounds, if prepared as mixtures in a positional scanning format [22], would require the preparation of only three sets of 30 mixtures each (each mixture being made up of 900 compounds). Assuming that each separate mixture can yield distinguishable signals from an assay of interest, clearly an exceptional time and handling savings is possible with such a library as compared with the individual compound array.

B. Positional Scanning Deconvolution Method

The positional scanning (PS) approach [22] is illustrated in the generic representation shown in Fig. 1. It involves the screening of separate, single defined position synthetic combinatorial libraries (SCLs) to identify individually the most important functionalities at each position of diversity within a library. A complete PS-SCL having three positions of diversity consists of three sublibraries (designated OXX, XOX, and XXO), each of which has a single defined functionality at one position represented as O and a mixture of functionalities represented as X at each of the other two positions. The generic library shown contains 64 compounds ($4 \times 4 \times 4 = 64$). The pooling of each sublibrary, which contain the same 64 compounds (4 mixtures of 16 compounds), would vary on the basis of the functionality at the defined position in that sublibrary. The structure of individual compounds can be determined from such a screening because each compound is present in one and only one mixture of each sublibrary. In theory, if only one compound was active in the library, activity corresponding to that compound would be found in the one mixture of each sublibrary containing that compound. When considered in concert, the defined functionality in each mixture can then be used to identify the individual active compound responsible for the activity within the mixture. In practice, the activity found is generally due not to the activity of a single active compound but to the sum of activities of families of homologous compounds within the active mixture. PS-SCLs are generally prepared by the reagent mixture approach [23] (Fig. 2b). Although the synthesis of PS-SCLs is possible with the divide, couple, and recombine (DCR) approach [24–26] (Fig. 2a), in reality the effort involved makes the synthesis much more tedious than the reagent mixture approach. Freier and coworkers [27,28] have presented an excellent discussion of the theoretical and experimental aspects of iterative and positional scanning deconvolution.

C. Solid-Phase Synthesis of Heterocyclic Compounds Using Amino Acids and Peptides as Templates

Substituted heterocyclic compounds offer a high degree of structural diversity and have proven to be broadly and economically useful as therapeutic agents. The development of strategies for the synthesis of heterocyclic compounds on the solid phase is expanding as a greater understanding of how to carry out such reactions successfully is attained. We have reviewed the current status of heterocyclic combinatorially libraries [20].

Because of their versatility and availability, amino acids have been used extensively for the synthesis of heterocyclic compounds. They have a wide range of available functional groups in protected form, which facilitates

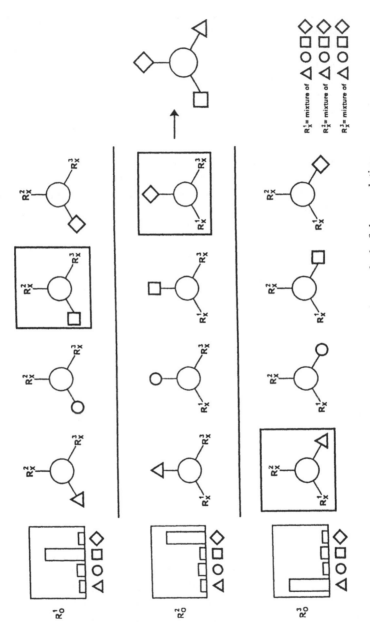

Figure 1 Schematic representation of the positional scanning method of deconvolution.

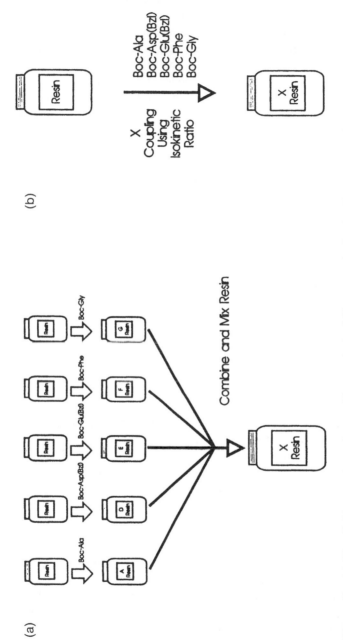

Figure 2 (a) The divide, couple, and recombine method of library synthesis. (b) The reagent mixture approach.

synthetic operations. Their activation, protection, and deprotection are well documented, and they are commercially available in enantiomeric forms.

A variety of heterocyclic compounds have been derived from the condensation of aldehydes with α-amino acids. Imines 1 (Fig. 3) are often used as intermediates in organic synthesis and are the starting point for chemical reactions such as cycloadditions, condensation reactions, and nucleophilic additions. The formation of imines via condensation of amines with aldehydes was first adopted for the reductive alkylation of resin-bound amino acids [29–31]. Imines have now been used as synthetic intermediates in the generation of a range of heterocyclic combinatorial libraries.

1. Isoquinolines and Derivatives

Isoquinoline derivatives are an important family of natural products. They have diverse biological activities and are used, for example, as bronchodilators, skeletal muscle relaxants, and antiseptics. The solid-phase synthesis of a 43,000-compound tetrahydroisoquinoline 2 combinatorial library has been reported by Griffith et al. [32]. The library was synthesized by a three-step procedure. An imine was formed by reacting a substituted benzaldehyde with a methylbenzhydrylamine (MBHA) resin-bound amino acid. Imine formation was driven to completion using trimethylorthoformate as a dehydrating reagent. Treatment of the imine with homophthalic anhydride provided the desired tetrahydroisoquinoline (Fig. 3a).

2. 4-Thiazolidinones and 4-Metathiazanones

Holmes et al. [33] reported the solution and polymer-supported synthesis of 4-thiazolidinones 3 and 4-metathiazanones 4 derived from amino acids. A three-component condensation of an amino acid ester or a resin-bound amino acid (glycine, alanine, β-alanine, phenylalanine, and valine), an aldehyde (benzaldehyde, o-tolualdehyde, m-tolualdehyde, p-tolualdehyde and 3-pyridine carboxaldehyde), and an α-mercapto carboxylic acid led to the formation of five- and six-membered heterocycles (Fig. 3b).

3. β-Carbolines

First described by Kaljuste and Unden [34], the reaction of polymer-bound tryptophan with a variety of aldehydes and ketones under Pictet–Spengler conditions [35] has also been reported by Mayer et al. [36]. The β-carboline derivatives 5 exhibited significant bioactivity. Following incorporation of 9-fluorenylmethoxycarbonyl (Fmoc)-tryptophan and deprotection, the condensation with an aldehyde or ketone under acidic conditions provided the β-carbolines in high yield and purity. The same strategy has been published

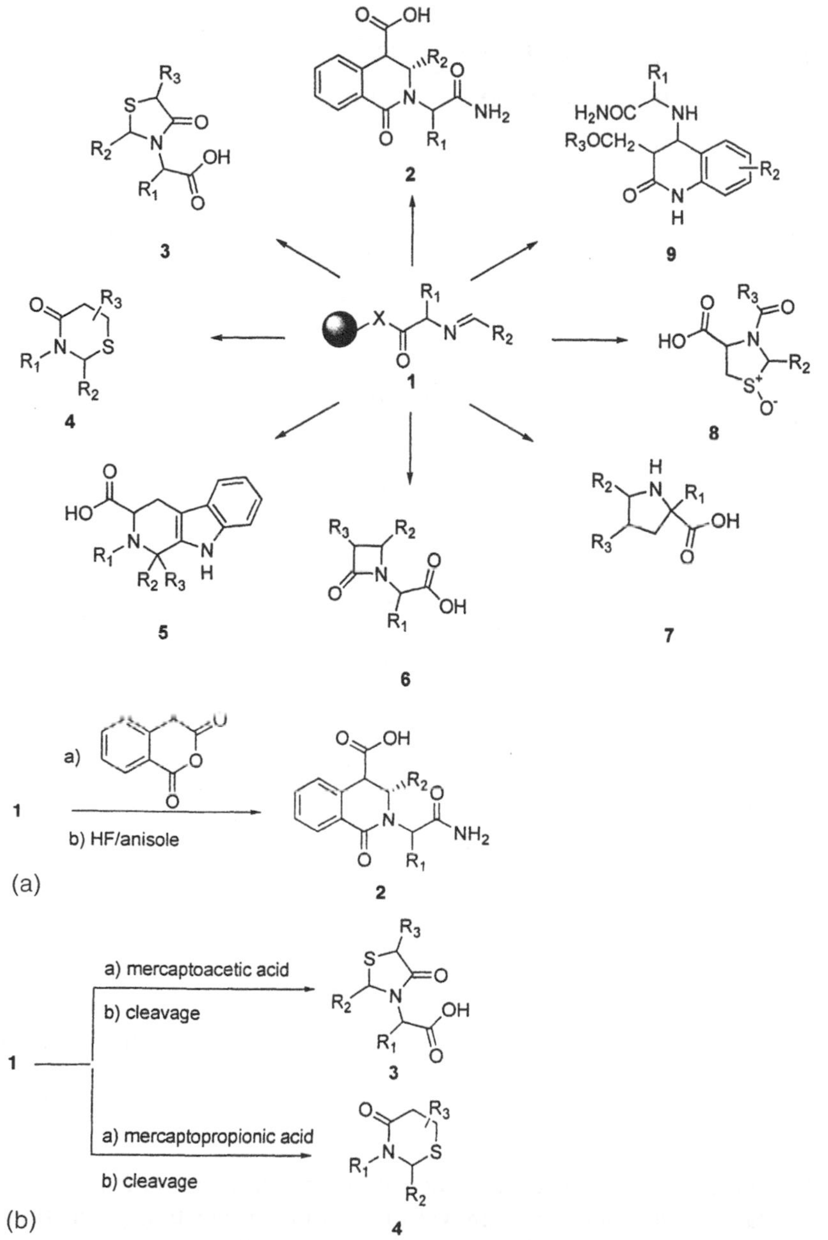

Figure 3 Solid-phase synthesis of heterocycles from amino acids and peptides through imine formation. Synthesis of (a) isoquinolines **2**, (b) thiazolidinones **3** and methathiazanones **4**, (c) β-carbolines **5**, (d) β-lactams **6**, (e) pyrrolidines **7**, (f) thiazolidines **8**, and (g) quinolinones **9**.

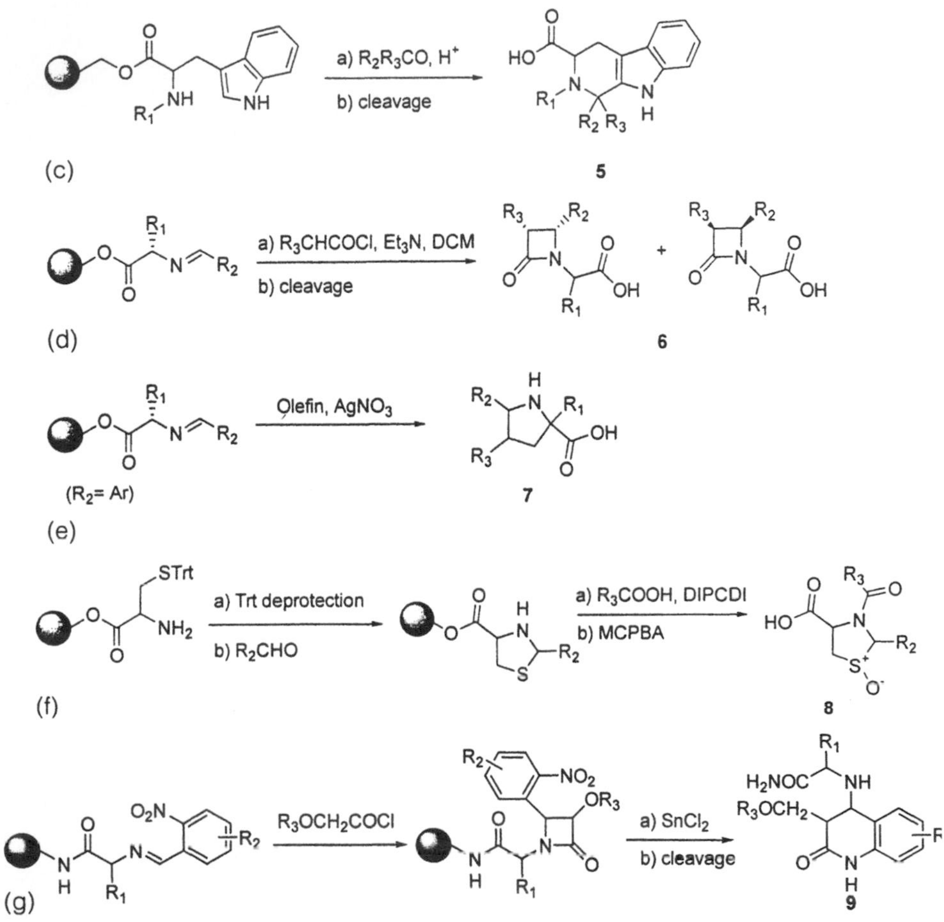

(c)

5

(d)

6

(e)

(R₂= Ar)

7

(f)

8

(g)

9

Figure 3 Continued

by Yang and Guo [37], with the only difference being the use of *t*-butylox-ycarbonyl (Boc)-tryptophan (Fig. 3c).

4. β-Lactams

Ruhland et al. [38] reported the combinatorial synthesis of structurally diverse β-lactams **6**. Following removal of the Fmoc protecting group from resin-bound amino acids, a 10- to 15-fold excess of alkyl, aromatic, or α,β-unsaturated aldehydes in a 1:1 mixture of trimethylorthoformate in dichloromethane (DCM) was added to yield resin-bound imines. The [2+2] cycloaddition occurred by addition of acid chlorides to a DCM suspension of the imine resins in the presence of triethylamine at 0°C. The reaction mixture was allowed to warm to room temperature with agitation for 16 h. The products were cleaved from the Sasrin support by treatment with a solution of 3% (v/v) trifluoroacetic acid (TFA) in DCM for 45 min (Fig. 3d).

5. Pyrrolidines

Murphy et al. [39] reported the synthesis of pyrrolidine **7** combinatorial libraries. Starting from polystyrene resin-bound amino acids, the α-amino ester was condensed with aromatic and heteroaromatic aldehydes in neat trimethylorthoformate to afford the resin-bound aryl imine. Pyrrolidine and pyrroline derivatives were obtained through cycloaddition of the 1,3-dipoles azomethine ylides to olefin and acetylene dipolarophiles. A library of 500 compounds was reported. The screening of this library for in vivo inhibition of angiotensin-converting enzyme (ACE) led to the identification of 1-(3'-mercapto-2'-(S)-methyl-1'-oxopropyl)-5-phenyl-2,4-pyrrolidinedicarboxylic acid 4-methyl ester as a potent ACE inhibitor that incorporates the mercaptoisobutyryl side chain (Fig. 3e).

6. Thiazolidines

Patek et al. [40] reported an elegant method for the synthesis of a number of N-acyl thiazolidines **8** via the condensation of aldehydes and ketones with β-mercaptoalkylamines. Following attachment of cysteine using an ester linkage to TentaGel S OH resin and simultaneous deprotection, condensation with an aldehyde resulted in imine formation and subsequent cyclization to the thiazolidine. It was found that acylation of the thiazolidine increases the stability of the ring to acidolysis. The factors influencing the stability of acylated thiazolidine heterocycles have been studied by Mutter et al. [41,42]. To increase the stability of the ring, the thioether was oxidized to the sulfoxide using m-chloroperbenzoic acid (Fig. 3f).

7. Quinolinones

An elegant method for the solid-phase synthesis of quinolinone **9** derivatives has been developed by Pei et al. [43]. Using the "teabag" technology, the synthesis of a 4-amino-3,4-dihydro-2(1H)-quinolinone library was carried out through the rearrangement of β-lactam intermediates on the solid phase. The condensation of o-nitrobenzaldehyde with a resin-bound amino acid yielded an imine that, following [2+2] cycloaddition with ketenes, afforded the β-lactams (Fig. 3g).

D. Heterocyclic Compounds Derived from Dipeptides or Dipeptidomimetics

Linear peptides and peptidomimetics have been extensively used and will continue to be employed as starting materials for the generation of heterocyclic compounds on the solid phase. As will be shown, many different structures have been prepared starting from dipeptidomimetics.

Multicomponent condensations such as the Ugi reaction [44] and the Biginelli condensation [45] are especially useful for the creation of diverse chemical libraries on the solid phase. Four-component condensations have been reviewed by Mjalli and Toyonaga [46] for the synthesis on the solid phase of small-ring heterocycles [47]. For example, the one-pot condensation of an amine (derived from amino acids) and an aldehyde, followed by the addition of an isocyanide and a carboxylic acid, provides a dipeptidomimetic N-alkyl-N-acyl-α-amino amide **10** that can serve as a useful starting point for the synthesis of imidazoles **11** and pyrroles **12**, which are pharmaceutically useful compounds (Fig. 4).

1. Pyrroles and Derivatives

The synthesis of pentasubstituted pyrroles has been reported by Mjalli and Toyonaga [46] using a multicomponent condensation. The treatment of the intermediate **10** with neat acetic anhydride or isobutylchloroformate and triethylamine in toluene, followed by the addition of a series of acetylenic esters, provided the polymer-bound pentasubstituted pyrroles **12** (Fig. 4). The reaction proceeded by in situ cyclization of the intermediate via [3+2]

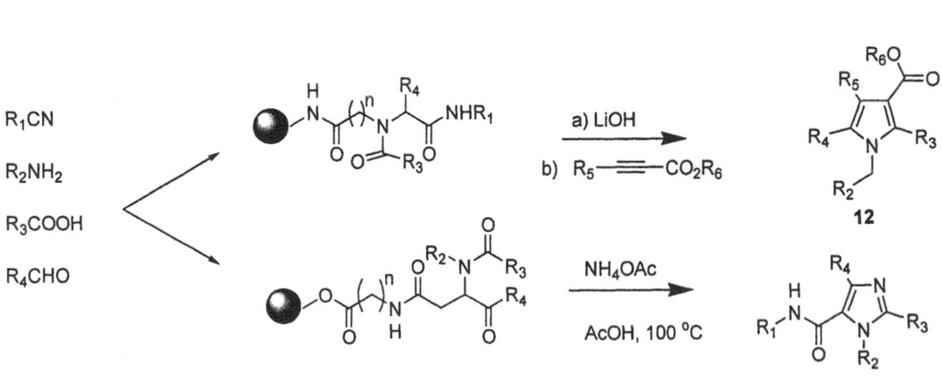

Figure 4 Multicomponent condensations (the resin can be attached to any of the four functionalities).

cycloaddition with a variety of alkynes. An isomeric mixture of pyrroles in an approximately 4:1 ratio was obtained after release of the product from the resin with 20% TFA–DCM. The desired products were obtained in overall yield of 35–75% over eight steps, illustrating the power of solid-phase synthesis.

2. Tetrahydroisoquinolines and Tetrahydroimidazopyridines

Meutermans and Alewood [48] reported the solid-phase synthesis of tetrahydroisoquinolines **13** and dihydroisoquinolines **13a** using the Bischler–Napieralski reaction (Fig. 5). The polystyrene resin-bound deprotected L-3,4-dimethoxyphenylalanine was acylated with acetic acid derivatives using N-[(1H-benzotriazol-1-yl)(dimethylamino)methylene]-N-methylmethanaminium hexafluorophosphate N-oxide (HBTU) as a coupling reagent. The product obtained was then treated with phosphorus oxychloride under optimized conditions to afford a Bischler–Napieralski cyclization. Hutchins and Chapman [49] reported the synthesis of tetrahydroisoquinolines **13b** and 4,5,6,7-tetrahydro-3H-imidazol[4,5-c]pyridines **14** via cyclocondensation of the appropriate dipeptidomimetic with various aldehydes (Fig. 6).

3. Isoxazoles and Isoxazolines

The synthesis of a series of isoxazoles **15** and isoxazolines **15a** on the solid phase has been reported by Pei and Moos [50] via [3 + 2] cycloaddition of alkenes and alkynes with highly reactive nitrile oxides. The cycloaddition reactions of resin-bound peptoids were carried out in toluene at 100°C or in DCM–H$_2$O at room temperature, depending on the precursors of the nitrile oxides. Benzaldehyde oxime and various nitroalkyl compounds were selected as nitrile oxide precursors. The nitrile oxides were generated in situ by reacting the nitroalkyl compounds with phenyl isocyanate and triethylamine or by oxidizing the oximes with sodium hypochlorite in the presence of triethylamine (Fig. 7).

4. Piperazine Derivatives

Goff and Zuckermann [51] reported the synthesis of 2-oxopiperazine **16** by intramolecular Michael addition on the solid phase. The coupling of resin-bound unsaturated dipeptoids with a variety of Fmoc-L-amino acids, N,N′-diisopropylcarbodiimide (DIPCDI), and 1-hydroxybenzotriazole (HOBt) affords tripeptoids, which after treatment with 20% piperidine in DMF, were acylated with benzoyl chloride–Et$_3$N in 1,2-dichloroethane. Following treatment with 95% TFA in H$_2$O, a diastereomeric ratio of monoketopiperazines **16** was obtained (Fig. 8). Rather than benzoyl chloride, phenylisocyanate or bromoacetic acid and an amine could be used.

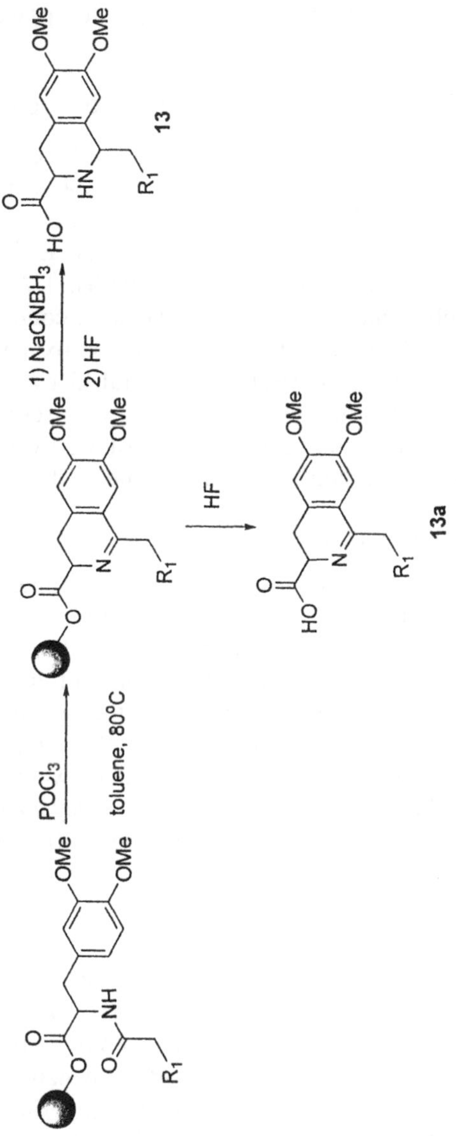

Figure 5 Synthesis of dihydro- and tetrahydroisoquinolines (**13a** and **13**).

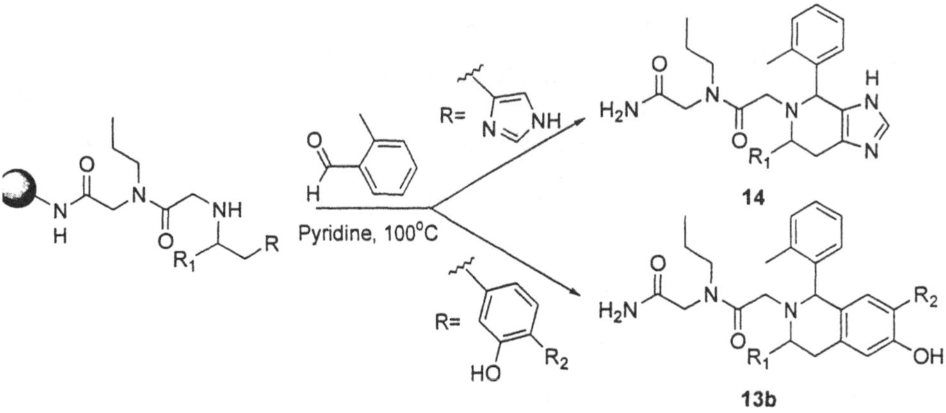

Figure 6 Synthesis of tetrahydroisoquinolines **13b** and tetrahydro-3*H*-imidazo-[4,5-*c*]pyridines **14**.

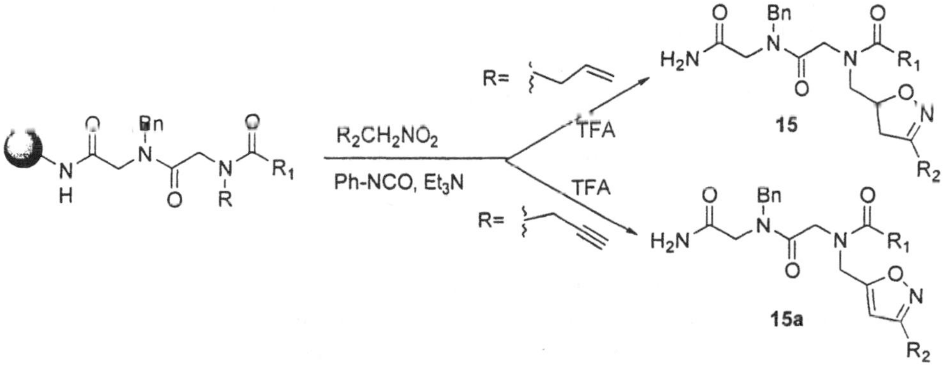

Figure 7 Synthesis of isoxazoles **15** and isoxazolines **15a**.

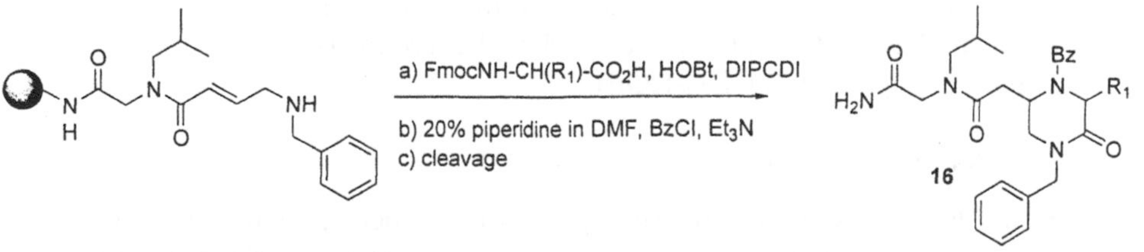

Figure 8 Synthesis of oxopiperazines **16**.

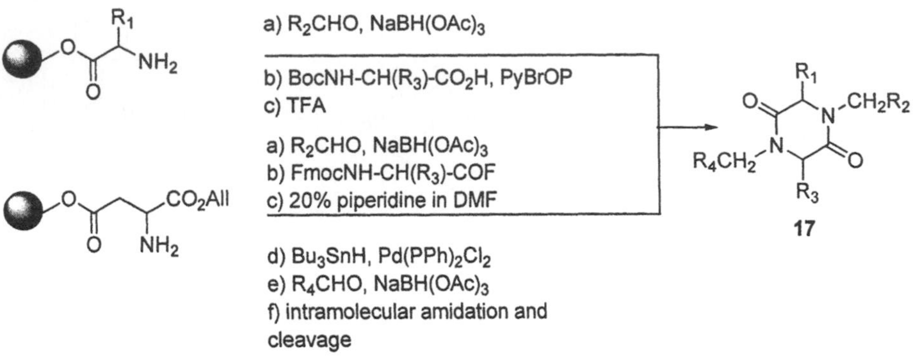

a) R_2CHO, NaBH(OAc)$_3$

b) BocNH-CH(R_3)-CO$_2$H, PyBrOP
c) TFA

a) R_2CHO, NaBH(OAc)$_3$
b) FmocNH-CH(R_3)-COF
c) 20% piperidine in DMF

d) Bu$_3$SnH, Pd(PPh)$_2$Cl$_2$
e) R_4CHO, NaBH(OAc)$_3$
f) intramolecular amidation and
cleavage

Figure 9   Synthesis of diketopiperazines **17**.

Gordon and Steele [52] developed a strategy for the solid-phase synthesis of diketopiperazines (DKPa) **17** based on reductive alkylation of a
support-bound amino acid with an aldehyde. The resulting secondary amine
was then coupled with Boc-protected amino acids employing bromotrispyr-
rollidino-phosphonium hexafluorophosphate (PyBroP) as the activating
agent. Following TFA treatment, cyclization to obtain the desired product
was accomplished by heating at reflux in toluene. Using the DCR approach,
a total of 1000 compounds were synthesized. Several DKPs were identified
that had significant biological activity, including affinity for the neurokinin-
2 receptors.

A similar approach to the solid-phase synthesis of diketopiperazine **17**
has been published by Krchňák et al. [53] (Fig. 9). This involved the coupling of the side-chain carboxyl group of Asp, Glu, or Ida to the resin.
Following deprotection of the α-amino group, the first position of diversity
was introduced by reductive alkylation of the polymer-supported amino
group using a variety of aldehydes in trimethylorthoformate with sodium
triacetoxyborohydride as the reducing agent. The resulting secondary amine
was then acylated with an amino acid. Following removal of the N_α-pro-
tecting group, a second reductive alkylation was accomplished employing
the same conditions. The desired dioxypiperazine ring was easily obtained
upon mild activation of the free carboxylic group using the conventional
coupling reagents DIPCDI–HOBt.

5.   Benzodiazepines

Benzodiazepines are an important group of therapeutic agents [54–56]. One
of the most interesting aspects of the benzodiazepines is their striking anticonvulsant properties in a variety of experimental models of epilepsy [54].

Virtually all known 1,4-benzodiazepines have the same profile of broad pharmacological activity, including antianxiety, sedative–hypnotic, anticonvulsant, muscle-relaxing actions and tranquilizing properties. The benzodiazepines are also well absorbed after oral administration. Many benzodiazepine drugs bind extensively to plasma and tissue proteins. The use of benzodiazepine in therapeutic applications has increased exponentially over the past 20 years. Benzodiazepines are now the most commonly prescribed group of drugs.

Starting from a support-bound N-alkylated glycine, Goff and Zuckermann [57] synthesized 1,4-benzodiazepine-2,5-diones **18**. Following acylation of the N-alkylated glycine (peptoids) with bromoacetic acid and displacement of the bromine with an α-amino ester, the resulting secondary amine was acylated with a substituted o-azidobenzoyl chloride. Treatment with tributylphosphine gave the iminophosphorane, which, upon heating at 125°C, cyclized to afford the resin-bound benzodiazepines. The desired product was removed from the resin by treatment with TFA (Fig. 10).

Dewitt et al. [58,59] reacted 2-aminobenzophenone imine with commercially available α-amino acids on Wang resin. The transimidation was followed by intramolecular cleavage leading to the desired benzodiazepine. Starting with five different amino acids and eight different benzophenone imines, 40 benzodiazepines **19** (Fig. 11) were produced in 9–63% yield (2–25 mg of material).

Moroder et al. [60] developed a new method for the generation of benzodiazepines from peptide precursors. A series of N_α-(2-aminobenzoyl)-N-alkylamino acid peptides were synthesized. The general tendency of these peptides to undergo acid-catalyzed cyclization to 1,4-benzodiazepine-2,5-diones was investigated.

6. Hydantoins and Derivatives

Dewitt et al. [58,59] also reported the synthesis of 40 hydantoins **20** using the Diversomer approach following the condensation of a variety of isocyanates to Wang resin–bound amino acids. Concomitant cyclization and cleavage leading to the hydantoin product occurred on heating the resin-bound urea in 6 M HCl (Fig. 11).

A hydantoin library of 800 compounds has been reported by Dressmann et al. [61]. In this approach, 20 different amino acids and over 80 primary amines were incorporated. Selected amino acids were attached via their N-terminus to hydroxymethyl polystyrene resin by using a carbamate linker. This enabled generation of the free acid resin-bound intermediates, which could then be converted to their corresponding amides by standard carbodiimide coupling reactions and excess primary amine. Concomitant

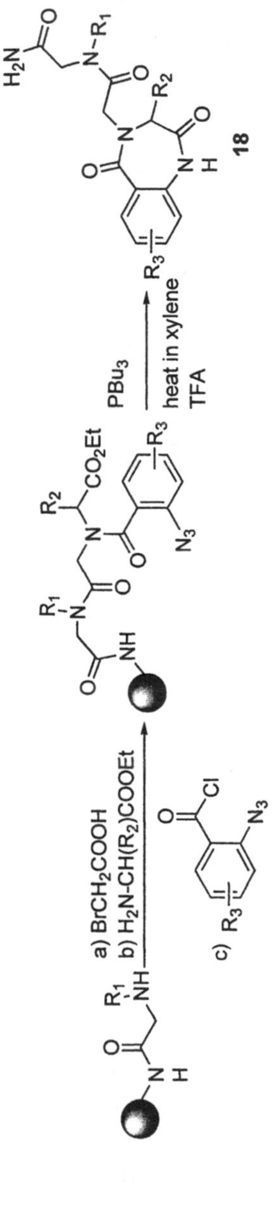

Figure 10 Synthesis of benzodiazepinediones **18**.

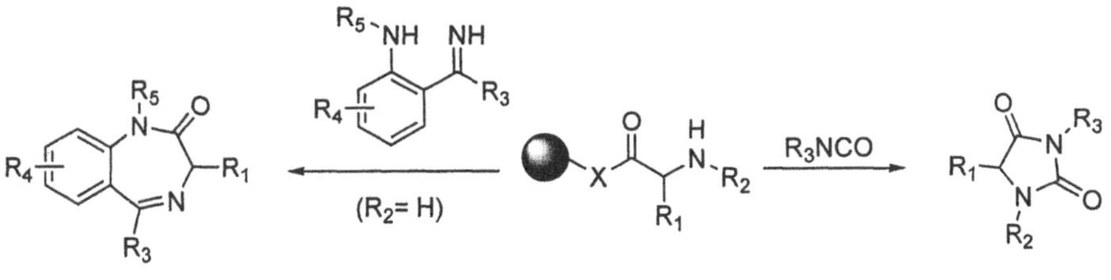

Figure 11 Synthesis of benzodiazepinones **19** and hydantoins **20**.

cyclization and cleavage occurred after treatment of the resin-bound intermediate with excess triethylamine in MeOH under reflux for 48 h.

7. Diazepines

The solid-phase synthesis of 1,3,4,7-tetrasubstituted perhydro-1,4-diazepine-2,5-diones **21** has been reported by Nefzi et al. [62,63]. The synthetic approach is illustrated in Fig. 12. The solid-phase synthesis of diazepine derivatives is initiated by reduction with NaBH$_3$CN in 1% AcOH in DMF of the enamine formed by the reaction product between an aldehyde and the α-amino group of the p-methylbenzhydrylamine resin-bound aspartic acid. Satisfactory results for the coupling of an Fmoc amino acid to the resulting secondary amine were obtained using double coupling with N-[(dimethyl-amino)-1H-1,2,3-triazolo[4,5-b]pyridin-1-ylmethylene]-N-methylmethanaminium hexafluorophosphate N-oxide (HATU). This coupling step depends strongly on the incoming amino acid. Good yields were obtained with Phe and Met(O), whereas low yields were obtained with hindered amino acids such as Val. Once the dipeptide has been formed, the Fmoc protecting group is removed and a second reductive alkylation is carried out under the same conditions. Following tBu cleavage, the thermodynamically favorable coupling of the resulting secondary amine to the side chain of aspartic acid was readily accomplished in the presence of HATU. By employing this strategy, 40 different diazepines with high-performance liquid chromatographic (HPLC) purities ranging from 15 to 87% were prepared.

8. Thiomorpholinones

The solid-phase synthesis of 2,4,5-trisubstituted thiomorpholine-3-ones **22** from a resin-bound protected cysteine has been reported by Nefzi et al.

21

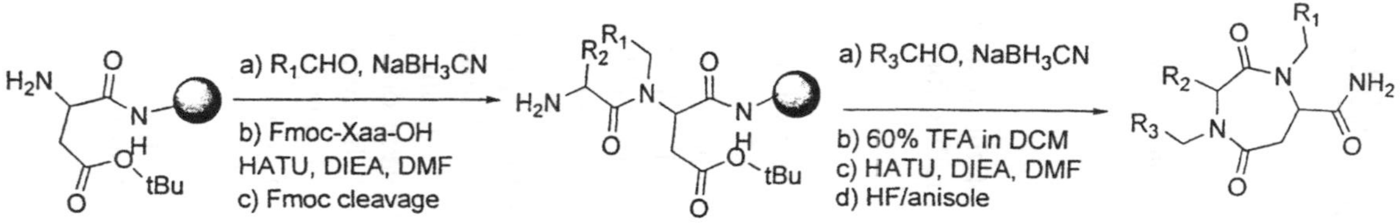

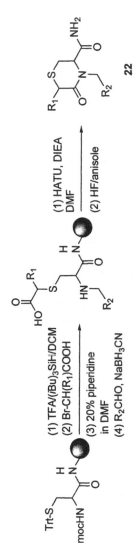

Figure 12 Synthesis of diazepines **21**.

22

Figure 13 Synthesis of thiomorpholinones **22**.

[62,64] (Fig. 13). Starting from MBHA resin, N_α-Fmoc-S-trityl-L-cysteine is coupled in the presence of DIPCDI and hydroxybenzotriazole (HOBt). Following cleavage of the trityl (Trt) group with 5% TFA in DCM in the presence of 5% of (iBu)$_3$SiH, the resin-bound Fmoc-cysteine was treated with a range of different α-bromo-α-alkyl carboxylic acids in DMF in the presence of N-methylmorpholine (NMM).

Poor purity was obtained for bulky R_1 groups such as phenyl or isopropyl; however, excellent results were obtained with bromoacetic acid (R_1 = H), 2-bromopropionic acid (R_1 = Me), and 2-bromovaleric acid (R_1 = Et). Following Fmoc removal with 20% piperidine in DMF, reductive alkylation of the free amine occurred in the presence of an aldehyde and sodium cyanoborohydrate (NaBH$_3$CN). The formation of thiomorpholinone occurred via intramolecular amidation with HATU as the coupling reagent.

E. Positional Scanning Solid-Phase Synthesis of Mixture-Based Organic and Heterocyclic Libraries from Amino Acids and Linear Peptides

1. Linear Ureas

A linear urea library **23** (Fig. 14) has been prepared. Reaction of a resin-bound amino acid with an individual preformed isocyanate affords the linear urea in good yield [62]. The isocyanate is generated by slowly adding an amine to a solution of triphosgene in anhydrous DCM in the presence of N,N-diisopropylethylamine (DIEA). The condensation of the isocyanate with the resin-bound amino acids yields the linear urea. In order to increase the number of compounds, selective N-alkylation was performed on the resin-bound amino acid. Following the individual synthesis of all control compounds in which the individual building blocks were varied while fixing the other position, a library of 125,000 linear N,N'-disubstituted ureas was prepared. This library was tested for opioid activity in mu, delta, and kappa opioid and sigma receptor binding assays. Following deconvolution of this

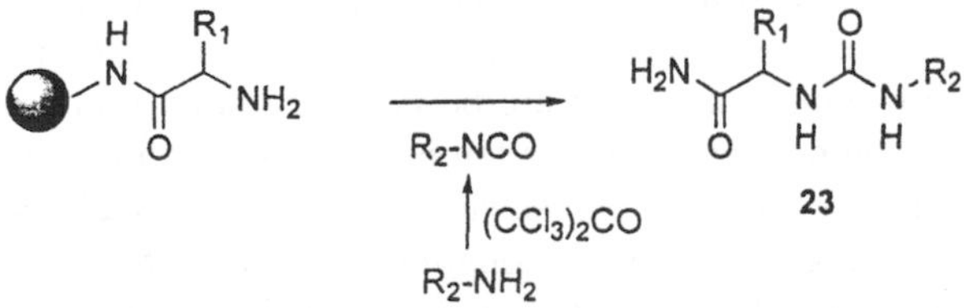

Figure 14 Synthesis of linear ureas **23** from resin-bound amino acids.

library, individual ureas having nanomolar affinities at the mu and sigma receptors were found.

2. Polyamines

The solid-phase construction of trisubstituted diethylenetriamines **24** was carried out starting from *p*-methylbenzhydrylamine (MBHA) resin-bound N_α-*tert*-butyloxycarbonyl (Boc) amino acids. Following removal of the Boc group with trifluoroacetic acid in dichloromethane, the resulting amine salt was neutralized and the resulting primary amine was then protected using triphenylmethyl chloride (TrtCl). The secondary amide could then be selectively alkylated in the presence of lithium *t*-butoxide and an alkylating reagent (methyl iodide or benzyl bromide). Following Trt deprotection, a second amino acid is coupled to the resin-bound *N*-alkyl amino acids. It was found that the alkylation of the amide resin linkage dramatically increases the acid sensitivity of the MBHA resin-bound peptide. Cleavage can be avoided through the subsequent use of Fmoc amino acid chemistry. Following Fmoc removal and *N*-acylation of the resin-bound dipeptide, exhaustive reduction of the amide bonds was achieved by using borane in tetrahydrofuran [63]. The desired products were obtained after cleavage of the resin-bound triamine with anhydrous HF (Fig. 15). We have successfully used peralkylation and/or exhaustive reduction of amide groups to generate compounds having completely different physicochemical properties from the starting peptides (peralkylated peptides, reduced peptides, and reduced peralkylated peptides) [65–69].

Following the strategy described before in conjunction with the teabag method, 29 different amino acids at the R_1 position, 28 different amino acids at the R_2 position, and 40 different carboxylic acids at the R_3 position were used to generate 97 different *N*-benzyltriamines and 97 different *N*-methyltriamines in which individual building blocks were varied while the remaining two positions were fixed. Modifications occurring to the amino acid side chains during the *N*-alkylation and reduction steps have been carefully studied. During the *N*-alkylation with methyl iodide and benzyl bromide, the protected N_ε-amine of lysine was alkylated, and after reduction the Boc protecting group was reduced, yielding the corresponding *N,N*-dimethyl and *N*-benzyl,*N*-methyl polyamines, respectively. These individual compounds served as controls for the synthesis of a mixture-based polyamine library.

The resin-bound polyamine was used as a template for the solid-phase synthesis of a range of heterocyclic compounds, such as cyclic ureas **25** and cyclic thioureas **26**. This work exemplifies our ongoing efforts on the solid-phase assembly of individual acyclic and heterocyclic compounds and combinatorial libraries using short peptides as starting materials.

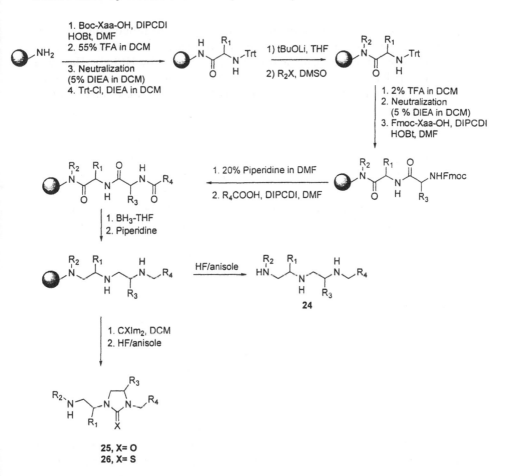

Figure 15 Synthesis of positional scanning heterocyclic libraries (diethyltriamines 24, cyclic ureas 25, and cyclic thioureas 26) derived from dipeptides.

3. Cyclic Ureas and Thioureas

Many biologically active compounds contain cyclic ureas, including inhibitors of human immunodeficiency virus (HIV) protease and HIV replication [70]. Kim et al. [71] presented an illustration of the synthesis of oligomeric cyclic ureas as nonnatural biopolymers. Applying the "libraries from libraries" [72] concept, triamines [65] such as those described earlier were used as templates for the generation of different heterocyclic compounds such as cyclic ureas, cyclic thioureas, and bicyclic guanidines [65]. The cyclizations to obtain the five-membered ring cyclic ureas and cyclic thioureas were performed using carbonyldiimidazole and thiocarbonyldiimidazole (Fig. 15). Following cleavage of the resin with anhydrous HF, the desired products were obtained in good yields and high purity (>80% by HPLC). The cyclization step has also been successfully carried out using triphosgene and thiophosgene. The reversed phase (RP) HPLC and LC mass spectra of the benzylated cyclic urea **25a** obtained from phenylalanine, valine, and phenyl acetic acid are shown in Fig. 16.

Cyclic urea **25** and thiourea **26** libraries were assayed for their ability to inhibit growth of *Candida albicans*, which is one of the most common opportunistic fungi responsible for infections and the one most frequently associated with HIV-positive patients. Following the screening, four sets of individual compounds were prepared. Each set corresponded to all possible combinations of the building blocks defining the most active mixtures at each position of the given SCL. A total of 8 *N*-benzyl amido cyclic thioureas, 12 *N*-methyl amido cyclic thioureas, and 12 *N*-benzyl amido cyclic ureas were synthesized and assayed in a manner similar to the SCLs. Greater activities were for the *N*-benzylated compounds than for the *N*-methylated compounds (minimal inhibitory concentrations of the most active compounds varied from 8 to 64 μg/mL and 64 to 125 μg/mL, respectively). Representative active compounds from the benzylated cyclic ureas were derived from L-asparagine, L-norleucine, and cyclohexylacetic acid and from D-cyclohexylalanine, L-norleucine, and *p*-dimethylaminobenzoic acid.

4.   Bicyclic Guanidines

A template similar to that used for the synthesis of cyclic ureas and thioureas is applied for the synthesis of bicyclic guanidines [65], with the exception that the resin-bound amide is not alkylated (Fig. 17). Treatment of a resin-bound *N*-acyl dipeptide with diborane affords a triamine having three available secondary amine functionalities. Following cyclization with thiocarbonyldiimidazole, the presence of a third secondary amine allows the reaction to proceed through highly active intermediates to afford after HF cleavage the desired bicyclic guanidines **27** in good yield and high purity [65]. By using 49 amino acids for the first position of diversity, 51 amino acids for the second, and 41 carboxylic acids for the third, a library of 102,459 (49 × 51 × 41) compounds was synthesized in the positional scanning format. The screening of this library in a radioreceptor assay selective for the kappa opiate receptor led to the identification of individual compounds showing excellent binding affinity (median inhibitory concentrations less than 50 nM) (Table 1). The library was screened at 4 μg/mL, and mixtures were incubated for 2.5 h at 25°C with 3 nM [^{3}H]U69,593 in a total volume of 0.65 mL guinea pig brain homogenate. Guinea pig cortices and

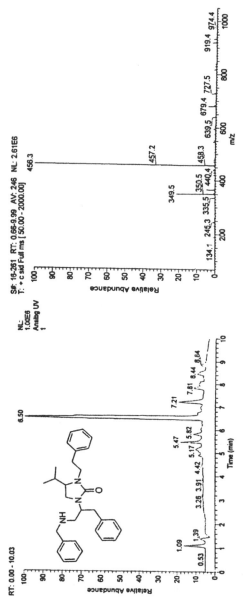

Figure 16 RP-HPLC and mass spectrum of a representative cyclic urea.

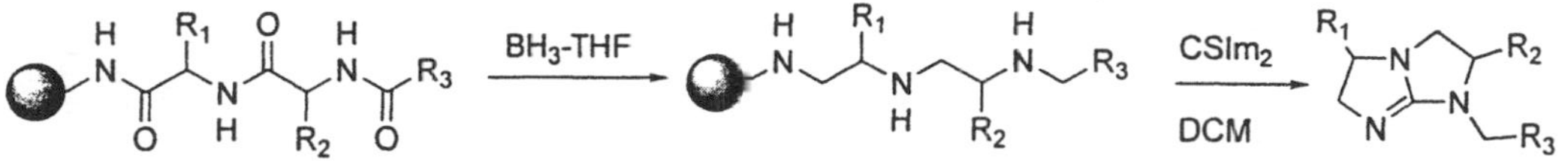

Figure 17 Synthesis of bicyclic guanidines 27 from reduced acylated dipeptides.

Table 1 Activity of Several Individual Bicyclic Guanidines Following
Deconvolution and Synthesis

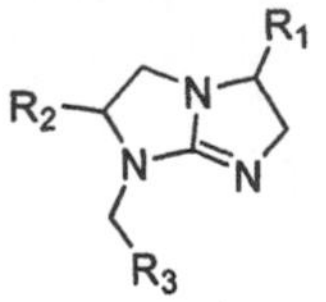

R_1	R_2	R_3	IC_{50} (nM)
Methyl	p-Methoxybenzyl	Cyclohexylpropyl	39
Methyl	p-Methoxybenzyl	1-Adamantylethyl	238
Cyclohexyl	p-Methoxybenzyl	1-Adamantylethyl	341
Methyl	p-Methoxybenzyl	Cyclohexylpropyl	502
Cyclohexyl	p-Methoxybenzyl	Cyclohexylpropyl	547

cerebella were homogenized in 40 mL of buffer A. Homogenates were centrifuged and incubated as described before. Each assays tube contained 0.5
mL of membrane suspension and 3 nM tritiated U69,593 in a total volume
of 0.65 mL. Assay tubes were incubated for 2.5 h at 25°C. The assay was
terminated and filtration and counting were performed as already described.
Unlabeled U50,488 was used as a competitor to generate a standard curve
and determine nonspecific binding.

5. Hydantoins and Thiohydantoins

Reaction of the N-terminal amino group of resin-bound dipeptide with phosgene (or triphosgene) leads to an intermediate isocyanate that reacts intra-
molecularly to form the five-membered ring hydantoin **28** (Fig. 18). The
solid-phase synthesis of numerous individual hydantoins was successfully
carried out, prompting us to design a library containing 5832 hydantoins.
The library was composed of 72 mixtures (R_1 position) each containing 81
individual compounds (R_2 position). In order to increase both the number
and class of available compounds, an additional library that was selectively
alkylated at two positions was prepared. In this library, the bond of the resin-
bound amide was first alkylated using LiOtBu and various alkylating agents,
yielding secondary amides bearing methyl, ethyl, allyl, benzyl, and 2-
naphthylmethyl groups. Following removal of the trityl protecting group and
neutralization, a second amino acid was coupled and the Fmoc protecting
group removed with 20% piperidine in DMF. After formation of the hydan-

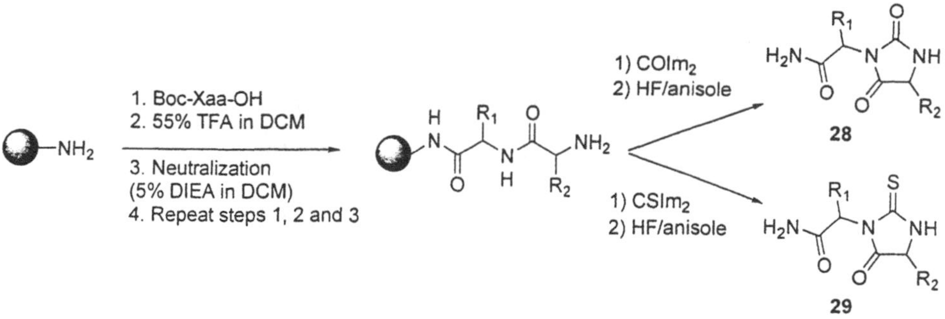

Figure 18 Synthesis of hydantoins **28** and thiohydantoins **29** from resin-bound dipeptides.

toin ring as described earlier, a second selective *N*-alkylation was performed using sodium cyclopentyldienylide as the base. The dialkylated hydantoin library synthesized was derived from 54 and 60 amino acids in the R_1 and R_3 positions, respectively, whereas R_2 was H, Me, and Bn and R_4 was Me, Et, All, and Bn. Furthermore, a positional scanning deconvolution format was used by synthesizing two distinct libraries, $O_1O_2X_3X_4$ and $X_1X_2O_3O_4$, that had two mixture (X) and two defined (O) positions each. The RP-HPLC and LC mass spectra of hydantoin **28a** obtained from phenylalanine and valine are shown in Fig. 19.

The library was tested in a competitive binding asay at the σ opioid receptor against [^{3}H]pentazocine. Initial screening revealed the importance of nonalkylated or benzylated hydrophobic amino acids in the R_1–R_2 positions [median inhibitory concentration (IC_{50}) in the μM range] and of benzylated basic amino acids in the R_3–R_4 positions (IC_{50}s in the 100 nM range). Benzylated 7-aminoheptanoic acid was included in the latter list as it was drastically different from the other hits and could therefore lead to interesting results in terms of structure–activity relationships (SARs). From these results, 12 individual hydantoins were synthesized and tested. These analogues showed improved binding affinities compared with the mixtures, as two of the individual dialkylated hydantoins had IC_{50}s of approximately 60 nM (Table 2) [62].

Similarly, thiohydantoins **29** were prepared by using thiophosgene or thiocarbonyldiimidazole as the reagent. A combinatorial library containing 6364 individual thiohydantoins was then assembled in the positional scanning format. For this library, 74 and 86 amino acids were used in the R_1 and R_3 positions, respectively.

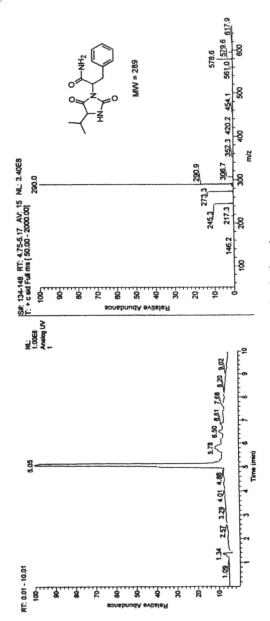

Figure 19 RP-HPLC and mass spectrum of a representative hydantoin.

Table 2   Activity of Several Individual Hydantoins Following Deconvolution and Synthesis

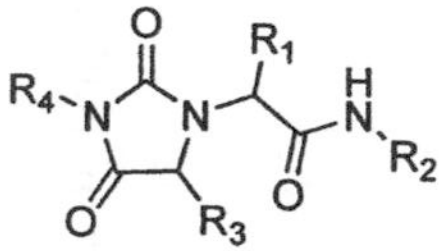

R_1	L-Ile	L-Nva	L-Nva	L-Ile	D-Cha	D-Cha	L-Ile	D-Cha	L-Nva	L-Nva	D-Cha	L-Ile
R_2	Bn	Bn	Bn	Bn	H	H	Bn	H	Bn	Bn	H	Bn
R_3	L-Lys	D-Lys	L-Lys	D-Lys	L-Lys	D-Lys	L-Orn	L-Orn	L-Orn	7-Aha	7-Aha	7-Aha
R_4	Bn	Bn	Bn	Bn	Bn	Bn	Bn	Bn	Bn	BN	Bn	Bn
IC_{50} (nM)	62	64	80	125	189	254	333	406	435	2417	3630	4615

III. CONCLUSION

The use of mixtures of compounds in the positional scanning combinatorial library format versus the use of individual compound arrays is clearly cost beneficial, allowing universities, research institutes, and small biotechnology companies to carry out the initial stages of the drug discovery process. One disadvantage of the use of mixtures is that a compound with the desired activity of interest could be overlooked. Although this is a legitimate concern, in practical terms we believe that the extreme cost and time savings make the incorporation of mixtures in the positional scanning format a powerful alternative to the use of large compound arrays. In fact, one would be best served by using both.

Extensive studies of the use of mixtures carried out by this laboratory and others have enabled the rapid, cost-effective identification of extremely active, highly specific individual compounds as reviewed in this chapter. There appears to be widespread skepticism that mixtures in combination with the positional scanning approach are effective in identifying active heterocycles, although they are widely accepted as methods for identifying peptides. However, these methods clearly are extremely effective and broadly applicable, as has been shown in this chapter and in many other studies. There is also nothing inherently unique about peptides or other oligomers that permits their successful use in these formats as compared with heterocycles.

Thus, the balance between classical high-throughput screening methods that have been used for individual compounds over the past 40 years and the use of mixture-based combinatorial libraries is tied to the balance between the need for perhaps overly complete and sometimes unnecessary data acquisition and the pragmatic and rapid gathering of compound information for lead development.

Combinatorial chemistry, in all its manifestations over the past 6 to 7 years, has fundamentally changed basic research and drug discovery. When coupled with other approaches such as computer-assisted design and molecular biology, combinatorial chemistry can be expected to enhance and continually increase the speed and thoroughness of drug discovery.

ACKNOWLEDGMENTS

The authors' work was funded in part by National Science Foundation grant CHE-9520142.

REFERENCES

1. RB Merrifield. J Am Chem Soc 85:2149–2154, 1963.
2. RB Merrifield. Science 232:341–347, 1986.
3. J Eichler, JR Appel, SE Blondelle, CT Dooley, B Dörner, JM Ostresh, E Pérez-Payá, C Pinilla, RA Houghten. Med Res Rev 15:481–496, 1995.
4. JM Stewart, JD Young. Solid-Phase Peptide Synthesis. 2nd ed. Rockford, IL: Pierce Chemical Company, 1984.
5. E Atherton, RC Sheppard. Solid-Phase Peptide Synthesis—A Practical Approach. Oxford: IRL Press, 1989.
6. HM Geysen, SJ Barteling, RH Meloen. Proc Natl Acad Sci U S A 82:178–182, 1985.
7. RA Houghten. Proc Natl Acad Sci U S A 82:5131–5135, 1985.
8. A Giannis, T Kolter. Angew Chem Int Ed Engl 32:1244–1267, 1993.
9. RMJ Liskamp. Angew Chem Int Ed Engl 33:305–307, 1994.
10. B Dörner, GM Husar, JM Ostresh, RA Houghten. Bioorg Med Chem 4:709–715, 1996.
11. JS Fruchtel, G Jung. Angew Chem Int Ed Engl 35:17–42, 1996.
12. JS Fruchtel, G Jung. In: G Jung, ed. Combinatorial Peptide and Nonpeptide Libraries. Weinheim: Verlag Chemie, 1996, pp 19–78.
13. SP Douglas, DM Whitfield, JJ Krepinsky. J Am Chem Soc 117:2116–2117, 1995.
14. M Schuster, P Wang, JC Paulson, C-H Wong. J Am Chem Soc 116:1135–1136, 1994.
15. JM Frecht, C Schuerch. J Am Chem Soc 492–496, 1971.
16. CC Leznoff. Acc Chem Res 11:327–333, 1978.
17. CC Leznoff, W Sywanyk. J Org Chem 42:3203–3205, 1977.
18. JI Crowley, H Rapoport. Acc Chem Res 9:135–144, 1976.
19. BA Bunin, JA Ellman. J Am Chem Soc 114:10997–10998, 1992.
20. A Nefzi, JM Ostresh, RA Houghten. Chem Rev 97:449–472, 1997.
21. FJ Sardina, H Rapoport. Chem Rev 96:1825–1872, 1996.
22. C Pinilla, JR Appel, P Blanc, RA Houghten. Biotechniques 13:901–905, 1992.
23. JM Ostresh, JH Winkle, VT Hamashin, RA Houghten. Biopolymers 34:1681–1689, 1994.
24. RA Houghten, C Pinilla, SE Blondelle, JR Appel, CT Dooley, JH Cuervo. Nature 354:84–86, 1991.
25. The Economist. 14 (December):91–92, 1991.
26. KS Lam, SE Salmon, EM Hersh, VJ Hruby, WM Kazmierski, RJ Knapp. Nature 354:82–84, 1991.
27. DAM Konings, JR Wyatt, DJ Ecker, SM Freier. J Med Chem 39:2710–2719, 1996.
28. L Wilson-Lingardo, PW Davis, DJ Ecker, N Hebert, O Acevedo, K Sprankle, T Brennan, L Schwarcz, SM Freier, JR Wyatt. J Med Chem 39:2720–2726, 1996.
29. AK Szardenings, TS Burkoth, GC Look, DA Campbell. J Org Chem 61:6720–6724, 1996.

30. PT Ho, D Chang, JWX Zhong, GF Musso. Peptide Res 6:10–12, 1993.
31. DW Gordon, J Steele. Biomed Chem Lett 5:47–50, 1995.
32. MC Griffith, CT Dooley, RA Houghten, JS Kiely. In: IM Chaiken, KD Janda, eds. Molecular Diversity and Combinatorial Chemistry: Libraries and Drug Discovery. Washington, DC: American Chemical Society, 1996, pp 50–57.
33. C Holmes, J Chinn, G Look, E Gordon, M Gallop. J Org Chem 60:7328–7333, 1995.
34. AA MacDonald, SH DeWitt, EM Hogan, R Ramage. Tetrahedron Lett 37:4815–4818, 1996.
35. A Pictet, T Spengler. Chem Ber 44:2030–2036, 1911.
36. JP Mayer, D Bankaitis-Davis, J Zhang, G Beaton, K Bjergarde, CM Anderson, BA Goodman, CJ Herrera. Tetrahedron Lett 37:5633–5636, 1996.
37. L Yang, L Guo. Tetrahedron Lett 37:5041–5044, 1996.
38. B Ruhland, A Bhandari, EM Gordon, MA Gallop. J Am Chem Soc 118:253–254, 1996.
39. MM Murphy, JR Schullek, EM Gordon, MA Gallop. J Am Chem Soc 117:7029–7030, 1995.
40. M Patek, B Drake, M Lebl. Tetrahedron Lett 36:2227–2230, 1995.
41. M Mutter, A Nefzi, T Sato, X Sun, F Wahl, T Woehr. Peptide Res 8:145–153, 1995.
42. T Woehr, F Wahl, A Nefzi, B Rohwedder, T Sato, X Sun, M Mutter. J Am Chem Soc 118:9218–9227, 1996.
43. Y Pei, RA Houghten, JS Kiely. Tetrahedron Lett 38:3349–3352, 1997.
44. A DomLing, I Ugi. Angew Chem Int Ed Engl 32:563–567, 1993.
45. P Biginelli. Gazz Chim Ital 23:360–366, 1893.
46. AMM Mjalli, BE Toyonaga. In: IM Chaiken, KD Janda, eds. Molecular Diversity and Combinatorial Chemistry: Libraries and Drug Discovery. Washington, DC: American Chemical Society, 1996, pp 70–80.
47. C Zhang, EJ Moran, TF Woiwode, KM Short, AMM Mjalli. Tetrahedron Lett 37:751–754, 1996.
48. W Meutermans, P Alewood. Tetrahedron Lett 36:7709–7712, 1995.
49. S Hutchins, K Chapman. Tetrahedron Lett 37:4865–4868, 1996.
50. Y Pei, WH Moos. Tetrahedron Lett 35:5825–5828, 1994.
51. D Goff, R Zuckermann. Tetrahedron Lett 37:6247–6250, 1996.
52. D Gordon, J Steele. Biomed Chem Lett 5:47–50, 1995.
53. V Krchňák, AS Weichsel, D Cabel, M Lebl. In: IM Chaiken, KD Janada, eds. Molecular Diversity and Combinatorial Chemistry: Libraries and Drug Discovery. Washington, DC: American Chemical Society, 1996, pp 99–117.
54. Benzodiazepines Divided: A Multidisciplinary Review. New York: Wiley, 1983.
55. LH Sternbach. J Med Chem 22:1–6, 1979.
56. MG Bock, RM DiPardo, BE Evans, KE Rittle, WL Whitter, DF Veber, PS Anderson, RM Freidinger. J Med Chem 32:13–16, 1989.
57. DA Goff, RN Zuckermann. J Org Chem 60:5744–5745, 1995.
58. SH DeWitt, JS Kiely, CJ Stankovic, MC Schroeder, DMR Cody, MR Pavia. Proc Natl Acad Sci U S A 90:6909–6913, 1993.
59. SH DeWitt, AW Czarnik. Acc Chem Res 29:114–122, 1996.

60. L Moroder, J Lutz, F Grams, S Rudolph-Bohner, G Ospay, M Goodman, W Kolbeck. Biopolymers 38:295–300, 1996.
61. B Dressman, L Spangle, S Kaldor. Tetrahedron Lett 37:937–940, 1996.
62. A Nefzi, CT Dooley, JM Ostresh, RA Houghten. Biomed Chem Lett 8:2273–2278, 1998.
63. A Nefzi, JM Ostresh, RA Houghten. Tetrahedron Lett 38:4943–4946, 1997.
64. A Nefzi, M Giulianotti, RA Houghten. Tetrahedron Lett 39:3671–3674, 1998.
65. JM Ostresh, CC Schoner, VT Hamashin, A Nefzi, J-P Meyer, RA Houghten. J Org Chem 63:8622–8623, 1998.
66. B Dörner, JM Ostresh, SE Blondelle, CT Dooley, RA Houghten. In: A. Abell, ed. Advances in Amino Acid Mimetics and Peptidomimetics. Greenwich, CT: JAI Press, 1997, pp 109–125.
67. JM Ostresh, B Dörner, RA Houghten. In: S. Cabilly, ed. Combinatorial Peptide Library Protocols. Totowa, NJ: Humana Press, 1998, pp 41–49.
68. B Dörner, JM Ostresh, GM Husar, RA Houghten. Methods Mol Cell Biol 6:35–40, 1996.
69. B Dörner, GM Husar, JM Ostresh, RA Houghten. Bioorg Med Chem 4:709–715, 1996.
70. PYS Lam, PK Jadhav, CJ Eyermann, CN Hodge, Y Ru, LT Bacheler, JL Meek, MJ Otto, MM Rayner, YN Wong, C-H Chang, PC Weber, DA Jackson, TR Sharpe, S Erickson-Viitanen. Science 263:380–384, 1994.
71. JM Kim, TE Wilson, TC Norman, PG Schultz. Tetrahedron Lett 37:5309–5312, 1996.
72. JM Ostresh, GM Husar, SE Blondelle, B Dörner, PA Weber, RA Houghten. Proc Natl Acad Sci U S A 91:11138–11142, 1994.

16

Solid-Phase Synthesis of Pseudopeptides and Oligomeric Peptide Backbone Mimetics

Gilles Guichard
Institut de Biologie Moléculaire et Cellulaire, Strasbourg, France

I. INTRODUCTION

The development of peptide drugs and potential applications of peptides as immunogens or as immunomodulators are often impaired by their high susceptibility to enzymatic degradation and their rapid clearance from the circulation. Moreover, the bioactive conformations of peptides are often poorly defined as a result of their inherent flexibility. With the aim of circumventing these problems, intense research has been focused in the past 20 years on chemical modifications of the peptide backbone [1], i.e., "backbone chemistry." Enhanced biological lifetime and stabilization of secondary structures can be achieved through the incorporation of D-amino acids, N-alkyl-α-amino acids, or other nonnatural constrained amino acid derivatives. Alternatively, the introduction of nonhydrolyzable peptide bond surrogates (denoted ψ[ ] according to Spatola's nomenclature for pseudopeptides [1]) or the replacement of the $C^{\alpha}H$ group by a nitrogen (azapeptides) also proved useful in converting biologically active peptides into more stable molecules with similar or enhanced biological activity, with improved selectivity, or with antagonist properties [1–7]. Methylene amino ψ[CH$_2$NH] [8,9], retro-inverso ψ[NHCO] [10–13], (E)- or (Z)-alkene ψ[CH=CH] [14,15], carba ψ[CH$_2$CH$_2$] [16,17], hydroxyethylene ψ[CHOHCH$_2$] [18–20], methyleneoxy ψ[CH$_2$O] [21–24], and ketomethylene ψ[COCH$_2$] [25,26], are the most popular amide bond replacements found in the literature. Although most synthetic routes to these modifications have been developed in solution, successful approaches to the solid-phase synthesis of ψ[NHCO] [27] and ψ[CH$_2$NH] [28] pseudopeptides were proposed as early as 1983 and 1987, respectively, allowing the rapid generation of large and diverse sets of pseudopeptides for biological testing. Since these pioneering works, the synthesis of several other peptide bond surrogates has been made accessible on solid supports, including (E)-alkene [29], aza(−azXaa-) [30–32], iminoaza (-Xaa ψ[CH=NH]azXbb-) [32], reduced aza (-Xaa ψ[CH$_2$NH]azXbb-) [32], carbaza (-Xaa ψ[CH$_2$CH$_2$]azXbb-) [33], N-hydroxyamide ψ[CONOH] [34], and depsipeptide ψ[COO] [35]. In addition, important synthetic efforts have been made to synthesize on the solid phase protease inhibitors incorporating stable transition state mimics such as hydroxyethylene, hydroxyethyleneamine [36–38], and diamino diol [39].

An exciting new focus of backbone chemistry is the creation of novel oligomeric compounds with specific biological functions [40] and/or with well-defined secondary structures [41,42]. An important step forward was made with the demonstration that β- and γ-peptides [40–47] and chiral oligo-N-substituted glycines (NSGs) also called peptoids [48,49] can form stable folded structures including helical and pleated-sheet structures in solution or in the solid state. Furthermore, oligomeric scaffolds with improved pharmacokinetics compared with peptides would be of major interest for new lead discovery. Synthesis and screening of a library of peptoids led to the identification of nanomolar ligands for the α_1-adrenergic receptor or the μ-opiate receptor [50]. The combinatorial approach also allowed the discovery of efficient cationic peptoid reagents for gene delivery [51]. Shultz and coworkers [52] have developed high-affinity ligands for the integrin GPIIb/IIIa based on linear and cyclic oligocarbamates. Human immunodeficiency virus type 1 (HIV-1) Tat-derived oligocarbamates and oligourea that bind the transactivation responsive region (TAR) RNA have been identified by Rama and coworkers [53,54]. For rapid screening of such potentially bioactive oligomers, there is a need for highly efficient syntheses of these molecules on solid supports from appropriate monomeric or submonomeric precursors. Since early reports on oligo-N-substituted glycines [55,56], the syntheses of a variety of unnatural oligomers including oligocarbamates [57], oligoureas [58,59], oligothioureas [60], β-peptides [61,62], β-peptoids [63], ureapeptoids [64], oligosulfonamides [65,66], and depsides [35] have been successfully achieved on solid supports.

II. METHYLENE AMINO ISOSTERES ψ[CH$_2$NH]

In the past 15 years, peptide analogues containing the reduced amide bond ψ(CH$_2$—NH) have been extensively investigated in biological and structural studies. Early syntheses of ψ(CH$_2$—NH) dipeptides involved a nucleophilic substitution reaction [67] or direct reduction of the amide bond in dipeptides with borane [68]. Reduction of protected endothiopeptides [69] with Raney nickel has also been proposed [70]. However, the preferred method for the introduction of the reduced peptide bond is reductive amination of N-protected α-amino aldehydes. Numerous reducing agents including H$_2$ in the presence of metal catalysts [71], zinc–acetic acid [72], borane–pyridine complex [73], sodium cyanoborohydride (NaBH$_3$CN) [8,74], and sodium triacetoxyborohydride [NaBH(OAc)$_3$] [75] have been evaluated for reductive amination of aldehydes and ketones. A common procedure for the preparation of ψ[CH$_2$NH] peptides involves the use of NaBH$_3$CN in acidified methanol [8,9]. However, NaBH(OAc)$_3$ in dichloroethane (DCE) or tetrahydrofuran (THF) is an interesting alternative because the reaction proceeds more rapidly and gives consistently better yields. The adaptation of this method to solid-phase technology was reported in 1987 by Sasaki and Coy [28,76]. In this case, the reduced amide bond is formed in situ on the resin (Scheme 1), which significantly improves the speed of synthesis as no extensive workup and purification are needed. The reaction usually proceeds by addition of a solution of freshly prepared N-protected α-amino aldehyde in DMF containing 1% acetic acid followed by addition of a solution of NaBH$_3$CN in DMF to the free amino group of the resin-bound peptide [28]. DMF is used instead of MeOH to allow greater swelling of the resin. The reaction can be monitored either by the ninhydrin test, as the secondary amine formed during the reductive amination produces only a weak coloration upon heating, or by the chloranil test (see Chapter 6).

Han and Chorev [77] reported a novel one-pot reductive alkylation of amines by S-ethyl thioesters mediated by triethylsilane and NaBH(OAc)$_3$ in the presence of Pd/C. This novel procedure is particularly attractive when the aldehyde is not stable enough to allow isolation. However, it is limited to solution-phase synthesis.

A. *N*-Protected-α-Amino Aldehydes

Numerous procedures have been reported for the synthesis of N-protected-α-amino aldehydes [78]. The N-protected α-amino aldehydes can be prepared either by oxidation of the corresponding N-protected β-amino alcohols with oxidants such as pyridinium dichromate [79,80] or SO$_3$–pyridine–dimethyl sulfoxide (DMSO) [81] or under Swern conditions [DMSO–oxalyl chloride–N,N-diisopropylethylamine (DIEA)] [81–83], by reduction of esters with diisobutyl aluminum hydride (DIBAL) [84], by reduction of N,N-disubstituted amides [85–87], by reduction of urethane-protected N-carboxy-

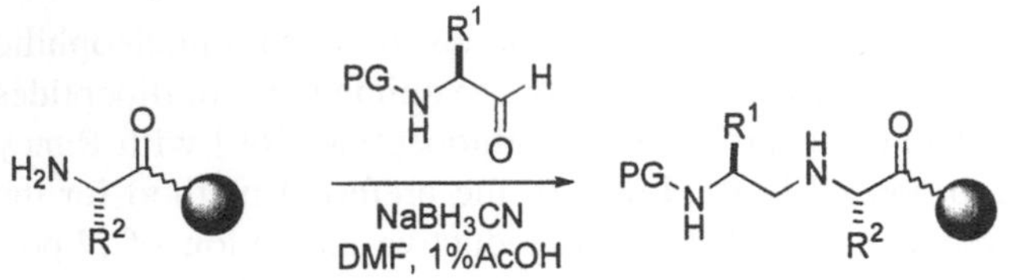

Scheme 1 Synthesis of ψ[CH$_2$NH] isosteres by reductive amination of N-protected α-aminoaldehydes on a solid support. PG, protecting group (Boc or Fmoc).

anhydrides (UNCAs) with LiAlH(O*t*Bu)$_3$ [88], or by reduction of thioesters with Et$_3$SiH in the presence of Pd/C [89,90].

In 1983, Fehrentz and Castro [86] described an efficient synthesis of optically pure N-tert-butyloxycarbonyl (Boc)-α-amino aldehydes **3** by reduction of the corresponding N,O-dimethylhydroxamates **1** (Weinreb amide [91]) with LiAlH$_4$ at 0°C (Scheme 2). Overreduction to an alcohol is precluded by the formation of an intermediate lithium salt **2** that upon hydrolysis affords the desired aldehyde. Yields are generally excellent and purification on silica gel is not necessary.

The method has been widely used and extended to the preparation of N-9-fluorenylmethoxycarbonyl (Fmoc)-α-amino aldehydes in good yields [83,92–94]. Fmoc- and Boc-β-amino aldehydes, which are key intermediates in the synthesis of carba [16,17] and carbaza peptides [33] (see later), are also prepared in good yield from the corresponding hydroxamates [95]. In this case, the Weinreb amide can be obtained in high yield by direct Wolff rearrangement of a diazomethylketone with N,O-dimethylhydroxylamine.

Protected aspartic and glutamic acid aldehydes, however, are difficult to obtain by the Fehrentz and Castro procedure because carboxylic esters are not stable under these hydride reduction conditions. Interestingly, Fmoc-

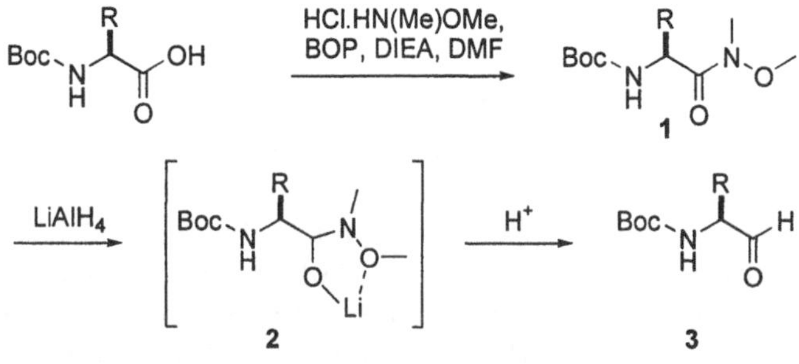

Scheme 2 Synthesis of N-Boc-α-amino aldehydes by reduction of their corresponding N,O-dimethylhydroxamates [86].

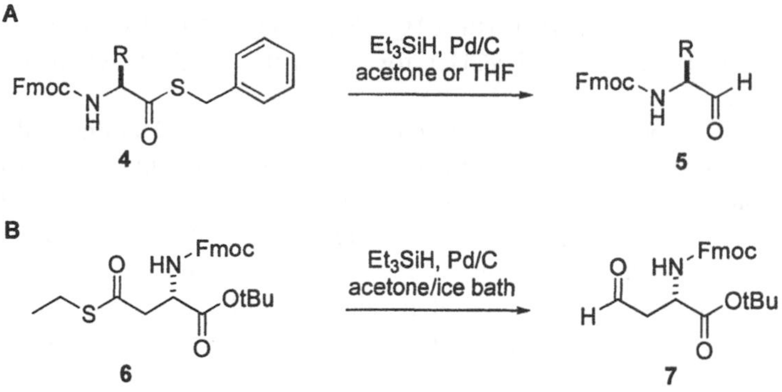

A

4   Et₃SiH, Pd/C   5

B

6   7

Scheme 3 Synthesis of *N*-Fmoc–protected amino aldehydes by reduction of thioesters [77,90].

Glu(O*t*Bu)-H was prepared in good yield (82%) by conducting the reaction at −78°C [83].

Reduction of *S*-ethyl [89] and *S*-benzyl thioesters [90] by triethylsilane in acetone in the presence of Pd/C as a catalyst also provides aldehydes in good yield. Ho and Ngu [90] have prepared *N*-Fmoc-α-amino aldehydes **5** from the corresponding *S*-benzyl thioesters **4** (Scheme 3), and *N*-protected Fmoc-Asp(O*t*Bu)-H, for example, could be obtained in 75% yield under these essentially neutral conditions. Fmoc-Glu(O*t*Bu)-H and Boc-Glu(O*c*Hx)-H have also been prepared by this procedure although in somewhat lower yield (54 and 42%, respectively) (G. Guichard, unpublished data). Recently, *tert*-butyl 2(*S*)-4-oxo-9-fluorenyl-methoxycarbonylamino-butyrate **7** was prepared in excellent yield (93%) by reduction of the corresponding *S*-ethylthioester **6** in acetone at a temperature kept below 20°C [77].

The Swern oxidation of *N*-Fmoc–protected β-amino alcohols **8** (Scheme 4) and the reduction of Weinreb amides prepared from Fmoc-α-amino acids with LiAlH₄ at −78°C were compared [83]. Both approaches afforded comparable synthetic yields, typically between 70 and 90% of the

8   DMSO/(COCl)₂/DIEA   5

Scheme 4 Synthesis of *N*-Fmoc–protected amino aldehydes by Swern oxidation of *N*-protected β-amino alcohols [83].

desired aldehyde **5** [R = *i*Pr, *i*Bu, Bn, CH₂O*t*Bu, (CH₂)₂COO*t*Bu] except in the case of methionine (R = CH₂SCH₃), for which the yield of Swern oxidation was low.

Stability of *N*-protected and particularly Boc-protected amino aldehydes **3** is rather limited. They are configurationally labile and may racemize even with rapid chromatography on silica gel [80]. Thus, they should be prepared just before use and storage for a long period of time, even under argon, should be avoided. However, as reported by Ho and Hgu [90], *N*-Fmoc-α-amino aldehydes **5** obtained after purification by flash chromatography on silica gel and eluted with organic solvents in the presence of 0.1% pyridine undergo little or insignificant racemization.

B. General Procedures for the Preparation of N-Protected Amino Aldehydes

1. Preparation of the N,O-Dimethyl Hydroxamates [86]

To a stirred solution of the *N*-Fmoc– or Boc-protected α-amino acid (10 mmol) and Benzotriazol-1-yloxytris(dimethylamino) phosphonium hexafluorophosphate (BOP) (10 mmol, 4.42 g) in DMF (30 mL) at room temperature are successively added HCl·HN(Me)OMe (11 mmol, 1.09 g) and DIEA (25 mmol, 4.36 mL). After stirring for 2 h, ethyl acetate (100 mL) is added followed by a saturated sodium bicarbonate solution (60 mL). The organic phase is then washed with saturated sodium bicarbonate solution (2 × 60 mL), water (60 mL), 1 M potassium hydrogen sulfate (2 × 60 mL), and saturated NaCl solution; dried over magnesium sulfate; and concentrated in vacuo. The crude product can be purified by column chromatography to give the pure *N,O*-dimethylhydroxamate.

2. Synthesis of N-Protected α-Amino Aldehydes 3 and 5 by Reduction of Their Corresponding N,O-Dimethylhydroxamate [86]

The *N,O*-dimethyl hydroxamate (3 mmol) is dissolved in 25 mL of anhydrous THF. The reaction mixture is cooled down to 0°C and LiAlH₄ (6 mmol, 228 mg) is added portionwise. After stirring the reaction for 30 min, ethyl acetate (70 mL) is added first, followed by a 1 M potassium hydrogen sulfate solution (50 mL), and the mixture is stirred for 30 min. The organic layer is then collected, washed with 1 M potassium hydrogen sulfate (2 × 50 mL) and a saturated NaCl solution, and dried over magnesium sulfate. The solvent is removed in vacuo to give the *N*-protected α-amino aldehyde. The α-amino aldehyde obtained by this method can be used without further purification as the thin-layer chromatographic (TLC) analysis shows only small amounts of impurities. If arginine aldehydes are required, the use of N^α-Boc-$N^\varepsilon,N^\varepsilon$-bisbenzyloxycarbonyl arginine avoids the formation of a cyclic aminal and allows the preparation of the aldehyde in good yield (two steps, 68%) [92].

3. Preparation of S-Benzyl Thioesters [90]

To a cold 0°C solution of N-Fmoc– or Boc-protected α-amino acid (10 mmol) and benzyl mercaptan (20 mmol, 2.48 g) in dichloromethane (30 mL) are successively added 4-(dimethylamino)pyridine (DMAP) (1 mmol, 122 mg) and 1,3-dicyclohexylcarbodiimide (DCC) (10.5 mmol, 2.17 g). After the reaction mixture is stirred overnight and the solvent removed in vacuo, ethyl acetate (100 mL) and a 1 N potassium hydrogen sulfate solution are added. The organic layer is washed with 1 N potassium hydrogen sulfate (2 × 60 mL) and saturated NaCl solution (60 mL), dried over magnesium sulfate, and filtered. The solvent is removed in vacuo and the S-benzyl thioester is recovered as a white solid after precipitation with ether–hexane (1:1, v/v).

4. Synthesis of the N-Protected α-Amino Aldehyde by Reduction of S-Benzyl Thioester [90]

The S-benzyl thioester (5 mmol) is dissolved in 25 mL of anhydrous THF or acetone containing 10% Pd/C as a catalyst (25% by weight) and the reaction mixture is cooled down to 0°C. A solution of triethylsilane (20 mmol, 2.32 g) in THF or acetone (5 mL) is then slowly added. After the reaction is stirred for 2 h at 10–20°C, the catalyst is removed by filtration and the solvent is evaporated in vacuo. The residue is purified by flash chromatography with 20–30% (v/v) ethyl acetate in hexane as eluant in the presence of 0.1% (v/v) pyridine. The solvent is finally removed in vacuo to leave the pure N-protected α-amino aldehyde.

5. Preparation of N-Protected β-Amino Alcohols [96]

To a cold (−15°C) solution of N-Fmoc– or Boc-protected α-amino acid (10 mmol) in 1,2-dimethoxyethane (10 mL) or THF (10 mL) are successively added NMM (1.11 mL, 10 mmol) and ethyl chloroformate (1.39 mL, 10 mmol). The reaction mixture is stirred at −15°C for 1 min and the precipitated NMM·HCl is rapidly filtered. An aqueous solution (5 mL) of NaBH$_4$ (570 mg, 15 mmol) is then added to the filtrate. After the evolution of gas is stopped, 250 mL of water is added at once. If the β-amino alcohol precipitates, it is filtered, washed with water and hexane, and dried under high vacuum. Otherwise, it is extracted with AcOEt and the organic layer is washed with 1 N potassium hydrogen sulfate (2 × 60 mL), saturated NaCl solution (60 mL), dried over magnesium sulfate, and filtered. The solvent is removed in vacuo to yield the pure β-amino alcohol.

6. Swern Oxidation of Fmoc-β-Amino Alcohol 8 [83]

To a cold −60°C solution of oxalyl chloride (26 μL, 0.3 mmol) in CH_2Cl_2 (0.5 mL), DMSO (43 μL, 0.6 mmol) is added under N_2 and the solution is stirred for 10 min. A solution of **8** (0.2 mmol) in CH_2Cl_2 (1 mL) is then added dropwise with a syringe. The reaction mixture is stirred at −60°C under N_2 for 15 min and DIEA (209 μL, 1.2 mmol) is added. The reaction is then allowed to warm to room temperature and water (5 mL) is added to the reaction mixture. The mixture is stirred at room temperature for an additional 10 min and extracted twice with CH_2Cl_2 (60 mL). The organic layer is washed with 1 N HCl solution and saturated NaCl solution, dried over anhydrous $MgSO_4$, and concentrated in vacuo. Flash chromatography yields the pure aldehyde **5**.

C. Procedure for the Synthesis of the Reduced Amide Bond on Solid Support [28,76]

After Boc or Fmoc deprotection, the resin is washed with DMF and a nihydrin test is performed. Aldehyde **3** or **5** (2.5 equiv.) is dissolved in 2 mL of DMF containing 1% acetic acid and the solution is added to the resin. After 1 min of agitation under nitrogen bubbling, a solution of $NaBH_3CN$ (3 equiv.) in 2 mL of DMF containing 1% acetic acid is added in one portion to the resin and the resin mixture is agitated with nitrogen bubbling. Typically, when N-Boc α-amino aldehyde **3** is used, the reaction is complete within 1 h as determined by the ninhydrin test. Otherwise, the reductive amination step is repeated using 3 equiv. of both N-protected α-amino aldehyde and the reducing agent as long as a positive ninhydrin test is obtained. All the couplings following the reduced peptide bond formation are performed using a twofold excess of N-Boc-α-amino acid and activation reagent and a single coupling procedure. After the last deprotection step, the peptide is cleaved from the resin using standard cleavage protocols.

Notes:

    1.   N-Fmoc α-amino aldehydes **5** have been found to be less reactive than the corresponding N-Boc α-amino aldehydes **3**. Thus, a larger excess of aldehydes and longer reaction times might be necessary when using N-Fmoc α-amino aldehydes **5** (G. Guichard, unpublished results).

2. Because of its high reactivity compared with other amino aldehydes, *N*-protected glycinal should not be used in large excess. Further reaction of the aldehyde in excess with the secondary amino group of the reduced peptide bond and subsequent derivatization have been observed (G. Guichard, unpublished results, and Ref. 93).

3. The protection of the secondary amine formed during the reductive alkylation is not mandatory as it has been shown not to undergo significant side reactions during the acylation step with various amino acids. However, *N*-protected glycine has been reported to react at this site, causing subsequent derivatization. Hence, a minimum excess of glycine derivative should be used for the coupling step.

4. Ho et al. [97] cautioned about possible diastereoisomer formation during solid-phase synthesis of reduced peptides. They found that ineffective trapping of the intermediate imine might be responsible for racemization. In order to keep the level of diastereoisomers as low as possible (<2%), portionwise addition of NaBH$_3$CN should be precluded. The authors also suggest combining solutions of aldehyde and NaBH$_3$CN before adding to the resin [97].

III. RETRO-INVERSO PEPTIDOMIMETICS

Retro-inverso modification of biologically active peptides is one of the most widely used peptidomimetic approaches to designing new bioactive molecules with increased stability. Synthetic methodology, its potential applications in drug design, and conformational properties have been fully and regularly reviewed over the past 20 years [10–13]. The concept of retro-isomers in cyclic peptides was recognized by Prelog and Gerlach [98] in 1964 and later by Shemyakin et al. [99]. Extension to linear peptides has been hampered by the noncomplementarity of end groups between the two isomers. The introduction of *gem*-diaminoalkyl residues and C-2 substituted malonyl derivatives was proposed as one solution to the end group problem [10]. Alternatively, incorporation of these non–amino acid residues within a peptide sequence led to the creation of a novel class of peptidomimetics, referred to as partially modified retro-inverso pseudopeptides (PMRIs), in which only one or several peptide bonds are reversed [10,11,13]. The synthetic methodology leading to these analogues has been developed extensively by Goodman and coworkers and applied by many groups to the design of peptide hormone analogues [enkephalins, dermorphine, bradykinin, gastrin, cholecystokinin (CCK), substance P, somatostatin, thymopentin [10], sweeteners [10], protease inhibitors [10], major histocompatibility complex (MHC) ligands [100–102] and RGD mimetics [103]. Solid-phase approaches to PMRIs have also been reported and will be detailed in this section.

Finally, we shall emphasize the successful developments of retro-inverso (all-D-retro) and end group–modified retro-inverso peptides in several biological areas including hormones [104,105], antibody recognition [106–109], peptide vaccines [110], MHC class II presentation [111,112], protein surface and receptor mimicry [113–115], antibacterial and channel-forming peptides such as mellitin and cecropin A [116,117], intracellular delivery systems [118], and nuclear import [119].

A. Partially Modified Retro-Inverso (PMRI) Isomers

As shown in Fig. 1, the retro-inverso modification ψ[NH—CO] requires the replacement of two successive amino acid residues by a 2-substituted malonate derivative and a *gem*-diamino alkyl residue. Hence, Xaa-ψ[NHCO]-Xbb- can be termed -gXaa-mXbb- ("g" and "m" symbolize *gem*-diaminoalkyl and malonyl residues, respectively).

For the reversal of additional peptide bonds, amino acids of opposite configuration at $C^{\alpha}H$ are inserted between the malonyl and the *gem*-diamino alkyl residues.

The 2-substituted malonic acid monoester **13** required for the synthesis of PMRI peptides can be prepared by alkylation of malonate diesters **9** with alkyl halides and subsequent hemisaponification. Alternatively, reductive alkylation of Meldrum's acid (2,2-dimethyl-1,3-dioxane-4,6-dione) **11** with the

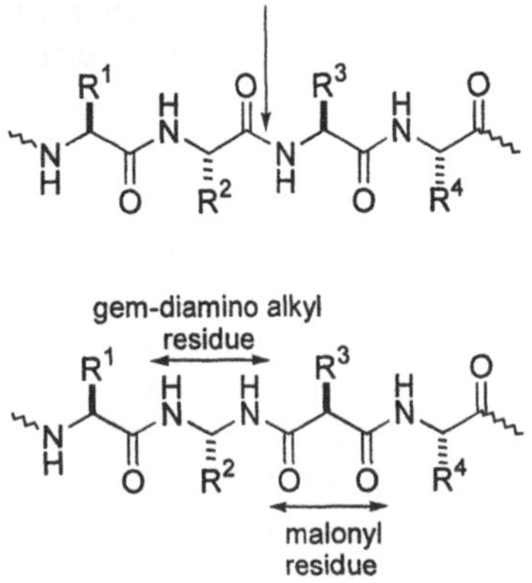

Figure 1   The parent peptide and a PMRI isomer.

appropriate aldehyde or ketone and borane-dimethyl amine complex [120] followed by ring opening in the presence of alcohol (Scheme 5) provides **13** [121,122]. One inherent problem in PMRI synthesis is the configurational instability of 2-monosubstituted malonyl residues under both neutral and basic conditions because of the fast tautomeric equilibrium between the keto and the coplanar enolic form. The incorporation of malonyl derivatives as a racemic mixture generates peptide diastereoisomers that might be separated and purified by reversed-phase high-performance liquid chromatography (RP-HPLC). However, isomerization at the resulting malonamide residue always occurs. The configurational stability at the 2-malonamide residue in the resulting peptides is profoundly influenced by the nature of adjacent side chains and by the conformational behavior of the peptide (by preventing a coplanar arrangement and/or deprotonation, steric hindrance and conformational strain may enhance the configurational stability at C-2). Depending on the peptide sequence and on the nature of the malonyl derivative, measured half-lives for isomerization vary between minutes and days. In some studies, configurational assignment of the resulting diastereoisomers has been proposed based on HPLC retention time or nuclear magnetic resonance (NMR) chemical shifts considerations as well as NOE measurements [11].

2-Methyl-2-alkyl-malonyl residues have been utilized by Dalpozzo and Laurita [123] to fix the configuration at the C-2 malonamide residue. Optically pure 2-fluoro-2-alkyl malonamide residues have also been reported [124].

Initially, *gem*-diamines were prepared from N-protected α-amino azide **14**. Intermediate isocyanates **15** obtained by Curtius rearrangement of **14** were immediately trapped with a primary alcohol to afford the corresponding N,N'-diacylated-1,1-diaminoalkane derivatives **16**, which were further deprotected to the monoacylated *gem*-diamine before coupling to carboxylic acids (Scheme 6) [125–128].

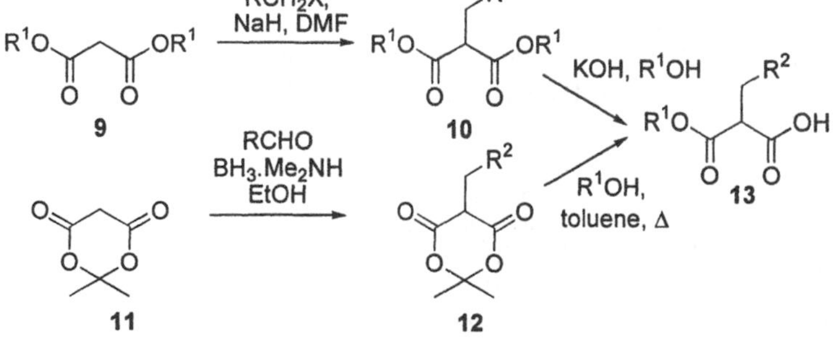

Scheme 5 Synthesis of C-2–substituted malonic acid monoester.

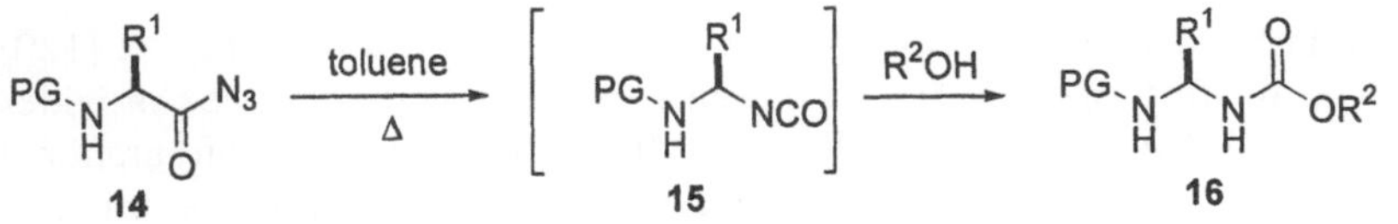

Scheme 6 Preparation of *N,N'*-diacylated-1,1-diamino alkane via Curtius rearrangement of *N*-acyl α-amino acylazide. (PG = Boc, Z, Boc-Xaa, Z-Xaa, Boc-peptide, or Z-peptide.)

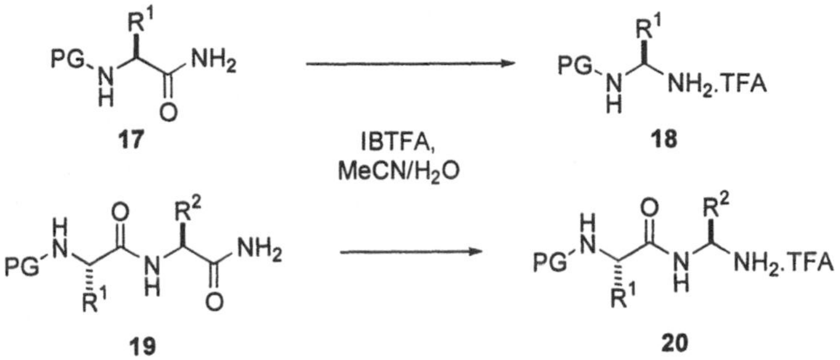

Scheme 7 IBTFA treatment of amides **17** and **19**.

However, optically pure *N*-acyl-1,1-diamino alkanes are most frequently and easily obtained through a Hofmann rearrangement (Scheme 7) by treatment of their corresponding carboxamide precursor with a mild oxidant such as iodobenzene bis(trifluoroacetate) (IBTFA) in a water–acetonitrile mixture [129,130]. Because *gem*-diamines **18** bearing a urethane-type protecting group tend to decompose under these conditions as a result of their greater sensitivity to hydrolysis as compared with the corresponding monoacylated *gem*-diamine [131], the reaction is performed on *N*-protected dipeptide carboxamide **19** to afford the corresponding *gem*-diaminoalkyl derivatives **20** in good yield [129].

Alternatively, IBTFA treatment of the malonylaminoacyl amide **21** allows the direct synthesis of a PMRI unit (**22**) (Fig. 2).

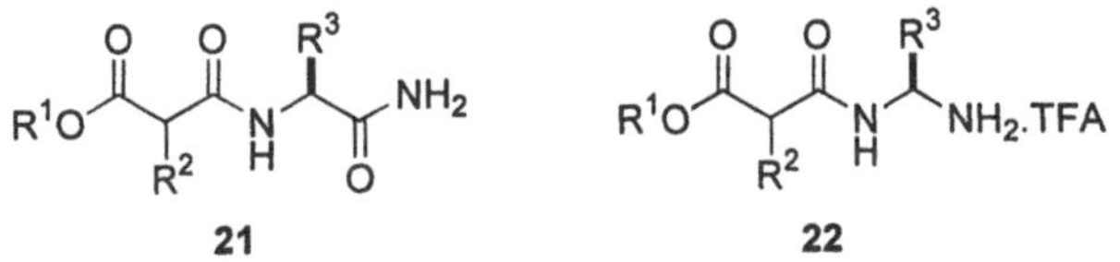

Figure 2 Malonylaminoacyl amide **21** and corresponding PMRI unit **22**.

In 1983, Pessi et al. [27,132] reported an efficient method for incorporating a retro-inverso amide bond into a growing peptide chain on a polyamide-type resin. The 2-alkylmalonyl-(D)-amino acid amide (**23**) prepared by classical synthesis in solution was coupled to the free amino group of the resin-bound peptide and the carboxamide function was then converted to the corresponding amine by treatment of the resin-bound peptide with IBTFA in DMF–H_2O. Elongation of the peptide chain was finally performed using a conventional solid-phase peptide synthesis procedure (Scheme 8). Cleavage of the peptide and reduction of oxidized methionine afforded the

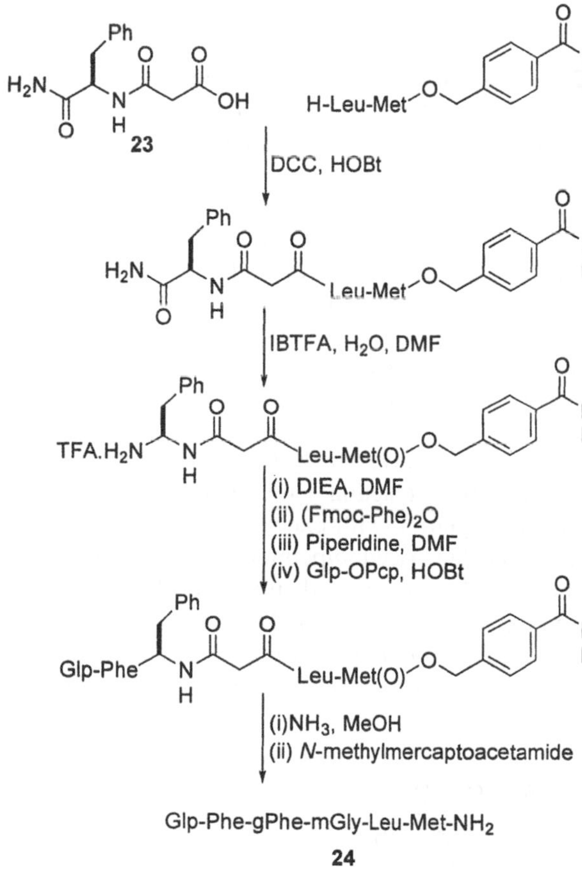

Glp-Phe-gPhe-mGly-Leu-Met-NH$_2$

24

Scheme 8 Solid-phase synthesis of the PMRI analogue of substance P: [Glp[6], gPhe[8], mGly[9]]SP(6–11) [27].

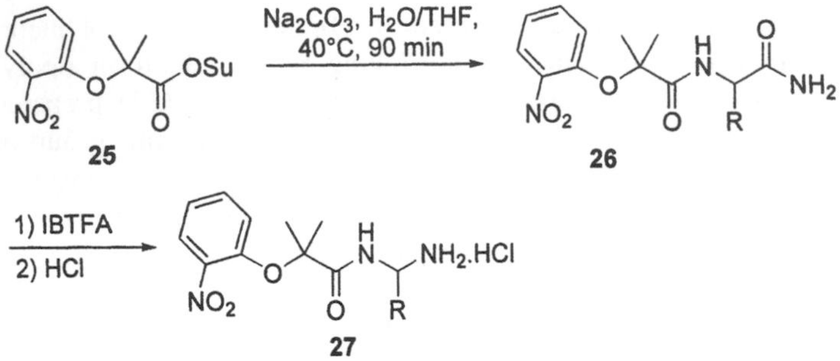

Scheme 9 Synthesis of stable Mnp-monoprotected *gem*-diamino alkyl residues
27 [122].

substance P analogue **24**. PMRI analogues of bradykinin [133] and of a
influenza nucleoprotein cytotoxic T-lymphocyte (CTL) epitope [100] have
also been synthesized successfully by a similar procedure.

Verdini and coworkers [122] have introduced the 2-methyl-2-(2′-ni-
tro)phenoxy-propionyl (Mnp) group for the preparation of stable monoprotected *gem*-diamino alkyl residues **27** via IBTFA treatment of the carboxamide precursor **26**. The precursor **26** is prepared by coupling amino
α-carboxamide and 2-methyl-2-(2′-nitro)phenoxy-propionic acid as its *O*-
succinimidyl ester **25**. Hydrochloride salts, such as **27**, are generally stable,
crystalline compounds that can be stored at room temperature for several
months. Once deprotonated, **27** is stable enough for acylation reactions
(Scheme 9) [122].

Further coupling of these derivatives with 2-substituted malonic acid
monoester **13** using DCC–HOBt or reaction with Meldrum's acid derivatives **12** after in situ trimethylsilylation with excess *N,O*-bistrimethylsilyla-
cetamide (BSA) gives the protected pseudodipeptide unit **28**, which can be
coupled to the growing peptide chain on a solid support (Scheme 10) [122].

Direct anchoring of Mnp-gIle-(*R,S*)-mLeu-OH (**28**) to hydroxyl resin
was reported by Verdini and coworkers [134] during the synthesis of the *C*-
terminal PMRI analogue of neurotensin [gIle,[12] (*R,S*)-mLeu[13]]NT(8–13)

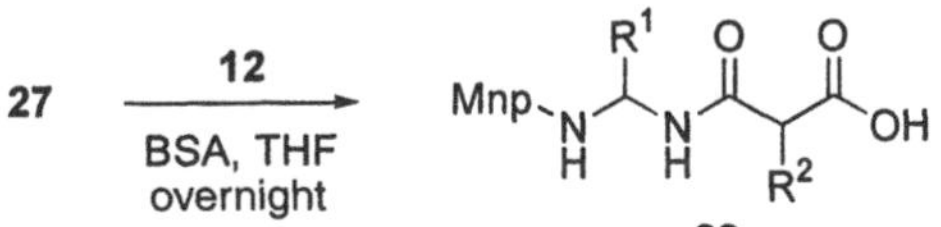

Scheme 10 Synthesis of Mnp-protected PMRI dipeptide unit **28** [122].

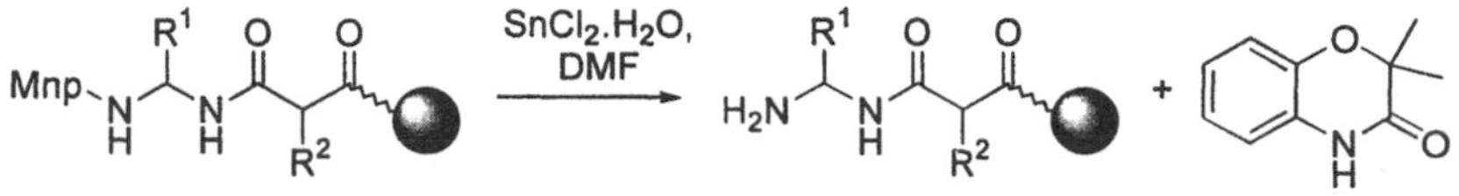

Scheme 11 Deprotection of Mnp-protected PMRI-peptide resin **31** [122].

(**29**). Attempts to anchor **28** with DCC–HOBt or BOP either alone or in the presence of DMAP did not result in a coupling yield greater than 15%. However, activation of **27** with carbonyl bis-(N-methylimidazolium) dichloride (**30**) in CH_2Cl_2 afforded higher coupling yields (50–70%). Generally, the Mnp protecting group is selectively and completely removed by two successive treatments of the PMRI-peptide resin **31** (Scheme 11) with a 2 M solution of $SnCl_2 \cdot H_2O$ in DMF (15 min + 45 min) [122]. However, during the synthesis of **29**, longer reaction times (2–3 h) were needed for complete Mnp removal. The 3,4-dihydro-2,2-dimethyl-2H-1,4-benzoxazin-3-one **32** that is formed during the reaction is removed by standard DMF washes. The mild Mnp cleavage conditions were found to leave tert-butyl– and benzyl-based side chains unaffected and thus to be compatible with Fmoc/tBu chemistry [122,134].

B. General Procedure for the Incorporation of the Retro-Inverso Modification on a Solid Support [27]

A solution of the 2-alkylmalonyl-(D)-amino acid amide (3 equiv.), DIC (3 equiv.), and HOBt (3 equiv.) in DMF (minimum volume) is added to the amino group of the peptide-bound resin and allowed to react overnight with agitation. The resin is washed with DMF and checked by the ninhydrin test. If necessary, a second coupling is performed. The resin is then washed and equilibrated three times with DMF–H_2O (3:1) prior to the addition of a solution of IBTFA (4 equiv.) in DMF. The mixture is allowed to react for 1 h. After intermediate DMF–H_2O (3:1) washings, the reaction step is repeated overnight with 4 equiv. of IBTFA. The resin is then washed and reequilibrated with DMF and neutralized with DIEA. The couplings of N-protected α-amino acids following the incorporation of the pseudopeptide unit and the cleavage of the peptide from the resin are conducted using standard procedures. Because reaction with IBTFA is performed in mixed aqueous organic media, Pessi et al. [27], used a polyamide-type resin that is freely solvated under these conditions. IBTFA treatment has also been performed on p-benzyloxybenzylalcohol-polystyrene-1% divinylbenzene in DMF–H_2O (9:1) mixture for 4 h. This solid-phase procedure has some limitations in the case of a peptide containing IBTFA-sensitive side chains. Protection of Asn and Gln residues is, of course, mandatory. Oxidation-sensitive side chains of Trp and Tyr residues must be appropriately protected. The side chain of methionine, which is oxidized to a sulfoxide, has to be reduced with N-methylmercaptoacetamide at the end of the synthesis. Cysteine derivatives are either deprotected or oxidized during IBTFA treatment and therefore should not be used.

C. Total Retro-Inverso (or All-D-Retro) Isomers

Prelog and Gerlach [98] originally postulated the topochemical equivalence between cyclic peptides and their retro-inverso (retro-all-D or retro-enantio) isomer. A few years later, Shemyakin et al. [99], taking advantage of the topochemical equivalence between a cyclic peptide and the cyclo retro-enantiomer, successfully applied the concept to gramicidin S, with the retro-enantiomer displaying antimicrobial activity similar to that of the parent peptide. Further extension of this principle to linear peptides was proposed by Shemyakin et al. [99] and Goodman et al. [125]. The retro-inverso isomer **34** of a linear peptide **33** is obtained through an inversion of the chirality at each α-carbon, i.e., replacement of L-amino acids by D-amino acids, and a reversal of the direction of the backbone amide bonds (Fig. 3). This results, theoretically, in maintenance of the side-chain topochemistry. A similar relationship exists between the enantiomer of the parent peptide and its retro

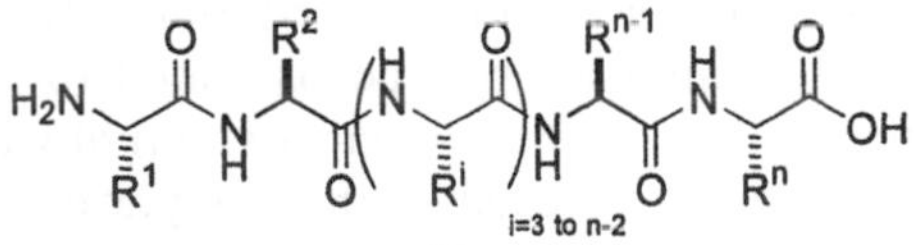

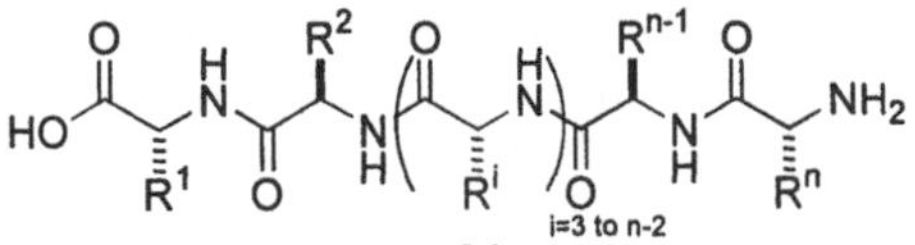

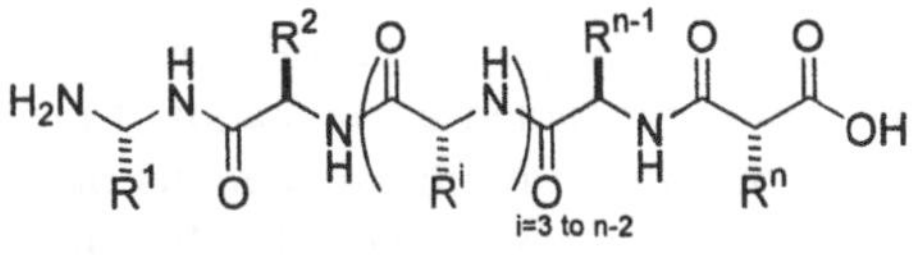

Figure 3 Structures of an L-peptide (**33**), the retro-inverso isomer (**34**), and the end group–modified retro-inverso isomer (**35**).

isomer. However, in such linear peptides, the two pairs of topochemically related isomers do not share end group and charge complementarity. Various approaches exist to solve this end group problem, depending on the importance of both terminal residues in the biological activity and on the degree of complementarity required [10]. For example, to avoid undesirable ionic interactions, N- and C-terminal can be simply acetylated and carboxamidated. Alternatively, replacement of N- and C-terminal ends by desamino and/or decarboxy residues has been described. However, the closest similarity between both isomers is achieved by replacing the N- and C-terminal amino acids by a *gem*-diamino alkyl residue and a 2-substituted malonate derivative, respectively (**35**) (Fig. 3). All these analogues, which are most closely related to the parent peptide, are referred to as end group–modified retro-inverso isomers [10].

We have extensively investigated the extent of antigenic mimicry achievable with retro-inverso peptides and evaluated their potential in the development of peptide-based vaccines [106,110]. We designed and synthesized using Fmoc/*t*Bu chemistry an end group–modified retro-inverso analogue **38** of the foot-and-mouth disease virus (FMDV) major antigenic site located within residues 141–159 of the viral VP1 protein (Fig. 4) [110,135].

In order to mimic the C-terminal end of the parent peptide (**37**), we introduced a 2-isobutyl malonyl derivative in place of the original C-terminal leucine residue. The malonyl residue was incorporated in the peptide chain as a *tert*-butyl monoester (**40**) obtained by alcoholysis of the 2,2-dimethyl-5-isobutyl-1,3-dioxane-4,6-dione (**39**) (Fig. 5).

The resulting diastereoisomers were separated and purified by RP-HPLC. The two isomers **38a** and **38b** were identified according to their HPLC retention times (the isomer retained for a shorter time was referred to as **38a**). We found that a single inoculation of **38b** elicits high levels of neutralizing antibodies that persist longer than those induced against the corresponding L-peptide and confers substantial protection in guinea pigs challenged with the cognate virus. The rates of isomerization of **38a** and **38b** at 37°C and pH 7.0 were followed by HPLC. After 21 h, no change was observed in the HPLC profile of **38a**. A slow equilibrium was observed

Ac-C→G→S→G→V→R→G→D→F→G→S→L→A→P→R→V→A→R→Q→Leu-OH   **37**

NH₂←D-cys←G←s←G←v←r←G←d←f←G←s←l←a←p←r←v←a←r←q←(R,S)mLeu-OH   **38**

Figure 4 Amino acid sequences of the retro-inverso analogue (**38**) of the FMDV FP variant VP1(141–159) peptide (**37**). The arrows indicate the sense of the peptide bond along the backbone. Lowercase letters indicate D-amino acid residues; m, malonate [110,135].

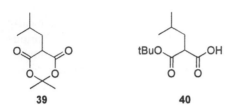

39 **40**

Figure 5 The 2-isobutyl malonyl derivative **40** used in the synthesis of **38** was prepared by alcoholysis of 5-isobutyl-1,3-dioxane-4,6-dione **39** [135].

after 21 h for **38b**, leading to about a 60:40 (**38b/38a**) equilibrium mixture of diastereoisomers [135].

1. Procedure for the Preparation of the End-to-End Retro-Inverso Isomer (**38**) [135]

a. *2,2-Dimethyl-5-isobutyl-1,3-dioxane-4,6-dione* (**39**) [101,120]. To a solution of Meldrum's acid **11** (5.76 g, 40 mmol) in MeOH (50 mL) is added a solution of borane-dimethylamine (2.36 g, 40 mmol). After stirring for 5 min, isobutyraldehyde (7.2 mL, 80 mmol) is carefully added. After 10 min, another portion of isobutyraldehyde (3.6 mL, 40 mmol) is added and the solution is stirred for another 30 min. Ice water (200 mL) is then added, and the mixture is acidified with HCl. The precipitate that is formed is filtered, washed with water, and dried in vacuo over phosphorus pentoxide. White solid. Yield 4.64 g (58%); R_f (CHCl$_3$–MeOH–AcOH 120:10:5) 0.85; melting point (m.p.) 69–70°C; ^{1}H NMR (200 MHz, CDCl$_3$) δ 1.14 (d, 6 H), 1.58 (m, 1 H), 1.74 (s, 3 H), 2.00 (m, 2 H); 3.44 (m, 1 H).

b. *(R,S)-2-Isobutylmalonic Acid Mono-tert-butyl Ester* (**40**). *tert*-Butyl alcohol (22.5 mL, 235 mmol) is added to a solution of **39** (9.42 g, 47 mmol) in toluene (10 mL) and the reaction is heated at 55°C for 96 h. The solvent is removed under reduced pressure, and the residue is purified by flash chromatography (*i*PrOH–CH$_2$Cl$_2$ 2:98). Oil. Yield 3.6 g (35%). ^{1}H NMR (200 MHz, CDCl$_3$) δ 0.96 (d, J = 6.4 Hz, 6 H), 1.47 (s, 9 H), 1.52-1.84 (m, 3 H), 3.36 (t, J = 7.6 Hz, 1 H).

c. *Solid-Phase Peptide Synthesis.* Assembly of the peptide chain is carried out on an automated peptide synthesizer on a Rink amide resin (25 μmol scale) using standard Fmoc chemistry. The sequence is inverted with respect to that of the parent peptide. *N*-Fmoc–protected D-α-amino acids are used instead of the L-amino acids in the parent sequence and are incorporated according to standard BOP–HOBt coupling procedures. A double coupling is performed systematically. In the case of amino acids containing a second chiral center (such as Ile or Thr), a D-allo derivative should be used in order to maintain the correct stereochemistry at the β-carbon. Furthermore, it has to be noticed that in sequences containing a proline residue, the spatial orientation of the proline ring is not maintained in the retro-inverso isomer as compared with the parent peptide. After removal of the last Fmoc protecting group, a solution of **40** (5 equiv.), BOP (5 equiv.), and HOBt (5 equiv.) in DMF (2 mL) is added to the resin. After addition of DIEA (15 equiv.), the mixture is nitrogen bubbled for 1 h. This acylation step is repeated once and stirring is continued as long as a positive ninhydrin test is obtained. The retro-inverso peptide resin is cleaved with reagent K for 2 h, and the peptide is collected in a tube filled with cold Et$_2$O. After centrifugation, the pellets are washed twice with cold ether. After the last centrifugation, the peptide is dissolved in 10% acetic acid in water and lyophilized. The two diastereoisomers **38a** and **38b** (respectively of the crude product as determined by HPLC) were separated and purified by RP-HPLC. **38a**: HPLC $R_t = 10.59$ min (5–65% B, 20 min, A: 0.1% aqueous TFA, B: 0.08% TFA in MeCN), matrix-assisted laser desorption–ionization mass spectrometry (MALDI-MS): 2073.7 [M+H$^+$]. **38b**: HPLC $R_t = 11.67$ min (5–65% B, 20 min, A: 0.1% aqueous TFA, B: 0.08% TFA in MeCN). MALDI-MS: 2073.9 [M+H$^+$].

IV. AZA, IMINOAZA, REDUCED AZA, AND CARBAZA PEPTIDES

Azapeptides are a unique class of pseudopeptides in which a nitrogen atom is substituted for the C$^\alpha$H (Fig. 6) [136]. These analogues have been extensively used as enzyme inhibitors (ACE [137], serine protease [138] and hormone analogues [luteinizing hormone–releasing hormone (LHRH) [139], oxytocin, enkephalin] [136]. In the past few years, several groups have in-

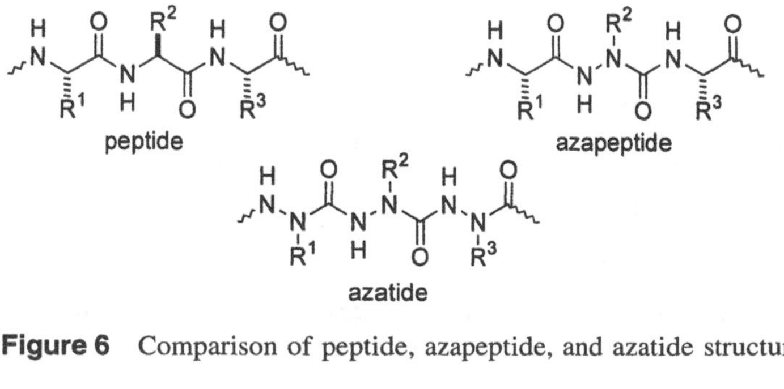

Figure 6 Comparison of peptide, azapeptide, and azatide structures.

vestigated possible synthetic routes to azapeptides on a solid support [30–32,140].

Incorporation of an aza residue into a peptide chain requires the synthesis of an acylsemicarbazide moiety and hence is a combination of hydrazine and peptide chemistry. Various strategies evaluated so far on the solid phase are as follows:

1. Reaction of carbazic acid chloride or activated carbamates derived from *N*-protected *N'*-alkyl hydrazines (**41**) (Fig. 7) with polymer-supported amines. In initial studies, activated carbamates obtained by reaction of (**41**) [R = $(CH_2)_2CONH_2$] with bis(2,4-dinitrophenyl)carbonate (**43**) and *p*-nitrophenylchloroformate, respectively, were found to be poorly reactive [30]. Reaction of (**42**) (R = H, Me, Bn, *i*Bu, *i*Pr) with bis(pentafluorophenyl)carbonate (**44**) and DMAP afforded a reactive intermediate that was used for the synthesis of "pure" azapeptides (also termed azatides) (see Fig. 6) [141]. However, the analogous intermediate obtained by starting from **41** was not reactive [141]. Marraud and coworkers [32] reported the successful use of bis(trichloromethyl)carbonate (triphosgene) (**45**) for clean and fast conversion of (**47**) to the corresponding carbazic acid chloride Boc-azLeu-Cl (**48**), which, upon reaction with the amino group of a resin-bound peptide (**49**), afforded the corresponding polymer-supported hydrazinocarbonyl peptide-PAM (**50**) [32]. Attempts to acylate **50** by an activated ester, by the preformed symmetrical anhydride, or by using TBTU, BOP, or DCC–HOBt as coupling agent were unsuccessful. However, the mixed anhydride procedure at low temperature gave the

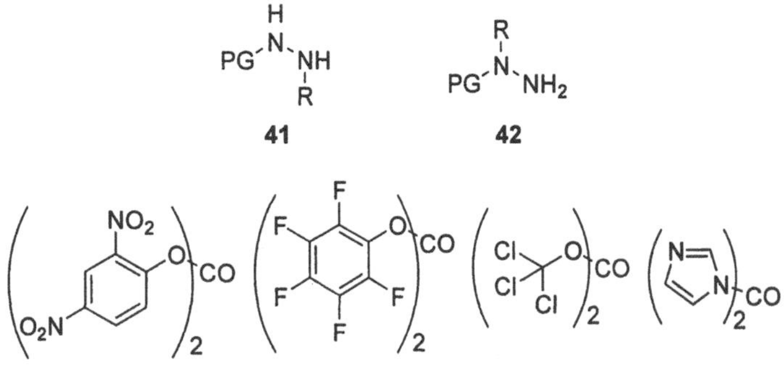

Figure 7 *N*-protected alkyl hydrazines (**41** and **42**) and carbonylating reagents (**43–46**) used in the synthesis of azapeptides and azatides.

corresponding azapeptide **51** in good yield after cleavage with TMSOTf–TFA (Scheme 12).

2. Reaction of **41** with a polymer-bound activated carbamate or isocyanate. In the original approach developed by Gray et al. [30], the free amino group of the resin-bound peptide was treated with **43** in DMF in the presence of N-methylmorpholine. After 15 min, the ninhydrin test was negative, indicating the absence of primary amine. Infrared (IR) analysis of the resin swollen in DMSO showed a characteristic absorption band at 2250 cm^{-1} attributable to an isocyanate group, suggesting that the initially formed active urethane had undergone an elimination. As examined by IR, addition of **41** [R = (CH$_2$)$_2$CONH$_2$] resulted in the disappearance of the isocyanate signal. However, in a related study, rearrangement of the isocyanate to hydantoin was found to occur up to 85% [31]. This side reaction can be reduced by treatment of the polymer-bound amine with **43** in the absence of N-methylmorpholine (NMM). Other carbonylating reagents including N,N'-carbonyldi-

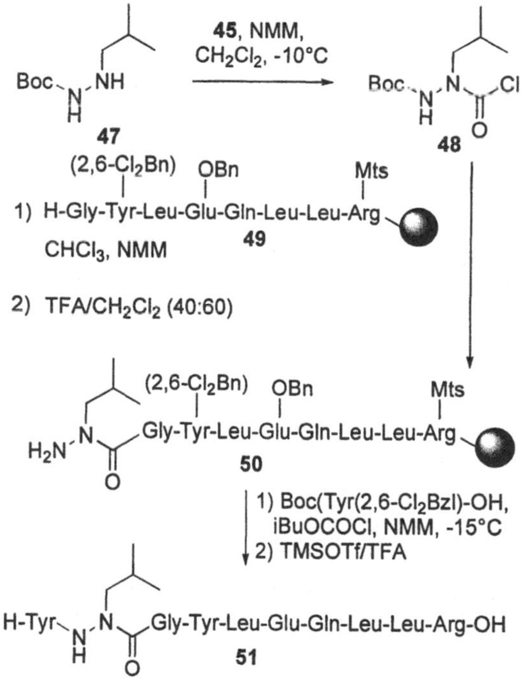

Scheme 12 Solid-phase synthesis of the azapeptide **51** [32].

imidazole (CDI) (**46**) and triphosgene (**45**) have also been used for activation of solid-supported amines (see later) [33].

3. Coupling of a preformed aza tripeptide. This method is more time consuming and may involve partial racemization at one residue [30,140].

The aza replacement also provided a framework for the construction of novel backbone modifications including iminoaza (semicarbazone) [32], reduced aza (semicarbazide) [32], and carbaza [33]. The semicarbazone motif is incorporated in good yield by condensation of the *N*-protected amino aldehyde Boc-Tyr(2,6-Cl$_2$Bn)-H (**52**) to the polymer-supported hydrazinocarbonyl peptide-PAM (**50**) at room temperature in the presence of sodium acetate and molecular sieves. Final cleavage with TMSOTf–TFA afforded the crude iminoaza peptide (**54**) in 45% yield. It is noteworthy that the semicarbazone moiety can be considered as a *cis* planar dipeptide isostere stabilized by intramolecular NH···N interaction closing a five-membered cycle. Alternatively, the polymer-supported semicarbazone (**53**) is easily reduced in situ by NaBH$_3$CN to give the corresponding reduced azapeptide (**55**) after final cleavage (Scheme 13) [32].

Limal et al. [33] reported the synthesis of the carbaza-peptide (**56**) (in 40% overall yield after HPLC purification) by coupling *N*-Boc or *N*-Fmoc *N'*-alkyl-propylenediamine derivatives (**57**) on solid-supported amines activated with triphosgene or CDI (Fig. 8). Reaction with triphosgene was found to be faster than with CDI and hydantoin formation was kept at a low level (<3%) when no tertiary base (DIEA) was added during the coupling procedure.

A. General Procedure for the Solid-Phase Synthesis of Aza, Iminoaza, and Reduced Azapeptide [32]

a. Boc-NH-NH-iBu (**47**). Isobutyraldehyde (Aldrich, 0.91 mL, 10 mmol) is added at room temperature over 10 min to *tert*-butyl carbazate (Aldrich, 1.32 g, 10 mmol) in THF (10 mL). After 3 h, the hydrazone is filtered off and hydrogenated over 10% Pd/C in isopropanol. Silica gel chromatography (AcOEt–petroleum ether 50:50) gives **47** as a colorless oil (75% yield). ^{1}H-NMR (CDCl$_3$, ppm): 0.93 (d, $J = 6.7$ Hz, 6 H, Me), 1.46 (s, 9 H, Me), 2.66 (d, $J = 6.9$ Hz, 2 H, CH$_2$), 1.73 (m, 1 H, CH), 3.30 (m, 1 H, CH$_2$), 5.80 (*br* d, 1 H, NH).

b. Boc-azLeu-Cl (**48**). To a cold $-10°C$ solution of **47** (0.94 g, 5 mmol) and triphosgene (**45**) (0.49 g, 1.67 mmol) in 5 mL of CHCl$_3$, NMM (0.55 mL, 5 mmol) is added dropwise under N$_2$ with vigorous stirring, and the mixture is maintained at $-10°C$ for 1 h. This intermediate is not isolated but is prepared just before use.

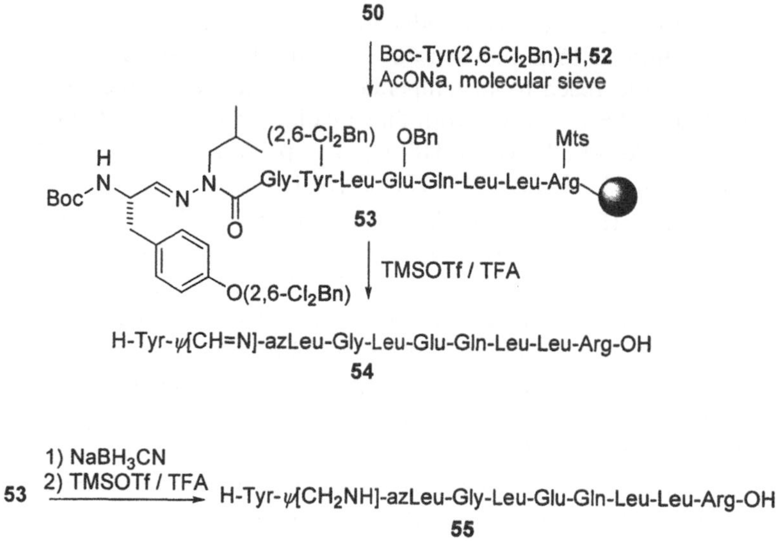
50
Boc-Tyr(2,6-Cl2Bn)-H,52
AcONa, molecular sieve
(2,6-Cl2Bn)
OBn
Mts
Gly-Tyr-Leu-Glu-Gln-Leu-Leu-Arg
Boc
53
O(2,6-Cl2Bn)
TMSOTf / TFA
H-Tyr-ψ[CH=N]-azLeu-Gly-Leu-Glu-Gln-Leu-Leu-Arg-OH
54
1) NaBH3CN
2) TMSOTf / TFA
53
H-Tyr-ψ[CH2NH]-azLeu-Gly-Leu-Glu-Gln-Leu-Leu-Arg-OH
55

Scheme 13 Solid-phase synthesis of iminoaza and reduced azapeptides [32].

c. Polymer-Supported Hydrazinocarbonyl Peptide (**50**). The octapeptide PAM resin (**49**) (0.2 mmol) prepared using standard Boc chemistry is introduced at $-10°C$ under N_2 into a moderately stirred solution of **48** (5 mmol) in $CHCl_3$. Then NMM (5 mmol) is added dropwise; the mixture is kept at $-10°C$ for 1 h and allowed to reach room temperature overnight. The Boc group is cleaved by 40% TFA in CH_2Cl_2 to give **50**, which is not characterized.

d. Azapeptide (**51**). To a stirred solution of Boc-Tyr(2,6-Cl_2Bn)-OH (0.22 g, 0.5 mmol) in CH_2Cl_2 (1 mL), NMM (5 μL, 0.5 mmol) and *i*buOCOCl

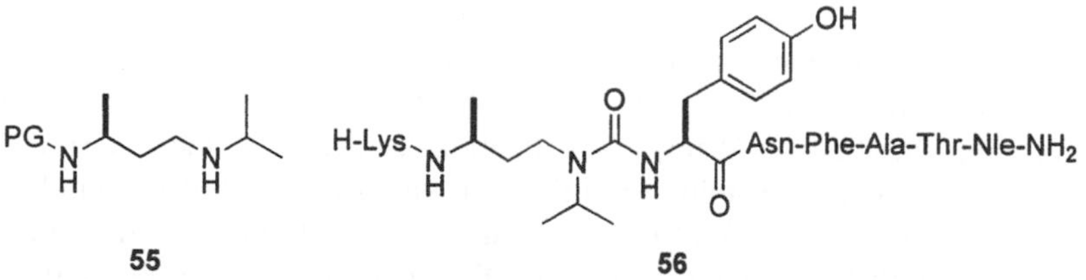
PG
H-Lys
Asn-Phe-Ala-Thr-Nle-NH2
OH
55
56

Figure 8 *N*-protected alkyl propylene diamine **55** used as building block in the synthesis of carbazapeptide **56** [33].

(65 μL, 0.5 mmol) are successively added at $-18°C$, and stirring is maintained for an additional 15 min. Then **50** (0.1 mmol) is introduced and the
mixture is allowed to reach room temperature with gentle stirring for 5 h.
Standard TMSOTf–TFA cleavage with thioanisole and 1,2-ethanedithiol affords crude **51** (yield 50% based on HPLC). The peptide is purified by RP-
HPLC and lyophilized. MALDI-MS: 1268.3 [M+H$^+$].

e. Iminoazapeptide (**54**). Boc-Tyr(2,6-Cl$_2$Bn)-H (**52**) (2.12 g, 0.5 mmol)
(prepared as described in the previous section dealing with N-protected α-
amino aldehydes) is added to **50** in DMF–CH$_2$Cl$_2$ (1:3) in the presence of
sodium acetate (0.164 g, 2 mmol) and molecular sieves. The beads are agitated for 5 h at room temperature and the coupling procedure is repeated
once. Cleavage of the resulting iminoazapeptide resin (**53**) as described for
51 affords crude **54** (45% yield), which is further purified and lyophilized.
MALDI-MS: 1252.60 [M+H$^+$].

f. Reduced Azapeptide (**55**). On-resin reduction of the semicarbazone
moiety in (**53**) (300 mg, 0.204 mmol) is carried out for 12 h in DMF–
CH$_2$Cl$_2$ by NaBH$_3$CN. Cleavage of the resulting reduced azapeptide resin as
described for **51** affords crude **55** (45% yield), which is further purified and
lyophilized. MALDI-MS: 1255.45 [M+H$^+$].

V. *N*-HYDROXY-AMIDE ANALOGUES ψ[CON(OH)]

So far, the influence of N-hydroxylation of the peptide backbone in biologically relevant peptides has been addressed only in rare studies. The N-
hydroxamide moiety is a strong proton donor that has been found to chelate
metal cations in natural siderophores. Conformational perturbations induced
by N-hydroxylation of model pseudodipeptides have been carefully studied
in solution (IR spectroscopy and ^{1}H NMR) and in the solid state (X-ray
diffraction) by Marraud et al. [142,143]. Until recently, incorporation of the
ψ[CON(OH)] peptide bond surrogate was carried out only in solution. However, with the aim of designing and evaluating a new class of MHC class I
ligands, Bianco et al. [34] have developed a method for the incorporation
of an N-hydroxy-glycine residue on a solid support. The N-hydroxy amide
ligands exemplified by H-Ser-Ile-Ile-ψ[CO-N(OH)]-Gly-Phe-Glu-Lys-Leu
(**58**) were synthesized on Wang resin. Peptide assembly of the C-terminal
part prior to the N-hydroxamide modification was performed using standard
Fmoc/tBu chemistry. The free amino group of the polymer-bound peptide
was acylated with bromoacetic acid and DIC in DMF to give **59**. O-Ben-
zylhydroxylamine hydrochloride (50 equiv.) was neutralized with DIEA in
DMF and added to **59**. Nucleophilic substitution of **59** required 12 days for

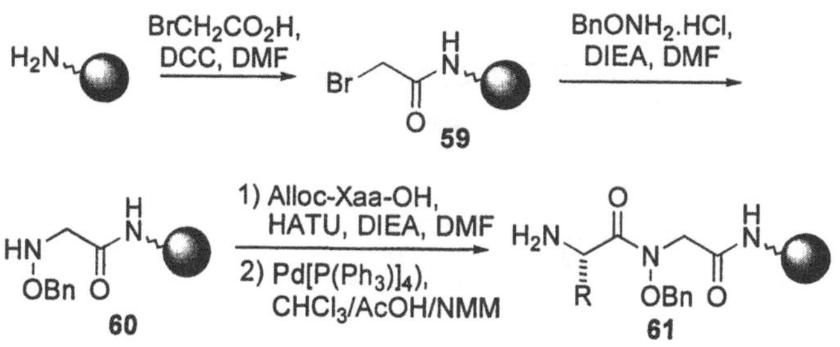

Scheme 14 Solid-phase synthesis of N-hydroxy peptides [34].

completion. The resulting **60** was reacted further with N-Alloc-protected α-amino acids using the uronium salt HATU as a coupling reagent. Attempts to use N-Fmoc amino acids were not successful as Fmoc deprotection with piperidine or DBU leads to a complex mixture of by-products (not characterized). Deprotection of the Alloc protecting group with Pd(Ph$_3$)$_4$ yielded the O-benzylhydroxamide-resin **61** (Scheme 14). Stepwise synthesis with Alloc α-amino acids or fragment condensation led to the fully protected peptide-resin, which was cleaved with 90% TFA in CH$_2$Cl$_2$ containing 5% triisopropylsilane. Finally the O-benzyl group was removed in solution by catalytic hydrogenolysis with Pd/C in McOH. Pseudopeptides were recovered in fair yield after RP-HPLC (20–65%).

VI. (E)-ALKENE PEPTIDE BOND ISOSTERE

The (E)-ψ[CH$_2$=CH$_2$] peptide bond isostere is a rigid nonhydrolyzable moiety that effectively mimics the amide bond structure (the $C^{\alpha i}$–$C^{\alpha i+1}$ distance is nearly equal to that found in peptides) [144,145] and has been incorporated in a number of biologically active peptides. Although numerous synthetic routes have been developed in solution, synthesis of (E)-ψ[CH$_2$=CH$_2$] replacement on a solid support has been achieved only recently by Wipf and Henninger [29]. The synthesis is based on the SN$_2$'-opening of N-protected alkenylazeridines [146]. The required polymer-bound alkenylaziridine **65** was synthesized using a solid-phase Wadsworth–Emmons reaction. The diethyl phosphonoacetate resin **63** (Scheme 15) was prepared by coupling the diethylphosphonoacetic acid **62** to a Wang resin (~ 0.81 mmol/g). Reaction of **63** with aldehyde **64** provided **65**. The 2-nitrobenzenesulfonyl (Ns) group was used as an N-protecting group in **64**. It can be removed selectively by treatment with thiophenoxide [147]. The resin was then swollen at room

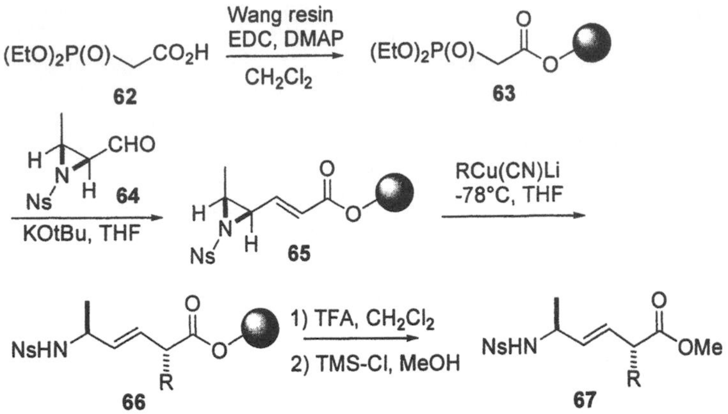

Scheme 15 Solid-phase synthesis of (*E*)-alkene isosteres [29].

temperature in THF and cooled to −78°C. A cold −78°C solution of the preformed alkylcyanocuprate was added to the resin and after 1 h the reaction mixture was quenched to give the corresponding pseudopeptide resin **66**. Cleavage from the resin and esterification afforded the protected dipeptides **67** in excellent purity and good yield (55–74%). Selective Ns deprotection of **66** and standard amino acid coupling afforded the expected pseudotripeptides after final cleavage and esterification.

VII. HYDROXYETHYLENEAMINO PEPTIDOMIMETICS

In the past few years, several groups have successfully developed efficient and original solid-phase synthetic routes for the preparation of enzyme inhibitor libraries. In the case of aspartyl protease inhibitors, stable isosteres mimicking the tetrahedral intermediate, i.e., hydroxyethylene, hydroxyethyleneamine [36–38], and diamino diol [39], have been successfully incorporated. In 1992, Alewood et al. [36] reported an efficient solid-phase synthesis of hydroxyethylene amino ψ[CH(OH)CH$_2$NH] pseudopeptides. Reaction of the free amine of the polymer-bound peptide **69** with *N*-Boc−protected α-amino-bromomethylketone **68** (obtained by treatment of the corresponding diazomethylketone with HBr in Et$_2$O) followed by reduction with NaBH$_4$ in THF yielded the *N*-protected hydroxyethyleneamino pseudopeptide-resin **70** (Scheme 16). Standard peptide elongation and HF cleavage afforded Ac-Ser-Leu-Asn-Phe ψ[CH(OH)CH$_2$NH]Pro-Ile-Val-OH (**71**) in 52% overall yield as a 4:6 (*S/R*) mixture of alcohol epimers. Alternatively, solid-phase reduc-

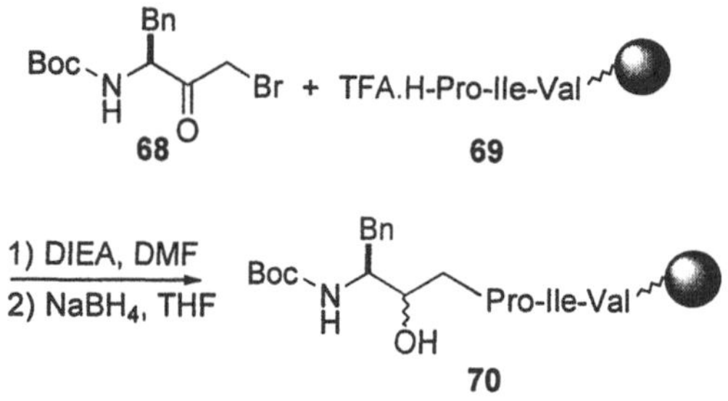

Scheme 16 Solid-phase synthesis of hydroxyethyleneamine pseudopeptides [36].

tive amination of α-hydroxyaldehyde **72** (Fig. 9) on a solid support has also yielded ψ[CH(OH)CH$_2$NH] isosteres [148].

Later, Kick and Ellmann [37,38] reported a powerful solid-phase approach to introduce functional diversity in hydroxyethylamine and hydroxyethylurea-based peptidomimetics. In the latest development of this work, the key intermediates **73a,b** (Nos = 4-nitrobenzenesulfonyl) provided a framework for the construction of the highly functionalized isosteres **74a,b** [38] (Fig. 10).

It is noteworthy that resin anchoring proceeds through the secondary alcohol of the starting scaffold. "It is the only invariant part of the inhibitor

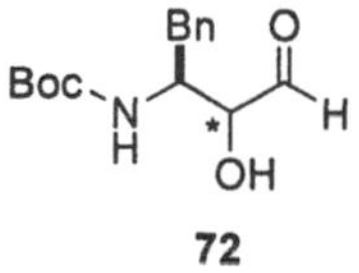

Figure 9 α-hydroxyaldehyde used in the solid phase synthesis of ψ [CH(OH)CH$_2$NH] isosteres [148].

X, Y = CO, CONH, SO$_2$

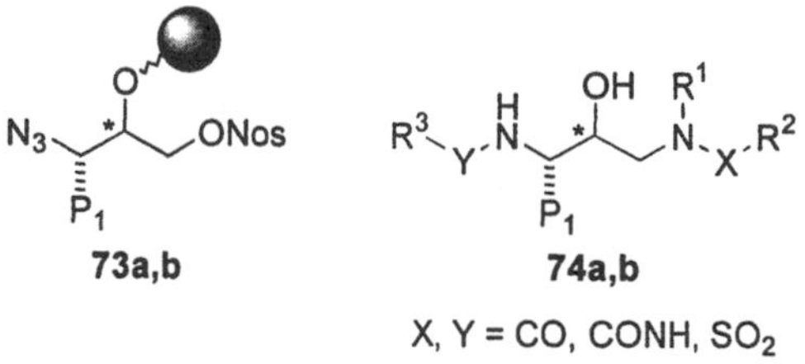

Figure 10 Key intermediate **73a,b** used by Ellman's group in the synthesis of aspartyl protease inhibitor structures **74a,b** [38].

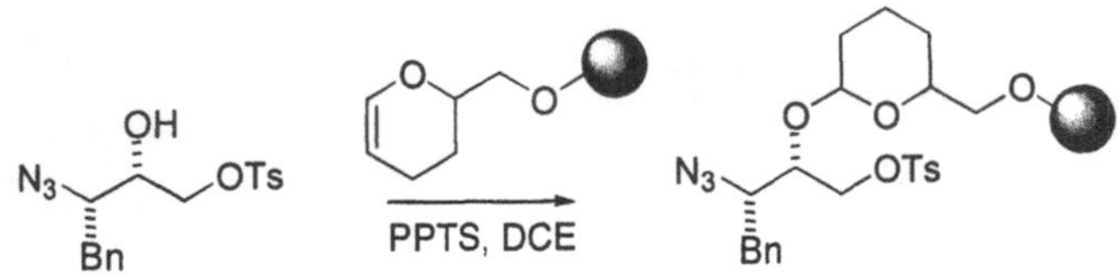

Scheme 17 Attachment of starting scaffold **75** to dihydropyran-functionalized polystyrene [37].

structure and allows diversity to be displayed at all variable sites of the inhibitor'' [38]. In early studies, the related intermediate **76** (Scheme 17) was obtained by anchoring **75** on a dihydropyran-functionalized polystyrene [149] support with pyridinium *p*-toluenesulfonate (PPTS) in 1,2-dichloroethane (DCE) [37].

In order to introduce diversity at the P1 position, intermediate **73a** was constructed from the corresponding polymer-bound pyrrolidine amide **80** (Scheme 18). Anchoring of pyrrolidine amide **78** (MMT = mono-*p*-methoxytrityl), which was prepared in four steps (76% overall yield) from commercially available (*S*)-methyl ester isopropylideneglycerate (**77a**) to (benzyloxy)benzyl bromide resin **79** (prepared from Wang resin by treatment with carbon tetrabromide and triphenylphosphine [150]), proceeded in 84% loading yield with minimal racemization (<2%) [151].

Addition of Grignard reagent to **80** provided ketone **81**, thus allowing the introduction of the P1 functionality (Scheme 19). Chelation-controlled reduction of **81** with Zn(BH$_4$)$_2$ afforded **82** with diastereoselectivities ranging from 90:10 to 80:20 depending on the P1 side chain. The high diastereo-

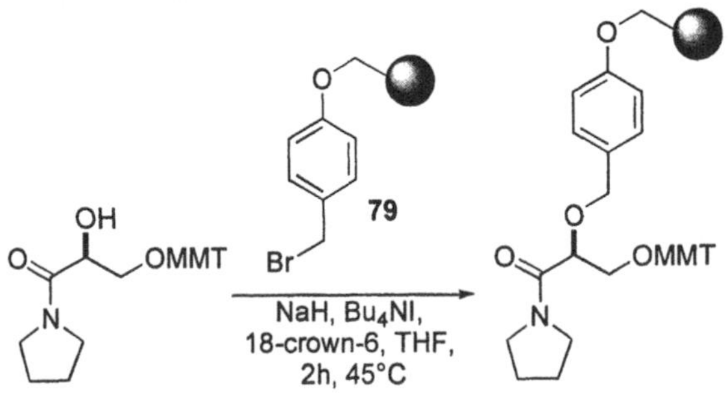

Scheme 18 Attachment of starting scaffold **78** to (benzyloxy)benzylbromide resin [38].

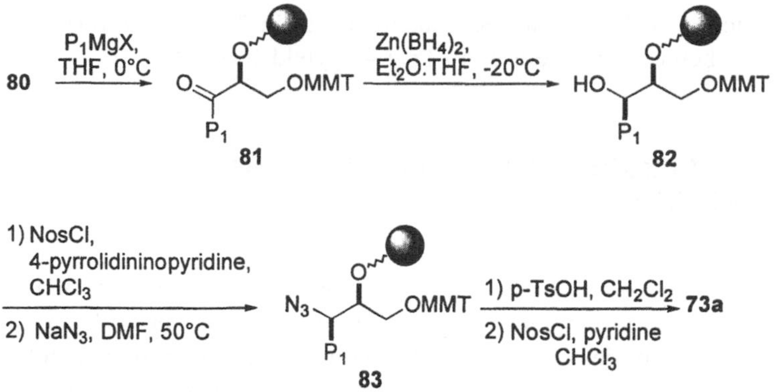

Scheme 19 Synthesis of polymer-bound key intermediate **73a** from polymer-bound pyrrolidine amide **80** [38].

selectivity observed was very much dependent on the choice of the linker: p-alkoxybenzyl versus tetrahydropyranyl. Secondary alcohol **82** was then converted to azide **83** through activation with 4-nitrobenzenesulfonyl-chloride and 4-pyrrolidinopyridine as a catalyst. Deprotection of the MMT group using 1% p-toluenesulfonic acid (p-TsOH) in CH$_2$Cl$_2$ and activation of the resulting alcohol with 4-nitrobenzenesulfonyl chloride afforded **73a**. The corresponding cleavage product was recovered in 70% overall yield (based on the original loading of **80**) after column chromatography.

A similar synthetic route led to **73b** starting from (R)-methyl ester isopropylideneglycerate (**77b**). However, in this case non−chelation-controlled reduction with L-Selectride resulted in poor diastereoselectivity (<60:40). Replacement of the MMT protecting group by the less hindered ethoxy-ethyl (EE) restored high diastereoselectivity (90:10).

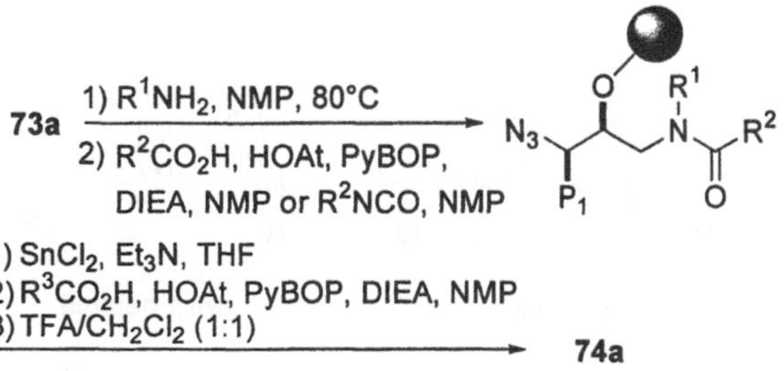

Scheme 20 Introduction of R^1, R^2 and R^3 functionalities in **73a** [38].

Introduction of R^1, R^2, and R^3 functionalities in **73a** was performed as depicted in Scheme 20 to yield **74a** (45–64% yield from **80**).

VIII. OLIGO-*N*-(SUBSTITUTED)GLYCINES (NSGs) OR "PEPTOIDS"

"Peptoids" or oligo-*N*-(substituted)glycines were originally reported by Simon et al. [55] as a new class of polymers designed for drug discovery. They differ from peptides in that the side chains on the α-carbon are shifted to the nitrogen atom (Fig. 11). As revealed by alignment of the peptide and peptoid sequences, the carbonyl groups and the side chains do not present the same overall position. However, it has been suggested that a closer similarity with the peptide sequence might be achieved by using the corresponding retro-peptoid sequence.

Peptoids have been shown to be resistant to enzymatic degradation, and combinatorial approaches led to the identification of high-affinity peptoid ligands for several important receptors [50,51]. Peptoids built exclusively of chiral *N*-substituted glycines with aromatic side chains were found to form stable helical conformations in aqueous solution [48,49].

Approaches to the synthesis of peptoids are summarized in Scheme 21. In the original method, Simon et al. [55] reported a monomer approach analogous to peptide synthesis in which the appropriate Fmoc-protected *N*-substituted glycines (**84**) were activated and coupled to the secondary amino group of the resin-bound peptoid chain to yield **85** (Scheme 21). Several synthetic routes to *N*-Fmoc–protected precursors (**84**) that were utilized in this early work are summarized in Scheme 22.

This approach was reinvestigated by Liskamp's group during the synthesis of peptoids and retro-peptoid analogues of Leu-enkephalin and substance P. In this study, *N*-substituted glycine precursors were obtained by alkylation of appropriate primary amines with ethyl bromoacetate. Subse-

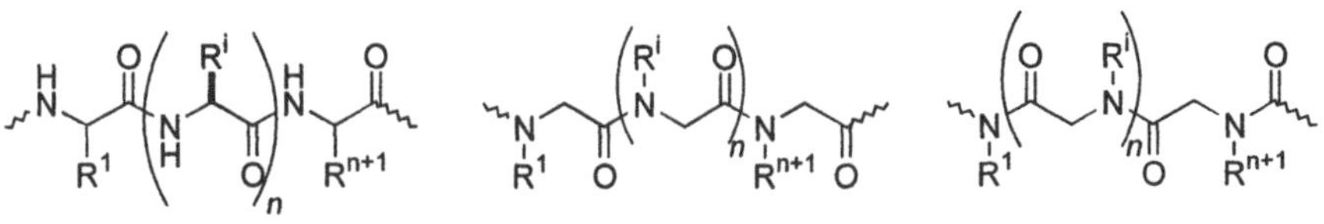

peptide	peptoid	retro-peptoid

Figure 11 Schematic representation of peptide, peptoid, and retro-peptoid oligomeric structures.

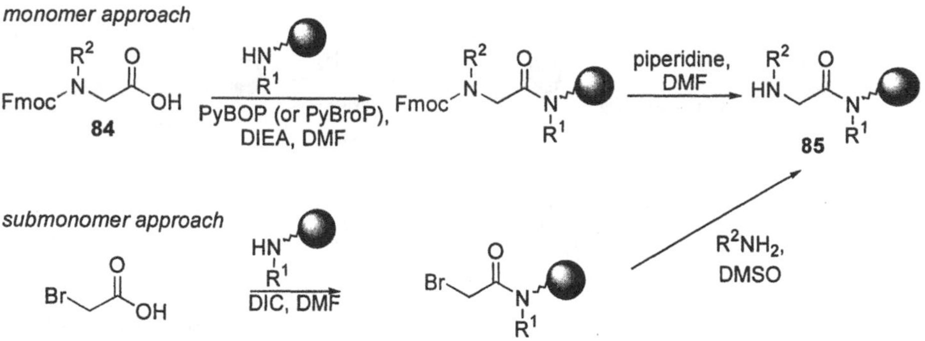

monomer approach

submonomer approach

Scheme 21 Solid-phase monomer [55] and submonomer [56] approach to the synthesis of peptoids.

quent saponification and Fmoc protection afforded the desired monomers (**84**) in 27–94% overall yield [152].

The access to peptoids was simplified with the solid-phase submonomer approach (Scheme 21) reported by Zuckerman et al. [56]. As shown in Scheme 21, the first step consists of acylation of the resin-bound amine with bromoacetic acid and DIC as the coupling agent, and in a second step, the side chain is introduced by nucleophilic substitution of the halide with an excess of a primary amine. This method, which allows the preparation of a wide variety of oligomers, has been applied successfully to the generation of diverse combinatorial peptoid libraries [50]. Furthermore, *N*-alkyl glycine residues containing peptide or peptoid–peptide hybrids can be read-

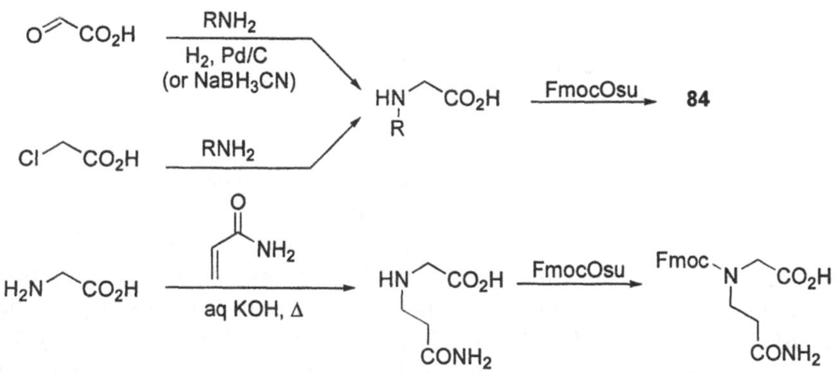

Scheme 22 Synthetic routes to Fmoc-protected *N*-alkyl glycines **84** and to Fmoc-NgIn-OH [55].

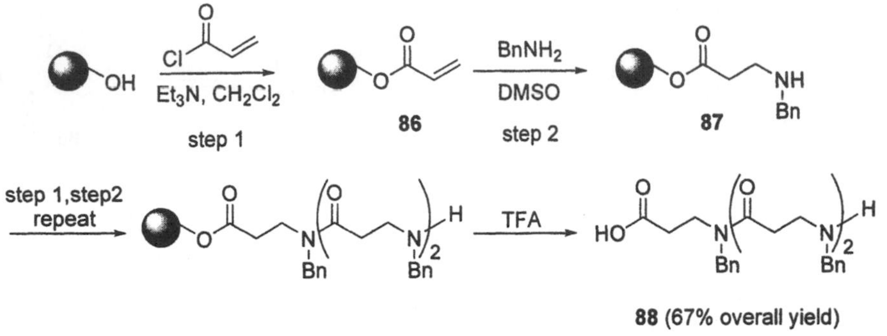

Scheme 23   Solid-phase synthesis of β-peptoid [63].

ily prepared by simple combination of peptoid and peptide chemistry [153,154]. Src homology 3 (SH3) domains bind specifically proline-rich sequences. Designing and screening a library in which key proline residues are substituted by N-substituted glycines yielded a ligand that binds the Grb2 SH3 domain with 100 times greater affinity [153].

A solid-phase submonomer approach to N-substituted β-aminopropionic acid oligomers or β-peptoids has been developed by Hamper et al. [63]. It is based on a simple two-step acylation and Michael addition reaction sequence. Treatment of Wang resin with 2 equiv. of acryloyl chloride in the presence of triethylamine in excess afforded the corresponding acrylate resin **86** (Scheme 23) [63]. Michael addition of a 6- to 10-fold excess of a given primary amine in DMSO afforded polymer-bound N-substituted β-alanines (**87**). Trimeric N-benzyl-β-aminopropionic acid (**88**) was prepared in 67% overall yield by repetition of this two-step sequence.

A.   Procedure for Peptoid Synthesis by the Submonomer Approach [56]

Peptoid oligomers are synthesized on a Rink amide resin (50 μmol scale) to avoid diketopiperazine formation. In order to suppress the formation of diketopiperazine during the synthesis of peptoids with a C-terminal carboxylic group, one can use the 2-chloro tritylchloride resin [155]. Following Fmoc removal, the resin is bromoacylated by successively adding a solution of bromoacetic acid (83 mg, 600 μmol, 12 equiv.) in DMF (830 μL) and 200 μL of DIC (103 μL, 660 μmol, 13 equiv.) in DMF (170 μL) to the resin. The reaction mixture is shaken for 30 min at room temperature. A double coupling is performed systematically. The resin is then filtered and washed three times with 2 mL of DMF. The nucleophilic substitution step is performed by adding a 2.5 M solution of the primary amine (2 mmol, 40 equiv.) in DMSO to the resin-bound bromoacetic acid. The mixture is allowed to react for 2 h with agitation. Optimization of the displacement reactions is performed by varying the amine concentration from 0.25 to 2.5 M. Alternatively, the reaction can be performed at 35°C for 40 min. The resin is then washed four times with 2 mL of DMF. After the last coupling step, the resin is cleaved with 95% TFA in water (10 mL) for 20 min at room temperature and lyophilized.

IX. OLIGOCARBAMATES

The oligocarbamate backbone was introduced in 1993 by Cho et al. [156], who demonstrated that libraries of oligocarbamates can be generated and screened for receptor binding. They developed an efficient solid-phase synthetic approach based on N-Fmoc–protected p-nitrophenyl carbonate monomers **89**. Carbonates **89** are easily prepared by reacting N-protected amino alcohol **8** (obtained either by reduction of N-Fmoc–protected amino acids or by N-Fmoc protection of the free amino alcohol) with p-nitrophenyl chloroformate in pyridine–CH_2Cl_2 (Scheme 24). Typically, overall yields for this two-step sequence are generally high and range from 50 to 80%. This general method allows the introduction of high diversity in side-chain functionality because carbonates **89** are virtually accessible from any amino alcohol including hydroxyproline and aminoindanol derivatives [57,156].

Stepwise solid-phase assembly of linear oligocarbamates was performed on Rink amide or amino acid ester–Wang resin. Typically, **89** (3–5 equiv.) was coupled to the polymer-supported amine in the presence of HOBt (10 equiv.) and DIEA (3–5 equiv.) for 4 h at room temperature. Coupling yields were greater than 99% per step as monitored by a quantitative ninhydrin test. The N-Fmoc protecting group was removed using 20% piperidine in N-methylpyrrolidinone (NMP). At the end of the synthesis, cleavage from the resin was accomplished with 5% triethylsilane in TFA (2 h) and the crude oligocarbamate was precipitated in a mixture of cold *tert*-butyl

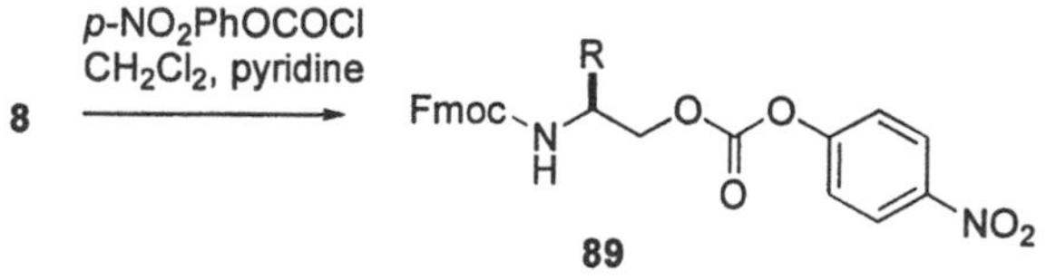

Scheme 24 Preparation of N-protected p-nitrophenyl carbonate monomers **89** [57,156].

methyl ether and hexane, purified by RP-HPLC, and lyophilized as exemplified by **90** (Fig. 12).

This procedure was adapted for the synthesis of N-alkyl oligocarbamates [157]. Prior to the coupling of carbonate **89**, the polymer-supported free amine was acylated with a carboxylic acid monomer and the newly formed amide bond was subsequently reduced with a 1 M solution of borane in THF for 1 h at 50°C. Washing the resin and quenching excess reagent were performed with 0.06 M DBU in MeOH–NMP (1:9). Coupling of **89** (5 equiv.) to the resulting secondary amine in THF in the presence of HOBt (10 equiv.) and DIEA (11 equiv.) at 50°C for 5 h was shown to proceed in greater than 96% yield. Repetition of these steps led to N-alkyloligocarbamates in good yield (70–90%) and purity [157].

Schultz and coworkers [57] reported the synthesis of cyclic oligocarbamates. Screening of libraries of linear and cyclic oligocarbamates for integrin GPIIb/IIIa binding led to the discovery of new high-affinity ligands as exemplified by **90** [median inhibitory concentration (IC_{50}) = 13 nM] and **91** (IC_{50} = 3.9 nM). Cyclization of oligocarbamates was performed on a solid support by reaction of the thiol of a C-terminal cysteine carbamate

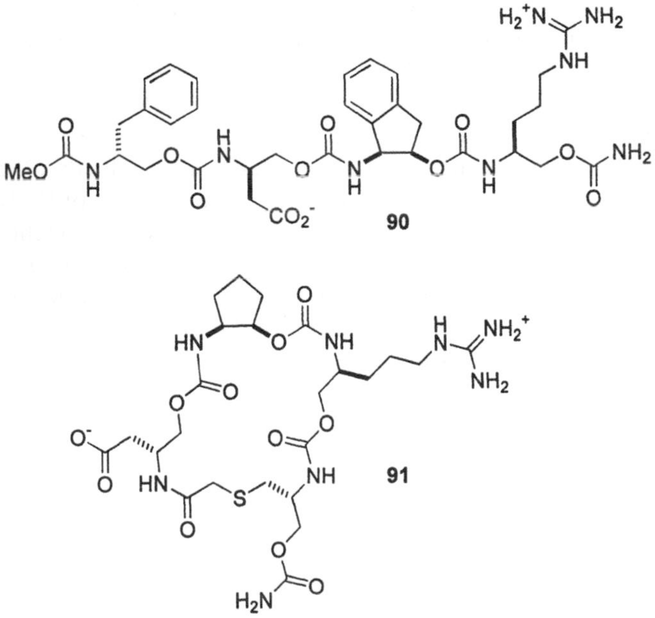

Figure 12   Linear and cyclic oligocarbamates with high affinity for integrin GPIIb/IIIa [57].

with a bromoacetylated residue. The trityl group that was used as a temporary protection was removed with 2% TFA–2% triisopropylsilane in CH$_2$Cl$_2$ (5 × 10 min). However, the choice of a linker proved to be a major problem in accessing cyclic oligocarbamate. The Rink amide resin could no longer be used because under conditions for Trt deprotection, the oligocarbamate was cleaved from the resin. Use of the 5-(4'-aminomethyl-3',5'-dimethoxyphenoxy)valeric acid (PAL) handle [158] also proved unsatisfactory because extensive oligocarbamate cleavage (90%) occurred. Instead, the use of the 4-(4'-aminomethyl-3'methoxyphenoxy)–butyric acid–p-methyl-benzhydrylamine (MBHA) resin (**94**) prepared from [4-(4'-hydroxymethyl-3'-methoxyphenoxy]butyric acid (HMPB)-MBHA resin (**92**) allowed selective Trt removal with less than 5% oligocarbamate cleavage (Scheme 25).

On-resin cyclization was performed with 25% DIEA in NMP for 24 h at 40°C under nitrogen. Cleavage and purification as described earlier for linear carbamates gave the expected cyclic carbamates.

Cyclization mediated by amide bond formation has been reported by Warrass et al. [159]. The linear oligocarbamate was assembled as described

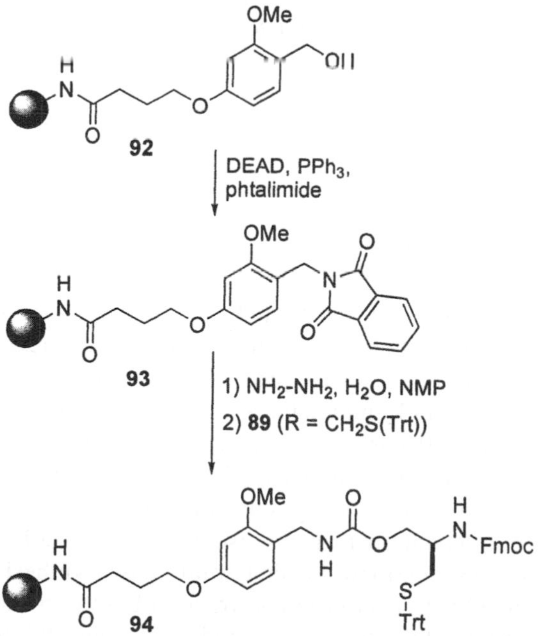

Scheme 25 Preparation of polymer-bound cysteine carbamate **94** for the synthesis of cyclic oligocarbamates [57].

by Schultz and coworkers [57] on a 2-chloro-tritylchloride resin substituted with an α-amino acid. At the end of the synthesis and after removal of the last Fmoc group, treatment of the resin with hexafluoro-2-propanol in DCM provided the side-chain protected oligocarbamate. After removal of the solvent, end-to-end cyclization was performed at high dilution (0.001 M) with TBTU–HOBt in CH$_2$Cl$_2$ and the side chains were finally removed with TFA. The deprotected cyclic oligocarbamates were recovered in 53–74% yield and high priority (70–98%).

A. General Procedure for the Synthesis of N-Fmoc– Protected p-Nitrophenyl Carbonates 89 [57,156]

To the appropriate N-Fmoc aminoalcohol (10 mmol) is added pyridine (11 mmol) and 50 mL of CH$_2$Cl$_2$ (or THF). A solution of p-nitrophenyl chloroformate (20 mmol) in CH$_2$Cl$_2$ (or THF) is added dropwise to the reaction mixture. The Fmoc-amino alcohol dissolves after addition of the chloroformate solution. The reaction mixture is stirred for at least 24 h. The mixture is diluted to 150 mL with CH$_2$Cl$_2$ and transferred to a separatory funnel. For reactions performed in THF, the THF is removed in vacuo and the residue is dissolved in CH$_2$Cl$_2$. The organic layer is washed with 1.0 M sodium bisulfate (3 × 75 mL) and 1.0 M sodium carbonate (5 × 100 mL). The organic layer is dried and the solvents are removed in vacuo. The crude product is purified by silica gel chromatography (CH$_2$Cl$_2$/hexanes 9:1, then CH$_2$Cl$_2$). Poor solubility of the Fmoc derivatives sometimes necessitates dry loading of the product on the silica gel column.

X. OLIGOUREAS AND THIOUREAS

Solid-phase syntheses and applications of substituted ureas and thioureas have gained increasing attention. Substituted ureas are essential components of biologically active compounds such as HIV protease inhibitors [160], CCK-B receptor antagonists [161], and endothelin antagonists [162]. Oligoureas have also been used as scaffolds for the creation of artificial β-sheets [163–166] and as peptide backbone mimetics [58,59,167]. Standard procedures for the formation of unsymmetrically substituted ureas involve the reaction of amines with carbonylation reagents, isocyanates, or carbamates. In 1995, Burgess et al. [58] reported the first solid-phase synthesis of oligourea on a solid support [167]. This synthetic route is based on the use of diaminoethane-derived monophthalimide-protected isocyanates **97** as activated monomers. Monophthalimide-protected diamine precursors **96** (Scheme 26) were prepared from N-Boc (or Fmoc)-protected α-amino acids in three steps (56–76% yield). The corresponding *N*-protected β-amino alcohol **95** was converted to the phthalimide derivative under Mitsunobu conditions and deprotection of the Boc (or Fmoc) protecting group afforded **96**. Treatment of **96** with phosgene or triphosgene as a carbonylating reagent yielded the corresponding isocyanates **97** (Scheme 26). Synthesis of oligourea was initiated by coupling the isocyanate monomer to the free amino group of the Rink amide resin for 5 h. The phthalimide group was subsequently removed by treatment of the resin with 60% hydrazine hydrate in DMF. Repetition of this coupling–deprotection cycle and cleavage from the resin gave the oligomers in 17–46% isolated yield. Similarly, *N*-protected isothiocyanates prepared from the *N*-Boc–monoprotected diamines by reaction with DCC and carbon disulfide have been introduced for the preparation of oligomeric thioureas on solid supports. It is noteworthy that long reaction times are usually required for complete reaction (2–4 days at 45°C) [60].

Alternatively, Schultz and coworkers [59] have proposed an approach related to that of Burgess which utilizes azido 4-nitrophenyl carbamates **98** as activated monomers. Carbamates **98** were prepared in four steps from alcohol **95**. Mesylation of **95** followed by azide displacement afforded the *N*-Boc–protected azide in high yield (80–90%). Boc deprotection and treatment of the resulting free amine with *p*-nitrophenyl chloroformate in the presence of pyridine in THF provided **98** (50–90% for the two steps). Solid phase urea bond formation was performed on a Rink amide resin by coupling **98** (5 equiv.) in CH$_2$Cl$_2$ in the presence of DIEA (7 equiv.) for 4 h at room temperature. Support-bound azide **99** was reduced in less than 2 h using

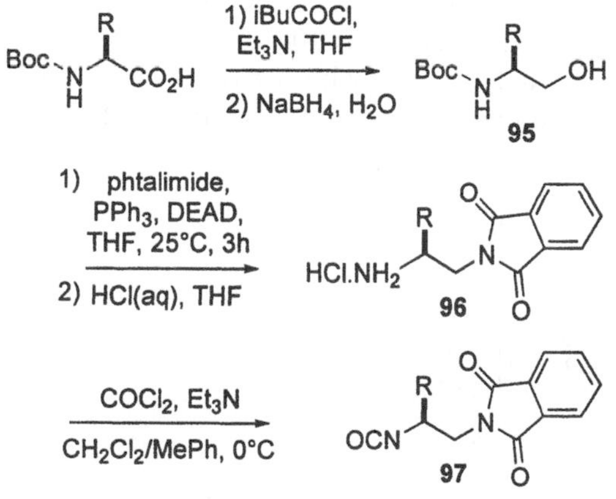

Scheme 26 Synthesis of diamino-ethane-derived monophthalimide-protected isocyanates **97** [58,167].

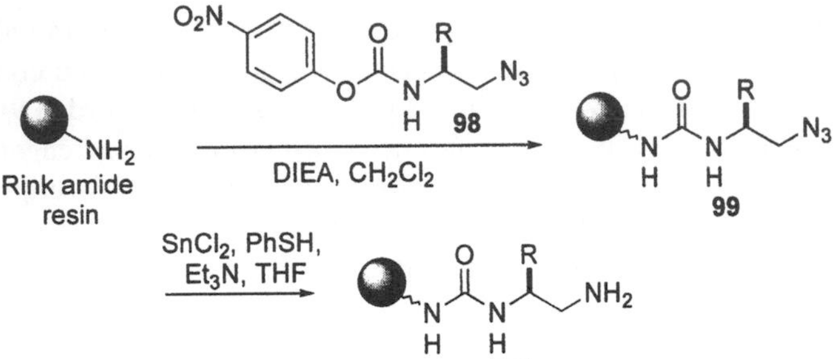

Scheme 27 Solid-phase synthesis of oligoureas using azido 4-nitrophenyl carbamates **98** as activated monomers [59].

SnCl$_2$ (5 equiv.), thiophenol (20 equiv.), and triethylamine (25 equiv.) in THF (5 mL per mmol resin) to give **100** (Scheme 27).

Four oligomers up to the tetramer (exemplified by **101**, Fig. 13) have been synthesized by this method. Purity of the crude material was high (75–94%) and yields after HPLC purification were in the 54–76% range.

Liskamp and coworkers [64] also reported the use of *p*-nitrophenyl carbamate derivatives obtained by reaction of Boc-protected *N*-substituted ethylenediamines with *p*-nitrophenyl chloroformate for the synthesis of ureapeptoids.

Guichard et al. [168] have reported the preparation of optically active *O*-succinimidyl-ethylcarbamate derivatives **105** and their use as activated monomers in the solid-phase synthesis of oligoureas. Boc- and Fmoc-protected β-amino acids **102** were first converted to the corresponding acyl azides **103** (Scheme 28). Intermediate isocyanates **104** obtained by Curtius rearrangement of **103** were immediately trapped with *N*-hydroxysuccinimide in the presence of pyridine to afford the corresponding carbamates **105** (51–

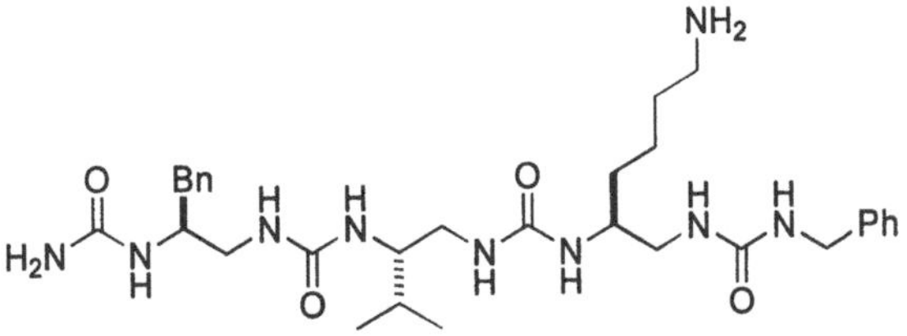

Figure 13 Example of oligourea synthesized using azido 4-nitrophenyl carbamates **98** [59].

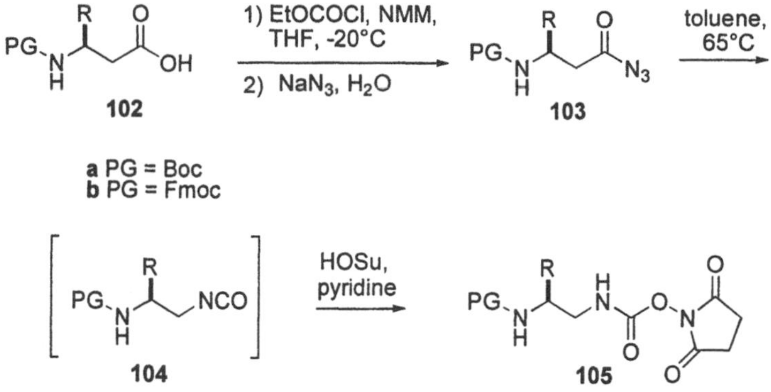

Scheme 28 Preparation of *O*-succinimidyl-ethylcarbamate derivatives **105** from corresponding β-amino acids **102** [168].

86%) as stable, crystalline products. It is noteworthy that this reaction sequence is completed within 1 h and that the mild conditions employed are compatible with the use of a number of functionalized side chains. In solution, carbamates **105** react readily with primary and secondary amino groups at ambient temperature to give the corresponding ureas (79–89%).

Solid-phase synthesis of the hexamer **106** (Fig. 14) was performed on Rink amide resin by coupling *O*-succinimidyl-(9*H*-fluoren-9-ylmethoxycarbonylamino)-ethyl carbamate derivatives **105b** (5 equiv.) with DIEA in DMF for 2 × 90 min. The Fmoc group was removed under standard conditions with 20% piperidine in DMF. After cleavage, the purity of the crude product was about 63% based on HPLC. Lyophilization and HPLC purification yielded **106** in 42% overall yield. Subsequently, it was found that the use of a weaker base such as NMM instead of DIEA led to significant improvement in overall yield (G. Guichard, unpublished data).

A. Procedure for the Solid-Phase Synthesis of Oligourea from *O*-Succinimidyl-(9*H*-fluoren-9-ylmethoxycarbonylamino)-Ethyl Carbamate Derivatives (105b)

1. General Procedure for the Preparation of *O*-Succinimidyl Carbamates **105b**

The *N*-Fmoc–protected β-amino acid **102b** (10 mmol) is dissolved in THF (30 mL) under Ar and cooled to −20°C. After addition of EtOCOCl (11 mmol) and NMM (11 mmol, 1.1 equiv.), the mixture is stirred at −20°C for 20 min. The resulting white suspension is allowed to warm up to −5°C and

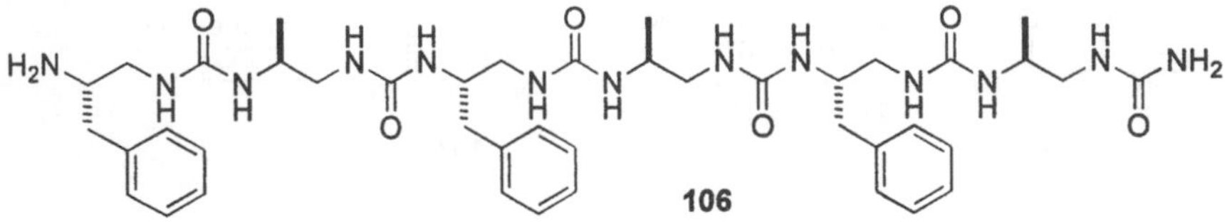

Figure 14 Oligourea prepared with monomers **105b**.

is treated with an aqueous solution (5 mL) of NaN$_3$ (25 mmol). The mixture is stirred for 5 min, diluted with EtOAc, washed with saturated NaCl, dried over MgSO$_4$, and concentrated under reduced pressure to give the acyl azide **103b**, which is used without further purification. Toluene is added under Ar and the resulting solution is heated to 65°C with stirring. After the gas evolution has stopped ($\sim$ 10 min), N-hydroxysuccinimide (10 mmol) and pyridine (10 mmol) are added. The mixture is stirred for 5 min at 65°C and then cooled to room temperature. In most cases the title compound crystallizes from the toluene solution and is collected by filtration. Recrystallization from toluene affords the pure O-succinimidyl carbamate **105b**. Otherwise the solvent is removed in vacuo and the residue is purified by recrystallization from the appropriate solvent.

2. Procedure for the Synthesis of **106**

Assembly of the oligomeric chain is carried out on an automated peptide synthesizer on Rink amide resin (50 μmol scale) using Fmoc-protected monomer **105b**. The monomer (4 equiv.) is dissolved in DMF (1 mL) and added to the resin. Then DIEA (8 equiv.) is added and the mixture is agitated with nitrogen bubbling for 90 min. Double coupling is performed systematically and the reaction is monitored using 2,4,6-trinitrobenzenesulfonic acid (TNBS) [169]. The Fmoc group is removed with 20% piperidine in DMF using a standard peptide synthesis procedure. After the removal of the last Fmoc protecting group, the resin is washed and dried prior to TFA–H$_2$O (95:5) cleavage (2 h). After concentration in vacuo, the crude material is diluted with H$_2$O–MeCN, lyophilized, and purified by RP-HPLC to give **106** (47%). MALDI-MS: 842.93 [M+H$^+$].

XI. OLIGOSULFONAMIDES

The growing interest in sulfonamides as peptide bond isosteres is motivated by their increased resistance to enzymatic degradation, by their structural similarity to the tetrahedral transition state intermediate, and thus by their

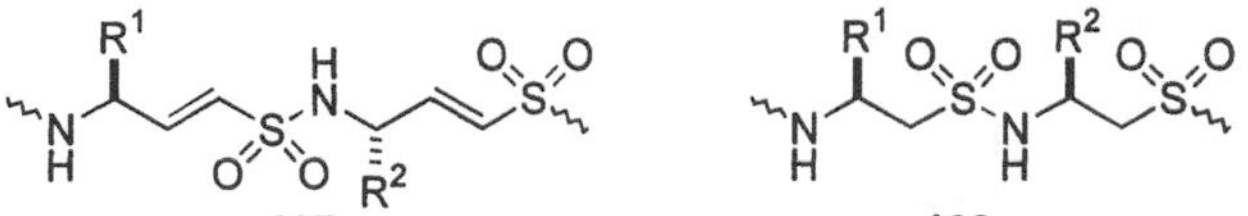

Figure 15 Vinylogous sulfonamidopeptides **107** and β-sulfonamidopeptides **108**.

potential utility in the development of protease inhibitors. Moreover, the sulfonamide group contains both hydrogen bond donors and acceptors (respectively stronger and weaker than in a normal peptide bond) [170] and hence is particularly suitable for the design of oligomers with potential folded structures, i.e., vinylogous sulfonamidopeptides (**107**) [65,171,172] and β-sulfonamidopeptides (**108**) [66,170,173–175] (Fig. 15).

Vinylogous sulfonamidopeptides, also termed vs-peptides (**107**), were introduced in 1994 by Gennari et al. [171]. A solid-phase approach employing N-Boc-vinylogous amino sulfonyl chloride monomers [Boc-vsXaa-Cl (**112**)] and generation of vs-peptide libraries were subsequently reported by the same group [65,172]. Wittig–Horner–Emmons reaction of N-Boc α-amino aldehydes **3** with methyl- or ethyldiethylphosphorylmethanesulfonate (**109**) afforded α,β-unsaturated sulfonates (**110**) (Scheme 29) with complete E-selectivity in good yield (75–85%). Treatment of **110** with tetrabutylammonium iodide in refluxing acetone gave the corresponding sulfonate salt (**111**), which was activated with SO_2Cl_2–PPh_3 in CH_2Cl_2 to the corresponding sulfonyl chloride **112** (85–90%).

Solid-phase synthesis of vs-peptides was performed starting from either Boc-Gly-TentaGel resin or TentaGel resin functionalized with a photoactivable linker (**113**, Novabiochem) (Fig. 16).

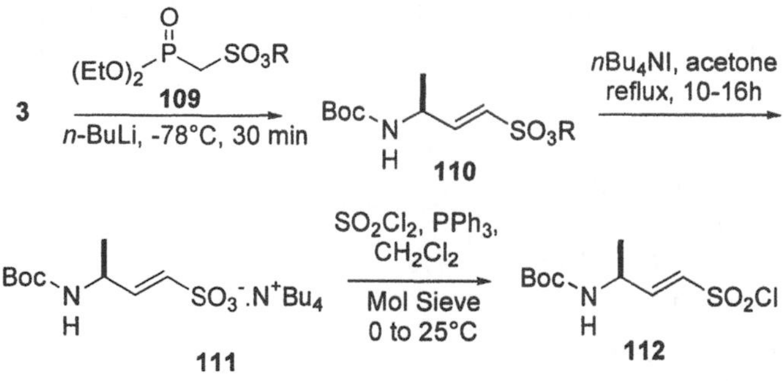

Scheme 29 Synthesis of N-Boc-vinylogous amino sulfonyl chloride monomers **112** [171].

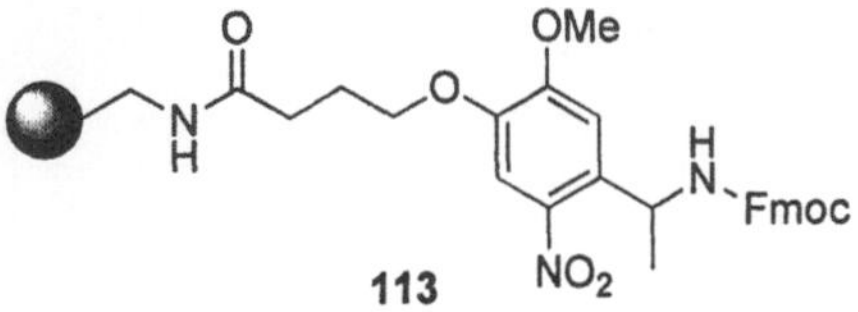

Figure 16 Photocleavable resin used for the synthesis of vs-peptides [65].

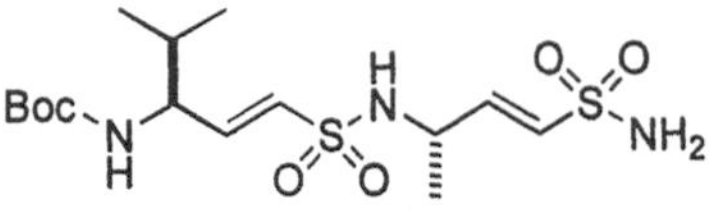

Figure 17 vs-dipeptides **114** synthesized on solid support [65].

Following Boc deprotection with 25% TFA in CH$_2$Cl$_2$ and subsequent neutralization, **112** (2 equiv.) was coupled to the polymer-bound amine in the presence of NMM (1 equiv.) (DBU, which was used initially, led to partial decomposition of sulfonyl chloride) and DMAP (0.5 equiv.) for 24 h. Because the ninhydrin test did not give satisfactory results, all coupling reactions were monitored with bromophenol blue. Cleavage of the resulting vs-peptide resin was best achieved when the photocleavable linker was employed [65]. Swelling the resin in MeOH and irradiating for 1–4 days at 354 nm afforded vs-dipeptide **114** (Fig. 17) in 86% yield after purification. Taking advantage of the acidic character of the sulfonamide NH (pK_a = 11–12) compared with BocNH (pK_a = 15–16), Gennari et al. [65] found that the sulfonamide group could be selectively alkylated on a solid support by using Cs$_2$CO$_3$ as a base in DMF.

Two different approaches to the synthesis of β-sulfonopeptides on a solid support have been reported to date. Gennari and coworkers and Liskamp's group used *N*-Fmoc– and *N*-Boc–protected β-sulfonyl chloride

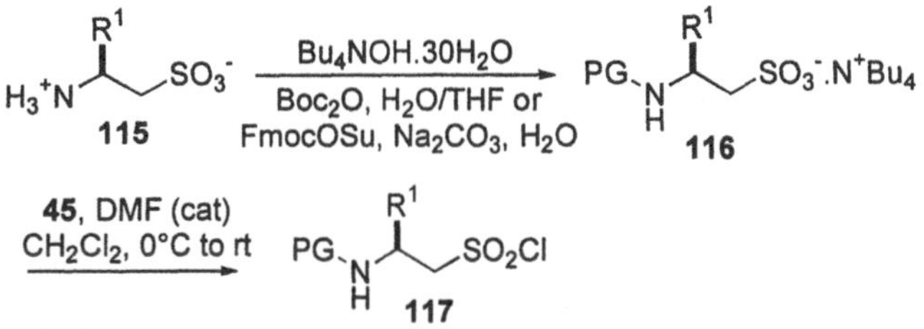

Scheme 30 Preparation of *N*-protected β-sulfonyl chloride **117**.

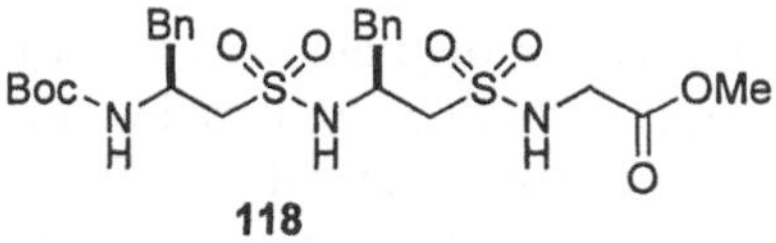

Figure 18 β-sulfonamidodipeptide synthesized on solid support [66].

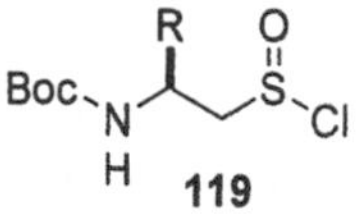

Figure 19 β-substituted sulfinyl chloride **119** [175].

(**117**) (Scheme 30) prepared in two steps from the corresponding enantio-merically pure substituted taurines (**115**), synthesized from N-protected α-amino acids according to Higashiura et al. [176]) [66,170,174].

These monomers (2 equiv.) were efficiently coupled to the amine on the solid support in the presence of 1-methoxy-2-methyl-1-trimethysilylox-ypropene (1 equiv.) and DMAP (0.5 equiv.). β-Sulfonamidopeptide **118** (Fig. 18) was synthesized starting from Boc-Gly-Tentagel resin in 70% yield [66].

Alternatively, Liskamp's group [175] initially reported the use of β-substituted sulfinyl chlorides **119** (Fig. 19) as activated monomers. NMM-mediated coupling to the polymer-bound amine led to the corresponding sulfinamide, which was further oxidized (OsO$_4$/N-methylmorpholine-N-ox-ide) to sulfonamide on the solid support [66]. However, the yield of the oxidation step is highly dependent on the sequence of the peptidosulfon-amide. Hence the former method utilizing β-sulfonyl chloride derivatives **117** is preferred [174].

XII. β-PEPTIDES

Peptides consisting exclusively of β- or γ-amino acids (ω-amino acids) have emerged as a promising new class of nonnatural oligomers (foldamers) that are able to fold into well-defined secondary structures [41–47]. So far, three different helical secondary structures and two turn motifs [177–181] as well as a parallel [177,179] and an antiparallel [179,182] sheet structure have been identified by two-dimensional NMR spectroscopy, circular dichroism (CD), and/or X-ray diffraction studies. In addition, cyclo-β³-tetrapeptides have been found to form nanotubes in the solid state [183] and have been used as transmembrane ion channels [184]. All these studies have demonstrated that by choosing the right substitution pattern of β-amino acids, it is possible to design β-peptides with well-defined folding propensities. For example, β-peptides containing either β^3-amino acids (120) (Fig. 20), β^2-amino acids (121) of (S) configuration [the (R/S) is sometimes misleading because in homochiral series, β^3Val and β^3Ser are of R configuration], or disubstituted $\beta^{2,3}$-amino acids (122) [with (S,S) configuration] exist as left-handed 3_1 helices (H-bonded ring size of 14 atoms) in MeOH solution [the helix handedness being opposite for peptides built from amino acids of (R) configuration] [177,178,180]. Alternatively, mixed β^2–β^3 peptides with alternating substitution pattern were found to fold in an unusual 12/10/12 helical structure with a central 10-membered and two terminal 12-membered rings [178]. It is worth mentioning that in X-ray studies of β-peptides made of rotationally restricted amino acids, trans-2-aminocyclopentanecarboxylic acid (trans-ACPC) (123) peptides fold into a 2.5_1 helix (H-bonded ring size of 12 atoms) that differs from the 3_1 helix adopted by the related trans-2-aminocyclohexanecarboxylic acid peptides (trans-ACHC) (124) [181].

In addition, the high resistance of β-peptides to proteolytic digestion suggests that these compounds may be attractive candidates for new drug discovery [185]. Nonapeptide analogues containing β-homoalanine oligomers have been reported as novel nonnatural MHC class I (HLA-B27) ligands [186]. Furthermore, Seebach and coworkers [187] have designed and synthesized a bioactive cyclo-β-tetrapeptide analogue of somatostatin, demonstrating for the first time that β-peptides can be used as peptidomimetics.

To facilitate the search for new types of secondary structures and for bioactive peptides, several groups have synthesized β-peptides (including β^2, β^3, and disubstituted $\beta^{2,2}$) on solid supports by Fmoc chemistry [61,62,180,188,189].

Numerous methods are available for the synthesis of β-amino acids and have been reviewed [190–192]. For the synthesis and structural studies

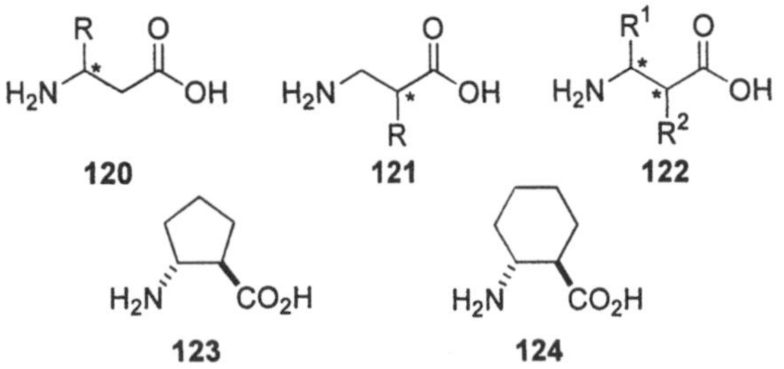

Figure 20 Example of β-amino acids with various substitution patterns incorporated into β-peptides.

of β-peptides, Seebach and coworkers [62,178,189,193,194] have prepared a variety of β-amino acids (including β^3-, β^2-, $\beta^{2,3}$-, $\beta^{2,2}$-, and $\beta^{3,3}$-amino acids) with different substitution pattern and with the side chains of natural amino acids. Enantiomerically pure N-Fmoc–protected β^3-amino acids **102b** (Scheme 31) were prepared via the Wolff rearrangement of diazoketones **125** prepared from the corresponding commercially available N-Fmoc-amino acids by reaction of their mixed anhydrides with diazomethane (**Caution:** the generation and handling of diazomethane require special precautions) [62,189]. In previously reported procedures, the diazoketones were decomposed in THF containing 10% H_2O with catalytic amounts of CF_3COOAg added as a homogeneous solution in Et_3N [177,193]. However, significant loss of the Fmoc protecting group was observed under these conditions [62]. Optimized procedures for the silver-catalyzed Wolff rearrangement involve the use of NMM in place of Et_3N [62], sonication and no tertiary base [195], or thermal decomposition in the absence of a tertiary base [196].

Two strategies, based on the use of either amino acids **102b** or diazoketones **125** as monomers, have been reported for the solid-phase synthesis of β^3-peptides [61,62]. Anchoring of the first residue on the resin was performed either by esterification of the *ortho*-chlorotrityl chloride resin with **102b** or by acylation of the 4-(benzyloxy)benzyl alcohol resin (Wang resin) with the ketene intermediate from the Wolff rearrangement of **125**. β-Peptides were then synthesized using conventional solid-phase peptide synthesis and BOP/HOBt as the coupling agent. Because the Kaiser ninhydrin test for free amino β^3-peptide resins is not reliable, all coupling reactions were monitored using 2,4,6-trinitrobenzenesulfonic acid (TNBS) [62,169]. After removal of the last Fmoc protecting group, cleavage from the resin (TFA–H_2O, 95:5), and precipitation, the crude β-peptides were recovered in good yields (45–99%) and purity (45–95%). For example, the purity of crude **126** (Fig. 21) was 94% as determined by HPLC [62]. Under these conditions, β^3-peptides of up to 15 amino acids have been synthesized. In some cases, incomplete Fmoc deprotection in the last steps of the synthesis was observed when using 20% piperidine in DMF. The use of 1,8-diazabicyclo[5.4.0]undec-7-ene resulted in significant improvements. Alternatively, β-

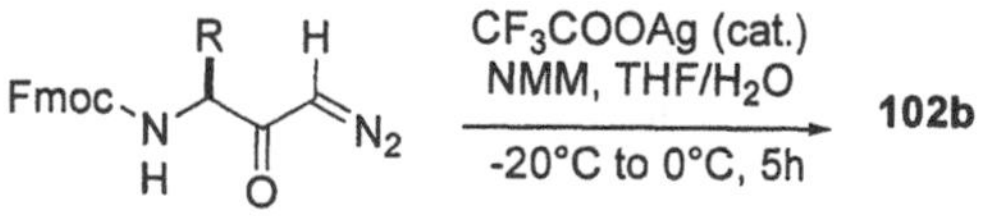

Scheme 31 Wolff rearrangement of diazoketone **125** to give the N-Fmoc–protected β^3-amino acid **102b** [62].

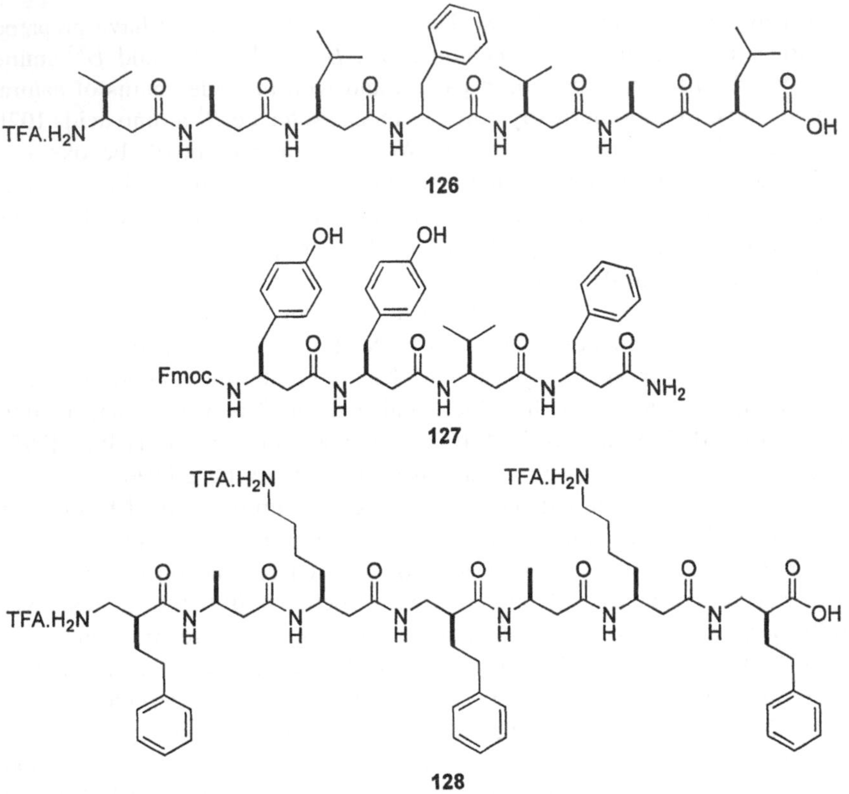

126

127

128

Figure 21 β^3- and mixed β^2–β^3 peptides synthesized on solid support [61, 62,188].

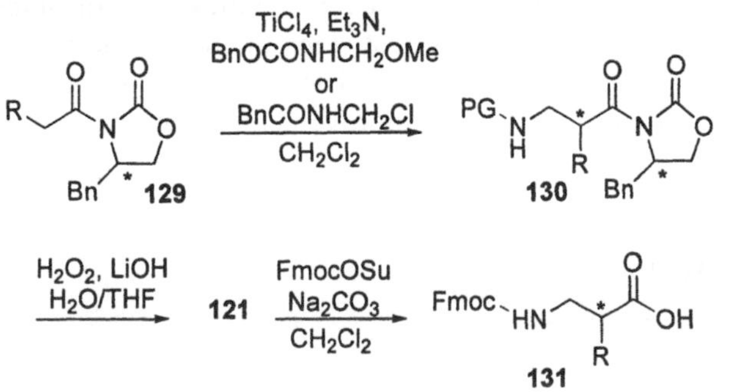

Scheme 32 Synthesis of enantiomerically pure N-protected β^2-amino acids **131** [178,194,198].

peptide bond formation via sequential Arndt–Eistert homologation on the solid phase has been described [61,62]. The tetrapeptide **127** was prepared in 60% yield and 90% purity by this method [62]. This approach, however, may be limited to the synthesis of short β-peptides.

β^2-Amino acid derivatives (**131**) (Scheme 32) required for synthesis of β^2-peptides have been prepared using the methodology of Evans et al. [197]. The (S)- and (R)-2-(aminomethyl)alkanoic acids (**121**) are constructed by asymmetric aminomethylation of optically pure (R)- and (S)-N-acyl-4-phenylmethyl)oxazolidin-2-ones (**129**) through TiCl$_4$-enolates with (benzoyl-amino)methylchloride or benzyl N-(methoxymethyl)carbamate. Hydrolytic removal of the auxiliary led to the N-protected (benzoyl or Z) amino acid [178,188,194,198]. Deprotection yielded the free amino acid (**121**), which was converted to the N-Fmoc–protected derivative **131**. As an example of the use of β^2-amino acid, the mixed β^2–β^3 peptide (**128**) was synthesized in high purity (80% as determined by HPLC of the crude product) and 33% yield after purification [188].

A. General Procedure for the Synthesis of N-Fmoc–Protected β^3-Amino Acids (102b) [62,188]

1. Synthesis of diazoketone 125

To a cold 20°C solution of Fmoc-α-amino acid (40 mmol) in THF (50 mL) under Ar are successively added i-BuOCOCl (5.71 mL, 44 mmol) and NMM (4.81 mL, 44 mmol). After 20 min of stirring at −20°C, the resulting white suspension is filtered rapidly and a solution of CH$_2$N$_2$ in Et$_2$O is added to the filtrate (**CAUTION**: the generation and handling of diazomethane require special precautions). Stirring is continued for 3 h as the mixture is allowed to warm to room temperature. Excess CH$_2$N$_2$ is destroyed by vigorous stirring. The mixture is diluted with Et$_2$O and washed with saturated NaHCO$_3$ solution, water, 1 M KHSO$_4$, and saturated NaCl solution. The organic phase is dried over MgSO$_4$ and concentrated in vacuo. Recrystallization in AcOEt–Et$_2$O–hexane and/or flash chromatography (AcOEt–hexane) yields **125** (60–93%), generally as a yellow solid and sometimes as a yellow oil [Fmoc-Ser(tBu)CHN$_2$, Fmoc-Thr(tBu)-CHN$_2$].

2. Wolff Rearrangement of 125

To a cold −20°C solution of **125** (34 mmol) in THF containing 10% H$_2$O (100 mL) is added with exclusion of light a solution of silver benzoate (856 mg, 3.74 mmol) in NMM (9.3 mL, 85 mmol) and the resulting mixture is allowed to warm to 0°C in 5 h in the dark. It is diluted with saturated NaHCO$_3$ solution (500 mL) and extracted with Et$_2$O (100 mL). The aqueous phase is acidified to pH 2–3 with 1 M KHSO$_4$ at 0°C and extracted with AcOEt (300 mL). The organic layer is dried over MgSO$_4$ and concentrated in vacuo. Recrystallization from CHCl$_3$–hexane or flash chromatography (AcOEt–hexane–AcOH) yields **102b** (47–81%).

ACKNOWLEDGMENT

The author greatly acknowledge Dr. J. P. Briand and Dr. A. Bianco for helpful discussions and suggestions.

NOTE ADDED IN PROOF

A general solid-phase approach to ketone-based cysteine protease inhibitors has been reported by Ellman's group [199]. The synthesis of support-bound α-chloro hydrazones key intermediates was achieved by attachment of chloromethylketones onto a support-bound hydrazine. The solid-phase synthesis of depsides and depsipeptides has been successfully achieved [200]. The hydroxyl group of hydroxy acids was protected as tetrahydropyranyl ether and couplings to the resin bound chain were performed with DIC/DMAP.

REFERENCES

1. AF Spatola. In: B Weinstein, ed. Chemistry and Biochemistry of Amino Acids, Peptides and Proteins. Vol 7. New York: Marcel Dekker, 1983, pp 267–357.
2. JL Fauchère. Adv Drug Res 15:29–69, 1986.
3. AS Dutta. Adv Drug Res 21:145–286, 1991.
4. JL Fauchère, C Thurieau. Adv Drug Res 23:128–159, 1992.
5. A Giannis, T Kolter. Angew Chem Int Ed Engl 32:1244–1267, 1993.
6. J Gante. Angew Chem Int Ed Engl 33:1699–1720, 1994.
7. TK Sawyer. In: MD Taylor, GL Amidorn, eds. Peptide-Based Drug Design Controlling Transport and Metabolism. New York: Oxford University Press, 1995, pp 332–422.
8. M Szelke, B Leckie, A Hallett, DM Jones, J Sueiras, B Atrash, AF Lever. Nature 299:555–557, 1982.
9. J Martinez, JP Bali, M Rodriguez, B Castro, R Magous, J Laur, MF Lignon. J Med Chem 28:1874–1879, 1985.
10. M Goodman, M Chorev. Acc Chem Res 12:1–7, 1979.
11. M Chorev, M Goodman. Acc Chem Res 26:266–273, 1993.
12. M Chorev, M Goodman. Trends Biotechnol 13:438–445, 1995.

13. MD Fletcher, MM Campbell. Chem Rev 98:763–796, 1998.
14. MM Hann, PG Sammes, PD Kennewell, JB Taylor. J Chem Soc Perkin Trans I 307–314, 1982.
15. DJ Kempf, XC Wang, SG Spanton. Int J Peptide Protein Res 38:237–241, 1991.
16. M Rodriguez, A Aumelas, J Martinez. Tetrahedron Lett 31:5153–5156, 1990.
17. M Rodriguez, A Heitz, J Martinez. Tetrahedron Lett 31:7319–7322, 1990.
18. M Holladay, DH Rich. Tetrahedron Lett 24:4401–4404, 1983.
19. BE Evans, KE Rittle, CF Homnick, JP Springer, J Hirshfield, DF Veber. J Org Chem 50:4615–4625, 1985.
20. SL Harbeson, DH Rich. J Med Chem 32:1378–1392, 1989.
21. RE Tenbrink. J Org Chem 52:418–422, 1987.
22. E Rubini, C Gilon, Z Selinger, M Chorev. Tetrahedron 42:6039–6045, 1986.
23. BH Norman, JS Kroin. J Org Chem 61:4990–4998, 1996.
24. NJ Anthony, RP Gomez, WJ Holtz, JS Murphy, RG Ball, TJ Lee. Tetrahedron Lett 36:3821–3824, 1995.
25. RG Almquist, WR Chao, ME Ellis, HL Johnson. J Med Chem 23:1392–1398, 1980.
26. MT García-López, R González-Muñiz, RJ Harto. Tetrahedron Lett 29:1577–1580, 1988.
27. A Pessi, M Pinori, AS Verdini, GC Viscomi. J Chem Soc Chem Commun 195–197, 1983.
28. Y Sasaki, DH Coy. Peptides 8:119–121, 1987.
29. P Wipf, TC Henninger. J Org Chem 62:1586 1587, 1997.
30. CJ Gray, M Quibell, N Baggett, T Hammerle. Int J Peptide Protein Res 40:351–362, 1992.
31. M Quibell, WG Turnell, TJ Johnson. J Chem Soc Perkin Trans 1 2843–2849, 1993.
32. C Frochot, R Vanderesse, A Driou, G Linden, M Marraud, MT Cung. Lett Peptide Sci 4:219–225, 1997.
33. D Limal, V Semetey, P Dalbon, M Jolivet, JP Briand. Tetrahedron Lett 40:2749–2752, 1999.
34. A Bianco, C Zabel, P Walden, G Jung. J Peptide Sci 4:471–478, 1998.
35. O Kuisle, E Quiñoá, R Riguera. Tetrahedron Lett 40:1203–1206, 1999.
36. PF Alewood, RI Brinkworth, RJ Dancer, B Garnham, A Jones, SBH Kent. Tetrahedron Lett 33:977–980, 1992.
37. EK Kick, JA Ellman. J Med Chem 38:1427–1430, 1995.
38. CE Lee, EK Kick, JA Ellman. J Am Chem Soc 120:9735–9747, 1998.
39. GT Wang, S Li, N Wideburg, GA Kraft, DL Kempf. J Med Chem 38:2995–3002, 1995.
40. RMJ Liskamp. Angew Chem Int Ed Engl 33:633–636, 1994.
41. SH Gellman. Acc Chem Res 31:173–180, 1998.
42. S Borman. Chem Eng News June 16:32–35, 1997.
43. D Seebach, JL Matthews. Chem Commun 2015–2022. 1997.
44. U Koert. Angew Chem Int Ed Engl 36:1836–1837, 1997.
45. A Banerjee, P Balaram. Curr Sci 73:1067–1077, 1997.

46. T Hintermann, K Gademann, B Jaun, D Seebach. Helv Chim Acta 81:983–1002, 1998.
47. S Hanessian, X Luo, R Schaum, S Michnick. J Am Chem Soc 120:8569–8570, 1998.
48. K Kirshenbaum, AE Barron, RA Goldsmith, P Armand, EK Bradley, KTV Truong, KA Dill, FE Cohen, RN Zuckermann. Proc Natl Acad Sci U S A 95: 4303–4308, 1998.
49. P Armand, K Kirshenbaum, RA Goldsmith, S Farr-Jones, AE Barron, KTV Truong, KA Dill, DA Mierke, FE Cohen, RN Zuckermann, EK Bradley. Proc Natl Acad Sci U S A 95:4309–4314, 1998.
50. RN Zuckermann, EJ Martin, DC Spellmeyer, GB Stauber, KR Shoemaker, JM Kerr, GM Figliozzi, DA Goff, MA Siani, RJ Simon, SC Banville, EG Brown, L Wang, LS Richter, WH Moos. J Med Chem 37:2678–2685, 1994.
51. JE Murphy, R Uno, JD Hammer, FE Cohen, V Dwarki, R Zuckermann. Proc Natl Acad Sci U S A 95:1517–1522, 1998.
52. CY Cho, RS Youngquist, SJ Paikoff, MH Beresini, AR Hebert, LT Berleau, CW Liu, DE Wemmer, T Keough, PG Schultz. J Am Chem Soc 120:7706–7718, 1998.
53. X Wang, I Huq, TM Rana. J Am Chem Soc 119:6444–6445, 1997.
54. N Tamilarasu, I Huq, TM Rana. J Am Chem Soc 121:1597–1598, 1999.
55. RJ Simon, RS Kania, RN Zuckermann, VD Huebner, VD, DA Jewell, S Banville, S Ng, L Wang, S Rosenberg, CK Marlowe, DC Spellmeyer, R Tan, AD Frankel, DV Santi, FE Cohen, PA Bartlett. Proc Natl Acad Sci U S A 89:9367–9371, 1992.
56. RN Zuckermann, JM Kerr, SBH Kent, WH Moos. J Am Chem Soc 114: 10646–10647, 1992.
57. CY Cho, RS Youngquist, SJ Paikoff, MH Beresini, AR Herbert, LT Berleau, CW Liu, DE Wemmer, T Keough, PG Schultz. J Am Chem Soc 120:7706–7718, 1998.
58. K Burgess, DS Linthicum, H Shin. Angew Chem Int Ed Engl 34:907–908, 1995.
59. JM Kim, Y Bi, SJ Paikoff, PG Schultz. Tetrahedron Lett 37:5305–5308, 1996.
60. J Smith, JL Liras, SE Schneider, EV Anslyn. J Org Chem 61:8811–8818, 1996.
61. RE Marti, KH Bleicher, KW Bair. Tetrahedron Lett 38:6145–6148, 1997.
62. G Guichard, S Abele, D Seebach. Helv Chim Acta 81:187–206, 1998.
63. BC Hamper, SA Kolodziej, AM Scates, RG Smith, E Cortez. J Org Chem 63:708–718, 1998.
64. JAW Kruijzer, DJ Lefeber, RMJ Liskamp. Tetrahedron Lett 38:5335–5338, 1997.
65. C Gennari, C Lonari, S Ressel, B Salom, U Piarulli, S Ceccarelli, A Mielgo. Eur J Org Chem 2437–2449, 1998.
66. M Gude, U Piarulli, D Potenza, B Salom, C Gennari. Tetrahedron Lett 37: 8589–8592, 1996.
67. E Atherton, HD Law, S Moore, DF Elliott, R Wade. J Chem Soc (C) 3393–3396, 1971.

68. H Oyamada, MA Ueki. Bull Chem Soc Jpn 60:267–271, 1987.
69. K Clausen, M Thorsen, SO Lawesson. Tetrahedron 37:3635–3639, 1981.
70. FS Guziec Jr, LM Wasmund. Tetrahedron Lett 31:23–26, 1990.
71. HE Johnson, DG Crosby. J Org Chem 27:2205–2207, 1962.
72. IV Micovic, MD Ivanovic, DM Piatak, VD Bojic. Synthesis 1043–1045, 1991.
73. MD Boman, IC Guch, M DiMare. J Org Chem 60:5995–5996, 1995.
74. RF Borch, MD Bernstein, HD Durst. J Am Chem Soc 93:2897–2904, 1971.
75. AF Abdel-Magid, KG Carson, BD Harris, CA Maryanoff, RD Shah. J Org Chem 61:3849–3862, 1996.
76. DH Coy, SJ Hocart, Y Sasaki. Tetrahedron 44:835–841, 1988.
77. Y Han, M Chorev. J Org Chem 64:1972–1978, 1999.
78. J Jurczak, A Golebiowski. Chem Rev 89:149–164, 1989.
79. CF Stanfield, JE Parker, P Kanellis. J Org Chem 46:4797–4798, 1981.
80. KE Rittle, CF Homnick, GS Ponticello, BE Evans. J Org Chem 47:3016–3018, 1982.
81. JR Luly, JF Dellaria, JJ Plattner, JL Doderquist, NA Yi. J Org Chem 52:1487–1492, 1987.
82. AJ Mancuso, D Swern. Synthesis 165–185, 1981.
83. JJ Wen, CM Crews. Tetrahedron Asym 9:1855–1858, 1998.
84. DH Rich, ET Sun, AS Boparai. J Org Chem 43:3624–3626, 1978.
85. J Zemlicka, M Murata. J Org Chem 41:3317–3321, 1976.
86. JA Fehrentz, B Castro. Synthesis 676–678, 1983.
87. WD Lubell, H Rapoport. J Am Chem Soc 109:236–239, 1987.
88. JA Fehrentz, C Pothion, JC Califano, A Loffet, J Martinez. Tetrahedron Lett 35:9031–9034, 1994.
89. T Fukayama, SC Lin. J Am Chem Soc 112:7050–7051, 1990.
90. PT Ho, KY Ngu. J Org Chem 58:2313–2316, 1993.
91. S Nahm, SM Weinreb. Tetrahedron Lett 22:3815–3818, 1981.
92. G Guichard, JP Briand, M Friede. Peptide Res 6:121–124, 1993.
93. JP Salvi, N Walchshofer, J Paris. Tetrahedron Lett 35:1181–1184, 1994.
94. JP Meyer, P Davis, KB Lee, F Porreca, HL Yamamura, VJ Hruby. J Med Chem 38:3462–3468, 1995.
95. D Limal, A Quesnel, JP Briand. Tetrahedron Lett 39:4239–4242, 1998.
96. M Rodriguez, M Llinares, S Doulut, A Heitz, J Martinez. Tetrahedron Lett 32:923–926, 1991.
97. PT Ho, D Chang, JWX Zhong, GF Musso. Peptide Res 6:10–12, 1993.
98. V Prelog, H Gerlach. Helv Chim Acta 47:2288–2302, 1964.
99. MM Shemyakin, A Ovchinnikov, VT Ivanov. Angew Chem Int Ed Engl 8:492–499, 1969.
100. H Dürr, M Goodman, G Jung. Angew Chem Int Ed Engl 31:785–787, 1992.
101. G Guichard, F Connan, R Graff, M Ostankovitch, S Muller, JG Guillet, J Choppin, JP Briand. J Med Chem 39:2030–2039, 1996.
102. M Ostankovitch, G Guichard, F Connan, S Muller, A Chaboissier, J Hoebeke, J Choppin, JP Briand, JG Guillet. J Immunol 161:200–208, 1998.

103. J Wermuth, SL Goodman, A Jonczyk, H Kessler. J Am Chem Soc 119:1328–1335, 1997.

104. M Ruvo, G Fassina. Int J Peptide Res 45:356–365, 1995.

105. M Mariotti, A Sisto. Eur Appl EP406931, 1991.

106. G Guichard, N Benkirane, G Zeder-Lutz, MHV Van Regenmortel, JP Briand, S Muller. Proc Natl Acad Sci U S A 91:9765–9769, 1994.

107. G Guichard, S Muller, MHV Van Regenmortel, JP Briand, P Mascagni, E Giralt. Trends Biotechnol 14:44–45, 1996.

108. A Comis, P Fischer, MI Tyler. PCT Int Appl WO 9405311, 1994.

109. A Verdoliva, M Ruvo, G Cassani, G Fassina. J Biol Chem 270:30422–30427, 1995.

110. JP Briand, N Benkirane, G Guichard, JFE Newman, MHV Van Regenmortel, F Brown, S Muller. Proc Natl Acad Sci U S A 94:12545–12550, 1997.

111. C Meziere, M Viguier, H Dumortier, R Lo-Man, C Leclerc, JG Guillet, JP Briand, S Muller. J Immunol 159:3230–3237, 1997.

112. A Comis, MI Tyler, P Fischer. PCT Int Appl WO 9523166, 1995.

113. BA Jameson, JM McDonnel, JC Marini, R Korngold. Nature 368:744–746, 1994.

114. JM McDonnell, AJ Beavil, GA Mackay, BA Jameson, R Korngold, HJ Gould, BJ Sutton. Nat Struct Biol 3:419–426, 1996.

115. JM McDonnel, D Fushman, SM Cahill, BJ Sutton, D Cowburn. J Am Chem Soc 119:5321–5328, 1997.

116. RB Merrifield, P Juvvadi, D Andreu, J Ubach, A Boman, HG Boman. Proc Natl Acad Sci U S A 92:3449–3453, 1995.

117. P Juvvadi, S Vuunam, RB Merrifield. J Am Chem Soc 118:8989–8997, 1996.

118. J Brugidou, C Legrand, J Mery, A Rabie. Biochem Biophys Res Commun 214:685–693, 1995.

119. ACS Saphire, SJ Bark, L Gerace. J Biol Chem 273:29764–29769, 1998.

120. DM Hrubowchak, FX Smith. Tetrahedron Lett 24:4951–4954, 1983.

121. M Chorev, E Rubini, C Gilon, U Wormser, Z Selinger. J Med Chem 26:129–135, 1983.

122. L Gazerro, M Pinori, AS Verdini. In: R Epton, ed. Innovation and Perspectives in Solid Phase Synthesis. Peptides, Polypeptides and Oligonucleotides. Macro-Organic Reagents and Catalysts. Birmingham: SPCC Ltd, 1990, pp 403–412.

123. A Dalpozzo, E Laurita. In: HLS Maia, ed. Peptides. Proceedings of the Twenty-Third European Peptide Symposium. Leiden: ESCOM, 1995, pp 714–715.

124. A Abouabdellah, JT Welch. Tetrahedron Assym 5:1005–1013, 1994.

125. M Chorev, CG Willson, M Goodman. J Am Chem Soc 99:8075–8076, 1977.

126. M Chorev, R Shavitz, M Goodman, S Minick, R Guillemin. Science 204:1210–1212, 1979.

127. N Chaturvedi, M Goodman, MC Bowers. Int J Peptide Protein Res 17:72–88, 1981.

128. M Chorev, M Goodman. Int J Peptide Protein Res 21:258–268, 1983.

129. AS Radhakrishna, ME Parham, RM Riggs, GM Loudon. J Org Chem 44: 1746–1747, 1979.

130. PV Pallai, M Goodman. J Chem Soc Chem Commun 280–281, 1982.

131. GM Loudon, MR Almond, JN Jacob. J Am Chem Soc 103:4508–4515, 1981.

132. A Pessi, M Pinori, AS Verdini, GC Viscomi. Eur Appl EP 097994, 1983.

133. F Bonelli, A Pessi, AS Verdini. Int J Peptide Res 24:553–556, 1984.

134. G Di Gregorio, M Pinori, AS Verdini. In: R Epton, ed. Innovation and Perspectives in Solid Phase Synthesis. Peptides, Polypeptides and Oligonucleotides. Andover Hampshire: Intercept, England, UK, 1992, pp 311–318.

135. S Muller, G Guichard, N Benkirane, Fred Brown, MHV Van Regenmortel, JP Briand. Peptide Res 8:138–144, 1995.

136. J Gante. Synthesis 405–413, 1989.

137. WJ Greenlee, ED Thorsett, JP Springer, AA Patchett, EH ULM, TC Vassil. Biochem Biophys Res Commun 122:791–797, 1984.

138. JC Powers, BF Gupton. Methods Enzymol 46:208–216, 1977.

139. AS Dutta, BJA Furr, MB Giles. J Chem Soc Perkin Trans I 379–388, 1979.

140. F Andre, M Marraud, T Tsouloufis, SJ Tzartos, G Boussard. J Peptide Sci 3: 429–441, 1997.

141. H Han, KD Janda. J Am Chem Soc 118:2539–2544, 1996.

142. M Marraud, V Dupont, V Grand, S Zerkout, A Lecoq, G Boussard, J Vidal, A Collet, A Aubry. Biopolymers 33:1135–1148, 1993.

143. V Dupont, A Lecoq, JP Mangeot, A Aubry, G Boussard, M Marraud. J Am Chem Soc 115:8898–8906, 1993.

144. MM Hahn, PG Sammes, PD Kennewell, JD Taylor. J Chem Soc Chem Commun 234–235, 1980.

145. P Wipf, TC Henninger, SJ Geib. J Org Chem 63:6088–6089, 1998.

146. P Wipf, PC Fritch. J Am Chem Soc 59:4875–4886, 1994.

147. T Fukuyama, CK Jow, M Cheung. Tetrahedron Lett 36:6373–6374, 1995.

148. D Tourwé, J Piron, P Defreyn, G Van Binst. Tetrahedron Lett 34:5499–5502, 1993.

149. LA Thompson, JA Ellman. Tetrahedron Lett 35:9333–9336, 1994.

150. K Ngu, DV Patel. Tetrahedron Lett 38:973–976, 1997.

151. B Altava, I Burguette, SV Luis. Tetrahedron 50:7535–7542, 1994.

152. JAW Kruijtzer, LJF Hofmeyer, W Heerma, Cversluis, RMJ Liskamp. Chem Eur J 4:1570–1580, 1998.

153. JT Nguyen, CW Turck, FE Cohen, RN Zuckerman, WA Lim. Science 282: 2088–2092, 1998.

154. S Ostergaard, A Holm. Mol Divers 3:17–27, 1997.

155. C Anne, MC Fournié-Zaluski, BP Roques, F Cornille. Tetrahedron Lett 39: 8973–8974, 1998.

156. CY Cho, EJ Moran, SR Cherry, JC Stephans, SPA Fodor, CL Adams, A Sundaram, JW Jacobs, PG Schultz. Science 261:1303–1305, 1993.

157. SJ Paikoff, TE Wilson, CY Cho, PG Schultz. Tetrahedron Lett 37:5653–5656, 1996.

158. F Albericio, N Kneib-Cordonier, S Biancalana, L Gera, RI Masada, D Hudson, G Barany. J Org Chem 55:3730–3743, 1990.

159. R Warrass, KH Wiesmüller, G Jung. Tetrahedron Lett 39:2715–2716, 1998.
160. PY Lam, PK Jadhav, CJ Eyermann, CN Hodge, Y Ru, LT Bacheler, JL Meek, MJ Otto, MM Rayner, YN Wong, CH Chang, PC Weber, DA Jackson, TR Sharpe, S Erickson-Viitanen. Science 263:380–384, 1994.
161. JL Castro, RG Ball, HB Broughton, MG Russell, D Rathbone, AP Watt, R Baker, KL Chapman, AE Fletcher, S Patel, AJ Smith, GR Marshall, W Ryecroft, VG Matassa. J Med Chem 39:842–849, 1996.
162. TW von Geldern, JA Kester, R Bal, JR Wu-Wong, W Chiou, DB Dixon, TJ Opgenorth. J Med Chem 39:968–981, 1996.
163. JS Nowick, EM Smith, GW Noronha. J Org Chem 60:7386–7387, 1995.
164. JS Nowick, S Mahrus, EM Smith, JW Ziller. J Am Chem Soc 118:1066–1072, 1996.
165. JS Nowick, DL Holmes, G Mackin, G Noronha, AJ Shaka, EM Smith. J Am Chem Soc 118:2764–2765, 1996.
166. DH Holmes, EM Smith, JS Nowick. J Am Chem Soc 119:7665–7669, 1997.
167. K Burgess, J Ibarzo, DS Linthicum, DH Russell, H Shin, A Shitangkoon, R Totani, AJ Zhang, J Am Chem Soc 119:1564, 1997.
168. G Guichard, V Semetey, C Didierjean, A Aubry, JP Briand, M Rodriguez. J Org Chem, 64:8702–8705, 1999.
169. WS Hancock, JE Battersby. Anal Biochem 71:260–264, 1976.
170. C Gennari, M Gude, D Potenza, U Piarulli. Chem Eur J 4:1924–1931, 1998.
171. C Gennari, B Salom, D Potenza, A Williams. Angew Chem Int Ed Engl 33:2067–2069, 1994.
172. C Gennari, HP Nestler, B Salom, WC Still. Angew Chem Int Ed Engl. 34:1763–1765, 1995.
173. WJ Moree, G van der Marel, RJ Liskamp. J Org Chem 60:5157–5169, 1995.
174. DBA de Bont, GDH Dijkstra, JAJ den Hartog, RMJ Liskamp. Bioorg Med Chem Lett 6:3035–3040, 1996.
175. DBA de Bont, WJ Moree, RMJ Liskamp. Bioorg Med Chem 4:667–672, 1996.
176. K Higashiura, H Morino, H Matsuura, Y Toyomaki, K Ienaga. J Chem Soc Perkin Trans I 1479–1481, 1989.
177. D Seebach, M Overhand, FNM Kühnle, B Martinoni, L Oberer, U Hommel, H Widmer. Helv Chim Acta 79:913–941, 1996.
178. D Seebach, S Abele, K Gademann, G Guichard, T Hinermann, B Jaun, JL Matthews, JV Schreiber, L Oberer, U Hommel, H Widmer. Helv Chim Acta 81:932–982, 1998.
179. D Seebach, S Abele, K Gademann, B Jaun. Angew Chem Int Ed Engl 38:1595–1597,1999.
180. DH Appella, LA Christianson, IL Karle, DR Powell, SH Gellman. J Am Chem Soc 118:13071–13072, 1997.
181. DH Appella, LA Christianson, DA Klein, DR Powell, X Huang, JJ Barchi Jr, SH Gellman. Nature 387:381–384, 1997.
182. S Krauthäuser, LA Christianson, DR Powell, SH Gellman. J Am Chem Soc 119:11719–11720, 1997.

183. D Seebach, JL Matthews, A Meden, T Wessels, C Baerlocher, LB McCusker. Helv Chim Acta 80:173–182, 1997.
184. TD Clark, LK Buehler, MR Ghadiri. J Am Chem Soc 120:651–656, 1998.
185. T Hintermann, D Seebach. Chimia 51:244–247, 1997.
186. S Poenaru, JR Lamas, G Folkers, JA López de Castro, D Seebach, D Rognan. J Med Chem 42:2318–2331, 1999.
187. K Gademann, M Ernst, D Hoyer, D Seebach. Angew Chem Int Ed Engl 38: 1223–1226, 1999.
188. S Abele, G Guichard, D Seebach. Helv Chim Acta 81:2141–2156, 1998.
189. D Seebach, S Abele, T Sifferlen, M Hängi, S Gruner, P Seiler. Helv Chim Acta 81:2218–2243, 1998.
190. DC Cole. Tetrahedron 50:9517–9582, 1994.
191. E Juaristi, ed. Enantioselective Synthesis of β-Amino Acids. New York: Wiley-VCH, 1997.
192. S Abele, D Seebach. Eur J Org Chem, in press.
193. J Podlech, D Seebach. Liebigs Ann 1217–1228, 1995.
194. T Hintermann, D Seebach. Synlett 437–438, 1997.
195. A Muller, C Vogt, N Sewald. Synthesis 837–841, 1998.
196. A Leggio, A Liguori, A Procopio, G Sindona. J Chem Soc Perkin Trans I 1969–1971, 1997.
197. DA Evans, F Urpi, TC Sommers, SC Clark, MT Bilodeau. J Am Chem Soc 112:8215–8216, 1990.
198. T Hintermann, D Seebach. Helv Chim Acta 81:2093–2126, 1998.
199. A Lee, L Huang, JA Ellman. J Am Chem Soc 121:9907–9914, 1999.
200. O Kuisle, E Quiñoá, R Riguera J Org Chem 64:8063–8075, 1999.

17
Instrumentation

Scott B. Daniels
PE Biosystems, Framingham, Massachusetts

I. INTRODUCTION

The use of an insoluble matrix facilitates the process of biopolymer synthesis. All intermediate purification steps that are required in solution-phase synthesis are eliminated by simply washing excess reagents and reaction by-products from the support-bound product.

Synthesis on solid supports was first developed by Merrifield [1] for the assembly of peptides. It has expanded to include many different applications including oligonucleotide, carbohydrate, and small-molecule assembly (see Chapters 11 and 14). The repetitive cycle of steps involved in the solid-phase synthesis of biopolymers can be performed manually using simple laboratory equipment or fully automated with sophisticated instrumentation. This chapter examines typical solid-phase reaction kinetics to identify factors that can improve the efficiency of both manual and automated synthesis. The hardware and software features of automated solid-phase instruments are also discussed. The focus of this discussion is not on particular commercial model synthesizers but on the basic principles of instrument operation. These considerations can assist in the design, purchase, or use of automated equipment for solid-phase synthesis. Most contrasting features have advantages and disadvantages and the proper choice of instrumentation depends on the synthetic needs of the user.

II. REACTION KINETICS

Before discussing the details of solid-phase instrumentation, it is important to understand the processes that are to be automated. The kinetics of the reactions provide valuable information that can be used to optimize the solid-phase assembly of biopolymers.

Figure 1 outlines the basic steps performed in most solid-phase biopolymer syntheses. A protected dimer attached to the solid support by a linker (A) is treated with a deblock reagent (D) to yield the unprotected dimer–support (B). The free protecting group or a protecting group–deblock reagent adduct (E) is released during this reaction. The unprotected dimer–support (B) is washed to remove the excess deblock reagent and any reaction by-products. It is then coupled with an activated protected monomer (F) to form the protected trimer–solid support (C).

The support is washed again, and these two reactions, deblocking and coupling, are repeated until the desired biopolymer–solid support has been assembled. The biopolymer is generally cleaved from the solid support for use in the desired application. Although other steps may be involved, depending on the type of biopolymer synthesized, deblocking and coupling are the main reactions in the assembly process.

The rate equations for the deblocking and coupling steps are shown in Eqs. (1) and (2), respectively. The letters correspond to the designations in Fig. 1 and k_d and k_c are the rate constants for deblocking and coupling, respectively.

$$A + D \xrightarrow{k_d} B + E \qquad \frac{d[B]}{dt} = \frac{d[E]}{dt} = k_d[A][D] \tag{1}$$

$$B + F \xrightarrow{k_c} C \qquad \frac{d[C]}{dt} = -\frac{d[F]}{dt} = k_c[B][F] \tag{2}$$

Although these equations are simplifications of the actual chemical processes, the kinetic information is instructive. The second-order rate equations indicate that the reaction rates are directly proportional to the concentration of the reactants and the magnitude of the rate constants. Efforts to maximize these factors can lead to reactions that are more efficient.

A.　Reaction Concentrations

Although the theoretical rate equations indicate that a higher concentration leads to a faster reaction, there are some practical limits to reactant concentration. The solubility of the reactant in the synthesis solvent restricts the upper limit of the concentration. There is also the possibility of the occurrence of undesired side reactions with highly concentrated or reactive reagents. This is particularly true for the deblock reaction. For instance, when higher concentrations of acid are used in the deblock reaction in oligonu-

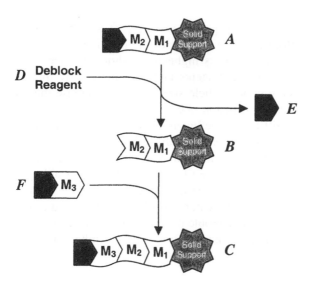

Figure 1 General synthesis scheme for the addition of one monomer in the biopolymer assembly process.

cleotide synthesis, more depurination of the adenosine residues can occur [2].

Increasing the effective concentration of available sites on the support can also increase the reaction rates. These "high-load" supports have been used successfully in solid-phase synthesis but are generally suited to the assembly of shorter chain biopolymers [3]. As the oligomer grows in length, the steric crowding within the high-load supports can adversely affect the efficiency of the synthesis.

The interaction of a liquid and a solid phase presents additional considerations that are not described by the preceding simple rate equations. Diffusion is important for the permeation of reagents and solvents into the reactive sites of the solid support matrix. The concentration of the reactants also has an effect on the diffusion of materials into the solid support. Diffusion is the random movement of molecules from an area of high concentration to one of lower concentration and can be described by Eq. (3).

$$\text{Diffusion rate} = D \frac{\Delta C}{x} \tag{3}$$

where D is the diffusion coefficient, ΔC is the concentration difference, and x is the diffusion distance.

A higher concentration of reagent delivered to the solid support will result in a higher ΔC between the solution and the interior of the solid support, where most of the functional groups reside. This will increase the rate of diffusion and therefore increase the rate of the reaction.

In addition, the rate of diffusion is higher over short distances. It is therefore important to distribute the reagents and solvents evenly through the solid support. A smaller support particle size will also decrease the diffusion distance and increase the diffusion rate.

The magnitude of the diffusion coefficient depends on the kinetic energy of the molecules and the permeability of the solid support. Any effect that increases the kinetic energy of the molecules, such as a higher temperature or stirring, will also increase the rate of diffusion. The permeability of the solid support can be enlarged by increasing the porosity of the matrix.

The diffusion of reagents into the common solid supports used in biopolymer synthesis is not typically a problem. Applications such as large-bead synthesis, however, may require modifications of the synthesis methods to speed the diffusion process [4].

B. Rate Constants

A reaction rate constant can be described by

$$k = CTe^{-\Delta G^{\neq}/RT} \tag{4}$$

where C and R are constants, T is the temperature, and $\Delta G^{\neq}$ is the free energy for activation [5].

The magnitude of a rate constant can be increased by elevating the temperature or decreasing the energy of activation. Higher temperatures lead to a faster reaction but also increase the rate of any other side reactions that might occur. Therefore, studies need to be performed before increasing the reaction temperature to ensure that it favors the formation of the desired product and not that of a by-product [6].

The free energy of activation can be decreased by using catalysts or reagents that are more reactive. Also, the solvent has some effect on the energy of activation and certain solvents may result in a lower energy transition state [5].

C. Reaction Times

The integrated form of the second-order rate equation for the coupling reaction is

$$t = \frac{1}{k([B_0] - [F_0])} \ln \frac{[F_0][B]}{[B_0][F]} \tag{5}$$

There is a logarithmic relationship between the reaction time and the extent

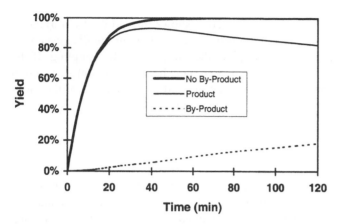

Figure 2   Product and by-product generation over the time course of a coupling reaction.

of the reaction. For example, if the reaction requires 30 min to reach 98% completion, it may take an additional 30 min or more to reach 99.9% completion. Therefore, it is the last several percent of a reaction that requires the greatest amount of time. A longer reaction period also allows more time for the accumulation of undesired by-products (i.e., racemized product).

Figure 2 shows a plot of concentration versus time for a second-order reaction. The solid bold line represents the formation of the desired product with no competing side reactions. The solid line represents the same reaction in which there is competing degradation of the product. The broken line describes the formation of this by-product. Figure 2 shows that there is a point in the time course of the reaction at about 40 min where the amount of desired product is maximized. If the reaction is carried out for a longer time, the amount of desired product decreases steadily.

D.   Reaction Monitoring

It is very useful to monitor the progress of the reactions to determine the completion of the process. It is difficult to monitor the reaction on the solid support in real time, but the disappearance of reactants or the generation of by-products in the solution can be followed to determine the status of the reaction.

Equation (1) for the deblock reaction indicates that the formation of the product (B) is equal to the liberation of the protecting group (E). If the accumulation of the protecting group can be monitored, then the extent of the reaction can be determined.

Equation (2) for the coupling reaction shows that product C formation is equal to activated monomer (F) destruction. Following the disappearance of monomer leads to information about the progress of the coupling. If the amount of activated monomer is much greater than the amount of reaction sites on the solid support, then the detection and quantification of the monomer disappearance may be difficult or inaccurate.

III. MANUAL SYNTHESIS

Manual solid-phase synthesis requires intervention by a chemist during the assembly of a biopolymer. A manual synthesis can vary from manual addition of all reagents and solvents to manual intervention at a single step (such as addition of the monomer).

A simple manual synthesis does not require extensive instrumentation —a reaction vessel, common laboratory glassware, a timer, and perhaps a stirrer or shaker suffice [7]. It is good training to do some manual syntheses when initially performing solid-phase reactions or when experimenting with new solid supports or reagents. Information may be obtained about the methods and details of solid-phase synthesis by manually performing and carefully observing the processes involved. Knowledge of the behavior of the solid support under different synthesis conditions can be valuable for a thorough understanding of how to automate the process properly.

Although automated solid-phase instruments are increasingly being used to construct biopolymers, manual synthesis is still used extensively with methods that are used to create molecular diversity (see Chapter 15). The "libraries" of molecules produced by these methods are often used for screening against biological targets.

The supports used for these methods include standard resins, polymers packaged in permeable "teabag"-style containers [8], derivatized plastic pins [9], and membranes [10] (see Chapter 1). The solid supports are typically combined for the common deblocking and washing steps and then separated to perform couplings on individual portions of the support.

IV. AUTOMATED SYNTHESIS

Automating the synthetic process was one of Merrifield's original intentions when he conceptualized solid-phase peptide chemistry. The first automated solid-phase synthesizer, which is now in the Smithsonian Institution in Washington, DC, was produced in Merrifield's laboratory at Rockefeller University by Stewart in 1965 [11]. Since that time, many automated solid-phase instruments have been developed [12]. Tables 1 through 3 list some of the more widely available automated solid-phase instruments for the synthesis of peptides, nucleic acids, and organic molecules, respectively, at the time of this publication.

Automated synthesis can run unattended and is less prone to human error than manual synthesis. As the biopolymer assembly process was automated, there was a tendency to produce longer sequences. This, in turn, required better synthetic methods, which has led to continual improvements in solid-phase chemistries and instrument design.

The goal of an automated peptide synthesis instrument is to produce the purest peptide in a reasonable amount of time with little waste of reagents and solvents. This can be achieved with efficient chemistries and reliable instrument hardware and software.

A. Chemistry

The basic success of an automated biopolmyer synthesis depends on the chemistry used for chain assembly. A well-understood and well-characterized chemical process has the best chance for success on an automated instrument. Fortunately, many of the reactions used in peptide and oligonucleotide synthesis are well characterized and approach 100% stepwise yields with minimal side reactions. Advances in this field are continually resulting in higher quality products.

To obtain a relatively pure product in a process with many synthetic steps, all reactions must be very efficient. This is especially true for longer sequences that can become lost in a mixture of by-products (mainly deletion sequences) if the stepwise efficiencies are not very close to 100%. Figure 3 shows the yield of desired biopolymer versus the length of the sequence at various stepwise yields. Even at 99% stepwise efficiency, the final yield of a biopolymer 100 monomers in length (broken line) is less than 40%. As more undesired synthetic by-products are generated, it also becomes increasingly difficult to purify the desired product.

A successful synthesis also requires high-quality and stable reagents. The chemicals must remain stable on the instrument for at least the duration of the synthesis (and preferably longer so that multiple syntheses can be performed before changing reagents). If chemicals are not stable under standard conditions, special measures may be required, such as blanketing with inert gas, cooling, or dissolving reagents just prior to use.

The characteristics of the supports used in solid-phase synthesis are very important for automated synthesis. Supports that are mechanically unstable and tend to fracture and produce fine particles can either clog frit material in the reactor or pass through the frit material and clog fluidic lines

Table 1 Peptide Synthesis Systems

Manufacturer	Instrument	Chemistry	No. of reactors	Scale	Delivery method	Monomers	Comments
ABIMED	ESP221 synthesizer	Fmoc	3	To 0.1 mmol	Syringe	Predissolved	
	AMS 422 multiple peptide synthesizer	Fmoc	48	0.005–0.05 mmol	Gas pressure and syringe	Predissolved	
	ASP 222 Auto-Spot robot	Fmoc	400 spots	<50 nmol	Syringe	Predissolved	Bulk operations carried out manually
Advanced ChemTech www.peptide.com	Model 90 tabletop peptide synthesizer	Fmoc and tBoc	2	0.1–10 mmol	Gas pressure	Predissolved or point of use dissolution	
	Model 348 multiple peptide synthesizer	Fmoc and tBoc	48	0.005–0.15 mmol	Gas pressure	Predissolved	
	Model 396 multiple bimolecular synthesizer	Organic/peptide (Fmoc and tBoc)	8 to 96	0.005–1 mmol	Gas pressure	Predissolved	Temperature control
	Model 357 flexible bimolecular synthesizer	Organic/peptide (Fmoc and tBoc)	36	0.005–0.25 mmol	Gas pressure	Predissolved	Temperature control, combinatorial peptide libraries
	Model 400 mini pilot plant synthesizer	Organic/peptide (Fmoc and tBoc)	1	0.1–5 kg support	Gas pressure	Predissolved	
PE Applied Biosystems www2.perkin-elmer.com/lab	433A peptide synthesis system	Fmoc and tBoc	1	0.005–1.0 mmol	Gas pressure	Point of use dissolution	Conductivity or UV detector
PerSeptive Biosystems www.pbio.com	Pioneer peptide synthesis system	Fmoc	2	0.02–2 mmol	Pump	Point of use dissolution	UV detector
	Pioneer with multiple peptide synthesis (MPS) option	Fmoc	16, 32	0.025–0.100 mmol	Pump	Predissolved	
Protein Technologies, Inc.	SONATA/Pilot	Fmoc and tBoc	1	0.1–50 mmol	Gas pressure and pump	Predissolved	
	Model PS3	Fmoc and tBoc	1	0.1–0.25 mmol	Gas pressure	Point of use dissolution	
	SYMPHONY/multiplex peptide synthesizer	Fmoc and tBoc	12	To 0.1 mmol	Gas pressure	Predissolved	On-line cleavage
SHIMADZU www.shimadzu.com	PSSM-8 peptide synthesizer	Fmoc	8	0.005–0.05 mmol	Syringe	Predissolved	

Table 2 Nucleic Acid Synthesis Systems

Manufacturer	Instrument	Chemistry	No. of reactors	Scale	Delivery method	Trityl monitor	Comments
Amersham Pharmacia Biotech www.apbiotech.com	OligoPilot II DNA/RNA synthesizer	DNA, RNA	1	0.01–4 mmol	Pump	Yes	
	OligoProcess DNA/RNA synthesizer	DNA, RNA	1	10–100 mmol	Pump	Yes	Explosion proof
Beckman Instruments, Inc. www.beckman.com	Oligo 1000M DNA synthesizer	DNA	8	0.03–1 μmol	Gas pressure	Yes	
Distribio www.distribio.com	Polygen DNA synthesizer	DNA	10	0.01–1 μmol	Pump	Yes	
PE Applied Biosystems www2.perkin-elmer.com/ab	3948 nucleic acid synthesis and purification system	DNA	48	10–60 nmol	Gas pressure	No	On-line cleavage, purification, and optical density measurement
	394 high-throughput DNA/RNA synthesizer	DNA, RNA	4	0.04–10 μmol	Gas pressure	Yes	On-line cleavage
PerSeptive Biosystems www.pbio.com	Expedite 8909 nucleic acid synthesis system	DNA, RNA, PNA	2	0.05–15 μmol	Pump	Yes	Separates halogenated waste
	Expedite 8909 with multiple oligonucleotide synthesis system (MOSS) option	DNA	16	0.05–15 μmol	Pump	Yes	Separates halogenated waste

Table 3 Solid-Phase Organic Synthesis Systems

Manufacturer	Instrument	No. of reactors	Reactor volume (mL)	Inert atmosphere	Temperature control	Comments
Advanced ChemTech www.peptide.com	ReacTech organic synthesizer	40	8	Yes	Yes	
	Benchmark organic synthesis systems	8–96	8	Yes	Yes	
	LabTech organic synthesis systems	8–96	8	Yes	Optional	
Argonaut Technologies, Inc. www.argotech.com	Trident automated library synthesizer	192 (four 48-reactor cassettes)	4	Yes	Yes—different temperature on each reactor cassette	
	Nautilus 2400	24	5–23	Yes	Yes—individual vessel temperature control	
	Quest 210 organic synthesizers	20	5 or 10	Yes	Yes	Optional automated solvent wash module
	Quest 205 organic synthesizers	10	100	Yes	Yes—two different temperature zones	Optional automated solvent wash module
Bohdan Automation, Inc. www.bohdaninc.com	Solid-phase synthesis workstation	48	5–6	Yes	Optional off-line station	
Mettler-Toledo Myriad Ltd. www.mtmyriad.com	Myriad chemistry developer	4	0.03 or 3	Yes	Yes	
	Myriad personal synthesizer	24	0.3 or 3	Yes	Yes	
PE Applied Biosystems www2.perkin-elmer.com/lab	SOLARIS 530 organic synthesis system	48	7	Yes	Off-line incubation workstation	

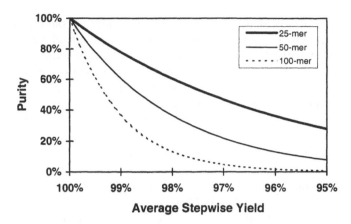

Figure 3 Purity versus average stepwise yields for the assembly of biopolymers of different lengths.

or cause harm to mechanical parts. Supports that swell greatly or collapse under synthesis conditions can also lead to problems with the distribution and flow of reagents and solvents.

The details of peptide and oligonucleotide chemistry are treated elsewhere in this volume and are not discussed herein except when related to instrument design.

1. Peptide Chemistry

Two principal protection schemes are used for peptide synthesis. One strategy incorporates the *tert*-butyloxycarbonyl (Boc) and the other the fluorenylmethyloxycarbonyl (Fmoc) group to protect the N^α position of the amino acid. The Boc protecting group is removed by concentrated acid [usually trifluoroacetic acid (TFA)] and the final peptide is cleaved from the support with a stronger acid such as hydrogen fluoride. The first automated solid-phase peptide synthesizers used Boc protection chemistry. The Fmoc protecting group is removed at each cycle by a basic reagent [usually 20% piperidine in dimethylformamide (DMF)] and the final peptide is released from the support with TFA.

The use of either protection scheme can produce peptides of high quality, but Fmoc chemistry has several advantages over Boc chemistry in terms of instrumentation and safety. The contents of the TFA reservoir typically used in Boc syntheses are generally under pressure and/or pumped to the reactor. A breach in the fluidic delivery system or bottle integrity could lead to a large spill of this corrosive acid. In addition, the postsynthetic cleavage of peptides constructed with Boc chemistry involves the use of extremely hazardous acidic reagents. Besides the important safety issues, a disadvantage of TFA is that its vapors can permeate through many fluorinated polymers, which are common materials used for tubing and valve components. The TFA vapors can then cause corrosion in the other parts of the system, which can result in instrument failure.

An early advantage of Boc-protected monomers was that they were much less expensive than the corresponding Fmoc-protected monomers. This price differential continues to narrow as Fmoc chemistry becomes more widely used. However, as chemical disposal becomes more costly, the higher costs associated with the disposal of halogenated TFA waste can become significant.

A major advantage of Fmoc chemistry for automated instrumentation is that the removal of the Fmoc group can be easily monitored by an ultraviolet (UV) or conductivity detector. This information can be useful for feedback monitoring of the synthesis (see Reaction Monitoring, Section IV.C.1).

The coupling reaction in peptide synthesis involves the reaction of an activated amino acid with a free amino group on the solid support. Activated amino acids are generally unstable in solution and must be prepared just before delivery to the solid support. The preparation of the activated amino acid solution ranges from simply dissolving a solid preactivated derivative to on-line reaction, isolation, and dissolution of the activated derivative.

Most current methods involve mixing an amino acid powder or solution with an activator, allowing time for the activation reaction, and then delivering the material to the solid support for the acylation reaction. This usually requires a separate vessel on the instrument for the activation reaction. The choice of activator depends on a variety of factors including difficulty of coupling, cost, and length of peptide (see Chapter 6).

2. Oligonucleotide Chemistry

The solid-phase assembly of oligonucleotides is predominantly performed using phosphoramidite chemistry (see Chapter 11). This is an excellent method for automated oligonucleotide synthesis because all reactions are highly efficient. The addition of a single monomer to the growing oligonucleotide chain can occur within 3 min with efficiencies greater than 99%. In addition, the reagent and monomer solutions are stable on the instrument for several weeks at room temperature.

The protecting group for the 5′ end of the oligonucleotide is the 4,4′-dimethoxytrityl (DMT) group. This protecting group is easily removed by dilute acid (typically dichloroacetic acid or trichloroacetic acid) to produce an orange-colored species. The quantity of color released can be determined with the aid of a simple visible light detector. This value is an indication of the extent of the previous coupling reaction and can be used to monitor the success of the synthesis.

The scale of an oligonucleotide synthesis is often 1 μmol or less (about 1000 times less than a typical peptide synthesis). At these small scales the instrument volumes (tubing, valves, etc.) become significant compared with the volume of the solid support. Most of the volume of reagents and solvents is used to flush the system and not the support. Therefore, minimizing the instrument system volumes maximizes the reagent and solvent efficiency during the synthesis.

The coupling of the phosphoramidites to the solid support is very moisture sensitive. Therefore, the coupling reagents need to be anhydrous and the instrument design should eliminate the possibility of contamination by water from the atmosphere. Molecular sieves can be added to the reagents to scavenge water. The sieves should be enclosed in a permeable membrane that allows free diffusion of the solution but prevents fines generated from the fragile material from clogging frits and harming system components.

The reagent reservoirs should be blanketed with a dry inert gas. Moisture can also contaminate solutions by slow diffusion through tubing walls. To reduce this potential problem, the system tubing exposed to the atmosphere should be as short as possible and should be flushed with dry wash solvent prior to the start of a synthesis.

B. Hardware

The hardware of an automated solid-phase synthesizer consists of all the electromechanical components of the system (reaction vessel, tubing, valves, pumps, detectors, power supplies, circuit boards, sensors, etc.). It is imperative that all these components are reliable and can operate without failure for many synthesis cycles before repair or replacement. Before using any component in an automated system, it should undergo compatibility and lifetime testing with the reagents that would be in contact during the synthesis.

However, some elements of an automated solid-phase instrument do experience more accelerated degradation. These items should be easily accessible to the user and replaced at appropriate intervals. Examples of some of these disposable elements are O-rings for sealing reagent bottles, filter frits for the reagents and reaction vessel, and the detector lamp.

1. Reaction Vessel

The heart of any solid-phase synthesizer is the reaction vessel that contains the solid support on which the biopolymer is assembled.

The configuration of the reaction vessel depends on the type of reagent delivery system employed by the solid-phase instrument. The main methods for solvent and reagent deliveries are batch and continuous flow. The batch method involves delivering the solution to the solid support, shaking to mix it thoroughly, and then draining the solution off to waste. Thoroughly washing the support requires multiple cycles of solvent delivery, mixing, and draining.

For the continuous-flow method, reagents and solvents are passed through a packed column of the solid support. This approach was first envisioned by Merrifield when he conceptualized solid-phase synthesis in 1959 [13]. However, the early solid supports were not rigid enough to be used effectively in a continuous-flow system. Each method has advantages and disadvantages. The washing provided by continuous flow is more efficient than that in the batch method, as seen in Fig. 4, where a dye is washed off a support using both methods. The success of continuous flow depends strongly on the properties of the solid support and the even distribution of reagents and solvents throughout the column.

Polystyrene solid supports that are not highly cross-linked are rather "soft." The compression of the support during continuous flow results in impedance to the free flow of reagents and solvents through the column. It is possible to use these supports in a continuous-flow system if they are mixed at a ratio of 1:4 with glass beads (100–200 mesh) [14]. However,

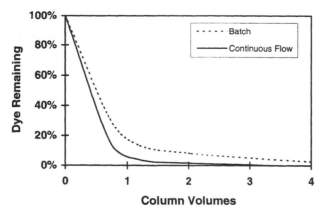

Figure 4 Elimination of dye from a solid support using batch and continuous-flow wash methods.

this is suitable only for relatively short peptides because the glass and polystyrene tend to separate as the synthesis progresses.

It was not until the introduction of more rigid solid supports that continuous-flow solid-phase peptide synthesis became practical. Initially, a support with polyacrylamide gel held within the pores of an inert macroporous matrix (diatomaceous earth) was used [15]. Although these fragile supports worked well for continuous-flow synthesis, careful handling was required to avoid generating fines that could clog the column frits and damage instrument components.

PEG-PS (Polyethylene glycol–polystyrene graft copolymer) supports are made from polystyrene with long polyethylene glycol molecules incorporated in the resin [16]. These supports provide a more rigid framework of more highly cross-linked polystyrene particles. This method retains the open channels necessary for rapid diffusion of reactants throughout the gel matrix so that little back pressure is produced under continuous-flow conditions. As continuous flow involves no mechanical shaking of the solid support, distribution of reagents and solvents depends mainly on a suitable synthesis column and diffusion. The column design for continuous flow must have a good height-to-length aspect ratio and an efficient inlet distributor.

For washing steps in which a relatively large quantity of solution is used compared with the volume of the column, even distribution is not too difficult. However, steps that require the delivery of a small quantity of reagents, such as monomer or activator during the coupling step, are susceptible to poor distribution. Even distribution of reagents during the coupling step can be accomplished by recycling the reactants through the column.

The recycling method has the added benefit that a small quantity (less than the column volume) of concentrated monomer can be initially passed as a band through the column. Figure 5 shows the relative concentration of the activated amino acid that the solid support is exposed to as the solution is recycled through the column. As previously discussed, the more concentrated reagent reacts more quickly with the solid support. As the solution is recycled through the column, it eventually mixes with the entire recycle volume to reach its equilibrium concentration, at which it is evenly distributed throughout the column.

The disadvantage of recycling is that it requires some additional system volume comprising tubing, a recirculation pump, and valves. If the recycle volume is large compared with the column volume, the equilibrium concentration of monomer may be too low for timely reaction of any remaining unreacted sites. Therefore, for continuous-flow systems it is important to minimize the volume of this recycle loop.

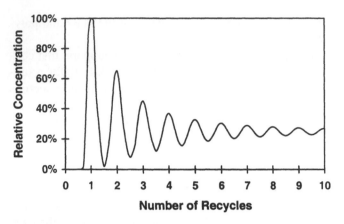

Figure 5   Relative concentration of activated amino acid in the solid support versus the number of passes through the column.

A batch reactor provides excellent distribution of reagents and solvents. The volume of a batch reactor is generally at least twice as large as the volume of the solid support in order to allow room for efficient mixing. It is important, however, that the extra area of the reactor vessel walls is efficiently washed so that solid support is not isolated at the side or top of the reactor. Most batch reactors are continuously agitated during reaction but continued shaking is not required once reagents are distributed (assuming diffusion is fast). Some batch synthesizers utilize some of the advantages of continuous-flow washing by delivering solvent to the top of the resin bed while simultaneously removing it from the bottom.

Batch reagent and solvent delivery is more practical for instruments with many reactors because continuous flow would require valves and a recirculation pump for each reactor.

2.   Delivery Systems

An automated solid-phase instrument requires systems to deliver the reagents, solvents, and activated monomers to the reaction vessel. The size and number of the reagent reservoirs are dependent on the chemistry, sequence, and scale of synthesis. The capacity of the reagents and solvent reservoirs should be sufficient to accommodate the desired scale range and multiple syntheses without refilling. The number of reagent reservoirs should be greater than required for the standard chemistries to allow the flexibility to perform specialized reactions.

a. Reagents and Solvents. Solutions can be transferred through an automated synthesizer by either pressure or positive displacement delivery. For the former, the solution source is pressurized and the valve path is opened to the solution destination. The solution then flows from the area of higher pressure to the area of lower pressure. Positive displacement delivery utilizes a mechanical pump that draws solution from the source and then pushes it to the destination. The source is often pressurized to assist in delivery of the solution to the pump.

The flow rate for pressurized delivery is determined by the difference in the source and destination pressures and any impedance to flow in the delivery path. The flow rates for a positive displacement device are generally quite accurate and reliable and should not change significantly with small variations in the blanketing pressure and the flow impedance of the system.

Positive displacement delivery can be used for either batch or continuous-flow modes, but pressurized delivery is suited mainly for the batch mode. In continuous-flow synthesis, the impedance of the flow path can change during the assembly process because of the swelling of the support in different reagents or as the biopolymer grows in length. This will have a large effect on the flow rate in pressurized delivery and can result in inadequate delivery of reagents or solvents.

Accurate flow rates for pressurized delivery depend on precise and reliable pressurization of the reagent reservoirs. In addition, the flow rates may need to be calibrated as often as every synthesis to ensure that they have not changed.

b. Monomers. Monomer solutions can be stored in reagent reservoirs on the instrument and delivered similarly to the other reagents and solvents. However, this is most practical if there is a small number of monomers and they are stable in solution (i.e., oligonucleotide synthesis). Otherwise, dissolution, activation, and transfer of powdered monomer derivatives may be required for coupling to the solid support.

The individual powdered monomer derivatives can be placed in cartridges or vials in a robotic device that delivers the activator to the monomer. This is achieved by either moving a delivery device to the desired monomer position or moving the desired monomer to a fixed delivery position. The monomer is then dissolved (usually by gas bubbling), withdrawn from the cartridge or vial, and transferred to the reaction vessel.

c. Reagent and Wash Volumes. The required wash and reagent volumes are not generally directly proportional to the scale of the synthesis. System volumes (tubing, valves, etc.), which do not change during scaling, become significant at lower scales compared with the volume of the reactor. Larger scales tend to push the limits of the mechanical device (i.e., higher flow rates and larger volumes). Adjustments to the cycle steps are often necessary to increase the duration of the steps to deliver a sufficient volume of solvent or reagent at a rate within the specifications of the instrument. The amount of solution required to wash a solid support effectively depends not only on the quantity of support but also on the physical characteristics of the polymer. A support that swells more in the synthesis solvents requires a larger amount of solvent to wash out the internal void volume contained in the support bed.

Determining the Void Volume of a Solid Support. The void volume of a support (V_{void}) dictates how much solvent is required to wash the support thoroughly. The void volume of a support can be described by two constants. The displacement constant ($K_{displacement}$) is the number of grams of support that displaces 1 mL of solvent. The spatial constant ($V_{spatial}$) is the volume that 1 g of support occupies in the solvent.

A high displacement constant indicates that the support has a high density and will require less wash solvent, whereas a high spatial requirement indicates that the support occupies a large volume and will require more wash solvent.

Use the following procedure to determine $K_{displacement}$ and $V_{spatial}$:

1. Place 7 mL of synthesis solvent [DMF, N-methylpyrrolidinone (NMP), acetonitrile (ACN), etc.] in a 10-mL graduated cylinder.
2. Add 1 g of solid support and wait for 15 min to allow the polymer to swell.
3. Tap the cylinder gently to pack the support and dislodge any trapped air bubbles.
4. Subtract 7 mL from the total volume in the cylinder. This value is $1/K_{displacement}$ in mL/g.
5. Read the volume that the support occupies (if the support floats in the solvent, read the volume at the bottom of the support and subtract it from the total volume in the cylinder). This volume is $V_{spatial}$ in mL/g.
6. These two constants can be used to estimate the resulting void volume of the support:

$$V_{void} = V_{spatial} - \frac{1}{K_{displacement}} \tag{6}$$

In general, as a synthesis progresses, the support swells and $V_{spatial}$ increases. However, as the biopolymer chain grows and fills the internal volume of the support, $K_{displacement}$ also increases. For more rigid supports that do not swell dramatically during the synthesis, these changes tend to cancel each other.

For supports that swell greatly during synthesis, the change in $V_{spatial}$ may not be balanced by the change in $K_{displacement}$. A higher value than the actual measured V_{void} should be used in scale-up calculations to ensure thorough washing of the support at the later stages of the synthesis.

Determining Wash Volumes. The void volume of the support and system volume of the synthesizer are useful in estimating the quantity of wash solvent needed to flush the reactor of excess reagents and their chemical by-products. The wash volume may be expressed by the following equation:

$$\text{Wash volume} = x \times V_{system} + y \times V_{support} \tag{7}$$

where the support volume ($V_{support}$) is the product of the void volume (V_{void}) and the quantity of support ($V_{support} \times$ g support). The size of the multiplication factors x and y depends on a variety of factors including the washing method (batch or continuous flow), the washout characteristics of the system components, and the thoroughness of washing required.

At larger scales, the system volume may become insignificant, whereas at small scales the system volume often exceeds $V_{support}$. In a manual synthesis, the system volume typically consists of the walls and cap of the reactor vessel; in an automated synthesizer, the system volume comprises the tubing and valves.

The multiplication factor x is chosen so that the system components are sufficiently washed. An instrument with narrow-bore tubing and zero dead volume valves is thoroughly washed using only two system volumes, whereas large-bore tubing and valves, which allow more mixing, may require larger wash volumes.

The multiplication factor y is chosen to ensure that the solid support is sufficiently washed. It usually ranges from 5 to 10, depending on a number of factors including the synthesis step, wash method, reactor design, and type of support. It is more critical at certain steps in a synthesis to wash the support more efficiently to ensure that none of the reagents from the previous step interferes with the next reaction. In the typical synthesis of both peptides and oligonucleotides, it is important to remove all traces of the deblock reagent prior to introduction of the protected monomers during the coupling steps. Any deblock reagent remaining during the coupling step could lead to premature deprotection of the N-terminus of the biopolymer and multiple insertions of the same monomer. As discussed earlier, the continuous-flow method of washing a solid support is generally more efficient than the batch method. To achieve the same low level of impurities, typically twice as much wash solvent is required in a batch synthesis as in a continuous-flow synthesis. Figure 4 shows the washout of dye from a support versus the volume of wash solvent. The figure illustrates that in using both batch and continuous-flow washing, more than 80% of the dye is removed with one column volume of solvent. However, to achieve a low level of dye, much more solvent is required in a batch wash.

The reactor design has a large effect on the amount of wash solvent required. In a batch synthesizer, the reactor should be appropriately sized so that it is large enough to allow efficient and easy mixing of the support. However, it should not be so large or oddly shaped that a large amount of solvent is required to wash the support from the walls of the vessel.

The design of the column is critical for the continuous-flow instrument to guarantee efficient washing. The flow distribution of solvents at the column inlet and the width of the column are important in providing even distribution of the solvents and equal washing of the solid support throughout the column. A narrower column is generally easier to wash efficiently but contributes more to the system pressure. A compromise between column width and height is usually necessary to accommodate reasonable system pressures and column geometries. Some solid supports may nonspecifically bind some of the synthesis reagents that are not easily removed by the wash solvent. Different washing strategies may need to be employed if all traces of the impurity are required to be removed before the next synthesis step (see Reaction Monitoring, Section IV.C.1).

Although it is relatively easy to measure system and support volumes, it is difficult to predict how efficiently and completely these volumes will be flushed with the wash volumes because of the factors described previously. When determining appropriate wash volumes, a good starting point can be determined using Eq. (7), but it is also important to test these wash volumes by measuring impurity levels after the wash step. This can be done by introducing a dye onto the solid support prior to the wash step and seeing if it is totally cleared from the support by the wash solvent (by eye or spectrophotometric measurement).

Monomer Excess. As previously described, the coupling rate is proportional to the concentration of monomer in solution. For efficient coupling to occur, the concentration of monomer needs to be as high as possible. The concentration may vary with the synthesis scale and can be adjusted by changing the amino acid excess factor. Equation (8) describes the relationship between the monomer excess (XS) and the scale of the synthesis.

$$XS = \text{conc}_{min} \times \left(\frac{V_{void}}{\text{loading}} + \frac{V_{system}}{\text{scale}} \right) + \text{fraction coupled} \tag{8}$$

where conc_{min} is the minimum desired concentration of monomer in mol/L, loading is the substitution of the support in mmol/g, V_{system} is the volume of the system elements that are included in the coupling reaction (i.e., tubing, pump, valves) in mL, scale is the total scale of the synthesis in mmol, and fraction coupled is the fraction of the coupling reaction that has been completed.

Equation (8) shows that as the loading and/or scale increases, the monomer required to maintain the minimum concentration decreases. However, as the void volume of the support or the system volume increases, a larger excess of monomer is needed to maintain the minimum concentration.

Figure 6 shows a plot of the scale versus the excess monomer at different system volumes. The support used in this example has a loading of 0.15 mmol/g and a void volume of 4.4 mL/g. The minimum concentration of monomer is 0.05 M and the coupling reaction is assumed to be complete (fraction coupled = 1).

For an ideal coupling reaction, where there is no system volume (broken line), the excess amino acid required to maintain the minimum concentration is independent of the synthesis scale. Batch reactions approach this ideal but there is usually some small system volume (i.e., filter frit volume). As illustrated in Fig. 6, the lower scale limit for an instrument with a V_{system} of 6 mL (solid bold line) is about 0.05 mmol with an eightfold excess of amino acid. Below this scale, the excess of amino acid required to maintain a minimum concentration increases sharply. Lowering the system volume to 2 mL (solid line) has a marked effect and about half as much amino acid is required to maintain the minimum concentration. A low system volume also extends the lower scale limit while still using a reasonable excess of amino acid.

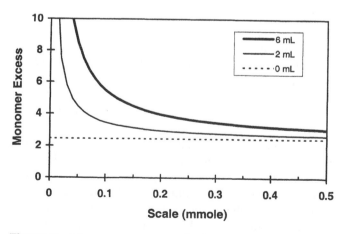

Figure 6   Monomer excess required to maintain a minimum concentration at different system volumes.

As the scale increases, the system volume compared with the support volume becomes insignificant and only a two- to threefold excess of monomer is required to maintain the minimum concentration.

A graph can be constructed using the preceding equation with the appropriate system volume and support to determine how great an excess of amino acid is required at the desired synthesis scale. For ideal synthesis conditions, the intersection of the lines from the synthesis scale and amino acid excess should be near or above the curve in the plot.

C. Software

The reliability of the software in an automated synthesizer is as important as that of the instrument hardware. The software controls most of the tasks of the chemist in the manual synthesis. It runs the instrument, issuing the commands telling how much, when, and where to deliver reagents. The main features of instrument software are the user interface, system control, and system monitoring.

The user interface of a biopolymer synthesizer should, at the minimum, allow the operator to input the sequence of the desired product. The interface can also be used to

Calculate reagent usage and waste generation.
Manage synthesis data and produce synthesis reports.
Display current system and synthesis statistics and data.
Allow modification of the synthesis steps by the user or by monitor feedback.

The ultimate control of an automated synthesizer is provided by a processor that translates the commands from the user into a series of signals to actuate the various electromechanical devices that are used to perform the synthesis. In addition to this processor, the system can include a computer workstation that performs all the user interface functions. Once the synthesis information is downloaded from the workstation, the instrument can independently perform the synthesis using its internal processor and the workstation can be used for synthesis data management or other applications.

The software can be used to monitor instrument sensors to determine whether the instrument is operating within the specifications. If a problem is detected, the software can halt the instrument and alert the user. If possible, the user can correct the problem and continue the synthesis or abort the synthesis and prevent waste of valuable reagents. Examples of some elements of an instrument that can be monitored are the gas pressure, fluid pressure, fluid levels, flow rate, and valve function.

1. Reaction Monitoring

A powerful feature of an automated synthesizer can be reaction monitoring. Reaction monitoring can provide valuable diagnostic and qualitative information about the synthesis process. If the monitoring occurs in real time, the software can provide several choices for feedback control based on the detector data to assist in assembly. Monitoring used in conjunction with other analytical techniques [e.g., high-performance liquid chromatography (HPLC), mass spectrometry (MS)] presents an effective means of measuring the success of a synthesis. The two main steps in oligomer assembly are the deblock and coupling steps. If these can be monitored, then the assembly process can be followed and optimized.

a. Deblock Monitoring. The detector is usually placed in the fluid stream after the reaction vessel so that the removal of the terminal protecting group can be followed. The removal of the Fmoc group for peptides can easily be monitored by a conductivity or UV detector and the removal of the DMT group in oligonucleotide synthesis can be monitored by a visible detector.

The detector output from a continuous-flow instrument can be constantly monitored. The trace of the output versus time can then be plotted to show a peak for the removal of the protecting group. The detector in a batch synthesizer can measure the amount of the protecting group removed in each portion of the deblock solution. In either case, the software can follow the deblock reaction and provide feedback to extend the reaction in the case of a slow removal of the protecting group. Slow protecting group removal may be attributed to poor accessibility to the growing biopolymer in the support matrix and thus the attachment of the subsequent monomer may also be difficult [17]. The software can then increase the coupling time and/or perform a recouple to drive the reaction to completion and produce a purer product.

Ideally, the magnitude of the detector output is proportional to the quantity of the protecting group removed from the biopolymer support and the coupling efficiency of the previous amino acid can be calculated. However, several factors reduce the accuracy of this relationship.

The correlation between the protecting group concentration and detector output may not be strictly linear. The detector output for more concentrated solutions could fall outside its linear range and thus be underestimated. For instance, in oligonucleotide synthesis, the DMT removal rate is different for the various monomers [18]. Typically, the DMT group is removed from a guanosine residue in the shortest period of time, resulting in a more intensely colored solution. Conversely, the DMT group is removed from the thymidine residue in the longest period of time, which results in a less intensely colored solution. Depending on the linearity range of the detector and the scale of the synthesis, the amount of DMT can be underestimated for guanosine residues whereas the thymidine residues will have the most accurate reading.

The magnitude of the detector output can be affected by the solvent. For instance, for the removal of the Fmoc group in peptide synthesis, the extinction coefficient of the dibenzofulvene–piperidine adduct is strongly dependent on the piperidine concentration in the solution [15]. The local concentration of piperidine as the adduct is eluted can vary within a single synthesis. If the deblock reaction is fast, then the adduct is eluted at the leading edge of the piperidine reagent and is mixed with the DMF in the column. In contrast, a slower deblock reaction results in adduct elution in a more concentrated solution of piperidine. The extinction coefficient of the dibenzofulvene–piperidine adduct was measured in various concentrations of piperidine in DMF. The extinction coefficient was about two times greater in a 5% piperidine solution in DMF than in a 20% piperidine solution in DMF. Therefore, equal peak areas for successive deblock reactions can be expected only under strictly equivalent experimental conditions.

Other reagents used during synthesis can be eluted and detected in the deblocking step. For example, in peptide synthesis, most triazole-based coupling reagents bind to the solid support and may be eluted during the piperidine wash [19]. These triazole-based reagents include HATU, N-[(dimethylamino)-1H-1,2,3-triazolo[4,5-b]pyridin-1-ylmethylene]-N-methylmethanaminium hexafluorophosphate N-oxide; HBTU, N-[(1H-benzotriazol-1-yl)(dimethylamino)methylene]-N-methylmethanaminium hexafluorophosphate N-oxide; HOAt, 1-hydroxy-7-azabenzotriazole; HOBt, 1-hydroxybenzotriazol; PyAOP, 7-azabenzotriazol-1-yloxytris(pyrrolidino)phosphonium hexafluorophosphate; PyBOP, benzotriazol-1-yloxytris(pyrrolidino)phosphonium hexafluorophosphate; and TBTU, N-[(1H-benzotriazol-1-yl)(dimethylamino)methylene]-N-methylmethanaminium tetrafluoroborate N-oxide. This results in a larger than expected Fmoc peak that does not correlate with the quantity of the protecting group removed.

The detector output following a poor coupling can be very large, because the triazole-based reagents also bind to the free amino groups that were not acylated by the incoming Fmoc-protected amino acid. Although this nonspecific binding does not interfere with successful synthesis of a peptide, it does interfere with the interpretation of detector data. The bound triazole reagent can be removed by a concentrated base [N,N-diisopropylethylamine (DIEA)] wash after the coupling. The Fmoc group is removed slowly ($t_{1/2}$ ~10 h) by DIEA [20]. Thus, a 1-min exposure of the Fmoc-protected support to neat DIEA will remove less than 0.2% of the Fmoc group. Premature Fmoc removal is nonproblematic in single coupling protocols because the Fmoc group is removed by the deblock reagent following the coupling step. In double coupling protocols, premature Fmoc removal following the first addition of amino acid results in undesired additional incorporation of this residue in the sequence. To avoid this undesired reaction, the base wash can be skipped when the next cycle is a recoupling cycle.

The detector output for the deblock reaction can be a useful tool for identifying slow reactions or gross coupling failures. Because of the inherent inaccuracies, it should not be used to measure the absolute success of the synthesis or to replace other analytical techniques (e.g., HPLC, MS) (see Chapter 19).

b. Coupling Monitoring. Deblock monitoring can give an indication of where problems may have occurred during a linear assembly, which can provide useful information if the synthesis needs to be repeated (where to perform recouples, etc.). Although there has been significant progress in the development of efficient coupling reagents, some coupling reactions remain difficult to perform. Difficult couplings can sometimes be predicted by examining the sequence. In peptide assembly, there are specific couplings (adjacent β-branched residues) or regions within a peptide (β-sheets or turns) that are known to cause coupling problems. In oligonucleotide synthesis, sequences rich in guanosine residues can be difficult to synthesize. However, in many cases (especially in peptide synthesis) it is difficult to anticipate where a difficult coupling might occur.

One solution would be to assume that every coupling is difficult and incorporate rigorous conditions to achieve nearly 100% coupling. This might include extended coupling times, double coupling, and coupling under varied conditions. However, this would greatly extend the time of the synthesis, consume a large amount of reagents and solvents, and generate unnecessary waste.

It is preferable to perform a coupling step for only the amount of time required for a complete reaction to avoid the accumulation of undesired byproducts. This can be achieved by monitoring the coupling reaction to determine when it is complete. In peptide synthesis, most monitoring strategies examine the amount of free amine remaining on the solid support. Of these techniques, the ninhydrin test is most commonly used [21] (see Chapter 1). Although this is a very sensitive method, it requires the destruction of a small amount of the solid support for each test. For a long peptide, this can significantly reduce the final yield of product. In addition, the ninhydrin test is labor intensive. Even with automated resin sampling, the results of the ninhydrin test are not usually known before the next amino acid is added. As with deblock monitoring, this provides valuable information for synthesizing the same peptide in the future but does not assist in the construction of the current peptide.

Several noninvasive monitoring methods that allow real-time determination of the coupling efficiency of the acylation reaction have been developed [22,23]. These methods use acid–base indicators that are colored when associated with the basic free amino groups on the solid support. Generally, when the color disappears there are no more free amines on the support and the coupling reaction is complete. For automated detection, these methods require special detectors and reaction vessels and can be quite insensitive in the region where the coupling is nearing completion.

Counterion distribution monitoring (CDM) can determine, in real time, when the coupling reaction is complete [24]. It uses an in-line UV detector and is most sensitive in the region where the coupling reaction is approaching completion.

In the CDM method, an acidic dye is partitioned between the basic free amino groups on the solid support and an added base in solution. The relationship between the percent coupling and amount of dye in solution is not linear. When the amount of added base is adjusted relative to the number of free amino groups on the support, most of the color is released into solution near the completion of the coupling reaction. Therefore, with this method, the region where the coupling is approaching completion is the most sensitive and provides an accurate estimate of the coupling efficiency.

As with the other monitoring methods, the CDM method has some major restrictions and limitations. It cannot be used with some of the most popular coupling chemistries that require a large amount of base (HATU, TBTU, HBTU, BOP). The dye also has a tendency to bind to polystyrene-based supports, leading to inaccurate results. CDM is most suited to a continuous-flow instrument where the solution is continuously monitored as the coupling reaction proceeds.

Currently, there is no technique for monitoring the coupling reaction of biopolymer synthesis that is without serious restrictions. Advances in this field along with more effective coupling reagents, software, and instrumentation that provide adequate feedback control will extend the limits of biopolymer synthesis.

D. Safety

The safety features of an automated solid-phase synthesizer are extremely important. These instruments often contain large volumes of hazardous chemicals, usually under pressure, and are routinely left unattended for long periods of time.

If there is a breach in the fluidic system, containment of spills and leaks is critical to limit exposure to dangerous reagents. Spills and leaks can be contained by the use of trays or overflow containers and doors. The capacity of the tray or overflow container should be at least the volume of the largest reagent reservoir in the system. Doors that cover the fluidic system help prevent the leak from endangering the user or going outside the range of the containment system.

The waste container should be placed in a secondary overflow tray. The instrument should either monitor the level of the waste container or provide an accurate value for waste generation so that the user can determine whether the waste capacity is sufficient at the start of the synthesis.

To protect the user from hazardous fumes, the instrument should either be operated in a hood or have provisions to be attached to a ventilation system.

If the instrument contains pressurized containers, the pressurization system needs to be monitored for an overpressure condition and have a relief valve which will vent the pressure before the containers fail.

Electrical safety is also imperative to prevent the possibility of electrical shock or fire. The electronics of the system should be isolated from the fluidics so that if a leak does occur, there is no possibility that they will be exposed to flammable solutions.

Additional safety measures may need to be implemented depending on the instrument design, the chemistry, and the environment.

E. Multiple Reactor Synthesizers

Multiple reactor solid-phase synthesizers provide a high throughput of synthetic biopolymers. The same considerations as for a single synthesis should be applied to a multiple synthesis. The reaction kinetics and reagent requirements are generally quite similar. However, a synthesizer with many reactors often has some major design differences as compared with an instrument that contains only a few reactors.

A major consideration with a multiple synthesizer is the speed and efficiency of delivery (and removal) of reagents. A 30-s wash step in a single synthesis may seem insignificant, but multiply that by 100 syntheses and the time for the wash step approaches 1 h. The delivery times are often reduced by lowering the scale of the individual syntheses and therefore the amount of reagent required or delivering simultaneously to multiple reactors.

A conventional single synthesis system utilizes valves and tubing to route the flow of reagents to the reactor. As the number of reactors increases, it becomes impractical to have valves and tubing for each reactor. To eliminate the need for this added hardware, multiple synthesizers are often based on robotic designs, where the reactors remain stationary in a rack and one or more robotic arms transfer reagents to the individual reactors. The common steps can be expedited by simultaneous delivery to many reactors using a multiple delivery device. The dissolution of monomers just prior to use also becomes impractical with a multiple synthesizer. The monomers are generally predissolved in the synthesis solvent and transferred to the reaction vessel. Measures to extend the lifetime of the monomers in solution include temperature control and blanketing with inert gas. The scale of the synthesis and efficiency of the synthesizer have a large affect on the reagent consumption and waste generation. As the scale increases and the efficiency decreases, the quantity of reagents and waste increases.

V. CONCLUSION

The field of solid-phase synthesis instrumentation is continually advancing. Improvements in synthesis reagents, reaction monitoring, and instrument hardware and software will extend the limits of the instrumentation. As synthesizer capabilities improve, there is the potential that more and more control will be taken from the user until the instrument becomes a "black box." It is important, however, to maintain an understanding of the principles of instrument operation and the chemistry that is being performed. The instrument is secondary to the chemistry but is an essential tool to help carry out the synthesis efficiently. The best instrument cannot improve ineffective chemistry and, conversely, a poorly designed instrument can compromise a very efficient chemical process. As long as the basic principles of reaction kinetics, fluid mechanics, and instrument safety are sustained, a solid-phase synthesizer can be used to its maximum potential and benefits.

REFERENCES

1. RB Merrifield, J Am Chem Soc 85:2149–2154, 1963.
2. T Tanaka, RL Letsinger. Nucleic Acids Res 10:3249–3260, 1982.
3. SA Kates, C Blackburn, BF McGuinness, GW Griffin, NA Solé, G Barany, F Albericio. Biopolym Peptide Sci 47:365, 1998.
4. WE Rapp. In: SR Wilson, AW Czarnik, eds. Combinatorial Chemistry. Synthesis and Application. New York: Wiley, 1997, pp 65–94.
5. FA Carey, RJ Sundberg. Advanced Organic Chemistry: Part A: Structure and Mechanisms. New York: Plenum, 1977, pp 127–139.
6. AK Rabinovich, JE Rivier. In: RS Hodgess, JA Smith, eds. Peptides: Chemistry, Structure and Biology, Proceedings of the 13th American Peptide Symposium. Leiden: ESCOM Science Publishers, 1994, pp 71–73.
7. JM Stewart, JD Young. Solid-Phase Peptide Synthesis. 2nd ed. Rockford, IL: Pierce Chemical Co., 1984, pp 71–95.

8.   C Pinilla, JR Appel, RA Houghten. In: I Lefkovits, ed. Immunology Methods Manual. London: Academic Press, 1997, pp 837–845.
9.   AM Bray. In: I Lefkovits, ed. Immunology Methods Manual. London: Academic Press, 1997, pp 809–816.
10. R Frank, S Hoffmann, M Kiess, H Lahmann, W Tegge, C Behn, H Gausepohl. In: G Jung, ed. Combinatorial Peptide and Nonpeptide Libraries: A Handbook. Weinheim: VCH, 1996, pp 363–386.
11. RB Merrifield, JM Stewart. Nature 207:522–523, 1965.
12. LE Cammish, SA Kates. In: W Chan, P White, eds. Fmoc Solid-Phase Peptide Synthesis: A Practical Approach. Oxford, Oxford University Press, in press.
13. JM Stewart, JD Young. Solid-Phase Peptide Synthesis. 2nd ed. Rockford, IL: Pierce Chemical Co., 1984, p. v.
14. SA Kates, LE Cammish, F Albericio. J Peptide Sci 53:682–684, 1999.
15. A Dryland, RC Sheppard. J Chem Soc Perkin Trans I, 125–137, 1986.
16. G Barany, F Albericio, SA Kates, M Kempe. In: JM Harris, S Zalipsky, eds. Chemistry and Biological Application of Polyethylene Glycol. ACS Symposium Series. Washington, DC: American Chemical Society Books, 1997, pp 239–264.
17. D Nalis, R Jacobs. In: CH Schneider, AN Eberle, eds. Peptides 1992: Proceedings of the Twenty-Second European Peptide Symposium. Leiden: ESCOM Science Publishers, 1993, pp 298–299.
18. T Atkinson, M Smith. In: Oligonucleotide Synthesis: A Practical Approach. Oxford: IRL Press, 1984, pp 49–50.
19. SB Daniels. Technical Bulletin PT927. PerSeptive Biosystems, Framingham, MA, 1997.
20. GB Fields, RL Noble. Int J Peptide Protein Res 35:181, 1990.
21. JM Stewart, JD Young. Solid-Phase Peptide Synthesis. 2nd ed. Rockford, IL: Pierce Chemical Co., 1984, pp 105–107.
22. V Krchnak, J Vagner, M Lebl. Int J Peptide Protein Res 32:415–416, 1988.
23. L Cameron, M Meldal, RC Sheppard. J Chem Soc Chem Commun 270–272, 1987.
24. SC Young, PD White, JW Davies, DEIA Owen, SA Salisbury, EJ Tremeer. Biochem Soc Trans 18:1311–1312, 1990.

18

The Purification of Synthetic Peptides

John S. McMurray
The University of Texas M. D. Anderson Cancer Center, Houston, Texas

The chemistry of modern solid-phase peptide synthesis [1] is well developed and is quite efficient, and aspects of this technique are described in this monograph as well as in several excellent reviews and books [2–6] A few recent developments that have contributed to the production of high-quality peptides are briefly outlined here. Several different types of resins possessing the *C*-terminal amino acid already anchored to the support are commercially available, and methods for low-racemization attachment of this residue are available. Coupling reactions using carbodiimides, phosphonium salts, and uronium salts typically proceed in very high yield. Side-chain protecting groups for amino acids such as arginine, histidine, asparagine, and glutamine have reduced side reactions attributable to these residues. New, improved deprotection and cleavage cocktails are being reported that produce less by-products. In spite of these advances, impurities such as truncated peptides, deletion sequences, peptides containing racemized residues, incompletely deprotected peptides, protecting group by-products, scavengers, and remnants of resins and linkers may be present in the crude product, thus necessitating purification steps.

A variety of purification methods are available to the peptide chemist, and these are often used in combination. Several reviews, chapters, and books are available that describe the purification of peptides (including but certainly not limited to Refs. 2–4, 7, and 8). This chapter is not meant to be an extensive review of these techniques. Rather, the techniques commonly employed in the author's laboratory will be described. As opposed to peptide fragments derived from protein digests, the purification of synthetic peptides will be discussed in this chapter. The emphasis will be on peptides synthesized on the 0.1–0.5 mmol scale.

Purification schemes depend on the perceived nature of the impurities. The actual experimental conditions depend on the nature of the peptide. For each technique discussed, the user is encouraged to get in touch with the manufacturers of the media, high-performance liquid chromatography (HPLC) columns, equipment, etc. to acquire technical notes, which can provide very useful information on topics such as preparing media, solvent conditions, and HPLC strategies.

Following solid-phase synthesis, cleavage from the resin, and precipitation in ether, the crude product is analyzed to provide information on purity and to assist in the development of a purification scheme. Weintraub and colleagues (see Chapter 19) discuss analytical techniques and provide an excellent reference. Most commonly, the crude product is analyzed by reverse-phase HPLC (RP-HPLC) using gradients of acetonitrile (ACN) in water [both containing 0.1% trifluoroacetic acid (TFA)]. A diode array detector is very useful for exploratory runs because peptides often have groups with characteristic signatures in the ultraviolet (UV)–visible range. For example, if tyrosine or tryptophans is included in the sequence, a peak centered at 275–280 nm in the UV spectrum will be present. The HPLC peaks that do not have absorbance at 280 nm obviously do not have Tyr or Trp in them. In our work on tyrosine kinase substrates [9], the HPLC was programmed to plot chromatograms at both 230 and 275 nm to monitor the presence of the carbonyl group of the peptide bond and the aromatic ring of tyrosine simultaneously. The 9-fluorenylmethoxycarbonyl (Fmoc) group also has a diagnostic UV absorbance spectrum, which we and others have utilized during HPLC in N-terminal protected peptides. For laboratories in which it is available, mass spectrometry (MS) of the crude peptide is invaluable for detecting the presence (or absence) of the desired peptide and characterizing side products such as deletion sequences or incompletely deprotected species. MS, either as a direct detector on the HPLC (LC-MS) or on fractions of the analytical HPLC run, provides information on the composition of individual peaks. This information can be used to determine which peak contains the desired peptide and the possible presence of coeluting contaminants. The composition of by-products such as failure sequences or incompletely deprotected peptides can help guide the development of alternative synthesis or deprotection strategies if necessary.

The initial HPLC analysis provides information on the purity of the peptide and the potential difficulty of purification. Sometimes the synthesis is so efficient that only one peak is present and only gel filtration chromatography is required to achieve purification. This is rare, however, and more stringent techniques such as ion exchange or preparative reverse-phase HPLC are usually required. If the main peak is well separated from impurities, then simple scaling up from analytical to preparative reverse-phase HPLC is all that is required to achieve purification. Sometimes more than one compound coelutes or there are several peaks close to the peak of interest. In these cases, other techniques such as gel filtration or ion-exchange chromatography may provide separations not possible with reverse-phase HPLC. In this chapter the techniques of gel filtration, ion exchange, and HPLC will be discussed and examples of each will be given.

I. GEL FILTRATION

Gel filtration is a chromatographic technique that separates materials by molecular size. The medium is typically a water-compatible, cross-linked polysaccharide or polyacrylamide gel that can be characterized as having pores. During migration down the column, smaller particles are retained in the pores for longer times than larger ones with the result that elution time is inversely proportional to the size of the molecule. Therefore, gel filtration is often used to desalt peptides after deprotection and cleavage from solid supports to remove acids, scavengers, protecting group remnants, and truncated sequences, usually smaller than the desired peptide, that were not removed by ether precipitation. This technique is limited, of course, and the separation of peptides having racemized residues or sequences differing by one or two residues from the desired product is not possible. Gel filtration can also be used to desalt peptides purified by reverse-phase HPLC or ion-exchange chromatography in which the mobile phase contains nonvolatile salts such as triethylammonium phosphate or NaCl.

Gel filtration chromatography requires a pump, a glass column fitted with a stopcock or valve to control flow, a detector, and a fraction collector. For low-pressure techniques in aqueous buffers such as gel filtration, a peristaltic pump is quite useful.

By varying the degree of cross-linking, gel filtration media can be manufactured that will separate different ranges of molecular sizes. Sephadex G-10, G-15, and G-25 from Pharmacia and Biogel P-2 and P-4 from BioRad are commonly used for peptide applications as these gels fractionate low-molecular-weight compounds.

Gel filtration media are supplied either as a dry powder or as a suspension in aqueous solution. In the former case, swelling the gel in the elution buffer is necessary. Swelling time depends on the medium and is available from the manufacturer. The lower Sephadex gels are swelled in 2–3 h, and swelling can be accelerated in a boiling water bath. After the swelling step, it is sometimes necessary to remove the smaller particles, i.e., the "fines," by suspending the medium in buffer, allowing it to settle, and then decanting the gel beads that remain suspended in the medium. A suspension of particles of uniform size will pack most evenly and give the best separation. Furthermore, failure to remove the fines can result in clogged filters on the bottom of the column, which may impair flow and cause high pressure. High pressure can cause fittings and tubing to fail and leaks to occur.

After swelling, the medium must be degassed. Degassing is necessary to remove dissolved gas that can form bubbles in the column after it is poured. These bubbles will cause channeling in the gel bed and poor separations will occur. We usually swell the resin in a side-arm Erlenmeyer flask and, after removing the fines, place a stopper on the flask and apply a vacuum through the side arm for 20–30 min with occasional stirring to dislodge any trapped gas bubbles. Never use a stir bar to agitate the medium, as this will cause fracturing of the particles, which can result in clogged frits on the column and give rise to impaired flow rates and high pressure in the system. To pour the column, about 1/10 of the column is filled with degassed buffer and the medium, suspended in buffer, is added. The stopcock at the column outlet is opened and solvent is allowed to drain. As buffer drains, the column packs. Suspended medium is added to the packing column in portions until the gel bed reaches the desired height. The size of the column depends on the amount of peptide that needs to be purified. Typically, a size of 2.5 × 100 cm is sufficient to purify peptides synthesized on a 0.1-mmol scale. Buffer is degassed in the same manner as the gel filtration medium before use. Several column volumes of buffer are pumped through the column to ensure complete packing and to equilibrate the gel in the eluant before it is used for chromatography. We often pump solvent through the column throughout the night.

The choice of buffer depends on the nature of the peptide. Volatile acids, bases, or salts are convenient mobile phase components in that they can be removed from the peptide product by lyophilization. Positively charged peptides are soluble at low pH, so acetic acid solutions are often employed. On the other hand, in our laboratory we work with head-to-tail cyclic peptides, which are insoluble in acidic solutions because of the side chains of aspartic acid and glutamic acid [9,10] and the lack of positively charged residues. We chose a dilute ammonia solution as the mobile phase, knowing that the pH would be high enough to dissolve the peptides and that all components of this buffer could be removed by lyophilization. Some researchers prefer more neutral buffers. For desalting runs prior to preparative reverse-phase HPLC, buffer composition is immaterial as long as the pH is below 7, because the components, volatile or not, will be separated from the peptide before it is eluted from the column.

The crude peptide is purified by dissolution in the elution buffer followed by applying the solution to the top of the column. For the 2.5 × 100 cm column, a sample volume of 5–10 mL is effective. We use columns fitted with a plunger so that the bottom of the plunger is located less than 1 cm above the top of the bed of resin. The inlet tube from the buffer reservoir is used to pump the sample onto the column. The sample container is rinsed with three 1-mL portions of buffer, which are also pumped onto the column. After the sample is completely transferred to the gel bed, the buffer inlet tube is replaced in the buffer, which is then pumped through the column. Typically, flow rates of 1 mL/min are used.

The eluant is passed through an optical detector to monitor the elution of the sample components. Again, observation wavelengths of 214–230 nm are useful in that they detect the carbonyl group of the amide bond. Tyrosine and tryptophan absorb at 280 nm, so this wavelength is useful for peptides having these residues.

Low-pressure chromatography typically requires hours, so a fraction collector is a necessity. It is best to configure the system so that a "tick mark" is made on the strip chart recorder during fraction changes. Fractions can be analyzed by analytical HPLC to determine which ones contain the peptide of interest. Fractions are then pooled and lyophilized. In cases in which preparative HPLC is the next step, the pooled fractions can be pumped directly onto the column.

The following is an example of how we used the ability of gel filtration to discriminate molecules of different sizes to study a side reaction in peptide synthesis.

A. Example: Separation of Oligomeric Peptides from a Cyclic Decapeptide

In a study of on-resin cyclization of peptides, gel filtration was used to separate oligomeric peptides from the desired cyclic peptide [10]. Using the Fmoc-based protection scheme, we synthesized a decapeptide, Fmoc-Tyr(tBu)-Ala-Ala-Arg(Mtr)-Gln-DPhe-Pro-Asp(OtBu)-Asn-Glu-ODmb (Dmb = 2,4-dimethoxybenzyl), attached to the support via the side chain of the C-terminal glutamic acid. Cyclization was achieved by removing the Dmb group followed by the N-terminal Fmoc group and then treating the peptidyl-resin with coupling agents. An undesired but major side reaction was the coupling of neighboring decapeptides resulting in the formation of oligomeric peptides of 20, 30, 40, or more amino acids. Gel filtration chromatography through a column of Sephadex G-25 in dilute NH_3 as the mobile phase afforded complete separation of the cyclic decamer from the oligo-meric by-products. We were able to quantitate the degree of oligomerization under various cyclization conditions by simply weighing the lyophilized powders. During the cyclization step racemization of the C-terminal Glu occurred, which could be quantitated by analytical reverse-phase HPLC. However, the racemized side product was not separated by gel filtration.

Procedure.   Approximately 120 g of Sephadex G-25 was swollen overnight in a 1-L side-arm flask in 600 mL of H_2O–concentrated NH_3, 200:1 (v/v). A size 12 neoprene stopper was placed on the top of the flask, a vacuum line was attached to the side arm, and intermittent swirling was carried out for 30 min. The same buffer (2 L) was degassed in the same fashion but in this case magnetic stirring was employed. A 2.5 × 100 cm glass column (Spectrum), fitted with a stopcock on the bottom, was filled to approximately 1/10 full with the buffer. A small amount of solvent was drained from the column to allow free flow of buffer through the outlet plumbing. The degassed gel was suspended by swirling and was poured into the column. The stopcock was opened and the buffer was allowed to drain. Suspended medium was added in portions to the draining column until a bed height of approximately 90 cm was obtained. The column was filled to the top with solvent and the plunger was fitted such that buffer flowed backward and no air remained in the plunger's tubing. Buffer was then pumped through the column overnight at 1 mL/min. Crude cyclic peptide (0.129 g) was dissolved in 10 mL of buffer, pumped onto the column, and washed in with the washing protocol described before. Solvent was pumped through the column at 1 mL/ min and 5-min fractions were taken. The first peak was lyophilized to give 40 mg of oligomers [fast atom bombardment-mass spectrometry (FAB)-MS 2387.0 (dimer), 3580.5 (trimer), and 4774.0 (tetramer)]. The second peak gave 90 mg of cyclic peptide containing 22% of the diastereomer with D-Glu [FAB-MS (M+H) expected, 1193.5; found, 1193.6]. Both oligomer and cyclic peptide gave correct amino acid analyses.

II.   ION-EXCHANGE CHROMATOGRAPHY, LOW PRESSURE

Ion-exchange chromatography is an excellent purification method that sep-arates compounds on the basis of ionic charge [11,12]. Peptides are typically charged and are therefore amenable to purification using this technique. A chromatographic column is packed with a stationary phase possessing charged groups such as carboxyl groups or diethylaminoethyl groups. The peptide to be purified is a counterion to the stationary phase. Buffer of increasing ionic strength is passed through the column, and the ions of the mobile phase displace the charged peptide, resulting in elution from the column. Increases or decreases of pH can also be used to displace the peptide from the column. Peptides of differing overall charge or isoelectric point (pI) can therefore be separated.

If the peptide is positively charged, then a stationary phase derivatized with negatively charged species such as the carboxymethyl (CM) or sulfo-propyl (SP) group is employed. For negatively charged peptides, supports with positively charged diethylaminoethyl (DEAE) or quaternary amines are used. CM and DEAE are considered weak ion exchangers, whereas SP and quaternary amines are strong exchangers. Atherton and Sheppard [2] describe some outstanding separations of peptides using low-pressure columns of CM-cellulose or DEAE-cellulose. CM- and DEAE-Sephadex resins are also available, as are similarly substituted polystyrene products (Dowex and Amberlite). These supports have a fairly high loading capacity and 1–2 g is quite sufficient for peptide separations on a 0.1-mmol scale.

Low-pressure ion-exchange chromatography requires essentially the same equipment as gel filtration: a peristaltic pump, a column, a detector, and a fraction collector. One uses a peristaltic pump and a gradient maker to deliver the mobile phase. Although Atherton and Sheppard [2] describe extensive washing of CM- or DEAE-cellulose with strong acid or base before use, in the case of the Sephadex resins we have found that simple swelling in the starting buffer and washing with several columns are sufficient preparation. The manufacturer's instructions should be consulted for swelling and equilibration of the resin. Because of the high capacity of ion-exchange resins, smaller columns than those used for preparative gel filtration are used. Columns of 1 cm diameter are adequate and are available from several manufacturers. The columns are packed similarly to gel filtration systems, and again several column volumes of starting buffer are used to equilibrate and pack the resin. Use of a plunger assembly on the top of the column is advantageous in low-pressure ion-exchange chromatography.

The choice of a buffer is important and is sequence dependent. Basic peptides can be purified via chromatography in an acidic buffer and can be eluted with a gradient of increasing NaCl or NH_4OAc. Simple gradients of ammonium acetate or ammonium bicarbonate have been reported [2]. The latter is useful for acidic peptides that are insoluble in acidic conditions. A gradient maker is employed in which the starting buffer is placed in a reservoir closest to the pump and the final buffer is located in a second reservoir connected to the first. These devices are available commercially. During the chromatographic run buffer is pumped out of the first reservoir into the column. Buffer from the second reservoir feeds into the first, resulting in increasing concentration of eluting salt over time.

Salts are a major component of the mobile phase in ion-exchange chromatography and must be removed before the peptide can be used. Although it is a time-consuming process, volatile salts such as ammonium bicarbonate or ammonium acetate can be eliminated by several lyphilizations from water. Nonvolatile salts can be removed by gel filtration chromatography or by reverse-phase HPLC. The following is an example from our laboratory in which low-pressure ion-exchange chromatography was a highly successful purification method.

A. Example: The Synthesis and Purification of (Tyr-Ala-Glu)$_n$

In a study of repeating tripeptides of the structure (Tyr-Ala-Glu)$_n$ [13], we used low-pressure DEAE-Sephadex chromatography for purification. Peptides were synthesized by coupling tripeptide blocks, Fmoc-Glu(OtBu)-Tyr(tBu)-Ala-OH, to polyamide resin using Fmoc-Asp(OtBu) as the C-terminal residue. After obtaining the desired length, the peptide was capped with Fmoc-Tyr(tBu)-Ala-OH, deprotected, and cleaved from the resin to give peptides of the formula (Tyr-Ala-Glu)(Tyr-Ala-Asp). (The C-terminal Asp was included to serve as a reference for amino acid analysis of these peptides.) Separation by ion exchange was chosen because each peptide would differ by at least one negative charge from deletion sequences. Syntheses were carried out on a 0.06-mmol scale in these studies. Crude products were desalted by gel filtration on a 2.5 × 100 cm column of Sephadex G-25 in aqueous NH$_3$ as described earlier.

Procedure. Sephadex A-25 (1 g) swollen overnight in 0.3 M NH$_4$HCO$_3$ was packed into a 1 × 10 cm column from Omnifit. The column was fitted with a plunger and was equilibrated in the same buffer. Desalted peptides were dissolved 0.3 M NH$_4$HCO$_3$ and were pumped onto the ion-exchange column. A gradient formed from 250 mL each of 0.3 M and 1.3 M NH$_4$HCO$_3$ was pumped through the column at 1 mL/min. Elution was monitored at 280 nm. The fractions containing the desired peptides were pooled and lyophilized. Water was added to the residues and lyophilized off several times to remove the NH$_4$HCO$_3$. Peptides were characterized by reverse-phase HPLC, amino acid analysis, and FAB-MS. In these experiments complete purification was achieved with only one pass down these columns.

III. ION-EXCHANGE HPLC

The principles of separation by ion-exchange HPLC [7,11,14,15] are essentially the same as in the case of low-pressure applications. A variety of ion-exchange HPLC columns are available. Strong anion exchange (SAX) columns typically consist of polyethyleneimine grafted onto silica. Strong cation exchange (SCX) columns consist of sulfonic acid groups grafted onto silica. As in the case of reverse-phase packings, care must be taken to keep the pH below 7–8 when using silica-based columns to avoid dissolution of the silica and removal of the ion-exchange groups. Ion-exchange HPLC columns packed with polymeric materials are also available that have the advantage of a greater pH range.

A. Example: Purification of Cyclo(Asp-Asn-Glu-Tyr-Ala-Phe-Tyr-Gln-D-Phe-Pro)

This peptide was synthesized by a solid phase–solution phase procedure [16] for use as a tyrosine kinase inhibitor. Analytical reverse-phase HPLC showed that the crude peptide was contaminated with several compounds that eluted very close to the main peak. In the electrospray ionization (ESI)-MS analysis, negative ion mode, the peptide was the main peak and both the singly charged and doubly charged species were present. A first-pass preparative HPLC provided only partial purification, so preparative ion exchange was used. A 2.5 × 25 cm TSK DEAE-5PW column was employed to exchange the negatively charged Asp and Glu residues. A gradient of NH_4OAc was used with the reasoning that this material mimics the peptide and the stationary phase. NH_4HCO_3, used in the low-pressure ion-exchange separation, is prone to bubble formation at high concentrations, which was avoided with NH_4OAc.

Crude peptide, 153 mg, was 50% pure as judged by RP-HPLC. It was first purified by HPLC using a 2.5 × 25 cm C_{18} column with a gradient of 1% min ACN in H_2O (0.1% TFA in both components). A discussion of preparative reverse-phase HPLC is given later.) Pooling of the best fractions gave only 50 mg of product, which was still contaminated with several materials. The ion-exchange HPLC solvents were A, 0.01 M NH_4OAc, and B, 1 M NH_4OAc. The TSK DEAE-5PW column was washed for 1 h with 1 M NH_4OAc and was equilibrated with 0.3 M NH_4OAc (30% B). The partially purified peptide was dissolved in 1 mL of H_2O with 2 drops of concentrated NH_3. It was loaded onto the column through a sample loop and 2 × 1 mL of buffer A was used to rinse the flask and transfer the contents quantitatively to the ion-exchange column. The gradient program was started, which was 30–70% B (0.3–1 M NH_4OAc)/70 min. Elution was monitored at 275 nm, taking advantage of the Tyr absorbance and avoiding the absorbance of the carbonyl group of the buffer salt.

Fractions containing only the cyclic peptide were pooled and desalted by pumping directly onto the reverse-phase column, pumping 10% ACN for

10 min, then running a gradient of 1% ACN/min. The appropriate fractions were collected, giving 44 mg of peptide that was 98% pure. [ESI-MS (M − H) expected, 1274.35; found, 1273.95. (M − 2H) expected, 636,68; found, 636.56.]

[Note: This synthesis was performed as a rush project under the maxim "time is money" and the diethylamine used to remove the Fmoc group from the linear precursor peptide was old and yellow colored. To save time, it was not distilled. The result was a very impure crude product that necessitated this extensive purification scheme. The conclusion from this exercise is that the maxim "garbage in, garbage out" takes precedence over "time is money."]

IV. REVERSE-PHASE HPLC

By far the most common technique for the purification of peptides is reverse-phase HPLC. This is a very powerful method that allows the separation of peptides from a variety of impurities, often including side products in which one of the amino acids has undergone partial racemization. Several papers and monographs should be consulted to learn basic principles of peptide HPLC, which include but are not limited to Refs. 7, 8, and 12–14. These tend to focus on analytical HPLC, but the principles are easily applied to preparative procedures and some discussion of the latter is included in several of the publications. In addition to these reviews and books, manufacturers of HPLC columns often publish technical notes that are excellent sources of information on methods of both analytical and preparative peptide HPLC.

As mentioned earlier, one performs analytical RP-HPLC on the peptide of interest and uses the retention time of the peptide to provide information about the concentration of organic modifier that is required to elute the peptide from the column. Typically C_4, C_8, and C_{18} columns are used, although phenyl or cyano columns are sometimes employed. Highly cross-linked polystyrene-co-divinylbenzene columns are also available. These have the advantage of being able to withstand a pH range of 1–14. In our laboratory, 4.6 × 250 mm columns for analytical HPLC are used, and we have settled on two "standard" gradients of ACN in H_2O. We use 0.1% TFA in both H_2O and ACN, but some laboratories reduce the TFA in the ACN to 0.085%, which provides a better baseline when monitoring elution in the carbonyl range <230 nm. The gradient called PEP1 goes from 10 to 50% ACN over 30 min at a flow rate of 1.5 mL/min (1.33% ACN/min). This is normally used for deprotected peptides. When chemistry of side chain–protected peptides, such as head-to-tail cyclization [9,15] or fragment condensation, is performed, a gradient of 10–80% ACN/50 min (1.4% min) is used (PEP2).

To scale up from an analytical run to a preparative run, we first calculate the concentration of ACN that elutes the peptide of interest and use that value to program a gradient for preparative HPLC. With the analytical HPLC used in our laboratory, the time for solvent to pass from the inlet to the detector is ~2 min. The retention time (r.t.) of the peptide of interest is used to calculate the concentration of ACN that is required to elute the desired product at a preparative scale:

For PEP1, (r.t.-2) × 1.3%/min + 10% = ACN % that "pushes" the peptide off the column.

For PEP2, (r.t.-2) × 1.4%/min + 10% = ACN % that "pushes" the peptide off the column.

For preparative HPLC we use 2.5 × 25 cm versions of the analytical column, typically C_{18}, and the same solvent system as for the analytical run. We then program our instrument to produce a gradient starting with 7% less ACN than the calculated elution percentage and ending at 8% greater. This gradient is run over 30 min for a 0.5%/min increase in ACN with a flow rate of 10 mL/min. The formula for converting the flow rate from an analytical chromatographic separation to a larger preparative column with the same gradient time is

$$F_P = F_A \times L_P/L_A \times d_P^2/d_A^2$$

in which F is the flow rate, L is the column length, d is the column diameter, subscript P denotes preparative, and subscript A denotes analytical [17]. For direct scaling of the analytical runs using 4.6 × 250 mm columns to the larger 25 × 250 mm columns, the flow rate would be 44 mL/min. Although we are not directly scaling up from analytical to preparative, we obtain excellent separations using our shallower gradients pumping at 10 mL/min. This lower flow rate consumes considerably less solvent, thus reducing the cost of the synthesis and the amount of chemical waste to be disposed of. Some manufacturers supply a series of gradients from curved to linear. In our laboratory we use linear gradients.

To decide which fractions to pool, samples of fractions under peaks are analyzed by analytical HPLC. Those that are considered pure are pooled and lypophilized. The advantage of the H_2O–ACN–TFA elution system, in addition to the sharp peaks and generally high resolution, is that this mobile phase system consists of volatile solvents and TFA that can be removed conveniently by lyophilization. In cases in which time is a factor, the solvents can be removed by careful rotary evaporation as well.

In addition to the simple H_2O–ACN–TFA system, other mobile phases have been found to be quite useful in preparative RP-HPLC. For example, gradients of ACN in aqueous triethylammonium phosphate at varying pH values have been described [18]. These systems produce rather effective purifications, but the products must be desalted by either rechromatography using H_2O–ACN–TFA RP-HPLC or gel filtration. Our laboratory typically works with acidic peptides containing several glutamic acid, aspartic acid, phosphotyrosine, and 4-carboxyphenylalanine residues. Typically H_2O–ACN–TFA gradients on C_8 or C_{18} columns are quite effective. As an alternative, 0.01 M NH_4OAc, pH 6.5, as the aqueous component of the mobile phase and a gradient of ACN or MeOH are employed [2]. The higher pH of this system ensures that the carboxyl groups are ionized, which increases the solubility of the peptide. In certain instances, complete separation of cyclic peptides in which one of the residues is partially racemized has been achieved. We found that reverse-phase HPLC purification of the repeating tripeptides mentioned earlier (Tyr-Xxx-Glu/Asp [13]) in the NH_4OAc system also resulted in excellent separations that were accomplished in about one sixth of the time for low-pressure ion-exchange chromatography.

Discussion of the instrumentation necessary for preparative HPLC will be limited. In general, a pair of pumps that are controlled by gradient-forming software or hardware is required. Flow rates tend to be high, so the pump heads must be able to deliver large volumes. Some laboratories use flows of 50 or 100 mL/min or greater. Our laboratory rarely pumps more than 10 mL/min for a synthesis on the 0.1–0.5 mmol scale. Our pump heads are capable of delivering 25 mL/min. The peptide is applied to the column using either a valve with a large sample loop or a third pump that can be activated with software and plumbed with appropriate valves to direct flow. In our laboratory we have two pumps, A and B, that deliver the aqueous buffer and the organic modifier, respectively; a third unit pumps directly onto the column. Its flow rate is identical to that of the running gradient. All three pumps are controlled by software. Column effluent is directed to a multiple-wavelength UV–visible detector that is set either to 230 or 275 nm, depending on the application. It has also been used at 310 and 410 nm to monitor N-terminal Fmoc groups and fluorescein isothiocyanate (FITC)-labeled peptides, respectively. Column effluent is collected in a fraction collector. At a flow rate of 10 mL/min, we collect either 0.5- or 1-min fractions, depending on the number of contaminants near the desired peak or the amount of peptide being purified.

The following are descriptions of two HPLC purifications that were performed in our laboratory. The purifications were challenging, but with imagination and persistence the peptides were purified.

A. Example: Purification of Biotinyl-ε-aminohexanoyl-Arg-Ala-Asp-Asn-Asp-Lys-Glu-Tyr(OPO$_3$H$_2$)-Leu-Val-Thr-NH$_2$

This peptide was synthesized on polyamide resin [19] using Fmoc chemistry. The peptide was phosphorylated by global phosphorylation on resin [20] and was capped with biotinyl-ε-aminohexanoyl-hydroxylsuccinimide. Both of these steps were considered risks. ESI-MS of the crude material indicated that the peptide was completely phosphorylated but that the biotinylation step was incomplete. Analytical RP-HPLC on a C$_{18}$ column using PEP1 indicated that both peptides were present in nearly equal proportions. The crude peptide (129 mg) was dissolved in 30 mL of H$_2$O. A few drops of concentrated NH$_3$ had to be added to effect complete dissolution. HOAc was added to adjust the pH to just under 7 (pH paper) and the solution was filtered through a 0.45-μm nylon filter. The sample was pumped onto two 2.5 × 25 cm Vydac C$_{18}$ 15–20 μm columns in series at 10 mL/min. Separation was achieved by running a gradient of 16–31% B/30 min at 10 mL/min (A = 0.1% TFA–H$_2$O; B = 0.1% TFA–ACN). The first 10 min of eluant was diverted to waste and fractions were taken thereafter.

Fractions under the two major peaks were collected to give 60 mg of the nonbiotinylated phosphopeptide and 86 mg of the biotinylated version. The combined weights of the two products were more than the weight of the starting material, which suggests that these peptides retain significant amounts of H$_2$O and/or TFA after lyophilization. No activity of the biotinylated phosphopeptide was reported at this stage, so the peptide was further purified. The HPLC columns were equilibrated with 5% ACN in 0.01 M NH$_4$OAc. The peptide was dissolved in 5 mL of H$_2$O, and again complete dissolution required a few drops of concentrated NH$_3$. The peptide solution was pumped onto the columns and a gradient of 5–25% ACN/40 min was run. The appropriate fractions were collected and evaporated on the rotary evaporator to remove ACN quickly. The columns were equilibrated with 16% ACN (0.1% TFA), and this material was rechromatographed with a gradient of 16–31% ACN/30 min. Although this was the third separation, impurities were still visible in this chromatographic run. Appropriate fractions were pooled and lyophilized to give 82 mg of a white, fluffy solid that was 94% pure by HPLC. ESI-MS (M−H) expected 1853.02, found 1852.43. (M−2H) expected 926.01, found 926.19. Following this purification strategy, biological activity of the peptide was exhibited.

B. Example: Purification of Phe-Val-Ala-Phe-Leu-Ala-Phe-Leu-Ala

This peptide was designed for use as a tyrosine kinase inhibitor. It was synthesized using Fmoc chemistry on polyamide resin [19], cleaved using TFA–phenol (95:5), and precipitated from Et_2O in the usual manner. As can be expected from a cursory examination of its sequence, this compound was insoluble in aqueous solution, even in the presence of the detergent Triton X-100. The crude peptide was also poorly soluble in dimethyl sulfoxide (DMSO). It seemed to be completely soluble only in neat TFA. Therefore analytical or preparative HPLC on a standard C_{18} column was not possible. However, polystyrene-divinylbenzene HPLC columns can withstand harsh acidic conditions and were chosen for the purification. Crude peptide (20 mg) was dissolved in 200 μL of neat TFA and injected onto an analytical polystyrene RP-HPLC column using the autoinjector of our analytical HPLC fitted with a fraction collector. The separation was carried out using our PEP2 gradient at 40°C. Pooling the middle fractions under the main peak gave 7.5 mg of the peptide [FAB-MS (M+H) expected, 998.4; found, 998.7] Amino acid analysis: Phe, 3.1; Val 1.0*; Leu, 2.3; Ala, 3.2.

ACKNOWLEDGMENT

This work was supported by the NCI, CA53617, and the Texas Higher Education Coordination Board, grant 073. I am grateful to Claire Lewis Edgemon and Dr. Nihal U. Obeyesekere for the synthetic chemistry and peptide purification.

REFERENCES

1. RB Merrifield. Solid-phase synthesis. I. The synthesis of a tetrapeptide. J Am Chem Soc 85:2149–2154, 1963.
2. E Atherton, RC Sheppard. Solid-phase Peptide Synthesis, a Practical Approach. Oxford: IRL Press, 1989.
3. J Stewart, J Young. Solid-Phase Peptide Synthesis. 2nd ed. Rockford, IL: Pierce Chemical Company, 1984.
4. MW Pennington, BM Dunn, eds. Peptide Synthesis Protocols. Totowa, NJ: Humana Press, 1994.
5. GB Fields, ed. Solid-Phase Peptide Synthesis Methods in Enzymology. New York, NY: Academic Press, 1997, p 289.
6. GB Fields, RL Noble. Solid-phase peptide synthesis utilizing 9-fluorenyl-methoxycarbonyl amino acids. Int J Peptide Protein Res 35:161–214, 1990.

7. CT Mant, RE Hodges, eds. High-Performance Liquid Chromatography of Peptides and Proteins: Separation, Analysis, and Conformation. Boca Raton, FL: CRC Press, 1991.

8. WS Handcock, ed. CRC Handbook for the Separation of Amino Acids, Peptides and Proteins. Boca Raton, FL: CRC Press, 1984.

9. JS McMurray, RJA Budde, DF Dyckes. Cyclic peptide substrates of pp60c-src. Synthesis and evaluation. Int J Peptide Protein Res 42:209–215, 1993.

10. JS McMurray, CA Lewis, NU Obeyesekere. The influence of solid support, solvent, and coupling reagent on the head-to-tail cyclization of resin bound peptides. Peptide Res 7:195–206, 1994.

11. G Choudhary, C Horvath. Ion-exchange chromatography. Methods Enzymol 270:47–82, 1996.

12. BM Dunn, MW Pennington, eds. Peptide Analysis Protocols. Methods in Molecular Biology. Vol 36 Totowa, NJ: Humana Press, 1994.

13. NU Obeyesekere, JN LaCroix, RJA Budde, DF Dyckes, JS McMurray. Solid-phase synthesis of (tyrosyl-glutamyl-alanyl)$_n$ by segment condensation. Int J Peptide Protein Res 43:118–216, 1994.

14. CT Mant, RH Hodges. Analysis of peptides by high performance liquid chromatography. Methods Enzymol 271:3–50, 1996.

15. CT Mant, LH Kondejewski, PJ Cachie, OD Monera, RH Hodges. Analysis of synthetic peptides by high performance liquid chromatography. Methods Enzymol 289:426–469, 1997.

16. JS McMurray, CA Lewis. The synthesis of cyclic peptides using Fmoc solid-phase chemistry and the linkage agent 4-hydroxymethyl-3-methoxyphenoxy butyric acid. Tetrahedron Lett 34:8059–8062, 1993.

17. Chromatography Columns and Supplies Catalog. Waters Corporation, 34 Maple St., Milford, MA 01757.

18. CA Hoeger, R Galyean, RA McClintock, JE Rivier. Practical aspects of preparative reversed-phase chromatography of synthetic peptides. CT Mant, RE Hodges, eds. High-Performance Liquid Chromatography of Peptides and Proteins: Separation, Analysis, and Conformation. Boca Raton, FL: CRC Press, pp 753–764.

19. JT Sparrow, NG Knieb-Cordonier, NU Obeyesekere, JS McMurray. A large pore polydimethylacrylamide resin for solid-phase peptide synthesis: Applications in FMOC chemistry. Peptide Res 9:297–304, 1996.

20. JW Perich. Efficient solid-phase synthesis of mixed Thr(P)-, Ser(P)-, and Tyr(P)-containing phosphopeptides by "global" "phosphite-triester" phosphorylation. Int J Peptide Protein Res 40:134–140, 1992.

19

Analysis of Synthetic Peptides

Nicholas P. Ambulos, Jr.
University of Maryland School of Medicine, Baltimore, Maryland

Lisa Bibbs
The Scripps Research Institute, La Jolla, California

Lynda F. Bonewald
The University of Texas Health Science Center at San Antonio, San Antonio, Texas

Steven A. Kates
Consensus Pharmaceuticals, Inc., Medford, Massachusetts

Ashok Khatri
Massachusetts General Hospital, Boston, Massachusetts

Katalin Fölkl Medzihradszky
University of California San Francisco, San Francisco, California

Susan T. Weintraub
The University of Texas Health Science Center at San Antonio, San Antonio, Texas

Synthetic peptides are being utilized increasingly in biochemical, molecular biological, and pharmacological research in addition to finding exciting new uses in drug discovery programs. Thus, sample purity and sequence integrity are both extremely important. The use of impure or incorrectly synthesized peptides can result in significant losses of time and resources. The Peptide Synthesis Research Group (PSRG) of the Association of Biomolecular Resource Facilities (ABRF) recommends that along with stringent quality control measures that are carried out during synthesis, each product should be characterized by independent biophysical methods, including, as a minimum, reversed-phase high-performance liquid chromatography (RP-HPLC) and mass spectrometry (MS) [electrospray ionization (ESI-MS) or matrix-assisted laser desorption–ionization (MALDI-MS)]. Furthermore, for accurate quantification of synthetic peptides, amino acid analysis (AAA) is the optimal method. Then, if discrepancies in a peptide sample are indicated by the analytical data, either repurification or resynthesis can be undertaken. Problems with a sequence can often be readily deduced by either Edman degradation analysis or tandem mass spectrometry. Capillary electrophoresis (CE) is also a powerful tool for further evaluating problematic peptides. In the sections that follow are descriptions of a number of the methods that are highly effective for evaluation of synthetic peptides, along with examples of their application for specific problems. The methods covered include analytical HPLC, CE, MS, AAA, and Edman degradation sequence analysis.

I. HIGH-PERFORMANCE LIQUID CHROMATOGRAPHY (HPLC)

A. Introduction

One of the most widely utilized methods for assessment of synthetic peptides is analytical HPLC. HPLC offers substantially higher resolution and faster separations than classical liquid chromatography. Presented here is a basic overview of HPLC as applied to analysis of synthetic peptides. There are a number of superb reviews on the subject that provide more detailed information on the theory and practice of HPLC [1–3].

Fractionation by HPLC results from interactions between the sample, the mobile phase (also called the buffer), and the stationary phase (i.e., the column packing). Analytical HPLC columns are usually either 3.2 or 4.6 mm in inner diameter (i.d.), with lengths from 10 to 30 cm. As a general rule, an increase in column length will lead to an increase in component resolution without significant peak broadening. Columns are normally purchased pre-filled with a solid packing material that has a well-defined particle size (usually 5 μm) and pore size (60 to 300 Å) as well as specific chemical groups on the surface for interaction with the analyte.

HPLC systems generally include the following components: an injector (usually a loop) for sample loading, pumps for the mobile phase (two pumps are most often employed for standard gradient separations), an in-line mixer to ensure efficient blending of the solvents during gradient formation, a column, and a detector [typically monitoring ultraviolet (UV) absorbance, in conjunction with either a computer data system or a strip-chart recorder]. A fraction collector is frequently connected to permit future re-use of the column effluent.

Separation of peptides by HPLC is accomplished using a variety of different strategies, essentially determined by the nature of the chemical groups on the surface of the stationary phase. There are two general categories of separation—those in which the target molecules bind to the column matrix and those in which no binding occurs. The latter is exemplified by gel filtration chromatography, where molecules are fractionated on the basis of size. All other commonly used chromatographic methods involve interaction of the target molecule with the column matrix. The ability of a particular method to effect separation is dependent on the degree of binding of the analytes and the nature of the eluant. Examples of these methods include reversed-phase, hydrophobic interaction, normal-phase, ion-exchange, and affinity chromatography.

B. Modes of Separation

1. Ion-Exchange Chromatography

Ion-exchange chromatography (IEX) separates molecules on the basis of their charge. IEX is often used to purify biologically active material because loading of the sample onto the column can be accomplished in most standard aqueous buffers, and elution conditions are mild enough to retain biological activity. IEX is a powerful tool for purification of peptides, proteins, antibodies, and oligonucleotides [4–6]. The bonded phase of the column packing contains either positively charged (anion exchange, AEX) or negatively charged (cation exchange, CEX) functional groups. These functional groups are further classified according to the level of pH dependence of the charge on the bonded surface. Weak ion-exchange groups tend to easily lose their charge as the pH of the mobile phase changes, whereas strong ion-exchange groups maintain their charge throughout most of the pH range (Table 1).

For IEX, samples are loaded onto the column in a dilute buffer (i.e., weak ionic strength), which allows electrostatic binding of the solutes to the charged functional groups. Samples are subsequently released from the col-

Table 1 Commonly Used Functional Groups for Ion-Exchange Chromatography

Ion-exchange system	Functional group
Weak anion exchange	Diethylaminoethyl
Strong anion exchange	Diethylmethylaminoethyl
Weak cation exchange	Carboxymethyl
Strong cation exchange	Sulfopropyl

umn by increasing the ionic strength of the buffer. Frequently, a salt gradient is employed for this purpose. At low salt concentrations, only the weakly bound solutes elute from the column; as the salt concentration is increased, the more highly charged molecules, which are more tightly bound to the column, are released. Alternatively, a gradient of either increasing or decreasing pH can be utilized, because changes in pH alter the charge on a molecule and specifically influence the electrostatic interaction between the column and the analyte. Knowledge of the pI of the molecule(s) of interest is very important in the selection of the appropriate ion-exchange column packing. In general, because peptides have a number of different groups that can become charged, at pH values above the pI, there will be an overall negative charge on the molecule, and thus an AEX column can be employed. On the other hand, at a pH below the pI, a peptide will exhibit a net positive charge and should be run on a CEX column. Once the column has been selected, a good starting point for the mobile phase is to use a buffer that is 1 to 1.5 pH units above (for AEX) or below (for CEX) the pI to allow adequate binding to the column. Increasing concentrations of NaCl can provide an adequate ionic strength gradient to elute most molecules. A typical gradient is 0 to 1 M NaCl over 100 column volumes. From these general conditions, separations can be optimized by varying factors such as the column surface chemistry, the pH, the gradient or even by adding organic modifiers such as acetonitrile or methanol for molecules that are hydrophobic or demonstrate poor solubility in aqueous systems.

2. Reversed-Phase Separations

Reversed-phase HPLC is the most frequently used method for analysis and purification of synthetic peptides [1,7,8]. Molecules are separated by RP-HPLC on the basis of differences in hydrophobicity. Reversed-phase column packings have a silica-based solid support with bonded alkyl groups, usually containing 4, 8, or 18 carbons; the longer hydrocarbon chain column packings (i.e., C_{18}) are generally more efficient for separation of peptides than the shorter chain materials. The highly nonpolar surface in a C_{18} column preferentially interacts with hydrophobic molecules (such as most peptides) if they are introduced in a polar mobile phase. Fractionation is then achieved by gradually changing to a more nonpolar buffer, generally through gradient addition of a water–miscible organic modifier, such as acetonitrile, methanol, or isopropanol. Enhancement of the interaction of a peptide with an RP column packing can be achieved by addition of an ion-pairing agent. Most synthetic peptides have one or more amino acid residues with a polar and/ or charged side chain. The ion-pairing agent is amphipathic, with a charged end that interacts with the polar or charged groups on the peptide and a hydrophobic end that interacts with the column to enhance binding. An ideal ion-pairing agent also has the property of being volatile, so it can be easily removed after HPLC analysis. Trifluoroacetic acid (TFA) is the most commonly used ion-pairing agent for silica-based RP-HPLC. Addition of TFA to the mobile phase has the added benefit of maintaining a low-pH environment during chromatography. This is important because for most RP column packing materials, at basic pH there can be irreversible stripping of the alkyl hydrocarbon chains from the silica base.

One critical decision in developing a reversed-phase method is selection of the rate (i.e., the slope) of the gradient to obtain efficient separation. The choice of the gradient depends largely on the hydrophobicity of the peptide and the nature of any contaminants that may be present. Often, a very shallow or near-isocratic gradient offers enhanced resolution. A good starting place for a binary gradient is 0.1% TFA in water for the A buffer and acetonitrile containing 0.085% TFA for the B buffer, increasing B from 0 to 100% over a 30-min period. The percentage of TFA is decreased over the course of the gradient in order to maintain a flat baseline in the UV trace. From these initial conditions, separations can be optimized, for example, by either increasing or decreasing the rate of the gradient, by changing the initial percentage of the B buffer, by varying the organic solvent, or by changing the ion-pairing agent.

Although reversed-phase HPLC is the most widely used method for chromatographic analysis of synthetic peptides, there are occasions when a technique called hydrophobic interaction chromatography (HIC) is preferable. HIC also uses hydrophobicity as a means of fractionating molecules. But, in contrast to RP-HPLC in which binding to the column packing is often strong enough to disrupt secondary and/or tertiary structure, interaction between peptides and HIC packing is much weaker [9]. HIC is often a recommended method for purification of biologically active peptides that must retain a higher order structure; HIC is also useful for analysis of peptides that are extremely hydrophobic and may bind irreversibly to a traditional RP column.

3. Column Maintenance

Resolution on analytical HPLC can be negatively affected by the accumulation on the column of strongly retained materials, such as proteins, lipids, or other hydrophobic substances. Filtering of samples and buffers as well as the addition of a guard column can help prevent accumulation of debris; however, thorough cleaning is beneficial and can often lengthen the useful life of an HPLC column. Silica-based RP columns can be cleaned using gradient elution with 0.1% aqueous TFA (solvent A) and either isopropanol/

0.1% TFA (solvent B) or acetonitrile/isopropanol (2:1) containing 0.1% TFA as an alternative solvent B. In some cases it may be necessary to wash a column with a stronger reagent mixture such as isopropanol/0.1 N HNO_3 (4:1) overnight at a low flow rate. Lipids or very hydrophobic compounds can frequently be removed by washing the column with nonpolar solvents such as chloroform or dichloromethane; however, intermediate use of a solvent such as isopropanol that is miscible with both the nonpolar solvents and standard RP-HPLC mobile phases must be employed. Washing with 0.1 N NaOH is useful for cleaning of IEX or HIC columns. Hydrophobic contaminants can be removed from these columns by washing with ethanol or methanol, and more strongly bound material may require washing with detergents or guanidine-HCl. In order to extend the life of an RP-HPLC column, it is best to store the column in a solvent such as methanol and avoid long-term exposure to mobile phases containing TFA.

4. Examples

To illustrate the analysis of peptides by the techniques described in this chapter, selected results from the 1998 study of the ABRF Peptide Synthesis Research Group will be presented. The goal of this study was to assess the quality control capabilities of participating laboratories. As part of this assessment, a synthetic peptide sample was distributed by the PSRG, and the laboratories were asked to verify whether the "requested" sequence of H-Asp-Glu-Gln-Glu-Ala-Leu-Asn-Arg-Ser-NH$_2$ (peptide **1**, 1059.5 daltons) was present. The peptide that was actually sent out for this part of the study had (by design) been synthesized in the reverse direction; in addition, the "requested" Ser residue had been replaced with a Thr and an Asp was incorporated instead of one of the Glu residues. As a consequence, the molecular mass of the test peptide was the same as the one requested, but, in reality, the sequence was H-Thr-Arg-Asn-Leu-Ala-Asp-Gln-Glu-Asp-NH$_2$ (peptide **2**). In preparation for the study, both **1** and **2** were analyzed by HPLC. Using a 1-mm C_{18} column and a standard acetonitrile/water/TFA mobile phase system at a flow rate of 50 μL/min, the peptides eluted very close to the solvent front. In order to gain better retention of the peptides, a shallower gradient was employed. The A buffer was 2% acetonitrile/0.1% aqueous TFA and the B buffer was 90% acetonitrile/0.09% TFA. Using a gradient of 2 to 12% B over 10 min, **1** eluted at 8.2 ± 0.2 min (mean ± SD for three injections) and **2** eluted at 8.3 ± 0.2 min (Fig. 1). Thus, under these conditions, it was not possible to distinguish between the two peptides when analyzed separately by HPLC. However, as illustrated in Fig. 2, when aliquots of each sample were mixed (using less of **1** as a marker), clear separation into two peaks was observed. Similar results were obtained when

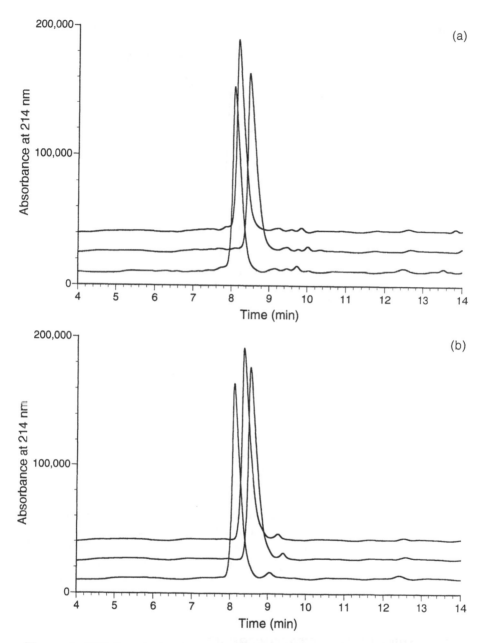

Figure 1 HPLC analysis of H-Asp-Glu-Gln-Glu-Ala-Leu-Asn-Arg-Ser-NH₂ (a, peptide **1**) and H-Thr-Arg-Asn-Leu-Ala-Asp-Gln-Glu-Asp-NH₂ (b, peptide **2**). HPLC, Michrom MAGIC 2002 (Michrom BioResources, Inc.); column, Michrom MAGIC MS C₁₈ (1 × 150 mm); buffer A, 2% acetonitrile/0.1% TFA; buffer B, 90% acetonitrile/0.09% TFA; gradient, 2 to 12% B over 10 min; flow rate, 50 μL/min; UV detection, 214 nm; shown are three successive injections of each peptide.

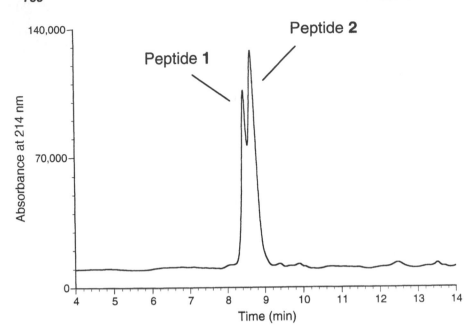

Figure 2 HPLC analysis of a mixture of H-Asp-Glu-Gln-Glu-Ala-Leu-Asn-Arg-Ser-NH₂ and H-Thr-Arg-Asn-Leu-Ala-Asp-Gln-Glu-Asp-NH₂. HPLC conditions were the same as in Fig. 1.

an isocratic gradient at 5% B was employed. These analyses clearly illustrate two important points related to HPLC analysis of synthetic peptides. First, for relatively small peptides, the lack of significant retention on a C_{18} column can make it difficult to achieve adequate component resolution. In addition, slight variations in retention time may be observed. As a consequence, it may not be possible to differentiate between individual peptides by HPLC behavior. Second, and more important, observation of a single HPLC peak does not guarantee that a sample is pure. Thus, analysis by additional techniques, as described below, is important for reliable characterization of synthetic peptides.

II. CAPILLARY ELECTROPHORESIS

A. Introduction

Capillary electrophoresis (CE) is a valuable tool for qualitative and quantitative analysis of a wide variety of biomolecules [10,11]. Separation by CE combines features from classical methods of polyacrylamide gel electrophoresis (PAGE) and HPLC. Although not as frequently employed, CE can provide a powerful complement to HPLC for analysis of synthetic peptides. A number of benefits are afforded by CE. The analysis requires only nanoliter volumes, thus preserving most of the sample for future use. In addition, CE can often separate impurities, such as truncated products or incompletely deprotected peptides, which can be difficult to fractionate by conventional analytical HPLC methods, especially for components that have a hydrophobicity similar to that of the desired product. Although various mass spectrometric techniques can be used to detect these products, peptide quantification by MS is difficult, at best. However, CE can be used to not only resolve but also to readily quantify most synthetic peptides. Although not trivial, CE can be employed for micropurification [12] in which individual components can either be isolated from a single run or pooled from multiple runs for further characterization by techniques such as amino acid analysis or N-terminal sequencing. CE can also be used in conjunction with MS, either on line or off line, for rigorous identification of synthesis impurities [13–16].

B. Separation Theory

Separation by CE is accomplished by applying a voltage across a buffer-filled glass silica capillary that spans two buffer reservoirs; separation of components is accomplished on the basis of their charge-to-mass ratio. A variety of factors can influence CE resolution, most notably electrophoretic migration and electroosmotic flow (EOF). Operational parameters such as the applied voltage, the viscosity of the buffer, and the length of the capillary as well as molecular characteristics such as the mass, the charge, and the shape of the analyte directly affect electrophoretic mobility. Electroosmotic flow, on the other hand, is related to the movement and velocity of the buffer within the capillary. Silanol (Si–OH) groups on the inner wall of the fused silica capillary interact with the solution within the capillary. These silanol groups can be readily converted to anions by the CE buffer and subsequently attract cations from the buffer to create a double layer of negatively and positively charged ions at the capillary wall. When a voltage is applied to the capillary, the cations interacting with the silanol groups begin to migrate toward the cathode, pulling water molecules along with them, thereby creating an electroosmotic flow. Buffers at high pH induce a greater degree of ionization of the silanol groups, making the capillary wall more negatively charged than at low pH. Therefore, at a higher pH, electroosmotic flow will be much greater than at neutral or low pH. Buffers in the neutral pH range are, thus, not ideally suited for CE because any positively charged analytes will tend to bind easily to the capillary wall surface and not fractionate cleanly from other components.

Use of a high-pH buffer can be quite effective in the separation of mixtures, in part because of the contributions of EOF. At high pH, cations move rapidly to the cathode as a result of both electrophoretic mobility and EOF. At the same time, uncharged analytes migrate at the velocity of the EOF and anions tend to move slowly toward the cathode. As a consequence, in CE, anion movement represents a balance in the competition between the EOF pulling these molecules to the cathode and the electrophoretic tendency of anions to migrate to the anode. In an environment where the charge of a peptide can be controlled by pH, EOF can greatly enhance the simultaneous separation of charged and uncharged molecules because they will migrate at different velocities. The ability to regulate the EOF thus gives CE greater flexibility in design of experimental methods to separate complex mixtures.

CE separations can also be influenced by the addition of buffer modifiers which affect resolution by changing the viscosity of the buffer, by altering the electroosmotic flow, or by modulating interactions between the analyte and the capillary wall [17–19]. Some of the common buffer modifiers are salts, ionic and nonionic surfactants, ion-pairing agents, and organic solvents. Binding of an analyte to the capillary wall can be minimized through addition of a divalent amine to the buffer or by using a capillary coated with a hydrophilic polymer. Neutral coated capillaries allow CE separation based primarily on electrophoretic mobility by greatly reducing EOF [20]. Using an amine-coated capillary, a reversed electroosmotic flow is observed because a strong cationic charge is present on the inner wall of the capillary.

C. Instrument Design

The length of a CE capillary generally ranges from 20 to 100 cm, with typical inner diameters of 50–75 μm. The outside of the glass capillary is coated with polyimide that renders the otherwise fragile glass pliable. A small region of the polyimide coating is burned off to expose a section of silica tubing that serves as a flow cell for UV absorbance detection.

The buffer reservoir at the inlet side of the capillary (where the sample is introduced) can be readily switched from the sample vial for injection to the buffer vial for the electrophoresis run. CE separations typically require high voltage, often as high as 500 V per cm length of the capillary. Because application of a voltage of this magnitude causes the temperature of the capillary to increase, resulting in band broadening, CE instruments provide a means for temperature control of the capillary, typically by using either liquid cooling or forced air convection.

There are several different ways to introduce a sample into the capillary. When either pressure on the inlet side or vacuum on the outlet side of the capillary is employed, the process is called hydrodynamic injection. The quantity of sample applied to the capillary in this manner can be controlled by changing either the length of time or the level of pressure or vacuum used for injection. When a voltage is utilized to introduce a sample into the capillary (electrokinetic injection), concentration of the sample can be obtained at the same time. Regulation of electrokinetic injection is accomplished by adjusting the length of time or the amount of the applied voltage. Each of these injection techniques has an associated disadvantage that must be considered when designing a CE analysis. When hydrodynamic injection is utilized, the sample is introduced into the capillary along with the buffer in which it is dissolved. Even though only nanoliter volumes are involved, there can be a profound impact on the EOF if the sample buffer contains salts or other buffer modifiers. With electrokinetic injection, on the other hand, a representative sampling of the components in a mixture may not be introduced into the capillary, depending on the charge of each analyte. Components with relatively high mobility will be loaded preferentially, causing significant sampling bias.

D. Applications

1. Free Solution in an Uncoated Capillary

In general, for analysis of a peptide with an unknown migration pattern, it is logical to start with an uncoated capillary at basic pH (0.05 M sodium borate buffer, pH 8.0, for example). Under these conditions, EOF will give a rapid initial separation of the sample components. Prior to the electrophoretic run, the capillary should be cleaned and conditioned. This is accomplished by rinsing the capillary in acid and base, as shown in Table 2. CE protocols often include these rinses to ensure that the capillary is prop-

Table 2 Conditioning Method for New Uncoated Capillary

Rinse solution	Pressure (psi)	Duration (min)
0.1 N NaOH	20	10
Water	20	2
1.0 N HCl	20	10
Water	20	2
0.05 M sodium borate, pH 8.0	20	10

erly conditioned for each run and to maintain reproducibility between runs. The peptide sample being analyzed is typically dissolved in the running buffer at a concentration of approximately 10 μg/mL and applied to the capillary hydrodynamically, as a plug, by a 5-s low-pressure injection. The separation is accomplished by applying a voltage in the range of 250 to 500 V/cm length of the capillary over a 10- to 30-min period, with UV monitoring at 214 nm.

A typical CE analysis sequence is as follows:

1. 0.1 N NaOH pressure rinse, 20 psi, 2 min
2. Water pressure rinse, 20 psi, 1 min
3. 1 N HCl pressure rinse, 20 psi, 2 min
4. Water pressure rinse, 20 psi, 1 min
5. 0.05 M sodium borate, pH 8.0, pressure rinse, 20 psi, 2 min
6. Pressure injection of sample, 1 psi, 5 s
7. Voltage separation in sodium borate, pH 8.0, 20 kV over 60 cm, 15 min

Use of this CE protocol should provide an initial indication of the purity of a peptide sample; however, changes in the quantity of sample injected, the field voltage applied, and the duration of the run may be necessary for more efficient separation in subsequent analyses. Although this method may yield significant information about contaminants present in a synthesis mixture, it is important to conduct one or more CE analyses at a lower pH (e.g., 0.025–0.05 M sodium phosphate buffer, pH 2.5) to determine whether any components comigrated at the higher pH.

2. Free Solution in a Neutral Coated Capillary

Standard buffers used for separation in a neutral coated capillary range from pH 3 to 8. Buffers outside this range can shorten the useful life of the capillary. The isoelectric point (pI) of the analyte of interest is the main determinant for choice of the pH of the buffer. Typical separations employ a buffer with a pH that is 1–2 units below the pI. Under these conditions, peptides generally exhibit a net positive charge. Citrate buffers (20 mM) ranging from pH 3 to 6 are commonly utilized for CE in a neutral coated capillary. It is sometimes advantageous to use a buffer with a pH above the pI (for example, 20 mM tricine, pH 8.0) in order to decrease CE analysis time; however, because most peptides exhibit a net negative charge at pH values above their pI, the electrode polarity must be reversed in order to obtain component fractionation. For CE in a neutral coated capillary, a relatively high electric field (500 V/cm) is generally employed because under these conditions there is little or no electroosmotic flow.

3. Free Solution in an Amine-Coated Capillary

The positively charged surface of an amine-coated capillary is ideal for separation of basic molecules that might ordinarily bind to the negatively charged silanol groups of the fused silica. With an amine coating, the EOF is generally of greater magnitude and in the opposite direction to the EOF in an uncoated capillary. The separation buffer should be selected on the basis of the "ionizability" of the peptide, with typical buffers including Tris pH 8.0, MES pH 6.0, and acetate pH 4.5.

4. Examples

The utility of CE as an analytical tool for characterization of synthetic peptides is nicely illustrated by analysis of the ABRF Peptide Synthesis Research Group 1998 peptides. As described earlier, these peptides were not easily resolved by HPLC. For analysis by CE, each peptide was separately dissolved at a concentration of approximately 10 μg/mL in 0.1% aqueous TFA, hydrodynamically injected into an uncoated fused silica capillary for 5 s at 1 psi, and electrophoresed using sodium borate, pH 8.0, as the run buffer at a field strength of 350 V/cm for 10 min, with UV monitoring at 214 nm. Triplicate analyses were performed. Overlaying the three runs for each of the peptides demonstrates the high degree of reproducibility of CE (Fig. 3a and b). When the two peptides were mixed and then hydrodynamically injected together under the same conditions, near baseline separation was observed (Fig. 3c). In order to obtain better resolution, the same peptide mixture was hydrodynamically injected into the capillary and fractionated using a pH 2.5 phosphate run buffer as a way to reduce the EOF. With this lower pH buffer, some additional separation was observed, although there was an order of magnitude decrease in sensitivity. To completely eliminate effects due to EOF, a neutral coated capillary and citrate buffer, pH 4.0, were then employed. The peptide mixture was once again hydrodynamically injected for 5 s at 5 psi and separated using a field strength of 350 V/cm. Under these conditions, the peptides were resolved by almost 5 min, although the sensitivity was still significantly reduced.

CE has not been widely utilized as a routine method in peptide synthesis laboratories. Analytical HPLC and mass spectrometry are generally sufficient for the majority of synthetic peptide samples. However, there are instances in which the greater flexibility in method development afforded by CE permits resolution of components that are not readily separated by other techniques (illustrated by the peptide rearrangement reported by Kates and Albericio [21]). Thus, CE should be considered as a powerful complement or even alternative to analytical HPLC for characterization of synthetic peptides.

(a)

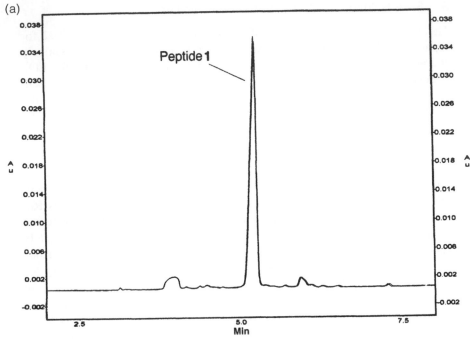

(b)

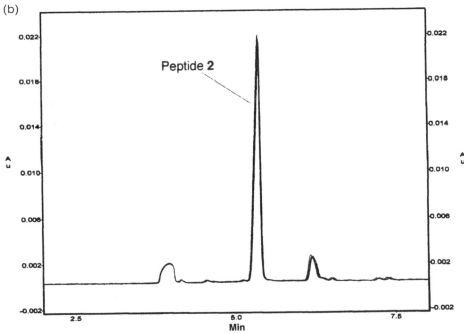

(c)

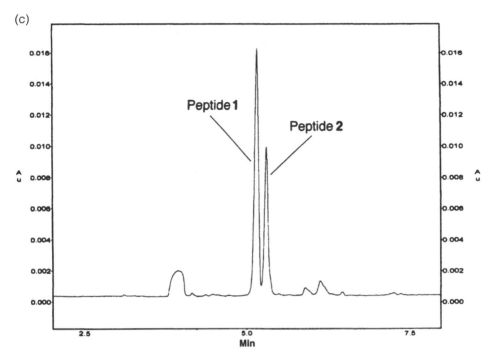

Figure 3 CE analysis of synthetic peptides. (a) H-Asp-Glu-Gln-Glu-Ala-Leu-Asn-Arg-Ser-NH$_2$; (b) H-Thr-Arg-Asn-Leu-Ala-Asp-Gln-Glu-Asp-NH$_2$; (c) a mixture of both peptides. Fractionation was obtained on a Beckman P/ACE-MDQ CE system, utilizing a 60-cm fused silica capillary in 50 mM borate buffer, pH 8.3. Samples were hydrodynamically injected for 5 s at 1 psi and separated at 20 kV for 10 min. The traces represent an overlay of 3 injections.

III. MASS SPECTROMETRY

A. Introduction

In the past decade, mass spectrometry (MS) has become the method of choice for quality control of synthetic peptides. Historically, plasma desorption (PD) and fast atom bombardment (FAB) were the first ionization methods used for the mass analysis of nonderivatized peptides. More recently, electrospray ionization (ESI) MS and matrix-assisted laser desorption ionization (MALDI) MS have found widespread utility for peptide analysis. Both of the latter methods yield protonated molecules and, thus, provide direct molecular weight information. As will be covered later, ESI can be employed with a variety of mass analyzers, including quadrupole, magnetic sector, ion trap, and time-of-flight (TOF) analyzers. On the other hand,

MALDI is most frequently coupled with TOF analysis. In the mass spectral characterization of peptides, sequence information can be obtained through decomposition and/or fragmentation of the ionized molecules. With the proper instrument configuration, ions produced by MALDI that fragment in flight (through a process called "postsource decay," PSD) can be focused and detected, thereby providing a facile means of obtaining sequence information for a wide range of molecular weight species. In ESI-MS, peptide fragmentation can be obtained through collisionally activated decomposition (CAD, also called collision-induced dissociation, CID). Analysis of the CAD products is most efficiently accomplished by a multistage instrument, such as a triple quadrupole, a double-focusing magnetic sector instrument, or an ion trap, after mass selection of the ion of interest.

B. MALDI

Most of the commercially available MALDI-TOF instruments are equipped with a nitrogen laser operating at 337 nm. To prepare a sample for MALDI-TOF analysis, a matrix absorbing at this wavelength is mixed and cocrystallized with each sample. Although many different matrices have been used successfully, the most widely recommended matrix for peptide analysis is 4-hydroxy-α-cyanocinnamic acid (HCCA). Other frequently used matrices that are suitable for analysis of larger peptides (>3000 daltons) are 2,5-dihydroxybenzoic acid (DHB) and 3,5-dimethoxy-4-hydroxycinnamic acid (sinapinic acid, SA). Matrix solutions can be prepared in many different ways. For example, HCCA and SA (both from Aldrich) can be used as saturated solutions in water–acetonitrile (7:3, v/v) that has been acidified with 0.1% aqueous TFA. Other preparations include dissolving 10 mg HCCA in 1 mL of ethanol–acetonitrile (1:1) or 10 mg of DHB (Fluka) in 1 mL of ethanol/water (1:1) (22). Alternatively, these three matrices can be purchased in solution, ready for immediate use (Hewlett-Packard, Palo Alto, CA). It is important to note that decomposition–fragmentation of peptides is both sequence and matrix dependent. Analysis in HCCA usually yields more fragmentation than in a "cooler" matrix (i.e., a matrix that does not induce as much fragmentation) such as DHB [22]. Substances that decompose readily under MALDI conditions, such as phospho-Ser, phospho-Thr, and phospho-His–containing peptides as well as sulfo- or sialylated glycopeptides, can be analyzed more successfully in neutral, two-component matrices [23].

For the analysis of synthetic peptides, it is recommended that the peptide solution be in the range 5 to 10 pmol/μL. Typically, 1 μL of the sample solution is added to 1μL of the matrix solution in a 0.5-mL polypropylene centrifuge tube (e.g., an Eppendorf tube); the mixture is then vortexed and centrifuged (Microfuge). This quantity is usually sufficient for four loadings on a MALDI target. To ensure accurate molecular weight determination, the MALDI-TOF instrument should be regularly calibrated with a standard peptide mixture, with the best results obtained by using standard peptides with $[M+H]^+$ values that bracket the mass range of interest. For example, an ~1 pmol/μL solution containing angiotensin II ($[M+H]^+$ m/z 1046.5) and adrenocorticotropic hormone (ACTH)-(18–39)-clip ($[M+H]^+$ m/z 2465.2) provides excellent calibration for peptides in the mass range of 1000 to 2500 daltons, with satisfactory results obtained up to m/z 3000. One complication is the fact that the matrices yield abundant cluster ions in the low mass range (<m/z 500) that may interfere with the detection of low-mass peptides. Because of the interference of these abundant background ions, during MALDI-TOF analyses only ions above m/z 400 are routinely monitored.

Different MALDI-TOF mass spectrometers provide different mass resolutions and mass accuracies. The simplest linear instruments are unable to resolve isotope clusters; thus, the $[M+H]^+$ ions observed represent average mass values. With this type of instrument, the error in mass measurement may be as great as 0.1 to 0.2% (1000–2000 ppm) with external calibration. Thus, the uncertainty for a peptide with $[M+H]^+$ at m/z 1000 is 1 or 2 daltons. Linear instruments fitted with delayed extraction or simple reflectron instruments are typically able to resolve isotope clusters up to ~m/z 2500–3000, thereby providing monoisotopic $[M+H]^+$ values with an accuracy of 0.05% or 500 ppm (0.5 dalton at m/z 1000). Reflectron instruments with delayed extraction operating at a resolution of about 10,000 can lower this error to below 100 ppm, or 0.1 dalton at m/z 1000. Although the use of internal calibration can provide significantly higher mass accuracy, it can be tricky to mix in just the right amount of standard.

Another point that must be mentioned is that in routine MALDI-TOF-MS analysis of peptides and proteins, the relative ion intensities observed do not necessarily reflect the true relative quantities of the components in a mixture. For example, Arg-containing peptides usually produce the most abundant signals in nonseparated tryptic digests and at the same time several other components may go undetected. In fact, ESI and MALDI analyses of the same sample can often yield dramatically different results, in part because of differential ionization and suppression effects. Another potential complication is that, as mentioned earlier, ionization by MALDI may trigger fragmentation. Sometimes this fragmentation occurs in the ion source, immediately upon ionization (so-called prompt fragmentation) and then the products appear as well-resolved ions—just as any other "real" component in the mixture. Peptides susceptible to this in-source fragmentation should be analyzed in a "cooler" matrix [23]. When the decomposition occurs after leaving the source (i.e., in the flight tube), the fragments can be detected only when a reflectron is used; these ions appear as wider, nonresolved, metastable peaks at slightly higher m/z values than appropriate for their real masses. Observation of such peaks in MALDI spectra may indicate the presence of structures that readily undergo decomposition. For example, during MALDI-TOF analysis, Ser/Thr-phosphorylated peptides routinely lose all or part of the phosphate group (-98 and -80), in contrast to peptides that contain phosphorylated Tyr, which exhibit a loss of only 80 mass units (occurring to a much lesser extent that in their Ser or Thr counterparts) [22]. Other commonly observed fragments include loss of a portion of the side chain from oxidized Met (-64) and cleavage of the peptide bond between Asp and Pro residues [24]. Switching to the linear mode will eliminate detection of these metastable ions, resulting in a simplified spectrum.

C. Electrospray Ionization

Electrospray ionization and MALDI represent major advances in the application of mass spectrometry to analysis of biomolecules. Although in many cases the two techniques are able to provide the same information, quite frequently they are complementary. For a variety of reasons, some samples are more successfully analyzed by MALDI and other by ESI. Thus, it is highly beneficial to have access to both types of instrument for characterization of synthetic peptides. In ESI-MS, the samples are introduced in solution at flow rates from less than 1 μL/min up to 1 mL/min, depending on the design of the ESI interface. Pure samples and simple mixtures are often analyzed by direct infusion; more complex mixtures (such as proteolytic digests) generally require either on-line or off-line HPLC fractionation prior to MS analysis.

ESI-MS as we use it today was introduced in the late 1980s by Fenn et al. [25]. In the ESI process, the analyte solution is forced through an extremely small capillary (either metal or fused silica) into a region of high electric field, producing a fine spray of droplets. Through a variety of different processes (which again depend on the interface design) the droplets diminish in size as the solvent evaporates, eventually leaving only charged analyte molecules with all solvent removed. A helpful presentation about ESI-MS can be found in *Introduction to Mass Spectrometry* by Watson [26]. A fundamental key to the success of ESI-MS for analysis of peptides and proteins is the presence of highly basic sites on residues such as Arg, Lys, and the amino terminus. By inclusion of acid in the electrospray solution, these sites become protonated, leading to a series of positively charged ions. (ESI-MS instruments can also be operated in the negative ion mode; however, this discussion will be limited to positive ion detection, as this is the mode most frequently utilized for characterization of synthetic peptides.

In practice, several different charge states of a peptide are usually detected at the same time. The actual masses of the charge states differ by 1 atomic mass unit (amu), but the peaks appear quite separated in the spectrum because the instrument measures mass divided by charge (m/z) as opposed to mass. This phenomenon has made it possible for ESI-MS to be used on relatively low-mass-range instruments (such as quadrupoles and ion traps) for analysis of proteins even up to 100,000 daltons. In nature, there is normally a sufficient number of basic residues in a protein so that the resulting distribution of charge states falls below m/z 2000. For example, horse heart apomyoglobin (16,952 daltons, containing 2 Arg and 19 Lys residues) is commonly utilized as a reference standard in ESI-MS. Under routine operating conditions, the predominant peaks detected center near the 20+ charge state, with the charge state distribution generally ranging from about 30+ [(16,952 + 30) ÷ 30] (m/z 566.1) to 10+ (m/z 1696.2). Fortunately, included with most ESI-MS data systems is a program that can mathematically deconvolute the array of charge states and readily provide the analyte molecular weight. This determination is much simpler for synthetic peptides because of their lower molecular weight, often yielding a 1+ ion ($[M+H]^+$) in the mass range $<m/z$ 2000.

To analyze a peptide by ESI-MS, the sample is usually dissolved in either acetonitrile/water or methanol/water (typically 50–70% organic) containing 0.5–1% acetic acid. Peptide concentrations from 0.5 to 5 pmol/μL are commonly used, depending on the specific instrument. For ESI-MS analysis, as for most other MS techniques, "more" is not better—that is, by staying at the lower concentration range it is possible to avoid problems with peptide aggregation and still minimize background interference. Because the analysis of synthetic peptides is not ordinarily sample limited, the easiest way to introduce the peptide sample is by direct infusion via a syringe pump or by loop injection. Spectral averaging can then be employed to enhance the quality of the acquired scans. In most cases, use of computer-assisted deconvolution is not necessary in the analysis of synthetic peptides because agreement with the expected sequence can be readily assessed by visual inspection of the spectrum. Any differences between the observed and predicted molecular weights can usually be quickly attributed to the presence of residual blocking groups and/or extra or missing residues.

For some electrospray interfaces, the presence of residual TFA in a sample can significantly interfere with ionization, resulting in "noisy," potentially uninterpretable spectra. In that case, it is beneficial to redissolve the sample without added TFA and relyophilize prior to analysis. Several cycles of solubilization–lyophilization may be necessary to eliminate interference from TFA. Peptide solubility can occasionally be a problem for ESI-MS. Routinely, the sample is first dissolved in either water or dilute acid

and then diluted appropriately into the ESI solvent mixture. For some peptides it may be necessary to start with an organic solvent and then make
subsequent dilutions. In the special case of highly hydrophobic peptides, a
combination of chloroform/methanol/5% acetic acid (5:5:1 or 2:4:2 v/v/v)
[27] has proved extremely useful.

D. Sequence Information

As mentioned earlier, sequence information can be obtained by MS after
either peptide decomposition in the flight tube in MALDI (PSD) or collisionally activated decomposition (CAD) in ESI-MS. A particular advantage
of this approach is that a peptide of interest can be characterized even if it
is in a mixture. In MALDI-PSD this is accomplished through use of an "ion
gate"—an electronic device for selecting an approximately 1% wide win-
dow around the desired m/z value. By using a strategy that involves alter-
ation of the reflectron voltage, the fragment ions formed in the free-flight
region can be focused and their masses accurately measured, yielding a PSD
spectrum. In ESI-MS/MS, the approach utilized depends on the instrument
configuration. When only one stage of mass analysis is available (e.g., with
a single quadrupole), "in-source CAD" (fragmentation generated in the electrospray interface) can be used to obtain sequence information about an
analyte. However, interpretation of the results is complicated by the likely
presence of multiple charge states and/or additional components. With tandem mass spectrometers, the ability to effect initial mass selection prior to
ion fragmentation significantly enhances the level of information that can
be obtained. For example, in a triple quadrupole MS, an ion of interest is
selectively transmitted through the first quadrupole analyzer (Q1), CAD
fragmentation is generated in Q2, and then mass analysis is conducted in
Q3. Similarly, in an ion-trap MS system, a peptide ion of interest can be
specifically retained while all other ions are ejected; subsequent CAD of the
trapped ion produces fragments that are then detected to yield the CAD
spectrum. The practical upper mass limit for this type of CAD (i.e., low-
energy CAD available in quadrupoles and ion traps as compared with high-
energy CAD in magnetic sector instruments) is around 3000 daltons. In most
cases, more extensive, yet interpretable sequence information can be obtained for 2+ (and sometimes 3+) ions. Thus, for characterization of a
synthetic peptide, mass selection of the 2+ ion followed by CAD is generally recommended over use of the 1+ ion.

Peptide fragments from PSD and CAD processes are usually formed
via peptide bond cleavages (i.e., between the backbone NH of one residue
and the carbonyl of the adjacent residue). Many of the same principles apply
for interpretation of ESI-MS/MS data as for MALDI-PSD. When the charge

is retained on the *N*-terminus, the ions produced are called **b** ions and are numbered sequentially starting at the *N*-terminus. Ions with charge retention at the *C*-terminus are analogously called **y** ions and are numbered from the *C*-terminus. It is important to keep in mind that although the **b**- and **y**-ion series are generally predominant in PSD and CAD spectra, alternate backbone cleavages are often detected, albeit to a significantly lesser extent. PSD and CAD sequence ions may fragment further, depending on their amino acid composition, losing, for example, CO (-28 from **b**-type ions only), NH_3 (-17), and H_2O (-18). In addition, immonium ions with a structure of $^+NH_2=CH-R$ may be detected, providing information about the individual amino acid residues in the peptide. Occasionally, these immonium ions produce further fragments that even more specifically indicate the presence of an amino acid by exhibiting a series of characteristic ions. A noteworthy example is the series of m/z 70, 87, 100, and 112 that is indicative of Arg [28]. Details about the nomenclature for peptide fragments can be found in *Methods in Enzymology* [29]. There is also a comprehensive article on the interpretation of peptide spectra generated by high-energy CAD [28] that contains useful tables and explanations that can also be helpful for PSD and low-energy CAD data. Computer software programs are available as well that can reliably predict the mass spectral fragmentation of peptides; examples include the "MS-Product" component in the Protein Prospector package of UCSF (*http://prospector.uosf.edu*), "BioWorks" available on Finnigan mass spectrometers, "MassLynx" by Micromass, "MacBioSpec" distributed with earlier Sciex instruments, and GPMAW (downloadable from the net at *http://welcome/to/gpmaw*).

E. Examples

As part of the 1998 ABRF PSRG study, the test peptides were analyzed by both MALDI-TOF-MS and ESI-MS. In Fig. 4 are the MALDI-TOF (a) and ESI-MS (b) spectra of peptide **2**. With MALDI-TOF, an intense ion at m/z 1060.7 was detected, representing $[M+H]^+$. When analyzed by ESI-MS, both basic sites on this peptide become protonated, yielding the ions at m/z 1060.6 ($[M+H]^+$) and m/z 530.9 ($[M+2H]^{2+}$), which were mathematically deconvoluted by the data system on the instrument to yield the calculated mass of the parent peptide (shown in the inset). So, at this point the results obtained by mass spectrometry would seem to indicate that the sample did contain the correct peptide. However, analysis by either MALDI-PSD (Fig. 5a) or ESI-MS/MS (Fig. 5b) revealed that there were problems with this peptide.

Shown in Table 3 is a list of the predicted **b** and **y** ions for the "requested" sequence (peptide **1**). As can be seen by comparison with the fig-

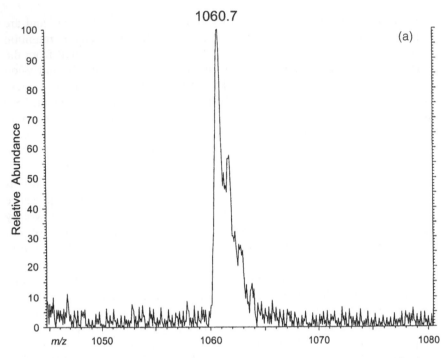

Figure 4a MALDI-TOF mass spectral analysis of H-Thr-Arg-Asn-Leu-Ala-Asp-Gln-Glu-Asp-NH$_2$. The spectrum was acquired on a TofSpec SE (Micromass) mass spectrometer equipped with a nitrogen laser and a reflectron, using a peptide solution of approximately 10 pmol/μL and HCCA (Hewlett-Packard) as the matrix.

ure, none of the predicted ions was detected in the PSD spectrum and only m/z 373.2 ("**b$_3$**") and 488.0 ("**y$_4$**") could conceivably be assigned in the ESI-MS/MS trace. An important point to note here is that the **b** and **y** ions are usually members of a series of ions. Although it is often observed that one series predominates over the other, depending on the sequence, commonly there are significant stretches of one or both of the series, with occasional breaks caused by specific residues. Inspection of the "requested" sequence shows that there was supposed to be an Arg near the C-terminus. Therefore, one would expect to see both series of ions in the PSD and CAD spectra because there should be a free amino terminus and a basic residue near the C-terminus. This was clearly not the case in the spectra shown in Fig. 5. In the ESI-MS/MS spectrum, from the major peaks in the 400 to 1100 m/z range, there is evidence that the peptide contains the sequence -Glu-Gln/Lys-Asp-Ala-, but no information is immediately obvious about

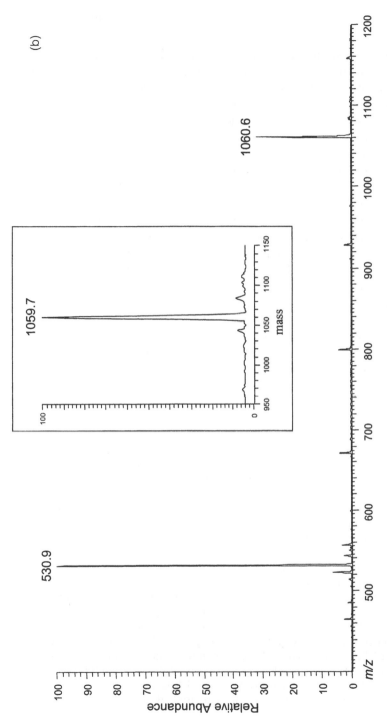

Figure 4b ESI mass spectral analysis of H-Thr-Arg-Asn-Leu-Ala-Asp-Asn-Gln-Glu-Asp-NH₂. The spectrum was acquired on a Finnigan LCQ ion trap mass spectrometer; the spectrum obtained by mathematical deconvolution is shown in the inset. The sample was dissolved at a concentration of approximately 0.5 pmol/μL in 50% acetonitrile containing 0.5% acetic acid and introduced into the mass spectrometer by direct infusion at a flow rate of 5 μL/min. Spectral averaging of 20 scans was employed. The small peaks in the spectrum were produced by "in-source" CAD of the peptide; peaks such as these are frequently, but not always observed in ESI-MS analysis of small peptides.

the direction of synthesis. From the immonium ion region in the MALDI-PSD spectrum, the presence of Arg (m/z 70, 87, 100, 112), Gln (m/z 84 and 101), Leu (m/z 86), and Thr (m/z 74) can be unambiguously determined. In addition, the ion at m/z 87 might indicate an Asn residue [28]. Then one can follow the trail of the N-terminal **b** ions accompanied by 17-dalton lower satellite ions due to NH_3 losses—a favored process when the fragment contains an Arg residue.

Actually deciphering the complete sequence from the MS fragmentation pattern is an extremely complex task that is beyond the scope of this chapter. However, verification of a proposed sequence should be possible, even for facilities just being introduced to peptide analysis by MS. In the case of the ABRF study described above, with the availability of MS/MS data it should have been obvious to the participating laboratories that the "requested" peptide had not been synthesized. With the results of this study as an example, it can be concluded that when other quality assurance mechanisms are in place in a peptide synthesis laboratory, molecular weight determination by either MALDI-TOF or ESI-MS can provide an excellent means of verifying the integrity of a synthetic product. When questions do arise about a specific peptide, MALDI-PSD and ESI-MS/MS then can permit rapid determination of sequence information that should yield valuable clues about the nature of a synthetic sequence.

---→

Figure 5 Sequence analysis of H-Thr-Arg-Asn-Leu-Ala-Asp-Gln-Glu-Asp-NH$_2$ by mass spectrometry. (a) MALDI-PSD spectrum acquired on the instrument described in the legend of Fig. 4. PSD was performed by making nine steps of the reflectron voltage; at each step the voltage was reduced to 75% of that in the previous step. Segments from each step were then stitched together to produce the complete spectrum. The spectral segments acquired for each step were obtained from separate positions on the sample target; data from 50–75 laser shots were collected at each position. Smoothing of the data was employed for the spectrum shown. As a result, the peaks represent "average mass" with no resolution between individual isotopes rather than monoisotopic mass. Calibration of each reflectron voltage step was accomplished using ACTH-(18–39)-clip peptide. (b) [see p. 776] ESI-MS/MS spectrum of m/z 530.9 acquired on the Finnigan LCQ. An ion isolation width of 2 was employed with collision energy of 35%. The sample was fractionated on a Michrom MAGIC 2002 HPLC (Michrom BioResources) fitted with a Michrom MAGIC MS C$_{18}$ column (1 × 150 mm); the eluant was introduced directly into the mass spectrometer. The mobile phase consisted of buffer A, 5% acetonitrile/0.1% TFA; buffer B, 90% acetonitrile/0.09% TFA; gradient, 5 to 65% B over 20 min; flow rate, 50 μL/min.

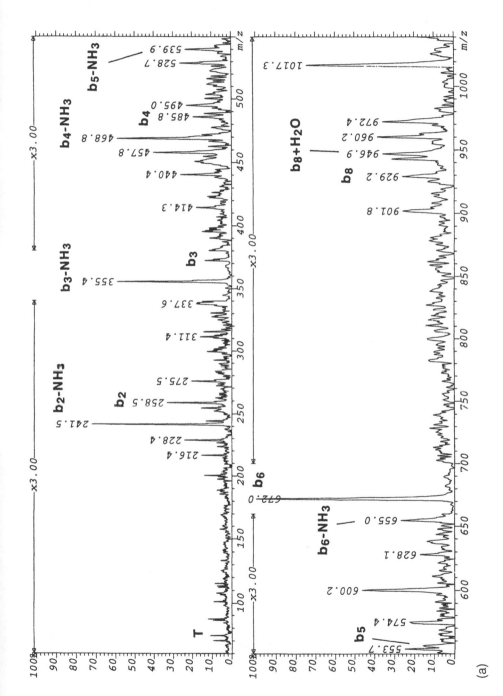

(a)

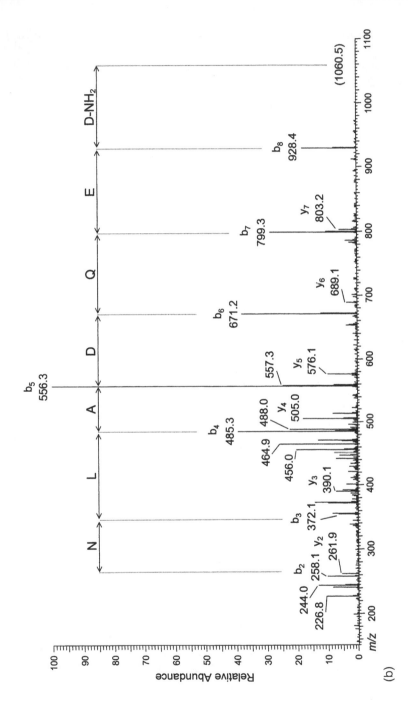

Figure 5 Continued

Table 3 Predicted MS/MS Fragmentation of H-Asp-Glu-Gln-Glu-Ala-Leu-Asn-Arg-Ser-NH$_2$[a]

b ions		Residue		y ions
116.1	1	Asp (D)	9	
245.1	2	Glu (E)	8	945.5
373.2	3	Gln (Q)	7	816.4
502.2	4	Glu (E)	6	688.4
573.2	5	Ala (A)	5	559.3
686.3	6	Leu (L)	4	488.3
800.4	7	Asn (N)	3	375.2
956.5	8	Arg (R)	2	261.2
1043.5	9	Ser (S)	1	105.1

[a]Calculation of the fragment m/z values utilizes the residue weights (molecular weight $-$ H$_2$O) of each amino acid as follows for a C-terminal amide peptide: **b** ions numerically equal the sum of the included residue weights plus 1 for the additional terminal H; y ions equal the sum of the residue weights plus 16 for the terminal NH$_2$ and 2 for two protons added during formation.

IV. AMINO ACID ANALYSIS

With the advances achieved in mass spectrometry instrumentation over the past decade, amino acid analysis (AAA) is less frequently utilized for the quality control of synthetic peptides than it was in the past. AAA remains, however, the method of choice for the quantitative analysis of synthetic peptides. Although in most cases AAA is conducted after cleavage of the peptide from the resin, a peptide that is still attached to the solid support can be analyzed as well. Analysis of a peptidyl resin is often facilitated by the fact that many polymeric supports incorporate an internal amino acid reference [30].

To perform AAA, a peptide must first be hydrolyzed in order to liberate the individual amino acids. To enhance the reliability of the results, each peptide sample should be analyzed in triplicate. Hydrolysis is accomplished under either vapor-phase or liquid-phase conditions, through use of 6 N HCl. For vapor-phase hydrolysis, a specially designed screw-capped reaction vessel that is fitted with an inert slide valve is utilized (supplied by vendors such as Pierce and Waters). The desalted, dry peptide is placed in an acid-washed glass tube (e.g., 6 × 50 mm). One or more of these tubes is then inserted into the reaction vessel. After addition of approximately 250 μL of 6 N HCl to the bottom of the reaction vessel, each vessel is purged with either argon or nitrogen and then a vacuum of approximately 100 mtorr is applied. The reaction vessels are then heated at 110°C for 24 h. Liquid-phase hydrolysis requires use of a glass tube that can be flame sealed and can withstand vacuum levels less than 100 mtorr. (Suitable tubes can be obtained from VWR or Scientific Products: Pyrex 9820-10, 75 × 10 mm rimless culture tubes.) The desalted, dry peptide is placed in the glass tube and approximately 100 μL of 6 N HCl is added. The tube is purged with argon or nitrogen, evacuated to less than 100 mtorr, and then flame sealed. As before, each tube is heated at 110°C for 24 h.

The amino acids must then be derivatized in order to be detected. Derivatization can be accomplished either before (precolumn) or after (postcolumn) separation by HPLC. At present, there are no clear-cut advantages of either method. The choice of derivatization technique depends largely on the instrumentation being used for AAA. After hydrolysis, there are specific sample handling procedures for each derivatization protocol; these procedures are described in detail in the references listed below for each method. The most commonly used postcolumn derivatization technique, developed by Moore and Stein in the late 1950s [31], initially uses ion-exchange chromatography to separate the amino acids. As they elute from the column, the amino acids are derivatized with ninhydrin and detected using an on-line visible wavelength detector. When postcolumn derivatization is employed, triplicate samples of approximately 1 μg of peptide should be hydrolyzed.

The precolumn technique that is most frequently employed today was developed during the early 1980s [32,33]. For this method, the classical Edman reagent phenylisothiocyanate (PITC) is used for amino acid derivatization after hydrolysis. Separation of the PTC amino acids is then accomplished by HPLC, with detection at 254 nm. Although standard C_{18} columns available from a variety of vendors are suitable for separation of the PTC-derivatized amino acids, there are specific columns that have been optimized for this purpose (e.g., Waters). Approximately 0.5 μg of peptide should be hydrolyzed for analyses using precolumn derivatization.

After hydrolysis, only 16 of the amino acids are routinely detectable. Asn and Gln are converted to Asp and Glu, respectively. Moreover, special procedures must be used to detect Cys and Trp [34,35]. In view of these potential complications, it is advisable to process a standard sample in parallel with the test peptide, regardless of which hydrolysis and derivatization techniques are being utilized.

An essential part of AAA is generation of a calibration curve, which is constructed from a plot of HPLC peak area versus amount for standard amino acids. Four points bracketing the range of interest should be used, and the resulting plot should be linear. For each amino acid, a response

Table 4   AAA Analysis of H-Thr-Arg-Asn-Leu-Ala-Asp-Gln-Glu-Asp-NH$_2$

Amino acid	Area	Response factor (RF)[a]	Quantity (pmol)[b]
Ala	2105	0.0587	123.6
Asx (D, N)	3601	0.1027	369.8
Glx (E, Q)	2650	0.0932	247.0
Thr	1250	0.1002	125.3
Leu	2095	0.0424	123.2
Arg	1600	0.0756	121.0
Number of residues	9		
Total amino acid (pmol)	1109.9		
Total peptide (pmol)	123.3		

[a]Calculated from the corresponding calibration curve according to the following equation:

$$RF = \text{quantity of standard (pmol)/peak area of standard}$$

[b]Calculated from the experimental data as follows:

$$\text{Quantity of peptide (pmol)} = \text{peak area of residue} \times RF$$

factor (quantity/area) is then calculated from the slope of the corresponding calibration curve. After analysis of a peptide hydrolysate, the amount of each amino acid present can be calculated using the appropriate response factor. An illustration of AAA is shown in Table 4 for peptide **2** (H-Thr-Arg-Asn-Leu-Ala-Asp-Gln-Glu-Asp-NH$_2$).

V.   EDMAN DEGRADATION SEQUENCE ANALYSIS

A.   Introduction

Edman degradation was originally developed for determination of the primary structure (i.e., amino acid sequence) of peptides and proteins. Sequence analysis is not regularly performed for quality control in routine peptide synthesis but is more often employed for problem solving. As described earlier in this chapter, efficient characterization of synthetic peptides can be readily obtained by a combination of RP-HPLC and mass spectrometry. Amino acid analysis is also valuable if MS is not available. If an incorrect mass or a discrepancy in the amino acid composition is found, one obvious alternative is to resynthesize the peptide. But, in order to deduce the cause of a failed synthesis, additional analyses must be performed. Both Edman degradation and tandem MS can be used to obtain sequence information about a "problem" peptide. However, interpretation of the results of MS sequence analysis is not always straightforward and a significant amount of experience is often needed to deduce a sequence de novo. In contrast, in Edman degradation, each residue is detected sequentially and identified by its characteristic HPLC elution behavior. Thus, it is readily possible to detect unambiguously substitutions, additions, and deletions in a peptide through use of Edman degradation [36,37]. Modification of an amino acid residue can also be determined by Edman degradation [38–40], although it may be necessary to use MS to gain insight into the exact chemical nature of the modified residue.

B. Chemistry

The strategy developed by Edman in 1950 [41] for chemical sequencing of peptides and proteins (summarized in Fig. 6) is still in use today. The process can be divided into three steps: coupling, cleavage, and conversion. Phenyl-isothiocyanate (PITC) is *coupled* to the amino-terminal amino acid of the peptide using a base such as trimethylamine (TMA) to form a phenylthio-carbamyl group (PTC-peptide). By-products are removed through a series of organic washes. *Cleavage* of the modified amino acid from the peptide is obtained by the addition of anhydrous trifluoroacetic acid (TFA) to give the anilinothiazolinone amino acid (ATZ-AA) and the peptide minus the amino-terminal amino acid. The released ATZ-AA is extracted from the peptide and *converted* to a more stable phenylthiohydantoin form (PTH-AA), which can be identified by RP-HPLC. The peptide is thus sequentially shortened by individual residues through repeated cycles in order to deduce the primary sequence.

C. Sample Preparation

Soluble synthetic peptides, multiple antigenic peptides (MAPs), or peptides still attached to resin can be sequenced by Edman degradation (refer to the review by Grant et al. [42] for more information about sequencing these three types of peptides). Because most sequences can provide accurate results for peptides in the 1–10 pmol range, quantity is not usually an issue for characterization of synthetic peptides. Although considerable losses can occur during the cycles of Edman degradation, most synthetic peptides of 15 to 20 residues can be completely sequenced to the carboxy-terminal amino acid. Peptide loss that occurs during sequencing is generally referred

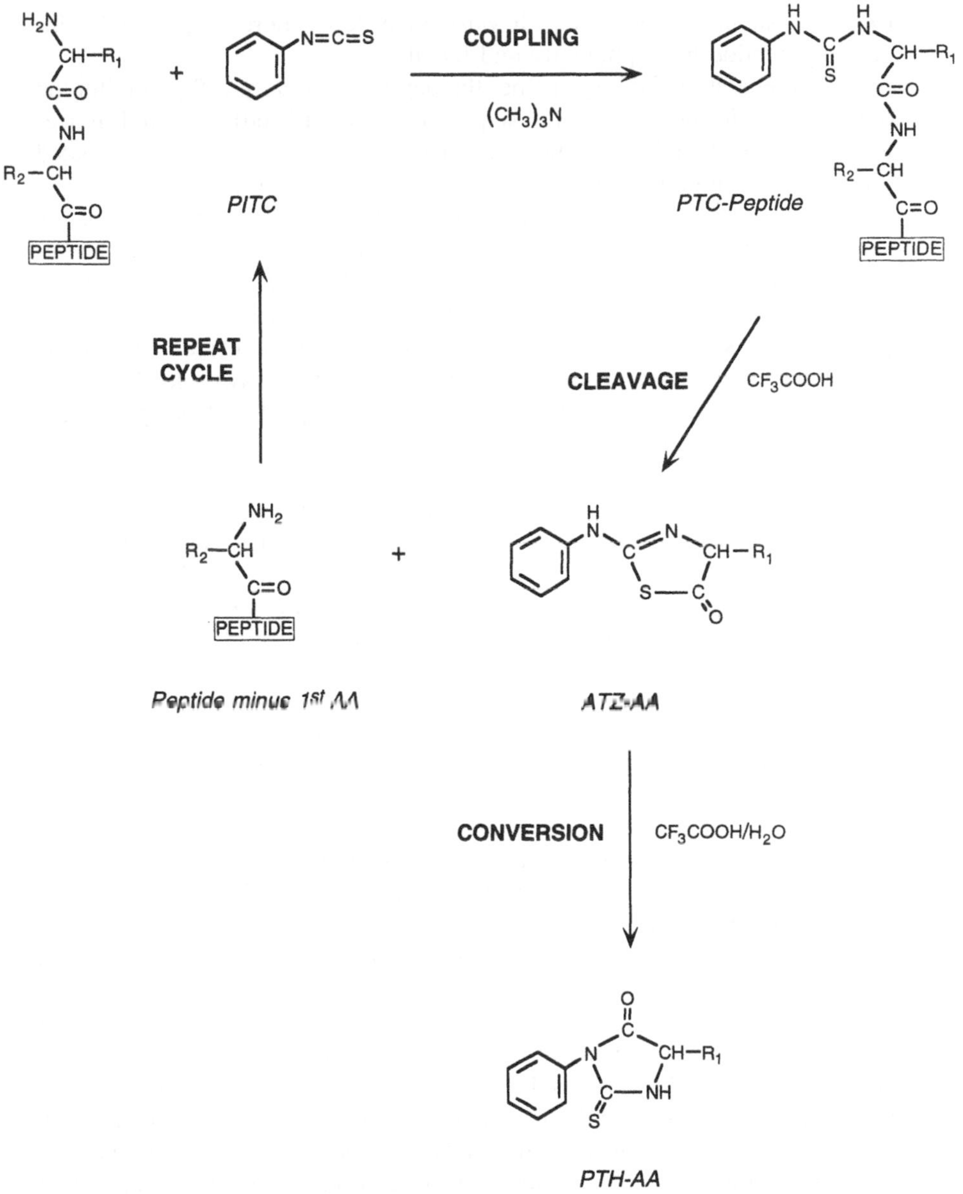

Figure 6 Edman degradation chemistry.

to as "washout." Problems with washout become more frequent with increasing peptide hydrophobicity and length.

In preparation for sequencing, the peptide (normally 100 pmol, but the optimal quantity depends on the sequencer being utilized) dissolved in acetonitrile or methanol is spotted onto a membrane [for example, a coated glass-fiber membrane (Biobrene; ABI) or a polyvinylidene difluoride (PVDF) membrane such as Immobilon (Millipore) or Problott (ABI)]; the sample is air dried and then the membrane is loaded directly into the sequence cartridge. If severe problems with washout are experienced, techniques for covalent coupling of a peptide to the membrane can be employed, provided that specific residues (e.g., an amino acid with a free amine on its side chain) are present in the sequence. However, a drawback of this approach is that amino acids that are covalently coupled will not produce a signal in the subsequent HPLC analysis.

MAPs, multiple antigenic peptides used for antibody production, are synthesized on a branched Lys core attached to a solid-phase support. These complex peptides are difficult to characterize because the product is heterogeneous. HPLC chromatograms of MAPs usually contain a large number of peaks and mass spectral analysis of an unfractionated preparation of MAPs is often ambiguous. However, sequence analysis based on Edman degradation can provide information that is representative of the average sequence of all the chains. If a "reasonable" sequence is obtained (i.e., if the predominant sequence indicated by Edman degradation is the sequence that was requested), the MAPs are often considered to be of sufficient quality for antibody production despite the fact that there may be sequence problems on a minority of the chains. The combination of HPLC (either on-line or off-line) with MS provides an alternative approach to characterization of MAPs [43]. After identification by MS, isolation of the desired product can then be accomplished by HPLC.

It is also possible to perform sequence analysis while a synthesis is in progress by loading a single bead or several beads directly into the sequencing chamber. Because a single bead of resin can hold 80–100 pmol of peptide, there is ample material for complete sequencing. One advantage of sequencing the beaded resin is that no washout occurs, in contrast to loading soluble peptide onto a membrane. Analysis of a bound peptide is particularly useful for assessing progress in the synthesis of peptides that have previously been difficult to produce. An important point to remember, however, is that the sequence deduced from one or a few beads may not be truly representative of the final cleaved and purified product. In order to sequence a resin-bound peptide, a few beads are suspended in an organic solvent such as methanol, acetonitrile, or 40% methanol in dichloromethane; 20 μL of the mixture is then deposited onto a TFA-treated glass fiber filter that is placed in the sequencer. Alternatively, a single bead can be selected by micromanipulation. In preparing the sample, if the resin is colorless and cannot be easily seen, staining with 0.1% bromphenol blue in methanol can be used to enhance visualization. When Fmoc chemistry is employed for the synthesis, the linkage between the peptide and the resin is acid stable. As a consequence, the bead(s) can be directly analyzed by Edman degradation because the TFA used during the conversion step will also remove the protecting groups on each amino acid, rendering the residues detectable by HPLC in the standard manner. However, if the synthesis is based on Boc chemistry or if other protecting groups are used, the level of TFA routinely employed for conversion is often not sufficient to remove the protecting groups. In that case, it is necessary to modify conditions for sequencing and for HPLC fractionation so that removal of protecting groups can be maximized and identification of any residual modified amino acids can be made.

D. *C*-Terminal Sequencing

A variety of different strategies for *C*-terminal sequencing have been reported, but derivatization of the *C*-terminus to a thiohydantoin [44,45] is the approach most widely utilized because it employs relatively efficient chemistry that is suitable for automation. Hewlett-Packard has designed an instrument (model HP241) that is capable of analysis by both Edman degradation and *C*-terminal sequencing, so that a sample can be sequenced from either terminus. Moreover, advances have been made in the methodology for *C*-terminal sequencing. In the past, a significant disadvantage of *C*-terminal sequencing was the inability to derivatize proline. Formation of the thiohydantoin requires the reaction of the peptide with a thiocyanate reagent in the presence of a carboxylic acid activating reagent. Although this process works well for the majority of the amino acids, proline has always been problematic, with the usual result that there is a halt in the *C*-terminal sequence analysis at this residue. However, it has been found [46] that proline-thiohydantoin can be efficiently produced using activation by acetyl chloride in TFA, followed by derivatization with ammonium thiocyanate at an elevated temperature (55°C instead of 30°C). This new chemical strategy should be easily automated in the near future.

Even though *C*-terminal sequencing has advanced significantly, the method of choice for sequence analysis of synthetic peptides is still Edman degradation. Although it is not ordinarily a serious disadvantage for synthetic peptides, 10 to 100 times more material is required for *C*-terminal sequencing than for Edman degradation. More important, however, a longer stretch of sequence can be deduced by Edman degradation than by currently available methods for *C*-terminal sequencing. The capacity to sequence synthetic peptides from the carboxy terminus does not add any particular advantage except in the case of peptides greater than 20–30 residues, which are too long for complete analysis by Edman degradation.

E. General Comments and Potential Problems

Edman degradation can be utilized only when a peptide has a free amino terminus. If the *N*-terminus is modified, for example, through acetylation, the peptide is considered "blocked." Likewise, if the peptide is cyclized head to tail via an amide linkage between the amino and carboxy termini, no sequence information can be obtained by chemical-based methods. Another issue that must be addressed is that free Cys is destroyed during Edman chemistry; Cys residues must be modified (such as by reduction and alkylation) in order to be detected by automated sequence analysis [47,48]. There are two separate problems with detection of Trp by Edman degradation. One is that PTH-Trp is recovered in low yields after derivatization and cleavage. Second, PTH-Trp often elutes near an Edman chemistry by-product, diphenylurea, adding uncertainty to the identification. Modification of HPLC conditions has proved useful for separation of PTH-Trp from this by-product [47]. Finally, quantification (but not identification) of PTH-Thr and PTH-Ser is often problematic because of variations in the number and amplitude of HPLC peaks from associated degradation products.

Modified or unusual amino acids such as hydroxyproline and phosphoserine and protected amino acids can also be identified by Edman degradation sequence analysis [38]. However, it is important to remember that in this method, identification of a residue is made solely on the basis of HPLC retention time. It is, therefore, essential to support the identification of a modified residue by comparison with authentic standards. Furthermore,

---

→

Figure 7 Edman degradation sequence analysis. (a) Chromatograms for each sequence analysis cycle; (b) [see p. 786] relative yield per cycle. The analysis was performed on a PE Biosystems cLC Procise 490 sequencer. Two microliters of an approximately 2 nM solution of the peptide in 0.2% TFA was applied to a PVDF membrane (0.4×0.4 mm) and allowed to air dry. Biobrene-T (2.5 μL of a solution containing Biobrene-T/methanol/0.2% TFA, 1:7:2, v/v/v; Biobrene-T from PE Biosystems) was then added on top of the dried peptide. Sequence analysis of the dry sample on the membrane was accomplished by means of instrument protocols provided by the manufacturer. A standard mixture containing 19 PTH-amino acids (all of the common amino acids except Cys) was utilized to determine the HPLC retention times for each derivatized residue.

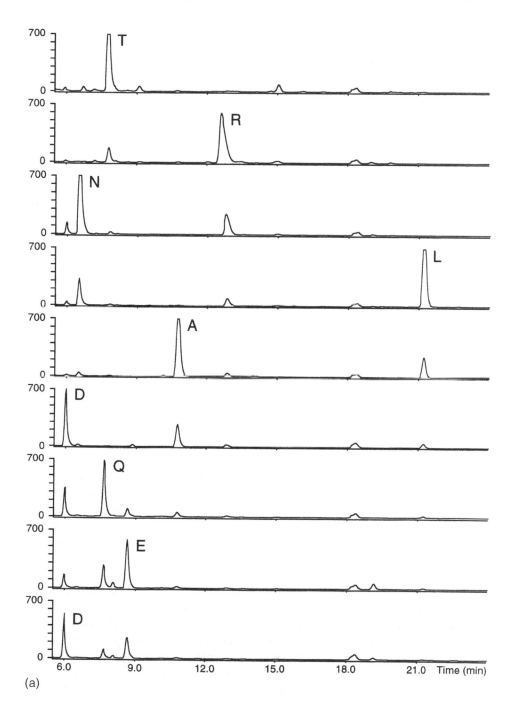

(a)

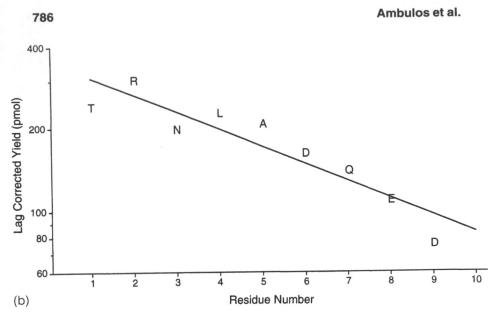

(b)

Figure 7   Continued

one must always be aware of the possibility that extraneous peaks related to sequencing artifacts can be misinterpreted as modified amino acids [42].

F.   Example

The power of Edman degradation sequence analysis for problem solving can be clearly seen in the analysis of the peptide that was presumed to have the sequence H-Asp-Glu-Gln-Glu-Ala-Leu-Asn-Arg-Ser-NH$_2$. Although discrepancies in the sequence were readily evident by mass spectrometry (as described earlier), it is unlikely that the actual sequence would have been deduced by tandem MS in most peptide synthesis laboratories. In contrast, using standard Edman degradation techniques it was very straightforward to determine that the sequence of the peptide was H-Thr-Arg-Asn-Leu-Ala-Asp-Gln-Glu-Asp-NH$_2$ (i.e., peptide **2**). As shown in Fig. 7a, intense, clearly identifiable peaks were detected in each analysis cycle. For every cycle after the first, a small peak from the amino acid in the previous cycle can be seen in addition to the peak for the amino acid in the current cycle. This phenomenon, usually called "lag," is recognizable in that lag peaks consistently decline in each subsequent cycle. The results presented in Fig. 7a illustrate the ability to sequence this peptide through the carboxy-terminal residue,

Asp. In Fig. 7b it can be seen that the percent yield for each amino acid decreased with each cycle; even though the yield for the carboxy-terminal amino acid was low, it was still sufficient for identification. Thus, the complete sequence of this peptide was readily obtained by Edman degradation.

In summary, Edman degradation sequence analysis of synthetic peptides can be very useful for identifying problems that occur during synthesis. However, because of time requirements in addition to the expense of the analysis, this method is not practical for routine use. As described earlier, the utilization of HPLC combined with mass spectrometry is more suitable for routine assessment of peptide purity and quality.

REFERENCES

1. KM Gooding, FE Regnier, eds. HPLC of Biological Macromolecules: Methods and Applications. New York: Dekker, 1990.
2. MTW Hearn, ed. HPLC of Proteins, Peptides and Polynucleotides: Contemporary Topics and Applications. New York: VCH, 1991.
3. CT Mant, RS Hodges, eds. HPLC of Peptides and Proteins: Separation, Analysis and Conformation. Boca Raton, FL: CRC Press, 1991.
4. M Dizdaroglu. J Chromatogr 364:223–229, 1983.
5. KY Kumagaye, M Takai, N Chino, T Kimura, S Sakakibara. J Chromatogr 327:327–332, 1985.
6. CT Mant, RS Hodges. J Chromatogr 327:147–155, 1985.
7. JG Dorsey, PJ Foley, WT Cooper, RA Barford, HG Barth. Anal Chem 64:353–389, 1992.
8. CT Mant, RS Hodges. Methods Enzymol 271:3–50, 1996.
9. JL Fausnaugh, E Pfannkoch, S Gupta, FE Regnier. Anal Biochem 137:464–472, 1984.
10. MV Novotny, KA Cobb, J Liu. Electrophoresis 11:735–749, 1990.
11. T Bergman, H Jornvall. In: R Hogue Angeletti, ed. Techniques in Protein Chemistry, III. San Diego: Academic Press, 1992, pp 129–134.
12. JW Kenny, JI Ohms, AJ Smith. In: RH Hogue Angeletti, ed. Techniques in Protein Chemistry, IV. San Diego: Academic Press, 1993, pp 363–370.
13. ED Lee, W Mueck, JD Henion, TR Covey. J Chromatogr 458:313–321, 1988.
14. ED Lee, W Mueck, JD Henion, TR Covey. Biomed Environ Mass Spectrom 18:844–850, 1989.
15. RD Smith, JA Olivares, NT Nguyen, HR Udseth. Anal Chem 60:436–441, 1988.
16. RD Smith, JA Loo, CG Edmonds, CJ Barinaga, HR Udseth. Anal Chem 62:882–899, 1990.
17. HE Schwartz, RP Palmieri, R Brown. In: P Camilleri, ed. Capillary Electrophoresis: Theory and Practice. Boca Raton, FL: CRC Press, 1993, p 201.

18. RM McCormick. In: JP Landers, ed. Boca Raton, FL: CRC Press, 1994, p 287.
19. RP Palmieri, JA Nolan. In: JP Landers, ed. Handbook of Capillary Electrophoresis. Boca Raton, FL: CRC Press, 1994, p 325.
20. RE Majors. LCGC 12:278–280, 1994.
21. SA Kates, F Albericio. Lett Peptide Sci 1:213–220, 1995.
22. RS Annan, SA Carr. Anal Chem 68:3413–3421, 1996.
23. JJ Gorman, BL Ferguson, TB Nguyen. Rapid Commun Mass Spectrom 10: 529–536, 1996.
24. W Yu, JE Vath, MC Huberty, SA Martin. Anal Chem 65:3015–3023, 1993.
25. JB Fenn, M Mann, CK Meng, SF Wong, CM Whitehouse. Science 246:64–71, 1989.
26. JT Watson. Introduction to Mass Spectrometry. 3rd ed. Philadelphia: Lippincott–Raven, 1997, pp 303–319.
27. PA Schindler, A Van Dorsselaer, AM Falick. Anal Biochem 213:256–263, 1993.
28. KF Medzihradszky, Al Burlingame. Methods: A Companion to Methods in Enzymology 6:284–303, 1994.
29. K Biemann. Methods Enzymol Appendix 5 193:886–887, 1990.
30. K Bláha, J Rudinger. Collect Czech Chem Commun 30:385, 1965.
31. S Moore, DH Spackman, WH Stein. Anal Chem 30:1185–1190, 1958.
32. BA Bidlingmeyer, SA Cohen, TL Tarvin. J Chromatogr 336:93–104, 1984.
33. RL Heinrikson, SC Meredith. Anal Biochem 136:65–74, 1984.
34. DJ Strydom, GE Tarr, YE Pan, RJ Paxton. In: RH Angeletti, ed. Techniques in Protein Chemistry III. New York: Academic Press, 1992, pp 261–274.
35. DJ Strydom, TT Andersen, I Apostol, JW Fox, RJ Paxton, JW Crabb. In: RH Angeletti, ed. Techniques in Protein Chemistry IV. New York: Academic Press, 1993, pp 279–288.
36. GB Fields, SA Carr, DR Marshak, AJ Smith, JT Stults, LC Williams, KR Williams, JD Young. In: RH Angeletti, ed. Techniques in Protein Chemistry IV. New York: Academic Press, 1993, pp 129–237.
37. GB Fields, RH Angeletti, SA Carr, AJ Smith, JT Stults, LC Williams, JD Young. In: JW Crabb, ed. Techniques in Protein Chemistry V. San Diego: Academic Press, 1994, pp 501–507.
38. MW Crankshaw, GR Grant. Identification of modified PTH-amino acids in protein sequence analysis. Presented at the Association of Biomolecular Resource Facilities, 1993.
39. RH Angeletti, LF Bonewald, GR Fields. Methods Enzymol 289:697–717, 1997.
40. LF Bonewald, L Bibbs, SA Kates, A Khatri, KF Medzihradszky, JS McMurray, ST Weintraub. J Peptide Res, in press.
41. P Edman. Acta Chem Scand 4:283–293, 1950.
42. GA Grant, MW Crankshaw, J Gorka. Methods Enzymol 298:395–419, 1997.
43. L Mints, R Hogue Angeletti, E Nieves. ABRF News 8:22–26, 1997.
44. NR Shenoy, JE Shively, JM Bailey. J Protein Chem 12:195–205, 1993.
45. CG Miller, JM Bailey. Am Biotech Lab 13:36–37, 1995.

46. K Hardeman, B Samyn, J Van Der Eycken, J Van Beeumen. Protein Sci 7: 1593–1602, 1998.
47. J Fernandez, K Stone. In: MC Flickinger, SW Drew, eds. Encyclopedia of Bioprocess Technology: Fermentation, Biocatalysis, and Bioseparation. New York: Wiley, 1999.
48. MW Hunkapillar. In: AS Brown, ed. Protein/Peptide Sequence Analysis: Current Methodologies. Boca Raton, FL: CRC Press, 1988, pp 87–117.

20

The Chromatographic Analysis of Combinatorial Arrays: Parallel HPLC and HPLC-MS

Vern de Biasi and Andrew Organ
SmithKline Beecham Pharmaceuticals, Essex, England

Combinatorial chemistry as a modality for the parallel synthesis of large numbers of potential pharmacophores constructed from a single template has now gained widespread acceptance and usage in the drug discovery process [1]. Because of the rapid evolution of the technology, there is, as demonstrated by the previous chapters in this book, a need for a continuous update and review of the science. The application of these massive parallel synthetic approaches has placed significant pressure on the analyst, who is required to provide the synthetic chemist with timely decision-making information associated with both the identity and purity of the products produced. It is therefore the intention of this chapter to highlight some of the pressures these automated synthetic technologies [2] cascade upon the analytical function and the strategy that we have developed to meet these requirements.

I. BACKGROUND

There are several established technologies for the application of both solution and solid-phase synthetic paradigms in the search for novel entities through the application of high-throughput synthesis, as detailed throughout this book. The increased implementation of these technologies has in turn forced a reexamination of the process of automated high-throughput analyses for the quality control of these compounds. Although the practice in some laboratories may vary, there is a general consensus that it is prudent to analyze either all the compounds synthesized or at least a statistically significant number for both purity and identity prior to performing biological screening. For this purpose high-performance liquid chromatography (HPLC) and high-performance liquid chromatography–mass spectrometry (HPLC-MS) are currently the methods of choice for characterizing both mixtures and single components synthesised by automated procedures.

Although efforts have been made to accelerate HPLC methods to match the rate of synthesis required, one of the underlying principles within the analytical laboratory undertaking this work is to have simple (generic) and thereby robust analytical methodology. The level of analysis undertaken within our laboratories has increased by at least an order of magnitude over the past 2 years. This increase appears set to rise at an even more dramatic rate as the level of automated instrumentation increases in the synthetic laboratory (Fig. 1).

In the current overall scheme of library design, synthesis, analysis, and biological screening the analyst is, in the majority of instances, in the position of being interdisposed between the chemist and the submission of the molecules for biological screening. It is therefore evident that any failure to maintain pace with any part of the synthetic process leads to the analysis becoming rate limiting. If the goal is to achieve 100% sampling of array or mixture products, then the analyst can be faced with in excess of several hundred HPLC or HPLC-MS assays per day. A delay in these analyses would in turn result in a backlog with a consequent delay in biological screening. Therefore in laboratories such as ours a great deal of time and effort has been invested in developing both robust and rapid methodologies to expedite the analysis of these compound arrays.

Over the period of evolution of combinatorial and automated synthetic procedures we have moved from the analysis of combinatorial libraries (typically 100–500 components) to the semiquantitative analysis of single products synthesized simultaneously in a parallel process. Analyses of libraries previously involved lengthy HPLC or HPLC-MS assays as the aim was then to prove the existence, or otherwise, of all components of the library under examination (Fig. 2). Such assays required in excess of 30 min to complete with lengthy manual processing of the resulting data.

Our present approach is to perform a large number of analyses to resolve the single desired product or other contaminants, use the target molecular weight to confirm the presence of the product, and generate the relative purity by either ultraviolet detection, evaporative light scattering detection, or some other detection protocol [3]. The samples for analysis are typically presented in 96-well microtiter plate (MTP) format and the result-

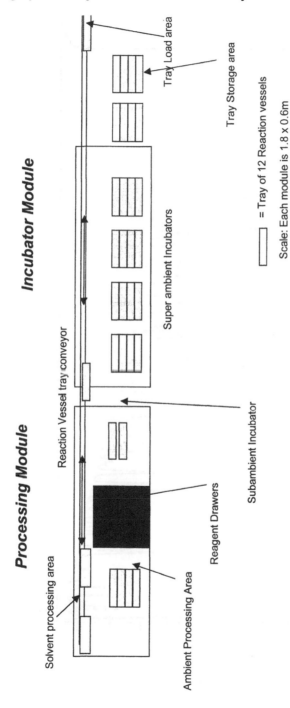

Figure 1 Schematic of the Myriad automated synthetic engine.

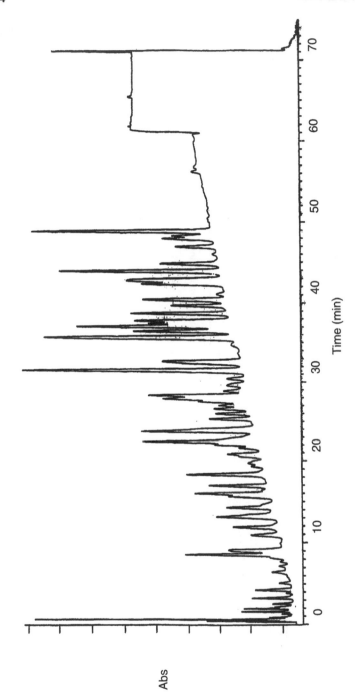

Figure 2 Example separation of a 100-component library by HPLC [Hypersil BDS, C$_8$ (3 μm 200 × 4.6 mm); water–acetonitrile + 0.1% TFA gradient, 0–100% over 60 min at a flow rate of 1 mL/min with UV detection at 220 nm].

ing relative purity data are used to make decisions prior to the biological screening of the components. As previously described, the robustness of the process is a key driver to ensure that there is no collapse in the material–information flow and hence no backlog. In an attempt to ensure this process, we have based our analytical methodologies on a generic approach. We have over time been able to modify and refine our methodologies to satisfy the demands of the increased rates of synthesis. However, there is a point at which further increases in analytical speed can compromise the robustness of the overall analytical process and as a consequence affect the timely delivery of data. It has therefore been our aim to develop an approach that can utilize the methodologies currently applied, maintain the robustness of the process, and also accelerate the throughput of compounds. It was evident that because we were applying generic methodologies to a wide range of compounds, the most facile way to increase throughput was to analyze multiple compounds simultaneously, namely by applying parallel separation and detection to the analysis of combinatorial arrays.

It is the intention of this chapter to illustrate the progress made to date through fast to ultrafast HPLC in a serial mode and then the development and application of parallel HPLC analysis to both of these separation modalities.

II. SEQUENTIAL ANALYSIS BY FAST AND ULTRAFAST HPLC

The development of separation methodologies to satisfy the throughput required for the rapid analysis of a wide variety of diverse molecules was based on generic separation methods with the application of fast HPLC. The main advantage of this technique, as its name implies, is the increased speed of separation, which arises from the use of columns of reduced length. Columns are typically of the order of 30 to 100 mm in length with internal diameter (i.d.) ranging from 2.1 to 4.6 mm. In order to achieve comparable column plate numbers, equivalent to those of the more conventional 250-mm-length columns, the particle size of the stationary phase is usually reduced to 3 or 3.5 μm. The columns are also operated at elevated flow rates typically ranging from 2 to 5 mL/min [4]. However, the combination of short columns and high volumetric flow rates can give rise to difficulties related to the high analyte elution speed, resulting in peak widths of only a few seconds. These difficulties are mainly confined to the response of the detector and data handling system. Thus, it is important to have a sufficiently fast response to represent the eluting peak accurately without contributing to the overall band broadening of the system [5]. It is also important to note that both the injection volume and concentration of the analyte must be reduced in order to obtain the same sample/stationary phase material ratio as in conventional chromatography; otherwise sample overloading will occur and an inaccurate representation of purity will be obtained. For the purpose of our work, fast HPLC methodology has been used with ultraviolet and mass spectrometric detection.

A. Experimental Conditions

Although a wide variety of HPLC equipment can be used for this work, we have developed our methodologies using an HP1100 HPLC system (Hewlett Packard, Waldbronn, Germany) with a binary pump, diode array detector, vacuum degasser, and column module. Sample introduction occurs via a Gilson 215 autosampler (Gilson, Middleton). Mass spectrometric detection utilized either a Micromass Platform II (Micromass, Manchester UK) or a Finnigan aQa (Finnigan Masslab, Manchester, UK). Data capture was achieved either through Mass Lynx 3.1 (Micromass) or MassLab 2.2 (Finnigan Masslab).

The initial separation work was performed on a Luna 2 C_{18} column 5 μm, 75 $\times$ 4.6 mm i.d. The following linear gradient profile was used: acetonitrile–water (5:95 v/v) containing trifluoroacetic acid (TFA) (0.1% v/v) at time zero to acetonitrile (100%) containing TFA (0.1% v/v) over 3.5 min. It was held at this composition for 1 min at a flow rate of 3.0 mL/min with a temperature of 40°C, giving an overall cycle time (injection to injection) slightly greater than 5.6 min.

The speed and throughput of the fast HPLC approach were increased by use of shorter columns with a reduced internal diameter, Zorbax SB C_{18} (3.5 μm, 30 $\times$ 2.1 mm i.d.) (Hewlett Packard, Waldbronn, Germany). These columns were used on the same instrumental hardware with the same mobile phase compositions at a flow rate of 2.0 mL/min with a temperature of 40°C. However, the linear gradient profile was completed in 2 min and held at this composition for 0.1 min, yielding an overall cycle time of 2.6 min.

The development of ultrafast HPLC methodology was carried out using a Symmetry C_{18} column (3.5 μm, 10 $\times$ 2.1 mm i.d.) (Waters, Milford, MA) with the same gradient profile, at the same flow rate and temperature, but chromatographed over 1 min, yielding an overall cycle time of 1.25 min.

Analyte detection in all these methods involved an ultraviolet diode array detector over the range 214–216 nm to minimize data file size. Mass spectrometric data were obtained following a split of the HPLC solvent stream of approximately 100:1. The mass spectrometer was tuned for optimal sensitivity and unit mass resolution. The mass range 180–800 daltons was scanned in 1 s and positive ion electrospray was used throughout. All solvents utilized were of HPLC grade.

B.   Results and Discussion

The fast HPLC development in our work has closely followed the availability and improvements in chromatographic stationary phase and column technology. The initial methodology, based on Luna material, was centered on a 5-μm particle size. These analyses with columns of 75 mm length and conventional 4.6 mm internal diameter gave analysis times of the order of 3.5 min (Fig. 3) with an overall cycle time, injection to injection, of almost 6 min. For such a cycle time the analysis of a 96-well microtiter plate (MTP) would be completed in 8 h, permitting the analysis of three MTPs per day.

It was difficult to improve throughput further with this column format. However, as the quality and robustness of columns packed with 3- or 3.5-μm stationary phase materials have improved, the materials have generally become the choice for this type of application. The potential of achieving the same number of theoretical plates with a shorter column length [5] permitted the development of faster analyses without significant loss or chromatographic resolution [6].

The development of this methodology was subsequently centered on Zorbax SB C$_{18}$ stationary phase material (3.5 μm, 30 $\times$ 2.1 mm) in a cartridge format. The reduced particle size helped to ensure that sufficient column efficiency was present for resolution of the potential components. The reduced length also permitted the use of volumetric flow rates (2 or 3 mL/min) in excess of those typically used for this column geometry. These combined features resulted in a 50% reduction in cycle time for the analyses with no apparent deleterious effect on the separation methodology (Fig. 4). Even with this fast cycle time, the complete QC of a 96-well MTP required in excess of 4 h.

Therefore, in an attempt to speed the analysis further, the column length was further reduced to 10 $\times$ 2.1 mm i.d. This approach gave adequate resolution of the components examined (Fig. 5), a reduced cycle time of 1.25 min, and no measurable decrease in column performance over the period of this work. However, the reduced stationary phase bed volume present in such columns resulted in a need to control the injected sample concentration to a degree not required with the 30 $\times$ 2.1 mm i.d. columns. This introduced an additional and thus far unnecessary step, which was felt to be deleterious to the overall QC process. Although it was practical to analyze samples in a 1-min time frame, there was an overall reduction of the efficiency of the complete QC process, which resulted in a reevaluation of the approach.

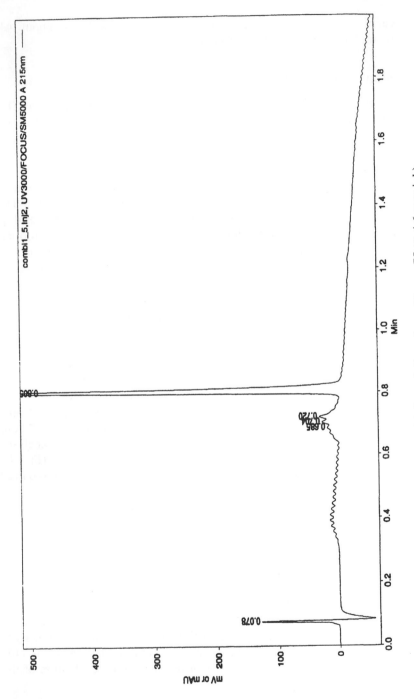

Figure 3 Typical analysis of an array sample with Luna 2 C$_{18}$ column (5 μm, 75 × 4.6 mm i.d.).

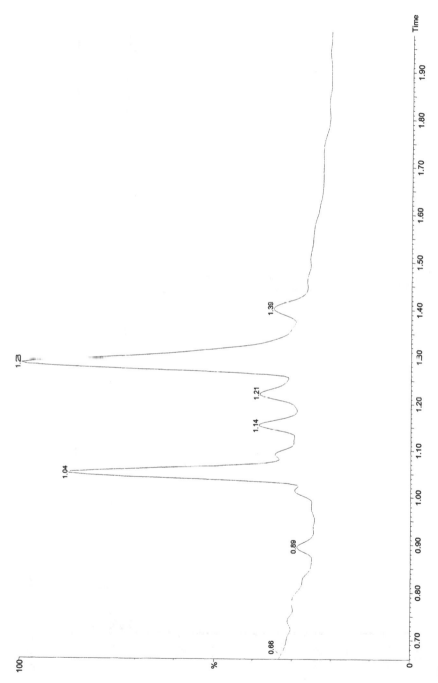

Figure 4 Typical analysis of an array sample with a Zorbax SB C$_{18}$ column (3.5 μm, 30 × 2.1 mm i.d.).

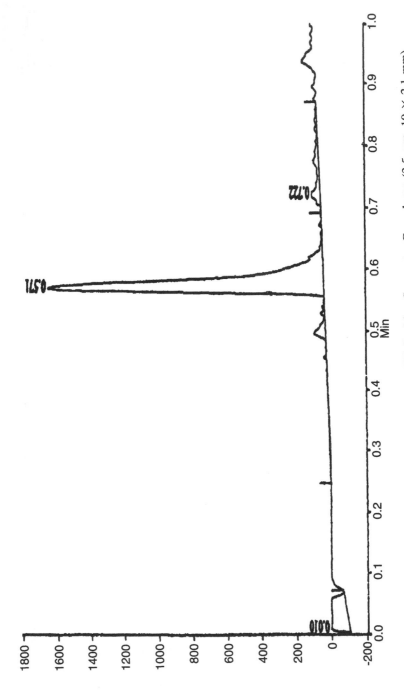

Figure 5   Analysis of combinatorial sample by ultrafast HPLC with a Symmetry C$_{18}$ column (3.5 μm, 10 × 2.1 mm).

To date, separation methodologies used in a combinatorial QC environment have been developed and refined to balance throughput and quality of analysis and maximize the robustness of the process. It is the case that in almost any analytical function, and especially in the combinatorial area, the emphasis on increased throughput while maintaining quality data is a key and constant driver. A reduction in column length and an increase in the volumetric flow rate were obvious routes to the reduction of chromatographic analysis time and subsequent increase in throughput. However, there was little advantage to be achieved in further reducing the analysis time with current equipment. Although this approach introduced improvements in relation to increasing throughput, fundamentally the same sequential analytical approach on a number of individual instruments is applied. Therefore, as sample numbers grow, the number of individual QC instruments required would increase. Currently, this has been the conventional approach as there has been no obvious alternative to meet these challenges. Unfortunately, with the sample number projections envisaged in the combinatorial area, it was believed that this process would not be efficient in the long term.

III. THE DEVELOPMENT OF PARALLEL HPLC

As we were applying the same generic method to the sequential analysis of these samples, in theory assays of a number of compounds, simultaneously incorporating the same analytical method, were envisioned. The application of parallel analysis to a number of other separation processes has been previously described [7–9]. Although these reports were not directly relevant to the analysis of combinatorial arrays, the authors have found obvious advantages to such an approach for their applications. The parallization of HPLC would also offer a significant advantage in terms of sample throughput and speed of analysis.

Unfortunately, automated parallel analytical separation systems were commercially not available at this time. However, in seeking to create the parallel strategy, we initiated studies on a manual-based injection system to validate the approach. It was thus possible to develop a manual parallel analytical separation system that allowed two columns to be used for the analysis of two samples concurrently. If this process could be automated and a greater number of analytes could be examined simultaneously by the same methodology, a quantal leap in productivity could be achieved. As previously discussed, the sample presentation typically occurs in an MTP format and so the simultaneous parallel analysis of a column or row from such a plate would seem a logical approach.

The availability of liquid handlers and the development of HPLC autosamplers, permitting the simultaneous manipulation of a number of individual analytes, led us to apply this technology to develop parallel HPLC and subsequently parallel HPLC-MS analyses.

A. Parallel Analysis Experimental Conditions

For parallel analysis an alternative set of instrumentation was utilized. The chromatographic system consisted of a Waters 600 pump (Waters, Milford, MA) with a Gilson 215 autosampler fitted with an 889 multiprobe (Gilson, Middleton,) and a vacuum degasser (Hewlett Packard, Waldbronn, Germany). The flow from the pump at 16 or 24 mL/min was delivered to an eight-way splitter (Valco, Switzerland) set to deliver nominally a flow of either 2 or 3 mL/min to each of the eight columns attached to the Gilson autosampler. The columns used were those previously described, Zorbax SB C_{18} (3.5 μm, 30 m $\times$ 2.1 mm i.d.). At present it has not been possible to develop a parallel UV detector, so we used eight individual UV detectors, from a variety of sources, each attached to an individual column, with detection set at 215 nm. An outline of the system is shown in Fig. 6. Data capture from the eight UV detectors was via a Waters 860 system (Waters, Milford, MA). The gradient profile was as used in the sequential analysis, previously described (Section II.A).

B. Results and Discussion

With the autosampler configuration of the Gilson 215/889 instrument, aspiration to a single column (eight samples) of a 96-well microtiter plate was possible (Fig. 7). These samples were then introduced into eight loops on the Rheodyne valves in one injection. The valves on the autosampler were then simultaneously rotated, from load to inject, and each sample was then introduced as a discrete entity into its own analytical column. At the point of the simultaneous valve rotation, the gradient was triggered, via contact closure from the autosampler, and the assays were initiated. By performing the analyses in this manner it was possible to complete the purity determinations for eight compounds in the same time required for a single injection in the sequential process, i.e., 2.6 min (Fig. 8). This in turn signaled the possibility of assaying a 96-well MTP in a total of 12 injections, or a total analysis time of approximately 30 min, compared with over 4 h achieved using the same methodology in a sequential process. Thus it would be possible to analyze over 46 MTPs, or greater than 4000 samples, per day on a single parallel HPLC instrument. We have also investigated this approach

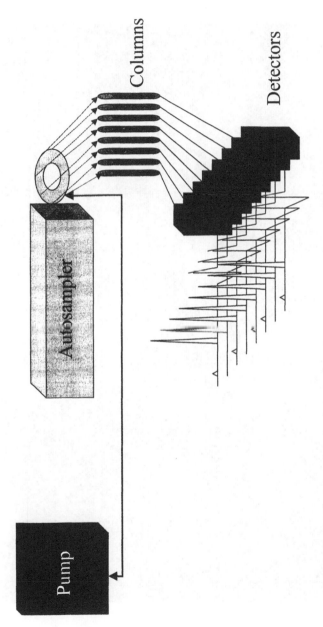

Figure 6   Diagrammatic representation of the eight-way parallel HPLC–UV system.

Figure 7 Simultaneous aspiration of a column of an MTP using Gilson 215 with an 889 multiprobe. (Reproduced by permission from Gilson.)

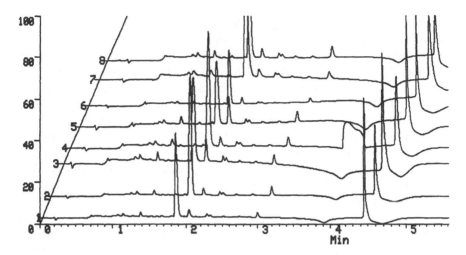

Figure 8 Parallel analysis of eight individual analytes using eight Zorbax SB columns.

using the developed ultrafast methodology, i.e., with the 10×2.1 mm i.d. column geometry and cycle time of 1.25 min. In that case, the analysis of a 96-well MTP has been completed in 15 min and permits over 96 MTPs, or greater than 9000 samples, to be analyzed in a 24-h period on a single instrument.

This initial parallelization of the purity analysis for combinatorial chemistry has demonstrated the considerable benefits of such an approach in terms of reduction of overall analytical time with no compromise of data quality. However, for a system that mirrors current practice in the majority of laboratories, and to obtain more information about the identity of the analyte, a parallel approach using MS detection must also be present.

IV. THE DEVELOPMENT OF PARALLEL HPLC-MS

The principle for the simultaneous introduction of more than one sample spray into the source of a mass spectrometer has been illustrated by others [10,11]. As is the case in the analysis of combinatorial arrays, the expected molecular weights of the compounds are known before analysis. Thus it should be possible to combine the downstream flows from the columns and introduce these into the mass spectrometer to identify the presence of the component of interest because the expected masses are known prior to injection. The obvious limitation of this approach is that it is essential that there is mass redundancy between adjoining wells of the MTP and this can be circumvented in the synthetic design process [12].

Initial work followed the approach adopted by Kassel and Zeng [11] and focused on a four-column parallel HPLC mode. We combined the four individual solvent streams, after splitting, to give a single sample stream entering the mass spectrometer (i.e., Finnigan aQa) (Fig. 9). Four mass chromatograms (upper four traces, Fig. 9) were obtained from the expected $[M+H]^+$ signals of the analytes (*m/z* 558, 482, 466, 452 daltons for four SB compounds). With the current Finnigan data system, only three (of the four) analog UV signals could be acquired simultaneously. This procedure worked sufficiently; however, the optimization of the process does require that an equal flow and signal are obtained from all four solvent streams. This is simply a "plumbing" issue, which may require some time to achieve because it is a trial-and-error approach. Therefore, the more solvent lines that are combined in this fashion, the more complex and involved the process.

From an analytical standpoint, we would ideally prefer to maintain discrete solvent lines even upon introduction into the mass spectrometer. This would ensure that continued data integrity throughout the analytical process could be maintained and no there would be constraints on sample adjacencies. We embarked on a technical collaboration with Micromass (UK) to develop and construct an interface into a mass spectrometer that would achieve our goal. The result of this collaboration is the construction of a novel four-channel multiplex electrospray liquid chromatography interface that has been used to analyze single components and quaternary mixtures by LC-MS as well as synthetic samples prepared by automated procedures.

The rapid LC methods employed for array analysis produce typical peak widths at half-height of less than 2 s. It was immediately apparent that the scan rates achievable on a radiofrequency (RF) quadrupole mass spectrometer would not be sufficient to generate high-quality mass spectra that could be used to confirm the presence of a desired chemical entity. Alternatively, commercial orthogonal time-of-flight (TOF) mass spectrometers can acquire full scan profile data at rates approaching 10 Hz so that rapid switching between up to four streams becomes feasible. Therefore this approach was followed in our development of the novel interface based on the Micromass LCT instrument.

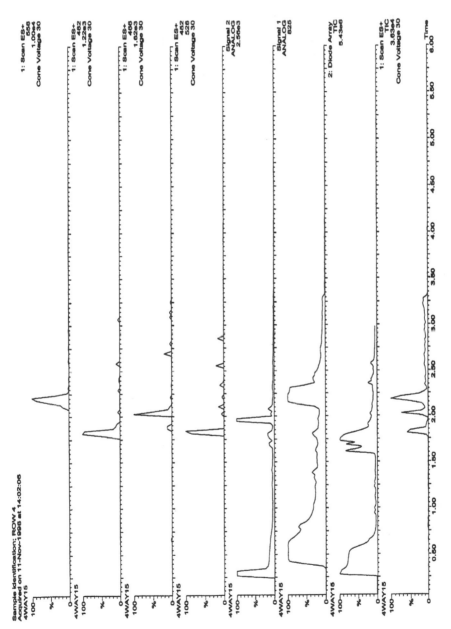

Figure 9 LC-MS data adopting a combined flow approach from four columns.

A. Parallel HPLC-MS Experimental Conditions

1. Liquid Chromatography

Liquid flow was provided by a Waters 600 quaternary pump (Waters Corp., Milford, MA). This flow was split four ways by use of a low-dead-volume Valco eight-way splitter (Valco, Switzerland). The separate streams were directed to the Rheodyne 7010 injection valves of a Gilson 889 liquid handler (Gilson, Middleton, WI). The Gilson 889 was mounted on a Gilson 215 multiple injection liquid handler. All samples were presented in polypropylene microtiter plates and an injection volume of 15 μL was employed throughout.

Separation was achieved with Zorbax SB C_{18} cartridges (3.5 μm, 30 $\times$ 2.1 mm). For this work a variety of UV detectors were used, each tuned to 215 nm, and the analog outputs were acquired by the mass spectrometer data system. The following gradient profile was employed using acetonitrile–trifluoroacetic acid (TFA, 0.1% v/v) as solvent A and water–trifluoroacetic acid (0.1% v/v) as solvent B: 5% A at time zero to 100% A at 1.5 min employing a linear gradient profile and held at 100% A until 2.0 min with a return to 5% A at 2.1 min. A flow rate of 8 mL/min was used, providing nominally 2 mL/min through each column. This flow was split 1:10 using three-way Valco low-dead-volume T pieces and the split streams were presented in separate lines to the novel four-way API electrospray device described in the following.

All solvents employed were HPLC grade and the test analytes, furosemide [molecular weight (MW) 331], reserpine (MW 609), triamterene (MW 254), and warfarin (MW 309), were all from Sigma Aldrich (Poole, Dorset, UK).

2. Mass Spectrometry

All mass spectrometric data were acquired on an LCT orthogonal TOF mass spectrometer (Micromass Ltd, Manchester, UK) fitted with a prototype version of a novel four-way multiplex API interface. Each liquid stream was sampled for 0.1 s with 0.1 s used to move to the adjacent sampling position. Mass spectra were acquired from 200 to 1000 daltons with a cycle time of 10 Hz. The instrument was operated under MassLynx software V3.3. The overall system layout is presented in Fig. 10 and the design of the multiplex head is shown in detail in Fig. 11.

B. Results and Discussion

The initial experiments with the system were conducted using four single-component standards, furosemide (MW 331), reserpine (MW 609), triam-

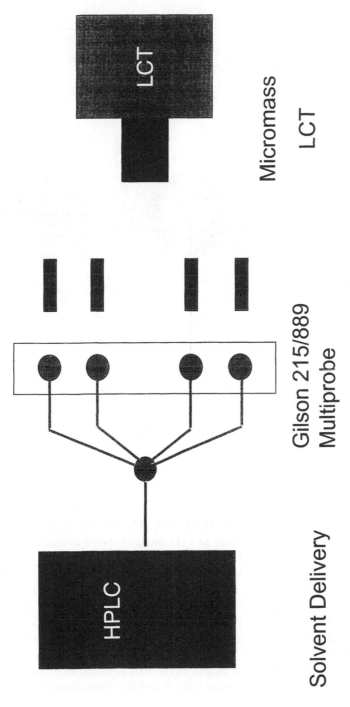

Figure 10 Schematic representation of four-way parallel HPLC-MS system.

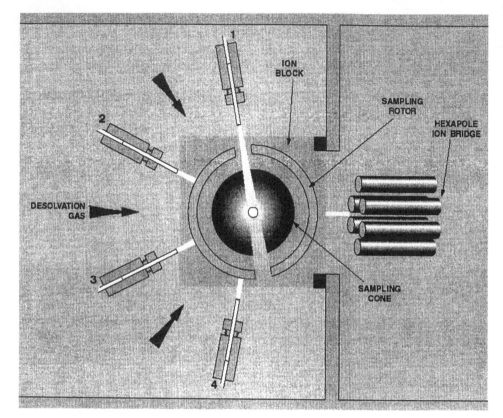

Figure 11   Design of multiplex head for LC-T mass spectrometer. (Reproduced by permission from Micromass.)

terene (MW 254), and warfarin (MW 309). The first set of data shows the mass chromatograms for the four M+H signals of the four standards injected simultaneously into the HPLC-MS system with an HPLC analysis time of 2.1 min (Fig. 12). This indicated that chromatographic integrity was preserved under the conditions of the high-throughput method employed. Mass spectral data were extracted from a single channel (sum of four spectra, LC line 3, Fig. 13) and it was evident that high-quality information was available. To evaluate the level of interchannel cross talk, the spectrum of furosemide (MW 331, injected in channel 1) was examined at a time point corresponding to elution of warfarin (MW 308, injected in channel 4) (Fig. 14). It was evident that there was little detectable interchannel cross talk in

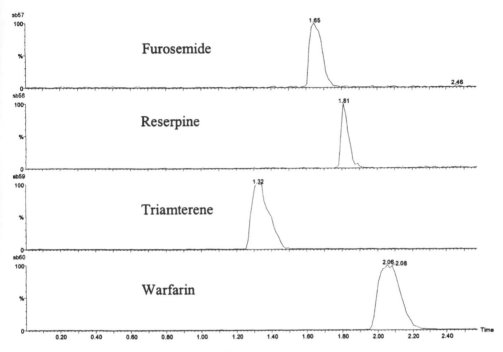

Figure 12 Mass chromatograms of the [M+H] signals for the four test compounds.

this experiment. From subsequent experiments, the interchannel cross talk was determined to be ~ 0.1 to 0.2% with this configuration.

The four individual components were then mixed and placed in each of the four wells. The four MTP wells were then analyzed simultaneously. One of these acquisitions is illustrated by the analog signal and extracted mass chromatograms presented in Fig. 15. These data indicate that even closely eluting components can still give unambiguous molecular weight information with this approach.

Since these initial experiments, we have applied this technology to interrogate compounds prepared by automated synthetic procedures employed in house at SB. Representative sets of four were analyzed, in parallel, from an MTP using the multiplex LC-MS probe. The data were compared with those obtained in serial mode using a single RF quadrupole instrument from mass analysis. The mass chromatogram for the four desired components is presented in Fig. 16 along with the relevant analog ultraviolet data. This information could then be used with the Diversity application (Micromass) to assign both relative purity and confirmatory mass, which we cur-

812

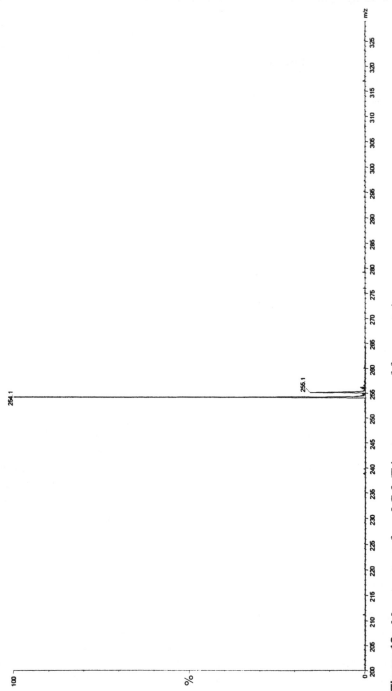

Figure 13 Mass spectrum from LC 3 (Triamterene; sum of four spectra).

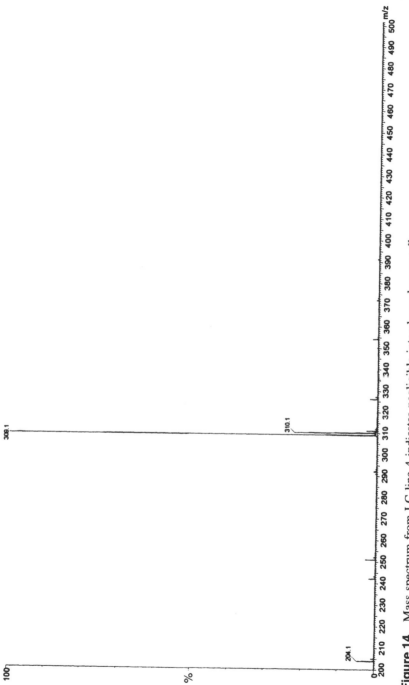

Figure 14 Mass spectrum from LC line 4 indicates negligible interchannel cross talk.

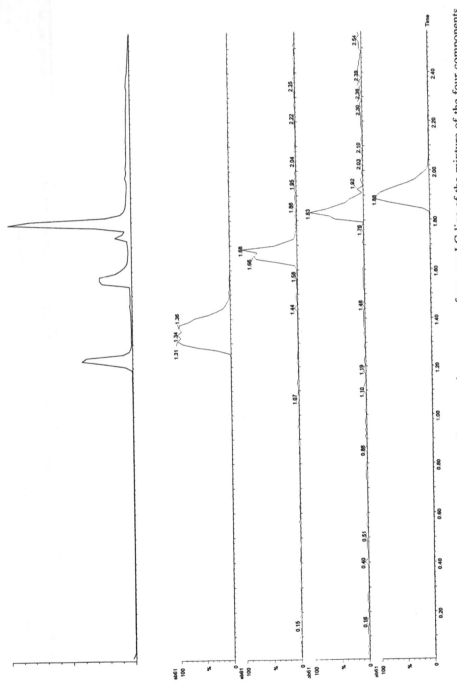

Figure 15 The analog signal and corresponding mass chromatograms from one LC line of the mixture of the four components.

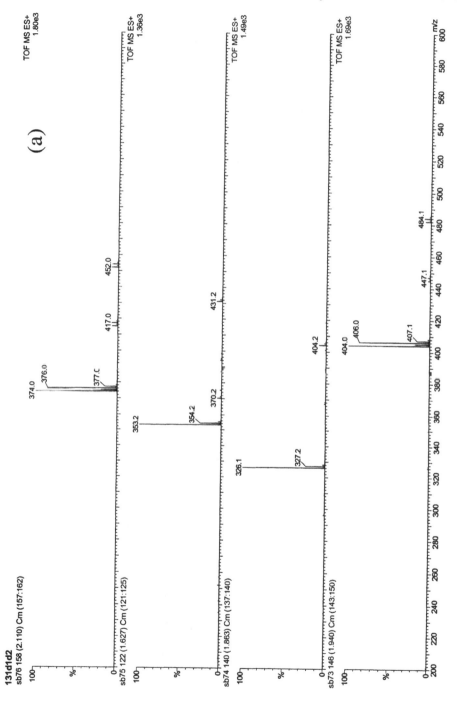

Figure 16 (a) Mass spectra of the four desired components. (b) [See p. 816.] An example UV chromatogram.

Figure 16 Continued.

rently practice. In our hands, it has been possible to extract high-quality mass spectral information from low-level synthetic by-products present as impurities at a typical level of 2% by UV area.

V. CONCLUSIONS

Quality control procedures for combinatorial libraries and arrays have traditionally been developed to keep pace with sample throughput. We have attempted to meet the challenge presented by the development and adoption of combinatorial technology and to build a process that is both fast and robust. The relative explosion of this chemical approach has in turn catalyzed the thinking of instrument vendors to satisfy the challenge of what is perceived to be the huge numbers of compounds that would require analysis. As with most new approaches, there has been a lag time before the nature of the analytical challenge is properly defined and also before the appropriate instrumentation becomes available. Our approach to the analytical problem has required a faster response from manufacturers for the development of products that were either nonstandard or not even conceived. As a result, we have been developing the technology in partnership with the vendors and especially with the development of parallel HPLC and HPLC-MS systems.

During the course of these developments we have taken our existing process and considered how to increase throughput without compromising any of its inherent strengths. In an area of high sample numbers, the advantages of parallel HPLC are manifest. It has been demonstrated that an eightfold increase in sample throughput can be achieved with a parallel HPLC approach. In addition, we have developed in conjunction with Micromass an interface that will allow us to combine the very demanding separation conditions with mass spectrometric detection that can also be performed in parallel. This development offers a potential paradigm shift in the way in which we determine the quality of both combinatorial arrays and libraries. In the current guise, the combination of these separation and detection technologies offers a fourfold increase in throughput. However, we can envisage that there is further potential to improve the throughput to mirror the current sample introduction technologies.

The analytical chemist has taken some time to see the light of parallelization. The reality is that in the analytical function within the high-throughput process, parallel analytical solutions to parallel synthesis now exist. The development of the technologies to achieve this and the learning curve we have gone through have enabled us to envisage carrying out both separation and detection in parallel for a whole range of applications, even outside the combinatorial area. From an analytical viewpoint, parallelization has only just begun!

ACKNOWLEDGMENTS

We would like to extend our thanks to our colleague Neville Haskins for his contribution to these developments. In addition, we express our gratitude to our other colleagues at SmithKline Beecham for their support in this work and our collaborators at Micromass, Robert Bateman, Kevin Giles, and Steven Preece, for their invaluable technical contribution to the design and construction of the MS interface and at Gilson, Henry O'Connell, for early access to the multiprobe injector.

REFERENCES

1. NK Terret, M Gardner, DW Gordon, RJ Kobylecki, J Steele. Eur J Chem 3: 1917, 1997.
2. N Hird, B MacLachlan. *ISLAR'98 Proceedings*, in press.
3. CF Kibbey. Mol Divers 1:247, 1996.
4. JB Li, J Morawski. LC-GC 16:468, 1998.
5. LR Snyder, JJ Kirkland. Introduction to Modern Liquid Chromatography, New York: Wiley Interscience, 1979, chap 5.
6. JH Knox. J Chromatogr Sci 15:352, 1977.
7. JM Guyot, G Michel, M Prost. Spectra 2000 156:62, 1991.
8. R Hashemi, S Meinitchouk, S Limbach. Labor Praxis 21:61, 1997.
9. F Gundlach. PI:WO 9835227.
10. R Bateman. 43rd ASMS Proceedings, p 98.
11. D Kassel, L Zeng. Anal Chem 70:4380, 1998.
12. I Hughes. J Med Chem 41:3804, 1998.

Index